Plant
Physiology

A TREATISE

EDITED BY

F. C. STEWARD

Department of Botany
Cornell University, Ithaca, New York

Volume II: Plants in Relation to
Water and Solutes

1959

 ACADEMIC PRESS, *New York and London*

Library of Congress Catalog Card Number: 59-7689

PRINTED IN THE UNITED STATES OF AMERICA

CONTRIBUTORS TO VOLUME II

T. A. BENNET-CLARK, *Department of Botany, University of London King's College, London, England*

O. BIDDULPH, *Department of Botany, State College of Washington, Pullman, Washington*

RUNAR COLLANDER, *Helsingfors Universitets, Botaniska Institution, Helsinki, Finland*

O. V. S. HEATH, *Imperial College of Science and Technology, Research Institute of Plant Physiology, Royal College of Science, London, England**

PAUL KRAMER, *Department of Botany, Duke University, Durham, North Carolina*

F. C. STEWARD, *Department of Botany, Cornell University, Ithaca, New York*

J. F. SUTCLIFFE, *Department of Botany, University of London King's College, London, England*

C. A. SWANSON, *Department of Botany, University of Ohio, Columbus, Ohio*

* Present address: Department of Horticulture, The University, Reading, England.

PREFACE TO VOLUME II

Through circumstances that were not foreseen, it has been found necessary to rearrange the subject matter of Volumes I and II of this treatise and to publish Volume II before Volume I. This makes it desirable to show the relationship of Volume II to the design of the whole work, which is outlined on pages ix and x, and to express in this volume certain general ideas which would have been superfluous if Volume I, with its Preface and Historical Introduction, had already appeared.

Until the end of the 19th century it was still possible for a comprehensive treatment of Plant Physiology to be written by one author, and the treatise of Pfeffer, in three volumes, reprinted in translation by the Oxford University Press in 1906, is the outstanding example. Since then there have appeared many textbooks designed for college or university students, and these vary greatly in scope and authority. At the other extreme, there is now appearing the encyclopedic work which is edited by Professor W. Ruhland and which is being printed partly in English and partly in German. In addition, the research specialist now has access to numerous reviews and summaries which are usually devoted to recent, often annual, advances in particular fields.

The aim of this treatise, some fifty or more years after Pfeffer, is to *say* what Plant Physiology is about and to do this in sufficient detail and with sufficient analysis of, and even extracts from, the ever expanding literature, so that each volume will be in large measure self-contained.

Plant physiologists will find that the treatment is sufficiently detailed to benefit their research in their own specialized fields and that the scope is broad enough to make reading of all portions of the work both stimulating and profitable. This treatise is, therefore, designed for the use of advanced and postgraduate students, teachers, research workers, and investigators in other fields of knowledge who need information about the present status of plant physiology. While such a synthesis of current knowledge is well justified by the great advances that have been made, especially in the last quarter of a century, its accomplishment requires the knowledge and mature experience of many authors who are aware of the trends in their often rapidly advancing fields of interest. Upon these authors, therefore, rests the quality and value of the work and to each the editor acknowledges his debt.

Although the treatise is now arranged in six volumes, each has been planned as a distinct unit and consists of a group of related chapters which, together, cover a major segment of the subject. Each chapter has been written by an authority in the field and analyzes the present status of its subject matter, giving pertinent references to the literature. The chief emphasis is on a synthesis of current knowledge, but consideration is also given to significant accomplishments of the past and, where possible, an insight into the problems of the future. Thus the reader may acquire an informed outlook on each topic.

While full advantage is taken of recent advances which accrue from the application of physical and biochemical techniques and the study of subcellular systems, the need to see the subject of plant physiology in terms of the morphology and organization of living plants is recognized throughout.

The treatise is intended not solely for use as a work of reference but is to be read by those who wish to obtain a reasoned analysis of the status and development of each subject which is discussed. Admittedly, and rightly, each chapter is affected by the author's own opinions, but, so far as possible in a work of this kind, an attempt has been made to achieve a measure of integration between the different chapters. Indexes enable information to be traced by reference to an author's name, to the plants used, or to the subject matter in question. For this volume the Index of Plant Names was prepared by Dr. W. J. Dress, and the Subject Index was compiled by Dr. H. Y. Mohan Ram. For this help the editor is grateful.

The separate acknowledgment of all those who have helped the authors and the Editor by permitting the inclusion of their published or unpublished material would be too great a task in a work of this kind. It should be understood, however, that both acknowledgment and thanks are conveyed by the form of citation in the text.

The Editor wishes especially to acknowledge the helpful cooperation of the personnel of Academic Press.

F. C. STEWARD

Cornell University
July, 1959

PLANT PHYSIOLOGY

The Plan of the Treatise

The treatise is planned in three main sections, as follows:

Section on *Cell Physiology and Problems Relating to Water and Solutes*

The purpose of this section is to present the properties of cells, their energy relations (Volume I) and behavior toward water and solutes with the closely related problems of the movement of solutes within the plant body and the economy of water in plants (Volume II).

The underlying theme of Volumes I and II is the basis of plant physiology in cell physiology.

Section on *Nutrition and Metabolism*

In this section the detailed facts and knowledge of nutrition and metabolism are presented, first with reference to the need for, and utilization of, inorganic nutrients (Volume III), and second with respect to the processes of organic nutrition (Volume IV). The treatment of organic nutrition leads to a reconsideration of photosynthesis and respiration at the level of organs and organisms. Volume IV describes the intermediary metabolism of carbon and nitrogenous compounds and presents a brief comparison of plants in health and in disease.

The main theme of Volumes III and IV is the nutrition, organic and inorganic, of plants and the biochemical steps by which these processes are achieved.

Section on *Growth and Development*

The purpose of the last section is to present the problems of plant physiology as seen through the analysis of growth and development, mainly with reference to flowering plants. This entails (Volume V) a reappraisal of the main events of growth and development from the standpoint of morphology and leads to a consideration of growth of cells and of organs. Tropisms and the role of hormones and the effects of synthetic growth regulators are discussed. In Volume VI the attention is focused upon the quantitative analysis of growth and development, the physiology of reproduction, the development of fruits and seeds, the problems of dormancy and perennation. The role of environmental factors in the control of growth and development merits separate treatment. Finally the problems of growth and development are examined from the standpoint of genetic control and from the interpretation of abnormal growth as seen in the formation of tumors. Throughout this treatment the controlling mechanisms of growth are evaluated.

Thus the last section of the work provides a synthesis of knowledge about plants since all their physiological processes converge upon growth and development.

The fulfillment of these objectives is possible only through the co-operation of many authors. The scope and treatment of individual chapters reflects the special interests of the contributors. While each volume is a complete unit, with its own table of contents and indexes, it is also an integral part of the whole plan.

Outline of the Plan

Section on *Cell Physiology and Problems Relating to Water and Solutes*

Volume IA. Cellular Organization and Respiration
Volume IB. Photosynthesis and Chemosynthesis
Volume II. Plants in Relation to Water and Solutes

Section on *Nutrition and Metabolism*

Volume III. Inorganic Nutrition of Plants
Volume IV. Organic Nutrition and Metabolism

Section on *Growth and Development*

Volume V. Analysis of Growth
Volume VI. The Physiology of Development

NOTE ON THE USE OF PLANT NAMES

The policy has been to identify by its scientific name, whenever possible, any plant mentioned by a vernacular name by the contributors to this work. In general, this has been done on the first occasion in each chapter when a vernacular name has been used. Particular care was taken to ensure the correct designation of plants mentioned in tables and figures which record actual observations. Sometimes, when reference has been made by an author to work done by others, it has not been possible to ascertain the exact identity of the plant material originally used, because the original workers did not identify their material except by generic or common name.

It should be unnecessary to state that the precise identification of plant material used in experimental work is as important for the enduring value of the work as the precise definition of any other variables in the work. "Warm" or "cold" would not usually be considered an acceptable substitute for a precisely stated temperature, nor could a general designation of "sugar" take the place of the precise molecular configuration of the substance used; "sunflower" and *"Helianthus"* are no more acceptable as plant names, considering how many diverse species are covered by either designation. Plant physiologists are becoming increasingly aware that different species of one genus (even different varieties or cultivars of one species) may differ in their physiological responses as well as in their external morphology, and that experimental plants should therefore be identified as precisely as possible if the observations made are to be verified by others.

On the assumption that such common names as lettuce and bean are well understood, it may appear pedantic to append the scientific names to them—but such an assumption cannot safely be made. Workers in the United States who use the unmodified word "bean" almost invariably are referring to some form of *Phaseolus vulgaris;* whereas in Britain *Vicia faba,* a plant of another genus entirely, might be implied. "Artichoke" is another such name that comes to mind, sometimes used for *Helianthus tuberosus* (properly, the Jerusalem artichoke), though the true artichoke is *Cynara scolymus.*

By the frequent interpolation of scientific names, consideration has also been given to the difficulties that any vernacular English name alone may present to a reader whose native tongue is not English. Even some American and most British botanists would be led into a misinterpretation of the identity of "yellow poplar," for instance, if this vernacular American name were not supplemented by its scientific equivalent *Liriodendron tulipifera,* for this is not a species of *Populus* as might be expected, but a member of the quite unrelated magnolia family.

When reference has been made to the work of another investigator who, in his published papers, has used a plant name not now accepted by the nomenclatural authorities followed in the present work, that name ordinarily has been included in parentheses, as a synonym, immediately after the accepted name. In a few instances, when it seemed expedient to employ a plant name as it was used by an original author, even though that name is not now recognized as the valid one, the valid name, preceded by the sign =, has been supplied in parentheses: e.g., *Betula verrucosa* (= *B. pendula*). Synonyms have occasionally been added elsewhere also, as in the case of a plant known and frequently reported upon in the literature under more than one name: e.g., *Pseudotsuga menziesii* (*P. taxifolia*); species of *Elodea* (*Anacharis*).

Having adopted these conventions, their implementation rested first with each contributor to this work; but all outstanding problems of nomenclature have been referred to Dr. W. J. Dress of the Bailey Hortorium, Cornell University. The authorities for the nomenclature employed in this work have been Bailey's *Hortus Second* and Bailey's *Manual of Cultivated Plants* for cultivated plants. For bacteria Bergey's *Manual of Determinative Bacteriology,* for fungi Ainsworth and Bisbee's *Dictionary of the Fungi* have been used as reference sources; other names have been checked where necessary against Engler's *Syllabus der Pflanzenfamilien.* Recent taxonomic monographs and floras have been consulted where necessary. Dr. Dress' work in ensuring consistency and accuracy in the use of plant names is deeply appreciated.

The Editor

CONTENTS

Chapter One

Chapter Two

CHAPTER SEVEN

Transpiration and the Water Economy of Plants
  by PAUL J. KRAMER 607

CONTENTS OF VOLUMES IA AND IB

PREAMBLE TO CHAPTER 1

From the first observations of protoplasm and of cells, special attention has been paid to their boundary surfaces and to the controlled entry and exit of substances which is an essential feature of the cell in life and the loss of which is an accompaniment of injury and death. Therefore, having treated cells, in Volume I, as organized units and having surveyed their ability to negotiate energy transformations through their metabolism, their ability to control the passage of substances across boundary surfaces, or membranes, needs now to be discussed as a very fundamental characteristic of cells. This is done in Chapter 1. While general principles of cell permeability are being sought, the great variety of cells and environments which are encountered in the plant kingdom needs to be recognized. Furthermore, it should also be recognized that cells which are in their most physiologically active, or growing, state possess properties additional to those of quiescent cells. Nevertheless, there is a vast body of information which relates to those passive permeability properties of cells that may influence the entry of solutes into, and their egress from, cells. These questions will now be discussed with special reference to the organization and properties of the boundary surfaces in cells, where these properties are believed to reside.

CHAPTER ONE

Cell Membranes: Their Resistance to Penetration and Their Capacity for Transport

Runar Collander

I. Introduction

Every living cell exists in a unique situation. On the one hand, its interior must be kept well isolated from its surroundings. This is imperative for many reasons. Imagine, for instance, a unicellular freshwater organism. If the surface layers were freely permeable to dissolved substances, then solutes could not be retained in the cell. No turgor pressure could be maintained in such a cell. Moreover, the sensitive vital machinery, with its multitude of enzymes requiring more or less constant hydrogen ion concentrations for their proper functioning, would be exposed to all the vagaries of the milieu. The orderly, integrated, and specific course of the life processes, so characteristic of all living organisms, would, under such conditions, be rendered impossible. Almost the same holds true of the cells of a multicellular plant: in order to be able to work successfully they must be effectively differentiated from their surroundings, that is, from the aqueous solution which bathes the cell walls.

On the other hand, every living cell must maintain a continuous exchange of substances with its surroundings: it needs oxygen for respiration; it must take up diverse nutrients; and it must get rid of the waste products of metabolism. Furthermore, in a multicellular plant numerous substances must be transported, say, from the photosynthesizing cells to the cells of the storage tissues, and from these to the meristematic cells and to others which consume organic foodstuffs.

Nature thus imposes two conflicting demands upon all living cells. From this point of view their life may be regarded as a perpetual cruise between Scylla and Charybdis. But how do they resolve this intricate situation? On the following pages we shall try to throw some light on this question.[1]

[1] For further particulars, and a more detailed bibliography, the reader is referred to the following comprehensive monographs: Stiles (157), Brooks and Brooks (25), Davson and Danielli (47), and W. Ruhland, ed., "Handbuch der Pflanzenphysiologie —Encyclopedia of Plant Physiology," Vol. 2. Springer, Berlin, 1956.

Living cells or, more properly speaking, the living protoplasts, are in fact isolated from their surroundings in such a way that the diffusion of numerous substances, outward and inward, is almost entirely prevented. At the same time the protoplasts are freely permeable to numerous other substances. Yet other substances behave in an intermediate manner in this respect. But we shall also find that the protoplasts have mechanisms at their disposal that make possible the active transport of certain substances, without regard to the direction in which these substances would tend to diffuse spontaneously.

This whole problem is often spoken of as the "permeability problem." This is not, however, a very appropriate designation, for it does not place the stress on the proper point. By far the most remarkable features are the very effective isolation of the protoplasts and also their striking capacity for active transport of substances even against concentration gradients.

II. The "Permeability Problem": A Historical Sketch

One of the very first to draw attention to the effective isolation of plant protoplasts was the Swiss botanist Carl Nägeli (114). As early as 1855, and thus only a few years after the recognition of protoplasm as the true site of the life processes, he published his observations on the impermeability of the protoplasm to the red, purple, or blue anthocyanins which occur dissolved in the cell sap of so many plants. In the same publication Nägeli described plasmolysis, a phenomenon that was to be important in later studies of the permeability properties of plant protoplasts. Nägeli's interpretation of this process was, on the whole, correct, for he realized that the plasmolytic contraction of the protoplast is due to an osmotic withdrawal of water from the cell sap. Only on a minor point was he mistaken, for he shared an idea then current that every osmotic process is composed of two opposing flows, endosmosis and exosmosis, which, depending on the qualities of the membrane and also on the composition of the solutions, bear a certain quantitative relation to each other.

Twelve years later Wilhelm Hofmeister (88) stressed that in many cases only an exchange of water between the protoplast and the surrounding medium is possible. He thus realized that, toward many solutions, protoplasts behave like semipermeable membranes. (The actual word "semipermeable," however, was only coined several years later by van't Hoff.)

Hugo de Vries found in 1871 that the protoplasts of the beet (*Beta vulgaris*) root, when plasmolyzed in a sodium chloride solution, for instance, will maintain their volume unaltered even for weeks. From this he concluded that these protoplasts must be considered practically impermeable to sodium chloride, for, if the protoplasts were even faintly permeable to the salt, during the long period of this experiment it would penetrate the protoplasts to an appreciable extent and thus raise their osmotic value. But this would have resulted in an increase of their volume, for exosmosis of sugar, which is the principal solute in the cell sap, could not be detected.

Even somewhat earlier, in the middle of the 1860's, Moritz Traube (162), a German-Jewish tradesman who in his spare time made several far-reaching scientific discoveries, had produced the first artificial semipermeable membranes and made preliminary studies on their properties. Wilhelm Pfeffer made a more thorough investigation of one of these, the copper ferrocyanide membrane. His studies, together with an abundance of new fundamental facts and ideas, were published in his book "Osmotische Unteruchungen," which appeared in 1877. Among other things, he pointed out that the high resistance of the protoplasts to diffusing substances is principally due to the extremely thin plasma membranes (*Plasmahaut*), the one surrounding the whole protoplast, the other situated at the interface between the protoplasm and the cell sap. In the same book (130), Pfeffer offered a general theory of the penetration of water and solutes through membranes of limited, or selective, permeability. He stressed that permeation does not depend only, as Traube had assumed, on the respective dimensions of the membrane pores and of the permeating molecules. Another important factor, according to Pfeffer, is the affinity of the diffusing substance for the membrane, i.e., its ability to dissolve in, or react reversibly with the membrane material. Moreover, if the membrane is provided with water-filled pores, the diffusion through these pores will also be influenced by the interfacial forces acting at the water-membrane interface. A permeability theory worked out eighty years ago was bound to be somewhat speculative; nevertheless we must admire the comprehensiveness of this early theory. It was, in fact, so general that most, if not all, of the permeability theories proposed later may be regarded, as special cases of the all-embracing theory of Pfeffer.

As early as the 1870's and 1880's observations showed that water is not the only substance which penetrates the living protoplasts readily. Thus de Vries, in his paper of 1871, showed that the protoplasts of the common beet, in a quite intact state, are rapidly penetrated by ammonia, as shown by the color change of the cells when they are immersed in very dilute ammoniacal solutions. In 1886, Pfeffer (131) observed that several basic dyes are also taken up rapidly by intact plant protoplasts and even accumulate in very high concentrations in the sap of many cells. He also found that retarding the vital activity of the cells, by low temperatures or by narcotics, had little effect upon the uptake of these dyes. Pfeffer therefore concluded, and rightly so, that the uptake of these dyes does not involve the intervention of an active transport mechanism. In the following year, Klebs (104) observed that if *Zygnema* sp. cells are put into a 10% glycerol solution, they will first undergo plasmolysis and then gradually recover. Obviously then glycerol penetrates the protoplasts with considerable rapidity, although by no means as rapidly as water. The analogous behavior of the epidermal cells of *Rhoeo discolor* toward glycerol and urea solutions was reported by de Vries in 1888 (170).

In spite of such scattered observations on the degree of permeability of plant protoplasts toward various solutes, it is scarcely an exaggeration to state that it was with the investigations of E. Overton from about 1890 onward that the permeability problem entered a new phase. Overton was the first to study cell permeability systematically and the first to state, explicitly, the nature of cell permeability and to distinguish this from the permeability properties of inanimate membranes.

Charles Ernest Overton, a distant relation of Charles Darwin, had been educated in Switzerland where, in the nineties he had a position as *Privatdozent* of biology at the University of Zurich. Preoccupied with investigations concerning the mechanism of heredity, he intended to study the influence of ethyl alcohol on the sex cells. First he sought to ascertain whether this substance can enter these cells. The literature contained no data on this point, and so Overton was compelled to decide the question experimentally and found that the protoplasts are about as permeable to alcohol as to water. This observation led Overton to study the permeability of the protoplasts toward other substances. Having discovered that the permeability of protoplasm to different solutes varies between very wide limits, Overton was gradually led to investigate what was the cause of this selective, or differential, permeability. With great tenacity Overton devoted the rest of his life to the elucidation of this problem and performed experiments with a very great variety of both animal and plant cells, testing as many different substances as possible. By 1899 he had made a great many experiments with more than 500 different substances.

Overton was obviously deeply impressed by the great uniformity with regard to permeability shown by cells which are very different in other respects; e.g., muscles and erythrocytes, on the one hand, root hairs and algal filaments, on the other. He found that the introduction of certain radicals, such as the hydroxyl, the carboxyl, or the amino group into a permeating molecule always greatly decreased its permeation power. Conversely, a lengthening of the carbon chain or an esterification of its carboxyl or hydroxyl groups enhanced the penetration power of a molecule. But how were these experimental results to be explained? In a paper read before the Naturforschende Gesellschaft in Zurich in 1898 and published the next year, Overton pointed out that there is a striking parallelism between the permeating power of different substances and their solubility in fats and fatlike solvents. However, it is not the absolute solubility in fats that is the decisive factor, but the relative solubility, i.e., the distribution, or partition, coefficient in the system fat-water. The smaller this coefficient, the more difficult is the passage of the substance through the protoplast. To explain these quite unexpected observations, Overton adopted Pfeffer's view that the diffusion resistance of the protoplast is, for the most part, located in the extremely thin plasma membranes. Overton now postulated that these membranes were impregnated with fatlike substances, or lipids.*

This was the fundamental theme of Overton's much discussed lipoidal theory of cell permeability, which immediately made the question of permeability a central problem in cell physiology. Numerous botanists, zoologists, medical men, pharmacologists, and chemists were attracted to the problem, but the progress in this field was not commensurate with the interest it aroused. On the contrary: during the first decades after the appearance of the lipoidal theory the discussion around it was characterized by a degree of confusion and passion rarely met in other fields of science. Many things contributed to this state of controversy.

In the first place, Overton unfortunately never published his experimental results in full, nor did he reply to the attacks on his theory. He was an extremely careful experimenter, but his magnum opus, intended to give a complete picture of the lipoidal theory and its experimental basis, was never completed, and only fragments of it were published (121–125). This made it difficult for other investigators to appreciate fully the firm foundation on which his views were based.

Furthermore, many of Overton's opponents failed to observe a fundamental distinction already made by Pfeffer and stressed again by Overton. From the very first Overton had emphasized that the interchange of substances between the cell and its surroundings comprises processes of two quite different kinds. According to Overton, this inter-

---

* Fats, the higher fatty acids, phosphatides, and sterols were formerly lumped together under the general name of lipoids, and this designation is often used even today, especially in the European literature. This group of substances is characterized by insolubility, or sparing solubility, in water and great solubility in "fat solvents," e.g., ether or chloroform. In accordance with the newer custom we shall here designate these substances as lipids.

change is partly regulated by the more or less passive resistance to diffusion of substances through the protoplasm. However, another part of the relations between the cell and its environment consists of an active transport of substances even in the reverse direction to that in which, from a purely physical standpoint, they would be expected to move. Thus fresh-water algae, for instance, take up salts from a very dilute solution, their natural living medium, and accumulate them in their cell sap. Such transport against the concentration gradient constitutes a certain amount of work done at the expense of energy released by the metabolic processes in the cell. Obviously the mechanism of such an active transport process must be fundamentally different from the mechanism of simple diffusion processes. In order to underline this difference, Overton coined a special term for this active transport, designating it as the adenoid (i.e., glandlike) activity of the cells. He stressed that the lipoidal theory aims solely at the elucidation of the passive permeability of the protoplasts, while their adenoid activity represents quite another problem.

Nevertheless, as already mentioned, many of Overton's opponents failed to discern, or at least to take into account, this fundamental distinction. Thus some of them pointed out that lipid-insoluble substances (e.g., sugars, amino acids, mineral salts) must also be taken up by the cells, and they thought that this fact alone would be enough to overthrow the lipoidal theory. Other writers looked upon the concept of adenoid activity as the manifestation of an obscure, unsound vitalism, and some even suspected that the adenoid activity doctrine had been invented as a mere screen to hide the shortcomings of the lipoidal theory. Today, however, there can no longer be any doubt concerning the general occurrence and great importance of active transport processes in plant cells. In fact, the concept of active transport constitutes an indispensable complement, not only to the lipoidal theory, but to every conceivable permeability theory in the proper sense of this word.

At all events, the lipoidal theory of cell permeability had to face a long-contested resistance before its validity, even in main outline, was generally recognized. Meanwhile several other permeability theories were energetically advocated. It will suffice, however, to mention here only three of these hypotheses.

(1) I. Traube, in a long series of papers, enthusiastically defended his "*Haftdruck*" or "retention pressure" theory, according to which the permeation of solutes depends on how great is their *Haftdruck* to the membrane substance as compared with that to water. Unfortunately, however, Traube was never able to make a clear statement of the true meaning of the word *Haftdruck*. Today, therefore, this theory is almost

forgotten. However, if we try to interpret it as benevolently as possible, we may perhaps take *Haftdruck* as a rough equivalent of "affinity" or "intermolecular attraction." Interpreted in this way, the theory of Traube is not very far from some modern views concerning the permeation mechanism.

(2) In order to account for the permeation of water, mineral salts, and other more or less lipid-insoluble substances, Nathansohn (115) postulated a plasma membrane composed of a mosaic of both lipids and proteins. On this view, lipid-soluble substances would enter the cell through the lipoidal parts of the membrane, while the lipid-insoluble substances would take the other route. This hypothesis explains why lipid-soluble substances are always able to penetrate the plasma membrane, while the uptake of others may vary considerably from time to time according to hypothetical changes in the protein part of the membrane. The objection may be raised to Nathansohn's theory that it is very speculative in nature. Besides, we now know that the uptake of mineral salts, for instance, is principally due to active transport, not to permeability in the strict sense of the word. Nevertheless the mosaic hypothesis, when accorded the best interpretation, may also perhaps be found to be not so very far from the truth.

(3) The ultrafiltration theory of W. Ruhland aroused still more discussion. In its original version (143, 144) it maintained that the permeation power of colloids depends on their state of dispersion: the greater their dispersion, the more readily they would, according to Ruhland, permeate into living cells. It will be noted that the ultrafiltration theory, in this form, said nothing about the principal point of the permeability problem, namely, how it comes about that living protoplasts are more or less impermeable to numerous noncolloidal substances. Moreover, if the permeation of highly dispersed colloids was due to their state of high dispersion, then it would be natural to conclude that all noncolloidal solutes, owing to their still greater dispersion, would have even more permeation power. Of course, such a conclusion was too absurd even in those days of confused thought in the field of cell permeability. It only shows that the ultrafiltration theory could scarcely be taken seriously before, in 1925, it appeared in an essentially new version.

The "new" ultrafiltration theory (146) was based on experiments carried out with the great, *Oscillatoria*-like "sulfur bacterium," *Beggiatoa mirabilis*. According to Ruhland and Hoffmann, its permeability toward nonelectrolytes and alkaloid salts was almost entirely dependent on the molecular volume of the permeating substance. Moreover, Ruhland and Hoffmann suggested that although **their** experiments had been made with a single organism, and indeed with a rather unusual one, the ultrafiltration theory would, nevertheless, prove applicable in its essentials to all living protoplasts.

The first impression was, of course, that the ultrafiltration theory differs so radically from the lipoidal theory that they would be mutually exclusive. But, as we shall see, this is not so: for it seems now that molecular size cannot be entirely neglected in a comprehensive understanding of the permeation processes.

From about 1925 it has been increasingly clear that qualitative observations on permeability are not enough, but that quantitative measurements, as exact as possible, are needed. Numerous such investigations have, in fact, been carried out during the last thirty years, and as a consequence the confusion that had characterized the field of cell permeability has begun to dissipate, enabling established facts and assumptions to be distinguished, although by no means all differences of opinion have been eliminated.

In what follows we will attempt, primarily, to find out which points concerning cell permeability have so far been established with reasonable certainty, and after this has been done, we will try to draw deductions from these facts. First of all, however, let us take a look at the active transport processes accomplished by plant protoplasts.

III. Active Transport Processes[2]

The occurrence of specific active transport processes brought about by the energy released in cell metabolism was recognized by such pioneers of cytophysiology as Pfeffer (132) and Overton (121). However, most of their successors had little appreciation of the fundamental significance of these phenomena. It is true that Höber, in his "Physikalische Chemie der Zelle und der Gewebe" (73), which appeared in numerous editions from 1902 onward, always gave considerable weight to the idea of active transport, but the very term "physiological permeability" used by him was not likely to make clear the radical difference between active transport and simple permeation. And so it happened that when, some thirty years ago, Hoagland, Steward, Lundegårdh, and others again stressed the importance of active transport, their views had, to a large extent, the charm of novelty. It is only during the last ten or fifteen years, however, that the central role of active transport has been universally recognized, so that this phenomenon has at last received the attention it deserves.

[2] For further information concerning active transport, the reader is referred to Volume VIII of the "Symposia of the Society for Experimental Biology" (159), which is in its entirety devoted to this problem.

A. PERMEABILITY AND ACTIVE TRANSPORT: TERMINOLOGY

In the preceding historical sketch, it was pointed out that many disputes concerning the "permeability" of living cells have arisen because this word has been used in quite different ways. Even today misunderstandings in this field may arise from a variable, ill-defined terminology.

Used in its widest sense the term "permeability" embraces the total interchange of substances between the cell and its surroundings. Permeation, taken in this sense, includes both active transport processes and diffusion. On the other hand, when used with their restricted meaning, the words permeation and permeability refer solely to those transfer processes in which the protoplast plays the passive role of a mere resistance to be overcome by the substance as it leaves or enters the cell. Permeability in this latter sense thus equals $1/R$, where R denotes the diffusion resistance of the cell.

Now, which terminology is to be preferred? Historically, the answer to this question is clear, for all the classical authors in this field, with Pfeffer and Overton leading the way, have used the words permeation, permeability, etc., in their restricted sense, and fortunately, it is more practical to retain this custom. For, if we should now decide to use these terms in a wider sense, we would be obliged to invent a new word for permeability in the narrower sense. On the other hand, permeability in its widest sense is synonymous with other terms such as transfer, exchange, uptake, absorption, flux, etc. Therefore, the present writer restricts the term permeability to its original meaning and speaks of the permeability of protoplasm in just the same sense as physicists and engineers speak of the permeability of inanimate membranes.

For active transport, several other names have also been used. One of the oldest is the term "adenoid activity" (121). In recent years the terms dynamic transfer, active secretion, metabolic, or nonosmotic transport have all come into prominence. Whereas nonosmotic transport should properly relate to the movement of water, it is largely a matter of taste which other term relates to the movement of solutes.

B. THE EXPERIMENTAL DISTINCTION BETWEEN PERMEATION AND ACTIVE TRANSPORT

Permeation processes always occur with the activity gradient or, in the case of ions, with the gradient of electrochemical potential. Hence in most cases permeation, like diffusion in general, tends to smooth out and finally to abolish any existing concentration differences.

On the other hand, neither the spontaneous, random thermal movements of the molecules nor the migration of ions under the influence of electrical potential gradients suffice for the active transport processes. These processes imply the intervention of mechanisms by which energy set free in cell metabolism is made available for transport, which may even create great concentration differences between the cell sap and the external medium.

It is not always easy, however, to decide whether a particular transfer process is to be regarded as metabolic or passive. The mere fact that some substance seems to be accumulated in the sap does not necessarily imply that the process is "metabolic" in the strict sense. Thus basic dyes and other weak bases are often apparently accumulated, even very efficiently, in cell saps, and yet the process may be regarded as passive, rather than active, because the entering solute is bound in the cell [see Bogen (15) who uses the term metaosmotic in these cases]. Moreover, the well-known Donnan equilibriums, also occurring in inanimate systems, result in an uneven distribution of diffusible ions on the two sides of a membrane, but nevertheless they involve no direct metabolic activity.

The following criteria may be used to decide whether a given process is to be considered as passive or active, i.e., as a case of "permeation" or "active transport."

(1) The rate of a permeation process is, in most cases, directly proportional to the concentration gradient, but in the case of active transport this is not so.

(2) Chemically similar substances often mutually depress the active uptake of one another, presumably because they compete for one and the same transport mechanism. In the case of simple permeation such competitions scarcely occur.

(3) As we shall see, the permeation power of different substances is correlated with their lipid solubility and molecular size. Now, if it is found that the penetration rates of two substances of about the same solubility and molecular size (e.g., two optical isomers) differ markedly, this is a strong indication that the uptake of at least one of them is not due to permeation alone.

(4) Elimination of free oxygen should, in aerobic organisms, depress or even prevent active transport. Permeability, on the other hand, seems scarcely to be affected by the absence of oxygen, except in so far as the membrane may be structurally affected thereby.

(5) Active transport may be reversibly reduced by narcosis, while the effects of narcotics on permeation processes are more complex.

(6) Substances known to inhibit certain enzymes, such as hydrocyanic acid, carbon monoxide, sodium azide, dinitrophenol, iodoacetate, and fluoride, have been successfully used to show that particular en-

zymes are involved, directly or indirectly, in certain absorption processes. A pronounced effect of enzyme inhibitors on permeation processes, although conceivable, is not very probable.

Thus there are numerous more or less suitable criteria which discriminate between passive permeability and active transport. While no criterion suffices alone, if used in combination the criteria will, in most cases, be conclusive.

There are, however, certain rather difficult cases. One of the most puzzling is that described as "facilitated diffusion" by Danielli (44). Facilitated diffusion occurs, as with simple diffusion, under the driving force of thermal agitation, and the equilibrium reached is the same as that achieved by simple permeation. However, facilitated diffusion occurs over a limited fraction of the cell surface (at the so-called active patches) only and at any one site is, like an enzymatic process, restricted by both structural and steric factors, so that only particular molecular species are concerned. The occurrence of facilitated diffusion in plant protoplasts is, however, still hypothetical.

C. ACTIVE TRANSPORT OF DIFFERENT SUBSTANCES

Active transport is most obvious in the case of strong electrolytes, i.e., in the case of ions. Active transport of ions is of such fundamental importance to the plant and is, at the same time, such a complicated phenomenon that it is treated in a special chapter of this treatise (see Chapter 4). In the present chapter we will concern ourselves mainly with the transport of essentially nonionized substances and of amino acids which, although strongly ionized, bear both positive and negative electric charges and are thus effectively electrically neutral (zwitter ions). Similarly, since activated movements of water are considered in Chapter 2 of this volume these will not be considered here.

1. Sugars

For a long time it has been surmised that sugars penetrate cells so slowly that they must resort to specific means to meet their requirements for these important nutrients.

Yeast cells afford an especially attractive object for studies on this problem, since they seem to be specialized for rapid uptake of sugar. It has been shown by Conway and Downey (40) and Rothstein (141), among others, that yeast cells are more or less impermeable to lactose, galactose, sorbose, and arabinose, while they will readily take up glucose, mannose, and fructose. Their ability to take up certain sugars but to reject others of similar lipid solubility and molecular size strongly suggests the existence of a special transport system for the sugars that are absorbed. The mechanism has been investigated especially by Rothstein and his co-workers (141). The starting point was the observation that uranyl ions (UO_2^{++}) even in low concentrations strongly inhibit glucose fermentation. Moreover, it could be shown that the uranyl ions do not enter the interior of the yeast cells but are bound to specific sites at the cell surface. Furthermore, uranyl was found to be an inhibitor of reactions specific to the uptake of hexoses, while it is without effect on all the metabolic reactions involved in the respiration and fermentation of other substrates. The action of uranyl must thus "be confined specifically to those reactions occurring at the cell surface which introduce sugars into the metabolic machine, without any effect on the integrity of the machine itself, which is presumably located inside the cell" (141). Probably it acts by chelating with adenosine triphosphate (or other phosphorylating agent), and thus prevents the phosphorylation of the sugar which normally proceeds at the cell surface in connection with active sugar uptake.

In the cells of higher plants, also an uptake of sugars is often readily observed when, for instance, pieces of leaf are floated on sugar solutions (175). The slowing down of the process by anaerobic conditions indicates that it occurs predominantly by active transport, although a slow sugar uptake by diffusion may occur simultaneously. A still more striking example of active sugar transfer is the copious secretion of sugars in nectaries. Less striking, but physiologically highly important, is the accumulation of sucrose in the sieve tubes of photosynthesizing leaves, an indispensable first step in the removal of assimilates from the leaves (176). Owing to technical difficulties, the mechanism of active sugar transfer has not yet been as thoroughly analyzed in higher plants as in yeasts, but a prevalent idea is that it depends on phosphorylation and dephosphorylation processes, while Gauch and Dugger (63) have also suggested that the sugar transport may depend on the formation of a boric acid-containing complex.

2. Amino Acids

Arisz and his collaborators (2) have studied the uptake and transport of amino acids and amino acid amides by two plant organs, namely the leaf parenchyma of *Vallisneria spiralis* and the leaf tentacles of *Drosera capensis*. In both cases absorption was found to be of the "active" type, as shown by its dependence on the oxygen supply and its inhibition by several enzyme inhibitors.

Another important series of investigations into the uptake of amino acids has been carried out by Gale and co-workers (62). They found that it cells of *Staphylococcus aureus* which are free of glutamic acid are suspended in a solution of glutamic acid, there is no accumulation of this substance within the cells as long as precautions are taken to exclude metabolic sources of energy. If, however, an energy-yielding substrate such as glucose is added, then rapid accumulation of glutamic acid takes place within the cells and the internal concentration may rise to as much as 400 times that of the external medium. The accumulation is abolished by any inhibitor which prevents the metabolism of glucose. Moreover, the rate of accumulation is independent of external concentration except for very low values of the latter. Finally, if cells rich in glutamic acid are washed and resuspended in water, there is only a very slow loss of glutamic acid to the external medium. There is thus a strong indication that the glutamic acid accumulation taking place in these cells is the result of active transport (cf. 112a).

3. Other Substances

For some twenty years it has been known that diatoms, when plasmolyzed in solutions of sugars, polyhydric alcohols, amides, etc., will deplasmolyze with much greater rapidity than other plant cells in the same situation. This has so far been interpreted as an indication of an exceptionally high permeability of the diatom protoplasts. However, according to Bogen and Follmann (19), the recovery from plasmolysis is largely prevented by enzyme inhibitors. This suggests that the normal deplasmolysis of the diatoms is to a corresponding extent due to active uptake of the solutes tested. A similar, although less striking, influence of metabolic inhibitors has been reported by the school of Bogen in other kinds of cells also. Bogen (16) therefore concludes that active transport is of much wider occurrence than has hitherto been assumed and that it occurs with a great variety of substances.

It seems that these claims will necessitate a critical re-examination of the question of the relative importance of active transport, on the one hand, and simple permeation, on the other. Until this has been done, the present writer is inclined to hold to the classical view that only slowly permeating substances are to any considerable extent subject to active transport. At any rate, it is obvious that the greater the permeability to a given substance, the greater also is the risk that this substance will leak out from the cell by diffusion and so nullify the results of the active accumulation process (cf. 85b).

IV. Diffusion in Inanimate Systems

It is not impossible that the penetration of water through some protoplasts is, to a certain extent, a filtration process, that is, a bulk flow of water through submicroscopic pores (cf. 128, 166). Mostly, however, permeation through plasma membranes is due predominantly or even solely, to diffusion.

To understand the diffusion of substances across plasma membranes one should examine what is known about diffusion in general and especially diffusion through artificial membranes which are better known as to structure and chemical composition than the plasma membranes.

A. DIFFUSION IN HOMOGENEOUS MEDIA

Diffusion is due to the spontaneous thermal movement of molecules and ions. Random movement of the particles leads ultimately to an even distribution within a single phase.

In 1855, Fick proposed the equation.

$$dS = -Da\frac{dv}{dx}dt$$

where dS is the quantity of substance which in the time dt passes across an area a in which dc/dx is the concentration gradient (i.e., the change in concentration with distance). D is a constant of proportionality and is called the diffusion coefficient. It gives the amount of substance diffusing across unit cross section in unit time when the concentration gradient is unity. The negative sign appears because diffusion takes place from a region of higher to one of lower concentration. The diffusion coefficient has the dimensions: area divided by time; it may thus be expressed, e.g., as square centimeters per second. A very exhaustive exposition of the application of Fick's law to cell physiological problems is that of Jacobs (96).

The Fick equation applies just as well to liquid and solid systems as to gaseous ones. In other respects, however, there are considerable differences between diffusion processes in these systems. Thus, in gaseous systems the molecules are subject to only small intermolecular forces and are therefore always free to move. In liquids and solids, on the other hand, there are considerable forces holding the molecules together in a more or less rigid framework. In such systems, therefore, a molecule can only diffuse if it has sufficient kinetic energy to overcome the forces holding it to its neighbors and if in addition it has sufficient energy to push other molecules out of the way. Each molecule will thus mostly vibrate about a mean position and only occasionally jump to a new one. The energy necessary for such a jump is called the "activation energy of the diffusion." Its magnitude will be the greater the stronger the bonds between the diffusing molecule and those of the medium and also the stronger the attraction of the molecules of the medium for each other, that is, the greater the viscosity of the medium. Thus, if the attraction is due to comparatively weak van der Waals' forces the diffusion will, other things being equal, be more rapid than when strong hydrogen bonds are involved. According to Danielli (42), the magnitude of $DM^{1/2}$ (where D is the diffusion coefficient and M the molecular weight of the diffusing substance) will be approximately constant for diffusion processes in aqueous solutions and in media of still lower viscosity, but this will not be the case in more viscous liquids nor in solids. On the other hand, the diffusion in gels is by no means as slow as one might be inclined to assume from the stiffness of the gel. This is due to the fact that the gel is traversed in all directions by innumerable water-filled channels in which most substances are able to diffuse almost as in ordinary water. Only when the diameter of the diffusing particles approaches the diameter of the channels is the diffusion rate considerably decreased.

B. DIFFUSION THROUGH MEMBRANES

Reference here may be made to the excellent monographs of Höber (75), Dean (48), and Pappenheimer (128).

In attempting to explain how a substance can penetrate a membrane by diffusion, we may have recourse to two quite different principles. If the membrane consists of a homogeneous layer of a water-immiscible substance, its permeability toward different substances is principally dependent on the extent to which these substances are soluble in the membrane substance. If, however, the membrane has a sievelike structure, similar to a filter, diffusion will take place, either exclusively or principally, across the channels which join the aqueous phases on the two sides of the membrane. Cases intermediate between these two also occur.

1. Homogeneous, Solventlike Membranes

If we have a membrane as a separate phase between two aqueous solutions, we may picture the permeation of a substance through this homogeneous membrane as composed of three successive steps: the permeator (the penetrating substance) has first to dissolve in the membrane substance on one side of the membrane, then to diffuse through the membrane to the other side, and, finally, to pass from the membrane phase into the second aqueous phase. The first process— the distribution of the permeator between the aqueous phase and the membrane phase—is obviously dependent on the magnitude of its distribution coefficient, k, defined by the equation

$$k = \frac{\text{conc. in membrane phase}}{\text{conc. in water}}$$

The rate of the second process—the diffusion across the membrane—is primarily dependent on the absolute difference in concentration of the permeator inside the two boundaries of the membrane. Substances having a distribution coefficient smaller than unity will therefore diffuse across the membrane in accordance with their distribution coefficient.*

Under many conditions the diffusion rate is thus almost directly proportional to the distribution coefficient of the permeator. Admittedly, the diffusion rate is also dependent on the viscosity of the membrane substance and on the molecular size of the permeator. However, even with great differences in molecular size, diffusion rates do not differ very much, the diffusion coefficients being, as we have seen, more or less inversely proportional to the square root of the molecular weight. On the other hand, the distribution coefficients of different substances often differ enormously. Their magnitude may therefore mostly be considered as the deciding factor.

A homogeneous membrane may, however, also be regarded as a "potential energy barrier" (42). This means that before a permeating molecule can enter the membrane phase from an aqueous solution it must acquire sufficient kinetic energy both to overcome the forces holding it to the water molecules and also to make a gap in the nonaqueous layer. Now, the more hydrophilic the permeator is—that is, the more firmly its molecules are attached to the water molecules—the greater, of course, is the kinetic energy required to bring about detachment from the aqueous phase. And since a large amount of kinetic energy is seldom available, the process of transition will be the slower the more hydrophilic is the permeator. When the permeating molecule has crossed the nonaqueous layer, it finally has to diffuse across the oil-water interface into the water. For hydrophilic molecules, however, this is easy.

It will be noted that the concept of the membrane as a potential energy barrier is by no means in conflict with the picture of the permeation process previously given. They are two different, but

* At this point for the sake of clarity, a warning to the reader is perhaps required. Although it was said above that substances having a low distribution coefficient will diffuse comparatively slowly across the membrane, this does not imply that the movements of the single permeating molecules are retarded but only that the number of these molecules in the membrane is reduced and that, in consequence of this, the quantity of the substance passing across the membrane in unit time is decreased. Perhaps it would therefore be better to speak of a decrease in the flux instead of a decrease in the diffusion rate.

equally justified, descriptions of one and the same reality—the one in the language of thermodynamics or the phase equilibrium theory, the other in that of molecular kinetics.

Strong electrolytes are very sparingly soluble in water-immiscible solvents. Homogeneous membranes consisting of such solvents are therefore almost impermeable to strong electrolytes. Nevertheless, aqueous solutions of strong electrolytes in contact with a homogeneous oil membrane which contains acidic or basic components, often give rise to electrical-potential differences. Thus, if we have a guaiacol membrane in contact with a potassium chloride solution, the potassium ion forms potassium guaiacolate and can diffuse across the guaiacol membrane as such, while the chloride ion and the hydrochloric acid formed by hydrolysis are practically insoluble and cannot diffuse in the oil layer. The guaiacol therefore behaves as a membrane permeable only to cations. Aniline, on the other hand, behaves as a membrane permeable only to anions, since chloride ions are soluble in aniline as aniline hydrochloride, and potassium ions and potassium hydroxide are not soluble in aniline. Since potential measurements are easy to make, such membrane potentials have been studied by numerous investigators. Dean (48) gives a good survey of the main results obtained.

2. Sievelike Membranes

Membranes exhibiting nearly all conceivable gradations of permeability may easily be prepared by pouring a suitable amount of a collodion solution on some suitable surface, such as glass or mercury, and allowing the solvent (alcohol-ether) to evaporate more or less completely. The permeability of the resulting film will be governed by the amount of solvent still present within it when it is immersed in water. Thus, if much solvent is present, the membrane will allow the passage of almost all kinds of molecules and ions, and perhaps even of some colloid particles. However, when the solvent is allowed to evaporate completely, a typical molecule-sieve, i.e., a membrane permeable only to the smallest molecules, results.

Table I contains some results obtained by Fujita (61) with two collodion membranes, the one (a) immersed in water some time before complete evaporation of the solvent and thus provided with relatively wide pores, the other (b) completely dried and therefore containing only very narrow pores. The permeability to urea is in both cases taken as equal to 1. As seen from the table, the permeation rates of different solutes are, in both cases, distinctly correlated with their molecular weights, but in a very different manner. The membrane with the wider pores is only slightly less permeable to the larger than to the smaller molecules. In fact, the expression $PM^{1/2}$ (where P is the apparent diffusion coefficient in the membrane) has in this case a fairly constant value, just as has the expression $DM^{1/2}$ in the case of so-called free diffusion. (Diffusion in, for instance, water is often called free diffusion, although, strictly speaking, this expression is correct only when applied to those gaseous systems in which intermolecular attraction is negligible.) On the other hand, in the experiments with the other membrane (b) the differences between the different solutes are much

TABLE I

PERMEABILITY (P) OF TWO COLLODION MEMBRANES TO NONELECTROLYTES[a,b]

Substances	M^c	Relative permeability (a)	(b)	$PM^{1/2}$ (a)	$PM^{1/2}$ (b)
Methyl alcohol	32	1.22	9.24	6.9	52.4
Acetone	58	1.11	7.08	8.5	53.9
Formamide	45	1.06	4.11	7.1	27.6
Ethyl alcohol	46	1.15	2.98	7.8	20.2
Propyl alcohol	60	1.00	1.03	7.7	8.0
Urea	60	1.00	1.00	7.7	7.7
Butyl alcohol	74	0.85	0.82	7.3	7.1
Ethylene glycol	62	0.80	0.27	6.3	2.1
Glycerol	92	0.81	0.22	7.7	2.1
Chloral hydrate	165	0.81	0.11	10.4	1.4
α-Monochlorohydrin	110	0.70	0.07	7.3	0.7
Glucose	180	0.54	0.00	7.2	0.0

[a]Fujita (61).
[b] One of the membranes (a) is provided with relatively wide pores, the other (b) with relatively narrow ones.
[c] M = molecular weight.

greater and the expression $PM^{1/2}$ is not even nearly constant but decreases rapidly with increasing molecular weight. The principal reason for this is probably the fact that the membrane (b) contains an assortment of pores of varying diameters and that only the widest of these pores will let the larger solute molecules through, while many more pores can be utilized by smaller molecules (cf. 75, 128).

On closer scrutiny of the results of Fujita, it will be found that, besides the molecular size, one or two other factors may also have influenced the permeation rates, especially in the case of membranes with narrow pores. Thus, acetone and butyl alcohol are seen to have penetrated the membrane (b) distinctly more rapidly than one would have expected from their molecular size, while formamide, urea, and ethylene glycol penetrate it more slowly. Now, the two first-named substances are decidedly surface-active and rather hydrophobic, while formamide and urea are surface-inactive and strongly hydrophilic. It is therefore probable that permeation through the narrow-pored collodion membrane is also influenced by adsorption and/or solubility phenomena which will favor the penetration of surface-active, hydrophobic solutes.

The classical copper ferrocyanide membrane is probably a more typical molecular sieve, for its permeability toward surface-active, hydrophobic substances seems not to be greater than toward surface-inactive ones.

Very interesting observations have been made on the penetration of nonelectrolytes into crystalline zeolites after the water normally present in the interstices within the crystals had been removed (9). It was found that zeolites possessing a fairly open kind of framework take up n-paraffins but not isoparaffins. Those of a slightly higher density do not take up either n- nor isoparaffins, except methane and ethane. Finally, the crystals with the narrowest interstices are only penetrated to a negligible extent by methane and ethane but take up molecular nitrogen and oxygen very rapidly. From these observations it can be concluded that the *shape* rather than the *volume* of the penetrating molecules exercises an influence upon the penetration power. The feature deciding the uptake or nonuptake is, above all, the cross section of the molecule, which is the same for all the n-paraffins, while the isoparaffin molecules have a somewhat greater cross section. Increasing the chain length does not materially influence the ultimate sorption equilibrium, but slows down the rate of intracrystalline diffusion. Such clear-cut results would scarcely have been possible if the interstices of a single zeolite crystal were not of very uniform diameter.

The permeability of sievelike membranes to ions has been extensively studied, especially by measuring the electrical potentials produced. Collodion membranes have proved especially useful in such studies (152). Slightly oxidized collodion (nitrocellulose) membranes contain negative charges arising from the ionization of scattered carboxyl groups situated in the pore walls. These fixed negative charges are combined with mobile ions of the opposite charge, say, sodium ions. Such a membrane will therefore be found permeable to sodium, and also to other, cations. At the same time it may be more or less impermeable to all anions, owing to the electrostatic repulsion exercised by the charges on the pore walls. The narrower the pores and the greater the density of the negative charges, of course, the more pronounced is this anion impermeability. The impermeability to anions makes the membranes impermeable to salts, too. However, the permeability of such a selectively cation-permeable membrane toward different cations varies considerably, depending on the dimensions of the ions in question. Thus, for instance, Michaelis (cf. 75) found the relative mobilities of some cations to be as shown.

	H	Rb	K	Na	Li
(a) In free solution	4.9	1.04	1.00	0.65	0.52
(b) Within a dried collodion membrane	42.5	2.8	1.00	0.14	0.048

These results are readily understandable when we remember that the hydration of the alkali cations decreases in the order $Li > Na > K > Rb$ and that the volume of the hydrated ions decreases in just the same order.

By impregnating collodion membranes with basic substances (alkaloids, basic dyes, basic proteins) they may be rendered electropositive. Such membranes are permeable to anions, but more or less impermeable to cations.

The selective ion permeability of the copper ferrocyanide membrane also depends upon the surface charge of the membrane and upon the charge, valency, and size of the diffusing ions (41). Thus the order of permeation power of some anions was found to be this: $Cl^- > Br^- > NO_3^- > I^- > IO_3^- > SO_4^{--} > oxalate^{--} > ferrocyanide^{---}$.

A detailed theory of the ion permeability of electrically charged membranes has been worked out by K. H. Meyer and Teorell (cf. 160).

3. Intermediate Cases

As already stated, even a "homogeneous" membrane, though operating principally as a selective solvent, will at the same time show just a slight trace of sieve action in so far as larger molecules will, other things being equal penetrate the membrane somewhat more slowly than smaller ones. On the other hand, sievelike membranes will seldom select the diffusing molecules solely according to their size: in most cases the result will also be more or less influenced by some kind of "affinity" between the permeating molecules and the membrane substance.

There are, however, other membranes which are still more clearly intermediate between the two main types so far discussed. Thus, by adding lipids to a solution of nitrocellulose in alcohol- ether, membranes may easily be prepared which combine the selective solvent properties of the lipid with the sieve action of the pure collodion membrane (174).

The artificial membranes so far mentioned have all been of a considerable thickness. The plasma membranes, on the other hand, are probably lipid-containing films only a few molecules thick. From the standpoint of the study of cell permeability, therefore, it appears to be of the greatest interest to gain some insight into the permeability properties of lipid films of the latter order of thickness. It is very regrettable that technical difficulties have so far largely prevented the study of the permeability properties of such membranes. As an example of the surprises possible in this field, it may be mentioned that it was only quite recently that the first reliable measurements of the resistance offered by fatty acid monolayers to the evaporation of water were carried out (1). According to this investigation such layers decrease evaporation by a factor of about 10^4, while in earlier measurements of this sort almost no resistance at all to the water evaporation had been found.

A film, a few molecules thick and consisting of arachinic acid and cadmium arachinate, has been found almost impermeable to ions but fairly permeable to H_2S and I_2 molecules (11). From the standpoint of cell physiology it is of the utmost importance that such investigations should be performed, quantitatively, with different types of films a few molecules thick and with a greater variety of permeators. To this end the films should, if possible, be spread between two aqueous solutions, though this would be extremely difficult.

V. Measurement and Quantitative Expression of Permeability and Active Transport

A. TERMINOLOGY

The terms permeation and permeability are often used synonymously, but this is not admissible. Permeation refers to a *process*, while permeability is a *property* of a membrane (cf. digestion—digestibility, etc.). This distinction is even clearer when we realize that the permeation rate always depends on two quite different factors, namely, (a) the permeability of the membrane and (b) the driving force of the process. Only too often the importance of the factor (b) has been underestimated or even neglected in relation to the factor (a). And yet it should be clear that each of them is essential to any actual permeation process.

Moreover, it is not desirable to speak of the permeability of a solute. The ability of a substance to penetrate a membrane is more properly called its permeation power.

B. Protoplasmic Permeability[3]

Imagine that a protoplast is put into a solution of a substance whose concentration at the surface of the protoplast is kept constant by continual mixing. Let us further assume that no active transport occurs and that the substance permeates relatively slowly through a thin surface layer of the protoplast, whilst the diffusion resistance in the interior of the protoplast is negligible. Obviously, then, the diffusion resistance of the surface layer alone controls the total permeation rate. By applying Fick's law we arrive at the equation

$$\frac{dS}{dt} = P \times A(C - c) \qquad (1)$$

where dS is the amount of substance entering the cell in time dt; A is the area of the cell surface; C, the equilibrium concentration of the permeator; and c, its concentration at any given time. P, the *permeability constant*, is a measure of the permeability of the protoplasmic surface layer. It may also be said to be a measure of the permeation power of the permeator in question. From this point of view it may be more logical to call it the *permeation constant*. Generally the equilibrium concentration (C) of the permeant in the cell sap is approximately equal to its constant concentration in the external solution. As far as our present knowledge goes, the exceptions to this rule are more apparent than real. Thus, in the case of basic dyes and other weak bases, which are often accumulated very effectively in the cell sap, it must be remembered that while the substance enters the cell as a free base, it is mostly not accumulated as such but is bound to an acidic constituent of the sap.

If V is the volume of the protoplast, $c = S/V$. If we assume that V and A remain constant during the permeation process, equation (1) can be integrated and becomes:

$$P = \frac{V}{At} \ln \frac{C}{C - c} \qquad (2)$$

The permeability constant, or permeation constant, P, has the dimensions length divided by time or, in the natural absolute CGS system, centimeters by seconds. The diffusion coefficient, on the other hand, as already stated, has the dimensions $l^2 \times t^{-1}$. The difference is

[3] For a more detailed mathematical treatment of the permeation processes, including methods of calculating permeability constants, see Rashevsky and Landahl (137), Jacobs (99), and especially the exhaustive treatise of Stadelmann (154).

due to the fact that the thickness of the diffusion-resisting surface layer is generally not known and must therefore be incorporated in the constant P.

In deriving equation (2) we have assumed, for the sake of simplicity, that the permeation rate is controlled solely by a thin surface of the protoplast. Now, strictly speaking, this assumption never holds true. First, because in most cases there is not one plasma membrane only, but two: one outer and one inner one. What is actually aimed at in the great majority of permeability experiments is thus to measure the sum of their diffusion resistances. Moreover, there is an unstirred layer of cell sap inside the protoplast and there is in most cases a cell wall outside it; it is self-evident that these must contribute more or less to the total resistance which the permeator has to overcome. The thicker and more impervious the cell wall, the bulkier the cell sap vacuole; and the smaller the diffusion resistance of the protoplast, the more, of course, will the experimentally found value of P differ from the true permeability of the protoplast.

There is, however, one means of correcting the experimentally measured P values to some extent, namely this: The total diffusion resistance of the cell is approximately equal to the sum of several successive resistance, viz., those of the cell wall, the plasma membranes, the mesoplasm, and the cell sap. Now, when the cell is killed, the diffusion resistance of the plasma membranes toward many substances is greatly reduced, while the other diffusion resistances will remain almost unaltered. The difference between the resistance of the cell in the living and dead state can therefore be used as an approximate measure of the resistance of the undamaged protoplasmic membranes:

$$R_{\text{plasma membranes}} \approx R_{\text{living cell}} - R_{\text{dead cell}}$$

or

$$\frac{1}{P_{\text{plasma membranes}}} \approx \frac{1}{P_{\text{living cell}}} - \frac{1}{P_{\text{dead cell}}}$$

Disregarding experimental errors, the permeability constants expressed in, say centimeters per second, are true measures of the protoplasmic permeability and thus can be directly compared.

For certain purposes, however, other measures of permeability may profitably be employed. Thus, the half-time of the permeation process, i.e., the time in which the initial concentration difference between the outer solution and the cell sap is reduced by half, may be found to serve as a measure of the rapidity of the permeation process. It should, however, be borne in mind that the half-times depend not only on the magnitude of the cell permeability but also on the dimensions of the cells. Thus, a *Micrococcus* cell only 1 μ in diameter will display half-saturation times 10,000 times smaller than a *Valonia* cell measuring 1 cm in diameter, provided the permeability constants are the same in the two cases. As a matter of fact, the great rapidity with which many interchanges occur between most plant cells and their environment is determined, to a greater extent than is often realized, by the very great relative surface of the majority of cells.

Finally, the water permeability may be expressed by a special water permeability constant indicating the amount of water that will pass across unit cell surface area in unit time per atmosphere of pressure difference. Such water permeability constants are not, of course, comparable with the permeation constants of solutes.

In the case of active transport processes there is hardly any other measure of their intensity than the flux, i.e., the amount of substance taken up in unit time per unit area of the cell surface.

C. EXPERIMENTAL METHODS OF MEASURING PERMEABILITY AND
   ACTIVE TRANSPORT

There exist a very great number of methods which have been used to determine the transfer of substances across plant protoplasts and also to measure their permeability. [For particulars and literature references, see (34).] But regrettably most of these methods are subject to sources of error which in most cases will render the results of such measurements more or less inaccurate and unreliable. One such quite general source of error has already been referred to, viz., the fact that the diffusion resistance of the cell wall and of the cell sap cannot be readily distinguished from that of the protoplasm, which thus seems greater than it really is. Especially in the case of rapidly penetrating substances such as, say, water, the errors caused in this way may be very great. Another point of general importance is the danger that the cells studied may have been damaged in some way so that their permeability has been pathologically increased or their power of active transport reduced. The gentlest possible handling of the cells is therefore necessary in carrying out such measurements.

The methods to be discussed will be grouped under five main headings: (1) Methods Involving Visible Changes within the Cells, (2) Analytical Methods, (3) Osmotic Methods, (4) Electrical Methods, and (5) Methods Based on Physiological Toxicological Effects.

1. Methods Involving Visible Changes within the Cells

a. Vital staining. Investigations into the permeability of living cells to dyes may seem very promising, for the penetration of dyes is often directly observable under the microscope. No wonder, therefore, that vital staining experiments have gained great popularity. In reality, however, as we shall see later on, the estimation of cell permeability on the basis of such experiments is far from easy. It is therefore really a pity that so many permeability studies have been based on this particular method.

b. *Color changes of indicators.* Acids and bases may produce color changes within cells containing either natural indicators (anthocyanins) or artificial ones (e.g., neutral red). It is, however, not easy to obtain quantitative results in this way, owing to the extent to which the cell sap may be buffered.

c. *Precipitation method.* Caffeine and other alkaloids, even in very low concentrations, cause precipitates within cell saps containing tannins. Up to now this method of studying permeability has only been used in a semiquantitative way.

All the above methods involve mechanisms of storage or binding of the solute after entry to a degree which affects the interpretation of the permeability per se.

2. Analytical Methods

In recent times direct analytical methods have been increasingly used for permeation studies. After immersion of cells or tissues in solutions of known composition, one can make an analysis either of the cells (or tissues) themselves, of the sap isolated from the cells, or of the immersion medium. The accuracy and reliability of the results obtained depend very much upon the nature of the experimental objects used.

Giant coenocytic vesicles, large enough to be handled singly, yield the most unambiguous results. Examples of the marine algae *Valonia* and *Halicystis* are described in this volume, Chapter 4. The isolation of samples of the sap from such cells is very simple. Even a perfusion of the living cell with solutions of known composition is possible. However, for many purposes the considerably smaller but more easily obtainable internodal cells of different members of the Characeae (*Nitella, Chara,* etc.) are equally suitable. In view of the slowness of diffusion it is often even an advantage that these cells, which generally have a thickness of about 0.2–1 mm and a length of about 2–10 cm, are not as bulky as the vesicles of *Valonia.* Moreover, the content of a single *Nitella* cell may well be enough for accurate microchemical determination. Sometimes it may be found most appropriate first to "saturate" the cell in a solution of the substance to be tested and then to follow its gradual outflux after the cell has been rinsed and transfered into pure water.

While a certain amount of work has been done with suspensions of bacteria, yeasts, or unicellular algae, the study of the permeability properties of such organisms is still more or less in its beginnings (see this volume, Chapter 4 for active transport in these organisms).

Cut disks of fleshy storage organs such as potato (*Solanum tuberosum*) tubers, beet roots, carrot (*Daucus carota* var. *sativa*) roots, etc., have been very extensively used for studies of both the loss and the absorption of solutes. The results, however, obtained with these systems relate more closely to active transport than to passive permeability and they are fully treated in this volume, Chapter 4. The same applies to excised roots and studies on various water plants.

Besides chemical analysis, several physical methods (spectroscopic, polarographic, conductrometric) have also been used for estimation of the often very small quantities of different substances taken up, or given off, by cells or tissues. In this connection the radioactive isotopes or tracers, of which a large assortment (e.g., C^{14}, Na^{24}, P^{32}, S^{35}, Cl^{36}, K^{42}, Ca^{45}, Fe^{55}, Zn^{65}, Br^{82}) is now available, deserve special mention. Their use in permeation studies is of great importance not only because the tracers can be determined, for instance with a Geiger counter, in extremely low concentrations, very exactly and yet comparatively easily. Still more important is the fact that the use of tracers enables us to follow the fate of certain atoms, ions, or molecules irrespective of the presence of other atoms, ions, or molecules of the same species. Thus a study of the exchange of, say, potassium ions against potassium ions, or of chlorine ions against chlorine ions, has been made possible by this method. Likewise and for the first time, it has proved possible to measure in a convincing manner the permeation power of single ionic species. Among the nonradioactive isotopes, H^2, N^{15}, and O^{18} are of greatest interest to the physiologist. Information concerning the isotope technique may be found in several handbooks (5, 37, 101a, 136, 138a, 148, 165).

3. Osmotic Methods

A feature common to the osmotic methods is that the entrance or exit of substances is not observed directly but only inferred from the consequential movement of water. The correctness of the osmotic permeation determinations thus rests on certain assumptions, the most important of which are the following: (a) the exchange of water between a cell and its bathing fluid is controlled practically solely by osmotic forces, not by imbibition phenomena, active water transport, etc. (b) Solutes taken up by the cells keep their osmotic water retaining power unchanged in the cell. (c) No exosmosis of normal cell constituents occurs during the experiment, nor is the osmotic value of the cell sap liable to changes due to metabolic processes.

General statements as to the justification of these assumptions are at present difficult to make, since the literature contains very conflicting evidence. Probably different types of cells behave very differently in this respect. Each case therefore requires appropriate control experiments.

In spite of this theoretical unreliability of the osmotic methods, they have played, and still play, a very important role in the exploration of the permeability phenomena. They are, in fact, of very general applicability and are among the most convenient to use. They may be divided into two groups: plasmolytic and nonplasmolytic methods.

The method of incipient (or, more properly, limiting) plasmolysis (*Grenzplasmolyse*) has been used by Overton (125), Fitting (58), Bärlund (8), and many others. Cells with osmotic concentrations as nearly equal as possible are put into a graded series of solutions, of increasing concentration, of the substance whose permeation is to be followed. The cells are then examined from time to time as to the appearance and disappearance of plasmolysis. Suppose that at a certain time the faintest traces of plasmolysis are observed in a 0.20 M solution, while 1 hour later the same state of limiting plasmolysis is found in a 0.25 M solution. In such a case it seems plausible to assume that during the 1 hour enough solute has entered the cells to raise the concentration of their sap from the equivalent of 0.20 M to 0.25 M. If the substance to be studied is so poorly soluble that a saturated solution of it does not cause any plasmolysis, it may be applied together with some nonpermeating substance (e.g., sugar or mannitol), the concentration of the latter being such that it would not alone plasmolyze the cells. This "method of partial pressures" is also applicable in the case of substances permeating so rapidly that they would not alone bring about a recognizable plasmolysis.

The plasmometric method of Höfler (78) has also proved very useful. In this case the cell to be studied is thoroughly plasmolyzed in a fairly concentrated solution of the substance to be tested. The volume of its protoplast is then evaluated from time to time by microscopic measurements. The more rapidly its volume is found to increase, the more rapidly the solute has entered the protoplast. This method requires only a single cell, but both the cell and its plasmolyzed protoplast should have geometrically simple forms to facilitate exact volume calculations.

If one wishes to compare the permeation powers of two substances, A and B, under strictly identical conditions, this can easily be done by a special modification of the plasmometric method. The cell is first plasmolyzed in an m molar solution of a nonpermeating sugar where the protoplast assumes the volume v. It is then transferred to a solution m molar as to sugar and n molar as to A. Here it is left until the concentration of A inside the protoplast has reached the same value as that outside it, as can be seen from the fact that its volume is again v. The external solution is now replaced by a solution which is m molar as to sugar and n molar as to B. Under these conditions A will diffuse out from the cell while B is at the same time diffusing in the opposite direction. Three alternatives are now conceivable: (a) $P_A = P_B$. The volume of the protoplast remains constant. (b) $P_A < P_B$. The volume first increases and then becomes equal to v. (c) $P_A > P_B$. The volume first decreases and then becomes equal to v.

When the protoplast shrinks away from the cell wall in plasmolysis, the original plasma membrane is probably often destroyed and a new osmotic membrane is formed. It would therefore not be surprising if the permeability of plasmolyzed protoplasts differed greatly from that of nonplasmolyzed ones. Experiments by Schmidt (149) and others have shown, however, that the difference in permeability is in most cases fairly slight. At any rate, it is fortunate that there are also some nonplasmolytic osmotic methods of permeability measurement. Thus, the changes in cell volume caused by osmotic withdrawal or uptake of water may be observed by watching the changes in length either of single cells or of tissue strips. Moreover, in *Beggiatoa mirabilis* and *Oscillatoria* spp. the withdrawal of water from the cells causes an easily observable inward bending (*Knickung*) of the lateral cell walls. Other nonplasmolytic osmotic methods have recently been described (102, 167).

4. Electrical Methods

The permeability of cells to ions may be studied by measuring the electrical conductivity of whole tissues or single cells. The first alternative is scarcely to be recommended, however, since it is difficult to ascertain to what extent the electric current passes through the protoplasts and to what extent it goes through the cell walls around the protoplasts. Analogous experiments with single cells of *Valonia*, *Nitella*, etc., are in this respect more reliable. It is even possible to insert an electrode into one such giant cell and then to measure directly the resistance across the protoplast.

5. Methods Based on Physiological or Toxicological Effects

Suppose it is found that a solution of a certain acid, A, rapidly kills cells of a given kind, while an equimolar solution of another acid, B, under the same conditions only exhibits harmful effects against cells of the same type after a longer time. This result might be explained by assuming that B does not enter the cells as rapidly as A does. But, of course, such a conclusion requires a rigorous examination of all the experimental circumstances before being definitely adopted. Moreover, truly quantitative permeability determinations are not easily attained in this way. However, experiments of this type are of special interest in so far as they may supply knowledge concerning the permeability of the outer plasma membrane, whose permeability properties are still very imperfectly known.

VI. Sites of Resistance to Penetration

A. LOCALIZATION OF PENETRATION RESISTANCE IN THE PROTOPLAST

Is penetration resistance uniform throughout the whole protoplast, or are there some zones of higher, and others of lower, resistance? This question was first put, and also answered, by Pfeffer (130, 132), who advanced the hypothesis that the resistance to permeation is principally due to two extremely thin plasma membranes, the one covering the outer surface of the protoplast, the other separating the cell sap from the protoplasm. The outer plasma membrane is now generally called the *plasmalemma* [Plowe (133)], while the inner one is called the *tonoplast* [de Vries (169)]. The bulk protoplasm between them is the *mesoplasm* (133).

The existence of an outer plasma membrane was postulated by Pfeffer on the basis of the following observation, among others. By treating protoplasts with dilute acid solutions it is possible to cause the outermost layers to assume a condition of rigor without destroying their initial impermeability toward certain dyes. Only after the outermost layer is ruptured will the dye enter through the tear and then rapidly spread throughout the interior of the protoplast.

The existence of an inner plasma membrane was first clearly demonstrated by de Vries (169), who was able to kill the bulk of the protoplast in such a way that its innermost layer still displayed the permeability properties of the intact protoplast virtually unchanged.

De Vries originally assumed that every plasma membrane was derived from a previous plasma membrane, just as a nucleus always originates from a previous one. Pfeffer demonstrated, however, that plasma membranes can form *de novo* when an algal filament is crushed in water, for the extruded spheres of protoplasm possess similar osmotic properties to the original cell. It is thus logical to conclude that a new plasma membrane is at once produced when the inner parts of the protoplasm are brought into direct contact with water. The following experiment of Pfeffer (132) is also very instructive. Small crystals of asparagine are introduced into the body of a myxomycete plasmodium immersed in a saturated solution of asparagine in water. If the bathing fluid is now diluted, the crystals imbedded in the protoplasm will dissolve in the water imbibed by the protoplasm. While this process is going on, sharply defined vacuoles form around the gradually vanishing crystals. These *de novo* produced "vacuoles" are found to behave osmotically just like natural ones.

The introduction of microinjection methods some forty years ago enabled new, and very fruitful, approaches to this problem. Thus Chambers made some impressive experiments consisting of the injection of ammonium chloride or sodium bicarbonate solutions into starfish eggs stained with neutral red. Jacobs (94) gives the following vivid descriptions of them: "Eggs were first stained and placed for a few moments in the ammonium chloride until the color of the intracellular indicator had visibly changed in the alkaline direction. Some of the same ammonium chloride solution was then injected into an individual egg. Immediately the indicator assumed its acid color at the point of injection and, what is more important, there was a rapid spread of the acid condition in all directions from the point of injection, which did not cease until the boundary of the egg had been reached. The last region of the egg in which the change from alkalinity to acidity occurred was therefore that which from the beginning of the experiment had been almost in contact with the solution in question. Experiments with the carbon dioxide-bicarbonate buffer system gave results which were entirely the same in principle except that the conditions of acidity and alkalinity were the reverse of those just described."

With plant cells, microinjection experiments are not so easily performed owing to the rigid cellulose wall which envelops the protoplast. Nevertheless, some very significant results have been achieved. The most important experiments were those of Plowe (133) concerning the behavior of certain dyes when injected into root hairs of *Trianea bogotensis*, epidermal cells of *Allium cepa*, and cells of the red alga *Griffithsia bornetiana*. All the dyes tested (Aniline Blue, Acid Fuchsin, bromocresol purple, phenol red) were such that they did not enter the cells when these were immersed in the dye solution. When injected into the central vacuole they spread in it but did not penetrate the protoplast. On the other hand, when injected into the mesoplasm itself, they were able to spread in the protoplast but unable to pass through the tonoplast into the vacuole or through the plasmalemma into the external solution. (In view of the fundamental importance of these experiments a repetition and extension of them seems desirable.)

These observations show that plasma membranes do exist and that they are less permeable, at least toward certain substances, than is the mesoplasm. There are still, however, quantitative problems to be solved. For, as Höfler (77) pointed out, if we put

$$R = R_L + R_M + R_T$$

where R denotes the permeation resistance of the whole protoplast, while R_L, R_M, and R_T give the resistances of plasmalemma, mesoplasm, and tonoplast, respectively, then it will of course be of great importance to designate the relative magnitudes of the three terms in question.

Let us begin with the permeation resistance of the mesoplasm. If, as is generally assumed, it contains plenty of both free and bound water, so that a coherent aqueous phase extends across the whole mesoplasm from the tonoplast to the plasmalemma, then it seems rather difficult to imagine that even the passage of strongly hydrophilic substances through the mesoplasm will be impeded much more than their diffusion through a water layer of corresponding thickness. Only in the case of extremely rapidly penetrating substances, such as water, might the diffusion resistance of the mesoplasm be of the same order of magnitude as that of the plasma membranes.

We now come to a more difficult question: How great is the penetration resistance of the plasmalemma as compared with that of the tonoplast?

In recent times it has often been claimed that the resistance of the plasmalemma is very much lower than that of the tonoplast and that many substances will therefore readily enter the bulk protoplasm from the ambient solution, although they do not reach the cell sap (3, 77). Or, as Höfler puts it: the "intrability" of the protoplast may be considerable, although its permeability is very low. Several quite different arguments have been advanced in favor of such views.

First, it is well known that the protoplast as a whole is almost impermeable to sugars, amino acids, and mineral salts—in fact, to many substances which play an important role in the metabolism of every cell. Now, it has often been thought that this paradox can be understood only if we assume that the great penetration resistance to substances of vital importance is located principally in the tonoplast alone, the plasmalemma being more or less freely permeable to the substances in question. It seems, however that those reasoning in this way have overlooked the fundamental fact that the uptake of substances into living protoplasts is not due to simple diffusion processes alone but is due, to a very large extent indeed, to active transport. In fact, it seems very probable that the plasmalemma is a site of effective transport activity, but this, of course, does not imply that its permeability must also be great.

Among the more direct observations which have been thought to indicate a relatively high permeability of the plasmalemma are those of Höfler (77, 80; cf. 153) on cap plasmolysis (*Kappenplasmolyse*). He found that the protoplasts of certain cells, when plasmolyzed in pure solutions of alkali chlorides, will sometimes swell considerably at their ends, thus indicating the fairly rapid entrance of the salt into the mesoplasm, although it does not enter the vacuole, which still remains strongly contracted. This is, no doubt, a very interesting phenomenon. On the other hand it occurs only occasionally, and so it does not prove that the plasmalemma is always, or even often, considerably more permeable to ions than is the tonoplast.

Moreover, according to Brooks (24) radioactive phosphate, sodium, and potassium ions are taken up considerably more rapidly into the protoplasm of *Nitella* cells than into their sap.

Finally, the observations concerning the so-called "apparent free space" in plant tissues have attracted great attention. By apparent free space is meant that portion of a tissue to which solutes apparently move "by free diffusion," i.e., without encountering any considerable resistance. In part, of course, this space, consists of cell walls and intercellular spaces, but according to the calculations of several recent investigators it is necessary to assume that the protoplasm, or at least parts of it, also belong to the apparent free space. If this assumption were correct, then it would follow that the plasmalemma must be readily permeable to ions, which only penetrate the tonoplast very slowly. As can be seen, for instance, from recent reports by Epstein (54, 55), Kramer (107), and Robertson (140), there are, in fact, several observations which appear to sustain such views. But, on the other hand, it may be permissible to doubt whether the evidence so far presented in favor of these ideas is convincing enough—to be a basis for such far-reaching conclusions as these, which would place the inclusions of the cytoplasm in virtual free contact with the changing composition of the external solution [cf. Levitt (110a)].

Such doubts seem the more justified when we remember the strong evidence suggesting that the diffusion resistance of the plasmalemma is of considerable magnitude. First, we have the analogy with the animal cells, in which the great diffusion resistance of the outer plasma membrane is in many cases extremely well established. Secondly, there is the general view that a condition for the undisturbed course of the life processes is that the living machinery should be effectively isolated from its environment, which, especially in the case of many unicellular organisms, may be subject to sudden changes in chemical composition. This requirement seems, in fact, to be well met, most protoplasts being remarkably resistant to lipid-insoluble poisons which, if they were to gain access to the interior of the protoplasts, could scarcely fail seriously to disturb the life processes. Thus, for instance, it is known that even wide fluctuations of the pH of the medium scarcely influence the rate of oxygen consumption of yeast. Moreover, Conway and Downey (40) have directly demonstrated that more or less lipid-insoluble acids (e.g., glyceric, malic, citric, and tartaric acid) can readily penetrate only a very restricted part of the yeast cell, the accessible region being probably identical with the cell wall alone. The plasmalemma of the cells of land plants may, on the whole, be somewhat more delicate (83), but in most cases, nevertheless, their protoplasts, too, show a fairly high degree of resistance to lipid-insoluble poisons, thus indicating that these substances can have almost no access to the mesoplasm across the intact plasmalemma. Finally, we have to remember that when plant cells are put into hypertonic solutions the protoplast, in the vast majority of cases, will contract as a whole, while, if the tonoplast alone were endowed with semipermeability, one would expect that only the vacuole would contract.

The problem of the permeability properties of the plasmalemma is thus, at present, rather puzzling in that suggestive arguments have been put forth in favor of, and also against, the existence of a great resistance to diffusion at the outer surface of the protoplast. The present writer inclines to the classical view, according to which the plasmalemma resembles the tonoplast in its diffusion resistance. But evidently it is not possible yet to answer this question definitely, least of all in a really quantitative manner. There is, of course, also the possibility that the location of the main osmotic barrier may vary in different kinds of cells and according to the solutes in question.

At all events, the present uncertainty concerning the permeability of the plasmalemma is one of the most serious gaps in our knowledge of the permeability properties of plant cells, because, for the functioning of the living machinery it must be far from insignificant whether the osmotic barrier is situated on the outer or the inner surface of the protoplasmic layer.

Concerning the tonoplast membrane, there seems to be fairly general agreement that its diffusion resistance toward hydrophilic substances is great. But here, too, reliable quantitative data are almost totally lacking.

A very special case is represented by the sieve tubes and latex vessels, which are believed to be devoid of tonoplasts [(60), page 84]. The same, of course, holds true of all cells lacking vacuoles.

B. PENETRATION RESISTANCE OF THE CELL WALL

On its way from the medium into the cell sap, or vice versa, a permeating molecule or ion has to overcome not only the resistance of the protoplast, including its membranes, but also that of the cell wall.

It is well known that the permeability of cutinized or suberized cell walls, even to water, is very restricted. On the other hand, strongly hydrophilic walls composed only of building materials such as cellulose, hemicelluloses, and pectins are readily permeable to the great majority of water-soluble substances. Nevertheless, such a wall represents an unstirred layer that by virtue of its thickness alone will offer a certain

TABLE II

PERMEABILITY OF THE CELL WALL OF *Nitella mucronata* TO NONELECTROLYTES[a,b]

Substances	M^c	$P \times 10^5$	$PM^{1/2} \times 10^4$
Methanol	32	75	43
Ethylene glycol	62	56	44
Glycerol	92	44	42
Tetraethylene glycol dimethyl ether	222	24	36
Sucrose	342	17	31
"Polyethylene glycol 400"	400	7.6	15
Raffinose	504	8.3	19
"Polyethylene glycol 600"	600	2.7	7
"Polyethylene glycol 1000"	1000	0.8	3

[a] The permeability constants (P) are given in centimeters per second, multiplied by 10^5.

[b] According to Collander (32) and some supplementary unpublished determinations.

[c] M = molecular weight.

resistance to diffusing molecules, which is more important as the diffusion resistance of the protoplast becomes smaller. At the same time, such a hydrophilic cell wall will function as a molecular sieve more or less impervious to all molecules whose molecular size exceeds a certain limit. As an example, the cell wall of *Nitella mucronata* may be taken (Table II). Although the values of the permeation constant stated in the table have not a high degree of accuracy, taken as a whole they show a fairly regular decrease of the permeation power with increasing molecular size. Moreover, the constancy of the $PM^{1/2}$ values at the beginning of the series and their rapid decrease at the end of it indicates how the sieve action of this cell wall gradually exerts its effect once the molecular weight exceeds a value of about 200 or 300. Finally, it should perhaps be pointed out that, although the permeability of the wall to such large-molecular substances as "polyethylene glycol 1000" may, at first sight, appear very small, it is nevertheless roughly 1000 times greater than the permeability of the protoplasts of *Nitella* toward, say, glycerol. Generally speaking, then, the hydrophilic cell walls will very seldom be an important hinderance to the uptake of solutes into plant cells.

For further particulars concerning the permeability of plant cell walls, the reader is referred to Brauner (21).

VII. Permeability to Nonelectrolytes

A survey of the permeability of cells to various substances may appropriately begin with nonelectrolytes. Members of this group are free from the complications which arise from variable ionization of the permeators, and the great variety of structure shown by these substances renders them especially suitable for studies of the relations between chemical and physical properties, on the one hand, and permeation power, on the other. Finally, the permeability of plant cells to nonelectrolytes may be said to be decidedly better known at present than their permeability to electrolytes.

In the following, we shall designate as nonelectrolytes all those substances which, at physiologically tolerable pH values, occur almost solely as nonionized molecules. Under the heading of nonelectrolytes we will thus also treat such extremely weak bases as urea, caffeine, and antipyrene, all of which have dissociation constants below 10^{-10}. Even the amino acids may conveniently be taken into consideration in this context inasmuch as they have no net electric charge.

A. OVERTON'S GENERAL PERMEABILITY RULES

As already stated, Overton, after studying the behavior of a great number of plant and animal cells toward several hundreds of chemical compounds, reached the conclusion that some general rules govern the relative permeation power of substances toward extremely different types of cells. These rules, which on the whole have been corroborated by later investigators, may be summarized as follows:

(1) Hydrocarbons, their halogen and nitro derivatives are all extremely lipophilic and have a very great permeation power.

(2) An increase in the number of hydroxyl groups tends simultaneously to decrease the lipid solubility and the permeation power. Thus, the monohydric alcohols permeate very rapidly, the dihydric ones considerably more slowly but still fairly rapidly, glycerol (a trihydric alcohol) rather slowly, erythritol (a tetrahydric alcohol) still more slowly. Finally, the hexahydric alcohols (e.g., mannitol), the

hexoses, and the corresponding di-and trisaccharides all have a scarcely detectable permeation power.

(3) A C=O group has qualitatively the same effect on the distribution and on the permeation as a C—OH group.

(4) A similar but stronger effect is exerted by the amino and especially by the —CONH₂ group. Thus the amides of the monobasic acids (e.g., acetamide) have about the same permeation power as the dihydric alcohols. Urea permeates rather slowly.

(5) Amino acids have an extremely low lipid solubility and permeate extremely slowly.

(6) An increase in the length of the carbon chain increases the lipid solubility and the permeation power. A similar effect is also obtained by substitution of the hydrogen atoms in the hydroxyl, carboxyl, or amino groups by alkyl, aryl, or acyl groups. Thus the permeation power increases strongly in the series glycerol < monoacetin < diacetin < triacetin and also in the series urea < methylurea < ethylurea < diethylurea < triethylurea.

(7) Substitution of oxygen atoms by sulfur atoms increases lipid solubility and permeation power. Thus, thiourea permeates somewhat more readily than does urea.

Overton claimed that these rules are valid for all kinds of cells. We shall see, however, that a few incontestable exceptions occur.

B. Lipid Solubility, Polarity, Hydrophilia, Hydrogen-Bonding Tendency

In view of the correlation found by Overton, and also by later investigators, between lipid solubility and permeation power the question of the relation of lipid solubility to other physicochemical properties and to the chemical constitution arises. Of course, when we are here speaking of lipid solubility, the relative lipid solubility alone is meant, i.e., the distribution coefficient lipid: water.

Little is actually known about the solvent properties of those lipids (phosphatides, sterols, etc.) which are commonly supposed to occur in the plasma membranes and whose solvent properties are therefore most relevant. It therefore seems best to consider the solvent properties of water-immiscible organic solvents in general: since these have so much in common, we shall be able in this way to conclude, indirectly, a great deal concerning the solvent properties of those more or less unknown cell lipoids in which we are primarily interested.

First, however, just a few words concerning two pairs of terms which are often used in this connection, namely, the terms polar and nonpolar, hydrophilic and hydrophobic. Polar compounds are character-

ized by a molecular structure which involves a shifting of one or more
electrons from their original position in the atom in such a way that the
molecule will show regions of positive or negative electrical charge.
The hydroxyl, carboxyl, and amino groups may be mentioned as examples of polar groups. The nonpolar compounds, on the other hand, of
which the hydrocarbons are the most typical representatives, have a
molecular structure which results from the sharing of electrons by the
atoms concerned without the production of regions of great electrical
dissimilarity. Now, on the whole, polar compounds tend to be more
soluble in water (which is itself highly polar) than in such nonpolar
or weakly polar organic solvents as hydrocarbons, ether, etc. Nonpolar
substances show the reverse relation.

Substances preferentially soluble in water are also called hydrophilic,
while those preferentially soluble in organic solvents are called hydrophobic,[4] lipophilic, or organophilic. The essential difference in nature
between these two groups of substances was virtually elucidated some
twenty years ago when the concept of hydrogen bonds (or hydrogen
bridges, as they are sometimes called) was advanced. It had for a long
time been known that water is in many respects a very peculiar liquid
(cf. 56). Thus, its vaporization heat, its surface tension, and its cohesion are unusually great. Furthermore, its boiling point is surprisingly high, as is at once realized by comparing it with that of the closest
analog of water, namely, hydrogen sulfide, which has a boiling point of
—60°C. These anomalies of water are understandable if we take into
account that the H_2O molecules in liquid water are not free but are
bound by hydrogen bonds into netlike complexes. Generally speaking,
the hydrogen bond is a chemical bond resulting from the attraction of
two electronegative atoms for a hydrogen atom. Only the most strongly
electronegative atoms are capable of forming hydrogen bonds, namely,
oxygen, nitrogen, chlorine, and fluorine atoms. Now, the hydrophilia
is just a manifestation of the tendency of a substance to form hydrogen
bonds: thanks to its hydroxyl, carboxyl, amino, or other similar groups,
it is able to form hydrogen bonds with the water molecules and thus
at the same time attracts, and is itself attracted by, water molecules.
On the other hand, molecules not capable of forming hydrogen bonds
are, as it were, squeezed out from water and other strongly hydrogen-
bonded liquids and thus they tend to accumulate in a neighboring nonaqueous liquid.

Now, it is the degree of hydrophilia, or hydrophobia, of the solutes

[4] Strictly speaking there exists no hydrophobia but only an absence of hydrophilia,
just as darkness is nothing but the absence of light. In many contexts the word
hydrophobia is nevertheless quite useful.

on the one hand and of the solvents on the other that is the main factors which governs the distribution of solutes between water and water-immiscible solvents. Here the old rule *similia similibus solvuntur* holds true. Thus, the more hydrophilic the solute and the more hydrophobic the nonaqueous solvent, the more completely the solute will accumulate in the aqueous phase. For instance, in the series propanol, propylene glycol, glycerol, the distribution coefficient in the solvent system ethyl ether-water falls off rapidly with increasing number of hydrophilic hydroxyl groups: $1.9 > 0.018 > 0.0007$. In the solvent system hexane-water, owing to the more strongly hydrophobic nature of the hexane, the distribution coefficients of the same substances are lower and fall off still more steeply, something like this: $0.04 > 0.0001 > 0.000001$.

Along with the hydrophilia yet another factor, namely, the acidity or basicity of solutes and solvents, may influence the distribution. But here the rule holds true that opposites attract each other, that is, acidic solutes tend to accumulate in basic solvents, while basic solutes have a stronger affinity for acidic solvents. In fact, even very low degrees of acidity or basicity may markedly influence the distribution. Thus, for instance, fatty acid amides have an appreciably greater affinity for alcohols than for ethers of a corresponding degree of hydrophilia.

Except for these two major factors, namely, the hydrophilia-hydrophobia and the acidity-basicity factors, the distribution is scarcely influenced by other more specific properties of the solvents. Thus, in spite of the many organic, water-immiscible solvents, they can all be classified into a consistent, readily surveyed system on the basis of these two criteria, viz., their varying hydrogen-bonding tendency, on the one hand, and their acidity or basicity, on the other. From the standpoint of cell permeability this has a very important practical consequence: Although almost nothing has so far been directly ascertained concerning the solvent properties of those lipids (phosphatides, sterols) which are most commonly supposed to occur in the plasma membranes, we are nevertheless in the position to predict, in a general way, the limits within which their solvent properties will vary.

C. PERMEABILITY OF THE INTERNODAL CELLS OF THE CHARACEAE

The results of Overton were published, as we have seen, in a semiquantitative form only. We will now turn to later, more truly quantitative, permeability measurements. In doing so, we will begin with the internodal cells of two characean plants, *Chara ceratophylla* (35) and *Nitella mucronata* (32), since these cells have been more thoroughly studied as regards their permeability to nonelectrolytes than have other types of plant cells. Moreover, the method of permeability measurement used in these studies was comparatively reliable, since it was based on quantitative microchemical determinations of the actual amounts of substances entering the cells, or leaking out from them, during appropriately chosen time intervals.

The authors of these works were anxious to decide to what extent the observed transfer of solutes may be regarded as due to simple per- meation processes and to what extent it may be due to active transport. To this end they first studied the concentration of the permeators in the cell sap after equilibrium had been reached. They found that in the cases studied the equilibrium concentration in the cell sap was about 90–100% of the concentration in the medium. Furthermore, the general course of the uptake was found to agree fairly with that pre- dicted for simple diffusion process. Finally, the transfer rates were found to be the same whether the solutes were entering the cells from outside or, on the contrary, were moving from the cell sap into the bathing fluid. All these findings strongly suggest that the transfer proc- esses so studied were not influenced appreciably by metabolism or by chemical combination of the permeators with any cell constituents. (This conclusion has been disputed by Bogen (16), but his arguments do not seem convincing.)

Figure 1 presents a survey of the principal results achieved with the *Nitella* cells and of their bearing on current permeability theories. To this end the permeability constants, P, multiplied by the molecular weights, M, raised to the power 1.5 have been plotted against the distri- bution coefficients in the system olive oil-water. The magnitude $PM^{1.5}$ has been chosen rather than P in order that the points which represent the different permeators shall fit a straight line as closely as possible. Besides, among the substances studied the variation in the magnitude of M is so slight (from 19 to 480) that, say, 1.2 or 1.8 used instead of 1.5 as exponent of M would not alter the picture very much. On the other hand, it will be noted that while the points representing molecules of medium or comparatively great molecular weight ($M = 60$–120 or 120–480, respectively) are fairly evenly distributed around the midline of Fig. 1, the smallest molecules ($M < 60$) are all situated above that line. Finally, it will be observed that the mid-line has a slope of about 53° to the abscissa. We see, then, that for molecules of a molecular weight between 60 and 480 the permeation constant P is roughly pro- portional to the expression $k^{1.3}/M^{1.5}$ (where k denotes the distribution coefficient olive oil:water) but the smallest molecules permeate some- what more readily than this formula implies.

With the cells of *Chara* and of a third characean plant, *Nitellopsis obtusulus*, virtually similar results were obtained, except that the

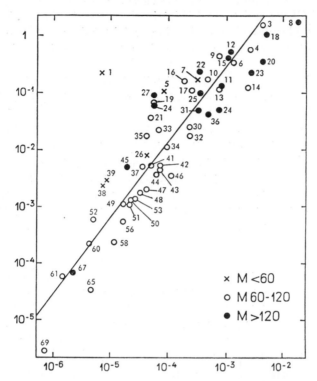

FIG. 1. Correlation between the permeation power of several nonelectrolytes toward *Nitella mucronata* cells on the one hand and their relative oil solubility and molecular weight on the other. Ordinate: $PM^{1.5}$; abscissa: distribution coefficient olive oil:water. The substances are: *1*, HDO; *3*, methyl acetate; *4*, *sec*-butanol; *5*, methanol; *6*, *n*-propanol; *7*, ethanol; *8*, paraldehyde; *9*, urethane; *10*, isopropanol; *11*, acetonylacetone; *12*, diethylene glycol monobutyl ether; *13*, dimethyl cyanamide; *14*, *tert*-butanol; *15*, glycerol diethyl ether; *16*, ethoxyethanol; *17*, methyl carbamate; *18*, triethyl citrate; *19*, methoxyethanol; *20*, triacetin; *21*, dimethylformamide; *22*, triethylene glycol diacetate; *23*, pyramidon; *24*, diethylene glycol monoethyl ether; *25*, caffeine; *26*, cyanamide; *27*, tetraethylene glycol dimethyl ether; *30*, methylpentanediol; *31*, antipyrene; *32*, isovaleramide; *33*, 1,6-hexanediol; *34*, *n*-butyramide; *35*, diethylene glycol monomethyl ether; *36*, trimethyl citrate; *37*, propionamide; *38*, formamide; *39*, acetamide; *41*, succinimide; *42*, glycerol monoethyl ether; *43*, *N,N*-diethylurea; *44*, 1,5-pentanediol; *45*, dipropylene glycol; *46*, glycerol monochlorohydrin; *47*, 1,3-butanediol; *48*, 2,3-butanediol; *49*, 1,2-propanediol; *50*, *N,N*-dimethylurea; *51*, 1,4-butanediol; *52*, ethylene glycol; *53*, glycerol monomethyl ether; *56*, ethylurea; *58*, thiourea; *60*, methylurea; *61*, urea; *65*, dicyanodiamide; *67*, hexamethylenetetraamine; *69*, glycerol. From Collander (32).

values of P were found to be more closely proportional to $k^{1.0}/M^{1.5}$ in the case of *Chara* and to $k^{1.15}/M^{1.5}$ in the case of *Nitellopsis*. In these cases, too, molecules of a molecular weight below 60 were found to have a greater permeation power than the expressions given would imply.

If $PM^{1.5}$ is plotted against the distribution coefficient ether:water instead of olive oil:water, the resulting graph will be fairly similar to Fig. 1, with the exception, however, that the points representing slightly basic substances (amines, amides) will be found lying too low. In this respect, then, olive oil, probably owing to the free oleic acid which it contains, is a decidedly better model substance than ether. Besides, by adding more oleic acid to the olive oil its solvent capacity toward slightly basic substances can be considerably increased.

The theoretical significance of these empirical results will be discussed below (Section XIII).

D. DIFFERENT PERMEABILITY TYPES

As already stated, Overton very strongly stressed the essential similarity, as to their permeability properties, between protoplasts of widely different origin, shape, and function. He did not fail to observe that there are also minor differences in this respect between different cells, but for reasons readily understood these differences did not interest him as much as the striking and unexpected resemblances between otherwise different cells. It was therefore only about 1930 that systematic investigations were started in order to reveal and, if possible, explain specific differences in permeability.

In the field of animal cells, such investigations have been successfully carried out by Jacobs (95, 98), who studied the permeability properties of the erythrocytes in different animal groups, detecting numerous interesting parallels between specific permeability properties, on the one hand, the taxonomic classification, on the other. These results are the more striking since they refer to cells with one and the same physiological function.

In the botanical field, "comparative" permeability studies were inaugurated in 1930 by Höfler (85a, 90b), who stressed the importance of establishing and comparing specific permeability series (*spezifische Permeabilitätsreihen*) for different plant protoplasts. In the following years his program was carried out by numerous workers, among whom Hofmeister (86), Marklund (112), and Elo (52) may be named. The results of these studies have been summarized by Höfler (82), who distinguished the following five permeability types.

Type I. The first type is called, by Höfler, the *Chara-Maianthemum*

Type. This denomination is derived from the names of two test objects, viz., the internodal cells of *Chara ceratophylla* and the subepidermal cells of the stem of *Maianthemum bifolium* [studied by Höfler (78)], which are considered to be characteristic of this type. The type might equally, however, be called the Main or Normal Type, for as a matter of fact the majority of plant protoplasts so far studied are of this type.

Table III gives a few, rather arbitrarily chosen examples illustrating the behavior of the representatives of this group. In spite of their very different nature (*Maianthemum* is a terrestrial flowering plant; *Chara*, a highly specialized green alga from brackish water; *Pylaiella litoralis*,

TABLE III

PERMEABILITY OF *Chara ceratophylla*, *Maianthemum bifolium*, AND *Pylaiella litoralis* CELLS TO SOME NONELECTROLYTES[a]

Substance	*Chara*[b]	*Maianthemum*[c]	*Pylaiella*[d]
Ethylene glycol	58	—	106
Thiourea	10	9.5	—
Methylurea	9.2	8.0	14
Lactamide	7.5	3.4	—
Urea	5.4	2.8	2.8
Glycerol	1.0	1.0	1.0
Malonamide	0.19	0.56	0.67
Erythritol	0.06	0.08	0.04

[a] The permeability to glycerol put equal to unity.
[b] According to Collander and Bärlund (35).
[c] According to Höfler (78).
[d] According to Marklund (112).

a marine brown alga) the similarity in their permeability is unmistakable.

Type II. The *Gentiana-Sturmiana* Type of Höfler is characterized by its unusually high permeability toward urea; urea penetrates these protoplasts more rapidly than does methylurea, while in "normal" cases the permeation rate of methylurea is about 2–5 times as great as that of urea. The last-mentioned behavior can be explained because the lipoid solubility of methylurea is some 2–5 times as great as that of urea. The more rapid permeation of urea, on the other hand, is probably due to some special feature, not yet exactly known, in the structure of the plasma membranes which allows the small, symmetrical urea molecules to penetrate the membrane more easily than the somewhat greater, asymmetric molecules of methylurea. So far it seems as if Type II is in all other respects very similar to Type I. Of the plant protoplasts hitherto studied, some 12% belong to Type II.

Type III. The *Rhoeo* Type owes its name to *Rhoeo discolor,* the epidermal cells of which were found by de Vries to constitute excellent material for plasmolytic permeability determinations. Bärlund (8) observed that these cells are relatively less permeable to all amides than are cells of the Normal Type. Thus, *Rhoeo* cells are less permeable to malonamide than to erythritol, less permeable to urea, thiourea, and methylurea than to glycerol, less permeable to formamide and acetamide than to ethylene glycol, while in all these respects cells of Type I behave in the contrary way. The *Rhoeo* Type could thus also be called the Amidophobic Type. Its peculiarities may easily be explained by assuming that the plasma membranes of this type contain lipoids which are less acidic and thus poorer solvents for the slightly basic amides than are the plasma membrane lipids of the Normal Type. There are not many representatives of this type as extreme as *Rhoeo* itself. Cases more or less intermediate between this type and Type I are, however, encountered.

Type IV (Beggiatoa Type). The permeability of *Beggiatoa mirabilis,* extensively studied by Ruhland and his co-workers differs strikingly from the Normal Type: lipid solubility seems in this case to have only a minor influence on the permeation power, while molecular size is the deciding factor. Moreover, the cells of *Beggiatoa* are extremely permeable to all solutes. An organism with permeability properties that resemble those of *Beggiatoa* is the blue-green alga *Oscillatoria* studied by Elo (52). It is not, however, as extreme in its behavior as *Beggiatoa.* Probably these two organisms are also taxonomically related, for *Beggiatoa,* which used to be classified amongst the bacteria, is now often considered to be a colorless member of the Oscillatoriales.

Type V (Diatom Type). Diatoms deplasmolyze with unusual rapidity in hypertonic solutions of several very poorly lipid-soluble substances. Their behavior thus resembles that of *Beggiatoa* and *Oscillatoria.* As may be seen from Table IV there is, nevertheless, a distinct difference between these two types. For in *Oscillatoria* the most striking feature is that even highly lipid-soluble compounds (e.g., propionamide) do not permeate very much faster than, say, glycerol, while the most outstanding property of the diatoms seems to be a considerable permeability to substances of extremely low lipoid solubility and of fairly large molecular size (erythritol, and even sucrose). Under heading IIIC it was mentioned that Bogen and Follman (19) maintain that the rapid deplasmolysis of diatom cells in hypertonic solutions is not due to an unusually great permeability but to active uptake of water and solutes. These authors, therefore, deny the very existence of a special Diatom Type of permeability.

The five permeability types outlined above are by no means sharply separated from each other. At least between Type I, on the one hand, and Types II and III, on the other, there is a quite continuous gradation. Types IV and V seem more isolated, however, according to our present state of knowledge.

It should also be noted that the permeability types discussed do not characterize certain plant species as such: different cells, say, epidermal and subepidermal cells, of one and the same plant may belong to distinctly different permeability types, and even the same cell may at different stages of its development represent different

TABLE IV

PERMEABILITY[a] OF TWO DIATOMS, *Licmophora oedipus* AND *Melosira* SP., AS COMPARED WITH THAT OF *Chara ceratophylla* AND *Elodea* (*Anacharis*) *densa* AS REPRESENTATIVES OF THE NORMAL TYPE AND WITH *Oscillatoria princeps*

Substance	Elodea[b]	Chara[c]	Licmophora[d]	Melosira[b]	Oscillatoria[d]
Propionamide	115	180	110	84	4.1
Acetamide	71	72	18	14	5.8
Ethylene glycol	51	58	7.2	11	5.4
Methylurea	12	9.2	2.6	3.4	3.2
Urea	4.8	5.4	1.1	1.2	1.0
Glycerol	1.0	1.0	1.0	1.0	1.0
Malonamide	0.61	0.19	1.0	0.5	0.2
Erythritol	0.11	0.06	0.8	0.4	0.04
Sucrose	0.02	—	0.5	0.2	—

[a] Permeability to glycerol put equal to unity.
[b] According to Marklund (112).
[c] According to Collander and Bärlund (35).
[d] According to Elo (52).

permeability types (112). Moreover, Hofmeister (87) found that in certain seasons of the year the subepidermal cells of *Ranunculus repens* behave as Type I, in other seasons as Type III. Finally, a simple rise in temperature increases the permeability toward different substances to very different extents and may thus cause an, at least apparent, transition from Type II to Type I.

In spite of all this, the recognition of separate permeability types seems decidedly useful, for it facilitates survey of the varying permeability properties of different cells. To the present writer it thus seems that Bogen (17) goes too far when he more or less denies the reality of the permeability types, attributing them to differences in active transport, etc.

Figure 2 presents a general view of the permeability differences between different kinds of plant protoplasts. In this graph each cell type is represented by a vertical line on which the permeation constants (expressed as centimeters per hour) of seven substances are denoted by points on a logarithmic scale. The cell types are arranged according to increasing permeability to erythritol. The sixteen cell types tested

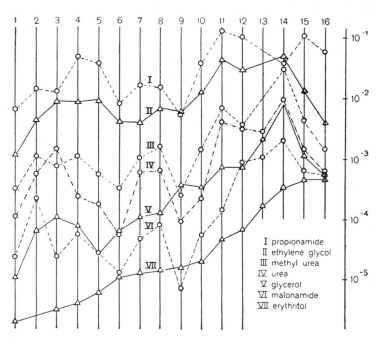

FIG. 2. Permeability of 16 different kinds of plant cells to some nonelectrolytes: *1*, leaf cells of *Plagiothecium denticulatum*; *2*, *Oedogonium* sp.; *3*, root cells of *Lemna minor*; *4*, *Pylaiella litoralis*; *5*, *Zygnema cyanosporum*; *6*, subepidermal cells of *Curcuma rubricaulis*; *7*, *Spirogyra* sp.; *8*, leaf cells of *Elodea* (*Anacharis*) *densa*; *9*, epidermal cells of *Rhoeo discolor*; *10*, epidermal cells of *Taraxacum pectinatiforme*; *11*, internodal cells of *Chara ceratophylla*; *12*, internodal cells of *Ceramium diaphanum*; *13*, *Escherichia paracoli*; *14*, *Oscillatoria princeps*; *15*, *Melosira* sp.; *16*, *Licmophora* sp. From Collander (29).

were intentionally chosen to represent types of plant cells as different as possible. Thus, such very different taxonomic groups as flowering plants, mosses, green algae, charophyta, zygophyta, diatoms, brown algae, red algae, blue-green algae, and bacteria are represented. Also the physiological character of the cells examined varies greatly. Therefore, this graph shows the range of variation in permeability which is encountered in plant cells in general. Of these sixteen cell types, numbers 4, 5, 7, 8, 11, 12, and 13 are seen to represent the Normal

Type; numbers 6 and 10 are intermediate between the Normal Type and the Amidophobic Type; numbers 1 and 2 represent the Normal Type except for a somewhat unusually high degree of amidophilia (malonamide has here a higher permeation rate than glycerol); number 3 belongs to the *Gentiana-Sturmiana* Type; number 9, to the *Rhoeo* Type; numbers 15 and 16, to the Diatom Type.

A general conclusion may be drawn from a collation of data like that given in Fig. 2. Although different plant cells do vary in their permeability properties, this variation cannot obscure the fundamental fact, already stressed by Overton, that there is nevertheless a very striking similarity between the permeability of all protoplasts, especially if we disregard a few extreme cases (*Beggiatoa*, *Oscillatoria*, diatoms). This conclusion would probably emerge even more clearly if the compilation were to comprise a wider range of permeators. Now it comprises only substances of low or moderate permeation power, since the plasmolytic method used in most of these determinations is not suited to give quantitative results with very rapidly permeating substances.

E. Some Especially Important Nonelectrolytes

1. Water

Water has sometimes been supposed to be the most hydrophilic of all substances. In reality, however, there are numerous substances (e.g., urea, polyhydric alcohols, sugars, amino acids, mineral salts) with a lipid:water distribution coefficient which is much smaller than that of water and which must therefore be considered to be more strongly hydrophilic even than water. Nevertheless water is, of course, strongly hydrophilic, and so it may seem surprising that, in spite of this, its permeation power is so great. This, and also the fundamental physiological role of water, makes it understandable that the permeability to water of both plant and animal cells has engaged the attention of such a large number of investigators. It is only a pity that by no means all of them have realized the great difficulties, both technical and theoretical, encountered in such work.

The chief technical difficulties arise from the extreme smallness of the diffusion resistance of protoplasm to water. The permeability to water is usually determined by watching the rate of plasmolysis or deplasmolysis. It may easily happen that the factor that controls the plasmolysis or deplasmolysis rate is not the water permeability of the protoplast, but rather the rapidity with which the plasmolyzing substance diffuses up to the outer surface of the protoplast or from this surface to the bulk of the bathing fluid. Particularly if the experiments are carried out, not with isolated cells, but with tissue sections, this source of error may be a very serious one. If we wish to measure the water permeability of the protoplast alone, the surrounding cell wall should, therefore, be as permeable as possible, not only to water, but also to the plasmolyzing substance. It is still better, however, to use isolated protoplasts so that the retarding influence of the cell wall is entirely eliminated. But even in such a case the assumption is not fully warranted that the rate of the osmotically induced contraction or expansion of the protoplast depends solely on the water permeability of the protoplast or the plasmatic membranes. For when water is withdrawn from, or taken up by, the protoplast, a layer of higher or lower concentration will be established just inside the protoplast, and this layer of different concentration will counteract the process under investigation. In view of these sources of error, which all tend to depress the water permeability values experimentally obtained below the correct value, it seems probable that most of the values recorded in the literature are actually too low, in some cases perhaps much too low.

Moreover, a quantitative comparison of water permeability and solute permeability is not easy. As stated before, the permeation of a solute across the plasma membrane may be said to be determined by the difference in concentration of that solute on the two sides of the membrane. But this does not hold true as regards the permeation of water. Imagine a 10% solution of, say, sodium chloride on one side of a semipermeable membrane and a 10% solution of sucrose on the other. These solutions contain the same concentration of water, and yet the diffusion pressure of water is nowhere near the same on the two sides of the membrane. Alternatively the permeation of water may be considered as due to differences in osmotic or hydrostatic pressure. A consequence of this has been that the water permeability has generally been expressed in length $\times$ time^{-1} $\times$ pressure^{-1} units. Such a procedure is in itself correct, but it has the regrettable drawback that the water permeability values expressed in this way are not comparable with other permeability values expressed in length $\times$ time^{-1} units.

Attempts have therefore been made to calculate permeation constants for water in virtually the same manner as for solutes. These attempts have for the most part been based on the assumption that a concentration gradient of solute determines a concentration gradient of water of equal magnitude but opposite sign and that under these conditions the diffusion of water molecules quantitatively obeys the same laws as does that of the solute molecules in a dilute solution. According to Jacobs (99) neither of these assumptions appears to be warranted. On the other hand, Bochsler (14) found that the permeation rate of water is proportional to the difference between the actual water concentration of the cell sap and its water concentration after osmotic equilibrium has been established.

There is, however, a simple method of determining the water permeability in exactly the same way as the permeability to solutes, namely to watch the permeation of isotopic water, either deuterium oxide or tritium oxide, which, when mixed with ordinary water, should behave essentially like any other solute. The theoretical drawback to this method, i.e., that heavy water permeates slightly more slowly than ordinary water, seems to be of no great significance.

TABLE V

PERMEABILITY OF SOME PLANT PROTOPLASTS TO WATER[a]

Cell type	Method	P_f	P_d	Authority for measurement
Fucus vesiculosus, egg cell	Osmosis	0.0008–0.001	—	(138b)
Allium cepa, parenchyma cell	Osmosis	0.002	—	(110b)
Salvinia natans, cell of water leaf	Plasmolysis	0.003	0.004[b]	(90b)
Nitella flexilis, internodal cell	Osmosis	0.04–0.11	—	(102)
N. mucronata, internodal cell	Isotope	—	9.0	(32)
Nitellopsis obtusulus, internodal cell	Osmosis	0.006	0.43[c]	(127)
N. obtusulus, internodal cell	Isotope	—	1.0	(171)
Vallisneria spiralis, epidermal cell	Plasmolysis	—	0.004[b]	(90b)
V. spiralis, mesophyll cell	Plasmolysis	—	0.016[b]	(90a)
Spirogyra sp.	Plasmolysis	—	0.011	(14)

[a] Permeability is expressed either in centimeters per hour per atmosphere (P_f) or in centimeters per hour (P_d). Temperature about 20°C.

[b] Value has been calculated by Bochsler (14).

[c] Value has been calculated by Wartiovaara (171).

According to experiments carried out with the isotope method (32), water permeates through the protoplasts of *Nitella* about 20,000 times more rapidly than urea and some 500,000 times more rapidly than glycerol. Thus water has, in fact, an extremely great permeation power in comparison with most other substances.

But the question of water permeability has also another aspect, for, as seen from Table V, the flow rate of water through a plant protoplast is usually even less than 0.1 mm per hour under the influence of a pressure difference of 1 atm. From this point of view, then, the actual resistance of protoplasm to the flow of water may be said to be very great. This statement seemingly contradicts that given previously, according to which water has an exceptionally great permeation power. This apparent contradiction is, however, readily resolved when we realize that the diffusion resistance of living protoplasts toward all substances is, in fact, very great. Although their permeability to water is much greater than their permeability to almost all other substances, their *absolute* permeability even to water is nevertheless rather low.

This is a fact which must not be forgotten when the course of the transpiration stream through parenchymatous tissues, e.g., in root and leaves, is being discussed. In streaming through such tissues, the water evidently has to choose between two alternative pathways, namely, (a) across the protoplasts, and (b) through the cell walls flanking the protoplasts. In reality both paths will, of course, always be used simultaneously, and it seems *a priori* clear that the intensity of each partial stream will be inversely proportional to the resistances encountered along each path. (Cf. Ohm's well-known law which governs the relative intensities of electric currents flowing in parallel.) Regrettably, however, the magnitudes of these resistances are not yet sufficiently known, and hence it is not yet possible to calculate the relative intensities of the two partial streams.

2. Oxygen

Free oxygen is one of the most rapidly penetrating substances. Its extremely rapid penetration through the protoplasm can be inferred from a simple observation made in connection with Engelmann's well-known bacterial method of demonstrating the evolution of oxygen in photosynthesizing cells. To this end a filament of, say, *Spirogyra* sp. is mounted in water containing positively aerotactic bacteria and covered with a glass slip whose edges are sealed to prevent the entrance of air. Now, if the preparation is exposed to light so that photosynthesis is started in the *Spirogyra* cells, the bacteria will not accumulate evenly around the photosynthesizing cells but almost exclusively just outside the individual chloroplasts. This indicates that the oxygen produced by the chloroplasts diffuses straight out from the cells without being so much retarded by the plasma membrane as to have time to spread to the sides before leaving the cells (Veijo Wartiovaara, private communication). There are also a few quantitative evaluations of the permeability of algal and bacterial cells to oxygen (137).

The great permeation power of oxygen is understandable in view of its considerable lipid solubility and the smallness of its molecules. Besides, all other gases, too, must have a very great permeation power, for they are all very lipid-soluble and/or have very small molecules.

3. Sugars and Amino Acids

These are among the most lipid-insoluble substances known. It is therefore only natural that plant protoplasts, when, for instance, plasmolytically studied should prove more or less impermeable to them. This applies even to such a relatively small-molecular substance as glycine. On the other hand, a continuous consumption of sugars and amino acids must be assumed in growing and respiring cells. But, as already pointed out in IIIC, the uptake of these substances seems largely to be due to processes of active transport.

VIII. Permeability to Salts and Ions

For several reasons it is much more difficult to study the permeation power of salts and ions than to measure that of nonelectrolytes.

Owing to the great electrical charges of the ions, measurable amounts of any single ionic species cannot be taken up or given off by a cell unless the ions in question are either accompanied by an equivalent amount of oppositely charged ions or exchanged for an equivalent amount of the same electrical sign.

Another very great difficulty is that the true permeability to ions, at least of most plant protoplasts, seem to be very small and that the transfer of ions through the plasma membranes is principally due, not to simple permeation, but to poorly known metabolic forces combined with electrical gradients of unknown sign and magnitude. While nonelectrolytes mostly tend to distribute themselves between the cell sap and the surrounding medium in such a manner that their final con-centrations are approximately the same in each solution, such a distribution almost never occurs in the case of ions (see this volume, Chapter 4). A discussion of salt uptake by plant cells in terms of simple permeation would therefore be not only difficult but misleading.

Therefore, and in spite of a vast literature in this field, our present knowledge of the permeability of plant protoplasts to salts and ions is still far from being as exactly established as our knowledge concerning the permeability to nonelectrolytes. In fact, although permeability constants to express quantitatively the permeability of plant protoplasts toward numerous nonelectrolytes have been calculated, no permeation constants for salts and only extremely few for ions are so far available.

The active transport of ions will be thoroughly treated in this volume, Chapter 4; here only the true permeability to salts and ions will concern us.

The first observations concerning the salt permeability of plant cells were all based on plasmolytic experiments. Many years ago de Vries (168) and Overton (121) found that concentrated solutions of such salts as sodium chloride or potassium nitrate produce a permanent plasmolysis. From this they concluded that the protoplasts are, in general, more or less impermeable to such salts. Later investigators, using a more refined plasmolytic technique have achieved somewhat more exact results. Thus Fitting (58), in his experiments with epidermal cells of *Rhoeo discolor*, was able quantitatively to measure the slow deplasmolysis of the cells in solutions of several alkali salts. Presuming that the rate of deplasmolysis is a measure of the penetration rate of the salt, he assumed that the penetration rate decreases in the order K > Na > Li as far as cations are concerned, while, in the case of the anions, sulfates penetrate considerably more slowly than chlorides, bromides, chlorates, and nitrates. In solutions of alkaline earth salts (chlorides and nitrates of Mg, Ca, Sr, and Ba) no deplasmolysis could be detected. The most rapidly penetrating salts seemed to have an initial permeation rate slightly less than that of glycerol; however, in a few hours the deplasmolysis rate decreased considerably, indicating, according to Fitting, a strong reduction in permeability to the salts in question. Similar results obtained with other cells were reported later by several other investigators. However, the exact interpretation of such seemingly simple experimental results is far from clear. Thus it is almost impossible to decide to what extent the results may have been influenced by factors such as a possible exosmosis of normal components of the cell sap or by active ion transport. Moreover, if an exchange of ions between the protoplast and the medium occurred, it would not, of course, be detectable by plasmolytic methods.

Evidently, then, the plasmolytic experiments must be supplemented, especially where ion permeability is concerned, with studies based either on chemical analyses or on the use of radioactive isotopes. A combination of chemical and radioactivity determinations is still better.

For a long time it has been known that the sap of some *Valonia* species contains high concentrations of potassium ions but only relatively low concentrations of sodium ions, while in the sea water from which the algae must have taken up all their constituents, the ratio between potassium and sodium is just the opposite. This is, in fact, a situation which is typical of a multitude of both plant and animal cells: with but few exceptions potassium dominates in the interior of the cells in spite of the frequent dominance of sodium in the fluids bathing the cells. Thus one or both of these ionic species is actively transported across the plasma membranes. At the same time, however, these membranes must be poorly permeable to the ions in question, for the greater the leakage occurring across them, the greater is, of course, the work required to maintain the constant ionic difference between the extra- and intracellular fluids.

With cells of *Nitella* and some other characeans, somewhat more detailed experiments have been carried out. Thus, for instance, some thirty years ago Hoagland *et al.* (72) found that when *Nitella* cells are placed in a dilute bromide solution a very slow exchange of bromide ions for chlorine ions occurs, so that within about 40 days a kind of steady state is attained. However, as pointed out by Steward and Millar (156), in these experiments a large population of *Nitella* cells was used, and so the probability exists that the bromide was actively taken up preferentially by the younger, still growing cells, while the chloride issued principally from more senescent ones. The exchange of bromide ions for chloride ions may thus in reality have occurred much more slowly than one would at first sight have been inclined to suppose. In other experiments (30), characean cells rich in potassium ions have been kept in solutions of rubidium salts. The exchange of rubidium for potassium was then found to occur at least a million times as slowly as an exchange of these cations by free diffusion through a water layer of the same thickness as the protoplasmic layer, or roughly a thousand million times as slowly as by diffusion through a water layer of the thickness of the plasma membrane. It thus seems fairly well established that cells of this type are almost impermeable to cations as well as to anions. In Section IV,B,2 it was stated that inanimate membranes may easily be prepared which are permeable to either cations or anions alone. Several animal protoplasts also seem to be more readily permeable either to positive or to negative ions. It may therefore be pointed out that among plants no clear examples of either cation- or anion-permeable protoplasts are so far known.

Besides, measurements of the electrical conductivity of living cells also show that their membranes are almost impermeable toward ions. Thus, for instance, the direct current resistance of *Nitella* cells has been found to be of the order of magnitude of 100,000 ohm per square centimeters. [Cf. Cole (28), Danielli (42), Blinks (13), and Umrath (164).]

More quantitative measurements of the permeability to ions have been carried out by Holm-Jensen *et al.* (89). Cells of some of the Characeae had been kept for several weeks in a constant milieu, and the authors therefore assumed that a steady state had been reached in which the loss of any ions by diffusion was balanced by active uptake of the same ion. When now one of the ions present in the medium is labeled by the addition of an infinitesimal amount of a radioactive isotope, without any appreciable alteration of the concentration of the medium, the protoplasm and the sap of the cells will gradually become more radioactive and the rate of increase of their activity can be used to estimate the rate of diffusion of the cation in question through the protoplasm. In this way the permeability of the protoplasts to potassium ions was found to be roughly 1×10^{-8} cm per second for *Nitellopsis* (= *Tolypellopsis*) and about 2×10^{-9} cm per second for *Nitella*, while the corresponding values for sodium ions were about 4×10^{-9} cm per second in the case of *Nitellopsis* and 0.4×10^{-9} cm per second in the case of *Nitella*. This means that the half-exchange times would be something like 20–400 days for these ions and cells. At any rate it is clear that these ion exchanges proceed exceedingly slowly, approximately as slowly as the permeation of sugars, for instance. On the other hand, Brooks (94) found a considerably more rapid uptake of radioactive potassium and sodium ions into the sap and especially into the protoplasm of *Nitella* cells. A critical repetition of these experiments therefore seems desirable.

Among experiments concerning the ion permeability of yeast cells, those of Conway and his associates [cf. Conway (39)] may be cited. They found that if resting yeast cells are shaken anaerobically in a solution containing labeled potassium ions, after some hours only a small percentage of the potassium ions from the external solution have mixed with the internal potassium ions. If, on the other hand, the yeast is shaken in the same solution in the presence of air, labeled potassium ions will enter more readily so that, after 1 hour, the mixing may have proceeded to some 40 to 50%, but this entry is almost entirely inhibited by cyanide. Especially in view of the small dimensions of the yeast cells their permeability toward potassium ions must thus be designated as very low. A high degree of impermeability of yeast cells to anions has likewise been reported (59).

There also exist a great number of studies on the ion exchange between excised roots or disks of various storage tissues, on the one hand, and aqueous salt solutions, on the other, but the proper interpretation of their results is no easy matter (cf. 54, 55, 107, 140, 158). In almost all these investigations there has been found a conspicuous dualism in regard to ion uptake: after an initially rapid uptake, or exchange, of ions, there follows a period of very much slower uptake or exchange. The first of these processes is largely independent of temperature, oxygen supply, and metabolic inhibitors, while the second step is strongly affected by all these factors. Hence it seems natural to assume that the uptake during the initial phase is essentially of the "passive" type, while the second one is metabolic in nature (cf. Chapter 4 in this volume). So far virtual agreement exists between most investigators in this field. But, as stated in Section VIA, the parts of a tissue to which strong electrolytes have free access during the first phase is still largely an open question: Is it to cell walls and intercellular spaces only, or to the cytoplasm or to parts of it, also? The answer to this question will prove of fundamental importance to the understanding of the influence of ions upon plant protoplasts.

Irrespective of whether the main diffusion resistance toward ions is located in both outer and inner plasma membranes or in the inner alone, it seems that almost all plant protoplasts possess a great, or even a very great, diffusion resistance toward ions. Thus, it is evident that active transport rather than diffusion is primarily involved in the entry of ions into cells (see Chapter 4 of this volume).

There may, however, exist a few exceptions to this general rule.

One such exception, and indeed a very outstanding one, is that displayed by the chief example of the ultrafiltration theory, *Beggiatoa mirabilis*, which seems to possess an extremely high permeability to salts (146, 147). But surely it is no mere chance that the cells of this organism are almost devoid of turgor: a certain degree of semipermeability is obviously indispensable for the development of turgor pressure.

Certain brown algae living in the tidal zone where they are exposed daily to great concentration changes are also characterized by an exceptionally high salt permeability (12), and the same seems to hold true for numerous diatoms (81). Perhaps the high salt permeability of these organisms is to be interpreted as an adaptation which makes it possible for them to endure the concentration changes in their natural habitats.

Concerning the ion permeability of bacterial cells there still exists some diversity of opinion. Thus, according to Roberts *et al.* (139) the cells of *Escherichia coli* are freely permeable to most, if not all, ions, so that equality of concentration with the ambient solution is reached within 5 minutes. On the other hand, several other investigations [Fischer (57); Hill (69); Collander (33); Mitchell and Moyle (112a)] clearly show the existence, in a variety of bacterial cells, of a peripheral osmotic barrier more or less impermeable to ions. It therefore seems probable that the relatively rapid uptake and release of solutes often observed in bacterial cells may depend in part on their exceptionally great surface:volume ratio and in part on more or less specific ion exchange mechanisms.

IX. Permeability to Acids and Bases

A. INTRODUCTION

The study of permeability to acids and bases is beset with special difficulties. First of all, it must be remembered that their hydrogen and hydroxyl ions, if present in too high concentrations, will display toxic effects. In all permeability experiments with acids or bases, it will therefore be especially important to ascertain whether the cells experimented with are still in a normal state at the end of the experiment. Otherwise there is the possibility that the observed penetration of acids or bases may be only a consequence of the protoplasts having lost their normal state of restricted permeability.

Another complication is that acids and bases may penetrate the cells as either (a) undissociated molecules, (b) ions, or (c) molecules and ions at the same time. In the preceding section we have seen that plant protoplasts are, in most cases at least, relatively impermeable to ions. We have thus to expect that weak or moderately strong acids and bases will penetrate exclusively, or at least preferentially, as undissociated molecules. Wide experience with numerous acids and bases shows that this is in reality so (151). This fact must, of course, be taken into account in calculating the permeability constants: the effective concentration gradient is not that of the total acid or base but that of the undissociated molecules alone. Evidently, then, the final equilibrium attained will be a function of the extra- and intracellular hydrogen ion concentrations, on the one hand, and of the dissociation constant of the penetrating acid or base, on the other (103). Thus, to take a specific example, ammonia molecules entering the acid sap of a cell will instantaneously be converted into ammonium salts of the acids present in the sap. These ammonium salts (i.e., the ammonium ions) are unable to penetrate the protoplast, while the ammonia molecules $(NH_3 + NH_4OH)$ penetrate it with extreme rapidity. The consequence of this will be that the ammonia is very effectively trapped, or accumulated, in the cell sap. This is not, of course, an example of the active transport capacity of the cells but only a case of diffusive or "metaosmotic" solute uptake in the sense of Bogen (15). By watching the rate of accumulation of ammonia it is therefore virtually possible to calculate its permeation constant provided the concentrations of the undissociated ammonia molecules outside and inside the cell are known.

B. ACIDS

The cell sap often contains lipid-insoluble acids such as oxalic, malic, tartaric, and citric acids in remarkably high concentrations. On the other hand, acetic, lactic, and pyruvic acids, although frequently arising in cell metabolism, are rarely found to accumulate in cells as free acids. It seems plausible to assume that this striking difference in the occurrence of the two groups of acids is due to differences relating to their permeation power: the acids of the second group possess a considerable lipid-solubility and have such a great permeation power that they simply cannot be kept accumulated in any cells whereas, on the other hand, the plasma membranes are practically impermeable to the acids of the first-mentioned group, which, therefore, if once accumulated in the cell sap, will remain there for unlimited periods of time.

One way of investigating the entrance of acids into plant cells is to observe the color changes brought about in cells which contain anthocyanins in their sap. Unfortunately, however, the anthocyanins of most cells are relatively insensible to slight increases in acidity (23, 49). Nevertheless, as was found by Jacobs (93), the pigment of the petal cells of *Symphytum peregrinum* is so sensitive to even a slight increase in the acidity of the sap that the entrance even of such an extremely weak acid as carbonic acid is clearly visible. With these cells Jacobs showed that carbonic acid or its anhydride CO_2 has a unique permeation power. Thus the cells "may be caused to develop an intracellular acidity in a solution of CO_2 in $M/2$ $NaHCO_3$, which has the alkaline reaction of pH 7.4, practically as rapidly as in a solution of CO_2 in distilled water, of pH 3.8; i.e., with an apparent hydrogen ion concentration 4,000 times as great, while in distilled water of pH 5.5–6.0 there is no change, although the apparent H-ion concentration is perhaps 100 times as great as in the first-mentioned solution." These seemingly paradoxical results are easily explained if the H-ions do not penetrate appreciably, but only the CO_2 (and H_2CO_3) molecules do so. Hydrogen sulfide is another weak acid which behaves toward living cells very much like carbonic acid (119).

An ingenious method of determining quantitatively the permeation power of both weak acids and weak bases was invented by Jacobs (97). Aqueous solutions of, say, ammonium acetate are always partially hydrolyzed into acetic acid and ammonia:

$$CH_3CONH_4 + H_2O \rightleftharpoons CH_3COOH + NH_4OH$$

Now, if we plasmolyze plant cells with an ammonium acetate solution, the salt as such will penetrate the protoplasts only very slowly. On the other hand, both acetic acid and ammonia molecules will enter very rapidly and will combine in the cell sap so that ammonium acetate is again produced. As a consequence of this, deplasmolysis will occur and its rate will depend primarily on the entrance rates of acetic acid and ammonia. This method has two great advantages. First, acid and base occur in such low concentrations that toxic effects are largely avoided. Secondly, although the acid and the base, if studied separately, would penetrate the cells so rapidly that exact measurements of their penetration rates would be extremely difficult, their joint penetration from the salt solution may nevertheless be readily measured owing to the very low concentration of the permeating molecules. From experiments carried out with this method it can be deduced that the permeation powers of the fatty acids increase in the order acetic < propionic < butyric < valeric acid. The influence of lipid solubility seems thus in this case more important than that of molecular size. The first member of this homologous series, formic acid, seems however to be exceptional.

Indoleacetic acid and other growth regulators of analogous composition constitute a physiologically important group of weak acids. Their physiological activity is, at least approximately, proportional to the concentration of the undissociated acid molecules in the external solution (120). The lipid solubility of all these substances is so great that they will readily penetrate all sorts of plant protoplasts. Also they should be accumulated from acid solutions in the more neutral or alkaline regions of the plant, such as, the cytoplasm and the content of the sieve tubes. According to Johnson and Bonner (100) phenoxyacetic acid, however, may be taken up, or bound, in several different ways.

C. Bases

Turning now to the permeation of bases, we shall find among them a state of affairs similar to that among the acids. Thus the extremely high permeability of all living protoplasts to carbonic acid and/or its anhydride has its counterpart in their behavior toward ammonia (NH_3 + NH_4OH). Just as living cells may turn more acid in an alkaline bicarbonate solution, so their pH may change in an alkaline direction when they are placed in a slightly acid solution of, say, ammonium chloride, owing to the very rapid penetration of the ammonium liberated by hydrolysis of the ammonium salt (92).

There are numerous other weak bases, such as alkylamines, alkanolamines, and the natural alkaloids, which behave in much the same manner toward living protoplasts as does ammonia (65, 122). A more quantitative comparison of the permeation powers of several such bases has been carried out by Äyräpää (6). His experiments were performed with bakers' yeast. The method was based on the color change caused by the penetration of the bases into yeast cells stained with neutral red. The time required to reach a certain standard color was used as an inverse measure of the penetration rate. His results indicate, as is seen from Fig. 3, that lipid solubility is a most important factor here. Thus even such very large alkaloid molecules as those of atropine (mol. wt., 289), cocaine (mol. wt., 303) and thebaine (mol. wt., 313) were found to penetrate the cells very rapidly. On the other hand, the very smallest molecules (ammonia, methylamine, hydrazine) showed permeation powers about 100–1000 times as great as those of somewhat larger molecules of corresponding lipid solubility. [Kral (105), using *Tradescantia* cells as his test object, has recently, with another technique, obtained results which are not quite in accord with those of Äyräpää.]

Like the strong acids (HCl, HNO_4, H_2SO_4, etc.), the strong bases

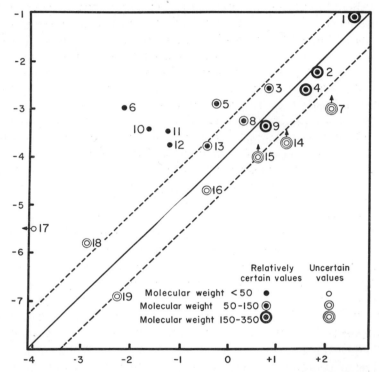

FIG. 3. Correlation between the permeation power of several bases toward yeast cells, on the one hand, and their relative ether solubility and molecular weight, on the other. Abscissa: log distribution coefficient ether:water; ordinate: log $PM^{1/2}$. The points are: *1*, diisoamylamine; *2*, sparteine; *3*, isoamylamine; *4*, novocaine; *5*, diethylamine; *6*, ammonia; *7*, cocaine; *8*, triethylamine; *9*, ephedrine; *10*, methylamine; *11*, dimethylamine; *12*, ethylamine; *13*, trimethylamine; *14*, thebaine; *15*, atropine; *16*, diethylethanolamine; *17*, hydrazine; *18*, ethanolamine; *19*, diethanolamine. From Äyräpää (6).

(NaOH, KOH, Ca(OH)$_2$, tetraalkyl ammonium bases, etc.) have almost no permeation power toward normal, uninjured protoplasts (23, 65).

X. Permeability to Dyes

Theoretically the permeability to dyes should already have been treated in the discussion devoted to the permeation of strong and weak electrolytes. However, the vital stains form such a peculiar group of permeators and have given rise to such a vast literature that their separate consideration seems appropriate.

It has already been pointed out that, although investigations into the permeability of living cells to dyes may at first sight seem very promising, they are in reality beset with numerous pitfalls the danger of which should not be underestimated [cf. Drawert (50)].

Thus most of the commercially obtainable dyestuffs contain large amounts of impurities. Many dyestuffs are in fact mixtures of several dyes differing from each other not only in chemical constitution but also in permeation power. A notable example of this is the much-used vital stain methylene blue. Pure methylene blue is tetramethylthionine, which is a strong base. It is in itself a poor vital stain. In aqueous solutions, however, especially at pH values above 9.0, it is gradually oxidized to trimethylthionine, which is a much weaker base and has a much greater permeation power. So it comes about that most vital staining with "methylene blue" is in fact due to impurities in the dyestuff used. The Biological Stain Commission (United States) has done much to facilitate the supply of stains of known chemical and physiological properties. (Cf. 38.)

Another drawback of many dyes, and especially of almost all basic dyes, is their great toxicity to living cells. In order to avoid damage to the cells, such dyes must therefore be used in very low concentrations. In layers less than, say, 1 mm thick such dilute stain solutions appear colorless. Obviously, then, the entrance of such dyes into cells of microscopic dimensions cannot be observed unless the dye is efficiently accumulated in the cell. This fact has given rise to endless discussions as to whether the nonappearance of a stain in cells is due to failure of the dye to penetrate the protoplasts or only to failure of the cells to bind the dye. There is, of course, no single answer to this question: each case requires special study.

Fluorescent dyes are often visible even if present in extremely low concentrations. In recent years therefore studies of their uptake have aroused much interest.

Exact measurements of the dye concentrations attained in cells of microscopic dimensions have hitherto proved very difficult. Only quite recently have instruments suitable for such measurements been devised (10). The possible intracellular reduction of dyes to colorless compounds, the strong adsorption of dyes by many cell walls, and the colloidal or semicolloidal nature of many dyes are other difficulties encountered in studies of the permeability of protoplasts to dyes.

It is not surprising, therefore, that the numerous attempts to test the permeability hypotheses by vital staining experiments have, on the whole, yielded rather disappointing results. "They have not only not as yet furnished a solution of the general problem of cell permeability but they have, if anything, been responsible for even greater confusion than existed before." These pessimistic words of a distinguished reviewer of this field (93) some thirty years ago are still difficult to refute.

Before discussing the details concerning the permeation of dyes, it should be pointed out that the permeation power and other physiological properties of dyes depend, to a very large extent, on their ionization, and the dyes may thus be divided into two main classes, i.e., basic and acid dyes.

When a dye is classified as "basic" this means that it is a salt of a colored base with some colorless acid, such as, for instance, hydrochloric or sulfuric acid. Most of the dye bases are weak bases with an ionization constant between about 10^{-3} and 10^{-8}. There are also, however, a few dye bases which may be called strong. The already mentioned tetramethylthionine or methylene blue base is an example of this.

Acid dyes, on the other hand, are salts of colored acids with colorless bases such as, for instance, sodium or potassium hydroxide. Among the dye acids some are weak acids which owe their acidic character to the presence of carboxyl or phenolic hydroxyl groups. The nitrophenols, phthaleins, and fluoresceins may be mentioned as examples of such dyes. On the other hand, numerous acid dyes are obtained by sulfonating basic dyes; they thus contain the —SOOH group and have the character of strong acids.

Numerous basic dyes are effectively accumulated by many, but not all, plant cells. The mechanism of this accumulation was outlined by Pfeffer in his fundamental investigation in the year 1886 (131) on the uptake of synthetic dyes by plant cells. Since then further details of the process have been filled in by McCutcheon and Lucké (111), Irwin (91), Jacobs (97), Kinzel (103), and others. There are two variants of this mechanism. The first of them is virtually identical with the mechanism of accumulation of ammonium ions already described (IX). The nonionized lipid-soluble dye base molecules enter the cells very rapidly. If the cell sap is acid enough, they form salts with the acids of the sap. These salts have practically no permeation power. Once formed, therefore, they cannot leave the cell again but are trapped there. New amounts of the dye base will continue, however, to enter until the concentration of the nonionized dye base is the same in the cell sap as in the surrounding solution. The upper limit of the accumulation brought about in this way is thus a function of the H-ion concentrations of the sap and of the ambient fluid, on the one hand, and of the ionization constant of the dye base on the other. [For particulars, see (103).]

Cells trapping basic dyes in this way, i.e., owing to their acid content alone, are said by Höfler and Schindler (84) to have "empty saps." There are, however, other kinds of cell saps too—"full saps" as Höfler called them. They contain other substances capable of binding dye bases. The chemical nature of these substances is known only in a few cases. So far it is not possible to state quantitatively the maximal accumulation capacity of such cells.

From what has been said above, it should be clear that the uptake of basic dyes differs from the more familiar uptake of, say, lipid-soluble nonelectrolytes only in one respect, namely, in that the base molecules do not remain free in the cell sap but are bound in some way. This binding of the dye bases does not prevent the calculation of their normal permeation constants. At least a minimum value for the permeation constant is readily obtainable if, as a first approximation, we assume that the concentration of the nonionized dye base molecules in the sap, during the time period in question, remains equal to zero. So far, however, very few permeation constants calculated in this way are to be found in the literature [e.g., Bartels (10)].

Turning now to the sulfonic acid dyes, we are confronted with substances which behave in quite another manner than the basic dyes. Thus, the sulfonic acid dyes are accumulated in comparatively few kinds of cells, and this accumulation is not due to a trap mechanism but to active transport (51). In most kinds of plant cells no entrance at all of the sulfonic acid dyes can be observed even in hours or days, in spite of the fact that most of these dyes are fairly nontoxic and can thus be used in high concentrations.

Is this nonentry of the sulfonic acid dyes caused by "lack of affinity" for these dyes on the part of the cell content or by true impermeability of the protoplasts toward them? That the latter alternative is the right one is shown, *inter alia*, by the following experiments (36): Cells of *Nitellopsis* were kept 1–2 weeks in fairly concentrated solutions of such dyes. At the end of the immersion period no traces of the dye could be detected in the isolated cell sap. At any rate the dye concentration in the sap was less than 1/1000 of its concentration in the bathing fluid. This cannot have been due to a continuous intracellular decolorization of the dye, because in special experiments with very small amounts of bathing fluid no decrease of its dye concentration could be detected. Nor can the absence of dye in the cell sap be explained by assuming that the dye is insoluble in the sap or that it is negatively adsorbed by some colloids of the sap, for analyses of the sap showed that it is essentially a not too concentrated aqueous salt solution containing only very low concentrations of organic substances.

The permeation power of the weak dye acids has not been much studied. It seems, however, possible that fluorescein, like the auxins, is accumulated in the sieve tubes. Such an accumulation, if really occurring, could well be the counterpart of the trapping of weak bases in acid saps, for the sieve-tube content is distinguished by such extremely high pH-values as 7.4–8.7 (176).

XI. Influence of External and Internal Factors on Permeability and Active Transport

A. INTRODUCTION

All properties of the living protoplast are subject to changes brought about by a great variety of internal and external factors. It is therefore only natural to assume that their permeability and their capacity for active transport also fluctuate according to existing conditions. The literature contains, in fact, an abundance of hints, assumptions, and statements concerning permeability changes in living protoplasts. Many authors seem, indeed, to regard this changeability as the most significant quality of protoplasmic permeability. It is also a very widespread opinion that these changes in the permeability of the protoplasts are of essential importance to the life processes occurring in the cells: an increase in permeability will, according to these views, promote the course of the life processes, while a decrease in permeability is supposed to slow them down. Experimental verification of these assumptions has not always been thought to be necessary, so that the permeability-change concept has largely served as a kind of magic formula considered adequate to solve even the most complex problems in a simple manner. Moreover, when verification has been attempted, the success achieved has often been about inversely proportional to the exactness of the methods used.

The craze for discovering permeability changes was understandable as long as it was thought that the exchange of substances between cells and their surroundings depended solely, or at least primarily, on the magnitude of cell permeability. But now, since it has been realized that active transport plays a major role in this process, it seems *a priori* more likely that cell activity will be controlled rather by the intensity of active transport than by variations in cell permeability. It also seems that the capacity for active transport is much more susceptible to both external and internal influences than is permeability proper.

B. INFLUENCE OF SINGLE FACTORS

1. Temperature

Temperature is one of those factors whose influence on both permeability and active transport is most indubitable and apparent. Indeed, the very fact that temperature is a measure of the mean kinetic energy of the molecules and ions makes it self-evident that all permeation and transport processes must of necessity be influenced by temperature.

The influence of temperature on chemical, physical and physiological processes is often expressed in terms of Q_{10}. This temperature coefficient, as it is called, is the ratio of the velocity constant of a process at a given temperature to its velocity constant at a temperature 10°C lower. Thus

$$Q_{10} = \frac{k_t}{k_{t-10}}$$

where k_t and k_{t-10} are the velocity constants of the process at the temperatures t and $t - 10$, respectively.

For a long time it has been known that the temperature coefficient of diffusion in aqueous solutions is generally of the magnitude 1.2–1.4, while that of chemical reactions is much higher, i.e., about 2–3. Now, the Q_{10} values of cell permeability are also very high, mostly between 2 and 5, sometimes even still higher. In the past this was frequently regarded as an indication of the "chemical nature" of the permeation process. This conclusion was, however, unwarranted, for we now know that diffusion processes in highly viscous media often show high temperature coefficients. It now seems more plausible to assume that the great influence of temperature on permeation rates might be, in the first place, an indication of a high viscosity of the plasma membranes. But the temperature coefficients of permeation are such complex quantities that a thorough analysis of them presents very great difficulties (173).

As is seen from Table VI, the temperature coefficients of permeation vary considerably, depending on the character of the permeator. Some time ago it was held that the coefficients were smaller in the case of rapidly permeating substances and greater in the case of slow permeators. According to Wartiovaara (173), however, it is more correct to say that the temperature coefficients of small molecules are, on the whole, lower than those of larger ones. Theoretically such a rule would not be surprising, since it seems comprehensible that the activation energy of the permeation process should increase with the bulkiness of the permeating molecules.

If the structure of the plasma membrane changes with the temperature, the temperature coefficient will, of course, be influenced by such changes. However, insofar as it is hitherto known, the value of the temperature coefficient is fairly constant within the whole temperature range so far studied, i.e., between about 0° and 30°C. It thus seems as though at least no sudden major structural changes occurred within these temperature limits.

TABLE VI

TEMPERATURE COEFFICIENTS (Q_{10} VALUES) OF THE PERMEATION OF DIFFERENT NON-ELECTROLYTES WITHIN CERTAIN TEMPERATURE INTERVALS AS COMPARED WITH THE PERMEATION CONSTANTS OF THE SAME SUBSTANCES AT 20°C[a]

Substance	Temperature interval (°C)			P (cm/hour)	Test object
	0–10°	10–20°	20–30°		
Urea	2.5	2.6	2.8	4×10^{-4}	*Chara ceratophylla*
Methanol	2.6			2.0	*Nitella mucronata*
Ethanol	2.7			2.0	*Nitella mucronata*
Propanol	2.7			3.3	*Nitella mucronata*
Ethylene glycol	2.3	2.9	3.5	0.010	*Nitellopsis obtusulus*
Methyl carbamate	3.2	2.9	—	0.75	*Nitellopsis obtusulus*
Glycerol	—	3.5	8	8×10^{-5}	*Nitellopsis obtusulus*
Tetraethylene glycol	—	4.9	—	5×10^{-4}	*Nitellopsis obtusulus*
Trimethyl citrate	—	5.5	—	0.05	*Nitellopsis obtusulus*
2,3-Butylene glycol	7.4	5.9	4.8	0.016	*Nitellopsis obtusulus*

[a] Mainly after Wartiovaara (173).

The temperature coefficient of active transport processes seems also to have values higher than 2.

2. Light

The literature contains numerous statements concerning permeability changes, especially permeability increases, caused by visible and invisible radiation [cf. Brauner (22)]. There are, however, many negative statements, and so this question is still somewhat obscure.

On the other hand, there can be no doubt concerning the favorable effect of light on the active uptake of salts by algae and other chlorophyll-containing cells. Probably this effect is, directly or indirectly, connected with photosynthesis.

3. H+ and OH- Ions

As already stated (Sections IX and X) the uptake of weak acids and bases is strongly influenced by the pH value of the ambient

solution. This is, however, primarily an influence on the ionization
of the penetrating substances rather than an influence on the cells themselves. On the other hand, if the proteins of the plasmalemma participate at all in the uptake of ions, the pH of the medium may be
expected to affect cell permeability, since it is well known that proteins
on the acid side of their isoelectric point will combine with anions,
while on the alkaline side cations are adsorbed. Experimental evidence
of such effects is, however, meager (55).

4. Other Ions

There are especially in the somewhat older literature, many statements that certain univalent cations such as Na+ and K+ tend to increase
cell permeability, while Ca++ and also some other bi- and trivalent
cations have the opposite effect. [Concerning the vast literature on "ion
antagonism," see the textbooks of Höber (76) and Heilbrunn (68).]
A more critical examination of the observations on which these ideas
are based shows, however, that this is not in all these cases the only, or
even the most probable, interpretation.

Thus, when it is found, for instance, that the uptake of bromide
occurs more readily from a postassium bromide solution than from
one of calcium bromide, this does not prove that the permeability of the
cell has been increased in the first-named solution or decreased in the
latter, for an even more plausible explanation is that when bromide ions
are actively taken up, the potassium ions are better suited as accompanying counterions than are calcium ions (see also Chapter 4 of this
volume).

Again, when it is found, for example, that the uptake of rubidium
ions is specifically depressed by potassium ions, and vice versa, this is
no indication of any permeability changes but is more plausibly interpreted as due to mutual competition of the two ions for one and
the same carrier (55). Such phenomena will be more fully treated in
Chapter 4 of this volume.

Finally, it should not be forgotten that pure solutions of, say, alkali
salts tend to injure living protoplasts and thus to decrease their normal
diffusion resistance, while calcium ions, for instance, are able to
counteract such harmful effects of the alkali ions. In such cases, then,
it may be somewhat questionable whether the permeability changes
observed should be interpreted as direct effects of the ions on the plasma
membrane or perhaps rather as symptoms of a general disturbance of
the living system.

It thus seems that there are not very many entirely unequivocal examples left of truly physiological and entirely reversible permeability

changes directly brought about by alkali or alkaline earth ions. But entirely to deny the existence of such influences would, no doubt, be an overstatement.

5. Oxygen

In Section III it was stated that a sufficient supply of free oxygen is, in many cases at least, an indispensable condition for effective active transport. Permeability proper, on the other hand, seems scarcely to be affected by anaerobic conditions, provided that they do not last so long a time that the life of the cell is jeopardized. It is often assumed that the existence of osmotic barriers in the protoplast depends on the continued activity of the cell. If this assumption holds true, plasma membranes of anaerobic cells will probably not persist under anaerobic conditions. At the same time, however, the life of the cells will probably be extinguished, and so it may be difficult to decide with certainty whether the abnormal increase in cell permeability is a direct consequence of the anaerobic conditions or whether it is a consequence of the death of the cells. At any rate, oxygen deficiency will tend to increase rather than decrease permeability at the same time as it checks active transport.

6. Anesthetics

We have already seen (Section IIIB) that active transport processes may be reversibly checked by anesthesia. However, in the case of true permeation processes the situation is much less clear, for while earlier investigators mostly assumed a similar decrease of permeability in the state of anesthesia, a reversible increase of permeability caused by anesthetics was later convincingly demonstrated in some cases (85b). It would seem then, that, depending perhaps on the special qualities of the anesthetics and on their concentration, both increase and decrease of permeability may occur. There is also the possibility that while the permeability toward some permeators is increased, that toward others is simultaneously decreased. In these respects our knowledge is still very imperfect (cf. 126).

7. Other Substances

The literature contains a multitude of statements concerning the alleged influence of different substances on permeability. Thus, to take a single example, the idea that auxins will increase the permeability to water and that they will in this way promote cell enlargement has found some supporters. The experimental evidence in favor of this hypothesis is, however, not very convincing (cf. 117). Other authors assume that auxins cause a nonosmotic, or active, water uptake, but this assumption, too, rests on a rather weak basis.

8. Nutritional State of the Cells

Hoagland and his associates (70) have shown that the salt-accumulation capacity of root cells is highly dependent on their nutritional state: the less salts and the more carbohydrates the roots contain, the greater is their capacity for salt uptake. Similar observations have since been made on other objects too. But a corresponding influence of the nutritional state on the permeability is not known.

9. Hydration and Dehydration

It is often assumed that a hydration of the plasma membranes will increase their permeability, while a dehydration of them will decrease it. Such an assumption seems in itself plausible, yet it must not be taken for granted that a hydration or dehydration of the bulk protoplasm will always be accompanied by a corresponding change in the osmotically determinative plasma membranes, which differ considerably both in chemical composition and in structure from the bulk protoplasm. Actually we know almost nothing about the hydration or dehydration of the plasma membranes.

10. Stimulation

The ion permeability of nerves and muscles is known to undergo sudden profound changes in connection with the passage of an excitation wave through them. It would therefore not be surprising if some similar change were to occur in plant cells concerned with the conduction of excitations or with the carrying out of sudden contractions. In fact, the sudden drop of turgor occurring, for instance, in the pulvini of *Mimosa pudica* and in the stamens of the *Cynareae* may well be interpreted as due to a sudden increase in permeability causing solutes and water to leak out from the cell sap through the protoplasm. The general occurrence of electrical responses in plants, more or less similar to those in animals, points in the same direction (164). From the standpoint of permeability these phenomena are, however, still rather imperfectly understood.

11. Seasonal Factors

As already mentioned (VIID), a seasonal change from one permeability type to another has in some cases been observed. Perhaps it will be found that the capacity for active transport is even more dependent on seasonal factors.

12. Frost and Drought-Hardening

According to Levitt (110), these cause a considerable increase in the permeability to polar substances, while the permeability to apolar substances remains unchanged.

13. Age

Young cells sometimes differ considerably as to permeability from middle-aged cells, just as these differ from senile ones (26). Probably changes in the capacity for active salt accumulation relative to age and development are even more common. However, as these points will be dealt with in greater detail in Chapter 4 of this volume, they need not concern us here.

14. Injury and Death

It was known even to the earliest students of cell permeability that the approximate impermeability of the protoplasts to anthocyanins, salts, sugars, etc., is characteristic of them only as long as they remain alive, while dead protoplasts are in most cases freely permeable to almost all dissolved substances. Indeed, the striking increase in permeability connected with death has become one of the most commonly used criteria for discriminating between living and dead protoplasts. However, as was pointed out by Overton (125), the increase in permeability does not always occur suddenly: under the influence of dilute formaldehyde solutions, for instance, it occurs gradually in such a way that the protoplasts first become permeable to relatively small molecules and ions and then successively also to larger and larger ones in roughly the following order: (a) alkali chlorides and nitrates, (b) sulfates, phosphates, and tartrates, (c) sucrose, (d) anthocyanins and tannins. Thus the plasma membranes, in such cases, assumed the properties of molecular sieves of gradually increasing pore size.

Osterhout (118) has carried out extensive investigations concerning permeability in relation to injury and death, but unfortunately the method used by him—measurements of the electrical conductivity of disks of the marine alga *Laminaria* sp.—was not very suited to reveal the finer details of the permeability changes in question.

XII. Synopsis of the Permeability Properties of Plant Protoplasts

On the preceding pages, a fairly detailed account of the permeability of plant protoplasts toward different substances has been given. In order to explain as completely as possible the permeability properties of the plasma membranes, we shall try, here, to summarize the principal evidence and submit a consistent picture of the problem as a whole.

While the aim is to present the pertinent facts in as unbiased a manner as possible, it is convenient to arrange the data according to two main aspects, namely, (a) correlations between lipid solubility and permeation power, and (b) correlations between molecular size and permeation power.

A. Lipid Solubility and Permeation Power

How close is the experimentally found correlation between lipid solubility and permeation power? A great difficulty when we try to answer this question is that we actually know very little about the plasma membrane lipids of plant cells—except what has been inferred from studies on protoplasmic permeability itself. This being so, there is but one way out of this dilemma, namely, to study the solvent properties of some more or less arbitrarily chosen lipoidal solvents, hoping that the solubilities of different substances in some of these solvents will be found to be at least reasonably well correlated with the permeation powers of these same substances.

While it may be argued that this is an empirical method of trial and error, like searching for a needle in a haystack, the situation is not quite as hopeless as it may at first appear. As already pointed out (Section VIIB), all kinds of lipids have certain important solvent properties in common. While we shall often have to speak of "lipid solubility" in a very general sense, we will at least try to be on our guard against drawing unwarranted conclusions from such a flexible concept as this.

Now, how close is the correlation between lipid solubility and permeation power? Or, in view of the fact that a major correlation between these two qualities is already generally recognized, it may be better to formulate the question in a slighly different way: What are the discrepancies found between lipoid solubility and permeation power?

The older literature especially contains numerous statements concerning such alleged discrepancies which are supposed to disprove the validity of the lipoidal theory. It is not possible here to examine each of them separately, but we will try to classify the most important of them into a few categories (a–d) which will be, at least briefly, discussed.

(a) It has been emphasized by numerous writers that living protoplasts must be able to take up several physiologically important, but lipid-insoluble substances, such as sugars, amino acids, and mineral salts, and it has been suggested that the uptake of these substances is in conflict with the lipoidal theory. In reality, however, this objection is erroneous for two different reasons. First, the often used classification of substances into lipid-soluble and lipid-insoluble ones is, of course, not quite correct, for theoretically all substances are endowed with a certain, although often extremely low, lipid solubility; therefore, no substances are absolutely unable to penetrate lipid membranes by diffusion. Still more important, however, is the fact that the uptake of most "lipid-insoluble" substances is largely due, as we have seen in Section IIIC, to the operation of active transport mechanisms. But such transport processes are superimposed upon permeability per se and thus do not vitiate the correlation between lipid solubility and permeation power proper.

Only in the case of very few, clearly aberrant kinds of cells (*Beggiatoa*, perhaps diatoms also) has a considerable permeability toward practically lipid-insoluble substances been established.

(b) It was claimed, a long time ago, that some basic dyes (methylene green, methyl green, thionine, Methyl Azure) are able to penetrate living cells with considerable rapidity in spite of being insoluble in lipoidal solvents (74, 142). But here we have to note that the alleged "insolubility in lipids" has not been very convincingly established. The method used was to shake an aqueous solution of the dye with a nonaqueous solvent and then simply to see whether the latter had become colored or not. Now there is, first, the possibility that some of the dye bases may be colorless and thus invisible in spite of the dye cations being colored. In such cases one may easily get the false impression that no dye has been taken up by the nonaqueous solvent. Secondly, it must be kept in mind that a distribution ratio of, say, 1:100 or even 1:1000 may be consistent with a fairly rapid penetration into cells. Such a slight lipid solubility is, however, easily overlooked. Finally there is the possibility that the dyes now in question might be almost insoluble in the neutral lipoidal solvents used in the tests but nevertheless distinctly soluble in lipoidal solvents of, for instance, a slightly acidic character (cf. 50, 116). A decision on this question is scarcely possible before the solubility properties of these dyes, and of other dyes too, have been more systematically and more exactly studied than hitherto.

(c) As pointed out in Section VIIC, it has been found that the permeation powers of nonelectrolytes toward *Nitella* cells are not a function of their lipid solubility alone, but also of their molecular size or some similar factor. The same has been found in several other cases, too. These discrepancies between lipid solubility and permeation power will be discussed below under the heading "Molecular Size and Permeation Power."

(d) Minor discrepancies still exist between the permeation powers of different substances and their solubilities in the lipoidal solvents so far tested, but these can be attributed, at least mainly, to experimental errors and also to the fact that the solvents so far tested (olive oil, etc.) are not identical with the plasma membrane lipids as regards their solvent properties.

B. MOLECULAR SIZE AND PERMEATION POWER

There is not a single nonelectrolyte or weak electrolyte known with a molecular weight below about 45, which does not penetrate cells rapidly. NH_3 (mol. wt., 17), H_2O (mol. wt., 18), HCN (mol. wt., 27), CO (mol. wt., 28), O_2 (mol. wt., 32), hydrazine (mol. wt., 32), methanol (mol. wt., 32), hydroxylamine (mol. wt., 33), H_2O_2 (mol. wt., 34), H_2S (mol. wt., 34), acetonitrile (mol. wt., 41), and CO_2 (mol. wt., 44) may be mentioned as examples of this. Now, most of these substances are fairly, or even strongly, lipid-soluble, too, so their permeation power may, to a considerable extent, be attributed to this fact. A comparison with other substances of similar solubility, however, reveals the fact that the substances mentioned, in all those cases when a quantitative comparison has been possible, penetrate plant protoplasts with a distinctly greater rapidity than would be expected if the permeation constants were proportional to, say $k/M^{1/2}$ or even to k/M, where k denotes the distribution coefficient lipid:water and M the molecular weight.

The influence of the molecular size is most strikingly shown when, within a homologous series, the general rule that the permeation power increases with increasing length of the carbon chain is broken in the case of the first member, or the first two members, of the series. Thus it has been found, in all the cases so far studied, that formamide has a greater permeation power than acetamide, while propionamide, again, permeates more readily than acetamide (31). Similarly ammonia, in at least some cases, seems to permeate more rapidly than methylamine, and this more rapidly than ethylamine, while amylamine, for instance, has a still greater permeation power (6). In the case of urea and the alkylureas, the situation is a little more complicated, for while most plant protoplasts are distinctly more permeable to methylurea than to urea, there exists a minority of cell types which display a considerably greater permeability to urea. On the other hand, in all cases so far studied, except *Beggiatoa*, ethylurea and dimethylurea have been found to have a greater permeation power than methylurea.

The influence of molecular size may also be evident when the permeation power increases throughout a homologous series with increasing length of the carbon chain. For, as pointed out by Wartiovaara (172), the increase in permeation power is not, in such cases, as rapid as the increase in, say, relative oil- or ether-solubility.

So far, we have here been concerned with relatively small molecules. In the case of molecules of medium or larger size, there are few even reasonably exact observations concerning possible correlations between molecular size and permeation power. As already stated, however, it seems that the permeation constants of nonelectrolytes of not too low molecular weight, toward *Nitella* cells, are inversely proportional approximately to the molecular weight raised to the power 1.5, provided, of course, that substances of similar lipid solubility are compared. So far, however, it is impossible to know how general the significance of this observation may be. Besides, even in the case of *Nitella* this rule holds true only as a first approximation. It may therefore well be that the factor actually concerned is neither the molecular weight nor the molecular volume but a hitherto unknown factor which is in some way correlated with them. Moreover, it should not be forgotten that even molecules as large as those of, for instance, neutral red (mol. wt., 252), atropine (mol. wt., 289), cocaine (mol. wt., 303), and thebaine (mol. wt., 313) are, in spite of their molecular size but in conformity with their great lipid solubility, endowed with a very great permeation power toward all the protoplasts yet studied in this respect.

C. RELATIVE IMPORTANCE OF MOLECULAR SIZE AND LIPID SOLUBILITY

Although this question cannot be decided unambiguously, the following points suggest that lipid solubility is relatively more important than molecular size. If all we know is the molecular weight, or molecular volume, of a substance, we cannot in most cases predict anything about its permeation power. Thus, substances of the molecular weight of, say, 50, 100, 200, or 400 may equally well belong to the very rapidly as to the very slowly permeating ones. On the other hand, if we know only the distribution coefficient ether:water or, still better, olive oil:water of a noncolloidal substance, we are at once able to anticipate, with reasonable certainty, the order of magnitude of its permeation power. At least from such a practical, or pragmatic, point of view, then, lipid solubility must be regarded as a decidedly more important criterion than molecular size.

This general statement may be supplemented by a more specific example. In Fig. 4 the permeation powers of 70 nonelectrolytes toward *Nitella mucronata* protoplasts are shown [cf. Collander (32), Table 3].

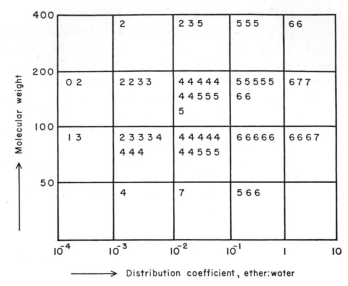

Fig. 4. Permeation powers of 70 nonelectrolytes toward *Nitella mucronata* proto-plasts as correlated with their molecular weights and ether:water distribution coeffi-cients. Each figure in the squares represents one permeator. For further details, see the text.

In this graph the relative ether solubilities, used as approximate measures of the lipid solubilities, increase from left to right on the abscissa, while the molecular weight increases upward on the axis of ordinates. In each square the permeation powers of the substances of corresponding molecular weight and ether solubility are indicated in groups or classes according to the scheme shown. (Every substance

	Permeation Constants Arranged in Classes							
P (cm/sec)	$<10^{-9}$	10^{-9}	10^{-8}	10^{-7}	10^{-6}	10^{-5}	10^{-4}	10^{-3}	10^{-2}
Class no.	0	1	2	3	4	5	6	7

is represented by one figure in the appropriate square.) On scrutiniz-ing Fig. 4 we find that the permeation powers increase both with in-creasing ether solubility and with decreasing molecular weight, but at least within the limits of the evidence available the correlation be-tween permeation power and ether solubility is distinctly closer than that between permeation and molecular weight. It is especially evident that when the distribution coefficients lie between, say, 0.1 and 1, or be-tween 1 and 10, even great variations in the molecular weight will not much influence the permeation power. It is also a remarkable fact that within each square the variability of the permeation power is fairly restricted, i.e., a given combination of solubility properties and molecular size always corresponds to a fairly constant permeation power.

The last-mentioned statement applies not only to *Nitella* cells but to plant protoplasts in general. Thus *if the lipid solubility and molecular size of a substance are known, its permeation power toward most plant protoplasts can be predicted with a high degree of certainty*. It seems appropriate to emphasize this fact, since it is not only theoretically interesting but also, in many cases, of considerable practical importance. There are, it is true, a few cell types (*Beggiatoa, Oscillatoria*, diatoms) whose behavior is somewhat divergent, but these are few indeed as compared with the large number of plant cells endowed with a more or less "normal" premeability.

XIII. Theory of Cell Permeability

A. INTRODUCTION

On the preceding pages an account of the empirically observable behavior of living plant protoplasts toward solutes and water has been given. Theories have so far been avoided in order not to obscure or distort facts. But it is now necessary to explain, as fully as possible, the mechanisms by which certain substances may permeate the plasma membranes, while other substances are, at the same time, more or less prevented from penetrating.

The monographs on permeability by, say, Stiles (157) or Jacobs (93) reveal that, some thirty years ago, a reviewer of the problem of cell permeability had to discuss many different, and often incompatible, permeability theories. No wonder, then, that Jacobs ended his account with the following remarks: "In conclusion, it may be emphasized that what is most needed in the field of cell permeability at the present day is facts. When sufficient accurate quantitative data covering a wide range of material . . . have become available, a satisfactory theory will follow as a matter of course. Until that time, speculations should be reduced to a minimum."

With the growth—both in extent and accuracy—of experimental evidence concerning permeability phenomena, it has now come about that most of the earlier bewildering array of permeability theories have simply died out, so that only two or three theories of cell permeability need now be considered. Even these theories are by no means sharply delimited but merge gradually into one main theory of cell permeability which has several modifications. This main theory embraces several principles referring to different aspects of the permeation process. The first principle is based on the empirical correlation between permeation power and lipid solubility. Let us call it the lipid-solubility principle. The second, which rests on the correlation between permeation power and molecular size, may be called the molecular-sieve or ultrafiltration principle. Both of these principles are now generally recognized, and the differences between the permeability doctrines of today are thus mainly that some investigators lay more stress on one principle, while others emphasize the other. Moreover, the principles themselves may be formulated and interpreted in somewhat different ways.

What we have to do now, therefore, is not to accept one of these principles and reject the other, but rather to find out how they may be combined so as to interpret actual examples.

B. THE LIPID-SOLUBILITY PRINCIPLE

The correlation between permeation power and lipid solubility, of course, invites the conclusion, already drawn by Overton, that the plasma membranes contain lipids and that substances pass across these membranes by dissolving in lipids. However, against this conclusion several objections (a–c) have been raised.

(a) It has been suggested that it would not be necessary to assume the occurence of lipids in the plasma membranes since, for instance, hydrophobic proteins would also favor the permeation of lipid-soluble, i.e., hydrophobic substances. In itself such an idea may seem plausible, however it should be noted that a differential permeability of protein membranes even remotely similar to that of living protoplasts has never been observed. So far, therefore, it seems most reasonable to ascribe the favored permeation of lipid-soluble substances through the protoplasts to the occurrence of some sort of lipids, or lipoproteins, in the plasma membranes.

(b) Lipid-soluble substances are often also surface-active. Some investigators have therefore assumed that the great permeation power of these substances is not primarily due to their solubility in the plasma membrane lipids, but rather to their adsorption at lipid-water interfaces in the membrane (150). However, the adsorption concept is scarcely applicable to the passage of permeating substances through the plasma membranes unless we assume the existence of water-filled pores in it. The predominance of such pores in the plasma membranes is, however, very questionable, except in a few cases such as *Beggiatoa*. (The pores of the plasma membranes will be more fully discussed below under the heading "The Molecular-Sieve Principle.")

Besides, the parallelism between adsorbability, i.e., surface activity, and lipid solubility is far from complete. The highest degree of surface activity will be attained if one end of the molecule is pronouncedly hydrophobic while the other end is strongly hydrophilic. The more hydrophobic the molecule as a whole, on the other hand, the greater is the lipid solubility. It should thus be possible to decide between lipid solubility and surface activity as permeation-promoting factors by comparing the permeation power of substances of equal lipid solubility but different surface activity. Such comparisons, however, seem not yet to have been attempted.

(c) Ullrich (163) has stressed that if the plasma membrane is only a few molecules thick and consists of strongly oriented molecules, it would not be correct to speak of the solubility of the permeators in the plasma membrane, since the prerequisite conditions for the validity of Henry's law will not be fulfilled. Bogen (17) has therefore proposed the use of the expression "the effect of intermolecular forces" instead of solubility. Theoretically this seems correct, for, as is well known, it is this strong tendency of hydrophilic substances to form intermolecular hydrogen bonds that tends to keep their molecules in an aqueous phase and prevent them from entering a lipoidal phase. On the other hand, retaining the old "solubility terminology" in this context is not only more convenient but is also factually justified, since the determination of the distribution coefficients is by far the most suitable way of measuring the integrated effect of all the interacting intermolecular forces. Hence we shall continue to speak of solubility in the plasma membrane lipids as a prerequisite for permeation. By this expression we only mean that permeating molecules probably diffuse through the plasma membranes, even if not in true solution in the plasma membrane lipids, nevertheless in a manner which may be most conveniently described as something very similar to a solution process. It may well be that this expression oversimplifies a very complex situation, but, if so, it is at present almost unavoidable, since the physical chemistry of today is unable to provide all of the knowledge necessary to interpret these processes adequately.

Now, what conclusions concerning the chemical nature of the plasma membrane lipids may be inferred from the permeability studies so far carried out?

As mentioned in Section VIIC, the permeation power of nonelectrolytes toward *Chara* cells varies proportionally to the first power of the olive oil:water distribution coefficient, while that toward *Nitella* cells is proportional to a slightly higher power of this coefficient. These observations may seem, at first sight, to indicate that the plasma membrane lipids in *Chara* are about equally as hydrophobic as olive oil, those of *Nitella* being somewhat more strongly hydrophobic. Probably, however, the situation is a little more complicated. The plasma membrane may consist, at least in part, of more or less rodlike lipid molecules orientated, with their hydrocarbon chains parallel, at right angles to the surface. If so, then there may exist in the plasma membrane a certain zone which behaves almost as a hydrocarbon layer, irrespective of how hydrophilic may be the end portions of these molecules. Probably it is the hydrocarbon-like layer that is the principal barrier slowing down the passage of hydrophilic substances across the plasma membrane. It thus seems possible that it is only this extremely thin layer within the plasma membranes of *Chara* and *Nitella* that is about as hydrophobic as, or somewhat more hydrophobic than, olive oil. The situation in other plant cell plasma membrane lipids is scarcely known even to this meager extent.

Another property of the plasma membrane lipids which may be ascertained by permeability studies is their acidity or basicity. In fact, we have already had cause to assume that the plasma membrane lipids of some plant cells are more acidic, as is seen from their preferential permeability to amides, while those of other cells seem to be more neutral (cf. VIID).

For the arguments favoring the view that phosphatides and sterols occur in the plasma membranes, see Section XIV,A,4.

C. THE MOLECULAR-SIEVE PRINCIPLE

If the permeability of the living protoplasts is compared with that of such artificial membranes as, for instance, the copper ferrocyanide or the collodion membrane, the contrast between them is found to be very striking indeed. By such comparisons we may, then, be tempted to classify the plasma membranes as homogeneous and nonporous. However, closer study of their permeability properties reveals features which render such a classification doubtful.

Danielli (42), applying the theory of activated diffusion to membrane problems, has made an attempt to derive equations which permit a conclusion as to whether a membrane is or is not homogeneous. According to Danielli, if (for slowly penetrating molecules) the quantity $PM^{1/2}e^{2500x/RT}$ when plotted against the oil:water distribution coefficient, or (for rapidly penetrating molecules) $PM^{1/2}$ when plotted against the distribution coefficient, gives a roughly linear relationship, the membrane is probably—as a first approximation—homogeneous. In these expressions P denotes the permeability constant, M the molecular weight, e the base of natural logarithms, x the number of nonpolar groups per molecule, R the gas constant, and T the absolute temperature. Danielli applied these tests to the results of Collander and Bärlund (35) on *Chara* and of Marklund (112) on other plant cells and concluded that the membranes of these cells are, as a first approximation, homogeneous lipid layers. It seems, however, that in applying his formulas he had overlooked some rather strongly deviating substances and that, if properly executed, the test would rather indicate some heterogeneity of the plasma membranes. It may be doubted, however, whether the test as applied is inherently reliable (177). Moreover, Danielli himself remarks that the test is difficult to apply in practice. The postulate of the approximately homogeneous character of the plasma membrane was accompanied by the reservation that there are certain small areas on the surface which have special properties and which permit certain types of molecules to enter and leave cells much more rapidly than can be accounted for by diffusion through a homogeneous lipid layer. It is thus unfair to attribute to Danielli such an oversimplification as that implied by the statement that the osmotic barrier of living cells is quite simply a homogeneous lipid layer.

In Section VIIC it was shown that the permeation power of nonelectrolytes toward *Nitella* protoplasts is inversely proportional approximately to the molecular weight raised to the power 1.5 (granted equality of lipid solubility). In itself this implies no more than a fairly high viscosity of the plasma membrane. What is more important, however, is that the very smallest molecules display a distinctly greater permeation power than would be expected on such an assumption. In other words, while the coefficients of free diffusion increase steadily with decreasing molecular size, the permeation power increases rather abruptly when the molecular weight decreases below about 50. This has been observed not only in *Nitella*, but for several other cells, too.

The simplest explanation for these observations is that the plasma membranes contain some sort of pores which will allow the smallest molecules to pass but are impervious to larger molecules. The plasma membranes thus seem to act simultaneously as selective solvents and as molecular sieves. This, admittedly rather general, view has been called the lipid-sieve hypothesis (134). It has been defended by the present author and his co-workers, among others (8, 29, 31, 32, 35, 52, 112).

However, if this view is accepted, great care should be taken in speaking about the assumed "pores" of the plasma membranes. Our familiar, macrophysical concepts are too crude to be applied here. In fact, the widespread opinion that the pores are permanent, water-filled channels traversing the membrane can scarcely be correct. For, as pointed out by Wartiovaara (172), if the permeating molecules could actually take either of two alternative routes, namely, solution in the lipids and movement through the pores, then the total permeation would equal the sum of the "lipid permeation" and the "pore permeation." If so, the smaller the permeation through the lipids, the more the pore permeation would predominate. In reality, however, this is not so: the pore permeation seems actually to be about as distinct in, say, the series of the monohydric alcohols as in that of the fatty acid amides, although the members of the first-named series have a much greater lipid solubility than those of the latter. Moreover, the great temperature coefficient values of the permeation processes also indicate that these processes do not consist of diffusion through permanent water filled pores.[5]

We are thus led to assume that in the great majority of plant plasma membranes there are no persistent, or mechanically fixed, water-filled pores. But, on the other hand, it really seems that there exist some kinds of plant protoplasts which must be assumed to be provided with such pores. This applies especially to *Beggiatoa*, whose permeability properties are difficult to explain without such an assumption. The pores of *Beggiatoa* must even be of considerable width, many of them letting through sucrose molecules, for instance. Danielli (42) has presented some calculations which indicate that either *Beggiatoa* cells possess an extremely small number of wide water-filled pores or else the diffusion-resisting layer is much thicker than are ordinary plasma membranes. The latter alternative seems decidedly more probable for, after all, the whole visible plasma layer and the whole cell wall should in this case also be understood as belonging to the "diffusion-resisting" layer. Moreover, Wartiovaara (172) has pointed out that the peculiar permeability properties mostly ascribed to the diatoms can be understood if it is assumed that about 3% of the area of their plasma membranes is of the *Beggiatoa* type, while the rest is of the usual type.

But how is the apparent sieve effect of the plasma membrane, i.e., its preferential permeability to the smallest molecules, to be explained in the case of membranes devoid of permanent, water-filled pores?

One possibility is that the osmotically determinative zone of the membrane is built up of regularly orientated, long lipid molecules, say, in a bimolecular layer (cf. Section XIV). Of course the plasma membrane may also contain layers of protein molecules etc., but these may perhaps be neglected here if we attempt only to obtain a simplified picture of what is responsible for the sieve effect. In a structure of this

[5] Ussing (166) has recently proposed a criterion by which he believes it possible to decide whether or not plasma membranes contain water-filled pores. If the water permeability of a cell membrane as measured by the rate of osmosis proves higher than its water permeability as measured by the rate of diffusion of isotopic water, then it seems to him appropriate to speak of a bulk flow of water through pores in the membrane. However, whether plant cell membranes are equipped with pores in Ussing's sense seems not yet to have been investigated.

kind, and owing to their incessant thermal vibration, the lipid molecules of the membrane will from time to time leave small gaps between each other. Each such gap will, of course, exist only for a very short time and then close again. These gaps will permit the passage of solute molecules more or less easily, depending principally on two factors, (a) their affinity for water, on the one hand, and to the plasma membrane substance on the other, and (b) the dimensions of the gaps and of the permeating molecules.

Factor a: It is clear that the more hydrophilic the permeating molecules, the more difficult will it be for them to become detached from the aqueous phase outside the plasma membrane and the greater the kinetic energy necessary to cause entry to the lipid phase. More precisely stated: the minimum energy required to enter the lipid phase is inversely proportional to the lipid:water distribution coefficient of the permeator. (Concerning the membrane as a "potential energy barrier" cf. Section IVB.)

Factor b: Even if the kinetic energy of the permeating molecule, at the very instant of contact with the membrane surface, equals or surpasses the minimum energy required, it is not certain that it will penetrate into the membrane. This will happen only if there is a hole of suitable dimensions precisely at the point of impact. Now, the occurrence of small holes is, of course, more likely than that of larger ones. This may perhaps suffice to explain the preferential permeability to small molecules. It is, however, conceivable that, strictly speaking, it is not the molecular size, or even the molecular volume, as such, that is decisive, because the steric properties of the permeating molecules may also be of importance.

Wartiovaara (172) has discussed in somewhat greater detail one such comparatively simple case—that of rod-shaped permeator molecules. He supposes that these molecules are able to penetrate the lipoidal membrane only if, on striking the surface, they are orientated approximately perpendicularly to its surface or if, before the hole has shut, they have time to achieve this orientation (cf. Fig. 5 which gives a schematic picture of what may happen). Starting from these premises, Wartiovaara reaches the conclusion that with increasing length of the permeating molecules, and thus also with increasing slowness of their reorientation, their orientation at the moment of striking the surface will concurrently become a more and more important factor reducing the probability that a molecule will actually penetrate. In other words: the greater the length of the permeator molecule the more exactly perpendicularly to the membrane surface must it be orientated in order to be able to penetrate. Wartiovaara found this theoretical postulate corroborated by experiments concerning the permeability of *Nitella* cells toward water and primary alcohols. In these experiments it was observed that, other things being equal, the chance for penetrating the plasma membranes is inversely proportional to the transverse moment of the inertia of the permeator molecule. (The magnitude of the transverse moment of the molecular inertia was calculated from the atomic weights and the mutual distances between the atoms in the molecule.) Quite similar results were obtained by Äyräpää (5) in his study of the permeability of yeast cells to ammonia and amines (cf. Section IXC).

In the case of molecules of a more complicated structure, no such quantitative calculations as those carried out by Wartiovaara and

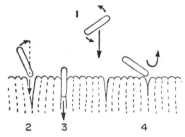

Fig. 5. Diagrammatic sketch of the penetration of rodlike solute molecules into a plasma membrane consisting of parallelwise orientated lipoid molecules. Molecule *1* is in free whirling motion in the aqueous medium surrounding the cell. Molecule *2* is just striking the plasma membrane and will, owing to its fairly favorable orientation and also to the existence of a hole in the membrane in the right place, probably be able to penetrate the membrane as molecule *3* is just doing. Molecule *4*, on the other hand, strikes the membrane in such an unfavorable orientation that it will be jostled back into the medium. From Wartiovaara (172).

Äyräpää have so far been possible. However, in experiments with *Nitella* (32) it has been found that richly branched, and thus bulkier, molecules, such as those of *tert*-butyl alcohol, triacetin, and trimethyl and triethyl citrate, all seem to permeate somewhat more slowly than their lipid solubility and molecular weight alone would lead one to expect.

To sum up, the old question of whether or not the plasma membranes are homogeneous now seems to be largely a question of definition, for the plasma membranes are intermediate between distinctly porous membranes, on the one hand and truly homogeneous ones, on the other. Their pronouncedly preferential permeability to the smallest molecules is suggestive of the molecular-sieve concept here. But, on the other hand, most plasma membranes are probably devoid of fixed and permanent water-filled pores and hence, from this point of view, they can behave as though homogeneous. This statement does not imply, however, that they are devoid of internal structure. On the contrary, some sort of anisotropic structure is almost certainly characteristic of these membranes and will, no doubt, influence their permeability properties. The details of this internal structure are, however, still obscure. Before they can be worked out, it will be necessary not only to submit the plasma membranes themselves and their permeability properties to a more thorough study, but also to investigate more closely the diffusion phenomena occurring in artificial membranes of molecular dimensions consisting of anisotropically orientated lipid molecules, so as to give a sound physicochemical basis for the understanding of the peculiar permeability properties of such systems.

D. SUCCESSIVE STEPS IN THE PERMEATION PROCESS

The permeation of a substance across a more or less homogeneous plasma membrane consists of three successive steps: the penetration through the water-plasma membrane interface, the diffusion through the interior of the plasma membrane, and the movement from the membrane into water again. Now, which of these steps will determine the rate of the over-all permeation process? This problem was first taken up by Danielli (42) but was later more thoroughly treated by Zwolinski et al. (177). According to them the rate-determining step is in some cases the diffusion within the membrane, in other cases the diffusion through the solution-membrane interface. The authors stress, however, that the data so far available are too inadequate to give a conclusive answer to this question.

These workers also calculated that the values of the diffusion coefficients for nonelectrolytes in the plasma membranes are of the order of 10,000 to 100,000 times as small as their values for diffusion in aqueous solutions. This is considered to be primarily due to the higher energies of activation for diffusion in the membranes. The diffusion coefficients as thus calculated "occupy an intermediate position in the spectrum of diffusion constants in solids and liquids which bespeak a semisolid structure for natural membranes." It seems, however, that these calculations, also, are all of a rather preliminary nature.

XIV. Other Evidence on the Composition and Structure of the Plasma Membranes

A. DIFFERENT LINES OF APPROACH

The concept of the plasma membranes which we have arrived at on the basis of our analysis of their permeability properties consists principally of the following points: (a) These membranes contain, as essential constituents, various lipids more or less resembling olive oil as regards their solvent properties. (b) The plasma membrane lipids are more acid in some cells, more neutral in others. (c) The structure of the plasma membranes is such that it enables them to behave to a slight extent as a kind of molecular sieve. (d) Permanent, water-filled pores can hardly be thought, however, to occur in the plasma membranes, except in a few exceptional kinds of cell. Finally, several active transport processes seem to imply the occurrence of certain enzymes in the plasma membranes.

Now this, of course, is only a very crude and a very imperfect picture, and it therefore seems necessary to examine whether there are any other lines of approach by which our concepts of the qualities of the plasma membranes could be supplemented and made more precise. One considerable difficulty in this connection is that these membranes are so extremely thin that they cannot be differentiated from the bulk protoplasm by microscopic means. Even the electron microscope has not so far revealed much of their structure, although some recent electron microscope measurements indicate that the tonoplasts of *Nitella* and *Vaucheria* have a thickness of about 70–100 A, while the plasmalemma is believed to be slightly thinner (64b). There are, however, some other modes of attack which have already contributed considerably to our knowledge in this field.

1. Micrurgical Evidence

Among the micrurgical experiments which have aided in elucidating the properties of the plasma membranes, those of Plowe (133) and of Chambers and Höfler (27) are especially noteworthy. In both cases the epidermal cells of *Allium cepa* were principally studied. Plowe found that when a strand is pulled out from the surface of the plasmolyzed protoplast by the microneedle, both mesoplasm and plasmalemma will at first follow the needle and cannot be distinguished from one another. Gradually, however, the mesoplasm rounds into droplets, while the plasmalemma persists as a slender thread connecting the droplets and forming a layer over each of them (Fig. 6). The conclusion that the outer layer and the inner cytoplasm are distinct seems inescapable, for although viscous fluids such as tar could no doubt also be pulled out into long, slender threads, there would in this case be no droplets on the thread. Tearing experiments with the microneedle also showed that both plasmalemma and tonoplast are elastic fluids, thus indicating that they do not consist solely of lipids but must also contain some fibrillar elements, probably proteins.

Chambers and Höfler (27) stress that the tonoplast membrane is a highly cohesive and extensible fluid film. Its cohesiveness is such that a glass microneedle can be passed readily through it without causing rupture. The membrane simply closes over the moving needle and remains intact. The highly fluid nature of the membrane is strikingly shown by the fact that a strand, extending to the tip of the needle, can be made, at its base, to slip along the surface of the tonoplast at right angles to the long axis of the strand. On coming into contact with the water-air interface, the tonoplast spreads out and disappears, leaving behind no appreciable trace of material. The lipoidal nature of the tonoplast is indicated *inter alia* by the observation that a droplet of liquid

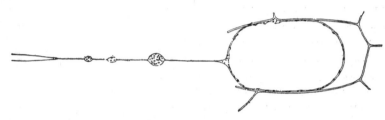

FIG. 6. Strand of protoplasm pulled out from the surface of a plasmolyzed protoplast by the microneedle. From Plowe (133).

paraffin or olive oil snaps on and adheres so strongly to the tonoplast that it cannot be removed without rupturing the membrane or carrying a portion of the tonoplast away with it.

2. Surface-Tension Measurements

Some important hints as to the composition and structure of the plasmalemma have also been gained from studies of its surface tension (66). The most remarkable finding is the smallness of this tension (about 0.1–2 dynes per centimeter) as compared with that of oil drops in water (about 8 dynes per centimeter). This shows at once that the cell surface cannot consist purely of lipids. On the other hand, Danielli and Harvey (45) have pointed out that the smallness of the tension at the cell surface is understandable if we assume that there is a layer of protein adsorbed on the lipid film. In fact such a layer would at the same time account both for the smallness of the surface tension and for the striking elasticity of the plasma membrane.

3. Chemical-Resistance Studies

Another method by which the outer plasma membrane may be studied is to watch its destruction by substances known not to penetrate

into the mesoplasm, and which therefore act solely on the plasmalemma. Except in the case of the red blood cells (cf. 46, 135) this approach has not so far been much used, but it is to be expected that
closer analysis of the ways in which the plasmalemma may be destroyed will add considerably to our knowledge of its original structure
and composition. Moreover, this is a method of long standing, for in
his classical paper of 1899 in which the lipoidal theory of permeability
was first enunciated, Overton already pointed out that the lipid impregnating the plasma membranes can scarcely be a fat in the strict
sense of the word, since algal filaments can be kept immersed for several days in dilute sodium carbonate solutions without showing any
sign of injury; by such treatment fats, if present in the plasmalemma,
would probably be saponified. The fact that numerous algal cells are
very resistent even to fairly concentrated sodium carbonate solutions
(pH about 11.8) has since been amply verified by Höfler (83), but the
cytological consequences of this fact are so far not quite clear since
there seems to be no comparable investigation revealing the saponifica-
tion rate of different fats under the influence of such solutions.

According to Ballentine and Parpart (7), neither tryptic nor peptic
enzymes affect the permeability of red blood cells, while lipolytic enzymes cause an increase of their permeability, presumably by splitting
off one of the long-chain fatty acids of the phospholipoids which are
present in their plasma membranes. Comprehensive experiments of
this kind with plant cells are still lacking. Mothes (113) incidentally
observed, however, that the tonoplasts of the alga *Sphaeroplea annulina*
are destroyed by papain. He therefore concluded that proteins constitute an essential part of them.

4. Comparisons with Blood-Cell Ghosts

When red blood cells are hemolyzed, i.e., caused to give off their
hemoglobin, very thin membranes are left which are supposed to be
plasma membranes, free of intracellular material. These so-called
"ghosts" constitute an easily obtainable and thus very attractive material for studies on the properties of plasma membranes. For instance,
the ghosts have been subjected to careful chemical analyses (cf. 129).
The results, show that they consist of relatively few substances—besides an unknown amount of water—namely, of lipids (both cholesterol
and different phosphatides) and proteins. The lipid:protein ratio shows
a remarkable constancy, varying only, in the case of the 17 different
mammals studied, between 1:1.4 and 1:1.8 by weight. This would
mean that there are some 70 lipid molecules for every protein molecule
present in the membrane. The bulk of the lipids is, however, not free

but occurs as lipoprotein complexes in which some of the lipid component (cholesterol) is more loosely, and some (phosphatides) more strongly, bound to the protein. Lipo-carbohydrate-protein complexes have also been observed in the ghosts.

The thickness of the ghost membranes has been evaluated in several different ways. The results are somewhat variable, but the most reliable values attained seem to be of the order of 50–100 A, corresponding to about 2–4 lipid molecules. Besides, the variation encountered may be due in part to the fact that some of the methods used will reveal the thickness of the lipid layer only, while others probably give the thickness of the total ghost membrane.

All these observations made on blood-cell ghosts are of considerable interest to the student of plant cell membranes, too. But, on the other hand, it should not be forgotten that the red blood cells are highly specialized cells, and thus what is true of their membranes may not necessarily be true of, for instance, plant cell membranes.

B. CONCLUSIONS

There exist, as we have seen, numerous observations which suggest that the plasma membranes of plant cells consist of lipids and proteins and/or lipoprotein complexes. But how are these components arranged in relation to each other?

Frey-Wyssling (60) assumes that the plasma membranes consist of globular protein molecules and that the interspaces between them are more or less filled with lipid molecules. The lipid filling is thought not only to explain the impermeability of the membrane toward lipid-insoluble substances but also to act as a stabilizer preventing the denaturation of the protein component. The question of the extent to which the lipid molecules are free or only loosely attached to the protein molecules and to what extent more or less stable lipoprotein complexes exist has not been definitely decided by him.

A more elaborate concept of plasma membrane structure has been presented by Danielli (42–44). According to him, the greater part of the plasma membrane consists of a roughly bimolecular lipid layer covered on both sides by adsorbed protein molecules, as shown in Fig. 7. The lipid molecules (at least the outermost of them, if there are more than two layers) are orientated perpendicularly to the membrane surface so that the hydrated polar groups are in the oil-water interfaces, while the hydrophobic tails are directed inward. The lipid layer has a fluid character, and thermal agitation will continually give rise to openings which are then shut again. How far the protein is mechanically superimposed upon the liquid, and how far the surface must be regarded as a specific protein-lipid complex, is undecided. The adsorbed protein molecules are regarded as denatured (possibly reversibly denatured). They "consist of polypeptide chains, or meshworks of such chains, lying in the plane of the interface [Fig. 8], with the hydrocarbon portions of the amino-acid residues dissolved in the lipid layer and the polar groups in the aqueous phase. There may be a further layer of globular protein adsorbed on to this primary layer. . . . The mechanical properties of the adsorbed polypeptide chains are probably largely responsible for the elasticity and relatively great mechanical strength of the plasma membrane. A single polypeptide chain may be 50 mμ or more in length and will be attached to similar chains by

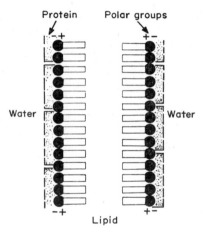

FIG. 7. Diagram of plasma membrane consisting of a lipid layer on each side covered with a layer of adsorbed protein molecules. From Harvey and Danielli (67).

hydrogen bonds and other linkages and will thus 'tie' the various parts of the membrane together, but, owing to the elastic properties inherent in such interlinked chains, without conferring upon the membrane an undesired brittleness." [Danielli (43), pp. 152–153.] Moreover, while the continuous lipid layer described above is supposed by Danielli to be the basic pattern of the plasma membrane, there may be relatively small areas, or patches, which permit abnormal permeation of special substances (e.g., glycerol in the case of certain erythrocytes).

In a long series of papers Bungenberg de Jong and his school [see Booij and Bungenberg de Jong (20)] have developed the hypothesis that the plasma membranes consist essentially of phosphatide double films which contain certain amounts of such substances as cholesterol and phosphatidic acid. Such a film will form a complex with protein and inorganic cations (Fig. 9). It will be noted that the coherence of such a film is secured not only by London-van der Waals' forces acting between the hydrocarbon chains but also by Coulomb forces between the ionized groups. Such a film will therefore have a considerable

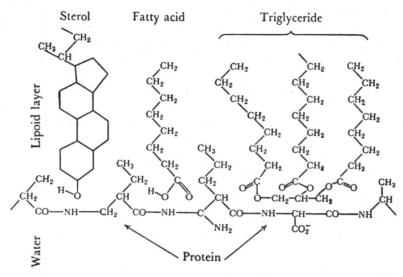

FIG. 8. Details of chemical structure of plasma membrane consisting of regularly orientated lipid molecules and covered with a layer of adsorbed protein molecules. From Danielli (42).

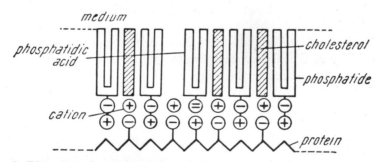

FIG. 9. Diagrammatic representation of plasma membrane consisting of phosphatides, phosphatidic acid, cholesterol, protein, and cations. From Booij and Bungenberg de Jong (20).

stability and will readily be formed spontaneously from constituents present in the cytoplasm. It is stressed by Booij and Bungenberg de Jong that although their models, when drawn on paper, may look very static, the molecules of the membrane will actually be in incessant thermic agitation. This will give rise to the appearance of smaller and larger holes between the carbon chains of the lipids. So it will be understood that such a membrane will act as a filter and as a selective solvent at the same time. The effect of temperature on the agitation of the membrane molecules also accounts for the high temperature coefficients of the permeation of large hydrophilic molecules. Moreover, variations in the amounts and properties of the membrane components will readily account for the variations found in the permeability of different cell types and there will be no difficulty in explaining permeability changes brought about by various substances. Booij and Bungenberg de Jong even suggest mechanisms enabling the active transfer of substances across such membranes.

In conclusion, it may be stated that while the details of the plasma membrane structure are still more or less hypothetical, the more gross features of their structure are now fairly well agreed upon. At least there exists a basis for continued research.

XV. Role of the Plasma Membranes in the Life of the Plant

During the two or three first decades of the present century there was a widespread tendency to think of living cells as if they were simply aqueous spaces isolated from their environment by selectively permeable, but inert, membranes. Today we realize that this "collodion bag concept" was a flagrant oversimplification. The principal defect of this mode of thought was, of course, that it totally neglected the active transport processes. That this really was a serious deficiency has become increasingly evident as the role of active transport has been appreciated.

It is also regrettable that there has been so much reference to the permeability of the plasma membranes, while their impermeability or, more properly speaking, the very great resistance they offer to the diffusion of many substances has been relatively neglected. And yet it is obvious that one of the most important functions of these membranes is to act as barriers which permit the undisturbed functioning of the highly susceptible living machinery by isolating it, on the one hand, from the cell sap and, on the other, from the fluctuating environment of the cell. It seems, in fact, plausible to assume that it is largely due to their effective isolation from the surrounding medium that living cells are able to inhabit such a variety of milieux displaying a very great variability as to pH, salt concentration, etc. The development of molds even in concentrated solutions of cupric sulfate may be cited as an impressive example of the remarkable independence of living cells of the chemical composition of their surroundings. Also the turgor of plant cells, so important for the normal elasticity of the plant body, would not be possible if the protoplasts were not almost impermeable to the salts, sugars, and other normal constituents of the cell sap and yet at the same time permeable to water. As was pointed out by Danielli, the effective isolation of the protoplasts also makes it conceivable that viruses are probably able to infect a plant only when they can reach the protoplasm through the surface of a damaged cell and that infection of healthy cells probably occurs via the protoplasmic connections between adjacent cells.

On the other hand, the isolation of the protoplasts is far from perfect. Thus there are numerous poisons which, either owing to their lipid solubility or to the smallness of their molecules, are able to penetrate all protoplasmic membranes. Ethyl alcohol, acetic acid, and also lactic acid may be cited as examples of toxic substances produced by living cells and endowed with such a great penetration power that no protoplast is able to prevent their entry. The same holds true of numerous poisons, less frequently met with under natural conditions, such as ether, chloroform, and other anesthetics, and also phenol, hydrocyanic acid, carbon monoxide, sulfur dioxide, hydrogen peroxide, formaldehyde, chlorine, bromine, iodine, and bichloride of mercury. (The last-mentioned substance is not a very strong electrolyte; its lipid solubility is considerable.) Also the highly lipid-soluble phenoxyacetic acids, etc., used for weed control, may be mentioned in this context. On the other hand, there are several poisons which penetrate protoplasts to a considerable extent only after destruction of the plasmalemma. Sodium and potassium hydroxide, hydrochloric, nitric, and sulfuric acids, potassium permanganate, and potassium chromate are examples of such substances.

If we then proceed to consider the uptake of nutrients by the cells, we observe that water, carbon dioxide, oxygen, and ammonia are examples of nutrients having such a great penetration power that their active uptake would seem superfluous. On the other hand, there is, as we have already seen, an ever-increasing body of evidence in favor of the view that salts especially, but probably also sugars and amino acids, are, preeminently, actively absorbed.

REFERENCE

1. Archer, R. J., and LaMer, V. K. The rate of evaporation of water through fatty acid monolayers. *J. Phys. Chem.* **59,** 200–208 (1955).
2. Arisz, W. H. Absorption and transport by the tentacles of *Drosera capensis. Acta Botan. Neerl.* **2,** 74–106 (1953).
3. Arisz, W. H. Permeability in plants. *Acta Physiol. et Pharmacol. Neerl.* **3,** 575–599 (1954).

4. Arisz, W. H. Significance of the symplasm theory for transport across the root. *Protoplasma* **46**, 5–62 (1956).

5. Aronoff, S. "Techniques of Radiobiochemistry." Iowa State College Press, Ames, Iowa, 1956.

6. Äyräpää, T. On the base permeability of yeast. *Physiol. Plantarum* **3**, 402–429 (1950).

7. Ballentine, R., and Parpart, A. K. The action of lipase on the red cell surface. *J. Cellular Comp. Physiol.* **16**, 49–54 (1940).

8. Bärlund, H. Permeabilitätsstudien an Epidermiszellen von *Rhoeo discolor*. *Acta Botan. Fennica* **5**, 1–117 (1929).

9. Barrer, M. R. Separations using zeolitic materials. *Discussions Faraday Soc.* **7**, 135–141 (1949).

10. Bartels, P. Quantitative mikrospektroskopische Untersuchungen der Speicherungs- und Permeabilitätsverhältnisse akridinorangegefärbter Zellen. *Planta* **44**, 341–369 (1954).

11. Beischer, D., and Oechsel, G. Ueber die Durchlässigkeit von Fettsäureaufbaufilmen für Moleküle und Ionen. *Z. Elektrochem.* **49**, 310–315 (1943).

12. Biebl, R. Zur Frage der Salzpermeabilität bei Braunalgen. *Protoplasma* **31**, 518–523 (1938).

13. Blinks, L. R. Some electrical properties of large plant cells. *In* "Electrochemistry in Biology and Medicine" (T. Shedlovsky, ed.), pp. 187–212. Wiley, New York, 1955.

14. Bochsler, A. Die Wasserpermeabilität des Protoplasmas auf Grund des Fickschen Diffusionsgesetzes. *Ber. schweiz. botan. Ges.* **58**, 73–122 (1948).

15. Bogen, H. J. Der Stoffaustausch der Zelle (und der Gewebe). Einführung, Begriffsbestimmung: Permeabilität, nichtdiosmotische Stoffaufnahme und -abgabe. *In* "Handbuch der Pflanzenphysiologic—Encyclopedia of Plant Physiology" (W. Ruhland, ed.), Vol. 2, pp. 116–124. Springer, Berlin, 1956.

16. Bogen, H. J. Die Aufnahme der Anelektrolyte. *In* "Handbuch der Pflanzenphysiologie—Encyclopedia of Plant Physiology" (W. Ruhland, ed.), Vol. 2, pp. 230–251. Springer, Berlin, 1956.

17. Bogen, H. J. Objekttypen der Permeabilität. *In* "Handbuch der Pflanzenphysiologie—Encyclopedia of Plant Physiology" (W. Ruhland, ed.), Vol. 2, pp. 426–438. Springer, Berlin, 1956.

18. Bogen, H. J. Die Theorie der Permeabilität. *In* "Handbuch der Pflanzenphysiologie—Encyclopedia of Plant Physiology" (W. Ruhland, ed.), Vol. 2, pp. 439–448. Springer, Berlin, 1956.

19. Bogen, H. J., and Follmann, G. Osmotische und nichtosmotische Stoffaufnahme bei Diatomeen. *Planta* **45**, 125–146 (1955).

20. Booij, H. L., and Bungenberg de Jong, H. G. Biocolloids and their interactions. *In* "Protoplasmatologia" (L. V. Heilbrunn and F. Weber, eds.), Vol. 1, Chapter 2, pp. 1–162. Springer, Vienna, 1956.

21. Brauner, L. Die Permeabilität der Zellwand. *In* "Handbuch der Pflanzenphysiologie—Encyclopedia of Plant Physiology" (W. Ruhland, ed.), Vol. 2, pp. 337–357. Springer, Berlin, 1956.

22. Brauner, L. Die Beeinflussung des Stoffaustausches durch das Licht. *In* "Handbuch der Pflanzenphysiologie—Encyclopedia of Plant Physiology" (W. Ruhland, ed.), Vol. 2, pp. 381–397. Springer, Berlin, 1956.

23. Brenner, W. Studien über die Empfindlichkeit und Permeabilität pflanzlicher

Protoplasten für Säuren und Basen. *Öfversigt Finska Vetenskaps-Soc. Förh.* A60 [4], 1–124 (1918).

24. Brooks, S. C. Penetration of radioactive isotopes P^{32}, Na^{24} and K^{42} into *Nitella*. *J. Cellular Comp. Physiol.* 38, 83–94 (1951).

25. Brooks, S. C., and Brooks, M. M. "The Permeability of Living Cells." Borntraeger, Berlin, 1941. (Reproduced by J. W. Edwards, Ann Arbor, Michigan.)

26. Bünning, E. Die Beeinflussung der Stoffaufnahme durch den physiologischen Zustand. *In* "Handbuch der Pflanzenphysiologie—Encyclopedia of Plant Physiology" (W. Ruhland, ed.), Vol. 2, pp. 418–425. Springer, Berlin, 1956.

27. Chambers, R., and Höfler, K. Micrurgical studies on the tonoplast of *Allium cepa*. *Protoplasma* 12, 338–355 (1931).

28. Cole, K. S. Permeability and impermeability of cell membranes for ions. *Cold Spring Harbor Symposia Quant. Biol.* 8, 110–122 (1940).

29. Collander, R. The permeability of plant protoplasts to nonelectrolytes. *Trans. Faraday Soc.* 33, 985–990 (1937).

30. Collander, R. Permeabilitätsstudien an Characeen. III. Die Aufnahme und Abgabe von Kationen. *Protoplasma* 33, 215–257 (1939).

31. Collander, R. The permeability of plant protoplasts to small molecules. *Physiol. Plantarum* 2, 300–311 (1949).

32. Collander, R. The permeability of *Nitella* cells to nonelectrolytes. *Physiol. Plantarum* 7, 420–445 (1954).

33. Collander, R. Permeability studies on luminous bacteria. *Protoplasma* 46, 123–142 (1956).

34. Collander, R. Methoden zur Messung des Stoffaustausches. *In* "Handbuch der Pflanzenphysiologie—Encyclopedia of Plant Physiology" (W. Ruhland, ed.), Vol. 2, pp. 196–217. Springer, Berlin, 1956.

35. Collander, R., and Bärlund, H. Permeabilitätsstudien an *Chara ceratophylla*. *Acta Botan. Fennica* 11, 1–114 (1933).

36. Collander, R., and Virtanen, E. Die Undurchlässigkeit pflanzlicher Protoplasten für Sulfosäurefarbstoffe. *Protoplasma* 31, 499–507 (1938).

37. Comar, C. L. "Radioisotopes in Biology and Agriculture." McGraw-Hill, New York, 1955.

38. Conn, H. J. "Biological Stains," 6th ed. Biotech Publications, Geneva, New York, 1953.

39. Conway, E. J. Some aspects of ion transport through membranes. *Symposia Soc. Exptl. Biol.* 8, 297–324 (1954).

40. Conway, E. J., and Downey, M. An outer metabolic region of the yeast cell. *Biochem. J.* 47, 347–355 (1950).

41. Craig, R. A., and Hartung, E. J. Studies in membrane permeability. *Trans. Faraday Soc.* 48, 964–969 (1952).

42. Danielli, J. F. *In* "The Permeability of Natural Membranes" (by H. Davson and J. F. Danielli), 2nd ed. Cambridge Univ. Press, London and New York, 1952.

43. Danielli, J. F. The cell surface and cell physiology. *In* "Cytology and Cell Physiology" (G. H. Bourne, ed.), 2nd ed., pp. 150–182. Oxford Univ. Press, London and New York, 1951.

44. Danielli, J. F. Morphological and molecular aspects of active transport. *Symposia Soc. Exptl. Biol.* 8, 502–516 (1954).

45. Danielli, J. F., and Harvey, E. N. The tension at the surface of mackerel egg

oil, with remarks on the nature of the cell surface. *J. Cellular Comp. Physiol.*
5, 483–494 (1935).

46. Davson, H. Haemolysis. *In* "The Permeability of Natural Membranes" (by
    H. Davson and J. F. Danielli), 2nd ed., pp. 249–263. Cambridge Univ. Press,
    London and New York, 1952.

47. Davson, H., and Danielli, J. F. "The Permeability of Natural Membranes,"
    2nd ed. Cambridge Univ. Press, London and New York, 1952.

48. Dean, R. B. The effects produced by diffusion in aqueous systems containing
    membranes. *Chem. Revs.* **41,** 503–523 (1947).

49. Drawert, H. Der pH-Wert des Zellsaftes. *In* "Handbuch der Pflanzenphys-
    iologie—Encyclopedia of Plant Physiology" (W. Ruhland, ed.), Vol. 1, pp.
    627–648. Springer, Berlin, 1955.

50. Drawert, H. Die Aufnahme der Farbstoffe. Vitalfärbung. *In* "Handbuch der
    Pflanzenphysiologie—Encyclopedia of Plant Physiology" (W. Ruhland, ed.),
    Vol. 2, pp. 252–289. Springer, Berlin, 1956.

51. Drawert, H., and Endlich, B. Der Einfluss von Sauerstoff und Atmungsgiften
    auf die Aufnahme und Speicherung saurer und basischer Farbstoffe durch die
    lebende Pflanzenzelle. *Protoplasma* **46,** 170–183 (1956).

52. Elo, J. E. Vergleichende Permeabilitätsstudien, besonders an niederen Pflanzen.
    Ann. botan. Soc. Zool. Botan. Fennicae Vanamo **8** [6], 1–108 (1937).

53. "Encyclopedia of Plant Physiology" (Handbuch der Pflanzenphysiologie,
    W. Ruhland, ed.), Vol. 2. Springer, Berlin, 1956.

54. Epstein, E. Mineral nutrition of plants: mechanisms of uptake and transport.
    Ann. Rev. Plant Physiol. **7,** 1–24 (1956).

55. Epstein, E. Uptake and ionic environment (including external pH). *In* "Hand-
    buch der Pflanzenphysiologie—Encyclopedia of Plant Physiology" (W. Ruhland,
    ed.), Vol. 2, pp. 398–408. Springer, Berlin, 1956.

56. Erichsen, L. von. Das Wasser, seine physikalischen und chemischen Eigen-
    schaften unter besonderer Berücksichtigung seiner physiologischen Bedeutung.
    In "Handbuch der Pflanzenphysiologie—Encyclopedia of Plant Physiology" (W.
    Ruhland, ed.), Vol. 1, pp. 168–193. Springer, Berlin, 1955.

57. Fischer, A. "Vorlesungen über Bakterien," 2nd ed. Fischer, Jena, 1903.

58. Fitting, H. Untersuchungen über die Aufnahme von Salzen in die lebende
    Zelle. *Jahrb. wiss. Botan.* **56,** 1–64 (1915).

59. Foulkes, E. C. Studies in cell permeability. *J. Gen Physiol.* **38,** 425–430 (1955).

60. Frey-Wyssling, A. Die submikroskopische Struktur des Cytoplasmas. *In* "Proto-
    plasmatologia" (L. V. Heilbrunn and F. Weber, eds.), Vol. II, Part A 2, pp.
    1–244. Springer, Vienna, 1955.

61. Fujita, A. Die Permeabilität der getrockneten Kollodiummembran für Nicht-
    elektrolyte. *Biochem. Z.* **170,** 18–29 (1926).

62. Gale, E. F. The accumulation of amino-acids within staphylococcal cells. *Sym-
    posia Soc. Exptl. Biol.* **8,** 242–253 (1954).

63. Gauch, H. G., and Dugger, Jr., W. M. The physiological action of boron in
    higher plants: a review and interpretation. *Contrib. Maryland Agr. Expt. Sta.*
    No. 2581 (1954).

64a. Goldacre, R. J. The folding and unfolding of protein molecules as a basis of
    osmotic work. *Intern. Rev. Cytol.* **1,** 135–164 (1952).

64b. Greenwood, A. D., Manton, I., and Clarke, B. Observations on the structure of
    the zoospores of *Vaucheria. J. Exptl. Botany* **8,** 71–86 (1957).

65. Harvey, E. N. Studies on the permeability of cells. *J. Exptl. Zool.* 10, 507–556 (1911).
66. Harvey, E. N. Tension at the cell surface. *In* "Protoplasmatologia" (L. V. Heilbrunn and F. Weber, eds.), Vol. II, Part E 5, pp. 1–30. Springer, Vienna, 1954.
67. Harvey, E. N., and Danielli, J. F. Properties of the cell surface. *Biol. Revs. Cambridge Phil. Soc.* 13, 319–341 (1938).
68. Heilbrunn, L. V. "An Outline of General Physiology," 3rd ed. Saunders, Philadelphia, 1952.
69. Hill, S. E. The effects of ammonia, of the fatty acids, and of their salts, on the luminescence of *Bacillus Fischeri. J. Cellular Comp. Physiol.* 1, 145–159 (1932).
70. Hoagland, D. R. Salt accumulation by plant cells, with special reference to metabolism and experiments on barley roots. *Cold Spring Harbor Symposia Quant. Biol.* 8, 181–194 (1940).
71. Hoagland, D. R., and Broyer, T. C. Accumulation of salts and permeability in plant cells. *J. Gen. Physiol.* 25, 865–880 (1942).
72. Hoagland, D. R., Hibbard, P. L., and Davis, A. R. The influence of light, temperature, and other conditions on the ability of *Nitella* cells to concentrate halogens in the cell sap. *J. Gen. Physiol.* 10, 121–146 (1926).
73. Höber, R. "Physikalische Chemie der Zelle und der Gewebe." Engelmann, Leipzig, 1902.
74. Höber, R. Die Durchlässigkeit der Zellen für Farbstoffe. *Biochem. Z.* 20, 56–99 (1909).
75. Höber, R. Membrane permeability to solutes in its relation to cellular physiology. *Physiol. Revs.* 16, 52–102 (1936).
76. Höber, R. "Physical Chemistry of Cells and Tissues." Blakiston, Philadelphia, 1945.
77. Höfler, K. Zur Tonoplastenfrage. *Protoplasma* 15, 462–477 (1932).
78. Höfler, K. Permeabilitätsstudien an Stengelzellen von *Majanthemum bifolium. Sitzber. Akad. Wiss. Wien. Math. naturwiss. Kl. Abt. I* 143, 213–264 (1934).
79. Höfler, K. Kappenplasmolyse und Salzpermeabilität. *Z. wiss. Mikroskop.* 51, 70–87 (1934).
80. Höfler, K. Kappenplasmolyse und Ionenantagonismus. *Protoplasma* 33, 545–578 (1939).
81. Höfler, K. Aus der Protoplasmatik der Diatomeen. *Ber. deutsch. botan. Ges.* 58, 97–120 (1940).
82. Höfler, K. Unsere derzeitige Kenntnis von den spezifischen Permeabilitätsreihen. *Ber. deutsch. botan. Ges.* 60, 179–200 (1942).
83. Höfler, K. Plasmolyse mit Natriumkarbonat. *Protoplasma* 60, 426–460 (1951).
84. Höfler, K., and Schindler, H. Volle und leere Zellsäfte bei Algen. *Protoplasma* 45, 173–193 (1955).
85a. Höfler, K., and Stiegler, A. Permeabilitätsverteilung in verschiedenen Geweben der Pflanze. *Protoplasma* 9, 469–512 (1930).
85b. Höfler, K., and Url, W. Kann man osmotische Werte plasmolytisch bestimmen? *Ber. deutsch. botan. Ges.* 70, 462–476 (1958).
85c. Höfler, K., and Weber, F. Die Wirkung der Äthernarkose auf die Harnstoffpermeabilität von Pflanzenzellen. *Jahrb. wiss. Botan.* 66, 643–737 (1926).
86. Hofmeister, L. Vergleichende Untersuchungen über spezifische Permeabilitätsreihen. *Bibliotheca Botan.* 113, 1–83 (1935).

87. Hofmeister, L. Verschiedene Permeabilitätsreihen bei einer und derselben Zellsorte von *Ranunculus repens*. *Jahrb. wiss. Botan.* **86**, 401–419 (1938).
88. Hofmeister, W. "Handbuch der physiologischen Botanik," Vol. I. Engelmann, Leipzig, 1867.
89. Holm-Jensen, I., Krogh, A., and Wartiovaara, V. Some experiments on the exchange of potassium and sodium between single cells of *Characeae* and the bathing fluid. *Acta Botan. Fennica* **36**, 1–22 (1944).
90a. Huber, B. Beiträge zur Kenntnis der Wasserpermeabilität des Protoplasmas. *Ber. deutsch. botan. Ges.* **51**, 53–64 (1933).
90b. Huber, B., and Höfler, K. Die Wasserpermeabilität des Protoplasmas. *Jahrb. wiss. Botan.* **73**, 351–511 (1930).
91. Irwin, M. Studies on penetration of dyes with glass electrode. *J. Gen. Physiol.* **14**, 19–29 (1930).
92. Jacobs, M. H. The influence of ammonium salts on cell reaction. *J. Gen. Physiol.* **5**, 181–188 (1922).
93. Jacobs, M. H. Permeability of the cell to diffusing substances. *In* "General Cytology" (E. V. Cowdry, ed.), pp. 97–164. Univ. Chicago Press, Chicago, Illinois, 1924.
94. Jacobs, M. H. Diffusion processes in non-living and living systems. *Proc. Am. Phil. Soc.* **70**, 167–186 (1931).
95. Jacobs, M. H. Osmotic hemolysis and zoological classification. *Proc. Am. Phil. Soc.* **70**, 363–370 (1931).
96. Jacobs, M. H. Diffusion processes. *Ergeb. Physiol. biol. Chem. u. exptl. Pharmakol.* **12**, 1–160 (1935).
97. Jacobs, M. H. Some aspects of cell permeability to weak electrolytes. *Cold Spring Harbor Symposia Quant. Biol.* **8**, 30–39 (1940).
98. Jacobs, M. H. Hemolysis and zoological relationship. *J. Exptl. Zool.* **113**, 277–300 (1950).
99. Jacobs, M. H. The measurement of cell permeability with particular reference to the erythrocyte. *In* "Modern Trends in Physiology and Biochemistry" (E. S. G. Barron, ed.), pp. 149–171. Academic Press, New York, 1952.
100. Johnson, M. P., and Bonner, J. The uptake of auxin by plant tissue. *Physiol. Plantarum* **9**, 102–118 (1956).
101. Jost, W. "Diffusion in Solids, Liquids, Gases." Academic Press, New York, 1952.
101a. Kamen, M. D. "Isotopic Tracers in Biology." 3rd ed. Academic Press, New York, 1957.
102. Kamiya, N., and Tazawa, M. Studies on water permeability of a single plant cell by means of transcellular osmosis. *Protoplasma* **46**, 394–436 (1956).
103. Kinzel, H. Theoretische Betrachtungen zur Ionenspeicherung basischer Vital-farbstoffe in leeren Zellsäften. *Protoplasma* **44**, 52–72 (1954).
104. Klebs, G. Beiträge zur Physiologie der Pflanzenzelle. *Ber. deut. botan. Ges.* **5**, 181–188 (1887).
105. Kral, F. Deplasmolyseverlauf in hydrolysierten Salzlösungen und Permeabilität des Plasmas für Essigsäure, Ammoniak und einige Amine. *Protoplasma* **46**, 481–522 (1956).
106. Kramer, P. J. Water content and water turnover in plant cells. *In* "Handbuch der Pflanzenphysiologie—Encyclopedia of Plant Physiology" (W. Ruhland, ed.), Vol. 1, pp. 194–222. Springer, Berlin, 1955.
107. Kramer, P. J. The uptake of salts by plant cells. *In* "Handbuch der Pflanzen-

physiologie—Encyclopedia of Plant Physiology" (W. Ruhland, ed.), Vol. 2, pp. 290–315. Springer, Berlin, 1956.

108. Kramer, P. J. The uptake of water by plant cells. *In* "Handbuch der Pflanzenphysiologie—Encyclopedia of Plant Physiology" (W. Ruhland, ed.), Vol. 2, pp. 316–336. Springer, Berlin, 1956.

109. Levitt, J. Further remarks on the thermodynamics of active (nonosmotic) water absorption. *Physiol. Plantarum* 6, 240–252 (1953).

110. Levitt, J. "The Hardiness of Plants." Academic Press, New York, 1956.

110a. Levitt, J. The significance of "Apparent Free Space" (A. F. S.) in ion absorption. *Physiol. Plantarum* 10, 882–888 (1957).

110b. Levitt, J., Scarth, G. W., and Gibbs, R. D. Water permeability of isolated protoplasts in relation to volume change. *Protoplasma* 26, 237–248 (1936).

111. McCutcheon, M., and Lucké, B. The mechanism of vital staining with basic dyes. *J. Gen. Physiol.* 6, 501–507 (1924).

112. Marklund, G. Vergleichende Permeabilitätsstudien an pflanzlichen Protoplasten. *Acta Botan. Fennica* 18, 1–110 (1936).

112a. Mitchell, P. A general theory of membrane transport from studies of bacteria. *Nature* 180, 134–136 (1957).

112b. Mitchell, P., and Moyle, J. Liberation and osmotic properties of the protoplasts of *Micrococcus lysodeikticus* and *Sarcina lutea*. *J. Gen. Microbiol.* 15, 512–520 (1956).

113. Mothes, K. Die Tonoplasten von *Sphaeroplea*. *Planta* 21, 486–510 (1933).

114. Nägeli, C., and Cramer, C. "Pflanzenphysiologische Untersuchungen," Vol. I, pp. 1–120. Schulthess, Zürich, 1855.

115. Nathansohn, A. Ueber die Regulation der Aufnahme anorganischer Salze durch die Knollen von *Dahlia*. *Jahrb. wiss. Botan.* 39, 607–644 (1904).

116. Nirenstein, E. Ueber das Wesen der Vitalfärbung. *Arch. Pflüger's ges. Physiol.* 179, 223–337 (1920).

117. Ordin, L., and Bonner, J. Permeability of *Avena* coleoptile sections to water measured by diffusion of deuterium hydroxide. *Plant Physiol.* 31, 53–57 (1956).

118. Osterhout, W. J. V. "Injury, recovery and death, in relation to conductivity and permeability." Lippincott, Philadelphia, 1922.

119. Osterhout, W. J. V. Is living protoplasm permeable to ions? *J. Gen. Physiol.* 8, 131–146 (1925).

120. Overbeek, J. van. Absorption and translocation of plant regulators. *Ann. Rev. Plant Physiol.* 7, 355–372 (1956).

121. Overton, E. Ueber die osmotischen Eigenschaften der lebenden Pflanzen—und Tierzelle. *Vierteljahrsschr. naturforsch. Ges. Zürich.* 40, 159–201 (1895).

122. Overton, E. Ueber die osmotischen Eigenschaften der Zelle in ihrer Bedeutung für die Toxikologie und Pharmakologie. *Vierteljahrsschr. naturforsch. Ges. Zürich.* 41, 383–406 (1896).

123. Overton, E. Ueber die allgemeinen osmotischen Eigenschaften der Zelle, ihre vermutlichen Ursachen und ihre Bedeutung für die Physiologie. *Vierteljahrsschr. naturforsch. Ges. Zürich.* 44, 88–135 (1899).

124. Overton, E. Studien über die Aufnahme der Anilinfarben durch die lebende Zelle. *Jahrb. wiss. Botan.* 34, 669–701 (1900).

125. Overton, E. Ueber den Mechanismus der Resorption und der Sekretion. *In* "Handbuch der Physiologie des Menschen" (W. Nagel, ed.), Vol. II, pp. 744–898. Vieweg, Braunschweig, 1907.

126. Paech, K., Wartiovaara, V., and Collander, R. Narkose and Narkotica. *In*

"Handbuch der Pflanzenphysiologie—Encyclopedia of Plant Physiology" (W. Ruhland, ed.), Vol. 2, pp. 779–791. Springer, Berlin, 1956.

127. Palva, P. Die Wasserpermeabilität der Zellen von *Tolypellopsia stelligera*. *Protoplasma* **32**, 265–271 (1939).

128. Pappenheimer, J. R. Passage of molecules through capillary walls. *Physiol. Revs.* **33**, 387–423 (1953).

129. Parpart, A. K., and Ballentine, R. Molecular anatomy of the red cell plasma membrane. *In* "Modern Trends in Physiology and Biochemistry" (E. S. G. Barron, ed.), pp. 135–148 (1952).

130. Pfeffer, W. "Osmotische Untersuchungen," 2nd ed. Engelmann, Leipzig, 1921.

131. Pfeffer, W. Ueber die Aufnahme von Anilinfarben in lebende Zellen. *Untersuch. aus dem Bot. Inst. Tübingen* **2**, 179–332 (1886).

132. Pfeffer, W. Zur Kenntnis der Plasmahaut und der Vacuolen. *Abhandl. sächs. Ges. Wiss.* **28**, 187–344 (1890).

133. Plowe, J. Q. Membranes in the plant cell. *Protoplasma* **12**, 196–240 (1931).

134. Poijärvi, L. A. P. Ueber die Basenpermeabilität pflanzlicher Zellen. *Acta Botan. Fennica* **4**, 1–102 (1928).

135. Ponder, E. Red cell structure and its breakdown. *In* "Protoplasmatologia" (L. V. Heilbrunn and F. Weber, eds.), Vol. X, Part 2, pp. 1–123. Springer, Vienna, 1955.

136. Radioactive isotopes and ionizing radiations in agriculture, physiology, and biochemistry. *Proc. Intern. Conf. Peaceful Uses Atomic Energy, United Nations, New York* **12** (1956).

137. Rashevsky, N., and Landahl, H. D. Permeability of cells, its nature and measurement from the point of view of mathematical biophysics. *Cold Spring Harbor Symposia Quant.* **8**, 9–16 (1940).

138a. Reports from the Second Isotope Conference Oxford. Butterworth's, London, 1954.

138b. Resühr, B. Hydratations—und Permeabilitässtudien an unbefruchteten *Fucus*—Eiern. *Protoplasma* **24**, 531–586 (1935).

139. Roberts, R. B., Cowie, D. B., Abelson, P. H., Bolton, E. T., and Britten, R. J. Studies of biosynthesis in *Escherichia coli. Carnegie Inst. Wash. Publ.* No. **607** (1955).

140. Robertson, R. N. The mechanism of absorption. *In* "Handbuch der Pflanzenphysiologie—Encyclopedia of Plant Physiology" (W. Ruhland, ed.), Vol. 2, pp. 449–467. Springer, Berlin, 1956.

141. Rothstein, A. The enzymology of the cell surface. *In* "Protoplasmatologia" (L. V. Heilbrunn and F. Weber, eds.), Vol. II, Part E 4, pp. 1–86. Springer, Wien, 1954.

142. Ruhland, W. Beiträge zur Kenntnis der Permeabilität der Plasmahaut. *Jahrb. wiss. Botan.* **46**, 1–54 (1908).

143. Ruhland, W. Studien über die Aufnahme von Kolloiden durch die pflanzliche Plasmahaut. *Jahrb. wiss. Botan.* **51**, 376–431 (1912).

144. Ruhland, W. Weitere Beiträge zur Kolloidchemie und physikalische Chemie der Zelle. *Jahrb. wiss. Botan.* **54**, 391–447 (1914).

145. Ruhland, W., and Heilmann, U. Ueber die Permeabilität von *Beggiatoa mirabilis* für Anelektrolyte bei Narkose mit den homologen Alkoholen C_1–C_9. *Planta* **39**, 91–120 (1951).

146. Ruhland, W., and Hoffman, C. Die Permeabilität von *Beggiatoa mirabilis*. *Planta* **1**, 1–83 (1925).

147. Ruhland, W., Ullrich, H., and Yamaha, G. Ueber den Durchtritt von Elektrolyten mit organischem Anion und einwertigem Kation in die Zellen von *Beggiatoa mirabilis. Planta* 12, 414–504 (1930).

148. Sacks, J. "Isotopic Tracers in Biochemistry and Physiology." McGraw-Hill, New York, 1953.

149. Schmidt, H. Plasmolyse und Permeabilität. *Jahrb. wiss. Botan.* 83, 470–512 (1936).

150. Schönfelder, S. Weitere Untersuchungen über die Permeabilität von *Beggiatoa mirabilis* nebst kritischen Ausführungen zum Gesamtproblem der Permeabilität. *Planta* 12, 414–504 (1930).

151. Simon, E. W., and Beevers, H. The effect of pH on the biological activities of weak acids and bases. *New Phytologist* 51, 163–197 (1952).

152. Sollner, K. Ion exchange membranes. *Ann. N. Y. Acad. Sci.* 57, 177–203 (1953).

153. Stadelmann, E. Plasmolyse und Deplasmolyse. *In* "Handbuch der Pflanzenphysiologie—Encyclopedia of Plant Physiology" (W. Ruhland, ed.), Vol. 2, pp. 71–115. Springer, Berlin, 1956.

154. Stadelmann, E. Mathematische Analyse experimenteller Ergebnisse: Gewinnung der Permeabilitätskonstanten, Stoffaufnahme—und—abgabewerte. *In* "Handbuch der Pflanzenphysiologie—Encyclopedia of Plant Physiology" (W. Ruhland, ed.), Vol. 2, pp. 139–195. Springer, Berlin, 1956.

155. Steward, F. C. The absorption and accumulation of solutes by living plant cells. *Protoplasma* 15, 29–58 (1932).

156. Steward, F. C., and Millar, F. K. Salt accumulation in plants: a reconsideration of the role of growth and metabolism. *Symposia Soc. Exptl. Biol.* 8, 367–406 (1954).

157. Stiles, W. Permeability. *New Phytologist Reprint* No. 13 (1924).

158. Sutcliffe, J. F. The exchangeability of potassium and bromide ions in cells of red beetroot tissue. *J. Exptl. Botany* 5, 313–326 (1954).

159. *Symposia Soc. Exptl. Biol.* 8 (1954).

160. Teorell, T. Transport processes and electrical phenomena in ionic membranes. *Progr. Biophys. and Biophys. Chem.* 3, 305–369 (1953).

161. Traube, I. A contribution to the theories on osmosis, solubility and narcosis. *Phil. Mag.* [6] 8, 704–715 (1904).

162. Traube, M. Experimente zur Theorie der Zellbildung und Endosmose. *Arch. Anat. Physiol. u. wiss. Med.* 87–128, 129–165 (1867).

163. Ullrich, H. Die Protoplasma-Permeabilität. *Naturwissenschaften* 35, 111–118 (1948).

164. Umrath, K. Elektrophysiologische Phänomene. "Handbuch der Pflanzenphysiologie—Encyclopedia of Plant Physiology" (W. Ruhland, ed.), Vol. 2, pp. 747–778. Springer, Berlin, 1956.

165. Ussing, H. H. Some aspects of the application of tracers in permeability studies. *Advances in Enzymol.* 13, 21–65 (1952).

166. Ussing, H. H. Membrane structure as revealed by permeability studies. *Proc. Symposium Colston Research Soc.* 7, 33–41 (1954).

167. Virgin, H. I. A new method for the determination of the turgor of the plant tissues. *Physiol. Plantarum* 8, 954–962 (1955).

168. Vries, H. de. Sur la perméabilité du protoplasme des betteraves rouges. *Arch. néerl. sci.* 6, 117–126 (1871).

169. Vries, H. de. Plasmolytische Studien über die Wand der Vakuolen. *Jahrb. wiss. Botan.* **16**, 465–598 (1885).

170. Vries, H. de. Ueber den isotonischen Coefficient des Glycerins. *Botan. Z.* **46**, 229–235, 245–253 (1888).

171. Wartiovaara, V. The permeability of *Tolypellopsis* cells for heavy water and methyl alcohol. *Acta Botan. Fennica* **34**, 1–22 (1944).

172. Wartiovaara, V. Zur Erklärung der Ultrafilterwirkung der Plasmahaut. *Physiol. Plantarum* **3**, 462–478 (1950).

173. Wartiovaara, V. Abhängigkeit des Stoffaustausches von der Temperatur. *In* "Handbuch der Pflanzenphysiologie—Encyclopedia of Plant Physiology" (W. Ruhland, ed.), Vol. 2, pp. 369–380. Springer, Berlin, 1956.

174. Weatherby, J. H. Permeability of the artificial phospholipid membrane. *J. Cellular Comp. Physiol.* **33**, 333–348 (1949).

175. Weatherley, P. E. On the uptake of sucrose and water by floating leaf disks under aerobic and anaerobic conditions. *New Phytologist* **54**, 13–28 (1935).

176. Ziegler, H. Untersuchungen über die Leitung und Sekretion der Assimilate. *Planta* **47**, 447–500 (1956).

177. Zwolinski, B. J., Eyring, H., and Reese, C. E. Diffusion and membrane permeability. I. *J. Phys. & Colloid Chem.* **53**, 1426–1453 (1949).

PREAMBLE TO CHAPTERS 2 AND 3

It is almost as difficult to conceive of the properties of cells and organs without water as it would be to conceive of biology without the unique properties of carbon. Chapters 2 and 3 are both concerned with the relations of plant cells to water.

Chapter 2 deals with the tendency of cells to gain water from, or lose water to, their immediate environment, and this historically important subject is discussed in terms of the general principles, physical and plant physiological which are involved.

Water absorption by plants is a concomitant of growth and as such is affected, or modified, by factors (nutritional, environmental, hormonal) that operate through their effects upon growth: these topics will find their place in subsequent chapters and volumes. However, in many situations in plants it is through the water relations of specialized "motor cells" that many reversible movements are determined, such as the rolling and unrolling of leaves, the action of pulvini in the posture of leaves, nyctitropic movements, and many examples of the responses of organs to external stimuli. Chapter 3 selects the classic example of the guard cells of the stomata for discussion as a special and very important case of the water relations of cells.

A later chapter in this volume (Chapter 7) considers the relations of plants to water from the standpoint of the plant as a whole in relation to its environment.

CHAPTER TWO

Water Relations of Cells

T. A. BENNET-CLARK

I. Introduction

The study of water economy of plants is a major preoccupation of physiologists and ecologists, and as water is the chief constituent of most tissues, every aspect of cell physiology is involved. One notes also that this interest of biologists provided in the first instance the data on which has been built up much of the early theory of the physical chemistry of dilute solutions.

It would probably be conceded that most of the work done before the middle of the nineteenth century is largely of antiquarian interest insofar as the water relations of cells are concerned. Stephen Hales's (30)

105

important work dealt with the integrated behavior of the whole plant rather than with the water relations of cells or tissues. He demonstrated the phenomena of root pressure and suction due to transpiration; in addition he introduced the technique of ringing as a device to detect the regions in which transport of water and solutes occurred.

The Abbé Nollet (47) is generally believed to have been the true discoverer of osmosis through membranes composed of animal bladder, but Dutrochet (25) was the first notable contributor. He discovered the processes that are described as osmosis and anomalous osmosis. One may note that in fact the so-called anomalous osmosis (in the sense that the requirements of an ideal osmotic system are not met) is, paradoxically, in all probability, much the commoner phenomenon. Dutrochet showed that if one encloses water inside an animal bladder and places this in an acid solution, water and acid diffuse into the pure water phase inside the bladder, generating an excess hydrostatic pressure. This process is termed nowadays negative anomalous osmosis.

With similar bladders, when sugar solution is enclosed, water diffuses in and sugar diffuses out. Dutrochet considered that the combination of *endosmose* of water and *exosmose* of sugar were essential components of the mechanism, which he thought would explain the absorption of water by plants. His views were of course criticized on the grounds that it was not possible to demonstrate the necessary *exosmose* from plants, and his views are today mainly of historical interest.

Dutrochet coined the word osmosis, the precise definition of which has since presented difficulty. It seems best now to use it in the sense of "diffusion through a membrane."[1]

The modern historical phase may be said to have begun with the work of Pfeffer (50), the success of which was due to the discovery by Traube in 1867 (62) of the relatively selective semipermeable membranes of copper ferrocyanide gel, coupled with Pfeffer's idea of precipitating such membranes in the walls of a porous pot.

Pfeffer's determinations of the osmotic pressures of sucrose solutions of a range of concentrations and at a range of temperatures are of classic importance both in the history of plant physiology and of physical chemistry. Within the limits of accuracy of the methods used and those available at that time, the proportionality of osmotic pressure and absolute temperature and of osmotic pressure and concentration were demonstrated and enabled van't Hoff to formulate a kinetic theory of dilute solutions and show the applicability of the gas law equation $PV = nRT$ (where n is the number of mols of solute in V liters), R

---

[1] There is no justification for the restriction made by some authors of the term osmosis to diffusion of solvent.

having the same value as the gas constant, P being the osmotic pressure in atmospheres, and T the absolute temperature.

From this work has also developed what we may call the classic view of the water relations of plant tissues. On this view each type of plant tissue element can be replaced effectively by a relatively simple non-living model. In the development of this picture many investigators have played a prominent part: De Vries, Dixon and Joly (24), Askenasy (3), Renner (51c), and Ursprung and Blum (65).

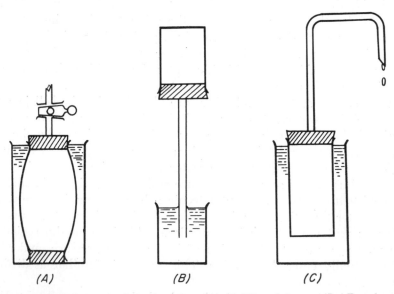

FIG. 1. Model of suggested mechanisms. (A) Rigidity of the turgid cell comparable to that of a parchment tube containing solution and immersed in solvent. (B) The transpiring shoot comparable to a porous pot filled with water; evaporation from the surface of the pot draws more water up from the reservoir. (C) Exudation and root pressure comparable to the development of pressure and outflow of solution when a porous pot with semipermeable membrane containing solution is placed in vessel of solvent.

The mechanical models usually adopted for the turgid parenchyma cell, the transpiring shoot, and the root system exhibiting guttation or root pressure are indicated by the diagrams of Fig. 1 which are probably familiar and broadly self-explanatory. These must now be subject to more detailed analysis.

II. Osmotic Quantities of Cells

The water relations of parenchymatous cells were first revealed by studies of the phenomenon of plasmolysis, and it simplifies discussion if we first consider the main functions for which we require quantitative estimates.

Before discussing the osmotic relations of cells, it is well to present simply the basis of the osmotic theory of solutions and to define the terms which will be used.

First consider three simple systems. In the first, a beaker of water is placed under a bell jar. Even under isothermal conditions the molecules, according to the kinetic theory, are not at rest and, at the boundary of the liquid, they have a net tendency to escape (fugacity), which is measured by the vapor pressure of the solvent. When the saturated vapor pressure, at the temperature in question, is reached, dynamic equilibrium is established between water molecules leaving and re-entering the liquid phase. This illustrates the familiar phenomena of vaporization and vapor pressure.

In the second system two beakers are placed under the bell jar; the one containing water, the other a solution. The vapor pressure of the solution is lower than that of the pure solvent, as shown by the fact that the temperature at boiling (when vapor pressure equals that of the surrounding atmosphere) is greater than in the case of the pure solvent. Thus, at the same temperature and above each liquid surface, there exists a pressure of water molecules; the one over the pure water being greater than that over the solution. In these circumstances water molecules will move through a membrane of air from the region of high vapor pressure to the region of lower vapor pressure, and some will eventually enter the solution. This process is familiar as distillation.

In the third system pure solvent and solution are not separated by a membrane of air, but by one endowed with special properties, namely that it is freely permeable to water (the solvent) while being completely impermeable to the solute. Under these circumstances water may pass from pure solvent to solution via two routes: (a) through a membrane of air under a vapor pressure gradient as in system 2, and (b) through the membrane. The observed facts are that in such circumstances water does pass across the membrane in the direction indicated, and this passage is called osmosis. Even without strict thermodynamic proof, however, it may be readily appreciated that if equilibrium were established via the vapor route by distillation, it would also be required to be established via the liquid route by osmosis. This is the basis of the familiar analogy between osmotic pressure and gas pressure, for the osmotic pressure of the solution has the dimensions of a "pressure lowering" and is the equivalent of a reduction in the vapor pressure of the solvent.

Pfeffer made his celebrated experiments with a particular membrane which approximated the ideal characteristics defined above (namely permeability to water but not to solutes). This membrane was the copper ferrocyanide membrane and the preferred solute was cane sugar. Somewhat unfortunately this type of membrane was called "semipermeable." With such a system Pfeffer measured the osmotic pressure of the solution by determining the superimposed pressure (with the dimensions of a positive hydrostatic pressure) which just stopped osmosis. Thus Pfeffer measured the positive pressure which was equal and opposite to the pressure deficit which caused water to enter the solution across the membrane. Having achieved a system which approximated the ideal, Pfeffer was able to measure the osmotic pressure and ascertain how it changed with temperature and concentration, and he quickly found that the osmotic pressure of the solution was proportional to the concentration of the solution ($(1/V)$ where V is the volume containing unit quantity of solute) and proportional to the absolute temperature (T_{abs}). When combined these two relations lead to an equation analogous to the ideal gas laws ($PV = KT$), and it was due to van't Hoff that it was appreciated that, if appropriate units were used (P in atmospheres, V in liters containing 1 gram molecule, T in absolute temperature), the constant of proportionality became equal to the gas constant; here $PV = RT$.

Understanding of this relationship, however, leads to a useful definition of the osmotic pressure of an ideal solution. This may be stated as follows: The osmotic pressure of a solution is equal to the gas pressure which the solute would exert if it were present as a gas, at that temperature, in a volume equal to the volume of the solution. This may be compared with the stricter definition of the physical chemists (cf. page 114).

It is important to recognize that in osmosis the solvent moves from pure solvent to solution, that osmotic pressure is measured by its equilibrant, or the pressure required to stop osmosis, that this pressure is a superimposed positive pressure, and that osmotic pressure has the dimensions of a *deficit of hydrostatic pressure*. Appreciation of these points will go far to eliminate confusion that otherwise often arises.

Pfeffer recognized in the work of de Vries that plant cells could be used to indicate solutions which had equal attraction for water in that they had equal ability to attract water from the cells. When different solutions were able to withdraw water from cells to the point where the tissue lost its "tonus," they were equal in their attraction for water and were said to be "isotonic," a term still used for solutions of equal osmotic pressure.

It was work with cells that first showed the unexpected osmotic pressure of salts with two or more radicals, and this in turn led Arrhenius to the basis of the ionic theory of solutions, since they behave as though they contain more ultimate particles than their molecular composition represents.

The water relations of parenchymatous cells were first revealed by studies of plasmolysis. They are most effectively dealt with by reference to three characteristic states in which such a cell may be found: (a) its normal state in a plant surrounded by other cells or the natural external medium, (b) the state induced by immersion in water or a solution of negligible osmotic pressure which causes the cell to swell, (c) the state of limiting plasmolysis when the protoplast starts to recede from the wall.

Water relations can be described then in terms of osmotic and diffusion pressures, hydrostatic pressures, and the opposite reaction to the hydrostatic pressure known commonly as wall pressure. These terms or conceptions will be discussed in greater detail.

A. TURGOR AND CELL STRUCTURE

The mature parenchymatous cell when immersed in water becomes turgid. Turgor is a state of rigidity resulting from distension of the cell by its liquid contents which are under pressure. Rigidity implies that work must be done to disturb the system and, as pointed out by Haines (29), if a cell or tissue is bent, stretching occurs only if the internal pressure keeps the inner and outer sides of the bending tissue apart. Otherwise, stretching is avoided by formation of a kink when the inner (concave) and outer (convex) walls approach.

Dixon pointed an analogy between a cell and an H-section girder. Rigidity depends on a combination of tension members, the flanges of the girder, and a compression member, the web which keeps the flanges apart. In the cell, the wall corresponds to the tension member and the excess hydrostatic pressure of the contents, to the compression member, and this excess hydrostatic pressure is a measure of the turgor. It seems appropriate that it should be termed the turgor pressure.

The *excess* hydrostatic pressure inside the cell i.e., the turgor pressure, is thus equal and opposite to the pressure set up by the elastic stretch of the cell wall. It is not possible to subdivide the wall pressure into a portion due to the cellulose wall and one due to the protoplast. As Haines has pointed out, the reaction between any two membranes such as wall and protoplast is not important for turgor or rigidity: it is the sum total of the tensions in the envelopes, which is balanced by the oppositely directed excess hydrostatic pressure that is significant.

Some confusion is possible as a result of different usages of the term turgor pressure by different authors. Burström (16) has suggested use of the term turgor pressure as meaning "the pressure acting on the cell wall [having] the nature of a diffusion pressure," and he equates it with the difference in osmotic values, actually in DPD's (diffusion pressure deficits) of the internal cell sap and the external solution. This matter of terminology arose from considerations of "turgor" in cells which are expanding in contradistinction to cells at equilibrium. It is clear that in such a cell there will be a turgor pressure (not in Burström's sense) which will be equal and opposite to the inwardly directed pressure due to the tension in the wall. In a cell which is not expanding and which is at equilbrium, the turgor pressure (older sense) will equal the excess hydrostatic pressure. In a cell which *is* expanding the excess hydrostatic pressure at a point in the wall is opposed by an equal and opposite reaction due to the elastic stretch of the wall, but this excess hydrostatic pressure is not equal to the mean hydrostatic pressure of the whole cell contents. As the cell is expanding, there will be a gradient of hydrostatic pressures from the surface of the cell inward, not, it is true, a very steep gradient, because it is exceptional to find very rapid expansion of plant cells.

Direct measurements of the hydrostatic pressure and osmotic pressures in a plant cell under any natural or experimentally applied conditions are usually not available, and recourse has been had to a range of indirect methods and deductions.

In order to clarify description it is necessary to define certain terms and consider in some detail the structural features of a typical parenchyma cell. This may be regarded as a protoplast containing a relatively large central vacuole or vacuoles. The boundary layer of this vacuole is termed the tonoplast, and several techniques show that the tonoplast has defined structure and mechanical strength. It is possible to dissect away the tonoplast with its enclosed solution from the rest of a plasmolyzed protoplast. Such protoplasts can be removed from plasmolyzed strips of epidermal cells, and, when touched with microdissection needles, the bulk of cytoplasm can often be pulled away from the tonoplast. The tonoplast has special permeability relations; water can readily pass through it, as shown by the volume changes when isolated tonoplasts are transferred from one solution to another.

It should also be pointed out that the protoplast as a whole is elastic and it compresses the tonoplast and vacuole. Doubtless the tonoplast itself is also a stretched elastic membrane. It sometimes happens that as a result of injury the cytoplasm which surrounds the tonoplast breaks off, and this is associated with a sudden expansion of tonoplast and vacuole as a result of release of the inwardly directed pressure due to the "cytoplasmic layer." In an experiment described by Konings-berger (39), this expansion due to release of cytoplasmic pressure was between 12 and 15%, and corresponded to an almost instantaneous reduction of internal osmotic pressure of 2 to 2.5 atm. The importance of this will be discussed briefly later.

Certain dyestuffs such as neutral red and other basic dyes pass through and accumulate in the vacuole when the internal pH is lower than the external. There is little or no information regarding the ability of an isolated tonoplast to accumulate or retain other solutes such as ions or sugars but the roughly constant volume shown by a tonoplast immersed in a saline or sugary plasmolyzing solution for several hours seems to indicate a rather low permeability and lack of activity of accumulation mechanism. The tonoplast dissected out in this manner is in the case of many cells quite a thick structure, and when it bursts a crumpled mass of protein and lipid remain. It should therefore not be regarded as only a bi- or paucimolecular lipid layer controlling permeability to polar solutes.

The status of the external layer of the protoplast presents even more difficulty than the tonoplast. It is commonly assumed that it possesses defined structure and permeability properties, and it is sometimes termed the plasmalemma. It has been regarded as lying inside the cell wall and pressed against this wall in a manner comparable to the pressure of an inner tube against the cover of a pneumatic tire. This analogy cannot be regarded as absolutely certain. The cellulose microfibrils of the cell wall appear to be formed within the cytoplasm; even, however, if they were formed outside or on the surface there is difficulty in understanding how the protoplast is not forced into the spaces between the fibrils as a result of the high excess internal hydrostatic pressure.

It may very well be that the effective outer surface of the protoplast is situated in the bulk of the wall, at least in certain types of cell or at certain stages of development. Analyses[2] have shown a considerable protein content of cell walls (20.8% in conifer cambium (1); 12% in cotton (*Gossypium*) primary wall (52)), but this hardly settles the question of the exact site of the plasmalemma.

After plasmolysis, in many tissues part of the protoplast frequently remains stuck to, or imbedded in, the cell wall, often appearing as a network (the so-called *wandständiges Cytoplasma*).

[2] Certain protoplasts adhere to certain walls and are not removed from them by plasmolysis. This is true of the radial walls of the endodermis to which the protoplast adheres and may also be true of meristematic cells which do not plasmolyze.

Bennett and Rideal (8) provide indirect evidence again for the view that the plasmalemma is immersed in, rather than lying alongside, the wall and pressed against it. They inserted a microelectrode (Ag-AgCl) into the vacuole and determined the capacitance and reactance between the inside electrode and a similar one outside the cell. On plasmolysis there was a marked decrease in impedance coupled with increase in capacitance. The simplest explanation was that plasmolysis caused an increase in the effective surface area of the plasmalemma and that the

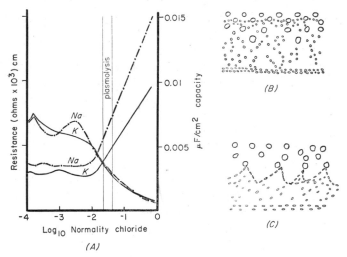

FIG. 2. (A) Resistance (descending curves, left ordinates) and capacitances (ascending curves, right ordinates) of a *Nitella* membrane measured between an electrode in the vacuole and one outside. Concentrations of the bathing solutions shown as abscissae. (B) Wall-protoplast organization suggested for the turgid cell with small external protoplast surface. The larger "circles" represent cross sections of cellulose microfibrils. (C) Wall-protoplast organization suggested for the plasmolyzed cell with large external protoplast area. From Bennett and Rideal (8).

plasmalemma constituted the insulation between the plates of the condenser. Their data and the picture presented of the association of cytoplasm and cell wall are illustrated in Fig. 2.

B. Osmosis

The water of the cell is clearly distributed between the phases vacuole, protoplast together with the various inclusions (mitochondria, plastids, nucleus), and the cell wall, which itself, as indicated above, may conceivably be, in part or in whole, a cytoplasmic inclusion. Water which is the major constituent of the cell is held in these various phases by a combination of osmotic and imbibitional forces.

The passage of water from one phase to another or to the environment depends on its activity or specific free energy. The earlier work on the physical chemistry of dilute solutions consisted of development of analogies with the kinetic theory of gases, and, arising from this, water relations of plant tissues have been expressed similarly rather than on a thermodynamic basis.

In the earlier plant physiological work, water movements in plants were regarded as correlated with osmotic pressure differences only. The importance of the hydrostatic pressures of cell contents, and consequently of the activity of water or the vapor pressure, was not at first recognized.

This history has left a legacy of terms, some of which have been used inaccurately and some of which are unnecessary. A brief and elementary statement has already been given of osmosis of water and osmotic pressure, but some recapitulation may add clarity.

Osmosis is strictly defined as diffusion across a membrane. It should be emphasized that it is generally applied to cases where there is a net flux across the membrane which is due to the activity (or vapor pressure) being different in the phases on opposite sides of the membrane —a circumstance which brings about an activity gradient in the membrane. Osmosis, or diffusion of water, may in principle be essentially similar to that of solutes. In the case of living cell membranes, it is clear that transfer of solutes is not a simple diffusion process such as may occur across a collodion membrane. It remains a matter of controversy whether the passage of water is simple osmosis or whether in addition so-called active, or nonosmotic passage, occurs.

The *osmotic pressure* of a solution is most rigidly defined [cf. Brønsted (12)] as the excess hydrostatic pressure which must be applied to the solution in order to make the chemical potential (activity) of the solvent in the solution equal to that of the pure solvent at the same temperature. This in effect is nearly the same as stating that it is the pressure that the solute would exert if it were a gas occupying the volume of the solution. As so defined, osmotic pressure is independent of the presence of any membrane—ideal or otherwise. Definitions of the type, "osmotic pressure is the maximum pressure which develops in a solution enclosed within a semipermeable membrane etc.," are quite unsatisfactory. The expression *osmotic potential* of a solution has been introduced quite unnecessarily on the grounds that the "solution is not exercising any pressure when it is in a beaker." Such an argument completely ignores the strict physicochemical definition of Brønsted.

There is perhaps a case to be argued for the use in regard to vacuolar contents of plant cells of the expression *osmotic value* (*osmotische Wert*), which has the meaning of "estimate of the osmotic pressure." As will be shown, the estimates obtainable by different methods differ quite considerably and all may be different from the actual, and at present unobtainable, real osmotic pressure of the vacuolar contents *in situ*. So that *plasmolytic osmotic value* or *cryoscopic osmotic value* describe the estimates obtained by plasmolytic or cryoscopic methods.

Incidentally a peculiar usage occurs in the literature more commonly in comparative animal physiological work than elsewhere: solutions are referred to as having *osmolar concentrations*. An osmolar concentration of 1.0 is equivalent to a solution having an osmotic pressure of 22.4 atm at 0°C.

In the literature, osmotic pressures or values are often cited in terms of molarities of sucrose or other osmoticum. Unfortunately, it is often impossible to discover whether the investigator has actually made up molar or molal solutions. A *molar* solution contains 1 gram-molecular weight in 1 liter of solution. A *molal* solution is 1 gram-molecular weight plus 1000 gm of water. It is simpler, when a series of concentrations are required, to make a standard molar solution and make fractional dilutions, as one weighing suffices. Half-strength molal sucrose is *not*, however, 0.5 molal! And one suspects that occasionally investigators have supposed that it is. Equally, of course, when osmotic pressures are cited it is not always made clear whether these are referred to 0°C, 20°C, or at the ambient temperature. Again when different solutes are used and solutions are described as isotonic, sometimes all that is really meant is that the solutions are of the same molarity. Equimolar solutions of a small molecule like urea and a large one like sucrose have very different osmotic pressures. Table I may be of value in illustrating these points.

The passage of water by diffusion from one phase to another depends on a free-energy difference. Water will diffuse from a phase having higher activity of water to one of lower activity or chemical potential. This was first recognized in relation to plant water relations in a general way by Renner (51c), who introduced the term *Saugkraft* as a measure of the reduction in activity of the water component of a phase. The *Saugkraft* of a solution at atmospheric pressure having an osmotic pressure (OP) of x atm was defined as x atm. Water will diffuse into such a solution through a semipermeable membrane from pure water and will similarly distill into it through a gas phase. This entry can be described as being due to the "diffusion pressure" difference of the water in the two phases. The diffusion pressure of water in the solution phase is clearly the lower, and the amount by which it is lower is termed the *diffusion pressure deficit* (DPD) by Meyer. The DPD is numerically equal to the *Saugkraft* (which has been translated into English, as suction force), but as dimensions are those of force/area the terms suction pressure, suction tension, or simply suction (*S*) are probably really more desirable than the earlier term *Saugkraft*.

Hydrostatic tensions can be treated algebraically as negative pressures (pressure deficits) so there seems little real objection to the terms *Saugdruck* or suction pressure, which are common in European literature and for which priority might be claimed. Little harm is done how-

TABLE I

OSMOTIC PRESSURES (OP) OF SOLUTIONS OF GIVEN MOLAL OR MOLAR
CONCENTRATIONS

Concen-tration	Molal sucrose		Molar sucrose at 20°C	Molar NaCl at 20°C	Molar KCl at 0°C
	OP at 0°C	OP at 20°C			
0.1	2.46	2.6	2.64	4.3	4.6
0.2	4.72	5.1	5.29	8.4	8.8
0.3	7.09	7.6	8.13	12.5	12.6
0.4	9.44	10.1	11.1	16.6	16.7
0.5	11.90	12.8	14.3	21.0	20.6
0.6	14.38	15.4	17.8	25.7	24.1
0.7	16.89	18.1	21.5	29.7	28.3
0.8	19.48	20.9	25.5	34.2	32.2
0.9	22.11	23.7	29.7	38.6	35.9
1.0	24.83	26.6	34.0	43.2	39.2
1.2			45.4		
1.5		41.8	65.8		
2.0		54.4	116.6	96.4	

ever if *S* and DPD are recognized as equivalent and convenient functions which are estimates of the vapor pressure deficit of the phase. It is on the whole convenient to be able to write equations containing the symbols for the osmotic pressure of a solution (*P*) and the excess hydrostatic pressure or turgor pressure or wall pressure (*W*) and the suction pressure or diffusion pressure deficit (*S*). By definition and axiom $S = P - W$.

In this account the symbol *S* will be used for the DPD or suction pressure. Incidentally one of the objections that has been voiced against the use of the term diffusion pressure deficit is that it required three letters! Levitt has introduced the term "*osmotic equivalent*" (symbol *E*), but it is in fact synonymous with DPD and the European terms *Saugkraft*, etc. By definition, Levitt's *E* might be expressed in energy units, being based on the relation $\Delta F = pV^1$; in practice, however, the same units as for DPD measurements are utilized: namely, liter-at-mosphere for free-energy changes.

The *turgor pressure* is the outwardly directed *excess* hydrostatic pressure which is equal and opposite to the inwardly directed reaction due to the stretched cell wall. Burström's suggestion that the term should be used to describe the difference in DPD of the cell contents and external medium has not proved acceptable. The DPD of the cell contents is the equivalent of what was at one time termed the "full suction pressure." The difference in DPD of cell and external medium (Burström's turgor pressure) is the equivalent of what was termed "net suction pressure" and can now be most conveniently termed the "DPD difference" $(S_i - S_e)$, where the subscripts i and e refer to the inside or contents of the cell and e the external medium. The confusion introduced by the terms full and net suction pressure is perhaps the best reason for dropping these terms and using instead DPD (or DPD difference, when necessary). For clarity and with the object of using a limited number of terms, we propose to use osmotic pressure; DPD; and turgor pressure, which equals the opposing reaction or wall pressure.

Comparable to the DPD is the estimate of the tenacity with which water is held in soils. This could be referred to as a DPD or suction pressure of the soil, but the soil physicist uses the reduction of activity or free energy of soil water over that of pure water in bulk. This reduction in free energy could be given in absolute units, but it is usual to follow Schofield's (54) suggestion and use the height of a water column in centimeters, but, as the range of free-energy differences (activities) is large, the logarithm of the height in centimeters is used and termed the pF of the soil: thus a suction of 1000 cm is termed pF 3 and is approximately 1 atm. In plant physiological terminology this would be described as a DPD or suction pressure of 1 atm.

III. Methods of Investigation

A. VAPOR PRESSURE LOWERING

The procedures that have been used in determinations of osmotic pressures of plant fluids (or culture solutions) have in general been based on properties of the solutions colligative with osmotic pressure, such as depression of vapor pressure or freezing point. These are based on Raoult's law that the vapor pressure lowering is proportional to the mole fraction of solute $(p_0 - p)/p = n_2/(n_1 - n_2)$, where p_0 and p are vapor pressure of solvent and solution respectively, and n_1 and n_2 are the number of molecules of solvent and solute.

When one can assume that the solvent vapor obeys the gas laws and that the partial molecular volume of solvent V_0 is independent of pressure, one obtains the relation

$$\text{osmotic pressure, } P = \frac{RT}{V_0} \ln \frac{p_0}{p}$$

Vapor pressure lowering methods, as applied in practice in plant physiological researches, have generally relied on comparison of the vapor pressure of experimental material with that of solutions of known osmotic pressure.

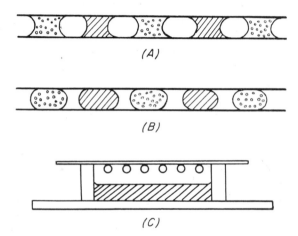

FIG. 3. Tubes filled as for the Barger method of comparing vapor pressures (osmotic pressures); known solution dotted; unknown, diagonal shading. (A) Uncoated tube. (B) Silicone-coated. (C) Ursprung's variant with series of known solutions in a small chamber above the unknown, which may be a solution or a tissue.

The method introduced by Barger (4), which has also been used in modified form by Ursprung and many others, depends on measuring the change in length of columns of experimental solution and standard solutions of known OP as shown in Fig. 3. The method is much simpler nowadays than when first introduced, as one can now coat the inside of capillaries with a silicone and achieve greater precision in filling the tubes.

Essentially the same principle is involved in Baldes and Johnston's method where two thermocouples are mounted so as to provide small loops into which are placed, respectively, the experimental solution and a drop of distilled water. The holder bearing the loops is placed in a moist chamber, and water condenses into the drop of solution; when a steady state is reached the thermocouples are connected to the galvanometer and the deflection is shown to be proportional to the OP of the solution. Calibration is based on the use of solutions of known OP.

Spanner (56) has devised an ingenious method which ultimately depends on the connection of vapor pressure and DPD or suction pressure. Use is made of the Peltier effect, or the cooling of a thermocouple when a current is passed through it. This cooling in Spanner's apparatus deposits dew on the junction, and as the dew point is dependent on the ambient water vapor pressure, an estimate of this last is obtained, and again calibration of galvanometer readings is effected against the atmosphere in equilibrium over solutions of known OP. In fact this method though potentially most valuable has been little used. It could in principle be used to determine DPD's of tissue *in situ* in the plant as microthermocouples were constructed. The Baldes and Johnston method is really applicable only to determine the DPD of extracted solutions, i.e., their osmotic pressures.

B. CRYOSCOPY

The commonest method in use is undoubtedly cryoscopy. At the freezing point of a solution (i.e., the temperature when ice separates), the vapor pressures of water of both liquid and solid phases are equal. The vapor pressure:temperature curves for dilute solutions and pure water are almost parallel, so $\Delta T / \Delta p =$ a constant K (where $\Delta T =$ freezing point depression and $\Delta p =$ the vapor pressure lowering), and substituting in the Raoult equation

$$\Delta T = K p_0 \frac{n_2}{n_1 + n_2}$$

K having the value 1.86°C per mol of an undissociated solute. The osmotic pressure at 0°C (P) is then given by the relation $P = \Delta T / 1.86 \times 22.4 = 12.04 \, \Delta T$.

In the practical procedure, the solution whose freezing point is to be determined is placed in a small tube jacketed by a larger one which is immersed in a freezing mixture. The temperature falls gradually, the solution being kept constantly stirred, and at the moment of crystallization or separating of ice the temperature rises as seen in Fig. 4. It is necessary to obtain the lowest temperature (T_1) and temperature after ice formation (T_2). As some supercooling usually occurs, the solution at the time when the temperature T_2 has been reached is more concentrated than in original solution. The correction suggested by Harris and Gortner (31) is based on the fact that approximately one eightieth of the water solidifies per degree of supercooling, so the true depression

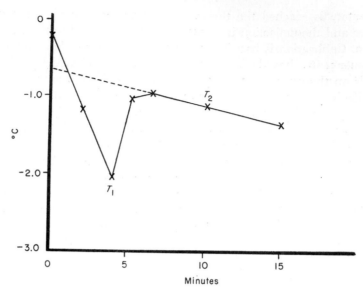

FIG. 4. Change of temperature with time when a solution is cooled. Ice forms at T_1. Extrapolation back from T_2 gives an estimate of the true freezing point.

ΔT is given by $\Delta T = T_2 - (T_2 - T_1)/80 \times T_2$. Supercooling errors can be frequently almost eliminated if one seeds the cooling solution with ice crystals. (The writer has also considered that seeding with silver iodide might act similarly.)

C. SAP EXTRACTION

These direct determinations of osmotic pressures by use of colligative properties can be applied to extracted samples of plant sap, and the value of the results is directly related to the certainty of origin of the sap. Where the sap is a natural exudate like that from hydathodes, the drip-tips of leaves, or cut vascular tissue, the authenticity of the fluid need not be questioned. The case of alleged vacuolar saps differs.

Sap can of course easily be pressed out of tissues, and its composition depends on both structural features and the previous treatment of the tissue. Where the cytoplasm and wall fraction of the tissue is of the order of, say, 5% of the total volume the totally extracted sap may nearly represent vacuolar sap. The major factors that would produce discrepancies are retention of solutes within the cell membranes, which would provide a too dilute expressed sap, or production of extra solute, either by breakdown of large molecules or release from adsorption following disorganization of the protoplasm, which provides a too concentrated supposed vacuolar sap. Retention of solute within the tissue certainly must occur if the normal semipermeability is unimpaired and if cells are not burst open, and it is usual to eliminate this source of error by killing at the low temperatures attained by liquid air or solid carbon dioxide. There is always a possibility that a small fraction of the total solute may be retained by adsorption on insoluble wall material. It is not strictly possible to evaluate the amounts lost by adsorption or gained by desorption; gains due to enzymatic hydrolyses are probably fairly easily avoided. There is, therefore, inevitably some lack of certainty that expressed vacuolar sap has the same composition or OP as it had in the tissue. The larger the vacuoles, the greater the likelihood of identity of the natural and expressed sap.

When, as in many tissues such as most meso- or xerophytic leaves, the relative bulks of vacuole and cytoplasm are more nearly equal, the origin of the expressed sap is distinctly conjectural.

D. PLASMOLYTIC METHODS

With certain simplifying assumptions, plasmolytic data have been used to evaluate osmotic pressures of vacuolar sap and, in certain cases, also the excess hydrostatic pressure of cell contents. The main plasmolytic methods for the determination of vacuolar osmotic pressure are (a) limiting plasmolysis (b) the plasmometric method.

In the limiting plasmolytic method it is not strictly possible to evaluate the OP of any one cell, as one requires a population of assumed similar cells. These are placed in a series of solutions of determined OP and the state of plasmolysis is recorded, the most satisfactory method being to record for each sample of the population the percentage of cells plasmolyzed in solutions of given osmotic pressures. The results can be plotted as in Fig. 5. This type of data clearly could be subjected to probit analysis, which would give a valid objective estimate of the mean OP of cells of the population.

Certain tissues have cells of relatively uniform OP, as the data in Fig. 5 show. In others there is a wide scatter of values. The plasmolytic value so obtained is the OP of the external solution which balances that of the vacuole when the protoplast is just about to shrink away from the wall; that is, when the turgor pressure is zero. As the fully turgid cell is larger, and thus more dilute, than the cell at limiting plasmolysis, the OP at full turgor P_t is given by the expression $P_t = P_l V_l / V_t$ (where P represents vacuolar OP; V, vacuolar volume; and the suffixes t, l, and p refer to the fully turgid, limit-plasmolyzed, and plasmolyzed conditions). V_t / V_l is termed the degree of turgor stretching, and this ratio can be evaluated by weighing or by volume measurements.

The plasmometric method is based directly on the assumed validity of the gas law relation, $PV = RT$. It is applicable when, by measurement, it is possible to obtain the volume of the cell vacuole initially and also after plasmolysis in a solution of known OP. The relation $P_t = P_p V_p / V_t$ holds if the gas laws are valid. In general the method is applicable with reasonable accuracy only in the case of cylindrical or spherical cells and only if the plasmolysis form is convex as in Fig. 6.

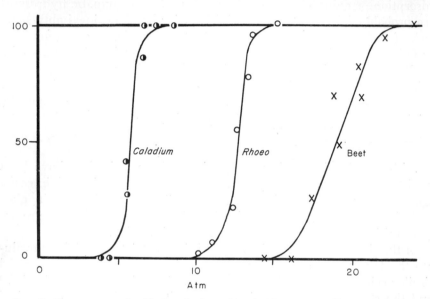

Fig. 5. Percentages of cells found plasmolyzed (shown as ordinates) in solutions of different osmotic pressures (shown as abscissas); *Caladium* petiole, *Rhoeo*, leaf, beet root. From Bennet-Clark *et al.* (7).

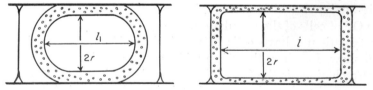

Fig. 6. Dimensions required when using the plasmometric method of estimating vacuolar osmotic pressure of a cylindrical cell.

It has been common to treat the plasmolyzed protoplast as a cylinder with two hemispherical ends, so that its volume is given as $(l - 2r)\pi r^2 + \frac{4}{3}\pi r^3$ [i.e., $(l - \frac{2}{3}r)\pi r^2$]. It must, however, be pointed out that the gas law relation, $PV =$ constant, is not strictly true with sucrose solutions of the concentrations found in many plant storage organs such as onion (*Allium cepa*) bulbs, and this inevitably introduces errors in this method. A correction factor, b, for the bulk of solute, giving the relation $P(V - b) = $ constant, can be applied and appropriate values for b determined by using a range of external osmotic pressures.

E. DPD DETERMINATION

In general and as indicated, the suction pressure, or DPD, of a cell or tissue is of far greater significance in relation to water movements than the OP, and unfortunately it is not readily determinable *in situ* in the living plant. Mature cells taken from land plants are generally found *not* to be in the fully turgid condition so that, on immersion in water, they swell and will ultimately come to a "quasi-equilibrium." In this condition it is assumed, on the classic view, that the activities of water inside and out are equal and that therefore the excess hydrostatic pressure inside equals the internal OP. This is not strictly a true equilibrium, as it is maintained by balanced activities of the tissue, which may include resorption of material leaking outward; gradual stretching, or growth, of individual cells may also occur.

The original suction pressure of such cells or tissue is obtained by finding an external solution the osmotic pressure of which just prevents this expansion of the tissue when immersed. More concentrated external solutions cause a shrinkage. The relation of tissue volume (or weight) to the external OP is of the type shown in Fig. 7. It will be seen that the DPD can be obtained by interpolation, and the external OP which causes no further shrinkage in volume can be considered as an estimate of the internal OP at which plasmolysis just commences. Tissue weight and volume changes have both been used as indicated in Fig. 7, and change in composition of the external solution has similarly been utilized, as in the refractometric method, where, for example, a sugar solution of given refractive index is treated with tissue; if the tissue has a higher DPD than the solution, water will be absorbed and the refractive index of the solution rises, and conversely. Since refractive indexes can be measured quickly and accurately, the method is useful, but like all methods it is subject to the uncertainties arising from absorption or release of solutes as between tissue and external solution. For this reason mannitol has frequently been chosen as the external solute in the hope that it might prove to be an "indifferent" one in the sense of not being consumed in metabolism or absorbed: definite proof of its "indifference" is, however, lacking.

The DPD (symbol S) of a portion of tissue, or of a cell, is given by the relation $S = P - W$ where W stands for the excess internal hydrostatic pressure (numerically equal to the wall pressure). The wall pressure of any given cell is altered directly the cell is cut from a massive tissue, if there is mutual compression of the cells in the intact tissue. This is certainly true of the tissue of herbaceous stems and of the stelar tissue of roots, and the values of suction pressure of such excised tissues cannot be treated as absolutely equivalent to the value *in situ* without some discussion and adjustment.

The suction pressure of the excised fragment tends to be larger than

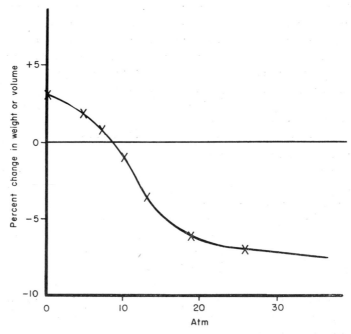

Fig. 7. Relation between change in tissue or cell volume or weight from its initial (or natural) value and osmotic pressure of the solution in which is it immersed.

it would have been *in situ* if the tissue *in situ* is compressed by surrounding tissues and, conversely, smaller if *in situ* it is subject to tension.

Spanner's method (see page 119) could be used to determine the actual suction pressure of tissue *in situ* in certain appropriate cases. Many records of suction pressures of tissues have therefore only a limited comparative value.

The DPD of a tissue can be obtained also by comparing its vapor pressure with those known solutions by Ursprung's modification of Barger's method (65) as illustrated in Fig. 3,c. A circular disk of leaf or other tissue is placed in the bottom of the glass ring sealed onto the slide in place of the solution as represented in the figure. The accuracy of equilibration can be tested by using a known solution in the base of the circular chamber and a series of known solutions in the capillaries in the air space.

Although published as a "new method" in 1930, it was really only a modification of the method used by Ursprung (64) and Renner (51c) in 1915 to arrive at an estimate of the DPD in the fern annulus or liverwort elators when the "springing-back" movements occurred. As presented in 1930 and as illustrated in Fig. 3,c, it is convenient in that five or more comparison solutions can be used in the capillaries and one can plot graphically the change in length of the liquid columns against their osmotic pressures and, by interpolation, get a satisfactory estimate of the DPD (and vapor pressure) of the material in the cup, which can, of course, be a solution, or plant tissue, soil sample, etc.

F. Turgor Pressure

If a cell, or group of cells, from a plant is subjected to mechanical pressure, it becomes deformed. The simplest case to consider is that of a

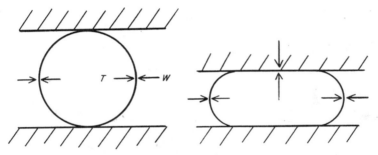

Fig. 8. Forces acting before and after application of pressure to a spherical cell by means of a pair of flat plates. T = turgor pressure; W = wall pressure.

spherical cell compressed by plates (Fig. 8). The pressure due to the plates at equilibrium is opposed by the excess hydrostatic pressure of the contents, and the wall pressure of the stretched portions of wall exposed to the atmosphere has been increased so that it is equal and opposite to this increased excess hydrostatic pressure.

When therefore, one gradually increases the pressure applied by the plates one should obtain theoretically a result of the type shown in Fig. 9. Increase in pressure will decrease the DPD of the vacuole until it equals zero, and since deformation of the cell must occur, each increment will cause outflow of water from the vacuole, but this will not be able to accumulate outside until the internal DPD equals zero.

After this, further increase in the excess hydrostatic pressure will produce a negative DPD; in other words, water will be pressed out and each successive pressure increment will press out smaller amounts; the actual amounts retained may be assumed to depend on the relation $PV = RT$.

This would provide a means of obtaining the internal excess hydrostatic pressure at full turgor (i.e., when the external environment was pure water), always assuming the experiment could be carried out.

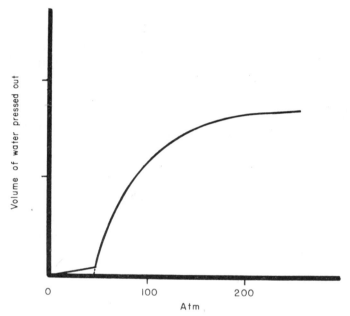

Fig. 9. Volumes of water theoretically pressed out (or decrease in cell volume) of tissue subjected to excess hydrostatic pressures shown as abscissas. See text.

It was in fact attempted by Dixon (21), who applied an external gas pressure to leafy shoots fixed as in Fig. 10. The gas pressure was raised in the hope that water would be pressed out of leaf parenchyma cells into the xylem (and, in principle, drip out of the cut end of the stem). The pressure needed to cause wilting was thought to be significant as an estimate of turgor pressure (i.e., excess hydrostatic pressure) of the leaf parenchyma.

The pressure-plate technique was used by Bennet-Clark and Bexon (5) to press a column of leaves carefully placed in a testing press between flat steel plates covered with pure tin foil. The output of water after successive pressure increments is shown in Fig. 11, and the point at which the curve cuts the zero ordinate was regarded as an estimate of the excess hydrostatic pressure of those leaf cells of lowest excess hydostatic pressure. It will be noted that the initial raising of the pressure produced by increasing the weight on the upper plate from zero to a value giving 20 atm pressure caused very little output of water as compared with volumes expressed after further pressure increments. This is related to the fact that the fully turgid leaf cells were only some 3% larger in volume than at limiting plasmolysis. The liquid pressed out with slowly increasing pressure on the plates proved to be water only very slightly contaminated by solutes which had probably come from cells split open or otherwise damaged by shearing (slipping movements at right angles to the applied pressure).

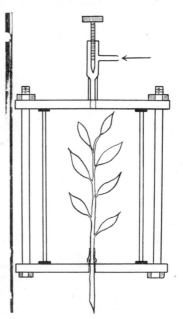

The phenomena associated with applications of pressure to plant tissues are perhaps worth further consideration. Crushing is the oldest method by which sap has been extracted; Dixon showed that the concentration of the first pressings from tissue was more dilute than in later pressings as a result of ultrafiltration of water; later Mason and Phillis showed that leaf tissue pressed carefully yielded a very dilute "extract." It seems, as pointed out above, that pressure does in fact cause ultrafiltration of water which one might legitimately expect would be contaminated by small permeating molecules whereas

FIG. 10. Dixon's method (21) of exposing a branch to an excess gas pressure. See also Haines (28a).

large molecules like anthocyans would be filtered off. This last is in fact observed.

Some cells of course burst and, when this happens, vacuolar solutes like anthocyans escape. In *Fagus sylvatica purpurea* the first burstings of cells were observed at applied excess hydrostatic pressure of just over 70 atm under careful loading conditions, but even at 150 atm a large proportion of the cells remained undamaged and those broken at lower pressures may have been torn or sheared open. This result is worth comparing with the bursting strength of *Nitella*, about 12 atm, as shown by Kamiya and Kuroda (see page 181); their results refer to end walls of *Nitella* cells and suggest a factor of safety of 2 as the internal excess hydrostatic pressure is around 8 atm. It must be remembered, however, that in normal circumstances the end wall is pressed against another cell and is thus subjected to less strain than are the side walls. In *Fagus* the safety factor appears certainly to be not less than 4 and probably well over 8 for most cells.

Some 40–50% of the total water is pressed out relatively readily, with little or no bursting of the cells. When the applied pressure is raised to over 150 atm, the tissue tends to disintegrate and cells

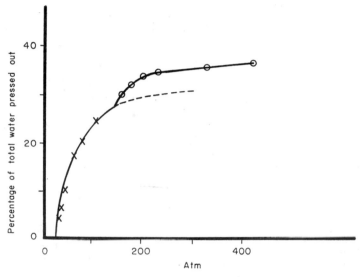

Fig. 11. Volume of juice exuded from leaves of *Fagus sylvatica purpurea* when subjected to pressures by the methods of Figs. 8 and 9. Colorless liquid of osmotic pressure less than 0.5 atm exuded during period is shown by curve —x—. The juice emerging after this was colored and had increasing osmotic pressure; volumes of colored juice are shown by curve —o—. From Bennet-Clark and Bexon (5).

separate from each other and are squeezed out sideways from between the plates used for application of pressure. This tears the cells open.

The tissue from which water has been expressed (providing this tearing or shearing has been avoided) retains about half the original total water and if killed, for example by freezing in liquid air, another 10–15% of water can be pressed out with very low pressures. Most of the remaining 30% can of course be pressed out readily if the tissue is retained in some sort of porous container. The behavior described emphasizes that the tissue water is held in two ways by the live tissue, about half being more readily pressed out than the remainder. This second half is, however, only "held" when the tissue is living and appears to be held in the cytoplasm. Microscopic examination shows that after the first half is pressed out the vacuoles have disappeared.

IV. Plasmolysis

A. GENERAL CONSIDERATIONS

The phenomenon of plasmolysis has been so much utilized in obtaining data of water relations of plant cells and of their permeability to solutes that it merits further detailed consideration.

In the process of plasmolysis water passes from the vacuole to an external solution of higher OP. If the plasmalemma were not present or if no barrier to diffusion occurred at the outer surface, all that would be expected to happen would be contraction of the vacuole. In normal plasmolysis this is however accompanied by tearing of the cytoplasm away from the wall. This behavior must be regarded as establishing the existence of a plasmalemma of limited permeability. The only alternative proposition would be a degree of mechanical structure in the cytoplasm causing its external surface to be adherent to the internal, a view for which there is as yet no evidence. There is no clear evidence as to whether the protoplast is drawn out of interfibrillar spaces of the wall or whether it is merely pushed off the inner wall surface to which it had adhered.

In the case of cells with rather thin walls, such as immature parenchyma (near a meristem) and leaf cells of many succulent plants, immersion in a hypertonic solution brings about a buckling of the wall plus protoplast (this is sometimes termed cytorrhysis); there is no tearing of protoplast away from the wall. This may be regarded as due in part to the protoplast impregnating the wall or being firmly attached to it and in part to the thinness and lack of rigidity of the wall. The phenomenon is also often observed with cells exposed to the air, such as those of moss leaves, terrestrial algae, and certain epidermal hairs, any of which might be expected to have relatively low water permeability.

Attempts have been made to obtain a quantitative estimate of what has been termed "adhesion pressure" of cytoplasm to cell wall by measuring the "limiting plasmolytic value" (i.e., OP of the external solution at limiting plasmolysis) when this point is approached by placing turgid cells in a range of solutions, and when approached by placing plasmolyzed cells in the same solutions. With beet root (*Beta vulgaris*), Buhmann (15) found values such as 19.6, 18.1, 18.9, atm on plasmolysis, but 3.3, 2.5, and 1.5 atm lower on deplasmolysis, and these latter figures were regarded as estimates of adhesion. With the same technique, similar, but much lower, values for adhesion were obtained by Currier (18), and in other cases these differences have not been observed.

It is important to recognize that the plasmolytic methods are based on an assumption that is not strictly correct, namely, that the cytoplasm per se exerts no pressure whatever on the vacuolar contents. Making this assumption, it is then supposed that at limiting plasmolysis (or in the case of an isolated protoplast) the internal and external osmotic pressures are equal.

In certain cases tonoplasts have been observed to "escape" from the cytoplasm which has become ruptured. The immediate consequence is sudden expansion (see page 112). In the experiment there described the sudden uptake of water caused the internal apparent OP to change from about 15.5 to 13 atm. Now if one assumes that the tonoplast pressure directed inwardly against the vacuole is negligible, one can give as an explanation that the original state of the isolated protoplast was that it possessed a DPD of 13 atm, that the vacuole contents had an OP of 15.5 atm, and that the DPD value was brought about by a turgor due to the cytoplasmic pressure of 2.5 atm. Release of this pressure caused an almost instantaneous uptake of water.

No doubt the cytoplasmic pressure varies from cell to cell and also with a variety of environmental conditions and internal so-called tonal factors. But it seems evident on the classical view of water relations that the DPD of an isolated or a plasmolyzed protoplast will always be smaller than the osmotic pressure. The plasmolytic value then is too low an estimate of the osmotic value.

The importance of this is that the discrepancies of plasmolytic values and osmotic pressures of extracted sap have called in question the classical view of cell water relations and suggested active water uptake.

In this connection it is important to note that the process of plasmolysis does not seem to cause any completely disruptive injury to cells. Red beet root (*Beta vulgaris*) tissue has been shown by Sutcliffe (58) to accumulate K^+ against a concentration gradient after plasmolysis. Disks of tissue having an internal potassium concentration of about 0.06 M before plasmolysis were immersed in solutions, all containing 0.02 M potassium chloride and made up to a series of osmotic pressures with different plasmolytica (Table II).

It will be noted that the reduction of turgor by raising the external OP to 5 atm promotes some extra K^+ uptake, but on plasmolysis there is some reduction which may be partly due to the increased leakage [see Table II (lower section)]. The more highly plasmolyzed cells[3]

[3] Calcium solutions showed some slight inhibition of accumulation.

again showed more active accumulation under these conditions where the ratio of external to internal K^+ was about 0.1. So the specialized and energy-requiring accumulation mechanism or "potassium pump" had clearly not been damaged by plasmolysis.

These commonly accepted views of the mechanism of plasmolysis have been called in question recently by Briggs and Robertson (11d) (cf. also Briggs 11a,b). Data relative to ion uptake by tissues (cf. Chapter 4) show that part of the cation uptake occurs very much more rapidly than the rest and suggest clearly the presence either of

TABLE II

ABSORPTION AND RELEASE OF K^+ BY BEET DISKS FROM MEDIA OF VARIOUS OSMOTIC PRESSURES (OP)[a]

Plasmolyzing agent	OP of tissue	OP of medium					
		0.9	5.0	10	15	20	30
K^+ absorbed[b]							
Sucrose	9.2	1.34	1.79	0.99	1.10	1.22	1.14
Glucose	8.3	1.48	1.82	0.95	1.22	1.06	1.11
$CaCl_2$	9.1	1.54	1.90	0.79	0.85	0.98	0.88
$MgSO_4$	8.9	1.51	1.68	0.96	1.19	1.29	1.05
K^+ released[c]							
Sucrose	8.4	0	0	0.34	0.16	0.12	0.06
Glucose	8.9	0.02	0.01	0.36	0.18	0.11	0.05

[a] From Sutcliffe (58).

[b] Milligrams absorbed per gram fresh weight during 6 hours by beet disks from media all containing 0.02 M KCl.

[c] Milligrams released per gram fresh weight during 6 hours at 7°C with media initially free of electrolytes.

separate mechanisms or of separate zones in tissues or cells which may be termed "free spaces" or "outer space" and a further and larger zone that can be termed conveniently "non-free space." Free space is the more readily accessible to penetrating or accumulating solutes. It is natural to equate these zones to intercellular or extracellular water, cytoplasm, and vacuole since much evidence exists of the different permeabilities of tonoplast and plasmalemma, the former being a much more impermeable barrier. From the data of Briggs et al. (11c), it is possible to obtain estimates of the volume of free space completely outside the cells and represented almost certainly by water in intercellular spaces, etc. This accounts for about 20% of the volume (200 ml/kg) in beetroot slices.

It is pointed out further that the cytoplasmic space almost certainly containing relatively nonmobile anions will act as a Donnan system. Detailed quantitative data of a suitable form are as yet very inadequate but it has been possible to estimate, from the concentrations of cations and anions in beet exchangeable at low temperatures, that the volume of the Donnan free space in beetroot slices is of the order of 20–30 ml per kilogram and the concentration of nonmobile Donnan anions is of the order of $0.5 N$ (0.5 equivalents per liter). The OP due to associated cation would thus be of the order of 11 atm (Briggs *et al.*, 11c). It is a not unreasonable assumption that OP's of vacuole and cytoplasm and also their HP's are equal.

It is clearly of the greatest importance for an understanding of the physiology of a plant cell or tissue that the distribution of materials between cytoplasm (including individual organelles of the cytoplasm), vacuole, and external medium should be clarified. Briggs (11a,b) expounds the view that the external boundary layer presents a relatively negligible barrier to diffusion in which the diffusion coefficients have the same order of magnitude as in an aqueous layer of the same dimensions.

The hydrostatic pressure within the cell wall maintaining turgor is due to two separate osmotic pressures of cytoplasm and vacuole regarded as probably equal in magnitude. The DPD's of these phases must clearly be equal except when water is passing from one to the other; the hydrostatic pressures must be equal unless the cytoplasm possesses some rigidity. The OP of vacuole is therefore equal to the OP of cytoplasm or is balanced by the sum of its OP coupled with any imbibitional or so-called "active" water secretion processes which might operate. The cytoplasmic OP in Brigg's view is determined by the Donnan system. An external plasmolyzing solution, on this view, brings about an increase in cytoplasmic OP by reason of the ready penetration of the solute from outside into the cytoplasm; this in turn causes osmotic flow of water from vacuole to cytoplasm which thus increases in volume.

There could be no plasmolysis at all if the cytoplasm were indeed strictly fluid. All that would happen would be "vacuolar contraction." To explain the shrinkage away from the wall which constitutes plasmolysis, it is necessary on this view to assume, as Briggs points out, that cytoplasm is elastic, such that when stretched (i.e. when the volume is increased) the hydrostatic pressure within the cytoplasm is increased.

In the plasmolyzed condition after equilibrium has been attained the DPD's of vacuole cytoplasm and external medium are all equal. In Briggs' view the OP of cytoplasm is larger than that of the external medium owing to the presence in it of the external solute which has freely diffused in, in addition to the Donnan ions. The actual amount of this excess of course is affected by the nature of the plasmolyticum as electrolyte ions will be redistributed by the Donnan system.

One could therefore in principle expect (cf. Briggs, 11a) that the limit plasmolyzing solution might have a higher OP than the vacuolar sap and that isotonic electrolyte and nonelectrolyte solutions might be in equilibrium with different vacuolar volumes.

Plasmolysis by salts such as KSCN does produce *Kappen-plasmolyse* which can take the extreme form of a tonoplast much shrunken and with the rest of the cell occupied by a much swollen cytoplasm (see Fig. 12). This on Briggs' view is due to the great extensibility of cytoplasm in presence of the ions of KSCN. Calcium chloride by contrast evokes nonextensibility of the cytoplasm. Considerable excess hydrostatic pressures of the cytoplasmic contents of the plasmolyzed protoplasts above those of external medium and vacuole have to be assumed on this view, comparable in magnitude to the original turgor pressure.

It would be important to consider how far this cytoplasmic extensibility and necessary mechanical structure is in agreement with the well established movements shown clearly in cine-films of living cells and often observed directly. There is also a possible difficulty in the high concentration of nonmobile Donnan anions of the order of $0.5 \, N$. Protein or peptide anions are rather unlikely to have equivalent weights much less than say 1200. The concentrations of complex phosphates are not very large. The amount of peptide anions is unlikely to be adequate as a source of Donnan anions which, it now seems probable, are provided by "pectic" material found in the cell wall.

Existing published data on changes in volume and apparent free space of cells subjected to differing degrees of plasmolysis are unfortunately inadequate. Direct observation and many analyses show that many ions pass very freely into an "apparent free space" which is probably cytoplasm. As it may well be necessary to assume a mechanical structure and elasticity of the cytoplasm in order to explain plasmolysis, it becomes impossible to adhere dogmatically to the view that the hydrostatic pressure within the cytoplasm necessarily equals that of the vacuole. It must be emphasized that although the external surface of the cytoplasm in certain tissues does appear to offer a relatively insignificant barrier to diffusion of certain ions, this is by no means generally or universally true. Some of the estimates of "free space" in tissues which are based on content of mannitol and on the quantity of mannitol which escapes freely after having been taken up by tissue most probably refer really to extracellular free space which

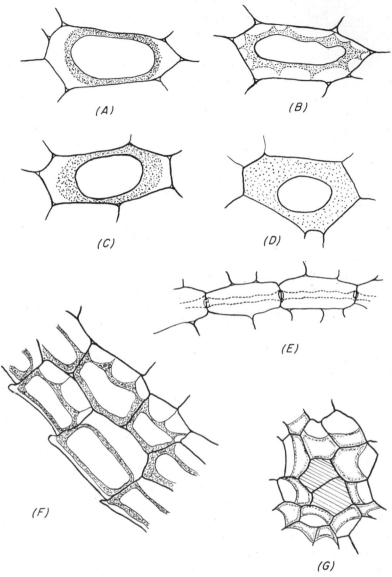

FIG. 12. Plasmolysis forms. (A) Convex plasmolysis. (B) Concave plasmolysis. (C) Cap plasmolysis. (D) Vacuolar contraction if space between wall and tonoplast is supposed to be living cytoplasm: this is termed tonoplast plasmolysis when one assumes that the cytoplasm is dead. (E) Plasmolysis of endodermal cells bearing a Casparian strip. (F) Plasmolysis of moss leaf cells showing contraction from the inner midrib sides of the cells. (G) Plasmolysis of cells near a necrotic area.

as shown by Briggs *et al.* (11c) is relatively large. The Donnan free space is small in volume and might be restricted to wall or cytoplasmic structures. The work of Bennett and Rideal (8) provided evidence for an effective barrier to diffusion at the plasmalemma in *Nitella,* and Walker's data (62b) similarly appear in better agreement with the older view that this outer membrane was responsible for maintenance of the high internal concentrations and OP of the cell. Cytoplasmic OP in any case must be controlled by the Donnan system modified possibly by "active" secretion processes at either or both surfaces, plasmalemma and tonoplast.

B. Plasmolysis Forms

It is well known that the manner of tearing away of protoplast from wall gives rise to a variety of so-called plasmolysis forms. Some distinctive types are illustrated in Fig. 12.

It has been supposed that the "form" is an index of the viscosity of the cytoplasm, the convex form being a sign of low viscosity. This type is produced by plasmolytica such as potassium thiocyanate and potassium nitrate, whereas the concave form is produced by bivalent cations, such as calcium chloride solutions. It is a little uncertain whether the primary cause is the higher viscosity of cytoplasm in the concave form or its adhesion to the wall, though this last may indeed be an expression of the high viscosity of such cytoplasm as does actually lie within the wall between polysaccharide microfibrils. The influence of ions on plasmolysis form and presumably on viscosity broadly follows lyotropic series, as expected, though exceptional behavior has frequently been reported. With the cation series Cs, K, Na, Li, Ca, Al, the cytoplasmic viscosity increases as one passes from left to right.

When multivalent cations like Al^{+++} are applied, it is sometimes impossible to obtain plasmolysis; this could be due to a viscosity within the wall so great that the protoplast cannot be pushed out. In these cases the cell as a whole crumples, the wall and protoplast being dragged inward by the contracting vacuole (cytorrhysis).

The adherence of the protoplast to the wall is evidently the reason for the special form of plasmolysis in the cells of the endodermis bearing a Casparian strip. In cells of some moss leaves (Fig. 12) the protoplast remains firmly attached to those parts of the cellulose walls which are relatively impermeable to water or solutes as a result presumably of cutinization; the internal walls are more permeable, which provides the observed plasmolysis form. The cells surrounding an injured or necrotic group frequently plasmolyze from the side distant from the injury; this is likely to be due to decrease in cell wall permeability near the injured cells. A possible cause may be impregnation by substances similar to "wilt toxins" (see Section V,C,3) which are released or formed by the injured cells and act as a wound-healing mechanism. The nature of this form of wound healing has not been studied in detail, but it seems possible that it may be related in mechanism to the tanning (hardening) of proteins by quinones which occurs in the anchoring of certain zoids.

It is incidentally a matter of note that in a favorite material for plasmolysis studies, a strip of onion bulb scale epidermis, it frequently happens that the cells close to the cut edge of the strip show convex plasmolysis (low viscosity or adhesion) and those in the center, or distant from the cut, show concave plasmolysis: this is attributable possibly to a shock effect on viscosity of the cytoplasm. It has been shown that shaking causes rather marked decreases in viscosity [Kahl (36)].

Certain reagents, like potassium thiocyanate, which are promoters of convex plasmolysis often cause the so-called cap plasmolysis (Fig. 12), which may develop into the condition of tonoplast plasmolysis in which the vacuole has become filled with cytoplasm much diluted with water. Little information exists at present as to whether in all cases the plasmolyticum has entered the cytoplasm, changing its swelling behavior.

C. ABNORMAL PLASMOLYSIS

In this category may be grouped the effects of certain "stimuli" which have been observed to cause cells to plasmolyze when mounted in hypotonic solutions, or even in water. This has been observed with leaves of *Elodea* (*Anacharis*) *canadensis*, root tips of *Zostera*, and corollas of various flowers, notably of the Boraginaceae, mounted in water. The phenomenon of vacuolar contraction is also presumably related to this so-called false plasmolysis. The precise molecular mechanism is completely unknown. Sudden loss of tonoplast permeability could in principle be expected to result in transfer of water from vacuole to cytoplasm. Study of the actual mechanism is impeded by inability to produce at will large numbers of cells showing these abnormal plasmolyses.

The marine diatom, *Ditylium brightwellii* might prove to be desirable research material in this connection: when this organism was transferred to hypotonic sea water, as shown by Gross (27), contraction of the protoplast occurred. Contraction was also produced as a result of a variety of other treatments which included darkening for 16 or more hours, treatment with isotonic but unbalanced saline solutions, and even mechanical shock. Gross considered that the contraction was similar to that which occurs when resting spores are formed.

Contraction also occurs when *Spirogyra* and other Conjugales form gametes. It is not known what happens to the vacuolar contents. There is no definite evidence to suggest that the vacuolar solutes are caused to disappear either by extrusion or condensation, leaving an internal solution ultimately isotonic with that outside, which incidentally is nearly pure water. There is no evidence that the semipermeable membrane has lost its normal semipermeability, which again could bring about equalization of activity of water in vacuole and external medium. Indeed, since the contracting gametes are caused to contract even more by external solutions of potassium chloride, it seems that part of the normal semipermeability is retained.

In the case of *Spirogyra longata* gamete contraction, Lloyd (43) suggested that work was done by contractile vacuoles "pumping out" the vacuole contents; this observation has not been followed up, nor was there information from Lloyd's work as to what was pumped out, whether water only or water together with solutes. The possibility should of course be envisaged that the contraction of the cytoplasm actually compresses the vacuolar contents and that the contracting gamete is turgid with compressed solution inside and, in place of a stretched cellulose wall, a stretched myosinlike membrane.

D. CONTRACTILE VACUOLES

Far more work has been done on animal than on plant contractile vacuoles, and, as there is every reason to suppose some similarity in mechanism and function, the following account is based on animal systems.

Sometimes contractile vacuoles occur in fixed positions as in the very complex apparatus of *Paramecium* and also in the apparently morphologically simpler *Chlamydomonas*, but frequently as in *Amoeba* they are described as "roving."

In operation, a vacuolar cycle consists of an expansion in volume (in *Paramecium*, a rapid entrance of liquid from feeder canals followed by slower further entrance of water), then contraction due to expulsion of liquid to the outside (in *Paramecium* again the contraction occurs in two phases, a slight volume reduction followed by complete expulsion of contents).

There is still considerable uncertainty regarding the source of the water which enters these vacuoles, and the composition of the liquid in the vacuoles is not known in detail. The rate of output is measurable from determinations of maximum size and frequency of cycles. Its value is dependent among other things on the OP of the external medium [Kitching (38a)], which suggests a function of osmoregulation for contractile vacuoles. In some organisms, for example the marine *Colthurnia*, there seems to be reason to think that water enters over the whole body surface and that its ejection takes place from the contractile vacuoles. As water enters the body surface more rapidly from a dilute than from concentrated solution, the body volume would tend to swell more in the more dilute external medium, and this is correlated with the greater rate of vacuolar output in dilute solutions.

When the vacuolar activity is inhibited by, say, 0.002 M cyanide, increase in body volume takes place, but this poisoning is reversible and, on washing, recovery of vacuolar activity and shrinkage begins. The problem of mechanism of entry of water into the vacuole remains unsettled; some authors believe that "active" water secretion into the vacuole occurs, others consider that secretion of a solute causes simple osmosis, or diffusion of water, into the vacuole along an activity or vapor pressure gradient. Kitching points out however that for *Colthurnia* placed in 25% sea water, in which the vacuole is very active, that solutes like urea or other products of catabolism could not be produced in adequate amounts to account for purely osmotic flow of water to the vacuoles; the animal's whole weight would be consumed in a few hours. This does not eliminate the possibility that salt might be secreted into the contractile vacuole. Accordingly there is here support without proof for the water-secretion hypothesis.

The mechanism of expulsion of the vacuole contents to the outside (systole) is also somewhat debatable. Active contraction of the vacuolar membrane has been suggested, but on the whole it seems more likely that increasing excess hydrostatic pressure of the whole body, or cell fluids, forces out the vacuolar liquid, and with fixed vacuoles the pore may well constitute a weak place which bursts.

It has been suggested that contractile vacuoles do not appear to occur in plant cells provided with cell walls. They occur in the naked beaks of *Chlamydomonas* cells, and possibly in the naked and contracted gametes of *Spirogyra*.

V. Water Movement

A. ACTIVE UPTAKE BY PARENCHYMA

Considerable controversy surrounds the question as to whether water movement in plant tissues is controlled solely by activity (vapor pressure) gradients or whether in addition to, or superposed upon, such movements there are also so-called nonosmotic movements, in which energy directly derived from metabolic processes is used to drive a molecular water pump.

In this connection it is important to recognize that so-called "active" or "nonosmotic" movement of solutes or water could be envisaged as being caused by an electric potential gradient, or possibly, as suggested by Goldacre (26) by the mechanical work performed by a contracting protein molecule of the myosin type, or finally by the operation of a carrier mechanism. In this last case the process is not "nonosmotic" so far as the carrier-substrate complex is concerned. This complex is supposed to diffuse normally along its activity gradient. The secretion mechanism is involved in either the formation or the breakdown of the carrier complex.

Active salt uptake by mechanisms as yet unknown are accepted as occurring in a wide range of both plant and animal tissues. Active water uptake by plant tissues is a matter of considerable controversy, although it tends to be accepted that it occurs, again by an unknown mechanism, in a variety of different animal tissues of which frog skin and many protozoa may be quoted as notable examples.

In plants, claims have been made of water secretion (active transport) into the vacuoles of certain parenchyma cells, into the vascular tissue of roots thus exhibiting root pressure, and by hydathodes.

Secretion of water into vacuoles was first suggested as a result of the finding that the osmotic value of certain parenchymatous tissues determined by plasmolysis was considerably larger than the osmotic pressure of their expressed sap determined cryoscopically. The osmotic pressure which just causes limiting plasmolysis has been generally assumed on the classical view to be equal to that of the vacuolar sap. It has already been pointed out that in fact the external balancing OP may be regarded as equal to the DPD of the vacuole but that as the vacuole is compressed by the cytoplasm, it must have a lower osmotic pressure than the vacuolar sap by an amount which is equal to the cytoplasmic pressure. As pointed out, this pressure may be of the order of 2–3 atm, which is larger than many people have expected.

It is, therefore, all the more striking that plasmolytic values actually obtained experimentally tend to be *higher* than the cryoscopically estimated osmotic pressures by amounts which vary from, say, 0.2 to 5 atm, and only rarely is equality found. A plasmolytic value *lower* than the cryoscopic, as it should be on the classical view, seems to be extremely rare.

Some representative results are given in Table III. Currier incidentally found that in certain beets showing a plasmolytically estimated osmotic pressure higher than the cryoscopic, i.e., a positive PCD (or plasmolytic-cryoscopic discrepancy), cylinders of tissue equilibrated in isotonic sucrose had an expressed sap (after freezing) of lower osmotic pressure than the external solution.

They confirmed the finding of Bennet-Clark *et al.* (7) that less than 50% of the cells are plasmolyzed in their own tissue juice; 50% plasmolysis would be expected on the classical view.

The suggested "explanation" was that wall pressure is balanced partly by the internal osmotic pressure and partly by an "active water pump" which operates to force water into the vacuole.

TABLE III
PLASMOLYTIC AND CRYOSCOPIC OSMOTIC "VALUES"[a]

Tissue	References	P_l	P_t	P_l deplasm.	OP	PCD
Root, beet (*Beta vulgaris*) A	(7)	23.4	22.6		15.5	7.1
Root, beet (*Beta vulgaris*) B		12.6	12.1		9.5	2.6
Petiole, *Begonia sempervirens*		8.0	7.7		5.3	2.4
Petiole, *Caladium bicolor*		5.5	5.3		5.8	0.5
Petiole, Caladium		6.4	6.2		5.8	0.4
Root, beet	(18)	16.2	15.7		10.7	5.3
Root, beet		12.6	12.3		12.0	0.3
Root, beet		17.9	17.5		16.1	1.4
Leaf, *Bergenia* sp.	(15)					
Epidermis		11.2		10.5		
Palisade		14.2		12.4		
Epidermis and palisade					9.2	1

[a] Limiting plasmolytic value $= P_l$:calculated osmotic pressure at full turgor $= P_t$. Limiting plasmolytic value approached by deplasmolysis $= P_l$ deplasm; cryoscopically determined osmotic pressure $=$ OP. Discrepancy $(P_t - OP) =$ PCD (plasmolytic-cryoscopic discrepancy).

This plasmolytic-cryoscopic discrepancy has been frequently observed, but a number of writers contend that all that is demonstrated by it is the inaccuracy of the methods.

When carefully controlled, the plasmolytic methods are unlikely to be subject to very great errors, but, as pointed out (page 130), they measure the DPD of the protoplast which must be *smaller* than the vacuolar osmotic pressure. Adhesion of protoplast to wall has been suggested as a cause of overestimation of the "plasmolytic osmotic value," but this is easily avoided by approaching limiting plasmolysis from the plasmolyzed condition.

Cryoscopic determinations of the osmotic pressure of expressed sap can give reasonably accurate results, but there is considerable uncertainty of the origin of the sap; the low values of the osmotic pressure have been ascribed to contamination of the vacuolar sap by water expressed from the cytoplasm [Currier (18)]. This implies that water is held in the living cytoplasm in some manner which is broken down at death; also that the volume of this releasable water is adequate to dilute the vacuolar sap and that this supposed release of water is not accompanied by a corresponding release of solutes.

It seems unlikely that the volume of cytoplasmic water accompanied by solutes is adequate in some of the tissues, like beet root, which have been studied. An element of doubt certainly exists. In leaf tissues of many thin-leafed plants, the volume of cytoplasmic water is a large fraction of the total leaf water, and there the cryoscopically determined osmotic pressure of an expressed sap may bear little direct relation to that of vacuolar sap.

The dual status of the water held in leaf cells was shown clearly in a study of the expression of leaf water (see also page 128). When the leaf was living, only a limited volume of water could be pressed out of leaves of *Fagus sylvatica* even at considerable pressures. When these same leaves were then killed by freezing in liquid air, a further large amount of water was pressed out at quite low pressure. Microscopic examination showed that pressing of the living leaf had squeezed the water out of the vacuoles and the cells appeared to retain the hydrated cytoplasm; there was no information regarding quantities or concentrations of solutes in this cytoplasm. Death, however, caused the release of water bound in the cytoplasm and also of solutes, which probably included both vacuolar and cytoplasmic solutes.

This ability of living cytoplasm to retain water against a considerable hydrostatic pressure is important as it shows that tissues with large proportionate cytoplasmic volumes will yield expressed saps which may be very different in composition from either vacuolar or cytoplasmic sap *in vivo*. This bound water also suggests that possibly cytoplasmic protein which might be similar to the "structure-proteins" of Banga and Szent-Györgyi (3a) may be important in plant cells in relation to accumulation processes, as has been suggested in animal tissues (cf. Chapter 4 of this volume).

Data of essentially similar type have been advanced by Bogen (9), who again suggests, in explanation, that part of the water uptake process in, for example, *Melosira* spp. is "active." In *M. varians* the plasmolytic osmotic value (P_l by our symbols, or O_g *osmotische Wert bei Grenzplasmolyse*) equals 0.280 M sucrose normally and 0.249 M sucrose in presence of 10^{-3} M azide, which would presumably stop active movement by interference with metabolism. The vapor pressure determination of vacuolar sap corresponded to that of 0.223 M sucrose. The plasmolytic value is thus reduced reversibly by about 1 atm by a respiratory inhibitor in low concentration and is about 1.5 atm above the osmotic pressure of the sap.

Here however, as in the case of parenchymatous tissues, there is inevitably some doubt regarding the equality of the concentration of expressed and vacuolar saps. The effect of azide could possibly be regarded as an interference with a "sugar pump" rather than a "water pump."

Another type of argument regarding the possibility of water secretion has been adduced by Levitt (40). If, in fact, a difference in hydrostatic pressure is maintained by such a process against continuous leakage by diffusion in the opposite direction, the work (W) done in maintaining the excess hydrostatic pressure (p) is given by

$$W = P \times a \times v \times t$$

where a is the area, v the velocity of water movement, and t the time. Also

$$W_1 = p \times A \times (P_w \times p)$$

where W_1 is the work done in 1 hour by 1 gm of tissue having a specific area A (cm^2 per gram) and P_w is the permeability as an estimate of the rate of leakage. The amount of leakage with an excess hydrostatic pressure (p) is $P_w p$. Controversy has thus been transferred to the magnitude of P_w, and it will be agreed that if P_w is big, only small excess hydrostatic pressures could be maintained. The work required to be done can be evaluated using the second equation given above; W_1 being given in ergs per gram per hour, p in dynes, A in cm^2 ($P_w p$) in centimeters per hour per atmosphere and atmospheres, respectively. Considering then the work necessary to maintain an "active pressure" of 5 atm, one obtains for beet root tissue:

	p	A	(P_w		p) =	W_1
Myers	5×10^6	600	0.7×10^{-4}		5 =	10.5×10^5
Levitt	5×10^6	240	20×10^{-4}		5 = 120	$\times 10^5$

The respiration rate observed (50 μl rising to 100 μl per gram per hour) gives 100×10^5 rising to 200×10^5 ergs per gram per hour. The energy requirements for active pressures of 2 atm, for example, are one-sixth those for 5 atm.

The thermodynamic argument can do no more than assign a maximum limit to the magnitude of any active water uptake. When permeabilities are relatively small as suggested by Myers' (and Höfler's) data, energy derived from metabolism is well in excess of the requirement for "driving a pump."

1. Bound Water

The concept of bound water has appeared in the literature from time to time with varying degrees of emphasis. Water molecules in a living organism (or in a relatively heterogeneous nonliving system) exist in an almost infinite range of states. Being dipoles, water molecules are attracted to charged ions and thus form an atmosphere around each ion. The attractive force depends on the square of the distance from the charged ion, and thus a range of attractions is found.

This also applies to the atmospheres of water molecules held by hydrogen bonding to the active groups of protein and carbohydrate molecules, as well as to those molecules attracted to ions of the proteins, etc. Consequently the amount of the so-called bound water which is found in any system depends on the method used to remove the water from the shells surrounding the binding sites.

Several methods have been used to obtain estimates of so-called bound water. One of these depends on the assumption that at some arbitrary temperature, say $-20°C$, the bound water will not freeze. The tissue or solution is frozen and is allowed to thaw out in a calorimeter, and the latent heat of melting of the tissue or solution is determined, hence an estimate of the ice content (or free water) of the material is obtained.

The cryoscopic method, which is applicable only to solutions, consists in dissolving a test solute (sugar) to a given known molarity in the solution and determining the freezing-point depression. If the observed depression exceeds the calculated depression, it is taken that part of the solvent was unavailable for solution and that the effective molarity of the sugar is greater than is to be expected from the weight of sugar used.

The fraction of the total cell water which is in the so-called bound state depends on the concentration of charged sites or sites at which hydrogen bonding can occur. The lower the absolute water content, the greater is the likelihood that importance should be attached to the extent of immobilization of water.

In view of the difficulties of arriving at adequate estimates of osmotic pressure of vacuole contents and of cytoplasm, it is hardly surprising that it has not proved possible to assign much certain significance to estimates that have been made of bound water of plant tissues.

Attempts have been made to correlate drought and frost resistance and bound-water content of expressed saps. The results are unconvincing. The origin of saps is presumably multiple since they may be derived from vacuole, cytoplasm, and wall, and the relative contributions made by these various sources are unlikely to be the same in susceptible, resistant, or hardened plants.

Levitt has determined "bound" water by the simple procedure of finding the dry weights of tissue (fungal mycelium, for example) at two temperatures. Rise in the temperature at which drying is effected drives off the more strongly bound water. This procedure enables one to obtain comparative figures rather simply. Levitt used it to show that *Aspergillus niger* grown in a range of solutions was more frost resistant and contained more bound water the higher the osmotic pressure of the culture medium.

In his 1956 survey of the general question of frost and drought resistance, Levitt (41), however, lists some thirteen studies in which no correlation between frost hardiness and bound-water content was found and some thirty-seven where such a correlation was found. This illustrates how indefinite these concepts really are.

B. ROOT PRESSURE AND EXUDATION

The other special case in which active uptake of water has been claimed (and also denied) is in the phenomenon of root pressure and exudation. The simplest "osmotic" explanation is indicated in diagrammatic form in Fig. 1. In order to substantiate this view it would be necessary to establish that the osmotic pressure of the exudate or xylem sap was in fact equal to either the excess hydrostatic pressure necessary to stop exudation (this is in fact what is strictly termed the root pressure) or, alternatively, that the xylem sap has the same osmotic pressure as that of the external medium which just stops exudation.

In practice completely unequivocal experimental results are remarkably difficult to obtain, but one entirely natural phenomenon is worth noting as a preliminary to the discussion of relevant work. This is the exudation of liquid from the so-called drip tips of certain leaves. This liquid, it should be explained, is not secreted by living cells in the leaf tip; it flows from rather large pores or water stomata behind which are plugs of loosely packed cells which constitute a tissue termed "epithem." The liquid is thus forced out by the excess hydrostatic pressure of the contents of the vascular system. This contrasts with the behavior of "active" hydathodes which consist of living and presumably secretory cells.

The liquid coming from drip tips or passive hydathodes frequently contains solutes. The encrusting saxifrages (e.g., *Saxifraga aizoön*) have a white deposit of calcium carbonate left on the surface after evaporation of water from the exudate; coleoptiles of grasses also excrete solutes. The most remarkable example to have been studied extensively is the tropical aroid *Colocasia antiquorum* which is capable of secreting up to 400 ml liquid per day. This liquid approaches pure water in composition; Dixon (23) cites the osmotic pressure determined cryoscopically as 0 atm and the specific conductivity as about 10^{-3} mho.

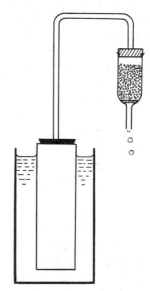

The mechanism of secretion of this water, as suggested by Dixon, is indicated by the diagrams of Fig. 13. The nonliving model contains a salt solution in the porous pot, absorbs water from the beaker, and secretes water from the end of the delivery tube; all that is required to keep the system running is a suitable absorbent (e.g., a mixed-bed exchange resin) in the bulb. In *Colocasia* "osmotic" entry of water into a solution in the xylem of the roots would provide root pressure and absorption of solute by living tissues en route to the leaves would provide the "active water secretion."

It is important to recognize that all so-called "active" or "nonosmotic" processes refer to the transport of one part only of a transport complex of unknown nature (transported material plus carrier). The transport of the *whole complex* is never "active." In the case of the active water transport of *Colocasia*, the suggested model, which incidentally is quite unproved, is a complex consisting of a solution inside a semipermeable membrane; i.e., a

FIG. 13. Model suggested to explain "secretion" of water by hydathodes of the passive (epithem) type. The porous pot with semipermeable membrane containing salt solution is immersed in water. Solution flows out over a mixed-bed ion exchange resin. See text.

solute which absorbs water from the external medium. This complex acts as an active water-transport mechanism by the assumed removal of solute from higher levels of the xylem, so that the material exuded is not the complex itself but the water component only.

It should be pointed out that no experimental evidence exists as to the nature of the carrier mechanism in *Colocasia*.

Other examples of exudation have been studied more extensively, such as the exudation from the cut stems of herbaceous plants. This

exudation is reduced and ultimately converted to a negative exudation
or intake of liquid as the osmotic pressure of the external liquid bathing
the roots is raised.

Van Overbeek (62a) determined the osmotic pressure of a mannitol
solution which just prevented exudation by decapitated tomato (*Lycopersicon esculentum*) plants. He then transferred the plants to water
and found that the osmotic pressure of the exudate was some 1–2 atm
lower than that of the balancing mannitol solution, from which he concluded that a "nonosmotic" component was involved. Van Overbeek

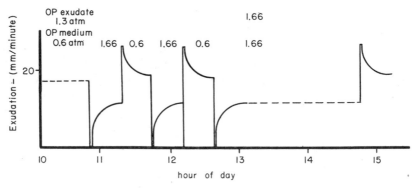

Fig. 14. Rate of exudation from decapitated tomato plants at times of day shown
as abscissas, during a series of changes of the external solution from Hoagland solution (OP, 0.6 atm) to Hoagland enriched with mannitol (OP, 1.66 atm). The osmotic pressures of exudation sap collected and of the medium are shown above the
graph. Based on Fig. 1 of Arisz *et al.* (2).

found further that HCN poisoning eliminated this nonosmotic component, which was regarded as a metabolically controlled water secretion. Van Overbeek exposed his root systems to the compensating mannitol solution for short periods of 5–15 minutes.

Arisz and co-workers (2) in a more extensive study also used
tomato plants. Measurements were made of the rate of bleeding every
30 seconds, and the root systems of the decapitated plants were placed
in a funnel containing solution to simplify rapid change of solutions.
Typical results are shown in Fig. 14. The plant was exposed first to
Hoagland solution of osmotic pressure 1.3 atm and transferred at the
times indicated by the vertical line to Hoagland solution plus mannitol
of OP 1.66 atm and then back to the original 1.3-atm solution at the
next vertical line.

The immediate effect of transfer to the more concentrated 1.66-atm
solution is a stoppage of exudation, but in about 5 minutes the rate has

risen to a equilibrium rate roughly half that on the original Hoagland solution. On transfer back to Hoagland solution the initial secretion rate is much increased but falls rapidly during about 5 minutes to approximately the original rate on Hoagland solution. The osmotic pressure of exuded sap in the first period on Hoagland solution was 1.3 atm, and when on Hoagland solution plus mannitol it rose to 1.66 atm. This building up of a higher concentration was supposed by Arisz and his associates to be due to continued salt secretion into the xylem at a constant rate coupled with the reduced water intake, due to the higher external osmotic pressure. Correspondingly, when the reverse change was made, increased exudation resulted from the higher internal concentration, and the increased inflow of water again diluted the internal solution.

It seems possible that van Overbeek's results have not allowed adequately for the rather rapid adjustment in concentration of the exudation sap revealed by this study. A truer compensating osmotic pressure of the medium, which is obtained only in the first 30 seconds after the change of medium, is 0.4–0.6 atm larger than that of the exudation sap. This difference in value is, in fact, quite large in relation to the actual osmotic pressure of the exuded sap and therefore of the root pressure. It is of course arguable that the exudation sap actually collected for cryoscopy is, like the exudate from *Colocasia* leaves, more dilute than the xylem contents of the roots, which actually determine the hydrostatic pressure that can be generated and which determine the magnitude of the compensating external osmotic pressure that is needed to stop inflow of water.

Other work on root water movements and root pressure is considered in dealing with water permeability.

C. WATER PERMEABILITY

1. Filtration or Osmotic Methods

Arguments regarding the mechanism of passage of water across plant membranes are important for the assistance they may give to an understanding of the structure of the membranes and of the possibility of control that membranes may exercise, at least when alive.

Earlier studies on water permeability of plant protoplasts have been reviewed by Stiles (57) and are quoted in the important paper by Huber and Höfler (34), which deserves further comment. As a measure of water permeability the units commonly chosen are volume of water transferred per unit area of surface per unit time per unit DPD difference: namely, $cm^3/cm^2/hour/atm$ or, simplifying, $cm/hour/atm$. Permeabilities are also quoted as quantity transferred (mols) per unit time per unit area per unit (molar) concentration difference.

Huber and Höfler determined the time course of plasmolysis of cells the geometrical shape of which made volume measurements relatively easy and used the plasmometric method which has been described (page 122). Typical results are shown in Fig. 15, in which ordinates

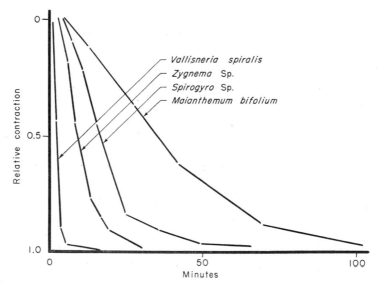

Fig. 15. Rates of contraction (or plasmolysis) of cells of *Vallisneria spiralis* leaf (*V*), *Zygnema* sp. (*Z*), *Spirogyra* sp. (*S*), *Maianthemum bifolium* (*M*). Ordinates show "relative contraction." If final degree of plasmolysis, V_p/V_t is written as G and degree of plasmolysis after other time periods is g, the relative contraction is $1\text{-}g/1\text{-}G$. From Huber and Höfler (34).

show the relative volume reductions based on the degree of plasmolysis (*Plasmolysegrad*), g, at times after immersion in plasmolyticum shown as abscissas. The final degree of plasmolysis can be written G, which in our former symbols equals V_p/V_o (V_p and V_o being the volumes at final plasmolysis equilibrium and original volume, respectively). G is thus equal to P_i/P_e, the ratio of the osmotic pressures of the internal cell sap to that of the external solution.

The permeability constant k is obtained from the relations:

$$\frac{dg}{dt} = -k\left(P_e - \frac{P_i}{g}\right)$$

integrating and rearranging gives

$$k = \frac{1}{P_e(t_2 - t_1)} \left(g_1 - g_2 + G \ln \frac{g_1 - G}{g_2 - G} \right)$$

where g_1 and g_2 are the degrees of plasmolysis at times t_1 and t_2.

These values of k provide relative permeabilities which are of some interest. Values of k found experimentally are shown in Table IV.

TABLE IV

VALUES OF k IN VARIOUS SPECIES AS FOUND EXPERIMENTALLY

Species	k value	Species	k value
Vallisneria spiralis	8.3	*Zygnema* 'velox'	6.0
Trichocolea tomentella	1.9	*Mougeotia* sp.	6.0
Spirogyra nitida	0.6	*Zygnema* sp.	0.8
Maianthemum bifolium	0.14	*Salvinia auriculata*	0.68

Some of the data were converted to absolute units ($cm^3/cm^2/hour/atm$) which require that the surface area be considered. Incidentally the assumption was made for this purpose that water passage through the plasmolyzing protoplasts occurred only through the hemispherical ends. The reason for this assumption is that when a long, plasmolyzed protoplast breaks into two bits, one with a large area and the other with a small area and presumably having the same internal osmotic pressures, the smaller protoplast shows in general the quicker further shrinkage. The conclusion drawn from the detailed data was that the water exchange in plasmolyzed *Salvinia* occurred very largely (*überwiegend*) through the free menisci and that the part of the protoplast pressed against the wall played a subordinate role in water transport. In *Salvinia* the permeability on this basis was found to be 33 μ per hour per atmosphere but would have been about 3 μ per hour per atmosphere if it had been expressed on the basis that the whole protoplast area is involved in water transport.

Levitt *et al.* (42) (and a number of other groups of workers) have used a permeability constant for water (p_w) on the basis

$$p_w = \frac{\text{change in volume}}{\text{average area} \times \text{average pressure difference} \times \text{time}}$$

The accuracy of assessment of volumes and areas depends on the nature of the cells and tissue. The pressure-difference determination may present more problems as evaluation of the DPD of the cell con-

tents is required. Levitt and co-workers examined the shrinkage of isolated onion protoplasts, which are easily obtained if a strip of epidermis
from an onion bulb scale is plasmolyzed and then cut with sharp
scissors and gently squeezed toward the cut edge.

The shrinkage of an exposed protoplast is incidentally often called
plasmorrhysis. Permeabilities, or rates of passage of water per unit
pressure difference, were determined both during plasmorrhysis and
deplasmorrhysis. Shrinkage of the protoplast and its accompanying dehydration were not accompanied by any very marked decrease in water
permeability; there was, however, a slight decrease. During deplasmorrhysis, however, the swelling was associated with a nearly four-
fold increase in permeability as the volume rose from a fifth of the
normal up to the normal volume. This, it seems, is due to mechanical
stretching that has damaged the cell surface. Indeed, it is easy to show
that the rate at which deplasmolysis or deplasmorrhysis occurs is of
fundamental importance, and in the work quoted, though this seems
to have been recognized, different rates of swelling and shrinking were
not examined.

The absolute value obtained for the water permeability of the isolated onion protoplast was 21 μ per atmosphere per hour, which may be
compared with Huber and Höfler's value of 33 μ per atmosphere per
hour for the *exposed* portions of a *Salvinia* protoplast, and these latter
workers suggested much lower values for the permeability of the part
of the protoplast pressed against the cell wall.

Other studies have provided a remarkable range of determined
values for permeability to water. Rosene (53), in a study of water uptake by individual root hairs of radish (*Raphanus sativus*), showed that
considerable variation in rate occurred. Micropotometers were used,
and rates of uptake between 0.2 and 3.1 μ per minute were observed.
The DPD differences providing this uptake were not recorded, but in
other experiments uptake was stopped by an external OP of about 5
atm. In other work, DPD's of root surfaces of the order of 2 atm have
been found, so the absolute permeabilities indicated by Rosene's work
might well be in the range between 5 and 60 μ per atmosphere per
hour for the slower and faster root hairs, respectively. These root hairs
were of course not subjected to the experimental treatment of plasmolysis which was involved in the other examples quoted.

Myers (45) determined the permeability of unplasmolyzed thin
slices of beet root tissue using essentially the relation

$$p_w = \frac{\text{volume change}}{\text{average area} \times \text{average PD} \times \text{time}}$$

Shrinkage of turgid slices was allowed to proceed in a nonplasmolyzing solution (9.5 atm). Water losses and suction-pressure changes of the tissue with time were obtained, and in addition p_w for similar tissue which had been plasmolyzed was determined using the standard plasmometric method.

p_w for plasmolyzed beet was 13 μ per atmosphere per hour, and for unplasmolyzed beet it was 0.7 μ per atmosphere per hour. This result agrees broadly with Huber and Höfler's contention that a protoplast pressed against a cell wall has much smaller conductivity for water than a naked or plasmolyzed portion of protoplast.

This reduced p_w of the unplasmolyzed tissue was regarded as due to the much reduced surface area of protoplast when pressed either against the cell wall or into the interfibrillar spaces. There are no quantitative data of dimensions of interfibrillar spaces for beet cell walls, but the pore volume in certain natural fibers like ramie (*Boehmeria nivea*) can be calculated from density determinations as about 0.1 of the total, and the cross-sectional area of the pores would then be expected to be about 0.07 of the total area. The ratio of permeabilities of plasmolyzed to unplasmolyzed cells 13.0:0.7 is thus of the same order as the probable free surface areas 1.0:0.07.

Before summarizing these data, the important study of *Nitella* by Kamiya and Tazawa (38) must be dealt with, as it has provided another extreme value for water permeability (p_w) namely 1110 μ per atmosphere per hour in contrast with 20 μ for various plasmolyzed cells or root hairs and 0.7 μ per atmosphere per hour for unplasmolyzed cells of beet.

A number of investigators had shown that it is simple to cause a current of water to flow through a *Nitella* internodal cell by immersing one end in water and the other in a solution of suitable osmotic pressure. The flow of water through the cell causes concentration of the vacuole contents at the exit end (which is incidentally readily seen if there is neutral-red staining of the vacuole). The water, it may be noted, appears to flow *through the vacuole*. The suggestions made from time to time that flow of materials or diffusion of materials may occur more readily through cell walls or protoplasts (a phenomenon termed transmeability) is clearly not applicable to this particular case.

Kamiya and Tazawa's experimental arrangements are indicated by the diagram of Fig. 16(A). When set up, both coompartments *A* and *B* are filled with tap water. On replacement of the water in *B* with 0.2 *M* sucrose, water flows from *A* into *B* through the *Nitella* cell as shown by the index bubble (cf. Fig. 16(A)). On replacing the sucrose with water again, there is a backflow and equilibrium is again re-established.

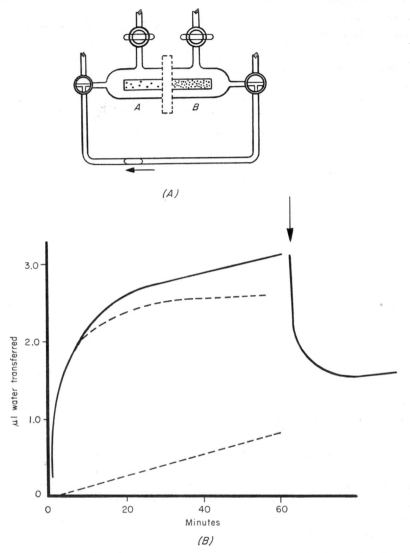

(A)

(B)

FIG. 16. Transcellular osmosis in *Nitella*. (A) Apparatus: left-hand chamber contains water; right, an osmoticum. With taps set as indicated, water flows through the cell, concentrating the vacuole contents toward the right; the index bubble goes to the left. (B) The graph shows the rate of flow through the cell. At the arrow the osmoticum and water bathing the ends of the cell were changed round so that flow through the cell was from right to left. From Kamiya and Tazawa (38).

The course of the forward flow of water through the cell into the sucrose characteristically does not come to an equilibrium when the cell solutes are polarized. Instead, after some 20–25 minutes, a steady state is reached in which a slow through-flow takes place. The depolarization is thought to be due to a stirring up of the solutes of the cell by protoplasmic circulation.

The kinetics of this through-flow are important. The initial rate of the transcellular osmosis is proportional to the difference in osmotic pressure at the two ends of the cell. So the rate or volume permeating per unit time (dv/dt) is given by $dv/dt = K\ (S_{en} - S_{ex})$ where S is the DPD difference causing a flow of water into the cell and the subscripts "en" and "ex" refer to the entrance and exit ends respectively.

And from the usual DPD-wall pressure relation

$$S_{en} = p_{en} - P_{en} - W$$
$$S_{ex} = p_{ex} - P_{ex} - W$$

where p_{en} and p_{ex} are the cell sap osmotic pressures at entrance and exit ends of the cell, P_{en} and P_{ex} are external medium osmotic pressures at the two ends, and W, the wall or turgor pressure, which is the same at both ends or virtually so. Hence

$$\frac{dv}{dt} = K[(p_{en} - p_{ex}) - (P_{en} - P_{ex})]$$

For the relatively most simple case, where water is supplied at the entrance end, P_{en} is zero and P_{ex} depends on the osmoticum used, it is then possible to derive the expression

$$K = \frac{1}{t} \times \frac{2.3 V_{en} V_{ex}}{p_i V_0 - P_{ex} V_{ex}} \times \log \left\{ \frac{p_i V_0 - P_{ex} V_{ex}}{P_{ex} V_{ex}} \left(\frac{p_i V_0}{p_i V_0 - P_{ex} V_{ex}} - e \frac{v}{V_{en}} \right) \right\}$$

where P_{en}, P_{ex}, V_{en}, V_{ex} are OPs and volumes of entrance and exit ends and p_i the internal OP and v the volume transported.

In the experiment illustrated in Fig. 16, the various constants had these values: $p_i = 6.35$ atm, $P_{ex} = 4.9$ atm, $V_{en} = 6.0$ μl, $V_{ex} = 5.35$ μl, $V_0 = V_{en} + V_{ex} = 11.35$ μl. The volume, v, transported in the first minute $= 0.79$ μl and time, $t = 1.0$ min.

$$\therefore K = 0.2 \ \mu\text{l/min/atm}$$

From the surface areas of the entrance and exit portions, the permeability in μl/mm^2/min/atm can be obtained.

However a complication exists, for it appears that the entrance permeability is different from the exit. This is shown by experiments in

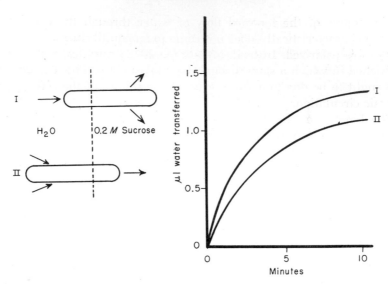

Fig. 17. (A) Osmosis in *Nitella* with lack of symmetry of intake and exit regions; *I*, small area of entry; *II*, large area of entry. (B) Rate of through flow of water with these two treatments. See text. From Kamiya and Tazawa (38).

which V_{en}, and thus A_{en}, the volume and area of the entrance end, is different from V_{ex} and A_{ex}. The flow-through of water is more rapid when $A_{en} < A_{ex}$ than when $A_{en} > A_{ex}$, as shown in Fig. 17. K is a permeability constant, thus $1/K$ can be treated as a resistance (R).

Hence
$$R_{en} = \frac{1}{R_{en}A_{en}} \quad \text{and} \quad R_{ex} = \frac{1}{k_{ex}A_{ex}}$$
$$R = R_{en} + R_{ex}$$
$$= \frac{k_{en}A_{en} + k_{ex}A_{ex}}{k_{en}k_{ex} \qquad A_{en}A_{ex}}$$

and if the ratio k_{en}/k_{ex} is called ρ

$$R = \frac{1}{k} = \frac{\rho A_{en} + A_{ex}}{\rho k_{ex}A_{en}A_{ex}}$$

In the experiment of Fig. 17, $A_{en} = 16.6$ mm², $A_{ex} = 42.8$ mm², $K = 0.15$ μl/min and when current was reversed, i.e., $A_{en} = 42.8$ mm² and $A_{ex} = 16.6$ mm², $K = 0.1$ μl/min. Substituting these values

$$k_{en} = 0.0184 \ \mu\text{l/mm}^2/\text{min/atm}$$
$$= 1104 \ \mu/\text{hour/atm}$$
$$k_{ex} = 0.0069 \ \mu\text{l/mm}^2/\text{min/atm}$$
$$= 414 \ \mu/\text{hour/atm}$$

It is worth while to summarize some of the data of water permeabilities that can be expressed in absolute units (see Table V).

Very little clear pattern emerges out of this surprisingly large range of values. The low permeability of a protoplast pressed against a cell wall in *Salvinia* or beet root is perhaps understandable as these walls are relatively thick and tough and have mechanical or retaining functions, in contrast to root hair cells where clearly the functional requirement is high permeability. One might expect high permeability to be achieved by a considerably looser texture of the walls, and the 20–100-fold greater permeability in root hairs might reasonably have been expected even if its precise molecular mechanism is still uncertain.

TABLE V

DATA ON WATER PERMEABILITY

	Site	μ/atm/hour	References
Beet root (*Beta vulgaris*)	Unplasmolyzed cells	0.7	(45)
	Plasmolyzed cells	13.0	(45)
Salvinia auriculata	Ends of plasmolyzed cells	33.	(34)
	Plasmolyzed cells[a]	3.	(34)
Arbacia	Unfertilized eggs	6.0	
	Fertilized eggs	12–24	
Fucus	Eggs	8–22	(51b)
Allium cepa	Epidermal cells	18–66	(18a)
	Isolated protoplast	21	(42)
Tolypellopsis		65	(51a)
Nitella	Outflow	420	(38)
	Inflow	1110	(38)

[a] If whole surface is regarded as permeable.

The increased permeability on plasmolysis in *Salvinia* or beet might on this basis reasonably be expected in cells or tissues where the cell wall is a dense structure, but one would not expect it in more permeable types of cell such as root hairs or, in general, in growing, enlarging cells.

The very high permeability of *Nitella* to water must at present be considered as defying comment. Clearly it is not merely a question of habitat; other fresh-water algae do not show such high permeabilities to water.

2. Isotope Exchange

The procedure has been to follow the uptake of "water" containing deuterium or tritium, i.e., DHO or THO, by tissue or, alternatively, the rate of outflow of these isotopic forms from previously treated tissue. The primary object of some of the published studies on plant tissues, in which these methods were used has been to compare water permeability of normal and auxin-treated tissues, and the data have not been expressed in a form to permit comparison with results just discussed. Rates of influx or efflux of DHO or THO could be utilized in calculating net fluxes imposed by a DHO concentration gradient, but that has not in fact been carried out for plant tissues. Exchange of DHO and H_2O would of course also be accompanied by exchange of D and D^+ for H and H^+ respectively.

Isotope exchange studies of water permeability are considered in relation to the information they have given regarding the influence of auxins (see pages 184, 185).

The use of D_2O (or DHO) or the tritium analogs for measurement of flux of water through a membrane of finite thickness yields results which do not agree with the observed rates of movement brought by an activity gradient due to the presence of dissolved substances on one side of the membrane.

Where a membrane separates water from a solution of given concentration, say 1 M, it can in general be considered that pores of the membrane are filled with water and a sharp concentration gradient occurs at the solution boundary. Osmosis, that is, a net flux of water, occurs because more molecules are ejected from the end of the pore adjacent to the solution phase in which activity or concentration of water is low, than are ejected into the pure water phase where the activity of water is high.

Where the membrane separates water, in which the D_2O concentration is virtually zero, from enriched water where the D_2O concentration is 1 M, the situation is different. The total water ($D_2O + H_2O$) activities on both sides are equal, and there will be no net flux of water as the numbers of collisions and rates of ejection of molecules will be the same from the two ends of the file. A D_2O molecule entering from the enriched side into the file may be ejected again by collisions onto its own side. This effect is more marked, the narrower and longer the file of molecules in the pore, and the transfer of D_2O may in fact be very small as compared with the flow of water which is brought about by an activity gradient. This was found by Hevesy and associates (32) in a study of water permeability of the frog skin, and an essentially comparable situation is present when fluxes of ions through membranes are investigated by the use of isotopically labeled ions [Hodgkin and Keynes (33)]. The relation between tracer flux and true flux depends on whether a net flux exists, as well as on the length of the column of particles in the membrane.

Differences in permeability as revealed by tracer fluxes and net flow due to a water activity gradient are probably not connected with the supposed difference between filtration permeability and diffusion permeability as has been suggested, at least in the case of most plant cell walls.

Filtration permeability refers to the bulk flow of liquid through narrow pores and it is defined by the Poiseuille relation:

$$\text{Flow rate} = \frac{\pi n r^4}{8\eta} \times \frac{dp}{dt}$$

where r is the pore radius; n, the number of pores in parallel; η, the viscosity; and dp/dt, the pressure gradient.

This relation is not applicable to conditions where the pore diameter is of the same order as molecular diameters. In such cases flow of water across the membrane takes place by diffusion. Where water is made to flow through a membrane separating two solutions, by a hydrostatic pressure difference, the solution adjacent to the membrane on the intake side becomes concentrated and on the outflow side diluted, so that a back diffusion occurs which is due to this concentration gradient and which is dependent on the thickness of the unstirred layer adjacent to the membrane. This back diffusion opposes the flow due to the hydrostatic pressure gradient.

This factor clearly must be considered where permeabilities are estimated by experimental application of a hydrostatic pressure gradient.

3. Permeability of Cell Walls

There is little direct quantitative information regarding the permeabilities of cell walls. In the integrated physiology of the plant in its natural condition, the permeability which is important is that of the wall cytoplasm complex.

When the cytoplasm is torn off or pushed out of the wall by plasmolysis, one can obtain estimates of the permeability of the cytoplasm, but not readily of the wall. Qualitative estimates of the permeability of the wall are fairly simple with dyestuffs. Thus the large molecules of Congo red fail to enter the space between plasmolyzed protoplast and cell wall of certain cells though acid fuchsin enters easily.

Results of that type give the impression that the plasmolyzed cell wall is relatively large-pored and permeable. Clearly very large differences in wall permeability are to be expected, ranging from the external cutinized walls of epidermal cells or suberized cork cells to the presumably very permeable walls of root hairs.

Indications of different permeabilities of different parts of the cell wall of certain cells are obtained from examination of plasmolysis forms. For example, endodermis cells show plasmolysis indicating the impermeability of the Casparian strip region, moss leaves show contraction of the protoplasts from the midrib sides of the cells, suggesting that the walls perpendicular to the surface are more permeable than the surface walls.

Certain "wilt-fungi" produce toxins which are believed to cause decreased wall permeability of leaf cells. These toxins are supposed to be carried up in the transpiration stream and to block the interfibrillar spaces of the cell walls adjacent to the vascular-bundle endings, thus reducing permeability and causing wilting. The effect can be imitated by certain substances of large molecular weight like inulin.

Determinations of these supposed altered wall resistances quantitatively would be of interest, but they are not readily achieved.

Qualitative demonstrations of wall conductivity were made by de Haan, Höfler, Huber, and others. Onion epidermal cells plasmolyzed in sucrose were put into water for 5–30 seconds and then into liquid paraffin. Deplasmolyis occurred as quickly in paraffin after a 30 seconds' wash in water as when they were left all the time in water. So it was argued that the space between plasmolyzed protoplast and wall equilibrated very rapidly with the external solution, thus showing high permeability of the cell wall.

On the other hand, a remarkable "change of plasmolyticum" effect is also observed (6). If onion epidermal cells plasmolyzed in sucrose are transferred to isotonic potassium chloride solution a very rapid shrinkage followed by an expansion of the protoplast occurs. This seems to be due in part to potassium chloride diffusing more rapidly inward than sucrose can diffuse out, but the length of time taken to establish the new equilibrium volume of the protoplast suggests that the walls are relatively impermeable to solutes. As, however, the final equilibrium volumes of the vacuole in solutions of potassium chloride and sucrose of the same osmotic pressure are *not* equal, the system is clearly complex. Potassium chloride of given osmotic pressure causes greater shrinkage of the vacuole than does sucrose of the same osmotic pressure.

It seems at present hardly practicable to obtain quantitative measures of wall permeability (as opposed to wall plus cytoplasm permeability) except in the cases of large and peculiar cells such as, for example, *Valonia* and *Nitella*.

4. Water Conductivity of Root Systems

The normal passage of water into land plants is through the root system. This flow is brought about by a gradient in DPD or activity of water on which may conceivably be superposed some so-called "active" water movement.

The gradient referred to is due in part to presence of solutes in xylem sap and in part to the fact that the liquid in the xylem may be subjected to a tension by the occurrence of active transpiration. Since the DPD or suction pressure (S) of contents of a cell or tissue element is given by $S = P_i - T$ (where P_i is osmotic pressure and T, the turgor pressure or excess hydrostatic pressure), when T is a tension or "negative pressure," S will be larger than P_i.

The driving force sending water from external medium to xylem is the pressure difference $S_x - S_0$ where S_x and S_0 are the DPD's of xylem and outside medium respectively. The conductivity of the tissue between the medium and xylem is expressible thus:

$$\text{conductivity} = \frac{\text{volume entering}}{(S_x - S_0)}$$

Among the earlier studies were those of Renner, in which he showed that the water uptake by transpiring rooted plants was considerably increased by cutting off the root system. It was fairly clear that the living root system offered a considerable resistance to the movement of water. A corollary of this is that the liquid in the vascular cylinder must have been in a state of tension.

When sections of a root are cut in order to use the Ursprung-Blum method of determining the DPD's of the series of cells from center to exterior (65), the release of the mutual tensions between neighboring cells causes the wall pressures of stelar cells to increase when the root is cut open, so that the measured DPD's of the stelar cells of the open root must be lower than the values they had when the root was intact and transpiring.

The actual gradient of DPD's in the intact root of a transpiring plant cannot be ascertained experimentally, but there must be a positive gradient from surface to xylem. The apparent jump in gradient at the position of the endodermis, as found by Ursprung and Blum, is almost certainly an experimental artifact produced as indicated above by release of mutual tensions on cutting.

Brouwer's (13) work represents one of the most extensive modern studies of root conductivity, following in the tradition of Renner, Brieger, Sierp, and Brewig.

Conductivities of different regions of a root were studied. (Brouwer's material was the root system of *Vicia faba* using the apparatus indicated by Fig. 18.) A single root was selected and threaded through rubber washers as illustrated; the remaining roots, forming the greater part of the root system, were enclosed in the "main chamber" where, by altering the osmotic pressure of the medium, an alteration of the DPD of the root system could be effected. The water conductivities of separate zones of the selected root were determined after an estimate of the internal DPD had been obtained by the standard method of opposing the entry of water by an external osmoticum. The rates of entry to a given zone were determined using a series of three or four external media of different osmotic pressure, and by interpolation the osmotic pressure producing no entry was obtained and treated as equal to the internal DPD (see Fig. 19(A)).

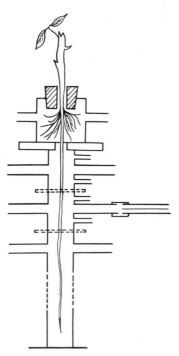

Fig. 18. Apparatus used in study of resistance to water flow through different zones of a root. The uppermost chamber contains most of the roots, and the osmoticum surrounding them can be changed. One root is threaded through rubber washers separating five chambers, the first two only are shown. They have inlet and outlet tubes for changing the osmoticum and a tube to which a capillary tube potometer is attached; one only is shown. From Brouwer (12a).

The uptakes of water given in mm³ per centimeter length of root are as in Table VI. The experimental plants have been treated in two ways; in one, the roots in the main compartment have been in water (Table VI, columns 2, 4, and 6), and in the other treatment, the main (uppermost) compartment has contained 2.5 atm solution (Table VI, columns, 3, 5, and 7).

It will be seen that the internal DPD's are not very markedly different in the different zones but the uptakes have a characteristic pattern: apical uptake much exceeds basal when the suction and uptakes are low, but with high suction and uptake there is a slightly greater uptake by basal than by apical regions. The resistance to water flow, in other words, of the basal regions is greatly reduced by an increase in DPD or suction. The resistance of the apical regions is not much affected. Similar results are illustrated in Fig. 19(B).

This effect on the water conductivity is found also when the DPD of the tissue is increased by application of an osmoticum outside the experimental root. The very large changes in conductivity are almost certainly controlled by activity of the protoplasts. It has, for example, been shown that respiratory poisons like hydrocyanic acid (HCN)

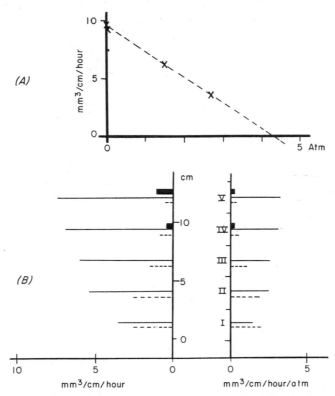

FIG. 19. (A) Rates of flow into a root zone shown as ordinates when exposed to external osmotica of osmotic pressures shown as abscissas. By extra- or interpolation, the osmotic pressure corresponding to zero flow is obtained.

(B) Left side: rates of flow (mm³/cm length/hour). Right side: conductivities (mm³/cm length/hour/atm). Values for the five different 2.5-cm zones are shown when DPD difference is 1.3 atm (------) and 2.5 atm (———). The blocks (shown ▬) give the values when the DPD difference is 2.5 atm in presence of 10^{-3} M KCN. From Brouwer (12b).

cause marked reduction of conductivity. Potassium cyanide at 5×10^{-3} M stopped water uptake immediately and at 10^{-3} M caused a gradual reduction to near zero over a period of 2–3 hr (see also the results given in Fig. 19B). As respiration is only slightly influenced by the lower concentrations of HCN, it is uncertain that its influence indicates the direct mediation of respiration. Similar results of lowered water conductivity by low oxygen concentration or respiratory poisons have been obtained by Rosene (53a) and others.

These effects of *increased* conductivities brought about by *increased* suctions are in contrast to the results of Brauner, who found lowered conductivity with increased dehydration of the membrane. In Brauner's work however very high osmotic pressures (over 100 atm) were used. In the root conductivity work, the state of hydration of membranes at the various relatively low suctions actually used must be nearly equal, but there is probably a turgor difference such that at high turgor (low suction) the protoplast may be forced further into

TABLE VI
WATER UPTAKE IN ROOTS OF BEANS (*Vicia faba*)

Zone	Main compartment[a]		DPD		Resistance[b]	
	Tap water	2.5-atm solution	H_2O	2.5 atm	Weak suction	Strong suction
Basal V	0.4	8.0	3.4	4.2	3.25	0.32
IV	0.2	7.0	2.0	4.9	2.60	0.34
III	1.2	6.0	1.6	2.6	0.92	0.42
II	3.6	5.0	1.3	2.5	0.47	0.51
Apical I	2.8	4.0	1.9	3.3	0.43	0.68

[a] In mm^3/cm/hour.
[b] Resistances (in atmospheres) required to give rate of 1 mm^3/cm/hour.

interfibrillar spaces of the wall and so bring about a reduced conductivity for water. Conductivity for, or uptake of, ions appears to be affected in a similar manner to water conductivity by alterations in DPD in the basal zones of the root.

These results may be compared with those of Bennett and Rideal (8) (see page 113 in this chapter) and Myers (45) (see page 150 in this chapter) which suggest that pressure of the external layers of the protoplast against, or into, the cell wall causes lowering of permeability. They also may be compared with figures for permeability in absolute units obtained by other workers. As the surface area of the experimental bean roots was not given one must make a reasonable estimate, say 30 mm^2 per 10-mm length. Extreme values for the conductivity are then about 4–80 μ per atmosphere per hour, which are comparable to conductivities found with other tissues in which also a wide range of values has been observed.

Difficulties of interpretation of water uptake phenomena of root systems are not confined to the problems presented by the effects of suction or cell turgor on the one hand, and the influence of metabolism on the other. Study of uptake and exudation from isolated, or cut, roots presents further problems.

Rosene determined intake and output from segments of onion root prepared as indicated in the diagram of Fig. 20. The observed quantities of uptake and output of liquid are indicated. The apical 4-cm segment showed no outflow from the apical end; the second segment showed outflow from both ends, and only in the oldest basal segment was there a flow inward from the upper basal end and through the vascular cylinder to an outflow at the lower end. In this last case, part at least of the vessel contents must be washed away by the flow of water downward under influence of gravity. It is common to observe in the younger segments more exudation from the basal than apical ends, which must mean that the later-formed vessels are closed at the lower (apical) end even though these ends are a considerable distance from the apex.

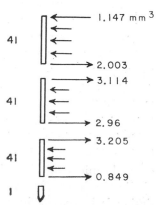

The critical experiment would consist of a through-flow of water from the upper (basal) cut end to the lower (apical) cut end, which would wash any solutes from the xylem (which may have been secreted by the innermost living cells) so effectively as to maintain the osmotic pressure in the xylem tract at the same value as the external medium. Any inflow then would be an indication of active water movement, and cessation of flow might mean that the flow was entirely passive diffusion. This experiment, however, probably cannot with certainty be achieved. Inevitably the water absorbed by cortical and outer stelar tissues will first

FIG. 20. Quantities of water exuded or absorbed from cut ends of segments of an onion root when absorbing water through potometers applied as indicated by the centrally placed inward pointing arrows. Lengths of the segments are indicated at the left. From Rosene (53).

flow into the youngest xylem elements, which possibly cannot be subjected to this through-flow, or the water may enter the phloem as suggested by James and Baker (35) and the exudation may be largely from the phloem.

There is no doubt whatever that the exudation processes of the root
system are closely controlled by secretory activity of the tissue and are
thus affected by metabolism. Equally it is not possible to define the
mechanisms involved or state which molecular or ionic species are subject to secretion or facilitated diffusion or whether any molecules can
pass through entirely without interference by metabolic or carrier
mechanisms.

The reduced permeability to water which is produced by anaerobic
conditions, or treatment with metabolic inhibitors like cyanide or azide,
could possibly be due to a change in the molecular pattern of those
layers of the tissue which present a barrier to diffusion. There is however no evidence whatever for such an effect. Where there exists clear
evidence of a carrier mechanism as in certain cases of ion uptake, these
mechanisms are inhibited by respiratory poisons, and the effects of such
interferences with metabolism on water uptake, or more strictly of the
water permeability of root systems, may indicate that part of the uptake occurs via a carrier system. It is worth pointing out that an active
carrier system might operate to transport a substance from one medium
to another in which its activity is higher than, equal to, or lower than
in the first medium. A truck may be driven by an engine uphill or
downhill; if it goes up, one knows for certain that it is being driven;
if downhill, one needs further data to know for certain whether it is
being driven or is freewheeling!

Periodicity in exudation is well known from the work of Grossenbacher (28) and others, which again shows the importance of active
protoplasmic control and its close correlation with endogenous metabolic rhythms. The amplitude of these fluctuations in the rate of exudation is increased, in certain cases, by treatment with 3-indolylacetic
acid, but the mechanism of this effect and of the actual observed
changes in rate is still unknown. Changes in permeability, or changes
in rate of secretion of either solutes or water, might conceivably be
involved.

5. Polar Permeability

The most striking case of polar permeability is undoubtedly that of
Nitella revealed in the transcellular permeation of water. Other cases
which must be considered for comparison are the polar permeability for
"filtration" of water through seed coats and the polarity of the water
loss, or transpiration, through insect cuticles, which at least superficially
seems to present a comparable problem.

The polarity of water flow through seed coats, first found by Denny

(19), has been more thoroughly studied by Brauner (11). Using seed coats of *Aesculus hippocastanum*, he showed that, under a pressure head of 70 cm Hg, water diffused through in the normal direction from outside to inside more rapidly than in the inverse direction. Typical results are given in Table VII.

TABLE VII

RATES OF DIFFUSION[a]

	Medium	Normal direction	Inverse direction
Aesculus hippocastanum: Var. I	Water	57	34
Var. II	Water	41	27
	0.0625 N K$_2$SO$_4$	38	38
	0.25 N K$_2$SO$_4$	30.5	31.1
Arachis hypogaea	Water[b]	152	100

[a] In microns per hour per atmosphere.
[b] Relative rates.

This polarity is correlated with the electrical asymmetry of the membrane, the nature of which for the case of *Aesculus* is illustrated in Fig. 21. The mobility of anions in the outer denser portion of the membrane is reduced to a greater extent than in the less dense inner portion and the resulting diffusion potential is therefore such that the outside surface is positive to the inside. A model membrane having the same elec-

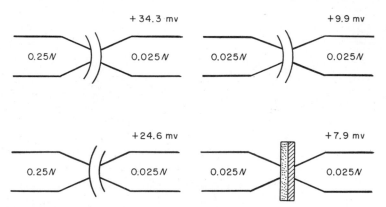

FIG. 21. Potential differences measured across a seed coat of *Aesculus hippocastanum* soaked in 0.25 N KCl and placed in contact with 0.25 N or 0.025 N KCl as indicated (convex side is the outside). The model membrane similarly soaked in 0.25 N KCl has an impermeable and a permeable layer placed in contact. From Brauner (11).

trical properties was made of two sheets of dense and permeable collodion. This electrical asymmetry causes the filtration to be accelerated in one direction and retarded in the other by a superposed electro-osmotic flow.

That this is the correct explanation of the polar permeability is substantiated by the results of treating the membrane with increasing concentrations of potassium sulfate (see Table VII) which reduce the zeta potentials of the liquid columns in the pores of the membrane. In the most concentrated, 0.25 N, there is some reversal of charge due to the high adsorbability of the SO_4^{--} ion and there is a correlated reversal of polarity of water permeability.

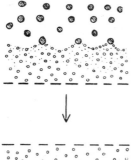
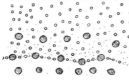

FIG. 22. Possible valve-type mechanisms for polar permeability in *Nitella*. Large circles (●) represent cell wall microfibrils; small circles and dots, cytoplasm; and the tonoplast is shown by dashed lines. On the upper (inflow) side, surface area and conductivity are larger than at the lower side (see text).

In principal the polarity of permeability of the protoplast of *Nitella* could be explained in a similar manner. An electroosmotic or other "active" water secretion from outside into the vacuole would have the effect of making the entrance permeability, k_{en}, greater than the exit permeability, k_{ex}, as the osmotic forces would be respectively assisted and retarded by the "active" forces. This cannot be regarded, however, as the only possible explanation at present, since in the *Nitella* experiment, unlike the seed coat studies, the system consists of two membranes in series, one of which is more permeable than the other. In addition to the asymmetry of the membranes, one having the protoplast on the outflow side of the membrane and the other having the cellulose wall on the outflow side, one membrane is bathed with water and the other is bathed with a solution of sucrose. It seems definitely improbable that the solution of sucrose will have reduced the permeability at the outflow side, in view of Brauner's findings, admittedly with quite different material. There remains a possibility of a sort of valve-like action. On the intake side the protoplast might have an effective larger surface area than on the outflow side as suggested by the diagrams of Fig. 22.

The possibility of this sort of control of water permeability, and indeed solute permeability, is suggested in Brouwer's work discussed on page 160. It is, however, very clear that the gradient in hydrostatic pressure along a *Nitella* cell through which water is being sucked at extremely low rates must be very small and the turgor pressures at the two ends must therefore be virtually equal. It seems, therefore, quite doubtful that the valve mechanisms of Fig. 22 could be operative. Active water secretion therefore appears as on the whole to be the more probable of the various possibilities.

D. ACTIVE WATER MOVEMENT: A SUMMARY

In foregoing sections part of the experimental evidence has been presented that has suggested that "active" water movement in plant tissues occurs. An attempt has been made to make clear the counter explanations according to which apparent water secretion has been regarded as either experimental artifact or expression of secretion of some solute. Three main lines of evidence can be summarized.

(1) The discrepancy between plasmolytic value and osmotic pressure of expressed sap. If this is to be discounted, the grounds must be the impossibility of obtaining an expressed vacuolar sap similar in osmotic pressure to the true internal vacuolar sap. A consequence of accepting this view is that all cryoscopic osmotic pressure data of natural tissues must be regarded as inherently unreliable.

(2) Secretion of water by hydathodes and development of root pressure undoubtedly occurs. It can be explained as due to secretion of solutes which "carry in" the water by reducing the activity of water on the secretory side of the membrane. There is really no method of providing experimental proof that the type of permeation process termed active movement does *not* occur in regard to any given cell constituent. It occurs by unknown method in respect of many ions and many nonelectrolytes (see Chapter 4 of this volume), and water cannot be excluded from the group of actively transported substances. Indeed it is almost unavoidable that part of its transport must be mediated by so-called "anomalous osmosis."

The transport of ions across a membrane, whether it is brought about by diffusion along an activity gradient, as in ordinary nonliving systems, or by a metabolic accumulatory process, results in a potential difference across the membrane. Experimentally observed potential differences across the boundary layers of plant cells vary from a few to about 100 mv.

A potential difference across a porous membrane must inevitably cause electroosmotic flow of water or, when flow is opposed, an excess hydrostatic pressure develops. The theory of electroosmotic flow and pressure was developed by Helmholtz and von Smoluchovski for the special cases of membranes whose pores could be regarded as large compared with molecular dimensions, and are given as

$$V = \frac{D\zeta}{4\pi\eta} \times \frac{l}{k}$$

$$P = \frac{D\zeta}{4\pi\eta} \times \frac{iW}{\lambda} \quad \text{where} \quad W = \frac{8\eta l}{\pi r^4}$$

where V is the volume transported per unit time, D the dielectric constant in the double layer, ζ the zeta potential, η the viscosity, and k the conductance within the pore which equals the bulk conductance. The pressure, P, depends on the reverse flow (U) down the middle of the pore, which is given by

$$U = \frac{P}{W} = \frac{\pi r^4 P}{8\eta l}$$

where r and l are respectively the radius and length of the pore. i is the current in the pore, which is more readily obtained than the field within the pore.

Both zeta potential and field are involved. The case of interest in regard to non osmotic water transport has not been analyzed mathematically, namely, the flow and pressure developed across a membrane of the type shown in Fig. 23,C, where the thickness of membrane and pore sizes are of molecular dimensions. Since membrane thickness is of the order of 50–100 Å (say 10^{-6} cm) and potential difference (PD) of the order of 10^{-2} to 10^{-1} volts, the field (possibly 10^4–10^5 volts per centimeter) may be considerable. The consequences and the meaning to be ascribed to a zeta potential in a system where pore diameters and lengths have molecular dimensions have not been examined theoretically and are not readily susceptible to experimental study.

The quantitative importance of electroosmosis in nonliving systems is indicated by experiments of Loeb and of Sollner in which membranes of oxidized collodion were used. Flow of water into an electrolyte solution of 0.01 M concentration (about 0.4 atm) was not prevented until the OP of the nonelectrolyte side of the membrane had been increased to over 7 atm.

The type of mechanism which is suggested in explanation of this anomalous osmosis in nonliving systems is indicated in Fig. 23. The case of electroosmosis of a solution is shown in Fig. 23,A, where a PD is applied by a battery and causes flow of solution in the direction shown by the dotted arrow, owing to the so-called slip of the electric double layer against the wall of the pore.

Positive anomalous osmosis is shown in Fig. 23,B, where the concentration of potassium chloride is greater at the left-hand side. This causes a membrane potential, E, to be developed at the surfaces of the narrow pores, which are relatively anion-impermeable, and a local current shown by the solid arrow flows, producing electroosmosis through the larger pores shown by a dotted arrow.

Negative anomalous osmosis occurs with a similar membrane supplied with lithium chloride at higher concentration on the left side. The

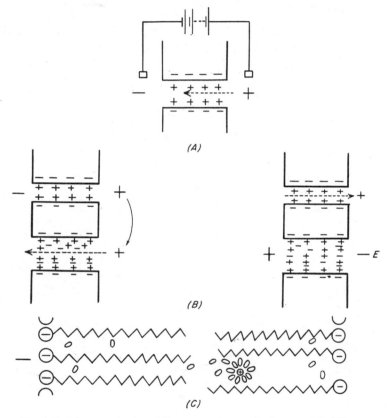

FIG. 23. (A) Electroosmosis with externally applied potential difference (PD) causing a current. (B) "Anomalous osmosis" explained as electroosmosis caused by currents due to diffusion potentials (see text). E = membrane potential. (C) The possible status of a hydrated ion and water molecules (shown by ellipses) in a bimolecular lipid film (see text).

membrane potential, E, is opposite (i.e., the dilute side is negative) because of the greater mobility of the anions, and electroosmotic flow occurs from the more concentrated to the more dilute side. The experimental findings of positive and negative anomalous osmosis are well established in nonliving systems. It is still debatable to what extent they may operate in living ones.

It is in fact, as pointed out on page 164, much more difficult to establish with certainty an "anomalous osmotic" flow or pressure which is positive than is the case with negative anomaly.

There remains the further possibility that anomalous osmosis is not involved in the postulated active water movement. Goldacre (26) suggested a scheme of contractile protein behavior which could in principle secrete water or salts. It is however a hypothesis on which it is remarkably difficult to obtain concrete evidence. A point which might be critical concerns the rhythmic movements of root hair vacuoles which Goldacre describes. These vacuolar movements are due, it is suggested, to ordered contraction and unfolding of the protein forming the vacuolar surface, and these cyclic changes are supposed to be associated with observed accumulation of neutral red. In fact the cytoplasmic circulation is stopped by 2,4-dinitrophenol, which does not inhibit neutral red uptake. This finding does not eliminate Goldacre's hypothesis though it does invalidate one of its supposed props.

Correlations between metabolism and water uptake, such as those observed in storage tissue disks for both water uptake and ion uptake (Chapter 4 in this volume), certainly predispose one toward interpretations of the water relations of cells which invoke metabolism and activated movements more than the classical interpretations imply.

(3) The most important new finding which might help to resolve the controversy is Kamiya's polar permeability, but even here the evidence, as pointed out, may prove to be equivocal.

Because of the equivocal nature of the evidence, we have in the general survey of water relations used the expression osmotic pressure of cell contents (P_i) as though the water flow was controlled solely by activity gradients of water. If in fact the hypothesis advanced to explain some of the anomalies in experimental data is correct, namely that active transport is also superposed, then flow will be due to $P_i + x$, where x is a secretion pressure brought about by unknown mechanism, which one would guess to be similar in general type to ion-uptake mechanisms.

VI. Osmotic Behavior of Tissues in Relation to Structure

The foregoing discussion of methods of study and of water relations of tissues will have shown the limitations imposed on the accuracy of information regarding activity of water in different parts of plants by the nature of the tissues and physicochemical methods available.

Some discussion is desirable of the role in the plant of the rather peculiar distribution of water supplies and of the state of hydrostatic pressure found in different tissues under various conditions.

A. MATURE TISSUES

It seems probable that the osmotic pressure of the vacuole contents of mature parenchyma is important chiefly in providing mechanical strength to the tissue. Turgor as pointed out (page 110) is broadly proportional to the excess hydrostatic pressure of cell contents. The rela-

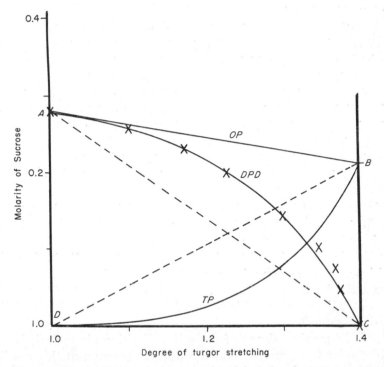

FIG. 24. Values of diffusion pressure deficit (DPD) of a *Nitella* cell at cell volumes shown as abscissas (×). The derived curves *OP* and *TP* give, respectively, values of vacuolar osmotic pressure calculated from volume and limiting plasmolytic value, and turgor pressure which is obtained by difference. The dashed lines, *DB* and *AC*, indicate the relationship that was assumed to exist between DPD or suction pressure and turgor pressure and cell volume before accurate measurements had been made. From Fig. 6 of Tamiya (59).

tion between this excess hydrostatic pressure or turgor pressure and cell volume is therefore of some interest.

Ursprung and Blum assumed that on expansion of a cell from the state of limiting or incipient plasmolysis the increase in turgor pressure was proportional to the increase in volume; so that plotting cell volume against turgor pressure as in Fig. 24, a straight line (the dashed line *DB*) would be obtained, and in earlier studies this was generally assumed. Oppenheimer (48), however, showed experimentally that the relationship is in fact hyperbolic. He showed in the case of pith cells of *Taraxacum officinale* that given increments of turgor pressure produce successively smaller increments of volume as the pressure increases. A similar state of affairs has been demonstrated for the cylindrical internodal cells of *Nitella* by Stow [quoted by Tamiya (59)]. These data are plotted in Fig. 24.

Calculations of the relationship between turgor pressure and extension of cells have been made, but in all cases simplifying assumptions have had to be made which reduce their practical value greatly. The two most simple cases are a sphere of radius, R, having a wall of thickness, r, and a cylinder of radius R and thickness r. In the sphere the tension, T, in the wall is given by $P = 2Tr/R$ and in the cylinder the tension is anisotropic, the transverse tension being $2Tr/R$ and the longitudinal being Tr/R.

As almost all cell walls are structurally anisotropic, the changes in shape resulting from tension in walls due to osmosis are not readily susceptible to calculation.

It is a matter of observation that the elasticity of different types of cells differs very markedly. In general the parenchymatous cells from the pith of stems and from bulky tuberous storage organs show considerable differences in volume in the fully turgid and limit-plasmolysis conditions as compared with mesophyll and epidermal cells of thin leaves and with the even more extreme behavior of cells of sclerophyllous and similar xeromorphic leaves. The contrast in behavior is indicated in Fig. 25.

In this figure the volume of the cell at limiting plasmolysis is taken as 100 (point 0 on curves) and the osmotic pressure at this condition is P. Turgor (TP) by definition is zero (point 0). Increase in volume of the vacuole results in reduction of osmotic pressure as indicated by line *PB*. If the vacuole volume is reduced, the osmotic pressure correspondingly rises as shown by *PC*.

Extensible cells may, in the fully turgid condition, be up to 40% larger in volume than at limit-plasmolysis; in this imaginary case a 40% extension similar to that found by Stow for *Nitella* has been shown. The course of change in DPD can be determined experimentally as the curve *PE*, and from this the turgor pressures at different degrees of extension can be calculated as shown by the curve 0*B*.

If such a cell were allowed to shrink to a smaller volume than that attained at limiting plasmolysis by letting water evaporate from the cell surface into the air, it is probable that the cell wall would cave in or become wrinkled, like a concertina, and that there would be little or no development of tension in the cell contents; this supposition, which has not been tested for the case of *Nitella*, is indicated by the line *0A* showing little deviation from the horizontal axis. The DPD-cell volume curve, *PD*, similarly indicates DPD values only very little in excess of

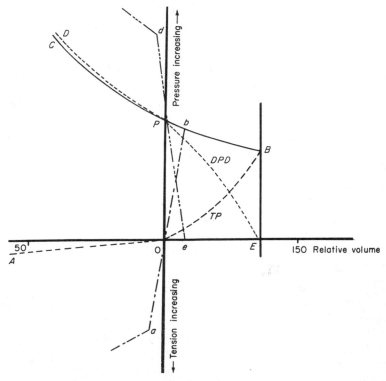

FIG. 25. The pressure-volume relations of different types of cell. The horizontal axis and abscissas show relative volumes, taking the volume of the cell at limiting plasmolysis as 100. The vertical axis shows relative pressures in atmosphere, taking the osmotic pressure at limiting plasmolysis as *P*. Osmotic pressures of contents are shown by continuous line. Turgor pressure and DPD (---) of a very extensible cell and turgor pressure (—·—) and DPD (—···—···) of a relatively nonextensible cell are also shown (see text).

the cell osmotic pressure. It is worth noting that the turgor pressure-extension curve is convex to the horizontal axis; in other words, the modulus of elasticity increases with extension so that extension is more difficult as the cell enlarges.

Thick-walled cells and mesophyll cells may be expected to behave differently. The maximum degree of turgor stretching may be only

2–5%, so the DPD cell volume curve will be of the type Pe which shows 10% turgor stretching, or may even be still steeper—nearly vertical—for thicker-walled cells. The turgor pressure-cell volume relation for 10% extension will then be $0b$. No experimental data appear to exist for such cases, but this is drawn concave to the extension axis, as it seems more probable that for walls showing limited elastic extension Hooke's law would be followed. Quite small reduction of water content below that attained at limiting plasmolysis might be expected to cause considerable rise in DPD and tension of cell contents as indicated by curves Pd and $0a$. That tensions are developed is indicated by the work of Chu (17). He used Ursprung's capillary method (page 124 and Fig. 3,C) to estimate the DPD's of leaves of a number of trees after they had been subjected to water losses of amounts insufficient to cause permanent injury or death. Some typical results are given in Table VIII. The relatively large tensions measured by Chu are probably reasonably reliable.

TABLE VIII

DEVELOPMENT OF TENSIONS IN LEAF CELLS

	Water deficit	DPD	OP	W (Turgor pressure)
Betula verrucosa (= *B. pendula*)	0	16.7	18.3	+ 1.6
	13.3	32.5	19.2	−13.3
	37.7	52.6	24.5	−28.1
Quercus pedunculata (= *Q. robur*)	0	16.4	18.9	+ 2.5
	11.8	26.2	19.6	− 6.6
	31.8	75.9	23.7	−49.2
Picea excelsa (= *P. abies*)				
July 5	0	12.7	14.3	+ 1.6
	20.7	29.7	27.6	− 2.1
March 14	0	18.8	25.5	+ 6.7
	29.8	145.6	46.3	−99.3

Still larger tensions were determined independently by Ursprung (64) and Renner (51c) in fern annulus cells and liverwort elators by essentially comparable methods. Sporangia and elators were exposed to atmospheres of different humidities in equilibrium with, and maintained constant by, a relatively large surface area of solution of suitably chosen osmotic pressure. As the external humidity is decreased the annulus cells and elators shrink more and more as indicated in Fig. 26. In the fern annulus the radius of the outer wall (r) decreases as the

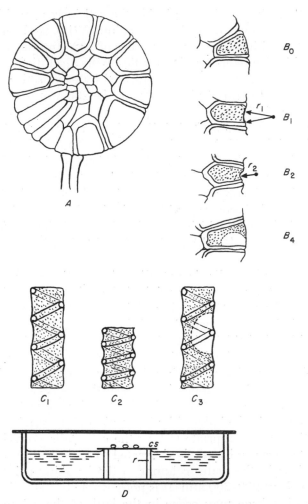

FIG. 26A. General schematic view of a fern sporangium.

B_0-B_4. Stages in the drying out of a single annulus cell. In the first stage of drying the convex outer wall becomes straight. At stage B_1 it is becoming convexly curved with radius r_1. With increase in tension the radius r_1 decreases (r_2 at stage B_2) and on breakage of the water column restoration of the original shape of the cell, B_4.

C_1-C_3. Stylized diagrams representing the contraction of a liverwort elator as at stage C_2. On breaking of the tensile water column there is almost instantaneous return to stage C_3. Owing to their length and irregularity of the thickening, shrinkage from C_1 to C_2 produces slow wriggling, and the sudden expansion from C_2 to C_3 a rapid irregular jerking.

D. Sporangium and elator movements were examined in this type of apparatus in which a glass ring, r, covered by a cover slip, cs, is sealed into a dish as shown. The atmosphere surrounding the sporangia (shown as circles on the cover slip) has a DPD controlled by the solution, the surface area of which is large.

tension in the column of water inside the cell increases; this draws in the side walls and tears open the sporangium. With external humidities in equilibrium with solutions of the order of 200 atm, the annulus cells remain in this shrunken condition, filled with a solution which is unlikely, even though somewhat concentrated by evaporation, to have an OP greater than 50 atm. As its DPD must be 200 atm the tension must be of the order of 150 atm.

Further evaporation and shrinkage is caused by increased DPD (lowered humidity) of the ambient atmosphere until the tensile water column breaks with the formation of a bubble of gas inside the cell. The elastic side walls immediately spring back to their original shape, the internal hydrostatic pressure becomes virtually atmospheric pressure, and the gas bubble becomes equilibrated with the external atmosphere. The external DPD at which these breakages of the tensile columns occur can be evaluated and have been found to be of the order of 250–350 atm.

Since the annulus cells do not all spring at the same moment the sporangium recovers toward its original shape in a series of jerks, each of which can throw out a number of spores. The elator mechanism is essentially similar, but, each elator being a long, narrow tubular cell armored with spiral thickenings, the elator shortens slightly during the drying-out process and when the water column breaks carries out a single sudden expansion which has the appearance of a "wriggle." The effect under natural conditions is to shake the spores loose. These annulus and elator movements brought about under natural conditions by evaporation can also be induced by immersion of sporangia or elators in sugar solutions of appropriate OP, as the cell walls are relatively impermeable to sugar and water is abstracted from the cells osmotically.

The tensions in the water columns at the breaking point have been found to be about 200–300 atm. It is not possible to state what precisely breaks. The breakage may occur in the bulk of the column, but it is perhaps more likely to be a break in the adhesion to the wall or a "tearing in" of the water menisci in the fabric of the wall. These tensile strengths are much below the theoretical value of 10,000 atm based on intermolecular attractive forces; however, they exceed considerably the values determined by cavitation techniques and by the Berthelot-tube method.

The Berthelot-tube method was used by Dixon and Joly (24) in the first determinations of the tensile strengths of plant saps. The procedure is to heat a sealed thick-walled glass tube containing water (or sap) and a small bubble of air. The expansion of the water exceeds that of the glass so that the bubble of air is forced into solution and the tube becomes completely filled with liquid. On cooling, the water column is subjected to a tension which can be calculated from the temperature and the thermal and elastic properties of glass and water. Finally the water column breaks with a characteristic click, and from the temperature at which this occurs the tension at breakage (or tensile strength) can be calculated. Temperley (60) has shown that Dixon and Joly's estimates of the tensions at the breaking points are almost certainly too high as they made the assumption that the pressure inside the tubes was atmospheric at the filling temperature; in fact, the tubes have to be heated about 5°C above the temperature at which the bubble seems to disappear in order to drive microscopic bubbles into solution, which corresponds to an excess hydrostatic pressure inside the tube of about 100 atm. Temperley found tensile strengths in the tubes of the order of 16–68 atm in different experiments. Here, as in the case of annulus cells, it is not clear whether cohesion of water or adhesion to glass or other suspended particles is the function which is being measured.

It is certainly quite clear that the DPD—volume curve Pd will show a break at some point, d, which may be determined by the breaking of the water column or, more commonly in leaf cells, by the caving-in or kinking of the cell wall (cf. Fig. 25).

The significance and importance of DPD gradients and of the diurnal and seasonal fluctuations which these undergo will be discussed in the chapters dealing with transpiration and transport of materials. One special case is, however, of particular importance, namely, the DPD and other osmotic relations of the water-absorbing cells of roots, root hairs, and other superficial cells. These cells are notably thin-walled and on being subjected to water losses by evaporation or water stress in the soil, the course of change in DPD with cell volume that would almost certainly always occur would be of the type shown by curve $0A$ in Fig. 25.

This is of importance in relation to the ability of plants to extract water from drying soils. Transpiration by a plant dries out the soil which is tapped by the root system, and a relatively well-defined condition termed "permanent wilting" can be established in which leaves which have wilted cannot be induced to recover by placing them in a saturated atmosphere. Water has to be added to the soil (see Chapter 7). The actual water content of different types of soil at the permanent-wilting point are markedly different, being smaller in soils of coarse texture or particle size. The DPD of soil at the permanent-wilting point has however in all cases a very similar value close to 15 atm. The relationship between soil DPD and water content is given by the curves of Fig. 27 for two types of soils. It will be noted that a given reduction in water content below the permanent-wilting point brings about a very large change in DPD. At approximately this water content a striking increase in resistance to the removal of water occurs.

If a plant can remove water from soil which is only slightly wetter than the permanent-wilting point, for example soil of DPD 14 atm, it follows that root hairs must have at least this effective osmotic pressure (or osmotic plus active absorption).

It is incidentally implicit in the results of many workers, or at least in their statements, that the permanent-wilting condition is an estimate of the plant's ability to extract water from soils and that the similarity

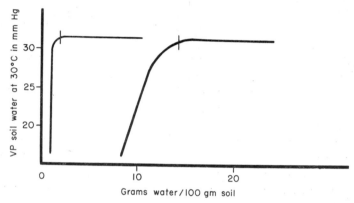

Fig. 27. DPD or water vapor pressures (VP) of soils having different water contents. The permanent-wilting region is indicated (see text).

of permanent-wilting point exhibited by different ecological types is very great.

This would appear to suggest that root cells are capable of a considerable degree of osmoregulation or control of their vacuolar osmotic pressure in relation to the DPD of the external medium, since it is quite certain that under conditions of favorable water supply, root hair osmotic pressures of 15 atm are unusual.

B. Osmoregulation in Plant Cells

The DPD of a plant cell can become adjusted to equality with that of the outside medium or with surrounding cells by change in the excess hydrostatic pressure of the cell contents, so that it has been commonly assumed that the specialized osmoregulatory devices which are met with in the animal kingdom are unnecessary and do not occur in plants.

Though osmoregulation in the sense of DPD adjustment probably usually takes place by change in turgor, it seems clear that changes in osmotic pressure also occur, but these are probably more usually correlated with a concentration difference of ions or other solutes across the protoplast rather than with water or turgor relations.

Under the experimental conditions to which plants are subjected when the permanent-wilting point of a soil is being determined, there may be a rapid build-up of high osmotic pressure up to or exceeding 15 atm, though it remains odd that there should be relatively slight specific variability between different plant types.

Halophytes are certainly in a quite special position in this respect. The osmotic pressures of cell contents in all cases studied have been above those of the ground water; in general changes in the salt content, and thus osmotic pressure of the external medium, are followed by changes in internal concentration, but the proportionality is not linear, as some of the data of Table IX show.

TABLE IX
OSMOTIC PRESSURE (OP) OF SOIL SOLUTIONS COMPARED WITH OSMOTIC PRESSURE OF PLANT CELLS

Plant	OP of soil solution	OP of plant cells
Triglochin maritimum	3.8	15.8
	5.8	15.8
	10.0	17.5
	18.2	19.5
Rhizophora mucronata	24.7	34.3
Avicennia marina	24.3	45.9

Several genera of mangroves do show a quite interesting feature: their seedlings which develop *in situ* on the parent plants have cell osmotic pressures well below those of the parent plant and commonly lower than the OP of the soil solution. This last result would seem to demand some specific "active" transport.

Parasites are in a situation comparable to halophytes, and it is generally stated that their tissues have higher osmotic pressures than those of their hosts. There would seem to be no special reason for this. Clearly the DPD's of parasite and host must be in equilibrium and actual haustoria might be expected to have a high turgor pressure and, therefore, a high osmotic pressure. There does not appear to have been any study of the osmotic relations at the actual zone of contact or infection.

An experimental treatment which is usually impossible of easy attainment is to change artificially the internal osmotic pressure of a cell and observe the consequences of this change. It has however been successfully accomplished by Kamiya and Kuroda (37). They caused "transcellular osmosis" of *Nitella* by placing one half of an internodal cell in water and the other in an osmoticum, then tying off the two

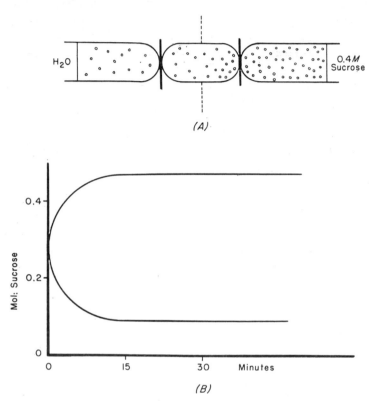

Fig. 28. (A) Experimental treatment of a *Nitella* cell through which water is flowing from water (at left) to 0.4 *M* sucrose at the right, concentrating the vacuole contents as indicated. The cell is then tied off with two ligatures. (B) The time courses of change in internal osmotic pressure of the intake end (lower curve) and outflow end (upper curve). From Kamiya and Kuroda (37).

ends. The tied-off daughter cells survive quite well, which, as Kamiya and Kuroda state, was, surprisingly enough, known to Dutrochet over a century ago.

The procedure is indicated by the diagram of Fig. 28. With water bathing the left half-cell and 0.4 *M* sucrose the right, the rate of change of osmotic pressure of the tied-off half ends was obtained by subjecting different cells to this treatment for 5, 10, 15, and 30 minutes, tying off, and determining the osmotic pressures by limiting plasmolysis. The changes are shown in Fig. 28,B, and it is clear that equilibrium is attained in about 15 minutes.

When the "concentrated" and "diluted" cells are returned to culture solution ($\frac{1}{100}$ Knop solution), the internal osmotic pressures slowly readjust themselves to the normal level. This readjustment will have been brought about, presumably, by passage of ions between vacuole and external medium. But a slower and less complete adjustment was also observed in cells suspended in moist air. The regulation was not conditioned by the turgor pressure, as it was found that if the high-concentration cell was placed in a sucrose solution which was just hypotonic to the internal solution, the osmotic pressure of the cell decreased so that it became plasmolyzed. Ultimately the internal osmotic pressure came down to the "normal" value. It would seem that the cells are stimulated to the adjustment of concentration by an abnormal concentration of some solute and that neither turgor pressure nor DPD provides the necessary osmotic control.

Incidentally this experimental treatment enabled a direct determination of the breaking strength of the cell wall to be obtained. Bursting took place when cells of internal OP 12–14 atm were put into water, the normal internal OP being about 6.2 atm. The walls always burst either at the ligature (an artificial weak place) or at the end wall. Evidently the end wall, which is not normally exposed to the outside, is mechanically weaker than are the side walls.

C. Osmotic Relations of Growing Cells

Water relations of growing cells will be dealt with in another chapter as part of the general discussion on physiology of growth. The expansion in volume of a growing vacuolated cell is largely brought about by the uptake of water, and continued expansion in growth clearly shows either a maintained DPD gradient between cell contents and the external medium or an "active" water uptake.

The actual entry of water would reduce the DPD gradient unless this entry were compensated by entry of solutes or reduction of wall pressure. As growth rates of expanding cells are maintained, it follows that one of these methods for keeping up a DPD gradient must be present. Both of the methods mentioned have been suggested by different investigators. As auxin affects the rate of cell expansion, its influence on water relations has been the subject of study and controversy which will receive further mention in Volume V on growth.

The enlargement of a meristematic cell is associated with increase in quantities of protein, cell wall material, and solutes, in addition to the gain in water. The relative amounts of the increases in these several constituents depends evidently on supplies of ions and of organic materials to the growing zone. In the case of root meristems these may be provided from the external medium or from the basal parts of the root. This can also, at least in experimental treatment, be true of shoot structures and of coleoptiles.

During the enlargement phase of a cell following its formation by a cell division, vacuoles develop and commonly coalesce to produce one main vacuole. Cytoplasm also increases in volume in all probability; certainly the protein content per cell rises. The vacuole volume:cytoplasm volume ratio increases greatly at the early stage of cell elongation. It would be of some interest to have data of the osmotic relations during this phase. However, that type of cell is least suited to yield experimental data of accuracy regarding osmotic quantities.

Expressed cell sap which could be used for cryoscopic determinations or for analysis is derived from both cytoplasm and vacuoles, and, as the vacuole cytoplasm ratio varies from near zero to, say, 20 as the cell enlarges, nothing can be deduced with certainty regarding vacuolar osmotic pressures.

Plasmolytic determinations by the method of limiting plasmolysis are unreliable both on account of adhesion of the protoplast to the thin walls, which may draw inward, and to the probably rapid accumulation, in many cases, of the plasmolyticum.

Pirson and Seidel (51) in a study of the developmental changes at the root apex of *Lemna minor* showed, by the plasmolytic method, that the apparent osmotic pressure dropped from 7.3 atm in the elongating zone to 5.6 atm immediately behind this zone. These data are presented in Fig. 29, together with times taken for recovery from plasmolysis in urea. These deplasmolysis times can be interpreted as evidence for a wide range of permeabilities in cells at this critical phase at the close of the period of expansion in size. They do not enable one to judge whether there are marked differences in the ability of cells of the different portions of the elongating root to accumulate the plasmolyticum, glucose. They emphasize the difficulties of interpretation.

In order to simplify the experimental situation, studies have been made of cell expansion of isolated portions of growing tissue, such as segments of grass coleoptiles, and slices of tissue, such as potato (*Solanum tuberosum*) or Jerusalem artichoke (*Helianthus tuberosus*) tuber. These will grow, absorbing water, and the rate or extent of growth is increased by treatment with auxins (3-indolylacetic acid and naphthylacetic acid). Some data for the expansion of artichoke slices are presented in Fig. 30. It seems now quite clear that the entry of water into

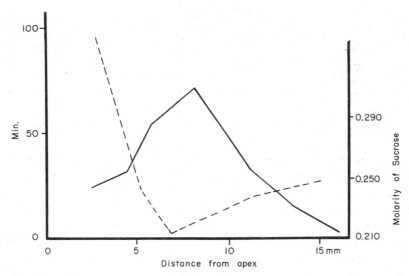

FIG. 29. Plasmolysis times (solid line) and osmotic pressures (dashed line) of cells at different distances (shown as abscissas) from the apex of the root of *Lemna minor*. From Pirson and Seidel (51).

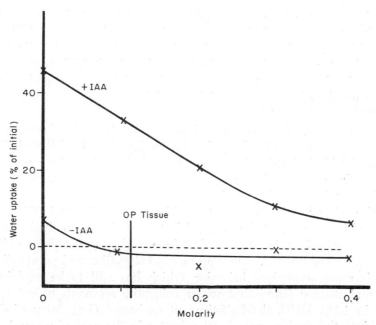

FIG. 30. Water uptake after 4 days by disks of Jerusalem artichoke (*Helianthus tuberosus*) in a range of osmotica with or without 3-indolylacetic acid (IAA). From Bonner *et al.* (10).

the expanding cells causes the dilution of vacuolar sap and reduction of osmotic pressure. The DPD of the turgid tissue before expansion must have been very close to that of the outside medium, water, namely zero; the rapid auxin-induced expansion occurs only under aerobic conditions, which at first suggested that "active" water secretion was involved [Bonner et al. (10)], but opinion has veered toward the view that in fact the DPD was caused to increase by reduction of wall pressure, so that while the internal osmotic pressure went down, the wall pressure went down to a slightly greater extent. This reduction of wall pressure, it should be pointed out, can be brought about in principle either by a change in modulus of elasticity caused by auxin or by a production of more wall material so that its area is increased.

In attempts to explain the mode of action of auxin, studies have been made of its effect on water permeability of cells. An effect on permeability alone could affect the rate but not the final length of extending cells. To obtain more extension one requires either a change in the cell wall, or a change in pressure with which water enters, brought about by "active" movement or by an increase in vacuolar osmotic pressure, possibly as a result of hydrolysis of reserve materials. The evidence as we have seen is against this last alternative.

The importance of the permeability studies is that they appear to show that the permeabilities of oat (*Avena sativa*) coleoptiles and potato disks, respectively, are quite high and that a metabolic component of water uptake is unlikely to be significant. This may be the case, but the data of Bonner et al. (10), in which auxin was shown to promote uptake of water by artichoke disks from a hypertonic solution, still have to be explained.

The permeability data are somewhat discordant. Koningsberger (39) used the plasmolytic method previously employed by Levitt et al. (42) and found with isolated onion protoplasts that, with a given pressure gradient, there was a slower rate of change of volume with auxin-treated than with control protoplasts.

On the other hand, Buffel (14), Thimann and Samuel (61), and Ordin and Bonner (49), who all used isotope exchange methods, found either no effect of auxin or somewhat increased apparent permeability due to auxin treatment in the case of oat coleoptiles and potato disks. There is some anomalous behavior which has still to be explained. Ordin and Bonner found that when coleoptile segments of oat were placed in 1.1% DHO, the half-time of exchange (i.e., entry of deuterium into the section) was 8–10 minutes in the case of normal sections and also of those treated with 5 mg per liter auxin, 3.0×10^{-4} M

KCN, or 5×10^{-5} M dinitrophenol; complete death of the sections also had almost no effect on the half-time of exchange. This suggests quite special water permeability relations of expanding growing cells, since, at least in ordinary, mature, tissue, death of the protoplast markedly changes permeability. There is also an anomaly in the results of Buffel, who found a half-time of exchange of DHO of about 5 minutes, but for osmotic outflow of water, the half-time was 55 minutes. There is at present no explanation; such a result could be caused by the presence of cytoplasmic or cell-wall water, which was more exchangeable with the external water than is the case with vacuolar water. This type of explanation appears, however, to be ruled out by the results of Ordin and Bonner, which showed that after about 2 hours the DHO concentration of the coleoptile segments had approached to about 96% of that outside.

D. WATER RELATIONS OF SPECIALIZED TISSUES

Three classes of tissue have somewhat specialized water relations. First, conducting tissues, xylem, phloem and possibly latex tubes, and certain types of parenchyma. Secondly, secretory structures such as active hydathodes, nectaries, salt glands. Thirdly, motor organs such as stomata, pulvini, and various structures concerned with dispersal of reproductive units.

1. Conducting Tissues

These are discussed in detail in Chapters 5 and 6 and consequently barely require mention here. The most important critical point concerns the magnitude of the tension to which columns of water containing air can be subjected without breaking.

2. Secretory Organs

Active secretory organs are in general recognized by their anatomical structure or by cytological reactions. Large nuclei or cells having a large proportion of their total volume occupied by so-called "dense" cytoplasm (cytoplasm with numerous granules) are characteristic of salt glands, nectaries, and also of hydathodes.

a. Salt glands. Little is known of the mechanism of action. Salt glands and nectaries have the special feature of polarity of activity, and the former have been the subject of a considerable amount of study. They occur in a number of halophytes, e.g., species of *Glaux, Limonium (Statice)*, and *Spartina*—to mention three that have been studied in some detail. The structures of these glands in *Limonium* and *Spartina* are shown in Fig. 31. Characteristic glandular cells with conspicuous nuclei and cytoplasm are present in each. The outer surfaces of the glands are armored with a thick cuticularized layer, and in *Limonium* the gland cells each have one pore rather less than 1 μ diameter in the

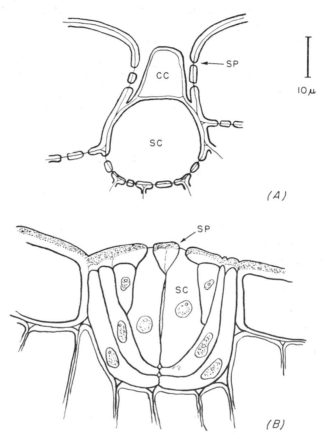

FIG. 31. (A) Salt glands of *Spartina townsendii; cc* = cap cell; *sc* = secretory cell; *sp* = secretory pore. (B) Salt gland of *Limonium latifolium; sc* = secretory cell; *sp* = secretory pore. Note the complex of cells adjoining the secretory cells. A, adapted from Arisz *et al.* (2) and B, from Skelding and Winterbotham (55a).

cuticle; the cellulose wall in this area is said to differ from that of the rest of the cell. In *Spartina* the two gland cells are cuticularized and pores occur in the cuticle of the neighboring epidermal cells.

Secretion by these glands in *Limonium* occurs both when the plant is intact and also from disks of leaf punched out and floated on water or salt solutions. The quantity of solution excreted by the leaves is very large; disks of leaf floating on water will secrete half their weight of solution in 24 hours. They have some 600–700 glands per square centimeter on each surface.

The secretion is stopped completely by anaerobic conditions and is impeded by an external osmotic pressure. The fluid secreted may be isotonic with the bathing fluid, and in general the greater part of the osmotic pressure of the secretion is due to sodium chloride.

It seems that these glands are highly turgid and force a salt solution out through pores under pressure. In *Spartina*, the salt solution appears to be forced out through neighboring nonglandular epidermal cells. The glands appear to be primarily desalting organs, and their activity reduces the salt content of mesophyll cells of leaves floating on water, so that ultimately the secretion becomes stopped.

b. Water glands. Water-secreting glands have received less study except in the case of the remarkable glands of certain carnivorous plants—*Utricularia*.

In *Utricularia*, the bladder entrance is closed by a door of specialized structure and the "setting of the trap" consists of a reduction in volume of water inside which causes the walls of the trap to cave inward. The strain on these walls is balanced by the tension (a slight tension) on the column of water in the trap. Roughly half the internal water is pumped out in the process of setting the trap. More water still can be taken out by placing the set trap in a solution of sugar or other osmoticum.

The pumping is assumed to be effected by the joint operations of the four-celled glands or trichomes on the inside of the bladder and the two-celled glands which accompany them on the outside. Nold (46) and Diannelidis (20) have each found electric potential differences ranging from 40 to 110 mv between internal liquid (positive) and external (negative). They suggest an electroosmotic flow of water from inside to outside as the mechanism of water secretion and, in favor, Diannelidis reports reduction of water flow and PD when the bladders are put into 0.1 M potassium nitrate. The critical experiment of artificially enhancing or reducing the metabolically maintained PD was unfortunately not done. This particular case is of interest as an apparently genuine water secretion in plants.

In many other water glands, the flow of water may conceivably be "active" or facilitated by metabolic processes; the conspicuous nuclei and dense granular cytoplasm perhaps suggests this. There is little clear evidence of water movements other than purely osmotic. However, there have been many postulations of "active" water secretion, for example, by the water glands on the leaves of many water plants, by the specialized glands on the scales of *Lathraea,* and by relatively unspecialized mesophyll. Compare, for example, Dixon and Barlee (22), Schmucker (55), and Wilson (63).

3. Motor Organs

The special behavior of motor organs certainly involves water relations of their constituent cells: in other words, changes in the internal excess hydrostatic pressure. These pressure changes are probably generally brought about by evaporation of water from the cell or tissue or by a change of osmotic pressure.

They will be considered in the appropriate chapters of this treatise. The more important types of movement may be recorded here: stomatal movements are, so far as is known, connected with changes of osmotic pressure of the guard cells. Because this mechanism is of such great importance, the mechanism of stomatal activity is discussed as a sequel to this chapter. Related also possibly to transpiration and gas-exchange control are the folding movements of grass leaves actuated by the hinge cells, which are enlarged cells of the upper epidermis. Their shrinkage, which causes rolling or folding of the leaf, is presumably due to loss of water by evaporation, but the osmotic and turgor pressures of folded shrunken and open hinge cells do not appear to have been compared.

Probably the most striking turgor movements are those which bring about nastic movements, and these are appropriately considered in another chapter.

REFERENCES

1. Allsopp, A., and Misra, P. The constitution of the cambium, of new wood; and the mature sapwood of ash, elm and Scotch pine. *Biochem. J.* 34, 1078–1084 (1940).
2. Arisz, W. H., Helder, R. J., and van Nie, R. Analysis of the exudation process in tomato plants. *J. Exptl. Botany* 2, 257–297 (1951).
2a. Arisz, W. H. The secretion of the salt glands of *Limonium latifolium. Acta. Botan. Neerl.* 4, 322 (1955).
3. Askenasy, E. Ueber das Saftsteigen. *Botan. Centr.* 62, 237–238 (1895).
3a. Banga, I., and Szent-Györgyi, A. Strukturproteine. *Enzymologia* 9, 111 (1940).
4. Barger, G. A microscopical method for determining molecular weights. *J. Chem. Soc.* 85, 286–324 (1904).
5. Bennet-Clark, T. A., and Bexon, D. Water relations of plant cells. II. *New Phytologist* 39, 337–361 (1940).
6. Bennet-Clark, T. A., and Bexon, D. Water relations of plant tissues. IV. *New Phytologist* 45, 5–17 (1946).
7. Bennet-Clark, T. A., Greenwood, A. D., and Barker, J. W. Water Relations of osmotic pressures of plant cells. *New Phytologist* 35, 277–291 (1936).
8. Bennett, M. C., and Rideal, E. Membrane behaviour in *Nitella. Proc. Roy. Soc.* B142, 483–496 (1954).

9. Bogen, H. J. Osmotische und nicht-osmotische Stoffaufnahme bei Diatomeen. *Planta* **45**, 125–146 (1955).

10. Bonner, J., Bandurski, R. S., and Millerd, A. Linkage of respiration to auxin-induced water uptake. *Physiol. Plantarum.* **6**, 511–522 (1953).

11. Brauner, L. Ueber polare Permeabilitat. *Ber. deut. botan. Ges.* **48**, 109–118 (1930).

11a. Briggs, G. E. Osmotic pressure of vacuolar sap by the plasmometric method. *New Phytologist* **56**, 305–324 (1957).

11b. Briggs, G. E. Some aspects of the free space in plant tissues. *New Phytologist* **56**, 258–260 (1957).

11c. Briggs, G. E., Hope, A. B., and Pitman, M. G. Exchangeable ions in beet disks at low temperature. *J. Exptl. Botany* **9**, 365–371 (1958).

11d. Briggs, G. E., and Robertson, R. N. Apparent free space. *Ann. Rev. Plant Physiol.* **8**, 11–30 (1957).

12. Brønsted, J. N. "Physical Chemistry." Chemical Publishing, New York, 1938.

12a. Brouwer, R. *Koninkl. Ned. Akad. Wetenschap. Proc.* **56C**, 106–115, Figure 1 (1953).

12b. Brouwer, R. *Koninkl. Ned. Akad. Wetenschap. Proc.* **57C**, 68–80, Figure 2 (1954).

13. Brouwer, R. The regulating influence of transpiration and suction tension on water and salt uptake of roots. *Acta Botan. Neerl.* **3**, 264–312 (1954).

14. Buffel, K. New techniques for comparative permeability studies on the oat coleoptile. *Mededel. Koninkl. Vlaam. Acad. Wetenschap. Belg.* **14**, No. 7 (1952).

15. Buhmann, A. Kritische Untersuchungen über Bestimmungen des osmotischen Wertes. *Protoplasma* **23**, 579–612 (1935).

16. Burström, H. A theoretical interpretation of turgor pressure. *Physiol. Plantarum* **1**, 57–64 (1948).

17. Chu, C.-R. Der Einfluss des Wassergehaltes der Blätter auf ihre Lebensfähigkeit. *Flora (Jena)* **130**, 384–437 (1936).

18. Currier, H. B. Water relations of root cells of *Beta. Am. J. Botany* **31**, 378–387 (1944).

18a. de Haan, I. Protoplasmaquellung und Wasserpermeabilität. *Rec. trav. botan. néerl.* **30**, 234 (1933).

19. Denny, F. E. Permeability of certain plant membranes to water. *Botan. Gaz.* **63**, 373–397 (1917).

20. Diannelidis, T. Beitrag zur Electrophysiologie pflanzlicher Drüsen. *Phyton Ann. rei botan.* **1**, 7–23 (1948).

21. Dixon, H. H. "Transpiration and the Ascent of Sap in Plants." Macmillan, London, 1914.

22. Dixon, H. H., and Barlee, J. S. Further experiments on transpiration into a saturated atmosphere. *Sci. Proc. Roy. Dublin Soc.* **22**, 211–222 (1940).

23. Dixon, H. H., and Dixon, G. J. Exudation of water from *Colocasia antiquorum. Sci. Proc. Roy. Dublin Soc.* **20**, 7–10 (1931).

24. Dixon, H. H., and Joly, J. On the ascent of sap. *Ann. Botany (London)* **8**, 468–470 (1894).

25. Dutrochet, R. H. J. Nouvelles observations sur l'endosmose. *Ann. chim. phys.* **35**, 393–400 (1872).

26. Goldacre, R. J. The folding and unfolding of protein molecules as a basis of osmotic work. *Intern. Rev. Cytol.* **1**, 135–164 (1952).

27. Gross, F. The osmotic relations of the diatom *Ditylum brightwelli*. *J. Marine Biol. Assoc. United Kingdom* **24**, 381–415 (1940).
28. Grossenbacher, K. A. Autonomic cycle of rate of exudation of plants. *Am. J. Botany* **26**, 107–109 (1939).
28a. Haines, F. M. Transpiration and pressure deficit. *Ann. Botany* **49**, 213 (1935).
29. Haines, F. M. An analysis of turgor and turgor pressure. *Ann. Botany (London)* **17**, 629–640 (1953).
30. Hales, S. "Vegetable Staticks," London (1738).
31. Harris, J. A., and Gortner, R. A. Notes on calculation of osmotic pressure. *Am. J. Botany* **1**, 75–78; **2**, 418–419 (1914).
32. Hevesy, G., Hofer, E., and Krogh, A. The permeability of the skin of the frog as determined by D_2O and H_2O. *Skand. Arch. Physiol.* **75**, 199–214 (1935).
33. Hodgkin, A. L., and Keynes, R. D. The K permeability of a giant nerve fibre. *J. Physiol. (London)* **128**, 61–88 (1955).
34. Huber, B., and Höfler, K. Die Wasserpermeabilität des Protoplasmas. *Jahrb. wiss. Botan.* **73**, 351–511 (1930).
35. James, W. O., and Baker, H. Sap pressure and movements of sap. *New Phytologist* **32**, 317–343 (1933).
36. Kahl, H. Uber den Einfluss von Schüttelbewegungen auf Struktur und Funktion des Plasmas. *Planta* **39**, 346–376 (1951).
37. Kamiya, N., and Kuroda, K. Artificial modification of the osmotic pressure of the plant cell. *Protoplasma* **46**, 423–436 (1956).
38. Kamiya, N., and Tazawa, M. Studies on the water permeabilities of a single plant cell. *Protoplasma* **46**, 394–422 (1956).
38a. Kitching, J. A. Contractile vacuoles. *Biol. Revs. Cambridge Phil. Soc.* **13**, 403.
39. Koningsberger, V. J. Over de primaire Werking van Groeistoffen. *Mededel. Koninkl. Vlaam. Acad. Wetenschap. Belg.* **9**, No. 13 (1947).
40. Levitt, J. The thermodynamics of active (non-osmotic) water absorption. *Plant Physiol.* **22**, 514–525 (1947).
41. Levitt, J. "The Hardiness of Plants." Academic Press, New York, 1956.
42. Levitt, J., Scarth, G. W., and Gibbs, R. D. Water permeability of isolated protoplasts in relation to volume change. *Protoplasma* **26**, 237–248 (1936).
43. Lloyd, F. E. Maturation and conjugation in *Spirogyra longata*. *Trans. Roy. Canadian Inst.* **15**, 151–193 (1926).
44. Meyer, B. S. The water relations of plant cells. *Botan. Rev.* **4**, 531–547 (1938).
45. Myers, G. M. P. The water permeability of unplasmolysed tissues. *J. Exptl. Botany* **2**, 129–144 (1951).
46. Nold, R. H. Die Funktion der Blase von *Utricularia*. *Botan. Cent. Beih.* **52A**, 415–448 (1934).
47. Nollet, M. Recherches sud les causes du bouillonment. *Mémoires de l'académie royale des sciences, des lettres et des beaux arts de Belgique* (1748).
48. Oppenheimer, H. R. Dehnbarkeit und Turgordehnung der Zellmembran. *Ber. deut. botan. Ges.* **48**, 192 (1930).
49. Ordin, L., and Bonner, J. Permeability of *Avena* Coleoptile sections to water measured by diffusion of deuterium hydroxide. *Plant Physiol.* **31**, 53–57.
50. Pfeffer, W. "Osmotische Untersuchungen: Studien zur Zellmechanik." Engelmann, Leipzig, 1877.
51. Pirson, A., and Seidel, F. Zell-und stoffwechselphysiologische Untersuchungen an der Wurzel von *Lemna minor*. *Planta* **38**, 431–473 (1950).

51a. Pulva, P. Die Wasserpermeabilität der Zellen von Tolypellopsis. *Protoplasma* **32**, 265 (1939).

51b. Resühr, B. Hydrations- und Permeabilitätstudien an unbefruchteten Fucus Eiern. *Protoplasma* **24**, 531 (1935).

51c. Renner, O. Theoretisches und Experimentelles zur Kohäsionstheorie der Wasserbewegung. *Jahrb. wiss. Botan.* **56**, 647 (1915).

52. Rollins, M. L. Some aspects of microscopy in cellulose research. *Anal. Chem.* **26**, 718–724 (1954).

53. Rosene, H. F. Water balance in onion root. *Plant Physiol.* **16**, 447–460 (1941).

53a. Rosene, H. F. Water absorption of root hairs. *Plant Physiol.* **18**, 588–607 (1943).

54. Schofield, R. K. The pF of water in soil. *Trans. Intern. Congr. Soil Sci. 3rd Congr. (Oxford)* **2**, 37–48 (1935).

55. Schmucker, T. Uber den Einfluss narkotischer Stoffe auf Transpiration. *Jahrb. wiss. Botan.* **68**, 771–800 (1928).

55a. Skelding, A. D., and Winterbotham, J. *New Phytologist* **38**, 69 (1939).

56. Spanner, D. C. The Peltier effect and its use in the measurement of suction pressure. *J. Exptl. Botany* **2**, 145–168 (1951).

57. Stiles, W. "Permeability." Wheldon and Wesley, London, 1924.

58. Sutcliffe, J. F. Salt uptake by plasmolysed tissues. *J. Exptl. Botany* **5**, 215 (1953).

59. Tamiya, H. Zur Theorie der Turgordehnung v.s.w. *Cytologia (Tokyo)* **8**, 542–562 (1938).

60. Temperley, H. N. V. The behaviour of water under hydrostatic tension. *Proc. Phys. Soc. (London)* **B58**, 436–443 (1946); **B59**, 199–208 (1947).

61. Thimann, K. V., and Samuel, E. W. The permeability of potato tissue to water. *Proc. Natl. Acad. Sci. U.S.* **41**, 1029–1033 (1955).

62. Traube, M. Experimente zur Theorie der Zellbildung. *Arch. Anat. Physiol. u. wiss. Med.* (1867).

62a. van Overbeek, J. Water uptake by excised root systems of the tomato due to non-osmotic forces. *Am. J. Botany* **31**, 265–269 (1942).

62b. Walker, N. A. Ion permeability of the plasmalemma of the plant cell. *Nature* **180**, 94–95 (1957).

63. Wilson, K. Water movement in submerged aquatic plants. *Ann. Botany (London)* **11**, 103 (1947).

64. Ursprung, A. Uber die Kohäsion des Wassers im Farn Annulus. *Ber. deut. botan. Ges.* **33**, 153 (1915).

65. Ursprung, A., and Blum, G. Zwei neue Saugkraft-Messmethoden. *Jahrb. wiss. Botan.* **72**, 254 (1930).

CHAPTER THREE

The Water Relations of Stomatal Cells and the Mechanisms of Stomatal Movement

O. V. S. Heath

I. Methods of Investigation

A medium-sized stoma measures perhaps $20 \times 15 \times 10$ μ overall; the largest are about three times as long and wide (Figs. 1–4). The possibilities of investigation of the mechanisms by which stomata open and close are therefore severely limited by their small size. Broadly, there are two possible experimental approaches. First, direct investigation of the physiology of individual guard cells and subsidiary cells, necessarily carried out under the microscope. Secondly, the more indirect approach of attempting to obtain evidence as to stomatal mechanism by following opening and closing responses to changes of external or internal factors. Methods of investigating the physiology of the individual cells include: (a) micromanipulation or surgery (see under Section II), (b) the determination of osmotic values by plasmolysis (Section III,A,2,a)—always difficult with guard cells owing to their small size and the invariable lack of anthocyanin in the vacuole, (c) determination of guard cell pH by means of indicators (Section III, A,2,a)—this unfortunately yields information as to the pH of the vacuolar sap, whereas the relevant pH is probably that in the cytoplasm, (d) visual or optical estimations of starch content (Sections

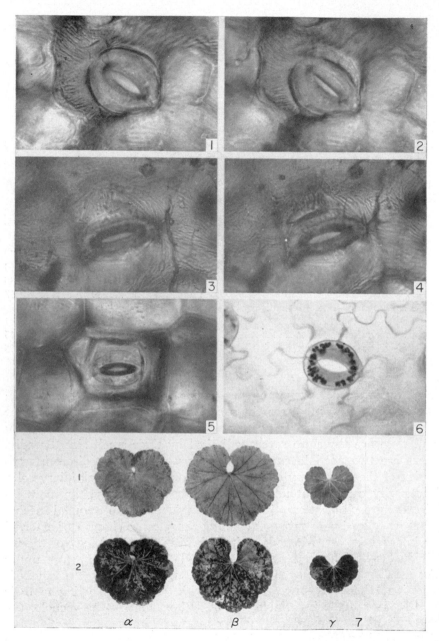

III,A,1,b; III,A,2,a), (e) microchemical tests, e.g., for sugar content or enzymes (Section III,A,2,a), etc. For investigating the responses of stomata to varying conditions the following methods are available.

A. DIRECT MEASUREMENT OF APERTURE

1. Direct Microscopic Observation on the Living Leaf

This provides a salutary check on any other method and is a good qualitative method in its own right. Although good images may be obtained using transmitted light from a condenser and a dry objective corrected for lack of cover glass (Figs. 1–5), the variation between different stomata on the same leaf makes it virtually impossible to measure a large enough number to give satisfactory mean values. Stålfelt (78) cuts out small pieces of leaf, mounts them in medicinal paraffin, and measures the apertures with an immersion objective.

2. Lloyd's Method

In this method (49) strips of epidermis are torn from the leaf and immediately plunged into absolute alcohol, which is alleged to fix the stomata in the condition of opening present on the living leaf. Since any tissue tensions between epidermis and mesophyll must be released on stripping, shrinkage or expansion of the former might be expected

---

FIG. 1. *Cyclamen persicum.* An open stoma on an intact leaf. Magnification: × 385. From Heath (20).

FIG. 2. The same stoma 1 minute after the upper guard cell was punctured and had collapsed to give a straight ventral wall, reducing the pore to half the aperture. No movement of the intact guard cell was seen. Magnification: × 385. From Heath (20).

FIG. 3. *Cyclamen persicum.* A partially open stoma on an intact leaf. Magnification: × 385. From Heath (20).

FIG. 4. The same stoma after the subsidiary cell on the upper side was punctured. Note further opening of the pore on that side. Magnification: × 385. From Heath (20).

FIG. 5. *Zebrina pendula (Tradescantia zebrina).* A partially open stoma on an intact leaf, showing the six cells of the stomatal apparatus developed within the contour of a single epidermal cell. Magnification: × 245. From Heath (20).

FIG. 6. *Pelargonium × hortorum.* Stoma on a "Lloyd's strip" transferred from absolute alcohol to iodine-phenol-KI reagent to stain the starch. The aperture is unaffected by the transfer. Magnification: × 530. From Heath (24).

FIG. 7. *Pelargonium × hortorum.* "Paired leaves" after infiltration with alcoholic gentian violet (pressed and dried before photographing). (1) Plus CO_2: single detached leaves enclosed under bell jars and illuminated for about 2 hours. (2) Minus CO_2: single detached leaves similarly enclosed and illuminated but in the presence of 2 N soda solution. γ Leaves are of a different strain from α and β leaves. In each pair injection is much heavier in the leaf minus CO_2. From Heath (25).

to occur. Nevertheless in the common cultivated geranium, *Pelargonium × hortorum* (sometimes, but incorrectly, referred to as *P. zonale*) Heath (24) has found by means of camera lucida drawings of stripped area and strip that no appreciable change of shape or size of the strip as a whole occurs; Lloyd's method thus gives valid estimates of numbers of stomata per unit area. In this species the stomata themselves open appreciably in the process of taking the strips (25) and the *absolute* apertures so obtained are thus of little value; comparison with porometer data (see below) shows, however, that Lloyd's method does serve to indicate trends of opening or closure. Although the labor of measurement can be deferred by storing the strips in absolute alcohol, it is not thereby reduced. A sample of 120 stomata of the cultivated geranium at medium apertures gave a standard error for pore area of about 5% of the mean (89); as the stomata close the variability increases enormously so that even larger samples are needed. Transfer of the strips to other organic liquids is liable to cause changes of aperture (Section II), while aqueous reagents cause complete closure; until recently it has therefore been necessary to store and examine the strips in absolute alcohol—a very poor medium for microscopy owing to its low refractive index. It has been found (24), however, that iodine and potassium iodide dissolved in phenol containing the minimum of water for liquefaction provide a reagent which causes no change of aperture in Lloyd's strips transferred from alcohol and which stains stomatal starch very well; this makes it now possible to observe starch content and aperture simultaneously in the same stoma (Fig. 6). The refractive index of this reagent is very high and excellent images are therefore obtainable.

B. INDIRECT METHODS

1. Infiltration Techniques

In a number of investigations (56, 76, and later workers), it was found that of a series of lipophile liquids which wet the cuticle some, such as medicinal paraffin, would only enter the leaf and inject the intercellular spaces through large and wide-open stomata whereas others, such as xylol, would enter through small and almost closed stomata. It seems likely that both the cohesion between the liquid and cuticle (or the surface tension at the liquid-solid interface) and the viscosity are involved in these effects; empirically such series as medicinal paraffin, absolute ethanol, benzol, and xylol can be used to grade leaves of a given species in respect of their stomatal aperture. More recent workers have used mixtures of two liquids in a series of graded proportions, thus providing a more sensitive test, e.g., Alvim and Havis (3) use ten mixtures of 0–90% Nujol with 100% to 10% n-dodecane. Another approach is to flood the leaf with a single liquid and estimate the fraction of the whole area that is injected. This was done by Heath (25) for the geranium using Williams' (85) modification of the infiltration technique, in which a strong solution of gentian violet in the absolute alcohol used makes the injected patches show clearly even in pressed dried leaves (Fig. 7). The leaves in each experiment were graded by eye in the order of the amount of injection and the ranked data were treated statistically by transformation to normal scores (11).

These infiltration methods are the best field methods available; they integrate the effects of large numbers of stomata and are far less laborious than is Lloyd's method.

Various methods have also been devised involving the injection of leaves with water under reduced pressure, the pressure, e.g., for half infiltration, being measured.

2. Porometer Methods

a. Viscous air-flow porometers. The name "porometer" was first used by Darwin and Pertz (8) for an instrument (see Fig. 8) which

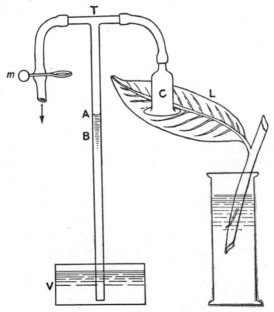

FIG. 8. The Darwin porometer. The cup, C, is cemented to a leaf, L; water is drawn from V up the vertical limb of T by suction applied at m and the rate of fall of the meniscus is then timed between A and B. After Darwin and Pertz (8).

made it possible, by timing the rate of fall of the water column between two marks *AB*, to measure the rate of flow of air under a known pressure difference, through the leaf into a small cup attached to its surface. This rate of flow was a function of the apertures of the stomata. With this porometer, and the numerous modifications of it made by later workers, the conductance or resistance of the stomata within the cup for viscous flow of air may be estimated, although as will be seen shortly certain other additional resistances in the leaf are also involved.

In the modification due to Knight (41) a "constant-pressure aspirator" (Fig. 9) provided the suction instead of a falling column of water.

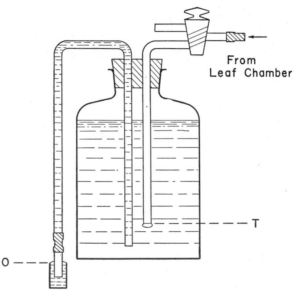

From
Leaf Chamber

Fig. 9. Constant-pressure aspirator. The reduction in pressure below atmospheric is due to the head of water from the bottom of the air-inlet tube, *T*, to the level of water overflow, *O*. After Knight (41).

The rate of air flow through the leaf was estimated by timing the emergence of bubbles in the aspirator. This was more convenient than resetting the water column for each reading as in the Darwin and Pertz porometer. In the "resistance porometer" of Gregory and Pearse (15) a further gain in convenience was obtained, the readings being in terms of two pressure deficits, one read on a manometer, from which the leaf resistance to air flow could be calculated. This apparatus was slightly modified by Heath (22) and more extensively by Spanner and Heath (75), whose instrument is shown in Fig. 10. A "gasometer" type of aspirator (*A*) provides air at a pressure of 5–10 cm of paraffin

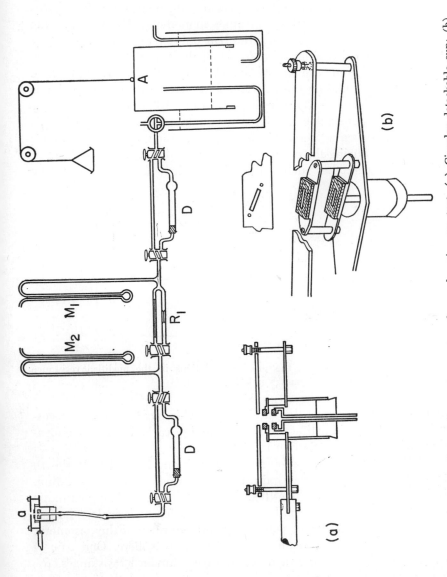

Fig. 10. Modified resistance porometer; for explanation see text. (a) Circular detachable cup; (b) rectangular detachable cup for use with wheat leaves. After Spanner and Heath (75).

above atmospheric. This causes a flow of air through a standard capillary resistance (R_1) and thence into the cup (a) and out through the leaf. Manometer M_1 indicates the pressure difference $(P_3\text{-}P_1)$ across the whole system due to the aspirator; manometer M_2 indicates the pressure difference $(P_2\text{-}P_1)$ across the leaf only; two drying tubes (D) protect the standard capillaries at R_1 when the porometer is not in use. The pressure relations are shown diagrammatically in Fig. 11, where the method of calculating the leaf resistance (R_2) in the same units as the standard resistance (R_1) is indicated. In Fig. 10 are shown two forms $(a$ and $b)$ of "detachable cup" for use with porometers: the leaf is lightly held between a "perspex" plate and an upper gelatine or rubber washer; except during the actual porometer readings, the lower part of the cup bearing another such washer is withdrawn, thus exposing the

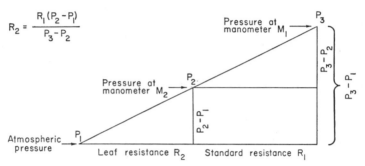

Fig. 11. Diagram illustrating the pressure relations in the modified resistance porometer (cf. Fig. 10). After Spanner and Heath (75).

stomata within the cup to the ambient air. This has been found necessary in order to avoid artifacts due to changing carbon dioxide concentrations within the cup (Section III,A,2,b).

The Wheatstone bridge porometer of Heath and Russell (33) shown in Fig. 12 represents a further gain in the ease and accuracy of reading. It is in fact a conventional Wheatstone bridge in which pneumatic pressures in a system of capillaries substitute for electrical potentials in conductors. The source of potential is an aspirator of the gasometer type and the "circuit" is completed by the atmosphere. One pair of arms of the bridge consists of a set of three standard resistances, R_F, any one of which can be selected at will, in series with the leaf resistance, R_L (shown conventionally without the necessary detachable cup); the other two arms consist of another standard resistance, R_f, in series with a high-precision needle valve R_v calibrated in terms of resistance. In place of the galvanometer as in a Wheatstone bridge circuit there is a sensitive differential manometer, D, containing Brodie's fluid and medicinal paraffin—this gives a sensitivity five times as great as does a water manometer, the difference of density of the two liquids being $1.03 - 0.83 = 0.20$. The tap, T, provides a "short" across the manometer for adjustment purposes. The instrument used by Heath and Russell (33) enabled a change in leaf resistance of 2.3% or less to be detected over a 10,000-fold range of leaf resistance, and the response time (for observing the existence of a barely detectable change of leaf resistance) varied from 1 second to 1 minute over that range.

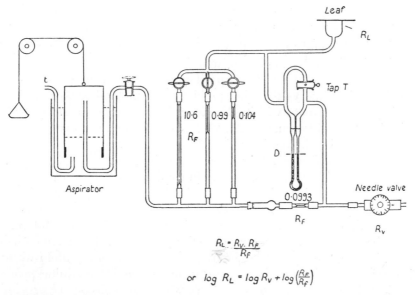

$$R_L = \frac{R_V \cdot R_F}{R_f}$$

$$\text{or } \log R_L = \log R_V + \log\left(\frac{R_F}{R_f}\right)$$

FIG. 12. The Wheatstone bridge porometer; for explanation see text. After Heath and Russell (33).

This porometer has the advantage that only one manometer need be observed, and that one is used as a null-point detector: the manometer range can therefore be small and hence high sensitivity can be combined with rapid response. Further, the aspirator pressure need not be constant so long as it does not change rapidly during a reading; a weighted rubber bladder can therefore be used for field work. One pair of arms of the bridge (R_F and R_L) may be arranged within a controlled-conditions cabinet at one temperature, while (for ease of manipulation) the other pair and the manometer may be outside at another.

Porometer methods have the advantage over such methods as that of Lloyd (49) (Section I,A,2) that they integrate the effects of large numbers of stomata—a cup of 0.5-cm² area may enclose from 5000 to 30,000. Unfortunately, the "leaf resistance" estimated by means of the instruments described above includes not only the resistance of the stomata within the cup to flow, but also that of the intercellular space system and of stomata outside the cup on both surfaces of the leaf. With changing stomatal aperture the distribution of air flow also changes so that varying proportions of these other resistances are included in the total measured. Thus, as the stomata open, the stomatal resistance becomes lower relative to that of the intercellular spaces; therefore a decreasing proportion of the air leaves (or enters) the leaf through stomata very remote from the cup, where the internal path is long, and hence relatively less of the intercellular space resistance is included in the total measured; under these conditions a large proportion of the air will pass straight through the leaf between the stomata within and immediately above the cup (if upper stomata are present). For a mathematical analysis of this problem and methods of estimating the resistance of stomata within the cup only, reference should be made to Heath (22) and Penman (62). The difficulty may, however, be circumvented by using leaves in which there are as many, or more, stomata on the surface immediately opposite to the cup (generally the upper surface) as within the cup itself. Under these circumstances very little of the air flows laterally through the leaf, since the vertical path is short. In wheat (*Triticum aestivum* var. 'Charter'), for instance, there are 1.3 times as many stomata on the upper surface and lateral flow may be neglected to a near approximation (35). The stomata within and above the cup may then be treated as two sets of parallel resistances in series and, if the short path through the intercellular spaces is neglected

$$R_{10,000} = R_{\text{Observed}} \left\{ \frac{n_U n_L}{n_U + n_L} \times \frac{10,000}{5690 \times 4310} \right\}$$

where $R_{10,000}$ is the resistance of 10,000 stomata within the cup, n_U and n_L are the numbers above and within the cup, $5690/4310 = 1.3$ and $5690 + 4310 = 10,000$. Even more favorable material is provided by such leaves as that of *Erythrina caffra* which have stomata on both surfaces (amphistomatous) and in which the internal structure completely prevents all lateral flow (52). Such leaves are called "heterobaric"; if they have stomata on one surface only ("hypostomatous" or "epistomatous"), they cannot be used with flow porometers (see page 206).

For any mathematical analysis of the "leaf resistance" measured, into components due to stomata within and outside the cup and intercellular space resistance (if this last is not neglected as above), a necessary assumption is that stomata within and outside the cup always have the same mean apertures. This has been found not to be the case unless they are exposed to similar conditions of carbon dioxide supply (Section III,A,2,b)—hence the need for detachable cups—and much early work has therefore been rendered suspect or invalidated. Especially is this so where porometer measurements with a permanently attached cup have been compared with transpiration or assimilation for the rest of the leaf.

b. Diffusion porometers. If stomatal apertures are considered in connection with their control of transpiration or assimilation, interest centers neither on their actual dimensions (Section I,A) nor on their resistance to mass flow (Section I,B,2,a) but on their diffusive resistance or conductance. Darwin (7) assumed that diffusion would vary directly as the area of the stomatal pore, while viscous flow would be proportional to the square of the area, as for a long capillary tube of circular cross section; he therefore used the square root of the porometer rate as an estimate of diffusive conductance, as did also Maskell (51). More recently several instruments have been devised which give a measure of stomatal aperture in terms of diffusion, and, making use of these, attempts have been made to calibrate experimentally the much more convenient viscous-resistance porometers in terms of diffusive resistance.

i. Hydrogen porometers. The diffusion porometer of Gregory and Armstrong (14) depends in principle upon the difference in the diffusion rates, through a perforated membrane, of hydrogen and air. A slightly modified apparatus used by Heath (unpublished) is shown in Fig. 13. When a reading is to be made the whole system is swept with pure hydrogen, which escapes through the trap (T), the stomata in the cup and the side arm (S)—the latter is important as efficient sweeping of the cup is essential. Hydrogen is also fed into the system from a small electrolytic cell at a rate controlled by a sliding resistance and indicated by a milliameter (MA). The side arm is then closed and taps (A, B) are turned to cut off the main hydrogen supply and the escape trap, leaving the electrolytic hydrogen being fed into a system closed except for the stomata in the cup area. Hydrogen is all the while diffusing out through the leaf, and air is diffusing into the cup, but at a slower rate. In the absence of the electrolytic hydrogen supply the pressure in the cup would therefore fall and in fact if this supply is insufficient to make up for the net loss of gas from the cup a fall of pressure is seen on manometer M. Conversely, too large a supply causes a rise of pressure and by trial and error a setting of the sliding re-

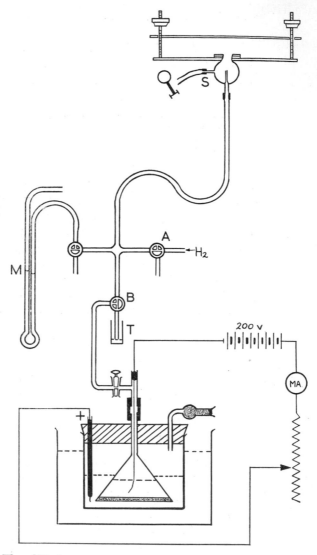

FIG. 13. The diffusion porometer of Gregory and Armstrong (modified); for explanation see text.

sistance is found such that *initially* on closing the system no change of pressure occurs. The meter then shows in terms of milliamperes the net diffusion out through the leaf and hence a measure of its diffusive conductance. Similar considerations with regard to the resistances of the intercellular spaces and the stomata outside the cup apply as with viscous-flow porometers; an approximate treatment of the theory is given by Penman (62). Early attempts to calibrate the resistance porometer in terms of diffusion using this apparatus [(14); and Heath, unpublished] were invalidated by the use of permanently attached cups; a later attempt with wheat stomata under similar conditions within and outside the cup (Milthorpe and Penman, unpublished) should be available soon.

The Spanner hydrogen porometer (74) makes use of a different physical phenomenon, first discovered by a Frenchman Dufour about 1872, viz., that when two gases at the same temperature interdiffuse, temperature differences arise where the gases mix. For hydrogen and air Spanner has observed temperature differences up to 5°C due to this cause, and as much as 8°C might be expected. This effect of course casts doubt upon the validity of the Gregory and Armstrong diffusion porometer measurements, since such temperature differences must occur in the cup and so affect the pressures observed; the magnitude of this source of error has not yet (1958) been reported. The Spanner porometer can only be used with amphistomatous leaves, wheat being particularly suitable. The most essential parts of the apparatus are shown in Fig. 14. To take a reading, a small jet of hydrogen at low pressure is blown against the upper surface of the leaf by opening the ball valve. Hydrogen diffuses through the leaf and out on the lower side, giving rise to a temperature wave by the Dufour effect where it meets the air. This wave is picked up by an exceedingly fine thermocouple (5 μ diameter wires) 2 mm from the leaf and the magnitude of the consequent "flick" of a quick-period galvanometer gives a measure of the quantity of hydrogen diffusing and hence of the diffusive conductance of the stomata. The leaf is held between two open glass rings and thus a detachable cup is not needed. Owing to the great rapidity of the reading, ambient temperature changes are not troublesome. In addition to the vertical intercellular space resistance the readings include the effect of a small external air path.

ii. Transpiration porometers. It is logical to include under "diffusion porometers" all those methods which give a measure of the transpiration from a small portion of a leaf with a standard "sink" substituted locally for the evaporating power of the ambient air. The use of such methods to estimate stomatal diffusive conductance assumes that the aqueous vapor at the surfaces of the transpiring mesophyll cells has saturation (or at least constant) pressure. They are therefore most unsuitable for any work involving different degrees of water strain; also for comparing the state of the stomata at different temperatures, unless a correction can be made for the resulting changes in vapor pressures at the mesophyll and the "sink" (III,C). On the other hand, for investigating the relations between stomatal diffusive resistance and assimilation they have the great advantage that the internal air paths concerned are much the same; they have the further advantage that they can be used with heterobaric hypostomatous leaves.

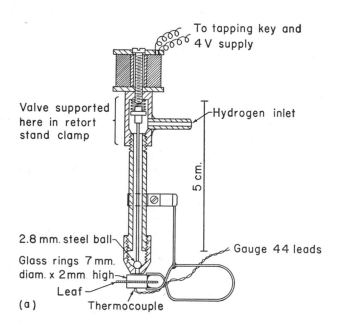

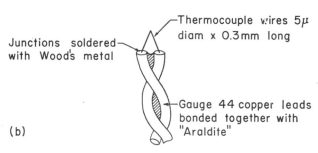

FIG. 14. (a) The Spanner hydrogen porometer; for explanation see text. (b) Enlarged view of the thermocouple. After Spanner (74).

The "horn hygroscope" of Darwin (6) consisted of a flat shaving of horn (cut across the grain) bearing a pointer; this was supported against the leaf and curled away from it as it absorbed moisture. A vast amount of useful information was collected by means of this apparently primitive instrument.

In the cobalt chloride method of Livingston and Shreve (48) a piece of filter paper impregnated with the salt is mounted between two color standards (one dark blue and the other pale blue), dried, and lightly clamped against the leaf surface. The reciprocal of the time taken for the paper to change from the color of the dark standard to that of the pale one is a measure of the rate of water loss from the area of leaf concerned. This provides the only available field method of estimating stomatal diffusive conductance, subject to the above-mentioned assumption.

In the "differential transpiration porometer" of Meidner and Spanner (52a) two small jets of air, at the same temperature but with slightly different constant humidities (e.g., 75% and 85% RH), are blown at the same speed against two neighboring portions of the lower leaf surface. This results in a difference in transpiration rate from the two areas concerned and hence in a differential cooling of the leaf; the temperature difference is measured on the other side of the leaf by means of thermojunctions arranged in contact with the upper surface immediately above the two jets. The air streams are commutated for taking a reading. The difference in the density of water vapor in the two jets is maintained constant for a given temperature, and a correction can be applied when effects of different temperatures are investigated. Hence the instrument does not give a measure of transpiration but of the *difference* in transpiration due to a practically constant difference in potential for diffusion; this is taken as a linear function of diffusive conductance. The method is very subject to the disturbing effects of small air currents and local ambient temperature changes; it can therefore be used only under carefully controlled conditions. It has been used to calibrate the Wheatstone bridge porometer in terms of diffusion and for the wheat leaf has yielded the relation

$$\log C_D = 0.37 \log C_V + \log K$$

where C_D and C_V are the diffusive and viscous conductances of the stomata, K is a constant for any individual leaf but varies from one to another, and the mean slope 0.37 showed a range from 0.32 to 0.39 for different leaves. Thus for the greatly elongated and narrow wheat stoma

$$C_D = K \sqrt[2.7]{C_V}$$

or almost a cube root relationship. For the much more nearly circular stomata of *Talinum triangulare* the corresponding relation (except at the smallest of apertures) was

$$\log C_D = 0.47 \log C_V + \log K$$

with a range for the slope from 0.43 to 0.52, i.e.,

$$C_D = k \sqrt[2.1]{C_V}$$

or very nearly a square root relation as used by Darwin (7) and
Maskell (51) (see I,B,2,b). Such a relation would be expected if the
stomatal pores were tubes, nearly circular in cross section and long
compared with their mean diameter (see page 203). The approximately
cube-root relation found for wheat may perhaps be explained by
assuming that the wheat stomatal pore is effectively a long tube of
very narrowly elliptical cross section in which the long axis is constant.
Viscous flow through a long capillary tube of elliptical cross section
varies as $2a^3b^3/(a^2 + b^2)$ where a and b are the semiaxes of the ellipse.
(If $a = b$, flow varies as a^4 or as the fourth power of the radius, as
assumed by Darwin, page 203). If b is very small compared with a, b^2
may be neglected in the denominator; if a is constant, flow varies as
b^3.

    c. Transpiration methods. A number of investigators have used
estimates of transpiration from whole leaves, shoots, or seedlings as an
indication of stomatal behavior, the same assumption being implicit as
for transpiration porometers. Frequently, no particular care is taken
to ensure that there is a standard "sink" for transpiration; seldom is
there any direct check as to stomatal movement. The work of Virgin
(81) using the extremely sensitive "corona hygrometer" (4) with carefully controlled ambient conditions has, however, yielded such consistent and interesting results (Section III,A,5) as to inspire confidence.

    d. Requirements for studying stomatal mechanisms. Although dif-
fusion methods (*b* and *c* above) have obvious advantages for the study
of the relation between stomatal movement and assimilation, they have
no special merit for the investigation of stomatal mechanism. For this
purpose the best measure would probably be the difference in turgor
between guard cell and subsidiary cell (cf. Section II), but no experimental method of measuring this directly has yet been devised.
When attention is confined to steady-state values (cf. III,A,3) almost
any measure of stomatal aperture is adequate if it may be expected that
higher values of the measure always correspond to wider (or alternatively to narrower) mean stomatal apertures and hence to more "open"
(or "closed") states of the mechanism. This may be assumed as a first
approximation with, e.g., log stomatal resistance to viscous flow (Section
III,A,3), though even here errors may be introduced by differences
in the scatter or standard deviation of aperture at the same mean value
[cf. (51)]. In dynamic studies, however, if rates of change of some
purely arbitrary measure, such as log (flow resistance) or transpiration

rate, are compared, the same observed rate of change may well correspond to very different rates of change of the actual turgor mechanism concerned when occurring in different parts of the range. For this and other reasons Heath and Orchard (31) suggest that some measure thought to bear at least an approximately linear relation to turgor difference should be used and they give reasons for supposing that cube root (viscous flow conductance) is such a measure as far as wheat stomata are concerned. Combining this with the relation between viscous and diffusive flow for wheat stomata found by Meidner and Spanner (Section I,B,2,b,ii) it appears that for this leaf the diffusive conductance of the stomata would also serve as a reasonable measure of changes in the stomatal mechanism.

A danger that must be borne in mind when carrying out dynamic studies arises from changes in rates of closure in darkness which have been found to result from "mechanical shock," as in fitting a porometer cup, and the conditions immediately following such "shock" (Section III,E,2).

II. The Motive Force

Although just over a century has elapsed since an account of the first experimental investigation of stomatal mechanism was published by von Mohl (55) almost the only definite statements that can be made are that the mechanism is operated by turgor changes and that, at least in many species, it is the turgor difference between the guard cells and subsidiary cells that decides the posture of the stoma. These facts were first shown by the experiments of von Mohl himself. Stomata of *Amaryllis* sp. were isolated from the subsidiary cells by making razor cuts on either side of the rows of stomata and then by further cuts parallel with the leaf surface, freeing small pieces of tissue which were floated on water or sugar solutions. Whereas on water the stomata invariably opened, closure could be caused by sugar solution, indicating a turgor mechanism. The immersion of intact leaves in water frequently caused closure, suggesting that here the increase of turgor of the subsidiary cells was greater than that of the guard cells.

On the basis of some experimentation but mainly of histological examination, Schwendener (67, 68) theorized extensively as to how turgor changes could operate various types of stomata; although judgment on some of the more elaborate arrangements of "hinges" may be suspended pending further experimental investigation, for simple stomata such as those of *Sprekelia* (*Amaryllis*) *formosissima* the generally accepted view based on his work appears reasonable, viz.; that in the guard cell the thinner and more extensible dorsal wall (remote from the pore) elongates under increased turgor—the thickened and relatively inextensible ventral wall (bordering the pore) cannot elongate appreciably and is therefore forced into a more semicircular shape by the expansion of the dorsal wall (Figs. 15 and 16). This is

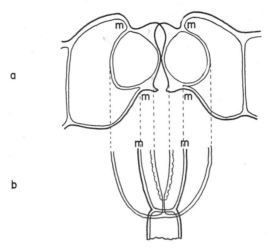

FIG. 15. Stoma of *Sprekelia* (*Amaryllis*) *formosissima:* (a) median transverse section; (b) surface view; dotted lines join corresponding parts of (a) and (b); *m*, points of attachment to subsidiary cells. ×540. After Schwendener (67).

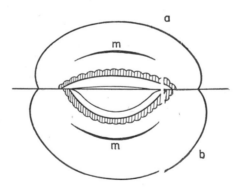

FIG. 16. Diagram of stoma of *Sprekelia* (*Amaryllis*) *formosissima* in surface view: (a) closed; (b) open; for explanation see text. After Schwendener (67).

rendered the more probable by the exhaustive work of Ziegenspeck (96, 97), who showed that in many species the micellae of cellulose run transversely to the long axis of the guard cell; since the walls would expand most readily at right angles to the micellae, this should ensure that the guard cells elongate rather than expand normally to the epidermis. This last is supposed to occur in the stomata of *Helleborus* species (see Schwendener's original figure (67), reproduced in almost every text book), but here change of shape rather than expansion of walls is concerned. In graminaceous stomata such as those of wheat (Figs. 17 and 18) the central part of each guard cell is narrow and very heavily thickened, only the dilated ends being sufficiently thin-walled to be capable of expansion. The swelling of these last, under increased

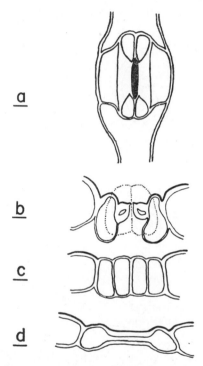

FIG. 17. Stoma of wheat; (a) surface view; (b) median transverse section; (c) transverse section near one end; (d) vertical longitudinal section of a guard cell. After Percivall, J. "The Wheat Plant." Duckworth, London, 1921.

turgor, causes the rod-like central portions to separate and open the stoma.

Nadel (58) produced evidence that another mechanism might be involved, operated by differential swelling of a guard cell wall of two or more layers. She found that stomata killed by "fixatives," on thin surface sections of *Citrus sinensis* and potato (*Solanum tuberosum*), could be made to show opening and closing movements when transferred from one organic liquid to another (e.g., xylol and alcohol). Since dead stomata could have no turgor she suggested that these move-

ments were due to differences in swelling of an outer (cutinized) layer
and inner (cellulose) layer of the wall and she cited Pringsheim's
opinion that even *in vivo* such swelling mechanisms might be important. Heath (20) made an experimental test of this possibility by
puncturing single guard cells of open stomata with a microneedle and
thus suddenly releasing the sap. If a wall mechanism were exerting an
opening force the stoma should open further on the punctured side,
whereas if a turgor mechanism alone were concerned the punctured
guard cell should immediately collapse, reducing the pore to half the

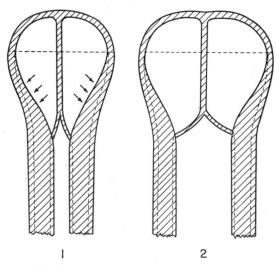

Fig. 18. Diagrams of wheat stoma: (1) in half-open state; (2) in wide-open state;
for explanation see text. After Schwendener (68).

aperture. This experiment was carried out many times, using leaves of
a monocotyledon and a dicotyledon (*Zebrina pendula* [*Tradescantia
zebrina*] and *Cyclamen persicum*), and in practically every case the
predicted result for a turgor mechanism was obtained (Figs. 1 and 2).

Heath (20) also demonstrated in the same two species that the degree
of stomatal opening depends upon the turgor difference between guard
cells and subsidiary cells: puncturing subsidiary cells next to partly
open stomata resulted in the guard cell concerned bulging into the
collapsed subsidiary cell and enlarging the pore on that side (Figs. 3
and 4). Rather more indirect evidence is provided by the frequently
observed stomatal opening which occurs in the early stages of rapid
wilting (44, 54, also see Fig. 27) and the closure which accompanies
rapid recovery from wilting (54). The opening may be attributed to

transpiration from the mesophyll causing the epidermis to lose turgor so rapidly that the guard cells bulge into the flaccid subsidiary cells and open the stoma; later, as the subsidiary cells, by virtue of their increased suction pressure, withdraw water from the guard cells, the stoma closes. The closure on recovery from wilting is presumably the converse process, the only source of supply of water to the guard cells being through the subsidiary cells. Further indirect evidence is available from osmotic pressure determinations by the plasmolytic method: Sayre (63) found that at night the guard cell osmotic value for closed stomata of *Rumex patientia* was always 1 or 2 atm less than that of the subsidiary cells, whereas for open stomata in the daytime it was sometimes as much as 5 atm higher (Section III,A,2,a).

III. Factors Controlling Turgor Difference between Guard Cells and Subsidiary Cells

The factors, both external and internal, known to cause stomatal responses are so numerous and diverse that it is not unreasonable to suppose that several mechanisms may be involved. The most important of these factors are: (a) light (intensity, quality, duration) and, closely associated with this, carbon dioxide concentration of the air; (b), water supply and evaporating power of the air, with the resulting internal factor of leaf water content; (c) temperature. Other known effective stimuli, most of them further removed from natural conditions, include lack of oxygen, wounding, mechanical shocks, electrical shocks, heat shocks, narcotics, acid and alkaline vapors, acids, alkalis, and neutral salts in solution. In some species also, the stomata have a diurnal rhythm of movement even under constant external conditions. It is probable that the movements in all these cases are produced by turgor changes, but the mechanisms may well differ and there is, for example, even evidence that several mechanisms are concerned in the response to changing light intensity (Section III,A,3).

A. Responses to Light and Carbon Dioxide

Hypotheses as to the mode of action of light in causing stomatal opening have tended to alternate between postulating a direct effect of light upon the guard cells and an indirect effect due to changes of carbon dioxide concentration in the substomatal cavities, brought about by photosynthesis and respiration in the mesophyll cells.

1. Postulated Direct Effects of Light Intensity

a. *Production of photosynthates in the guard cells*. The early hypotheses all dealt with direct light effects; thus von Mohl (55) supposed

that the chloroplasts in the guard cells produced osmotically active substances, under the influence of light and warmth, which brought about
stomatal opening by increasing turgor. No experimental evidence of
photosynthesis in the guard cells was obtained until nearly a hundred
years later (Section III,A,3,d).

b. Lloyd's starch ⇌ sugar hypothesis. Following up a suggestion made
by Kohl (43), Lloyd (49) developed the hypothesis that stomatal opening in light was due to the secretion in the guard cells of an enzyme
which hydrolyzed the starch in the plastids to sugar, thus increasing
the osmotic pressure and raising the turgor. This view was based on a
vast number of qualitative observations of stomatal starch content and
measurements of stomatal aperture as seen in epidermal strips (Section
I,A,2) taken at intervals throughout the day under natural conditions,
mostly for the desert plant *Verbena ciliata.* On the whole, stomatal
starch tended to be negatively correlated with aperture: the stomata
opened during the morning and starch content became very low; in
the afternoon starch increased and sometimes the stomata partially
closed; at night the stomata were closed and starch content was high.
(These diurnal changes are the converse of those seen in the mesophyll
and ordinary epidermal cells where starch increases by day and disappears at night; see also page 231.) Lloyd also carried out a few actual experiments, giving light and darkness at times of day other than
the normal, but in these the correlation of guard cell starch and aperture was not convincing; see Heath (24). In another experiment he
found that after 6 days in blue light the stomata remained devoid of
starch, but they still showed opening by day and closure at night.

Other investigators and especially Loftfield (50) observed similar
correlations in diurnal trends of stomatal apertures and starch contents
in other species. Loftfield found, however, that in alfalfa (*Medicago
sativa*) and *Rumex patientia* the correlation was much less complete,
mainly owing to the fact that midday closure was not accompanied by
changes in guard cell starch (see Section III,C).

c. The permeability hypothesis. Linsbauer (46) found that stomata
could open in carbon dioxide-free air (see Section III,A,2,b), which
had also been observed by Darwin (6) and Lloyd (49), showing that
opening did not necessarily depend on the photosynthetic production
of osmotically active substances in the guard cells. He suggested therefore that removal of carbon dioxide by photosynthesis was the immediate cause of opening and that this operated by changing the permeability of the guard cells to solutes. This was supported by Kissilew
(40), who found that indigo carmine penetrated more rapidly into the
guard cells when the stomata were closed than when they were open.

Various other workers, including later Linsbauer himself (47), found however, that guard cell permeability increased in light. Apart from the experimental uncertainties, permeability hypotheses suffer from the theoretical difficulties that solutes which are supposed to leak out of the guard cells with dark closure must go somewhere and must be regained for light-opening to occur. It may be noted that guard cells appear to be more or less isolated from the ordinary epidermal cells: no plasmodesmata have been observed between the two although these do occur in the outer walls of guard cells (37); the latter also long survive the ordinary epidermal cells when the leaf dies (17).

2. Postulated Indirect Effects of Light Intensity

a. Sayre's extension of the starch ⇌ sugar hypothesis. The hypothesis of an indirect light effect was first put forward in detail by Sayre (63) on the evidence from a comprehensive investigation using *Rumex patientia*. He observed the following diurnal changes to occur under natural illumination: In the morning the stomata opened (direct observation); the guard cell starch decreased (staining with iodine); "reducing sugar" increased (estimated by reduction of Fehling's solution in the guard cells); osmotic pressure in the guard cells increased from a night value of 12–14 atm to about 18–20 atm—the subsidiary cells remained at about 15 atm (plasmolytic method); the pH in the guard cells increased (infiltration with bromophenol blue). At night the converse changes occurred. The correlation between aperture and starch content was again somewhat incomplete, especially during midday closure (III,C). Unfortunately Sayre did not vary light experimentally, but he investigated the effects of pH on the stomata on detached epidermis, using dialyzed expressed sap as a buffer adjusted with acid or alkali. At pH 4.2–4.4 he found the widest stomatal opening, with decreased starch and increased "reducing sugar" and osmotic pressure; at higher or lower pH, closure was accompanied by the converse changes. He also found that on intact leaves closure could be caused even in light by acetic acid vapor and opening, even in darkness, by ammonia vapor.

Like Lloyd (49), Sayre thought that the plastids in the guard cells were capable of little if any photosynthesis and regarded them as probably amyloplasts. He called attention to their very pale green color, small size, and the way in which the large starch grains caused them to bulge (Fig. 6), unlike normal chloroplasts. He was unable to obtain microchemical reactions for chlorophyll (cf. page 222).

Sayre postulated that stomatal opening was initiated by light increasing the pH in the guard cells; this might be an indirect effect due to photosynthesis in the mesophyll cells removing carbon dioxide of respiration or a direct effect causing the oxidation of organic acids in the guard cells to carbon dioxide and water. The increase in pH was supposed to favor the hydrolysis of guard cell starch to sugar by diastase, resulting in increased osmotic pressure and stomatal opening. The converse changes were supposed to be caused by darkness. It should be emphasized that the observed effects could in fact have been due to a diurnal rhythm rather than to light and dark.

Sayre's hypothesis, as to an indirect light effect, was adopted by a succession of workers (1, 65, 72, 73), who obtained similar data covering a wide range of species, and for many years it was almost universally accepted. Following Hanes' (18) finding in plant tissues of the enzyme phosphorylase, catalyzing the reaction: starch $+$ inorganic phosphate $\rightleftharpoons$ glucose-1-phosphate with a pH-dependent equilibrium point, it was supposed by Small *et al.* (72) and Alvim (1) that this was the enzyme concerned in stomatal responses to light. Yin and Tung (95), by immersing epidermal strips in glucose-1-phosphate solution, obtained evidence that phosphorylase was present in the guard cells of tobacco (*Nicotiana tabacum*) and *Vicia faba*. It should be noted, however, that the above reaction involves no change in osmotic pressure, the total number of molecules in solution being unaltered; further reactions might result in hydrolysis of the glucose-1-phosphate and so raise the osmotic pressure, but energy would then be needed to re-form starch for stomatal closure to take place. As pointed out by Steward (personal communication) closure should therefore be impossible in the absence of oxygen, at least in darkness; it has been found by Heath and Orchard (31), however, the anaerobic conditions increase the closing responses of wheat stomata to "closing treatments" (Section III,A,2,d). It thus appears improbable that phosphorylase controls a major part of the mechanism.

Reviewing the literature, Heath (24) pointed out that "nearly all the direct evidence for an effect of light on starch contents of the guard cells is derived from the correlation of diurnal trends, an unsafe basis for a causal theory, and even these correlations . . . are not always satisfactory." His own work on the cultivated geranium using Lloyd's method and the iodine-phenol-KI reagent (I,A,2) gave qualitative evidence of a marked diurnal rhythm of guard cell starch but none of a light effect. Williams (88) obtained quantitative and significant evidence of this rhythm by measuring the area of plane projection of the guard cell starch grains stained as above (92); he also found an apparent effect of light in reducing starch under enclosed and very humid conditions, but this effect was absent for stomata freely exposed to the air. Later work (57, 91) suggests that this apparent light effect was really due to the reduction of carbon dioxide supply in light in the restricted volume of air.[1] If there is a light effect per se on stomatal starch, it is very small compared with effects of carbon dioxide, of diurnal rhythms, and especially of water strain (Section III,B). The existence of the carbon dioxide effect, however, implies that light can indirectly affect the starch ⇌ sugar balance by the photosynthetic removal of that gas, but this only becomes at all obvious with an artifically restricted external supply. Such a mechanism is at least not universal, for Heath (27) showed that the stomata of onion (*Allium cepa*), which are at all times free of starch, respond normally to light, and it appears that the starch ⇌ sugar balance is much more important in relation to water strain (Section III,B) and diurnal rhythms (Section III,D) than to the light response.

A great number of investigations (e.g., 63, 72, 73) have been concerned with the effects upon stomatal aperture and guard cell starch of floating epidermal strips or surface sections on buffer mixtures, salts, acids, and alkalis. Stomata under these circumstances have been found not to respond normally to light and, as pointed out by Heath (24), such experiments are not necessarily relevant to the question of the mechanism of the light response; they will not be discussed here.

b. The role of carbon dioxide. In the earliest investigations (e.g., 6) pure carbon dioxide or very high concentrations were used and the more interesting comparison between ordinary and carbon dioxide-free air was not explicitly made until Linsbauer (46) found that stomata tended to open in the latter whether in darkness or light. This entirely correct result was rendered suspect by the crude technique employed, and Paetz (60) came to the opposite conclusion. Freudenberger (12) passed air of known carbon dioxide concentrations (zero, 0.003, 0.015, 0.03, and 2.5%) through a porometer cup attached to leaves of *Canna* sp. and *Sanchezia* sp., and readings, taken at intervals of about an hour, showed generally an opening response to decrease of carbon dioxide and a closing response to increase. Owing to the concentrations being changed before steady-state readings were obtained, the results were purely qualitative.

Heath (23, 25), investigating the abnormally wide stomatal opening in light found inside permanently attached porometer cups (36, 85), discovered independently[2] that stomata of geranium (*Pelargonium* × *hortorum*) and of wheat (*Triticum aestivum*) opened markedly in response to lowering of the carbon dioxide content of the external air

---

[1] This work and the whole question of the starch ⇌ sugar hypothesis is discussed in detail by Heath (28).

[2] Freudenberger's paper was not available in England.

below the normal 0.03%, both in light and darkness. This was confirmed by the infiltration technique (Fig. 7) and by porometer experiments. Completely closed stomata of geranium, however, did not respond in darkness to changes of external carbon dioxide, showing that it was the concentration in the substomatal cavities that controlled opening. These findings were consistent with an indirect light effect operating through intercellular space carbon dioxide, but the importance of the external supply (in light) and especially of wind in maintaining the concentration close to the stomata were emphasized—wind would thus tend to cause closure and reduce the risk of excessive transpiration.

c. *Scarth's amphoteric colloid hypothesis.* Scarth (64) considered that the opening of stomata on illumination was too rapid to be due to hydrolysis of starch. He suggested that the removal of carbon dioxide by photosynthesis in the mesophyll raised the guard cell pH, which took amphoteric colloids in the vacuoles further from their isoelectric point so that they imbibed water; this imbibitional swelling drew water into the guard cells and caused stomatal opening. His only evidence to support this ingenious suggestion consisted of observations on the appearance of the vacuoles of the guard cells after staining with neutral red and treating with alkali. Alvim (1) was unable to confirm these observations. Later, Scarth (65) abandoned this hypothesis in favor of a scheme like that of Sayre (63).

d. *Williams' hypothesis—active water excretion from guard cells.* This hypothesis arose from an analogy between the movements of guard cells and of the pulvinules of *Mimosa pudica*, and from Weintraub's (83) claim to have observed contractile vacuoles in the pulvinules. For guard cells, Williams (90) postulated that: "(i) The important movement is a *closing* movement; opening being a return, probably osmotic, to a resting state. (ii) This closing is an 'active,' nonosmotic, energy-requiring transfer of water from guard-cell to neighbouring cells, possibly mediated by contractile structures of some type." He suggested that all light-induced opening movements were due to reduction of carbon dioxide concentration—carbon dioxide being supposed necessary for the functioning of the contractile vacuoles. This hypothesis lacked the support of any direct evidence as completely as did that of Scarth; in particular, contractile vacuoles had not been observed in guard cells. An experimental test was, however, available, for such "active" excretion of water from the guard cells should be prevented by lack of oxygen, as pointed out by Williams (90) himself. Heath and Orchard (31) therefore investigated the response of wheat stomata to four "closing treatments" (darkness, dry air, high carbon dioxide concentration, and nil or control) both in the presence and absence of oxygen. Rates of change of cube root (conductance) were used as data (Section I,B,2,d). So far from lack of oxygen preventing stomatal closure, it had the effect of significantly increasing the closing tendency when the "closing treatments" were applied. Thus Williams' hypothesis in its original form was disproved (for wheat stomata), but the results would be consistent with an "active" oxygen-requiring up-take of water by the guard cells during opening, and this might tend to be inhibited by carbon dioxide.

3. Postulated Direct and Indirect Light Effects: Interactions of Light Intensity and Carbon Dioxide

The demonstration that stomata opened in response to a reduction of the external carbon dioxide concentration below the normal (Section III,A,2,b) proved that light must to some extent at least control stomatal movement indirectly via the photosynthesis in the mesophyll cells and the carbon dioxide content of the intercellular spaces. Other possible mechanisms were: indirect effects transmitted by carbon dioxide gradients or in some other way from cell to cell rather than through the intercellular space atmosphere, and also direct light effects upon the guard cells themselves. These last could conceivably be independent of photosynthesis; they might however, be due to the guard cell chloroplasts reducing the internal carbon dioxide content or producing photosynthates active by osmosis or in some other way.

In a series of preliminary factorial experiments, Heath and Milthorpe (30) investigated the interactions of carbon dioxide concentration and rate of air flow at different light intensities, using the wheat leaf. Their most important and unexpected finding was that although the stomata opened markedly with reduction of external carbon dioxide concentration from 0.03% to 0.01%, further reduction to zero concentration caused no further opening; there were also indications of a direct effect of light upon the guard cells. These matters were further investigated by Heath and Russell (34, 35) and are discussed below.

a. Direct light effects. In an attempt to eliminate the influence of the changing carbon dioxide content of the substomatal cavities, Heath and Russell (34) conditioned the stomata within and immediately above the porometer cup by forcing carbon dioxide-free air through them between porometer readings. A pair of transparent rectangular chambers of about 1 cm² area was attached above and below part of a wheat leaf, and a flow of 250 ml per hour of moist carbon dioxide-free air was passed in under pressure, escaping laterally through the leaf to the exterior. Every 15 minutes the flow was interrupted and a Wheatstone bridge porometer reading (Section I,B,2,a) made, using the lower chamber as cup. The continuous line in Fig. 19 shows the effect of light intensity upon steady-state values of log (stomatal resistance) under these conditions—evidence for a direct light effect upon the stomata.

b. *Indirect light effects independent of carbon dioxide.* In the experiment under discussion a second pair of chambers was attached to the leaf 2 cm distant from the first pair and similarly treated except that the light intensity was maintained constant at the intermediate value [270 foot-candles (ft-c)]. The illumination of the zone between the two pairs of chambers was also kept constant, but at 800 ft-c. The intention was to have this second pair of chambers as a "control" and by using

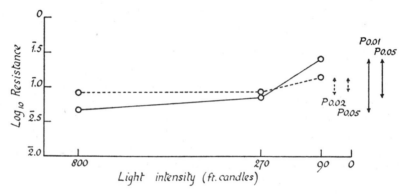

Fig. 19. Effect of light intensity on wheat stomata in carbon dioxide-free air. Continuous line: experimental area, light varied. Dotted line: control area at 270 foot-candles throughout, showing light effect transmitted across a 1.7-cm "light barrier" at 800 foot-candles. "Steady-state" values of log (stomatal resistance). Vertical lines indicate magnitudes of differences required to reach statistical significance. See text. After Heath and Russell (35).

"experimental-control" differences as porometer data to reduce the variation between different leaves. As shown by the broken line in Fig. 19, however, the stomata within the "control" chambers also showed a significant, though smaller, response to the variation of light intensity 2 cm further along the leaf. This is considered to be evidence for an indirect light effect transmitted from cell to cell and independent of carbon dioxide: the effect could not be conveyed in the intercellular spaces, which were swept with carbon dioxide-free air, nor would a carbon dioxide gradient be likely to be maintained across 2 cm of tissue illuminated with 800 ft-c. The mechanism of such a transmitted light effect is indeed completely unknown. The statistical evidence for its existence is good, though not overwhelming. If it be accepted, the possibility exists that the apparent direct effect mentioned (Section

III,A,3,a) was really entirely due to this indirect effect transmitted from other cells.

c. *Indirect light effects via carbon dioxide—the range of carbon dioxide concentration for stomatal responses.* In a much more ambitious factorial experiment, Heath and Russell (35) attempted to control the substomatal cavity carbon dioxide to various known concentrations by means of a "through" flow of air. An "over" flow of the desired air supply was swept at 8 liters per hour across the leaf surfaces, in a similar pair of chambers, and a portion of this was forced in through the stomata under 5-cm head of mercury. Figure 20 shows the results of the

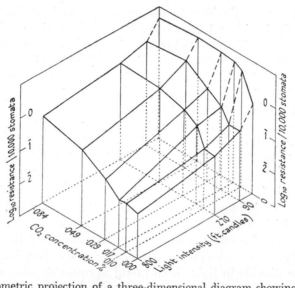

FIG. 20. Isometric projection of a three-dimensional diagram showing the relations between light intensity and carbon dioxide concentration, as affecting steady-state values of log (stomatal resistance) for wheat. After Heath and Russell (35).

experiment, with log (stomatal resistance) plotted vertically against the six carbon dioxide concentrations and three light intensities used. Incorporated in the same diagram are the results of a supplementary experiment carried out with the leaf in darkness. The interpretation which follows assumes that the results of the two experiments were comparable and that the desired concentrations were maintained at the permeable parts of the guard cells. For a discussion of the validity of this last assumption, reference must be made to the original paper.

The carbon dioxide response curves in Fig. 20, at light intensities above zero, *illustrate* the way in which changes of light intensity must control stomatal movement indirectly by changing the intercellular space carbon dioxide concentration. As this concentration is reduced the stomata open until about 0.01% is reached; beyond this point they fail to respond, but in fact very little further reduction due to photosynthesis is to be expected.[3] Thus the stomata respond appreciably only to concentrations down to about the minimum that can occur in the substomatal cavities under natural conditions. For darkness the curve found suggests that intercellular space carbon dioxide is almost without effect *above* 0.01% and hence that there is little or no control by this means; these are steady-state values, however, and the carbon dioxide concentration may affect rate of closure in darkness.

   d. Direct light effects—photosynthetic reduction of carbon dioxide content. The carbon dioxide response curve for darkness in Fig. 20 becomes virtually horizontal above 0.017% concentration; this indicates that the carbon dioxide-operated mechanism becomes saturated in darkness at about that external concentration, no doubt owing to the carbon dioxide of respiration in the guard cells supplemented by that of the adjacent epidermal cells. Under 90-ft-c illumination, however, saturation does not occur until 0.049% external concentration is reached and at 800 ft-c even 0.084% concentration apparently fails to saturate the mechanism, for the curve is still rising. This trend is attributed to photosynthesis in the guard cells, which rises with light intensity and makes higher external concentrations necessary to maintain saturation internally. This is thought to be the first experimental evidence that such photosynthesis plays an active part in the normal light responses of stomata, which was first postulated by von Mohl (Section III,A,1,a). Clearly however, the effect does not operate mainly by the production of osmotically active substances as he thought, for then increasing external carbon dioxide in light should tend to cause opening rather than closure.

   Since 1949 (2) evidence has accumulated that guard cell chloroplasts are in fact capable of slow photosynthesis (cf. page 215), though this evidence by itself does not demonstrate participation in the process of opening. In particular Shaw and Maclachlan (70, 71), using an autoradiographic technique, showed greater uptake of $C^{14}O_2$ in light than in darkness by guard cells on detached epidermis of several species (Figs. 21–24). They estimated that even the maximum rate of photosynthesis in the guard cells was about 50 times too low to account for the maximum rates of osmotic-pressure change observed during opening. They demonstrated the presence of chlorophylls a and b in the

[3] For wheat leaves at 900 ft-c and 25°C [the temperature of the experiments of Heath and Milthorpe (30) and Heath and Russell (34, 35)] Γ has been found to vary between 0.005% and 0.008% (see page 234) (28).

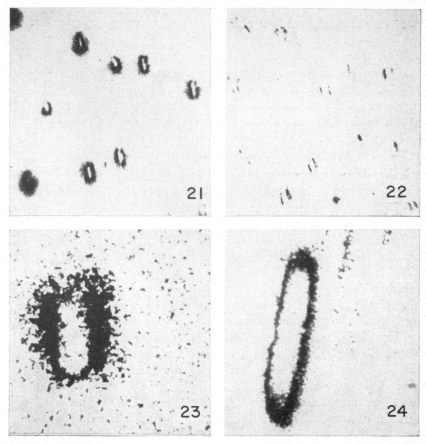

Fig. 21. Radiograph of onion leaf epidermis, after 3 hours in radioactive carbon dioxide in light. Magnification: × 63 (approx.). From Shaw and Maclachlan (71).

Fig. 22. Radiograph of onion leaf epidermis, after 3 hours in radioactive carbon dioxide in darkness. Magnification: × 63 (approx.). From Shaw and Maclachlan (71).

Fig. 23. Radiograph of a single onion stoma, as in Fig. 21. Magnification: × 300 (approx.). From Shaw and Maclachlan (71).

Fig. 24. Radiograph of single stoma of barley (*Hordeum vulgare*), after exposure of the epidermis to radioactive carbon dioxide for 3 hours in light. Magnification: × 300 (approx.). From Shaw and Maclachlan (71).

guard cells of *Tulipa gesneriana* but estimated the chlorophyll concentration in a guard cell chloroplast as about one-sixteenth that in a normal chloroplast of *Mnium*. These results support the view that the importance of photosynthesis in the guard cells lies in the removal of carbon dioxide rather than the production of carbohydrate, except perhaps for the long-term maintenance of osmotic pressure.

e. Direct light effects independent of carbon dioxide. The carbon dioxide-saturated parts of the curves in Fig. 20 (between 0.049% and 0.084% external concentration) show that even at these concentrations opening of the stomata occurred with increase of light intensity from zero to 90 ft-c and thence to 270 ft-c. Similarly a highly significant response to light occurred between 90 and 800 ft-c with zero and 0.010% external carbon dioxide; a light response was also found with zero carbon dioxide concentration by Heath and Russell (34) (Fig. 19) and with zero or 0.01% by Heath and Milthorpe (30). These light effects occurring at concentrations such that external carbon dioxide, and therefore presumably internal carbon dioxide also, was without effect suggest a component in the light response which operates independently of carbon dioxide. The simplest interpretation is a direct light effect upon the guard cells, but in view of the transmission found by Heath and Russell (34) (Section III,A,3,b) the possibility exists that the effect under discussion was entirely indirect and was transmitted from ordinary epidermal or even mesophyll cells.

Such a light effect might or might not depend on the presence of chlorophyll. Wilson (93) had found that light caused stomatal opening in etiolated sweet potato (*Ipomoea batatas*) leaves in which chlorophyll could not be detected by the fluorescence microscope in guard cells or mesophyll. The suggestion of a direct effect involving carotenoids rather than chlorophyll is also supported to some extent by some data of Liebig [(45); see also Section III,A,4]. Virgin (82) found that etiolated wheat seedlings did not begin to transpire rapidly (and by implication, open their stomata) in light until the time when rapid chlorophyll formation began—this, however, would coincide with the formation of carotenes.[4]

4. Responses to Light Quality

In order that investigations of stomatal responses to light of different wavelengths should yield unequivocal information bearing on the mechanism concerned, the number of quanta actually absorbed by the mesophyll or the guard cells (for indirect and direct effects respectively) should be estimated, for it is obvious that light reflected or transmitted cannot affect the stomata. Measurements of light reflection and transmission, and hence estimations of absolute absorption, do not appear to have been made in such investigations, but Liebig (45) did at least give some consideration to relative absorption of different wavelengths by the leaf pigments. It is also important that the intensities used at all wavelengths should be low, to ensure that the opening processes are

[4] See Addendum, page 727.

severely "limited" by light rather than by other factors, especially as Liebig (45) found that above a certain intensity stomatal aperture decreased again in blue and red, but not in green, light.

Using *Tradescantia fluminensis* (*T. viridis*), Liebig (45) compared at low intensities the amounts of light energy of red:blue and red:green necessary to give the same steady-state porometer readings. [This was the "compensation method" of Paetz (60).] Taking reciprocals of these intensities and calling that for red 100, Liebig obtained "efficiencies" (in terms of energy) of 100:42:168 for red:green:blue. She compared such efficiencies with calculated values for the relative numbers of quanta that a mixture of chlorophylls a and b, carotene and xanthophyll, in the proportions found in leaves would be expected to absorb, namely, red:green:blue :: 100:30:219. The efficiencies should, however, be corrected for the different energy contents of quanta of the wavelengths used; they then become red:green:blue :: 100:50:234. This reasonable agreement between relative efficiency and estimated relative absorption, with a tendency for higher efficiency of the blue light absorbed, not only suggests the importance of carotenoids but to some extent supports the hypothesis of light reactions not involving photosynthesis, for in photosynthesis blue light might be expected to be less efficient per quantum absorbed (10).

Liebig (45) also found that if the upper surface of the leaf was illuminated instead of the lower one, much more light energy was necessary to produce a given opening of the lower stomata, the ratios being 9–10×, 18×, and 5× for white, red, and green light respectively. She concluded from this that light was ineffective unless it actually reached the plastids of the guard cells. This is certainly an overstatement, for indirect light effects via the intercellular space carbon dioxide must occur, but her results do indicate that direct effects are very much more important.

5. *Responses to Duration of Light*

Virgin (81) made use of the change, caused by the transpiration of a group of wheat seedlings, in the relative humidity of a rapid air stream as a measure of their stomatal aperture. This was determined with the very sensitive "corona hygrometer" of Andersson *et al.* (4). In Fig. 25 it is seen that in intermittent light the stomata opened more rapidly than in continuous light of the same total quantity, but reached about the same final aperture. This was attributed to the rapid formation of osmotically active substance in the guard cells in light, with the resulting entry of water subject to a time lag. In another experiment (Fig. 26) it was found that the average value reached in intermittent light

of a given intensity was constant and independent of the frequency of
alternation of light and dark. Virgin concluded that under such conditions the rate of formation of osmotically active substance in light was
equal to the rate of "decomposition" in darkness and that "the reaction
is completely reversible at least within short periods of time." In view
of the effects upon opening and closure of assimilation and respiration
in the guard cells, however, the chemical pathways concerned in light
and darkness are almost certainly different and the "reaction" is most
unlikely to be a reversible one in the chemical sense.

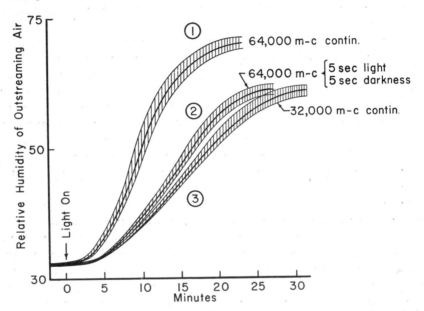

Fig. 25. Transpiration from wheat seedlings in continuous and intermittent light.
1. Continuous illumination 64,000 meter-candles (m-c). *2.* Intermittent illumination
(5 sec light, 5 sec darkness) 64,000 meter-candles. *3.* Continuous illumination, 32,000
meter-candles. Each series was repeated six times and all curves fell within the
shaded areas. After Virgin (81).

Stålfelt (77) postulated that stomatal opening obeyed a "product
law," the attainment of a given aperture being determined by the total
quantity of light received (i.e., intensity × time), indicating that the
principal reaction in light is photochemical. (In this connection the
high temperature coefficients found for light opening may be noted;
see p. 236.) Virgin (81) also concludes that in some of his data for continuous light the rates of increase of relative humidity conform to such
a law. It was found by Harms (19), however, that plotting time of
illumination against reciprocal of light intensity for a series of stomatal

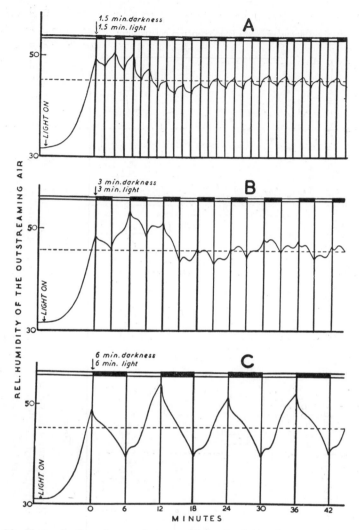

FIG. 26. Transpiration from wheat seedlings in intermittent light. Following darkness, continuous light was given in each series until the same transpiration rate was reached; then intermittent light with periods of: (A) 1.5 minutes; (B) 3 minutes; (C) 6 minutes. Light intensity, 64,000 meter-candles. See text. After Virgin (81).

apertures did not give straight lines $(i \times t = k)$ except for a very small range of aperture. The observation of different steady-state apertures at different light intensities (e.g., Fig. 20) would seem in any case to disprove a "product law," under which the stomata should ultimately open fully whatever the intensity.

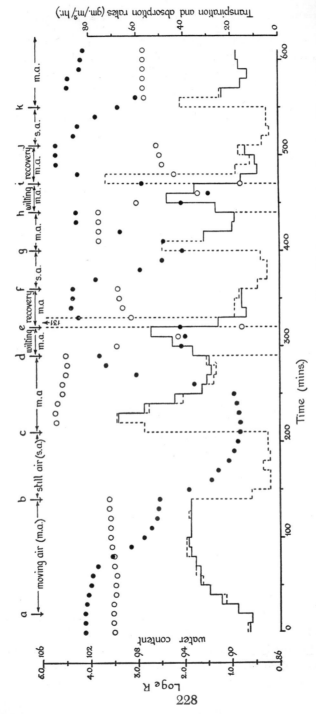

FIG. 27. Stomatal movement and leaf water content in wheat, as affected by still air and moving dry air, wilting (with no water supply to the leaf sheath), and recovery with the leaf sheath in water. Black circles = log (stomatal resistance); open circles = leaf water content percentage of initial value. Darkness until *a*, then light. See text. After Milthorpe and Spencer (54).

B. Responses to Water Supply, Evaporating Power of the Air, and Changing Leaf Water Content

Reference has already been made (Section II) to the preliminary stomatal opening which occurs when a leaf is rapidly wilted and the initial closure that occurs when water is again freely supplied to the

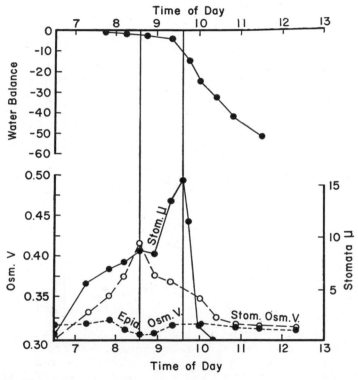

Fig. 28. *Vicia faba.* Time course of stomatal width, stomatal and epidermal "osmotic values" (Osm. V.) (by plasmolysis)* and leaf water deficit. Note (8:40–9:40 A.M.) the "passive" opening due to increasing water deficit, which must be accompanied and reduced by a "hydroactive" closing tendency since the osmotic value is falling. See text. After Stålfelt (79).

* This plasmolytic estimate of "osmotic value" should include any "nonosmotic" or "active" water-secreting force (cf. Chapter 2).

petiole of a wilted leaf. These stomatal movements are illustrated by Milthorpe and Spencer's (54) data shown in Fig. 27, where it is seen that the aperture of the stomata is largely independent of the *level* of leaf water content. In such cases, as stressed by these authors, it is the rate and direction of change in local water contents rather than the magnitude of the water deficit of the leaf as a whole that is important in affecting stomatal movement. Such movements are not observed with

230   O. V. S. HEATH

more gradual wilting or recovery from wilting, when equilibration of
suction pressure between guard cells and subsidiary cells is more nearly
approximated throughout; it is likely, therefore, that they are rare
under natural conditions. Nevertheless these effects must always
modify to an unknown extent the rates of closure on wilting and of

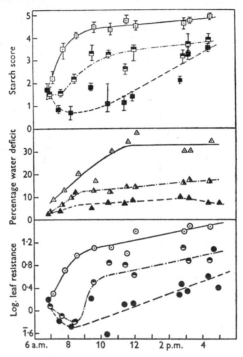

FIG. 29. *Chrysanthemum maximum.* Time course of stomatal movement, stomatal
starch content and leaf water deficit. Starch: plant A (watered throughout),
■ - - - ■; plant B (watered previous day), ▣ — . — . ▣; plant C (watered 2 days
before experiment), ⊡ ——— ⊡. Water deficit: A, ▲ — — — ▲; B, △ — . — . ▲;
C, △ ——— △. Log leaf resistance: A, ● — — — ●; B, ◐ — . — ◐; C, ☉ ——— ☉.
See text. After Yemm and Willis (94).

opening on recovery. Attempts such as those of Stålfelt (79) to separate
on a basis of whole-leaf water deficit "passive" and "hydroactive"
phases of stomatal movement, the latter due to chemical or physicochemical changes induced by water deficit, seem therefore foredoomed
to failure. Both these mechanisms may function concurrently and the
stomatal movement at any time will be the resultant of all the various
mechanisms operating—"passive" and "hydroactive" effects of water
content (Fig. 28), direct and indirect effects of light (Section III,A,3),
closing and opening effects of high temperatures (Section III,C), etc.

That severe water deficit does induce chemical changes in the guard cells leading to closure is indicated by the observation under such conditions of rapid and massive increases in stomatal starch content, and in some cases, decreases in osmotic pressure, by workers from Iljin (38) onward. Yemm and Willis (94) found with *Chrysanthemum maximum* grown outdoors in conditions inducing considerable water deficits, that opening in the early morning (presumably due to light) was followed by closure which occurred sooner, the greater the water deficit (Fig.

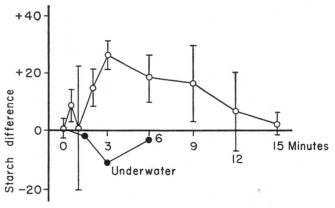

FIG. 30. *Pelargonium × hortorum*. Time course of stomatal starch content following stripping of nearby epidermis at time 0. Open circles = leaf in air; black circles = leaf under water. Fiducial limits represent 10% probability points. See text. After Williams and Barrett (91).

29). Under these conditions stomatal starch content was highly correlated with degree of closure (Fig. 29). It is a striking and unexplained fact that these changes in stomatal starch induced by wilting are the converse of those occurring in the mesophyll cells. There, starch is rapidly hydrolyzed on wilting (39), thus increasing the suction pressure (DPD) available for drawing up water from the soil. The great rapidity with which water strain, even if very localized, can cause increase in stomatal starch is indicated by the results of Williams and Barrett (91): when an epidermal strip was taken from a *Pelargonium × hortorum* leaf an increase in guard cell starch on adjacent parts of the leaf could be detected in a strip taken only 30 seconds later; this difference reached a maximum in strips taken after an interval of 3 minutes and then declined. If the leaf was immersed in water during the experiment, this effect was not found; this suggests that it was due to water loss from the torn edges of the epidermis (see Fig. 30).

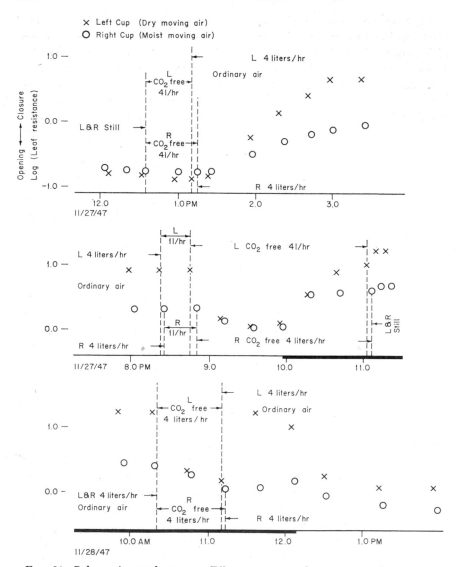

Fig. 31. *Pelargonium × hortorum.* Effects on stomatal movement of presence or absence of carbon dioxide in moving moist and dry air. Ordinates: log leaf resistance. Heavy lines on abscissas indicate periods of darkness. See text. After Heath (25).

One of the most interesting aspects of the effects of water strain upon stomatal movement is the apparent increased sensitivity to all other factors. This is well illustrated by the responses of wheat stomata to moving dry air following periods of still air, shown in Fig. 27. The closing effect of the moving air is here probably compounded of a carbon dioxide effect and a drying effect, but the responses are much more rapid after recovery from wilting than before it occurred. Similarly, even the application of dry air locally to a small area of the leaf (*Pelargonium*) has been found to accelerate closure due to carbon dioxide (25; see also Fig. 31) and to darkness (25, 86). Remarkably, however, the opening induced by light or carbon dioxide-free air was also found by these authors to be more rapid on portions of the leaf given dry conditions (Fig. 31). It is difficult to find a hypothesis to account for both more rapid closure *and* more rapid opening under dry conditions; some such effect as reduced rigidity of the cell walls would seem to be necessary. Williams' (86) suggestion of an increase of guard cell permeability to carbon dioxide under dry conditions is not adequate to account for the facts, since entry and loss of that gas are both involved in opening and in closing effects, e.g., entry during light opening or high carbon dioxide closure, loss during dark closure or low carbon dioxide opening.

As might be expected, different species show wide variation in the sensitivity of the stomata to conditions of high evaporating power. Thus *Pelargonium* $\times$ *hortorum* stomata show scarcely any closure in response to dry, as compared with moist, moving air (25), although responses to other factors are accelerated as noted above. On the other hand wheat stomata have been shown to close markedly when the evaporating power is increased[5] (30) and the same appears to be true of *Chrysanthemum maximum* (94).

C. RESPONSES TO TEMPERATURE; MIDDAY CLOSURE

Early experiments, such as those of Darwin (6), indicated that high temperatures (25–30°C) caused stomatal opening in some species but not others; in Darwin's experiments such apparent opening might have been due to the rise in vapor pressure with temperature (I,B,2,b,ii). Later Wilson (93), using permanently attached porometer cups, obtained indications that under outdoor conditions stomata of *Camellia japonica* and *Ligustrum japonicum* remained closed at about −4°C and opened wider with rise of mean daily air temperature up to about 25°C; the latter species also responded more rapidly to fluctuations of light (total radiation) at the higher temperatures. Loftfield (50) had found a very approximate temperature coefficient (Q_{10}) of 2 for the light-opening of stomata of alfalfa growing outdoors. Wilson (93) also made some experimental determinations with *Gossypium hirsutum* and *Nicotiana tabacum* at constant high light intensity and found opening with increasing temperature up to 25° or 30°C and some closure at 35°C.

[5] This may be in part an effect of water strain on internal CO_2 (99) (see Addendum, page 727).

It was observed by Loftfield (50) for many species, including onion (*Allium cepa*) and alfalfa, and confirmed by Sayre (63) for *Rumex patientia* that the stomata sometimes closed partially or completely at about midday, reopening later in the afternoon; such closure was not found to be accompanied by obvious correlative changes of guard cell starch (onion stomata are starch-free). Both authors attributed such midday closure to water strain, but the striking increases of stomatal starch that normally accompany wilting (Section III,B) cast doubt on such an explanation. Nutman (59) obtained evidence that the midday closure in *Coffea arabica* could occur in the absence of water strain (as when a whole tree was defoliated except for a single leaf) and that it was shown when the incident radiation on the leaf exceeded about 0.9 gm cal per square centimeter per minute. Immediate reopening could be obtained by shading the leaf concerned even if all other leaves on the tree were fully exposed and presumably maintaining the water deficit. He therefore attributed midday closure to supraoptimal light intensity (total radiation). Such closure was accompanied by a fall in apparent assimilation rates (i.e., the difference between assimilation and respiration), but in view of the effects of intercellular space carbon dioxide on stomatal aperture (Section III,A,2,b; III,A,3) doubt exists as to which was cause and which, effect.

Heath (27) found that temperatures above about 25°C tended to cause closure of onion stomata and interpreted this in terms of inter- cellular space and guard cell carbon dioxide contents. Like Parkin (61) he thought that onion guard cells contained no chloroplasts [shown to be incorrect by Shaw (69), *inter alia*], but he also pointed out the very high proportion of nongreen tissue in the onion leaf, which would be expected to cause an especially pronounced increase in intercellular space carbon dioxide concentration with rising temperature, owing to the higher Q_{10} for respiration than for assimilation. He suggested that the midday closure of onion stomata might be a high-temperature effect exerted via internal carbon dioxide, and he implied that the same might apply in *Coffea*. Later experimentation showed that higher tempera- tures do in fact result in higher carbon dioxide concentrations in the intercellular spaces of illuminated leaves. Miller and Burr (53) had found for whole plants and Gabrielsen (13) and Heath (21, 26) for detached leaves that photosynthesis under 1000–2500 ft-c illumination could reduce the carbon dioxide concentration of the air only to about 0.01%; this concentration, later denoted by the symbol Γ (gamma), was interpreted by Heath (26) as the *minimum* concentration in the intercellular spaces for the prevailing conditions of light and tempera- ture. Miller and Burr (53) had claimed that Γ was unaffected by temperature, but it was shown by Egle and Schenk (9) and Heath and Orchard (unpublished) to increase markedly with temperature in a wide range of species. Heath and Orchard (32) investigated the effects of leaf temperature on Γ for onion, *Coffea arabica*, and *Pelargonium* $\times$ *hortorum* leaves under 900 ft-c illumination; the results are plotted as $\log_e \Gamma$ in Fig. 32. For *Pelargonium*, which has not been observed to show midday closure, $\log_e \Gamma$ increased in a linear manner with temperature over the whole range; Γ itself increased by about 30% for each 5°C rise and reached a value of only 0.012% at 35°C. On the other hand, for onion and *C. arabica*, $\log_e \Gamma$ increased similarly to 30°C but

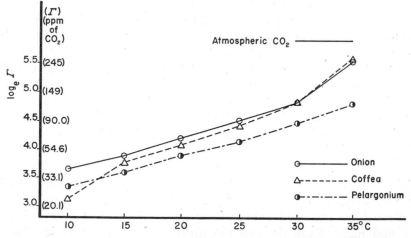

FIG. 32. Effect of temperature on $\log_e \Gamma$ for three species (*Allium cepa, Coffea arabica,* and *Pelargonium* $\times$ *hortorum*). For explanation see text. After Heath and Orchard (32).

then rose more steeply, Γ increasing from about 0.012% at 30° to about 0.025% at 35°C. An indirect estimate of stomatal diffusive conductance indicated that the onion stomata closed slightly from 25° to 30°C (coincident with the rise of the carbon dioxide concentration above 0.01%) and markedly from 30° to 35°C, whereas the *Pelargonium* stomata apparently opened somewhat with rising temperature up to 35°C.

Further support for the hypothesis that midday closure in onion is due to high temperature, operating by means of an increase in intercellular space carbon dioxide, was obtained by Heath and Meidner (29). Using the "differential transpiration porometer" of Meidner and Spanner (I,B,2,b,ii) they carried out a factorial experiment on the effect of leaf temperature, carbon dioxide, and light intensity upon responses of onion stomata. At either 300 or 900 ft-c an increase of temperature from 25° to 35°C caused marked stomatal closure if the leaf was sealed at the tip (as in nature); no such closure was observed with the leaf cavity open at the tip and swept with carbon dioxide-free air, and there were then even indications of greater opening at the higher temperature. Sweeping the leaf cavity carbon dioxide-free was without effect on the stomata at 25°C, when Γ would be about 0.01%. The opening effect of high temperature, when accumulation of carbon dioxide was prevented, was definitely established in a further experiment, and it thus appears that there are two effects of high temperature—a closing effect due to increased carbon dioxide concentration and an opening effect which may be independent of carbon dioxide or may perhaps be due to the lower solubility at high temperature of carbon dioxide in guard cell sap.[6] These results were all in terms of steady-state values of stomatal conductance, but with the leaf cavity swept as above the *rate* of opening from darkness under 300 ft-c illumination was also found to be higher at 36° than at 26°C. The temperature coefficient was 2.35 [cf. Loftfield's approximate value of 2 for alfalfa outdoors (Section III,C)] suggesting that the light-opening was mainly controlled by the rate of a "dark" chemical reaction even at such a high temperature range and low light intensity.

D. Autonomous Diurnal Rhythms

The stomata of several species have been observed to show a 24-hour cycle of movement even under constant conditions, and these rhythms have presumably been induced by the normal alternation of day and night. Examples are the diurnal rhythms found in *Prunus laurocerasus* by Maskell (51) and in the common cultivated geranium by Gregory and Pearse (16), both under constant illumination and both showing a minimum opening about or soon after midnight. The design of experiment used by Heath and Russell (35) with wheat made it possible to detect a highly significant diurnal effect, which is shown in Fig. 33; the data are averaged over all treatments of light and carbon dioxide, except for the open circles, which are not strictly comparable with the other points. A progressive reduction of steady-state opening from morning to afternoon is seen on each day. Such a "closing tendency" in the afternoon was observed in a different way by Darwin (6) using his horn hygroscope (Section I,B,II,b,ii); in several species less rapid opening and more rapid closure, in light and dark respectively, occurred in the afternoon than in the morning. Heath (24) suggested that periodic changes in starch $\rightleftharpoons$ sugar balance might provide the turgor changes for such autonomous diurnal rhythms

[6] Later estimates suggest that the change in solubility would have a very small effect.

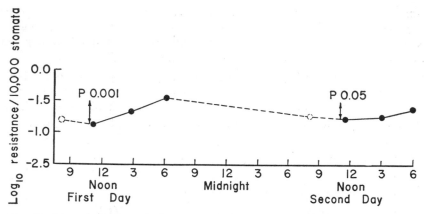

FIG. 33. Diurnal rhythm of wheat stomata in the experiment shown in Fig. 20. For explanation see text. After Heath and Russell (35).

of movement in view of the widespread occurrence of corresponding rhythms in stomatal starch content. It may be noted that for onion stomata, which are at all times starch-free, attempts to demonstrate autonomous diurnal rhythms have so far failed (Meidner and Heath, unpublished).

E. SHOCK EFFECTS, NARCOTICS, ETC.

1. Wounding

Stomatal responses to wounding of the leaf are surprisingly small or in some cases apparently nonexistent. Heath (20) found that when he punctured any one of the six cells which together make up the stomatal apparatus of *Zebrina pendula* (*Tradescantia zebrina*) (Fig. 5) closure, which was apparently irreversible, occurred within a few minutes. The response to puncturing other epidermal cells adjacent to the "unit" of six cells appeared to vary with the extent of the wounding but was in any case apparently confined to the single stomata concerned. Heath mentioned the possibility that these responses might be purely osmotic phenomena, due to the sap which remained in the punctured cell exerting a suction pressure (DPD) equal to its full osmotic pressure when the wall pressure was removed. However, in *Cyclamen persicum* no closing response to puncturing subsidiary or other epidermal cells was found and even when one guard cell was punctured the other was unaffected (Fig. 2). Williams (84) also found with the common geranium (*Pelargonium*) that cutting the main veins had no apparent effect on stomata in a porometer cup attached to the area served by these veins.

2. Mechanical Shocks

Early work with porometers (42) suggested that stomata underwent almost immediate partial closure with the "shock" due to attaching a porometer cup, later recovering and opening widely if illuminated. With the discovery of the sensitivity of stomata to subnormal carbon dioxide concentration (23, 25) this apparent shock effect was shown to be mainly due to the rather small normal aperture before the cup

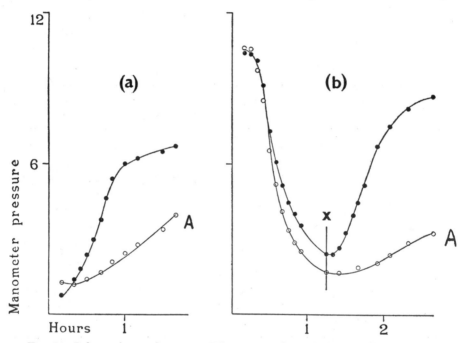

FIG. 34. *Pelargonium* × *hortorum*. Effects of mechanical shock, followed by darkness. (a) Two porometer cups were fitted and illuminated for about 2 hours; the area above one cup (black circles) was then given a sharp pressure and both were at once darkened (beginning of record). Note the more rapid closure in the stimulated area. (b) Opening in light the next day and closure following darkening (at X). Note persistence of the effect of stimulation. See text. After Williams (87).

was attached, the so-called "recovery" in light being the abnormally wide opening induced within the cup by the lowering of the carbon dioxide content (85). Williams found, however, with *Pelargonium* a very small, but apparently real, immediate closing response to mechanical shock (pressing the leaf above the cup) if the stomatal apertures were not abnormally wide; he interpreted this in terms of an increase in respiratory carbon dioxide, for Audus (5) had shown that mechanical stimulation of a *Pelargonium* leaf could increase the respiration more than 1.5 times, some of the effect persisting for more than 24 hours. Later Williams (87) found that such stimulation produced much more striking aftereffects as follows: When the shock stimulus was followed by darkening, much more rapid closure occurred than in a "control" unstimulated area; this difference in rate of dark closure persisted the next day (Fig. 34). When, however, the shock

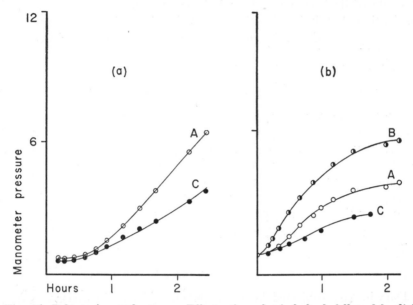

FIG. 35. *Pelargonium* × *hortorum*. Effects of mechanical shock followed by light. (a) As in Fig. 34 until the area over one cup was pressed; then illumination for a further 2 hours, followed by darkness (beginning of record). Note the less rapid closure in the stimulated area (*C*) than in the control (*A*). (b) Three cups were fitted and illuminated for about 2 hours; cup *A* was darkened without stimulation (beginning of record); the area above cup *B* received a shock and was darkened immediately (beginning of record) as in Fig. 34 (a); the area above cup *C* received a shock, was illuminated for a further 2 hours, and then darkened (beginning of record). Graph *C* has been displaced to coincide with *A* and *B*. Note less rapid closure in *C* and more rapid closure in *B* than in *A*. See text. After Williams (87).

was followed by a period of 1–2 hours' illumination, subsequent dark closure was slower than in the control area (Fig. 35). Although the more rapid dark closure due to "shock followed by darkness" may readily be explained in terms of increased respiration as above, the slower dark closure due to "shock followed by light," implying a *reduction* of respiration as a result of mechanical stimulation, does not readily fall into this picture. The interpretation is further complicated by the fact that under all cups, including the control areas (*A* in

Figs. 34 and 35), the leaf first experienced "shock followed by light" when the cups were fitted; Williams actually found in some cases that light or dark following a normal fitting of a porometer cup without further "shock" treatment altered the sensitivity to subsequent darkening as above. It thus appears that two successive "shock followed by light" treatments (C in Fig. 35) have a cumulative effect in reducing the rate of dark closure. The whole matter clearly merits further investigation, but it implies that procedure in fitting and illuminating porometer cups should be carefully standardized in any given experiment. This provides another example of the difficulty of observing the behavior of living organisms without altering it.

3. Heat Shock

Williams (84) investigated the transmission of stimuli, caused by burning the leaf edge of the common geranium, to porometer cups at different distances from the burn. Such stimuli traveled across the leaf at about 0.8 cm per minute and caused (generally) a slight closing movement, which appeared to be more or less irreversible especially in cups near to the burn. Cutting main veins isolated the area served by those veins from the effects of burns outside. Williams suggested that the effects were transmitted by a wound hormone traveling in the phloem.

4. Electric Shock

Darwin (6) subjected leaves to electric shocks by means of a Leclanché cell and Ruhmkorff coil, with two mercury electrodes touching the leaf surface; slight stimulation caused stomatal opening, but stronger stimulation caused closure which appeared again to be irreversible.

5. Narcotics and Other Poisons

Narcotics such as ether and chloroform cause stomatal closure which can be followed by opening on recovery from narcosis if an overdose is avoided (6, 66). This is consistent with the suggestion of "active" uptake of water being concerned in opening (III,A,2,d).

Stålfelt (80) investigated the effects of a relatively high concentration (0.01 $M = 0.065\%$) of sodium azide on Vicia faba leaves and leaf sections. He found the azide tended to prevent stomatal closure, or even caused opening, and concluded that this supported Williams' hypothesis of "active" water excretion from the guard cells (Section III,A,2,d). These results could, however, well be interpreted as due to greater susceptibility of the subsidiary cells than the guard cells to poisoning, causing "passive" opening. The guard cells are in general far more resistant to damage than other epidermal cells (Section III,A,1,c) and Stålfelt stated with reference to one of his experiments: "The leaf in azide quickly loses its turgor, droops and dries, but its stomata do not close"; he himself attributed this to excessive transpiration through the wide-open stomata. In this connection it is worthy of note that Mouravieff (57) found that all the epidermal cells on leaf sections of *Pelargonium* × *hortorum* were dead when examined after 4 hours in sodium azide, even at so low a concentration as 0.01%. He found that azide supplied from gelatine to "isolated stomata" (where the subsidiary cells could not be concerned) *prevented* the opening which normally occurred in light and absence of carbon dioxide; he considered this due to the blocking of water movement, which had been found by other workers to be an effect of azide on roots and storage tissues.

F. SUMMARY—THE MECHANISMS OF RESPONSE TO EXTERNAL AND INTERNAL FACTORS

The motive force operating stomatal movements is provided by turgor alone, turgor difference between the guard cells and subsidiary cells determining the degree of stomatal opening (Section II). The problems of stomatal mechanism thus relate: (a) to the mechanics of guard cell movement due to changing turgor, which have been little investigated since Schwendener's classic work (Section II) and (b) to the way in which various external and internal factors produce changes in the turgor difference between guard cells and subsidiary cells. The latter aspect only is considered in detail here.

1. Light and Carbon Dioxide

There is evidence of photosynthesis in the guard cells, which plays an active part in stomatal opening; it is, however, too slow to account directly, by production of carbohydrate, for the rapid increases in osmotic pressure which accompany light-opening and operates rather by reducing the guard cell carbon dioxide content (Section III,A,3,d). The existence of different steady-state stomatal openings at different light intensities (Figs. 19 and 20) disproves a "product law" (Section III,A,5), and they must be the resultants of at least two processes, one causing closure and the other, opening, e.g., the production of carbon dioxide by respiration and its removal by photosynthesis. These two must operate not only in the guard cells (direct effect) but also via the intercellular space atmosphere from the mesophyll (indirect effect). Guard cell photosynthesis, although slow, is apparently much more effective in controlling stomatal movement than that in the mesophyll, for with the lower stomata illuminated directly instead of through the leaf only 1/18 the intensity of red light is necessary to produce a given aperture (Section III,A,4). For mesophyll cells the ratio of assimilation to respiration is very high (of the order of 10 to 1) and this would be expected to give an "all-or-nothing" stomatal response to light except at the lowest light intensities. The graded response actually found over a wide range of intensities supports the above suggestion that a rather small proportion of the control is exerted by the intercellular space atmosphere and suggests that in the guard cells the effective ratio is much lower—largely due no doubt to the low rate of photosynthesis (Section III,A,3,d) but also to respiration in adjacent epidermal cells where these are free of chloroplasts (as in wheat). External carbon dioxide supply will control the stomata in a manner similar to the above indirect light effect and since it apparently operates from the substomatal cavities only [in *Pelargonium* (Section III,A,2,b)] the degree of such control must diminish as the stomata close.

The way in which changes of carbon dioxide concentration (whether due to light or external supply) control turgor differences is by no means certain. There is some evidence that low external carbon dioxide causes hydrolysis of guard cell starch (Section III,A,2,a) and hence that light should to some extent operate a starch $\rightleftharpoons$ sugar mechanism, though there is little evidence for a light effect on starch with an unrestricted supply of ordinary air to the leaf. At least the system is not universal, for some species of *Allium* have no starch in the guard cells (Section III,A,2,a); hydrolysis of soluble polysaccharides might be concerned here (page 236). Changes in starch $\rightleftharpoons$ sugar balance have been attributed to phosphorylase with a pH-dependent equilibrium, but the evidence is conflicting (III,A,2,a). The data collected for guard cell pH must all refer to the vacuoles, and the important value is the pH in the cytoplasm. Although it seems unlikely that carbon dioxide could directly affect guard cell pH, owing to the probable presence of buffer systems, it might do so by affecting organic acid formation. This would still leave unsolved the question of control of turgor difference by pH change, and here the evidence for the starch $\rightleftharpoons$ sugar balance (Section III,A,2,a), amphoteric colloid mechanism (Section III,A,2,c), or permeability mechanism (Section III,A,1,c) must all be considered unsatisfactory.

In addition to the four processes found to control stomatal movement through carbon dioxide (viz., assimilation and respiration in the guard cells and mesophyll cells) and the probable effect of the carbon dioxide contents of the subsidiary cells, there is evidence for an indirect light effect transmitted from cell to cell (Section III,A,3,b) and possibly a direct effect also (Section III,A,3,e), which are both independent of carbon dioxide. The mode of operation of these is quite unknown, though there is some slight evidence to suggest that light absorption by carotenoids may be concerned (Section III,A,4).

2. Water Supply, Evaporating Power of the Air, and Leaf Water Content

Water movements between guard cells and subsidiary cells lag appreciably behind changes in suction potential owing to the resistance to flow; this is indicated by the results of an experiment with intermittent light (Section III,A,5) and also by the preliminary opening found with rapid wilting and closure with rapid recovery (page 212). Such "passive" movements of stomata depend entirely on rate and direction of change of local water contents and may be virtually independent of the water deficit of the leaf as a whole (Section III,B). In the progressive closure found with more gradual wilting the guard cells must be losing turgor more rapidly than the subsidiary cells in spite of the above-mentioned resistance to water flow (which must nevertheless *reduce* the rate of closure); this may be attributed to the dramatic accumulation of guard cell starch, found by many workers to accompany wilting, which presumably indicates a rapid fall in sugar content and hence in osmotic pressure (Section III,B). That such starch formation may be appreciable within a minute of the beginning of water strain in the epidermis is suggested by the local increases in guard cell starch near the site of an epidermal strip (Section III,B). This suggests that "hydroactive" closing processes may be going on in the guard cells during the much longer period usually occupied by "passive" opening, which has also been shown by a fall in guard cell osmotic potential during such opening (Section III,B). How water loss from the guard cells causes so immediate a condensation of starch is not known, but these rapid "hydroactive" stomatal responses to water strain, coupled with the hydrolysis of starch caused in the other cells of the leaf (Section III,B), which increases the suction potential available for drawing up soil water, provide a system of obvious biological value.

Species vary greatly in the sensitivity of their stomata to high evaporating power when the leaf is provided with an unrestricted water supply. Even *Pelargonium* × *hortorum*, however, which shows hardly any closing response to moving dry air, responds more *rapidly* to other stimuli (light and dark, high or low carbon dioxide concentration) in dry air. Since both opening and closure are accelerated, a turgor hypothesis would seem inadequate, nor can a change in permeability to carbon dioxide fit all the observations; an increase, due to low humidity, in permeability to water or a decrease in rigidity of the guard cell walls would satisfy the requirements, but both seem rather improbable. (see Section III,B).

3. Temperature

High temperature exerts two opposing effects upon the stomata (for details see Section III,C):

(1) At a given light intensity the *minimum* intercellular space carbon dioxide concentration (Γ) increases with temperature. This increase has been found to be especially marked between 30° and 35°C in species (onion and coffee) which show midday closure of stomata, Γ rising nearly to the concentration in ordinary air. Midday closure is therefore thought to be a high-temperature effect operating through the substomatal cavity carbon dioxide. This is supported, for onion leaves, by the observation that raising the leaf temperature from 25° to 35°C caused closure, but not if the interior of the leaf was swept with carbon dioxide-free air. At other temperatures, and in other species, increased temperature must *tend* to cause closure if Γ is raised above the lowest concentration to which the stomata respond (e.g., about 0.01% for wheat).[7]

(2) If excessive accumulation of carbon dioxide in the intercellular spaces does not occur, raising the temperature causes opening. This opening effect of high temperature might be due to the lower solubility of carbon dioxide in the guard cell sap[8] or might be independent of carbon dioxide.

Under conditions such that high temperature causes opening, the rate of light opening from darkness is greater at higher temperature, with a Q_{10} of more than 2 even over a high temperature range (26° to 36°C) and at a moderate light intensity. This suggests that rate of light opening is mainly controlled by a "dark" chemical reaction, for example enzymic hydrolysis of polysaccharide.

4. Autonomous Diurnal Rhythms

In view of the observation of regular diurnal changes of stomatal starch content in many species it seems likely that both autonomous stomatal movements and "opening and closing tendencies" (Section III,D) are caused by changes in starch $\rightleftharpoons$ balance. Such a rhythm of hydrolysis in the morning and condensation in the evening can scarcely be induced by diurnal changes of water content, which would be expected to give a maximum starch content about the middle of the day

[7] See Addendum, page 727.
[8] See footnote to page 236.

and a minimum at night. It might therefore be suggested that it is due to the normal *diurnal* changes in carbon dioxide content of the guard cells. This would imply a light effect and there is, as noted elsewhere (Section III,A,2,a), little evidence of a light effect on stomatal starch except with a restricted external carbon dioxide supply—at least any immediate light effect would appear to be much smaller than that of normal diurnal rhythms. Effects upon the stomatal starch of temperature at normal levels do not appear to have been investigated—possibly this is the main factor concerned.

5. *Effects of Shock, Narcotics, and Other Factors*

The apparent irreversibility of the closure due to burning the leaf edge, electric shock, and (where it occurs) to wounding suggests that here perhaps an increase of permeability with a consequent loss of solutes may be involved. It is tempting to interpret the effects of mechanical shock in terms of carbon dioxide of respiration, but the recorded phenomena are complex and some are difficult to reconcile with this (Section III,E,2). Small doses of narcotics could affect an "active" water uptake mechanism, but Scarth and Shaw (66) have interpreted the effect of chloroform as a prevention of photosynthesis in the guard cells and it could equally well act by preventing photosynthesis in the mesophyll or, indeed, in a number of other ways—the use of poisons in so complex a system may be expected to provide evidence of the utmost ambiguity (Section III,E,5).

G. CONCLUSION

It may be hoped that this extremely condensed account of the evidence and hypotheses bearing on stomatal mechanism will serve to indicate the complexity of such an apparently simple matter as a turgor-operated cell movement and hearten research workers attracted to this field with the conviction that all is not yet discovered. Stomata seem to provide another example of the tendency found in many living organisms for a single system to be operated by a number of mechanisms which interact in an elaborate way to provide a sensitive control well adapted to the conditions normally encountered. Speculation as to the biological value of the various stomatal mechanisms provides an entertaining pastime but a sterile one unless it leads to experimentation, and it is reasonably certain that their interrelations will be elucidated only by systematic large-scale factorial experiments. Such experiments have already indicated that the stomatal mechanism can no longer be considered in terms of classical osmotic equilibriums, changed merely by condensation or hydrolysis of starch, but rather that steady states are concerned, the resultants of many different metabolic activities both

246 O. V. S. Heath

in the guard cells and elsewhere. Further progress will depend on a combination of indirect inference from the results of factorial experiments and such direct investigation of the metabolism of guard cells and subsidiary cells as may be made possible by the development of new microtechniques.

REFERENCES

1. Alvim, P. de T. Studies on the mechanism of stomatal behaviour. *Am. J. Botany* **36**, 781–791 (1949).
2. Alvim, P. de T. A atividade fotossintetica das celulas guardas. *Lilloa. Rev. botán. (Tucumán)* **19**, 5–10 (1949).
3. Alvim, P. de T., and Havis, J. R. An improved infiltration series for studying stomatal opening as illustrated with coffee. *Plant Physiol.* **29**, 97–98 (1954).
4. Andersson, N. E., Hertz, C. H., and Rufelt, H. A new fast recording hygrometer for plant transpiration measurements. *Physiol. Plantarum* **7**, 753–767 (1954).
5. Audus, L. J. Mechanical stimulation and respiration in the green leaf. II. Investigations on a number of angiospermic species. *New Phytologist* **38**, 284–288 (1939).
6. Darwin, F. Observations on stomata. *Phil. Trans. Roy. Soc. London* **B190**, 531–621 (1898).
7. Darwin, F. On the relation between transpiration and stomatal aperture. *Phil. Trans. Roy. Soc. London* **B207**, 413–437 (1915).
8. Darwin, F., and Pertz, D. F. M. On a new method of estimating the aperture of stomata. *Proc. Roy. Soc. London* **B84**, 136–154 (1911).
9. Egle, K., and Schenk, W. Der Einfluss der Temperatur auf die Lage des CO_2-Kompensationspunktes. *Planta* **43**, 83–97 (1953).
10. Emerson, R., and Lewis, C. M. The dependence of the quantum yield of *Chlorella* photosynthesis on wavelength of light. *Am. J. Botany* **30**, 165–178 (1943).
11. Fisher, R. A., and Yates, F. "Statistical Tables for Biological, Agricultural and Medical Research," 3rd ed. Hafner, New York, 1948.
12. Freudenberger, H. Die Reaktion der Schliesszellen auf Kohlansäure und Sauerstoffentzug. *Protoplasma* **35**, 15–54 (1940).
13. Gabrielsen, E. K. Threshold value of carbon dioxide concentration in photosynthesis of foliage leaves. *Nature* **161**, 138 (1948).
14. Gregory, F. G., and Armstrong, J. I. The diffusion porometer. *Proc. Roy. Soc. London* **B121**, 27–42 (1936).
15. Gregory, F. G., and Pearse, H. L. The resistance porometer and its application to the study of stomatal movement. *Proc. Roy. Soc.* **B114**, 477–493 (1934).
16. Gregory, F. G., and Pearse, H. L. The effect on the behaviour of stomata of alternating periods of light and darkness of short duration. *Ann. Botany (London)* [N.S.] **1**, 3–10 (1937).
17. Hagen, F. Zur Physiologie des Spaltöffnungsapparates. *Beitr. Allgem. Botan.* **1**, 260–291 (1918).
18. Hanes, C. S. Enzymatic synthesis of starch from glucose-1-phosphate. *Nature* **145**, 348 (1940).
19. Harms, H. Beziehungen zwischen Stomataweite, Lichtstärke und Lichtfarbe. *Planta* **25**, 155 (1936).
20. Heath, O. V. S. An experimental investigation of the mechanism of stomatal movement, with some preliminary observations upon the response of the guard cells to "shock." *New Phytologist* **37**, 385–395 (1938).
21. Heath, O. V. S. Experimental studies of the relation between carbon assimilation

and stomatal movement. I. Apparatus and technique. *Ann. Botany (London)* [N.S.] 3, 469–495 (1939).

22. Heath, O. V. S. Experimental studies of the relation between carbon assimilation and stomatal movement. II. The use of the resistance porometer in estimating stomatal aperture and diffusive resistance. Part 1. A critical study of the resistance porometer. With an appendix by H. L. Penman. *Ann. Botany (London)* [N.S.] 5, 455–500 (1941).

23. Heath, O. V. S. Studies in stomatal action. Control of stomatal movement by a reduction in the normal carbon dioxide content of the air. *Nature* 161, 179–181 (1948).

24. Heath, O. V. S. Studies in stomatal behaviour. II. The rôle of starch in the light response of stomata. Part 1. Review of literature, and experiments on the relation between aperture and starch content in the stomata of *Pelargonium zonale*. *New Phytologist* 48, 186–211 (1949).

25. Heath, O. V. S. Studies in stomatal behaviour. V. The rôle of carbon dioxide in the light response of stomata. Part 1. Investigation of the cause of abnormally wide stomatal opening within porometer cups. *J. Exptl. Botany* 1, 29–62 (1950).

26. Heath, O. V. S. Assimilation by green leaves with stomatal control eliminated. *Symposia Soc. Exptl. Biol.* 5, 94–114 (1951).

27. Heath, O. V. S. Studies in stomatal behaviour. II. The rôle of starch in the light response of stomata. Part 2. The light response of stomata of *Allium cepa* L., together with some preliminary observations on the temperature response. *New Phytologist* 51, 30–47 (1952).

28. Heath, O. V. S. Light and carbon dioxide in stomatal movements. *In* "Handbuch der Pflanzenphysiologie—Encyclopoedia of Plant Physiology" (W. Ruhland, ed.), Vol. 17/1, pp. 415–464. Springer, Berlin (1959).

29. Heath, O. V. S., and Meidner, H. Midday closure of stomata. Effects of carbon dioxide and temperature on stomata of *Allium cepa* L. *Nature* 180, 181–182 (1957).

30. Heath, O. V. S., and Milthorpe, F. L. Studies in stomatal behaviour. V. The rôle of carbon dioxide in the light response of stomata. Part 2. Preliminary experiments on the interrelations of light intensity, carbon dioxide concentration and rate of air flow in controlling the movement of wheat stomata. *J. Exptl. Botany* 1, 227–243 (1950).

31. Heath, O. V. S., and Orchard, B. Studies in stomatal behaviour. VII. Effects of anaerobic conditions upon stomatal movement—a test of Williams' hypothesis of stomatal mechanism. *J. Exptl. Botany* 7, 313–325 (1956).

32. Heath, O. V. S., and Orchard, B. Midday closure of stomata. Temperature effects on the minimum intercellular space carbon dioxide concentration. *Nature* 180, 180–181 (1957).

33. Heath, O. V. S., and Russell, J. The Wheatstone Bridge Porometer. *J. Exptl. Botany* 2, 111–116 (1951).

34. Heath, O. V. S., and Russell, J. Studies in stomatal behaviour. VI. An investigation of the light responses of wheat stomata with the attempted elimination of control by the mesophyll. Part 1. Effects of light independent of carbon dioxide and their transmission from one part of the leaf to another. *J. Exptl. Botany* 5, 1–15 (1954).

35. Heath, O. V. S., and Russell, J. Studies in stomatal behaviour. VI. An investigation of the light responses of wheat stomata with the attempted elimination of control by the mesophyll. Part 2. Interactions with external carbon dioxide, and general discussion. *J. Exptl. Botany* 5, 269–292 (1954).

36. Heath, O. V. S., and Williams, W. T. Studies in stomatal action. Adequacy of the porometer in the investigation of stomatal aperture. *Nature* 161, 178–179 (1948).

37. Huber, B., Kinder, E., Obmüller, E., and Ziegenspeck, H. Spaltöffnungs-Dünnstschnitte im Elektronenmikroskop. *Protoplasma* 46, 380–393 (1956).

38. Iljin, W. S. Die Regulierung der Spaltöffnungen im Zusammenhang mit Veränderung des osmotischen Druckes. *Botan. Centr. Beih. Abt. 1* A32, 15–35 (1914).

39. Iljin, W. S. Der Einfluss des Welkens auf den Ab- und Aufbau der Stärke in der Pflanze. *Planta* 10, 170–184 (1930).

40. Kissilew, N. Veränderung der Durchlässigkeit des Protoplasma der Schliesszellen im Zusammenhange mit stomatären Bewegungen. *Botan. Centr. Bieh. Abt. 1* A41, 287–308 (1925).

41. Knight, R. C. A convenient modification of the porometer. *New Phytologist* 14, 212–216 (1915).

42. Knight, R. C. On the use of the porometer in stomatal investigations. *Ann. Botany (London)* 30, 57–76 (1916).

43. Kohl, F. G. Über Assimilationsenergie and Spaltöffnungsmechanik. *Botan. Centr.* 64, 109–110 (1895). Abstracted by Lloyd (see Reference 49).

44. Laidlaw, C. G. P., and Knight, R. C. A description of a recording porometer and a note on stomatal behaviour during wilting. *Ann. Botany (London)* 30, 47–56 (1916).

45. Liebig, M. Untersuchungen über die Abhängigkeit der Spaltweite der Stomata von Intensität und Qualität der Strahlung. *Planta* 33, 206–257 (1942).

46. Linsbauer, K. Beiträge zur Kenntnis der Spaltöffnungsbewegungen. *Flora (Jena)* [N.F.] 9, 100–143 (1916).

47. Linsbauer, K. Weitere Beobachtungen an Spaltöffnungen. *Planta* 3, 527–561 (1927).

48. Livingston, B. E., and Shreve, E. B. Improvements in the method of determining the transpiring power of plant surfaces by hygrometric paper. *Plant World* 19, 287 (1916).

49. Lloyd, F. E. The physiology of stomata. *Carnegie Inst. Wash. Publ.* 82, 1–142 (1908).

50. Loftfield, J. V. G. The behaviour of stomata. *Carnegie Inst. Wash. Publ.* 314, 1–104 (1921).

51. Maskell, E. J. Experimental researches on vegetable assimilation. XVIII. The relation between stomatal opening and assimilation. A critical study of assimilation rates and porometer rates in leaves of cherry laurel. *Proc. Roy. Soc.* B102, 488–533 (1928).

52. Meidner, H. The determination of paths of air movement in leaves. *Physiol. Plantarum* 8, 930–935 (1955).

52a. Meidner, H. and Spanner, D. C. The differential transpiration porometer. *J. Exptl. Botany* (in the press).

53. Miller, E. S., and Burr, G. O. Carbon dioxide balance at high light intensities. *Plant Physiol.* 10, 93–114 (1935).

54. Milthorpe, F. L., and Spencer, E. J. Experimental studies of the factors controlling transpiration. III. The interrelations between transpiration rate, stomatal movement and leaf water content. *J. Exptl. Botany* 8, 414–437 (1957).

55. Mohl, H. von. Welche Ursachen bewirken die Erweiterung und Verengung der Spaltöffnungen? *Botan. Z.* 14, 697–704, 713–720 (1856).

56. Molisch, H. Das Offen- und Geschlossensein der Spaltöffnungen, veranschaulicht durch eine neue Methode (Infiltrationsmethode). Z. Botan. 4, 106–122 (1912).
57. Mouravieff, I. Action du CO_2 et de la lumière sur l'appareil stomatique séparé du mesophyll. II. Expériences avec les stomates maintenus sur les milieux complexes. Botaniste (Paris) 11, 195–212 (1956).
58. Nadel, M. On the influence of various liquid fixatives on stomatal behaviour. Palestine J. Botany Hort. Sci. 1, 22–42 (1935).
59. Nutman, F. J. Studies in the physiology of Coffea arabica. II. Stomatal movement in relation to photosynthesis under natural conditions. Ann. Botany (London) [N.S.] 1, 681–693 (1937).
60. Paetz, K. W. Untersuchungen über die Zusammenhänge zwischen stomatärer Öffnungsweite und bekannten Intensitäten bestimmter Spektralbezirke. Planta 10, 611–665 (1930).
61. Parkin, J. Contributions to our knowledge of the formation, storage and depletion of carbohydrates in monocotyledons. Phil. Trans. Roy. Soc. London B191, 35–79 (1899).
62. Penman, H. L. Theory of porometers used in the study of stomatal movements in leaves. Proc. Roy. Soc. London B130, 416–434 (1942).
63. Sayre, J. D. Physiology of stomata of Rumex patientia. Ohio J. Sci. 26, 233–266 (1926).
64. Scarth, G. W. The influence of H ion concentration on the turgor and movement of plant cells with special reference to stomatal behaviour. Proc. Intern. Congr. Plant Sci., 1st Congr., Ithaca, 1926 pp. 1151–1162 (1929).
65. Scarth, G. W. Mechanism of the action of light and other factors on stomatal movement. Plant Physiol. 7, 481–504 (1932).
66. Scarth, G. W., and Shaw, M. Stomatal movement and photosynthesis in Pelargonium. II. Effects of water deficit and of chloroform: photosynthesis in guard cells. Plant Physiol. 26, 581–597 (1951).
67. Schwendener, S. Über Bau und Mechanik der Spaltöffnungen. Monatsberichte Königlichen Akad. Wiss. Berlin. 46, 833–867 (1881).
68. Schwendener, S. Die Spaltöffnungen der Gramineen und Cyperaceen. Sitzber. kgl. preuss. Akad. Wiss. 6, 1–15 (1889).
69. Shaw, M. Chloroplasts in the stomata of Allium cepa L. New Phytologist 53, 344–348 (1954).
70. Shaw, M., and Maclachlan, G. A. Chlorophyll content and carbon dioxide uptake of stomatal cells. Nature 173, 29–30 (1954).
71. Shaw, M., and Maclachlan, G. A. The physiology of stomata. I. Carbon dioxide fixation in guard cells. Can. J. Botany 32, 784–794 (1954).
72. Small, J., Clarke, M. I., and Crosbie-Baird, J. pH phenomena in relation to stomatal opening: II–V. Proc. Roy. Soc. Edinburgh B61, 233–266 (1942).
73. Small, J., and Maxwell, K. M. pH phenomena in relation to stomatal opening. I. Coffea arabica and some other species. Protoplasma 32, 272–288 (1939).
74. Spanner, D. C. On a new method for measuring the stomatal aperture of leaves. J. Exptl. Botany 4, 283–295 (1953).
75. Spanner, D. C., and Heath, O. V. S. Experimental studies of the relation between carbon assimilation and stomatal movement. II. The use of the resistance porometer in estimating stomatal aperture and diffusive resistance. Part 2. Some sources of error in the use of the resistance porometer and some modifications of its design. Ann. Botany (London) [N.S.] 15, 319–331 (1951).
76. Stahl, E. Einige Versuche über Transpiration und Assimilation. Botan. Z. 52, 117–145 (1894).

77. Stålfelt, M. G. Die photischen Reaktionen im Spaltöffnungsmechanismus. *Flora* (*Jena*) [N.F.] 21, 236–272 (1927).

78. Stålfelt, M. G. Neuere Methoden zur Ermittlung des Öffnungszustandes der Stomata. "Handbuch der biologischen Arbeitsmethoden," E. Abderhalden, ed., Abt. 11, Teil 4, Heft 1, pp. 167–192. Urban and Schwarzenberg, Berlin and Vienna (1929).

79. Stålfelt, M. G. The stomata as a hydrophotic regulator of the water deficit of the plant. *Physiol. Plantarum* 8, 572–593 (1955).

80. Stålfelt, M. G. The water output of the guard cells of the stomata. *Physiol. Plantarum* 10, 752–773 (1957).

81. Virgin, H. I. Light-induced stomatal movements in wheat leaves recorded as transpiration. Experiments with the corona-hygrometer. *Physiol. Plantarum* 9, 280–303 (1956).

82. Virgin, H. I. Light-induced stomatal transpiration of etiolated wheat leaves as related to chlorophyll *a* content. *Physiol. Plantarum* 9, 482–493 (1956).

83. Weintraub, M. Leaf movements in *Mimosa pudica* L. *New Phytologist* 50, 357–382 (1952).

84. Williams, W. T. Studies in stomatal behaviour. I. Stomatal movement induced by heat-shock stimuli, and the transmission of such stimuli across the leaves of *Pelargonium zonale. Ann. Botany* (*London*) [N.S.] 12, 35–51 (1948).

85. Williams, W. T. Studies in stomatal behaviour. III. The sensitivity of stomata to mechanical shock. *Ann. Botany* (*London*) [N.S.] 13, 309–327 (1949).

86. Williams, W. T. Studies in stomatal behaviour. IV. The water–relations of the epidermis. *J. Exptl. Botany* 1, 114–131 (1950).

87. Williams, W. T. Studies in stomatal behaviour. III. The sensitivity of stomata to mechanical shock. Part 2. True shock-phenomena and their implications. *J. Exptl. Botany* 2, 86–95 (1951).

88. Williams, W. T. Studies in stomatal behaviour. II. The rôle of starch in the light response of stomata. Part 3. Quantitative relationships in Pelargonium. *J. Exptl. Botany* 3, 110–127 (1952).

89. Williams, W. T. Studies in stomatal behaviour. II. The rôle of starch in the light responses of stomata. Part 4. Variation under constant conditions. *J. Exptl. Botany* 3, 424–429 (1952).

90. Williams, W. T. A new theory of the mechanism of stomatal movement. *J. Exptl. Botany* 5, 343–352 (1954).

91. Williams, W. T., and Barrett, F. A. The effect of external factors on stomatal starch. *Physiol. Plantarum* 7, 298–311 (1954).

92. Williams, W. T., and Spencer, G. S. Quantitative estimation of stomatal starch. *Nature* 166, 34 (1950).

93. Wilson, C. C. The effect of some environmental factors on the movements of guard cells. *Plant Physiol.* 23, 5–37 (1948).

94. Yemm, E. W., and Willis, A. J. Stomatal movements and changes of carbohydrates in leaves of *Chrysanthemum maximum. New Phytologist* 53, 373–396 (1954).

95. Yin, H. C., and Tung, Y. T. Phosphorylase in guard cells. *Science* 108, 87–88 (1948).

96. Ziegenspeck, H. Die Micellierung der Turgezenzmechanismus. I. Die Spaltöffnungen. *Botan. Arch.* 39, 268–309 (1938).

97. Ziegenspeck, H. Das Vorkommen von Fila in radialer Anordnung in den Schliesszellen. *Protoplasma* 44, 385–388 (1955).

PREAMBLE TO CHAPTER 4

After the water relations that exist between cells and their environment have been discussed (Chapters 2 and 3), the regulatory mechanisms that determine the internal composition of cells, with respect
especially to inorganic ions, need to be considered.

In Chapter 4 this question is treated first as part of a larger biological
problem: namely, the regulation of the distinctive composition of the
body and cellular fluids (Part I). Answers to these problems will come
close to explaining some of the essential features of living systems which
maintain their internal composition distinctively different from that of
the environment.

These problems are also discussed in Chapter 4 (Part II) from the
standpoint of cells and against a background of knowledge of their
morphology their permeability properties, their respiration and ability
to do work—considerations which are treated in Volume I and in
Volume II, Chapter 1.

The distribution of dissolved salts throughout the higher plant body
is seen (in Part III) as a problem in which cells in one region absorb
salts released from another, so that the ability of salts to be absorbed
at the "sink" and released at the "source" may exert an influence upon
the movements which occur. The distribution of dissolved salts in the
plant body is discussed and seen as subject to factors that operate
through the integrated way in which the plant body grows. In this
treatment Chapter 4, Chapters 5 and 6, which deal with solute movement, and Chapter 7, which is concerned with water movement, should
all be considered together.

Chapter 4 of Volume II is concerned primarily with the role of salts
as solutes, not as nutrients, for the problems of inorganic plant nutrition
will be taken up in Volume III.

CHAPTER FOUR

Plants in Relation to Inorganic Salts

F. C. STEWARD AND J. F. SUTCLIFFE

Part I. Salt Absorption—General Considerations

A. INTRODUCTION

1. Active Secretion and Physical Diffusion in Biology

a. Unequal composition of extra and intracellular fluids in plants.
Plant cells are characteristically bounded by membranes—the cellulose
cell walls which are freely permeable to water and solutes and the
limiting boundary surfaces, or plasma membranes, which allow solutes
to pass only in a controlled or regulated manner. Although some animal
cells may ingest solid materials, or liquid droplets ("pyknosis"), plant
cells absorb solutes as ions, or molecules, in true solution. Equally

characteristically plant cells, by virtue of their elastic cellulose walls, can exist in a medium which is more dilute, often very much more dilute, than their own liquid contents, whereas animal cells are more usually bathed by solutions which are isotonic with their own contents. The greater total concentration of the solutes which comprise the internal fluids of plant cells is reflected also in many of the individual solutes; thus cells are said to "accumulate" solutes, and this process of "accumulation" is at the very core of the plant nutritional process and of plant cell physiology. By its aid plant cells, without recourse to motility or to ingestion of liquid or solid from the medium, may acquire and store amounts of solutes which are present only in a very much larger volume of the external medium. Two different ratios have been used to measure this absorptive or "accumulatory" power. The absorption ratio (225) is the amount of an absorbed solute per unit volume of tissue divided by its external concentration. The "factor of concentration," later called the accumulation ratio, as used for ions by Hoagland (90) and for salts by Osterhout (153), is the ratio of the internal concentration of the solute or ion to the external concentration. Often greater than unity, these ratios may approach infinity as cells may almost exhaust their external environment of some solutes. In this chapter this problem will be considered in detail.

The problem is not peculiar to plants, for it is but a part of a greater biological problem. The internal composition of cells and organisms is so regulated that they are set in, but are not continuous with, their environment, and the distinction between the composition of cells and of their ambient fluids is an essential factor in the maintenance of life. Unrestricted diffusion of solutes would be incompatible with the high degree of heterogeneity upon which life depends. In short, the principles of diffusion do not operate *in vivo* in a "free and unrestricted fashion"; death, and perhaps only death, is compatible with the attainment of true, physical, diffusion equilibria.

Plants contributed early, if indirectly, to the idea that the composition of the internal fluids of organisms may be very different from that of the fluids that bathe their cells, and to the knowledge that some constituents of the external solution may be enriched in the organism. This conclusion followed from the well-known recognition of certain chemical elements in the ashes of seaweeds. The very name "potash," from pot-ashes, signifies the enrichment of potassium in marine plants, which was the basis of the use of kelp as a potassium-furnishing fertilizer and derives from the method of obtaining potassium from the ashes of the burned seaweed. The elements bromine and iodine, present in sea water in much greater dilution than the element potassium, were also recognized in the composition of seaweeds. Since the main constituent of sea water is sodium chloride, these purely chemical observations could have led directly, and earlier in fact than they actually did, to the recognition of the problem that surrounds the accumulation, or selective absorption, of certain ions. This problem will need to be discussed at greater length below.

B. Historical Sketch: Trends of Thought at Different Periods

The trends of thought that have characterized salient periods in the investigation of this problem will now be indicated in general outline before the subject matter is discussed in detail.

1. The Classical Period

The classical period of cell physiology—ranging from the discovery of plasmolysis by Nägeli (1855), and its more complete recognition as an osmotic phenomenon through the works of De Vries and Pfeffer, leading to the emergence of an osmotic theory of solutions and of water relations in cells—needs no re-emphasis here; it will receive attention in other chapters. In this period, following the work on artificial membranes and beginning with the investigations of M. Traube (1867), attention tended to be focused upon membranes and upon the role of permeability in regulating transfer from one aqueous phase to another. The first concept of the ideal "semipermeable membrane"—permeable to solvent and completely impermeable to solutes—was manifestly not achieved by natural membranes, and this led to the use of special artificial systems which approximated to these conditions and permitted osmotic pressures to be measured, as in the special case of the copper ferrocyanide membrane and such solutes as cane sugar. This trend of investigation continues to the present day, for it is still a salient problem to interpret the structure of membranes *in vivo* from the knowledge gained from artificial membranes (dried collodion), or from oriented molecular films of fat and protein at water-air interfaces (see Volume IA, Chapter 1 and Chapter 1 of this volume).

Already in the works of De Vries there were suggestions that inorganic salts of potassium and sodium were responsible in no small measure for the osmotic value of the cell sap [see Table I, which is quoted from Stiles (223)].

From the work of De Vries in the 1880's, through the turn of the twentieth century and its first decade or so, came the recognition that individual ions might exist in the internal fluid in concentrations greater than those in the external solution. As may be seen from the account by Stiles (223), such observations followed from the early work of Meurer (1909) and Ruhland (1909), who used slices of such storage organs as the carrot (*Daucus carota* var. *sativa*) and the beet (*Beta vulgaris*) root, and they also followed from analyses of the internal fluids of aquatic organisms, such as those made by Nathansohn on *Codium* (1903), and by Meyer (1891), Wodehouse (1917), and later Crozier (1919) on *Valonia*. From this time to the present the problem has received attention which, at each period, has largely reflected, in the techniques adopted and the interpretations attempted, the current status of general and cell physiology as well as biochemistry.

TABLE I

Osmotically Active Constituents of Cell Sap[a]

	Osmotic pressures (as saltpeter values) of the various constituents in different species						
Constituent	*Heracleum sphondylium* leaf stalk	*Gunnera scabra* Young	*Gunnera scabra* Old	*Rheum officinale* stem	*Rheum hybridum* leaf stalk	*Crassula (Rochea) falcata* leaf	*Rosa "hybrida"* petal
Organic acids	0.020[b]	0.023[b]	0.028[b]	0.063[c]	0.124[c]	0.055[b]	0.023[d]
Potassium salts of organic acids	0.013	0.004	0.004	0.012	0.013	0.004	0.012
Glucose	0.152	0.026	0.021	0.085	0.052	0.030	0.218
Potassium chloride	—	0.062	0.090	—	—	—	—
Sodium chloride	0.014	—	—	—	—	0.015	—
Potassium phosphate	—	—	0.003	0.012	0.007	—	—
Total	0.199	0.115	0.146	0.172	0.196	0.104	0.253
Found saltpeter value of sap	0.22	0.12	0.16	0.20	0.22	0.13	0.27

[a] Data from DeVries; taken from Stiles (223).
[b] Principally malic acid and calculated as such.
[c] Principally oxalic acid and calculated as such.
[d] Not identified with certainty, but probably malic acid and calculated as such.

To the extent that solute accumulation received separate consideration in the classical period, it was regarded as an expression of special "physiological" permeability properties of the accumulating cells. Indeed one may summarize the knowledge to 1924 from the work by Stiles, entitled "Permeability" (223).

The first quarter of the twentieth century saw an awakened appreciation of the need to apply to biology the knowledge and the precise analytical methods of chemistry and interpretations based on equilibrium criteria. This trend was stimulated by such workers as Jacques

Loeb. The discovery and explanation of the Donnan effect in 1911 was also used to interpret data which a knowledge of permeability alone left obscure. An epitome of knowledge, as at the year 1924, may now be made as follows.

2. Summary as of 1924: after Stiles

By this time it was generally recognized that absorption of salts was an ionic phenomenon, with unequal intake of anion and cation frequently occurring, so that a measure of ion exchange was necessary to preserve electrical balance. The unequal intake of anion and cation by disks of carrot root and of beet root had been observed by Meurer in 1909, the cation absorption usually exceeding the anion (where the latter was chloride), but in the case of potassium nitrate the anion was absorbed in excess of the cation. This unequal absorption of ions was also recognized as a property of the living, as contrasted with the dead, tissue. Although unequal uptake of anion and cation had frequently been observed, there were already examples (notably that of Hoagland (84a) in 1919, concerned with the uptake of potassium and of chloride by barley) which indicated that the two ions of a salt could be absorbed in equivalent proportions. Indeed Redfern (166) also showed that the unequal intake of the two ions of a salt tended to become less apparent the more dilute the external solution from which the absorption occurred.

By 1919, Stiles and Kidd had introduced the concept of the absorption ratio to measure the degree of accumulation in the cell. This ratio measured the degree of accumulation of the absorbed salt or ion, *neglecting any amount of those same constituents that might have been present in the tissue at the outset.* Recalculating some of the earlier data from Nathansohn on *Codium*, and Meurer on dahlia (*Dahlia pinnata*) and carrot, on the absorption of sodium nitrate, ammonium nitrate, potassium chloride, and sodium chloride, Stiles and Kidd [see Stiles (223), p. 191] showed that the absorption ratios were relatively small and in fact were usually less than unity. This was partly because the absorption occurred from relatively strong solutions and partly because the conditions necessary to promote active salt accumulation were then not understood.

The use of dyes permitted some advance to be made. Generally speaking the basic dyes accumulated in the cells, whereas the acidic (usually sulfonic acid dyes were used) only reached concentrations inside the cells that were less than those in the medium. Redfern had shown for carrot tissue, by 1922, that the value of the absorption ratio for dyes varied very greatly with the external concentration; this led to the oft-quoted generalization that the absorption, or accumulation ratio, increases greatly with the *dilution* of the solute in the external solution. Chemical combination of these basic dyes with acid constituents in the cells may have played a much greater part in their supposed accumulation than was at that time realized. Nevertheless the absorption of dyes, by its ease of demonstration, gave an impetus to investigation of the general problem.

At this time the objective was to find some simple equilibrium concept to explain the experimental facts. For example, to quote Stiles [(233), page 191], "It will be observed that in scarcely any case does the absorption ratio approximate to unity, although having regard to the fact that thin slices of tissues were used and that the salts employed have fairly high coefficients of diffusion, *the equilibrium condition could not be far off at the end of the experiment.* This is also indicated by the value of the absorption ratio at the end of, say, four days, being practically identical with that of the end of six days in the case of the absorption of sodium nitrate by *Dahlia* tuber."

Later events were to confound this quotation at several points. In the first place, the coefficient of diffusion here in question was not that which prevails in water but in the tissue, and ions which are relatively freely diffusible in water may encounter far greater obstacles to diffusion in tissues. The arbitrary criterion that a diffusion equilibrium is established in 4–6 days ignores the effects of other variables than time, which were not understood when the quoted passage was written.

Work with dyes seemed to give some support to the equilibrium interpretation (166). Measurements of the changes that occurred in concentration of the external solution were made, followed by calculations of the concentration in the internal solution; this led to the idea that these relationships conformed to the adsorption isotherm, because on log/log plots the internal (i) and external (e) concentrations were related by approximately straight lines ($\log i - m \log e = \log k$ where $\log k$ is a constant). From such work, also extended to the absorption of alkali chlorides, the similarities between the processes of accumulation and adsorption were stressed, though without any assurance that the ions or solutes absorbed were in fact adsorbed on any specific colloid.

To Stiles and Kidd (226) belongs the credit for one of the first systematic investigations of the absorption process, even though much of their experimental approach was later supplanted by different methods. By 1919 they had investigated absorption from a variety of salt solutions by relatively thin disks of carrot root and potato (*Solanum tuberosum*) tuber tissue. The change in the external solution was followed by measurement of the electrical conductivity, a technique which was by that time in general favor. Stiles and Kidd were predisposed to see an equilibrium state which they felt was attained after a relatively few hours, and their experiments commonly lasted about 50 hours. From very dilute solutions there was not an actual uptake of salt, at least by potato disks, but rather a loss; albeit a loss of ions which was *reduced* in the presence of salts in comparison with that which occurred to distilled water. From this evidence an absorption was implied. Stiles and Kidd calculated the amount "absorbed," assumed it to be equally distributed throughout all the cells of the tissue mass, and then calculated an absorption ratio as they had defined it. After 52 hours of absorption by carrot root disks from potassium chloride, absorption ratios of 25 (at a dilution of $N/5000$) to 0.78 (at $N/10$) were calculated. This indicated again the now well-established fact that the absorption ratio *increases greatly with dilution*, and Stiles and Kidd concluded from the approximate approach to a steady state after 40 to 50 hours that they were at this time dealing with equilibrium conditions.

The brief epitome presented above, shows the extreme reluctance on the part of the plant physiologists of the day to recognize that the protoplasm of the living cell participates actively in the absorption process. Though it was evident that the properties of the living cell differ essentially from those of the dead cell, it was assumed that in some special manner the accumulation of the solute in the internal fluids of the cell was achieved by purely equilibrium mechanisms.

Although the work of this period focused attention on the problem, it contributed only a few ideas that were of lasting value. First, it emphasized the independent absorption of the individual ions. Second, it stressed that ions were rarely absorbed to the same concentration within and without the cell. Third, it showed that the relative absorption (absorption ratio) was greatest from the most dilute solution. Fourth, it emphasized that one ion could influence the relative absorption of another, so that such fast-moving ions as potassium and sodium could interfere with each other and certainly compete with the more slow-moving divalent ions. Thus the sequence of absorption from salts with a common anion, such as chloride, was given as $K > Na > Li > Ca > Mg$; while the absorption of anions from salts with a common cation was given as $NO_3 > Cl > SO_4$.

3. First Metabolic Concepts of Hoagland

Even by 1924, when the work on permeability by Stiles was issued, new experiments from Hoagland's laboratory were destined to change the trend of thought completely and to initiate what may now be regarded as the modern approach to this problem. Faced with the task of explaining the absorption of ions from the often very dilute soil solution, Hoagland had turned his attention to large coenocytic algae with which to investigate the essential phenomena, hoping thereby to obtain cells from which uncontaminated cell sap could be obtained for analysis. Also, he sought a nontoxic ion, not normally present in the cell sap, the absorption of which could be determined accurately. His choices were *Nitella clavata* for the cells and bromide for the ion, and his work led at once to a greater appreciation of the role of metabolic processes in the absorption and accumulation of ions. Contemporaneously other workers, notably of the Osterhout school, were giving increasing attention to other coenocytic structures, particularly marine members of the Siphonales, and the genus *Valonia* in particular. Thus, beginning in 1923, the need to supplement the familiar concepts of cell permeability and the knowledge of equilibria, in order that the absorption process *in vivo* could be explained, became increasingly evident; and this point of view was given clear expression between 1923 and 1929 by the writings of Hoagland and Davis (90, 91, 92).

4. First Views on the Role of Respiration

The knowledge that green cells respond to light by changed salt accumulation and cell composition led to the next development, which was the recognition that the role of metabolism was mediated largely through the process of respiration even in cells that were not green. This became apparent in varying degrees to different workers, and the part played by respiration was investigated from many different points of view. The importance of aeration of soils and water culture solutions together with work on slices of storage organs and upon excised roots, which demonstrated the effects of oxygen tension on salt uptake, quickly led to the view that respiration is the ultimate source of the energy which cells use to accumulate ions by a process which is essentially *not* an equilibrium; this point of view, developed from the earlier work of Hoagland, was expressed by Steward (201). Other workers visualized the role of respiration in somewhat different ways; notably Lundegårdh (123), who invoked a definite component of respiration which was the *consequence*, rather than the cause, of the uptake of anions (Lundegårdh and Burström, 130–132). However, by 1937 it could be stated (206): "The problem [that is, salt accumulation] is not one of equilibria. The transfer of salts from external solution to cell sap tends to increase their total energy and work must be done and energy expended. The living cells thus evade the limitations of true physicochemical equilibrium—a condition they rapidly approach when killed—by means of their own metabolism." As this point of view gathered momentum and became more general, such terms as "active transport," "active secretion," "dynamic transfer," etc. became prevalent and gave recognition to the fact that cells use energy of metabolic origin to perform work in the migration of water, solutes, and ions. This point of view was well documented by Höber (95) when the older work "Physical Chemistry of Cells and Tissues" was reissued under the same title in 1945.

5. Energy Coupling: Phosphate Bond Energy

The conversion of chemical energy into muscular work made it feasible to visualize that energy from the oxidative process of metabolism could also permit cells to do osmotic work. Although this general idea was already familiar in the role of the kidney, its application to vacuolated plant cells was slow. The works of Kalckar (107) and of Lippmann (119) furnished the evidence that phosphate bond energy could be the transferable currency through which energy changes *in vivo* were mediated. By the time that Höber's textbook was reissued it was possible to visualize phosphate bond energy being built into contracted, phosphorylated muscle proteins (e.g., myosin) and subsequently released on expansion and dephosphorylation. In muscle, movements of K^+ and Na^+, concomitant with stimulation and recovery, had been observed by Fenn [see Höber (95), pp. 467–469]. By 1944, as summarized by Hoagland (86), attention was being directed along similar lines with respect to the problems of salt accumulation in cells.

6. Indirect Nature of the Role of Respiration

While in general terms these ideas represented advances, more concrete problems arose. For example, every cell that can produce carbon dioxide is not necessarily able to accumulate salt. After the first realization of the importance of respiration, it was seen that the process of salt accumulation needed to be identified with even more intimate properties of cells which are capable of active metabolism, growth, and even development. Visible cytological events (protoplasmic streaming, starch-sugar changes, etc.) all indicated that the process of accumulation was one associated with active cells, which were using their metabolic energy in various special ways to do work.

7. The Role of Growth

The initial accumulation of solutes in cells proceeds simultaneously with their growth and, therefore, may be thought of as an integral part of growth itself. There are many ways, however, in which this relationship may subsequently be modified so that changes may be induced in the composition of cells, in response to changes in their environment, which may be less closely identified with their growth. The more intimate association of the primary processes of salt accumulation with the actual events that occurred during the growth of the cell was being prominently stressed by Steward in the period preceding 1935, by comparing different accumulating systems according to their capacity for further growth by cell division or enlargement (205). Since then this primary uptake of salts has been increasingly identified, in the growing cell, with that part of the metabolism which relates protein synthesis to respiration and, in turn, to salt accumulation.

8. Modern Trends

Even, however, to identify the primary process as one which is coextensive with the growth of the cell leaves the detailed mechanisms obscure, and here we come to modern aspects of the problem upon which finality is still hard to achieve. At various points in the period here briefly reviewed different ideas have emerged to explain the essential phenomena. All of the following ideas have been suggested from time to time.

(a) The functional membranes have been conceived in even more complex terms, combining the porous properties of molecular sieves with the solvent properties of more fluid membranes. The molecular sieves by their electrical properties greatly exaggerate the differences in mobility of ions in water as they move through water-filled pores in the ultramicroscopic membrane structure.

Porous membranes can be selectively permeable depending on the size of the pores and their electrical charge. Mosaic membranes composed of cation- and anion-permeable areas can also be made. Such mosaic membranes allow the diffusion of water, but not of salt, and allow ions to pass only by exchange.

(b) The behavior of complex molecules, like those of fat and proteins, in oriented molecular films at interfaces shows how the membrane structure may be so visualized that it combines the properties of a fluid, or solid, according to its degree of condensation, or compression. Complex laminated films of fats and proteins can now be visualized, embodying amphoteric substances, so that the mosaic type membrane with cation- and anion-permeable areas could well be visualized to occur in vivo.

(c) Ideas of ionic exchange, which restrict entry of ions to exchange for the issuing ions of carbonic acid, have been prominently emphasized (20b).

(d) Various devices that involve carrier molecules for ions in the cytoplasm, the properties of the carriers being amenable to reversible cycles, have been proposed. The carriers would enable the ions to serve to traverse the cytoplasm under the control of metabolism, circumventing the limitations of diffusion.

(e) Ideas which utilize the specificity of enzyme protein substrate complexes and which even go so far as to visualize specific and adaptive "permeases" for almost every entering solute have also been suggested (167).

(f) The view has been put forward (172, 175, 176) that mitochondria may be specifically identified with the transport of ions across membranes. This arises from the following observations:

(i) Mitochondria are the loci of respiratory enzymes and therefore are sites at which energy may be released.

(ii) There is some evidence in plants that ions may be associated with cytoplasmic particles.

(iii) In gastric mucosae and in kidney tubules mitochondria are observed to aggregate at the site of secretion (52).

But in spite of all these ideas, the primary need in the modern period is for experimental investigation which will stipulate the intimate characteristics of the vital machinery that are concerned in the processes of active transfer. Now, newer tools of radiobiochemistry; newer knowledge of enzyme systems, and the way they may be modified by inhibitors, and of the intermediates of metabolism; newer tools for the investigation of the growth of cells (tissue culture) and for their observation (phase and electron microscopy); methods for extracellular handling of cytoplasmic inclusions (e.g., mitochondria) and the investigation of their role in oxidative metabolism—all permit the older aspects of the problem to be re-examined in a new light. Today the inquiry into the mechanism of salt accumulation carries one into almost every aspect of cellular biology, and it is clear that its solution requires a synthesis of knowledge from them all. It is a paradox that to explain what may seem to be a specific feature of the vital machinery, one has almost to explain life itself. This, however, is the dilemma encountered in many biological problems which can only be satisfactorily investigated in organized systems. However, it is in the attempt to understand the problem as so conceived that the investigator and the student are brought to a fuller realization of the ultimate problems of plant physiology and of life.

On this note this brief historical sketch ends. The first optimistic inquirers embarked upon the investigation of salt accumulation and the absorptive process in cells on the assumption that this could be explained purely in terms of the relatively simple chemistry and physics of the day. As knowledge of the problem grew, it became increasingly identified with the vital machinery of the cells. The present task is to understand how this is coordinated to bring about movements of water and solutes that require the intervention of an active, working cellular machine and the extent to which this is dependent upon cells that can grow. For this we need to turn to the knowledge that has been gained through the investigation of typical systems. This will be done under heading E below and in Part II of this chapter, but, prior to this, it is instructive to note that similar problems arise in plants and animals.

C. ACTIVE TRANSFERS AND SOLUTE ACCUMULATION IN THE PLANT AND ANIMAL BODY

1. The Control of the Body Fluids of Aquatic Organisms

Problems and phenomena similar to those discussed above also occur in animals, but, in the much more complex organization of the animal body, the interpretations rendered have been more concerned with the morphology of special organs. However, to see the general biological problem in perspective, some examples of concentration differences that occur in the animal body will now be mentioned.

Through the gastrointestinal tract, the animal encloses within its body a region which is in virtual free connection with the external environment. For the internal gaseous phases of the plant body a similar situation prevails; however, this is less evident for the aqueous phases of the plant body, although it is conceivable that soil solutions may pass freely across the root cortex to the endodermal surface (193). Again, and unlike the higher plant, there are in the animal body extracellular fluids which together comprise a large part of its total weight. These fluids are represented by blood plasma and the interstitial fluids of the body. Whereas virtually all of the water content of plants may be regarded as intracellular, in the animal body about 50% of the water may be extracellular. Thus blood plasma and interstitial fluid represent, in effect, an internal aqueous environment against which the cells of the body maintain their special composition. Thus it is a problem to know how the composition of blood plasma and interstitial fluids are maintained distinct from the fluids which bathe the boundary surfaces of the organism and, secondly, to know how the intracellular fluids of the body are regulated against the composition of the blood plasma or the interstitial fluids.

In animals contact with the external environment may be made across the skin or the surfaces of stomach and intestines, the kidneys, and the lungs, or gills. Thus an animal body may be represented diagrammatically as in Fig. 1.

With this diagram in mind one may now visualize the remarkable range of events that the animal kingdom presents. In a marine teleost fish the composition of the extracellular fluids of the body, i.e., the blood plasma and the interstitial fluid, reflect in the general proportions

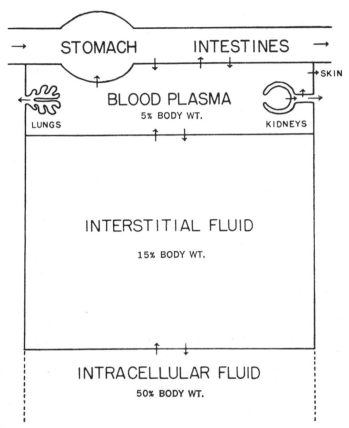

FIG. 1. Diagrammatic sketch of relations of intracellular and extracellular fluids in the animal body and with the external environment.

of the ions present, the composition of the external sea water; but the sea water is very much more concentrated than are the extracellular fluids. Thus, a bony fish keeps its internal fluids dilute relative to sea water: in essence, it conserves water and excludes salts. The composition of the cells, however, may be very different from that of the extracellular (intercellular) fluids of the body; for, whereas the latter reflects the relative composition of the external sea water, the former differs

266   F. C. STEWARD AND J. F. SUTCLIFFE

from it very greatly. This is chiefly indicated by the fact that the ionic composition of the intracellular fluids emphasizes potassium, magnesium, and phosphate; whereas the composition of the intercellular fluids contains mainly sodium and chloride (Fig. 2).

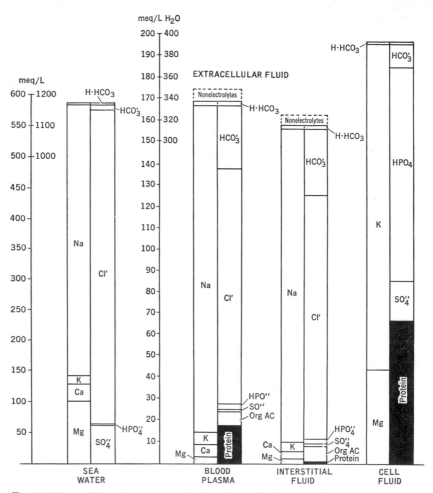

FIG. 2. Composition of external solutions, body fluids, and cellular fluids in marine animals. After Gamble (69).

Aquatic animals, however, exhibit a wide range of situations. The problem of a fresh-water fish is quite a different one, for its blood is much more concentrated than the medium with which the animal is bathed. This is achieved by a greater excretion of water than of salts. The striking adjustment that must ensue in a euryhaline fish, which can adjust itself upon transfer from sea water to fresh water, should be appreciated. Such an organism must switch from a situation in which it conserves water and excludes salt, to a situation in which it conserves and accumulates salt and excludes water.

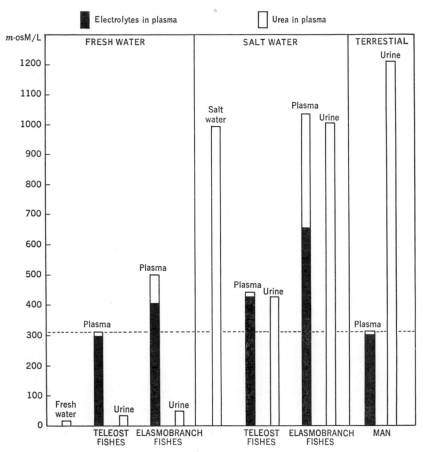

FIG. 3. Osmotic equilibria in the animal body. (m-osM = milliosmoles.) After Gamble (69).

Osmotic regulation in aquatic animals presents, however, other relevant features. The marine bony fish conserves water and excretes small quantities of concentrated urine. The fresh-water fish, on the contrary, conserves salts and excretes urine copiously, but this is more dilute than the blood. Thus osmotic regulation in the animal body requires controlled active movements of *water and/or solutes*. Also, the individual cells of the body maintain their distinctive composition in juxtaposition with an internal environment composed of the blood or the interstitial fluids.

Even a third facet may be added, for the sharks and rays differ from the bony fishes, in that they regulate their blood and urine at approximately the same osmotic value as the external sea water. While they thus evade the problems of controlled movement of water, the composition of the solutions in these organisms is far different from that of the external sea water, for these fishes utilize a solute, namely urea, to bridge the gap between the high salt concentration of the sea and the much lower salt concentration of the internal fluids. Marine elasmobranchs, therefore, substitute a mechanism which must control water and salts, as in the teleost fish, for one which relies largely upon control obtained by another solute, namely urea; this substance, freely circulating in the blood, maintains osmotic balance but is not itself allowed to leak into the sea water (Fig. 3).

In aquatic animals the urine may be more dilute than the blood, as in fresh-water fish; it may be approximately the same concentration as the blood and more dilute than sea water, as in the marine bony fish; or it may be approximately the same concentration as both blood and sea water, as in the elasmobranch fishes. In man, also, the kidney is regarded as performing osmotic work in the excretion of urine.

All these situations may be readily comprehended from the diagrams (Figs. 2 and 3) which are here reproduced from Gamble (69), and it becomes apparent that true physicochemical or diffusion equilibriums between the organism and its environment do not occur.

Faced with these problems in the animal, physiologists have sought explanations in part through the morphology of the cells and organs at the active centers, or through the intervention of the metabolism of the living cells which enables the organism to evade the limitations of true physicochemical equilibrium.

2. Terrestrial Plants and Animals Compared

The problems in the animal body have been set out above for comparison with the similar problems for the plant. Even when in soil, roots in effect grow in an aqueous medium, namely the soil solution. The extent, however, to which external solutions penetrate the tissues of the root and become continuous with the solution that bathes the protoplasts of the inner cortical cells is not as easy to specify. Although some cells of the root may be regarded as absorbing directly from the external solution, other cells of the plant body are differently situated. It is true that all cells should be regarded as bathed by a somewhat tenuous intercellular solution which perfuses the cell walls, but there is no obvious analog to the circulating fluid which is the blood of the animal.

The liquid contained in the xylem perfuses the permeable cellulose walls of the surrounding tissues and may well represent the interstitial fluid from which living cells obtain, or against which they must retain, their solutes. Also the fluid which emerges under certain situations from hydathodes may be regarded as continuous with that which perfuses the cell walls of the leaf tissue. However, these fluids represent a much smaller part of the total water content of the organism than the extracellular fluids of the animal body, and their exact composition is much more difficult to specify.

The animal body presents many other examples of the controlled movement of solutes across boundary membrane surfaces. The skin of the frog is a membrane across which oriented movements of both water and solutes occur. Red blood cells store potassium against the serum with which they are bathed, and muscle fibers maintain unequal potassium-sodium concentrations with respect to the lymph, and, indeed, as muscle contracts and expands these potassium-sodium concentrations change responsively. The salivary glands secrete a fluid which is hypotonic to blood plasma but has, as its most striking characteristic, a surplus of potassium. A secretory organ is thus one which transforms the composition of one fluid to that which is secreted, and generally this activity occurs in situations or structures which show enhanced metabolic activity. Brooks and Brooks (24) saw a general parallelism between metabolic activity and growth and the secretion of high internal concentrations of sodium and potassium instead of calcium, as shown by a hibernating animal in different stages of its development and activity.

The point is thus amply made that in nature "the principles of diffusion rarely operate in a free and unrestricted fashion," for the chemical composition of cells and organs is so regulated that the vital machinery may operate. Overton gave early recognition to this idea when he referred to "passive" permeability of cells to denote phenomena which he investigated through plasmolytic experiments, in contrast to what he called "adenoid activity" now more usually referred to as "active" or "physiological" permeability, in which it is recognized that the otherwise slow movement across boundary membranes may be enhanced in ways that are attributable to the physiological activity of the cells (cf., Chapter 1).

3. Some Examples of Active Secretion in Plants

The selective accumulation of weak electrolytes or nonelectrolytes in cells is a general problem in plants. Similar phenomena also exist and have been studied in bacteria. From the work of Gale (68) it is apparent that *Staphylococcus aureus* concentrates glutamic acid, and its cells may be rendered "deficient" in soluble nitrogen compounds (amino acids), so that they may subsequently absorb these compounds from solution. Such "deficient" cells absorb the basic amino acid lysine in apparent excess of the external concentration. This apparently occurs passively, e.g., by adsorption. Glutamic acid, however, is accumulated only if the cells have access to usable and uninhibited sources of metabolic energy.

Although the source of sugars, organic acids, and other organic solutes may often be endogenous rather than exogenous, their secretion to high concentrations into the more aqueous phases of the cells represents "accumulation" in the sense that they are retained even when such cells (e.g., beet in sterile solutions) are placed in direct external contact with water. De Vries showed that such solutes were retained by beet slices for a period as long as several days, but slices of storage tissue (potato tuber and red beet root) have also been leached aseptically with water for very much longer periods (up to 49 days) with only insignificant removal of their diffusible solutes (199).

Parasitic plants [as quoted by Stiles (223)] have osmotic concentrations which frequently exceed that of the host, and, according to Brooks and Brooks (24) galls and tumorous cells commonly accumulate more of certain solutes, e.g., potassium, than the normal cells of the host. Harley and his associates (79) have also shown that in mycorhiza, there is a relationship between the absorption of phosphate in the fungus and in the cells of the host. Also Arisz et al. (7) have investigated the secretion of salts by the glands on leaves of halophytes. These phenomena (discussed in Part III below) again emphasize the generality of the phenomena in question. However, where accumulation, or retention, of organic molecules is being considered, the problem appears strictly comparable to the accumulation of salts only *if the molecular or ionic species does not owe its "accumulation" or retention to the formation of a new and more indiffusible molecular species.*

4. Sites of Selective Accumulation in the Plant Body

In the typical flowering-plant body the problem of salt accumulation is much more diffuse than it is in the animal body and is more readily visualized in terms of the composition of cells than in terms of distinct levels of accumulation as shown, in animals, by the relative composition of intracellular fluids, the intercellular fluids, and the external solution.

Regarding a typical flowering plant body as at Fig. 4 the general situation is as follows.

At the level of the root hairs, the root has just passed out of its elongating phase. Its external surface, bounded by a piliferous layer but

FIG. 4. Diagrams showing internal structure of the plant body of a 2-year-old dicotyledon in relation to absorption and movements of solutes (drawings by M. H. Wilde).

A. Vertical section through plant axis (diagrammatic).

B. Median longitudinal section through shoot apex (diagrammatic).

C. Median longitudinal section through root apex (diagrammatic).

D. Transverse section portion of leaf.

E. Bundle endings in the leaf (apple), showing mesophyll cells and intercellular air spaces in the plane of the leaf surface.

F. Diagrammatic representation of the axis in a dicotyledon showing (a) segment of the axis in the primary region of the shoot, (b) segment of the axis in the secondary region of the shoot, and (c) segment of the root in the region of root hairs.

G. Obliquely tangential section in the cambial region of the stem, showing progressive maturation of secondary xylem elements (*Rhus typhina*).

H. Transverse section cambial region of stem (*Rhus typhina*).

I. Part of transverse section of a dicotyledonous root through the root hair region.

KEY: A. (C) cambium; (Co) cortex (periderm omitted); (E) epidermis; (LG) leaf gap; (LT) leaf trace; (Pet) petiole base; (Pi) pith; (Pr Ph) primary phloem; (Pr X) primary xylem; (S Ph) secondary phloem; (SX) secondary xylem.

B. and C. The above plus: (B) branch primordium; (End) endodermis; (I Met X) immature metaxylem; (I Pro Ph) immature protophloem; (I Pro X) immature protoxylem; (LP) leaf primordia; (Per) pericycle; (Proc) procambium (provascular tissue); (Pro Ph) protophloem; (Pro X) protoxylem; (RC) root cap; (RH) root hair.

D. The above plus: (M) spongy mesophyll; (Pal) palisade parenchyma; (Ph) phloem; (X) xylem.

E. (In S) intercellular space.

F. The above plus: (Pro XV) two vessels of the protoxylem, extending from root through stem into leaf surface.

G. The above plus: (IV) immature vessels of secondary xylem; (FI) fusiform cambial initials; (XV) mature vessels of secondary xylem with pitted secondary walls and tertiary thickening; (XVS) segment of immature secondary xylem vessel, transverse walls still intact.

H. The above plus: (Par) lignified wood parenchyma; (SP) simple perforation in oblique end wall of vessel segment.

I. The above plus: (End) endodermis with Casparian strips and passage cell (Pa C); (R Co) root cortex with air spaces.

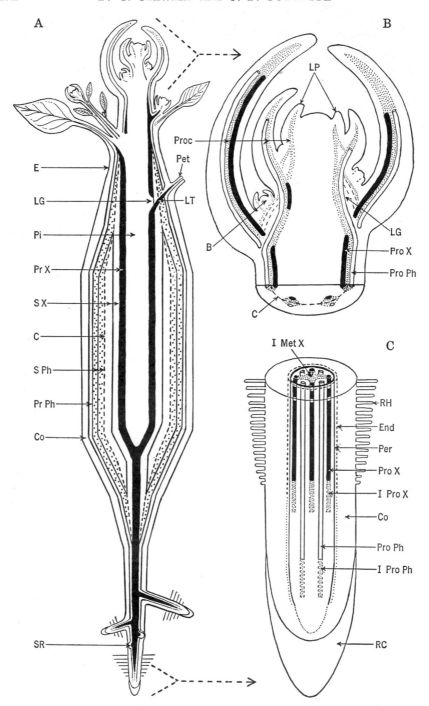

Fig. 4 (*Cont.*). For legend and key see page 271.

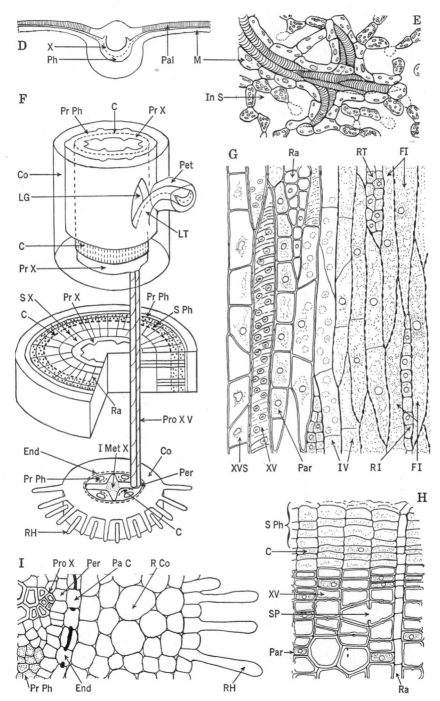

FIG. 4 (*Cont.*). For legend and key see page 271.

free of cuticle, is not clothed by the relatively impermeable layers (exodermis or periderm) which may later develop and, consequently, the living cells of the root cortex may be regarded as in free contact with the external solution, consisting of films of solution around colloidal soil particles or of an artificial culture medium. Each cell of the cortex and of the piliferous layer thus establishes, *independently*, its internal composition and maintains this against the intercellular solution with which it is bathed and against the competitive removal of solutes to other tissues of the plant body. At this very early stage in the development of the root the endodermis is already formed with well-defined Casparian strips.

In the region of mature tissue in the root, the endodermis may be regarded as a cylindrical sheath of living cells occupying the meshes of an impervious network of Casparian strips and relatively unbroken by passage cells occurring opposite protoxylem elements. This tissue thus separates the solution perfusing the cortical cell walls from the solution which, freely permeating the xylem elements, is contained within the stele, and it is thus in a position to control secretion into the stele and leakage outward from the xylem (see Fig. 4, I).

Without prejudging the case for movement of ions in either xylem or phloem, or the mechanism by which transfer to lateral organs is achieved, one may recognize that water and solutes must reach the shoot via the vascular strands and that, via the elongated elements of the bundle endings in the leaf, living parenchyma may again be conceived as bathed by a solution against which they maintain their internal solute concentrations; the extent to which these walls are flooded with solution will depend upon the interplay of evaporation from the leaf and supply via the vascular tissue (Fig. 4).

At the apex of shoot and root, the organized vascular tissues do not penetrate to every cell of the growing regions, but they terminate well behind, leaving a finite distance of unspecialized cells still to be traversed in order that the actively growing cells may receive both water and salts (Fig. 4). Also, in shoots special attention needs to be given to the accumulation of solutes in the growing cells of the cambium cylinder and the adjacent differentiating elements of xylem and phloem (see Part III of this chapter) and its role in the internal economy of the plant body (Fig. 4).

Obvious and important differences between the organization of monocotyledons and dicotyledons may also have a bearing on this discussion, especially the distribution of meristematic regions as potential centers of growth and accumulation of solutes. The absence of a vascular cambium in the monocotyledon is associated with a distinctive vascular anatomy which will influence the distribution of salts within the shoot. The presence of intercalary meristems at the bases of leaves of monocotyledons is associated with marked differences in their form and method of growth. This in turn may well affect the distribution of salts; in the leaf, for example, the younger cells at the base of a monocotyledonous leaf may accumulate more than those at the tip. Despite these different anatomical patterns there is, however, no reason to believe that the basic processes of ion accumulation in the cells are different in the two kinds of angiosperms.

Thus it is not enough to understand how each cell achieves its internal composition and maintains it against the liquid with which it is bathed, for it is also necessary to understand the relations which exist between the different tissues and the organs of the plant body, and the ability of one to contribute solutes to, or to be deprived of solutes by, another.

At the end of the translocation path, solutes are accumulated in cells, that is, at the "sink." Before they can be translocated they equally need to be released from cells at the "source." Therefore, a different problem, not encountered in single cells, arises. What stimuli determine the release of solutes at the source and their subsequent movement to, and accumulation in, the cells at the "sink"? To what extent will the distribution and rate of movement be determined by the facility with which solutes are released in the one place and stored in the other? This problem overlaps the material in Chapters 5 and 6 of this volume. Although old ideas, patterned somewhat on the analogy of the circulation of the blood, suggested that the inorganic ions as "unelaborated foodstuffs" moved upward with the water in the xylem and were circulated backward with "elaborated foodstuffs" in the tissues of the phloem, the problem can hardly be so simply conceived today.

To form an adequate picture of the plant body as an integrated nutritional system it should be possible to visualize the following events in sequence.

(a) The accumulation of ions in various regions of the root in relation to its morphology and manner of growth.

(b) The movement of ions from the cells in which they were initially accumulated into the tissues of the stele in which longitudinal movement over long distances occurs. Here the problem needs to be conceived in terms which give due regard to the anatomy of the roots in question and pay particular attention to the endodermis.

(c) Having penetrated the vascular tissues, the manner in which longitudinal movement occurs over long distances requires to be specified.

(d) The physical state of the ions, while in process of movement, needs to be specified. For example, are the ions in free solution in water? Do they move under gradients determined by their several concentrations or activities, or are they, as it were, swept along by a moving mass of water in elements which do not interpose impermeable protoplasmic barriers?

However, before these larger problems can be analyzed, it is essential to understand how the individual cell regulates its composition against the much more dilute external solution with which it is usually bathed.

D. Composition of the Media from Which Plant Cells Accumulate Solutes

1. Sea Water and Fresh Water

In different organisms, plant cells accumulate their salts from solutions of very different composition. The extremes are illustrated by sea water and fresh pond or river waters on the one hand, and typical culture and soil solutions on the other. A range of composition of such fluids is indicated in Tables II and III.

TABLE II

The Composition of Some Biologically Important Fluids

Fluid	Na	K	Ca	Mg	Cl	NO_3	SO_4	H_2PO_4
Composition of sea H_2O (gm equiv. per liter)	0.480	0.010	0.011	0.054	0.557		0.028	
Hoagland's solution: (mg equiv. per liter)		6	10	4		15	4	1
A sap of potato tuber (mg equiv. per liter)		69.7	13.0	42.1	15.8		20.1	10.0
Relative composition of sea H_2O (Na = 100)	100	3.6	3.9	12.1				
Relative composition of serums:								
Dogfish	100	4.6	2.7	2.5				
Dog	100	6.9	2.5	0.8				
Man	100	6.1	2.7	0.9				

First, sea water is relatively uniform in its composition wherever it occurs. The familiar idea that life originated in the sea gained credence because the ionic ratios between principal cations are very similar for sea water and blood serum, e.g., of the dogfish, the dog, and man (Table II). An exception is the element magnesium, which is more concentrated in sea water relative to sodium. Except for magnesium, blood serum resembles diluted sea water.

TABLE III
DISPLACED SOIL SOLUTIONS[a]

Soil	Water content	pH	NO$_3$	HCO$_3$	Cl	SO$_4$	PO$_4$	Ca	Mg	Na	K
7 A[b]		7.4	116	83	0	438	1.1	189	71	33	16
B[c]	16	7.6	1781	73	55	454	1.2	672	134	75	38
S[d]		7.2	1468	69	313	184	3.3	547	112	123	39
7 A: Apr. 30, 1923	10.7	7.4	174	83		655	1.1	283	106	49	24
Sept. 4, 1923	12.5	7.6	58	155		432	0.6	193	47	40	9
Apr. 28, 1924	14.2	7.6	222	142		571	0.6	296	67	52	11

The table column header "Parts per million in displaced solution" spans NO$_3$, HCO$_3$, Cl, SO$_4$, PO$_4$, Ca, Mg, Na, K.

[a] Reproduced from Hoagland (86); data from Burd and Martin (31). For similar comparisons on six other soils see the original papers.
[b] Cropped soils.
[c] Soils followed cropped one year; cultivated and irrigated, four years; merely watered, four years.
[d] Soils stored in air-dry condition brought to uniform water content.

TABLE IV
SAP COMPOSITION OF VARIOUS COENOCYTIC VESICLES—DATA TO 1936[a,b]

Ion	Sea H$_2$O	Valonia macrophysa (Bermuda)	Valonia ventricosa (Florida)	Valonia utricularis (Naples)		Halicystis ovalis (Monterey)	Halicystis osterhoutii (Bermuda)
Cl	0.580	0.597	0.608	0.419[c]	0.639	0.543[d]	0.603
Na	0.498	0.09	0.035	0.152	0.266	0.212[e]	0.557
K	0.012	0.5	0.576	0.291	0.372	0.318[e]	0.006

[a] For ultimate original sources of these data see Osterhout (153) and Collander (42).
[b] All data in equivalents per liter.
[c] Old data (138a) originally expressed as salt concentrations; recalculated in terms of ions.
[d] Chloride determined directly by Hollenberg (95a).
[e] K and Na ions expressed as percentage of chloride by Brooks (21), calculated values from Hollenberg's figure (95a).

However, *Valonia ventricosa*, a member of the Siphonales, contains a large central cavity, the sap of which has frequently been analyzed; this differs from the composition of sea water in a very remarkable way. While its total salt content is not greatly in excess of that in the sea, the relative proportions of the cations are quite different, as shown in Table IV.

Fresh pond water, and brackish water, in which certain *Nitella*

TABLE V

THE SAP COMPOSITION OF VARIOUS SPECIES OF *Nitella*
AS GROWN IN MEDIA OF DIFFERENT COMPOSITION[a,b]

Nitella species	Na	K	Mg	Ca	Cl
Nitella clavata sap[c]: Max.	86	59	22	19	107
Min.	40	44	11	12	101
Medium	1.2	0.05	3.0	1.3	1.0
Accumulation ratio (max.)	71.6	1180	7.3	15.3	107
N. flexilis sap[d]	63	58		17	131
Medium	0.17	0.03		0.22	0.04
Accumulation ratio	390	1900		77	3300
N. gracilis sap[d]	8	128		32	194
Medium	0.26	0.01		0.65	0.12
Accumulation ratio	31	13000		49	1600
N. hyalina sap[e]	97	52		8	184
Medium	43	0.9		2.4	50
Accumulation ratio	2.2	57		3.3	3.6

[a] Cited from Collander (42).
[b] All concentrations in milliequivalents per liter.
[c] Original data from Hoagland [see Collander (42), page 319].
[d] Original data from Collander [see (42), page 320].
[e] Original data from Collander [see (42), page 319].

species may grow, represent much more dilute solutions; and the plant in question contains in its large internodal cells a solution which has been analyzed (Table V). The table shows the principal ions which are accumulated. In fact, all ions present in the external medium, with the notable exception of nitrate, may be accumulated in these plants.

2. Soil and Culture Solutions

The fluid which bathes colloidal soil particles may be displaced by various means; this has been analyzed and shown to vary in composition with the season of the year and in response to the interplay of the removal of ions by plants and their resupply from the insoluble phases of the soil. Representative data of Burd and Martin (31), showing the range of composition in this soil solution, are given in Table III. Numerous artificial nutrient solutions have been devised, but one may take the composition of Hoagland's nutrient solution to represent a typical fluid in which higher plants will grow (Table II).

3. Interstitial Fluids

Typical analyses of the sap expressed by pressure from plant tissues reveal that potassium is the principal cation and chloride and organic acids are the principal anions (Table I). These are usually present in the internal fluids, e.g., of potato tuber and barley (*Hordeum vulgare*) root, in concentrations far higher than those of the typical nutrient solution or in the typical soil solution.

Only the cells at the external surface of the root are in direct contact with the external solution, and only indirect evidence may be cited to indicate the composition of the fluid which bathes the cell walls of the remainder of the plant body. Assuming that this fluid is similar in composition to the fluid present in the xylem, one may cite the known composition of xylem sap. An early attempt to determine the composition of the xylem sap, on the view that this fluid nourished the

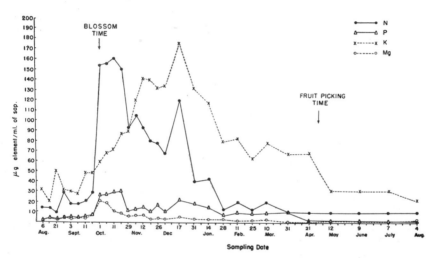

FIG. 5. Elementary composition of the xylem sap of the apple tree showing seasonal variation in New Zealand. From Bollard (16).

shoot, was that of Priestley and Wormall (164). The evaporated material from 100 liters of the bleeding sap of a vine, the shoot of which was cut at its base, yielded surprisingly little of either inorganic or organic solutes. Only 1.56 gm of dry matter per liter of this solution were obtained, and one-third of this was organic. Therefore, the salt content of this fluid was of the order of 1.0 gm of salt per liter. Calculating this in terms of potassium chloride, this solution would be of the order of 15 meq per liter, whereas the potassium chloride content in cells is commonly of the order of 100 meq per liter of expressed sap.

Similarly Wilson (247) analyzed the liquid that emerged from hydathodes under conditions of guttation and showed that it contained about 1030 ppm (parts per million) of solutes in the case of maize

(*Zea mays*) and 200–500 ppm in the case of timothy (*Phleum pratense*). The organic constituents contributed more to these concentrations than the inorganic constituents, so that the salt concentrations were low. A more elegant method of determining the composition of the fluid contained in the xylem is the analysis of the solution extruded by pressure from the vessels present in an isolated segment of woody stem (2, 11). This has been done for the so-called tracheal sap, analyzed in the autumn, after growth had subsided, and in May, after recovery had occurred during the winter and before extensive growth had occurred. These figures and the data of Fig. 5 taken from Bollard (16) (see Table VI) show that the liquid present in the xylem elements

TABLE VI

INORGANIC CONSTITUENTS IN TRACHEAL SAP OF PEAR (*Pyrus communis*)[a]

Constituent	Tracheal sap	
	Nov. 10 (ppm)	May 10 (ppm)
Ca	16.6	84.7
Mg	0.8	23.5
K	23.6	59.6
Fe	1.0	2.1
SO$_4$	8.3	31.8
Cl	3.2	4.5
PO$_4$	10.6	25.2
Totals	64.1	231.4

[a] From Anderssen (2).

contains ions at higher concentrations in the spring than in the autumn. This may imply some degree of ionic accumulation by secretion into the xylem sap, at least in the spring, the intensity of which is related to factors, e.g., blossoming, which reflect the general activity of the tree. The concentrations in the sap, however, are still far below the ionic concentrations that are normally obtained in the cells of higher plants. A still further measure of "accumulation" must occur at the point where the tracheal sap is imbibed by, or perfuses, the leaf parenchyma cell walls (see Fig. 4, E).

Thus the fluid inside the cell is typically maintained at higher concentrations of the principal inorganic ions than the bathing solution, and this is shown for the widest range of elements in those fresh-water organisms which grow in dilute solutions.

E. MATERIALS AND EXPERIMENTAL VARIABLES INVESTIGATED

1. Types of Plant Systems

An understanding of the accumulation process is required—first at the cell level, then at the organ level, and finally facing all the complexities of the integrated plant body. In unicells (such as *Chlorella*, yeast, and bacteria) the entire problem is encountered at the cellular level. The problem has been attacked with materials from higher plants of varying degrees of complexity. In many aquatic vascular plants, roots are rudimentary and either leaves or the entire shoot are the main absorbing organs. Such plants as *Vallisneria, Elodea (Anacharia), Lemna,* have therefore been used as experimental materials (see Part III page 434). However, the problem of salt accumulation at the cellular level has been investigated mainly with isolated tissues and organs of flowering plants and with certain coenocytic algae which, by their morphology, offer special advantages. In consequence there are far too few comparative data from different levels of the plant kingdom and not enough is known of the way these properties change during the development of the organism. Manifestly, cells of many storage organs at one period of development absorb and retain their solutes, while at others they release and supply them to other growing regions.

a. Some morphological concepts. The logical starting point for the investigation of nutrition and salt accumulation in the plant body is either a fertilized egg or a spore, for at these stages the individual plant is reduced to a single cell endowed with the ability to grow—spontaneously in the case of the spore, and usually only after fertilization in the case of the egg. Whereas eggs of fishes, or of many invertebrates, can be readily investigated in direct contact with external solutions, the egg as the starting point for the study of nutrition of higher plants presents much greater difficulty. Following fertilization, however, there is an intensive period of growth and development in which zygote absorbs from the fluids of the embryo sac and is nourished by the tissue of the nucellus. Growth of the endosperm, at the expense of the nucellus, often furnishes a mass of special reserve material, which is sometimes cellular and sometimes fluid, for the subsequent nourishment of the embryo (Fig. 6).

In albuminous seeds and fruits, the endosperm persists (as in the case of *Fagopyrum, Ricinus,* etc.) and in this way stores nutrients which are subsequently used upon germination. In exalbuminous seeds (e.g., as in legumes) the nutrient is mainly absorbed during the development of the embryo and prior to the germination of the seed, to be stored in such storage organs as cotyledons, hypocotyl, etc. In these circumstances, cotyledons often represent the first conspicuous absorbing organs of the embryo plant, as they are pushed out into the contents of the endosperm. Even where the embryo is embedded in a mass of endosperm at seed maturity (as in *Fagopyrum* (Fig. 6A)), it is often the

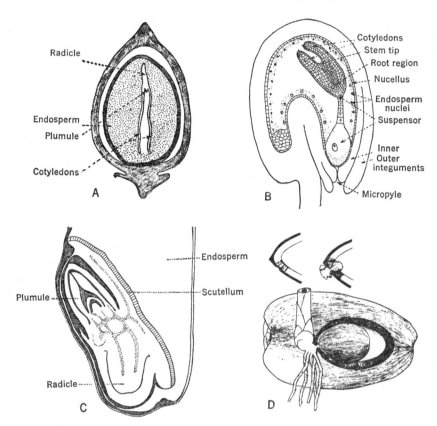

Fig. 6. Cotyledons as absorbing organs for embryos. A. Cotyledons embedded in endosperm of *Fagopyrum*. After Holman and Robbins (95b). B. *Capsella* embryo pushed into endosperm by the suspensor. After Holman and Robbins (95b). C. Embryo of *Triticum*, showing scutellum as an absorbing organ. After Woodhead (247a). D. *Cocos nucifera:* immature embryo embedded in solid endosperm; cotyledon growing into the endosperm cavity. After Kirkwood (110a).

cotyledons that absorb the nutrient from the endosperm and transmit it to the growing embryo on germination. In the familiar caryopsis of grasses the scutellum, whether this is regarded as a cotyledon or not, is such an absorbing organ. In certain cases (e.g., the coconut) Fig. 6D), the development of the embryo is long delayed, and the absorption of the impressive amounts of nutrients which are contained in the liquid contents of the nut (the liquid endosperm or coconut milk or coconut water) awaits the penetration into this cavity, and the growth, of a single cotyledonary structure.

Little is known about the early stages in embryonic development at which the accumulation of organic and inorganic solutes first occurs and whether this differs, in any essential respects, from the process as it has been investigated at later stages in the development. Nor is it known how much significance should be attached to the specialized heterotrophic nutrition of the young embryo in relation to its ability to obtain salts from the parent. In some cases, such as the embryo of orchids and those partially or wholly parasitic plants in which the embryo is often deficient in storage substances or storage organs, the embryos are dependent upon an external, rather than an internal, environment for their nutrition. Clearly more use could be made of such systems in the investigation of the processes of solute absorption.

With the elaboration of a photosynthetic system the processes of absorption and accumulation of organic solutes in cells of autotrophic plants has tended to be considered somewhat apart from the processes by which ions are absorbed. However, in such organisms as yeast and bacteria the absorption of ions, occurring concomitantly with the intake of sugars, or with their utilization at the cell surface, has compelled these functions to be regarded as complementary. Lack of a large aqueous sap vacuole in the microorganisms makes the parallel with the vacuolated parenchyma cells of higher plants somewhat less apparent, but the analogy with meristematic cells may well be closer.

b. Parenchyma cells. In higher plants the process of salt accumulation has been studied most frequently in mature parenchyma. This is because of their conspicuous vacuoles, which contain the bulk of the water of the cell and which through plasmolysis and deplasmolysis may be presumed to contain the bulk of the osmotically active dissolved material. Whereas the vacuole is the presumptive location of most of the osmotically active materials in parenchyma cells, the seat of the osmotic proporties lies in the protoplasmic boundary membrane surfaces. Therefore, brief reference is now necessary (a) to the plasma membranes and (b) to the vacuole, as principal features of the system involved in the process of salt accumulation.

i. Plasma membranes. Essential facts about the protoplasmic membranes may be summarized as follows. (For further details see Volume IA, Chapter 1, and Chapter 1 of this volume.)

Protoplasm and water are essentially immiscible though protoplasm itself has high water content. At surfaces of contact between protoplasm and other media, membranes invariably occur—their normal existence, and spontaneous formation when naked protoplasm is extruded or otherwise injured, are well known. The role of surface-active substances in the accumulation of membrane or film-forming substances at the interface is a cardinal feature of all views on membrane composition. While such purely physical phenomena no doubt underlie the spontaneous repair of membranes on injury, the surfaces of protoplasm seem to be bounded by morphological membranes which are thicker than mere interfaces, and they are the seat of mechanical properties, which can be demonstrated by microdissection, and of resistance to the passage of solutes, as can be shown by microinjection. The behavior of oriented molecular films of fats, fatty acids, and of protein at water-air interfaces and the formation of complex films by adsorption one upon another suggest how such plasma membranes might possess all the regularity of structure that is associated with a crystal and yet, according to the degree of compression in the film, possess the properties of either a liquid or a solid. Such regularity of structure is compatible with the ability of the membrane to behave as a system of water-filled pores toward the entry of ions, while it is not inconsistent with the ability of other molecules to penetrate the membrane by solution in its nonaqueous components.

The mechanical stability of the tonoplast has been amply demonstrated by the isolation of free, floating vacuoles. However, plasmolysis equally requires that the integrity of the outer protoplasmic surface membrane, which is a barrier to the plasmolyzing solute, be also recognized unless the cytoplasm has underlying structure which causes the protoplast to contract as a single unit. The outer membrane surface of contact with the medium, being the surface through which oxygen is absorbed and carbon dioxide emitted, may also be identified as a site of intense chemical reactions in the cell.

Dynamic interchanges between the cytoplasm and its boundary surface membrane may well be visualized—of the sort that occurs, for example, when a globular protein molecule is deposited on water, there to be unrolled, with its peptide chain along the water surface and its hydrophobic groups at right angles, and again rerolled if the molecule leaves the membrane and re-enters the cytoplasm. The passage of an ion across the cytoplasm and into the vacuole is obviously, therefore, a complex phenomenon. The ready appearance of an ion in the vacuole, or in the cytoplasm, is not evidence per se that the outer membrane presents little resistance to diffusion for, by selective storage in the inclusions of the cytoplasm, or physical binding to sites of opposite charge, the speed of movement across such an outer membrane may be accelerated.

There is every reason to believe, therefore, that both the outer membrane (plasmalemma) and the inner membrane (tonoplast) make their respective contributions to the passage of ions, but the actual rate of transfer across these barriers will be determined by the mechanism, in cytoplasm or vacuole, which causes them to be stored and not alone by the degree to which the membrane imposes barriers to the free diffusion of the ions in question.

ii. Vacuoles. Briefly, there are two views with respect to vacuoles. According to one, this region of the cell forms wherever there are local accumulations of soluble materials to attract water osmotically and, this having occurred, a new membrane is formed. The new or *de novo* formation of vacuoles in protoplasm by implantation of soluble matter, like crystals, or by various devices which submit cells to changed external environment, have been well known since the time of Pfeffer. Almost since that time there have been those who regarded vacuoles merely as the physical result of this local accumulation of soluble substance; whereas others [for the views of Went and the Dangeards, see Guilliermond (74)] have regarded it as an autonomous organelle of the cell, multiplying by division of its minute precursors in the form of plastid-like inclusions, or at least maintaining the continuity of some vacuole-forming substances. Thereafter vacuoles may fuse to produce the large vacuole of the mature cell (Fig. 7). To De Vries is traceable the concept that the tonoplast or boundary wall of the vacuole is on organelle especially endowed with the ability to secrete water and solutes internally, similar to the manner in which sugar may be secreted across the surface of the leucoplast except that, in the latter case, there is the internal formation of starch.[1]

The parallel between the vacuole and other cell organelles is an apt one. It seems undeniable that the local accumulation of solutes causes water to enter and thus form aqueous inclusions in cytoplasm, as at all

---

[1] Although we now know that glucose-1-phosphate and the enzyme phosphorylase will form, extracellularly, a product which is chemically similar to starch, it is equally true that this observation does not dispose of the organized plastid as a living, cellular inclusion concerned in the plant body with starch formation. There is a characteristic range of form of starch grains which require still other superimposed enzymes [the branching enzymes of Peat and Whelan (160)] for synthesis, and starch as it occurs biologically in the form of its grains is structurally still more complex, for in its laminations it incorporates still other constituents. It is easy to see in the leucoplast the source of the regulatory control of starch hydrolysis and synthesis; and it is evident that through the multiplication of these bodies by division and their transmission from cell to cell at mitosis these properties continue through and life cycle and may be transmitted in heredity.

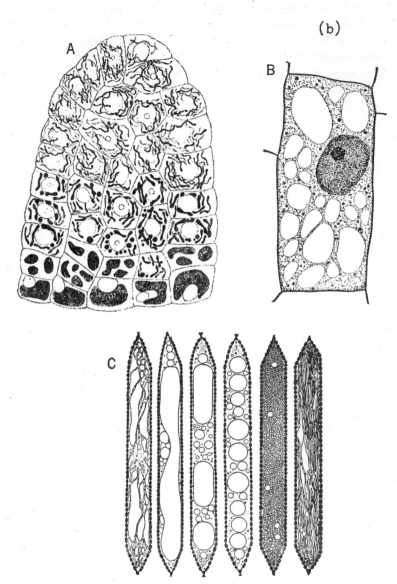

FIG. 7. The vacuome and its development. A. Development of the vacuome in rose (*Rosa*) petals. After Guilliermond (74). B. Vacuoles in a parenchyma cell. After Sharp (195). C. Various forms assumed by vacuoles in fusiform cambium cells of locust tree. After Bailey (9a).

surfaces of contact between water and protoplasmic membranes. It is equally clear, however, that *in their development* the vacuoles are preceded by minute cellular inclusions, rich in basophilic substances and that the vacuole as it occurs *normally* is the outgrowth, by swelling, of these organelles. At maturity, or full expansion, such aqueous sap vacuoles, which may occupy almost the entire volume of the cell, are not only to be regarded as surrounded by the whole cytoplasm as a complex membrane system, but they are also limited at the vacuole-cytoplasm surface by the tonoplast. This membrane is a remnant of the inclusion within which the vacuole is formed and is also the most stable of the known cytoplasmic boundary surfaces. An original experiment of Seifriz (194), since repeated by Chambers and Höfler (36) and Plowe (162) among others, shows how isolated vacuoles stripped free of the bulk of the cytoplasm may be obtained freely suspended in water and bounded by the tonoplast and apparently no other membrane. Such systems have been observed to retain their solutes and to respond osmotically to changes in the external medium, but the extent to which they metabolize is unknown.

Thus the vacuole itself has a long and complex developmental history, from the first-formed minute colloidal deeply-staining bodies, which pass through recognizable morphological stages as they swell and absorb water until, by coalescence, the large central vacuoles of mature parenchyma cells appear.

These events have been described in detail by the French school of cytologists, who regard the whole vacuole system—the vacuome—as an organelle of the cell. The vacuome passes through recognizable and complex morphological changes as cells, e.g., of the cambium, successively become active and undergo rest (Fig. 7, C). It is, therefore, an oversimplification to regard a vacuole merely as a more or less accidental internal drop of water, for it has a morphological history all its own (Fig. 7). In the sense that all protoplasm, even of meristematic cells, contains the precursors of vacuoles, all cytoplasm is vacuolated. Thus the solutes ultimately reside in vacuoles; the machinery to build up *de novo* the concentrations they contain, and to maintain these across the boundary surface of the protoplasm, must obviously reside in part, if not wholly, in the cytoplasm within which this organelle also developed.

c. Coenocytic vesicles. To demonstrate accumulation of ions in a fluid enclosed within a layer of cytoplasm, the situation in certain coenocytic structures has been exploited. In notable examples drawn from the Siphonales and from the Charales, as in species of *Valonia* (Fig. 8, a and b), and in the internodal cells of species of *Nitella* (Fig.

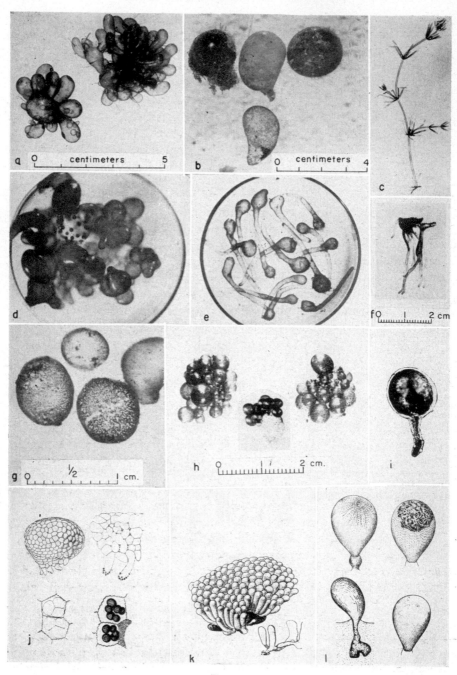

Fig. 8.

8, c) and of *Chara*, it is possible to obtain the liquid from the central cavity uncontaminated with cytoplasmic constituents. Since much later reference will be made to the analysis of these fluids, it is pertinent to refer to the morphological nature of the sap-filled cavity.

i. The sap cavity of coenocytes. In the first place, these cavities are only comparable to normal vacuoles of parenchyma cells in the sense that they are enclosed by a cytoplasmic sheath. Since the cytoplasm is multinucleate, these structures represent, not single, but multiple protoplasts. This is especially true of the much investigated vesicles of *Valonia*, in which the thin sheath of cytoplasm, within the complex laminated cellulose wall, contains nuclei which are not further distant from each other than they would be in most cellular tissues (56). The cytoplasm contains other inclusions, notably chloroplasts, and in some species, particularly *Valonia ventricosa*, this protoplasmic mass readily fragments into its individual protoplasts, forming separate sporelings (aplanospores) (Fig. 8, d, g) which can grow in the cell cavity, and from which, by special methods, the entire organism (Fig. 8, h) has been cultured (207).

These considerations raise the question whether the central cavity of these large vesicles, which in the case of *Valonia ventricosa* (Fig. 8, b) may contain as much as 30 ml of fluid in a single vesicle, is the true homolog of the vacuole of parenchyma cells. It seems that the solution in these large vesicles, which is almost entirely composed of potassium and sodium chloride, is much more comparable to an intercellular fluid such as that which is surrounded by a sheath of living cells (the endodermis) in the stele of higher plants. On this view the true homologs of the vacuole of higher plants would be more minute cytological inclusions in the cytoplasm of *Valonia*, where the primary process of accumulation may occur, to be followed by a later secretion into the central sap cavity. The *V. ventricosa* vesicles can only be grown from

---

FIG. 8. Various coenocytic vesicles which contain sap. (a) *Valonia macrophysa*. (b) *V. ventricosa*. (c) *Nitella flexilis*. (d) *V. ocellata:* vesicles in a petri dish showing aplanospores. (e) *V. ocellata:* germinated aplanospores in petri dish. (f) *V. ocellata:* showing branched rhizoidal processes. (g) *V. ventricosa:* vesicles with aplanospores. (h) *V. ventricosa:* aplanospores grown for months on marble in sea water. (i) *V. ventricosa:* aplanospore with rhizoidal processes which penetrate into calcareous substratum. (j) *Dictyosphaeria* sp.: diagram showing septate thallus and aplanospores within a compartment of the thallus. (k) *V. utricularis:* diagram, palisadelike mass of vesicles branched at the base. (l) *Halicystis ovalis:* diagram showing vesicle as growth of one year and areas from which swarmers are shed. (a) and (b) from Steward and Martin (215); (c) Hoagland (86); (d–i) from Steward (207); (j) after Borgeson, from Fritsch (67); (k) Oltmanns (147); (l) from Kuckuck (114).

aplanospores, which send rhizoidal processes (Fig. 8, g and i) into the substratum (e.g., marble) on which they are sown. This also raises the question whether these processes and the thin-walled hapteron cells may not be the functional absorbing organs, the ions being subsequently secreted into the vesicle. Protoplasmic streaming in the hapteron cells, which is never seen in the vesicle, again suggests that these cells are more physiologically active than the main vesicle. Such conceptions of *Valonia* do not rob these organisms of interest, but they do place them in a different relationship to the whole problem.

The Valoniaceae and related plants present a graded series of morphological situations. *Dictyosphaeria* (Fig. 8, j), encloses (within a sheath of cells in which the individual protoplasts are separated by septa) a similar sap cavity, which may contain fluid and accumulate ions in the manner found for *Codium* [cf. Brooks (22)]. In this case the external sea water is separated from this central cavity, not by a single cytoplasmic layer, but by an envelope divided by septa into compartments, each of which contains a multinucleate protoplast. Indeed within single cavities of this structure, sporelings may develop. When the vesicular structure of *Dictyosphaeria* becomes larger, it may eventually break, and the central cavity then becomes continuous with sea water without, in this case, causing the protoplasts of the envelope to perish. Thus, these large vesicles are only comparable to single cells in that the entire structures happen to be enclosed within one cellulose wall: the true homologs of the parenchyma cell are probably to be found in the *individual protoplasts* into which the structure of this colony, or coenobium, readily fragments.

After much earlier investigation upon *Valonia* species, it was possible to show that the first accumulation of radioactive ions occurred *in the cytoplasm*, possibly in, or on, inclusions in the cytoplasm, and it occurred only secondarily in the central sap cavity (23). Evidence bearing on this point was also obtained by Hoagland and Broyer (89) with *Nitella* (see also page 328).

ii. Rates of growth of coenocytic vesicles. In spite of the attractive feature that the size of these organisms permits uncontaminated sap to be extracted, they have some disadvantages. The vesicles grow relatively slowly (215), and the sap once formed is not normally subject to rapid change in composition. The exact age of the largest vesicles of *Valonia ventricosa* is not known, but, since the vesicles *in situ* are commonly much encrusted with epiphytes, their slow growth and longevity seems certain.

Thus, while these vesicles are suitable for the demonstration of the composition of the sap *after* absorption has occurred, they are not as satisfactory for the study of the accumulation in progress, since this occurs slowly under natural conditions that are difficult to duplicate. The interest in *Valonia*, therefore, centers mainly around the remarkable discrimination between K^+ and Na^+, rather than the total accumulation of salt. There is some evidence, however, that discrimination between K^+ and Na^+ declines with age (215); this could be interpreted in various ways. Either the vesicles may, increasingly with age, permit Na^+ to enter passively, by diffusion and exchange, or there may be an active excretion of Na^+ which decreases with age of the vesicle. Other secondary changes may occur, or be induced (e.g., as in the presence of NH_4^+) but these are superimposed upon the sap composition as it developes during growth. Wherever such induced changes result in movements of ions which are consistent with physical diffusion, they are, however, of little significance in relation to the problem in question. What may be called the "primary process" by which solutes are accumulated *de novo* differs from these "secondary" or "induced" changes in the important respect that by its very nature, the former is irreversible whereas the latter are, to some extent at least, reversible.

d. The use of tissue aggregates. As investigators depart from the use of single cells, or coenocytes, new problems arise. Tissue masses, made up of single cells too small for individual manipulation or analysis, have principally been obtained from such massive plant storage organs as the tubers of the potato, or the Jerusalem artichoke (*Helianthus tuberosus*), the enlarged stem of kohlrabi (*Brassica caulorapa*), or the storage roots of such plants as carrot, beet, dahlia, radish (*Raphanus sativus*) or they have been obtained by the use of organs such as excised roots. These experimental materials introduce obvious problems, due to the morphology of each organ: the surface volume relations of the experimental tissue, as determined by the thickness of the experimentally cut disks or cylinders; the extent to which the observed behavior is modified by the prior treatment which prepares the tissue for experiment, or by its response to the factors inherent in the transfer of the tissue from its situation in the natural organ, or environment, to the solutions in which it is immersed for experimental observation. Other features of the use of such tissue masses now need to be mentioned.

i. Diffusion in cell masses. When tissue aggregates are exposed to experimental solutions, the solute in question first moves by diffusion up to the protoplasts which it eventually enters. Where solutes enter the protoplast with relative difficulty, the principal channels for such diffusion into the tissue will be the liquid films which permeate the cell walls and in effect lie outside, and between, the protoplasts of adjacent cells. Thus, unless there are active mechanisms promoting rapid transit across unspecialized parenchyma cells, it may be asked how far diffusion *alone* will suffice and how far the uptake may be limited by the restricted channels which are available for diffusion.

Diffusion of solutes across thin slices of physiologically inactive storage tissue is indeed slower than in water alone, because the principal barriers to free diffusion are the living protoplasts. When induced to contract and eventually to withdraw from the cell wall (as by the use of higher osmotic pressures in the permeating medium), thus allowing water to infiltrate, the effective channel for diffusion is increased. It can be shown that, after correction for the viscosity of the medium, the amount of solute that diffuses across a thin slice of unspecialized tissue is related to the effective area for diffusion, which is represented by the liquid that lies between adjacent protoplasts. Also from work of this sort (200) the apparent diffusion coefficients of the solutes that diffuse through such tissue masses have been calculated and shown to be the order of 1/100 of that which prevails in water for nonelectrolytes, and it is even less for some electrolytes, e.g., phosphate. Electrolytes are impeded by the charges on the cell walls or cell surfaces. Whereas the diffusion coefficient for glucose in water is of the order of 10^{-5} cm^2 per second, the corresponding figure for diffusion across slices of plant tissue is of the order of 10^{-7} cm^2 per second. Smaller molecules, such as those of pentose sugars, diffuse more rapidly through tissue than larger molecules such as hexoses and disaccharides, but the differences between them are exaggerated in comparison with their respective diffusion rates in water. This is a familiar effect in diffusion through collodion membranes, where movement occurs through pores of near molecular dimensions.

A similar problem faced Osterhout in 1921 (148) in the use of the electrical conductivity of cells and tissues to measure permeability. Osterhout reached the conclusion that at least one-quarter of the direct current conducted by *Laminaria* sp. tissue passed around, rather than through, the cells in question. In fact, the disks of *Laminaria* frond are anatomically so complex that there may well have been still other complicating factors.

Diffusion studies (200) with tissue slices showed quite clearly that a solute could traverse a layer 0.33–0.66 mm thick by diffusion, without appreciably entering the cells of which it was composed. The paradox presented by the observation that a particular solute (bromide) could do this, although it was known to accumulate in other cells (*Nitella*), led to the eventual understanding of the physiological conditions necessary for the solute to accumulate. The diffusion experiments had been made in such a way that the tissue, though exposed to solution and also in contact with air, was in a closed system *without supplementary aeration.* This led to the investigation of the role of aeration in the accumulation of potassium bromide in thin submerged disks of potato tissue. The work on diffusion of solutes across layers of cells also raised other problems now to be considered.

Where extensive movements of solutes must occur across regions which are not endowed with special tissues for longitudinal transport (see Fig. 4, C and I), as e.g., across the root cortex, from the bundle ends to parenchyma cells in leaves, from the latest vascular strand of the axis to the dividing cells of the apex of shoot and root, in the deposition and removal of solutes from parenchyma in storage organs, etc., one may well need to invoke some more active mechanism of transfer than the very low rates of physical diffusion, which are compatible with the apparent diffusion coefficient of solutes through membranes of isolated, physiologically inactive cells. Indeed various workers [Arisz (4); Crafts and Broyer (51)] have revived the idea that movement of salts may occur from cell to cell through parenchyma in the cytoplasm via the protoplasmic connections. Such a movement could be an active, or a passive, one (see also Part III of this chapter).

ii. The barrier to diffusion: "Apparent free space" in tissues. On the classical view, the cytoplasm is bounded by the plasmalemma and by the tonoplast, and both these regions represent barriers to free diffusion. While the stability of the tonoplast is more clearly established, many facts of microinjection and microdissection indicate the existence of an outer membrane also, although cytological demonstration of a plasmalemma in vacuolated cells is not always possible. Acid dyes, for example, will perfuse the protoplast when they are injected into it, but will not traverse the undamaged plasmalemma if supplied in the external medium. If the outer membrane is destroyed or the cell placed in the condition which Pfeffer termed "rigor," then such penetration occurs.

Although this position seemed well established, a recent concept has arisen of "apparent free space" which signifies the volume in a cell, or tissue, into which solutes (ions or molecules) can apparently penetrate freely (20a). This concept appears in its most extreme form when it is suggested that entering solutes penetrate and perfuse the cytoplasm until they encounter a main diffusion barrier at the vacuolar surface. The idea originates from observations on the amounts of salts absorbed by, and easily lost from, tissues under conditions of reduced metabolism. The "apparent free space" is calculated on the assumption that the ions absorbed passively remain at the same concentration within the tissue as they are in the medium. However, it is evident that these ions (especially cations) in fact become concentrated within the space thus accessible to them, if only by adsorption and by Donnan equilibrium effects, so that the volume of tissue actually penetrated is invariably smaller than the "apparent free space" as calculated. The rapid initial uptake of cations which occurs when a tissue low in salt is transferred to a solution containing ions is a well-established phenomenon. Inasmuch as it involves diffusion of salt into intercellular spaces, and adsorption onto cell wall constituents, this presents no problem. However, more cations may be absorbed passively than can be easily accounted for in this way [(229); see also page 393], and it has been concluded that the base-exchange capacity of at least a part of the cytoplasm is also concerned. Thus, even in the passive entry of ions into the tissues, different sites which hold them with different degrees of tenacity are involved. This conception suggests that cations may reach the binding sites of the cytoplasm without the intervention of metabolic energy.

The passive movement of anions across the plasmalemma is less clearly established. The amounts of an anion, such as chloride or bromide absorbed under conditions of inhibited metabolism are very low, but measurable [e.g., Butler (32) and Sutcliffe (229)]. Most, if not all of this, can be accounted for in the intercellular spaces and cell walls of the tissue. Even if penetration of the outer membrane is possible, only a small number of anions would be expected to move into the cytoplasm owing to the predominance of indiffusible anions already present and to resultant Donnan effects. In support of the idea that passive absorption of anions does not occur, Arisz (6) has obtained indirect evidence with *Vallisneria spiralis* which may suggest that transport of chloride both across the plasmalemma and tonoplast requires an expenditure of metabolic energy.

iii. Thin disks. From the considerations described above obvious advantages accrue from the use of thin disks cut from tissue masses; such disks may place the maximum number of cells in direct contact with the external medium. Potato tuber tissue disks of the order of 1 mm thick represent 8–10 cells, and ideally even thinner disks should be used. Owing to difficulties of manipulation, disks thinner than about $\frac{1}{3}$ mm are not feasible. A great deal of work has been done with disks about 1 mm thick, but, by the use of disks of graduated thickness, one may extrapolate to the behavior of cells at the tissue surface, as will later be described. The surface cells are the seat of distinctive phenomena which bear upon their ability to accumulate (see Fig. 9, A and B).

e. Excised roots. Recognizing the interplay between the photosynthetic regions of the shoot and the accumulation of salts by the root, investigators have had recourse to excised roots from plants grown in water culture. Such work has intrinsic interest because of the role of the root as the absorbing organ for the intact plant. However, these organs when isolated also present certain complications. Even though it is retained in its culture medium, the subsequent behavior of an excised root is profoundly modified by the removal of the shoot. This is not attributable only to the removal of the sugar supply, which can be furnished externally under aseptic conditions. Excised roots, especially of monocotyledons, rapidly cease to grow, and to retain them in the continually growing state requires the familiar but special techniques of organ culture. These points need to be borne in mind in the interpretation of any experiments on excised roots where the impact of the processes of growth and development on the process of salt accumulation is in question.

The techniques which were devised by Hoagland and Broyer (87) for the investigation of salt accumulation in excised roots have special advantages. In this technique the roots were obtained from plants (barley) of known nutritional status, a condition which was achieved by allowing them to deplete a limited volume of external nutrient solution. In this way, and in the normal competition between shoot and root, the shoot depletes the root of its absorbed inorganic salts, so that, *for a strictly limited period* and under specified environmental conditions, the excised root has a marked ability for renewed salt absorption. These "low-salt roots," which have also been shown to be roots high in organic solutes, particularly sugar, have therefore the great advantage that their ability to absorb and accumulate ions is controllable by the conditions that apply during their growth on the plant. However, the consequences of this pretreatment during growth, for the cells that will eventually absorb, need to be more fully recognized (see also Fig. 39). Furthermore, such root systems represent extremely heterogeneous cell populations. Along each root axis there is the graded transition from the organized root apex, with its strictly meristematic regions confined to a mere fraction of a millimeter of root length, through procambial and elongating regions to that of the mature tissues. Where roots branch, new centers of meristematic activity arise in the pericycle. In response to this gradation in growth, there is a graded activity in salt absorption (163, 220). An additional complication is the possibility that absorbed salt may subsequently be exuded from the stele and the cut surfaces, since this is well known to occur when the cut surface is in contact with air (73, 89, 125). While investigations with excised root systems have shed much light on the behavior of the root as an absorbing organ, it is difficult to interpret results obtained with this

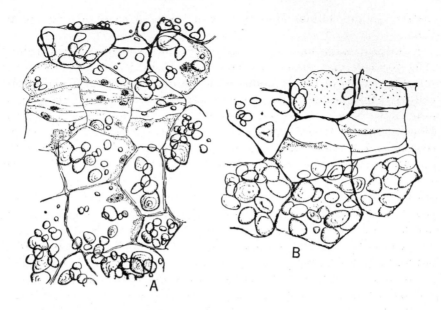

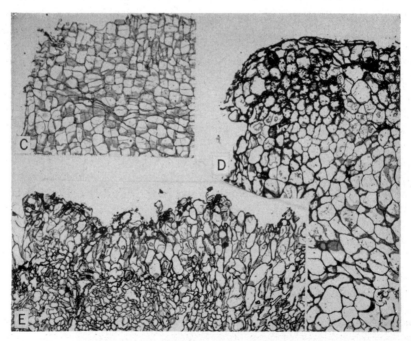

Fig. 9.

material at the cellular level because of the complexity of their organization.

f. Aseptic tissue and organ culture. With keen insight Haberlandt (75) foresaw that many problems of cell physiology would become solvable if cells of higher plants could be individually cultured aseptically and thus brought into a state for investigation in which they could also grow. This ultimate goal has not, even yet, been fully achieved. However, much is now known about the continuous culture of certain root tips and about the continuous culture of tissue explanted from certain storage organs, notably organs of secondary origin such as the carrot root and the Jerusalem artichoke tuber. In such cases it is now possible to submit tissue to the conditions in which they will either enlarge without dividing or grow indefinitely, in liquid media, by cell division (Fig. 9, C, D, and E). In some cases (explants of carrot and artichoke) the cultured tissue grows to form a more or less randomly proliferating mass, and in others, excised roots of tomato (*Lycopersicon esculentum*), it forms a continuously growing organ culture, i.e., a typical root system, by the prolonged activity of the explanted tip. Even so all roots do not grow with equal ease. Whereas some dicotyledonous roots can be cultured indefinitely, excised monocotyledonous roots, even when given the full advantage of known treatments that are successful for the culture of root tips in other plants, still fail to grow continuously. The most successful monocotyledon root culture yet achieved is that of Roberts and Street (169).

More applications of these systems to the study of salt accumulation can obviously be made. However, Haberlandt's ultimate goal of a system composed of free-floating cells from higher plants, which can grow singly and divide, is still to be realized; although progress in this direction by the modification of standard tissue culture techniques is now being obtained (Fig. 9, F and G).

Such a free-floating population of angiosperm cells, whose growth by division or cell expansion could be controlled by known nutrients,

---

FIG. 9. Cells of storage tissue in relation to salt accumulation.

A. Cells at surface of potato disk in air showing starch disappearance, divisions parallel to the surface and effects penetrating several layers from the surface. From Steward *et al.* (222).

B. Cells at surface of potato disk in aerated solution — events shown at (A) compressed into much narrower zone of tissue. From Steward *et al.* (222).

C. Section of carrot phloem explant from carrot root. From Steward *et al.* (213).

D. Section of carrot phloem explant from carrot root after growth in basal medium plus casein hydrolyzate, showing cell enlargement (213).

E. Section of carrot phloem explant from carrot root after growth in basal medium plus coconut milk showing cell divisions (213).

growth factors, and hormones, would represent a great advance in the systems now available for the investigation of salt absorption and accumulation. Figure 9 shows the varied responses of parenchyma cells under different conditions which affect their growth (Steward, unpublished observations).

   g. Vacuole sap and expressed sap from tissue aggregates. In the use of cell aggregates, the determination of the actual concentrations of solutes in the internal solutions presents a problem. Analyses of the expressed sap have been made on the implicit assumption that this represents both the bulk of the water of the cell and of the dissolved substances. Plausible as this assumption may be, it should still

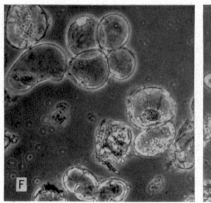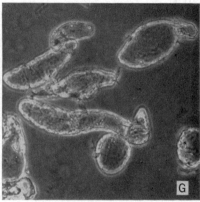

Fig. 9 (*cont.*) F. Cultured free-floating cells from potato tuber. Magnification: ×225, phase contrast.

   G. Cultured free-floating cells from carrot phloem. Magnification: × 225, phase contrast.

be recognized that there may be inclusions in the cytoplasm having solutions at different concentrations from that of the vacuole and, when the sap is expressed, these fluids may mingle.

   From the earliest investigations of sap expression, e.g., by H. H. Dixon as recorded in his well-known work (55), it is known that sap varies in its composition (osmotic pressure) with the speed of expression, and according as the tissue is, or is not, killed prior to the application of the pressure. Complete extraction of water and solutes by pressure was thus recognized to occur only after the permeability properties of the tissue are destroyed—usually by freezing. As so obtained, expressed sap merely indicates the average composition of the total solutes dissolved in the total water of the cell.

   A more novel concept arose from the experiments and ideas of Bennet-Clark and Bexon (13) and of Mason and Phillis (137). These workers reinvestigated the techniques of sap expression in relation to the composition of the issuing fluid. Leaf tissue, with its very regularly oriented palisade cells, could easily be subjected to pressure, in gradual increments, in such a way that liquid could be extruded without breaking the living cell by shearing forces: thus water, relatively free from solutes, could be expelled. Hence the concept emerged that the aqueous vacuole of the parenchyma cell is much less the *sole* repository of the dissolved substances than the classical, simple osmotic view of the cell would imply. It is true that these experiments encounter certain difficulties of interpretation. If water is caused to leave the vacuole under the influence of an osmotic gradient, and if this is done without detriment to the boundary membranes, there is no question that the solutes in the vacuole are retained. If the water is induced to leave the cell by a superimposed gas or fluid pressure, and if this is achieved without mechanical damage to the membranes of the cell, there seems also to be no reason why the solutes in the vacuole should not again be retained so that the issuing liquid approaches pure water, as in fact it does. Nevertheless, the re-examination of vacuole composition in the light of controlled experiments on sap expression, has been interpreted to mean that about 70% of the water in the cell is contained in the cytoplasm, and that the cytoplasmic sap is about five times more concentrated than the vacuolar sap with respect to certain salts. On this view, access to cytoplasmic sap is only obtained when the membrane system is destroyed by shearing forces due to the applied pressure or by killing the cell by other means. However, more usually the cytoplasm of vacuolated cells represents only a small part (10%) of the whole cell volume.

There seems to be no unequivocal way, at present, by which the exact location and concentration of dissolved substances in cells may be ascertained from the evidence of sap expression alone. The existence of various phases in the cell, each with its membrane system and each preserving a measure of independent composition, cannot be denied. The possibility that in cytoplasm there are aqueous inclusions which are able, in a more intimate way than the main vacuole, to participate in the initial process of salt accumulation phenomena also cannot be denied. The best that can be said at present is that *complete* sap expression, with *complete* destruction of the cell membranes, yields evidence of the *average* composition of the *total* solutes of the cell dissolved in the *total water*. Where the vacuole volume is large and the cytoplasmic volume is small, as in most parenchyma cells, there seems little reason to doubt that the expressed sap concentration is a near approximation to vacuolar sap (29).

2. Sap Composition of Plants in Relation to the Habitat

a. *Coenocytes. The concept of "normal" sap composition.* Some striking examples of aquatic plants which, though growing in media of the same composition (i.e., the sea), yet have very different composition have led to the view that the composition of the sap is a characteristic, even a specific, feature of plants. These examples have also suggested that the composition of the sap is more constant than in fact it is and, by underestimating the variations which normally occur, has diverted attention from the knowledge that can be gained by a study of the causes of that variation. Attention will here be focused on the coenocytic vesicles of the Valoniaceae and cells of the Characeae, about which most is known.

Several distinct species of *Valonia* have been investigated from time to time. First the Mediterranean species *V. utricularis* (Fig. 8, k) and two species best known from the West Indies, namely *V. macrophysa* (Fig. 8, a) and *V. ventricosa* (Fig. 8, b). A much less widely known species is *V. ocellata* (Fig. 8, d), to which reference will also be made. As commonly collected in the West Indies, *Valonia* sheds neither zoospores nor gametes, but in *Halicystis* (Fig. 8, l), a vesicle is produced annually and acts as a reproductive structure which eventually floats away. So similar are the vesicles of *Halicystis* and *Valonia ventricosa* that they were first confused at Bermuda, but later the floating specimens were distinguished as *Halicystis osterhoutii*. *Halicystis ovalis* is a species which occurs on the Pacific coast of the United States. These plants, therefore, present an array in which one may compare the composition of the sap from very similar vesicles; this has often been done. Table IV is compiled from figures that were widely quoted by Osterhout and others up to 1936. In Table IV attention is confined to the ions, potassium, sodium, and chloride, which comprise the bulk of the salts in these saps.

Certain features are immediately evident. In all cases the species of *Valonia* accumulate potassium from sea water in preference to sodium, though to a degree which varies with the species. Whereas the saps of *H. ovalis* resembles *V. utricularis*, *H. osterhoutii* has a higher concentration of sodium than of potassium. It is, however, necessary to ask how far these sap compositions, as determined, really are characteristic of the species in question.

To obtain these so-called "normal" sap compositions, the usual technique was to collect, from a sufficient number of vesicles, a single composite sap sample and submit this to the standard procedures of quantitative analysis. This was done by Meyer in 1891 for *V. utricu-*

laris, and few subsequent analyses on this plant have been reported. The favored figures, often quoted by Osterhout, for the composition of the sap of *V. macrophysa* at Bermuda were obtained from a collection of sap made by Crozier in 1919 and submitted to Van der Pyl for analysis. Between 1919 and 1936 these figures were accepted, almost without question, as being representative of *V. macrophysa* as it exists at Bermuda. When *V. macrophysa* was collected in other areas, such as Naples, Italy, the Dry Tortugas, Florida, etc., the composition of the organism in that locality was characterized, in a similar way, by the analysis of a single composite sap sample. Also the composition of *V. ventricosa* became known from relatively few collections, and the favored analyses for this organism, up to 1936, were those often cited by Osterhout and attributed by him to Cooper and Blinks (48).

TABLE VII

Composition of the Sap of *Valonia ventricosa* and *V. macrophysa* at Dry Tortugas, Florida[a,b]

Species	No. of samples	Ion	Mean	Mean ± 2 S.D.	Mean ± 2 S.D. of Mean
V. macrophysa	30	K	0.509	0.569–0.449	0.520–0.498
		Na	0.113	0.167–0.059	0.123–0.103
		Cl	0.624	0.647–0.601	0.628–0.620
V. ventricosa	62	K	0.591	0.621–0.561	0.595–0.587
		Na	0.043	0.067–0.019	0.046–0.040
		Cl	0.628	0.650–0.606	0.631–0.625

[a] From Steward and Martin (215).
[b] Sap concentrations in grams equivalents per liter.

b. Variation in the sap of Valonia spp. Despite the relative constancy of the composition of the sea with respect to potassium, sodium, and chloride, it is hardly credible that the composition of any plant in its natural habitat should be constant. Thus, in any attempt to characterize the organism by the ions accumulated in its aqueous fluids, one needs to derive both the average or mean sap composition and the *range of variation* which the organism exhibits. This can only be done by analyzing as wide a variety of samples as possible from a range of habitats and then calculating the mean sap composition and the standard deviation of the population so investigated. Steward and Martin (215), at the Dry Tortugas, analyzed as many as 62 samples of sap of *V. ventricosa* for potassium, sodium, and chloride, and as many as 30 analyses of sap were obtained for *V. macrophysa.* The essential data obtained are reproduced as Table VII.

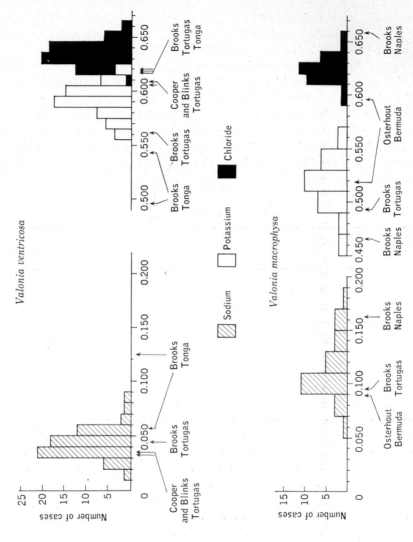

FIG. 10. A. Range of sap composition of *Valonia*. (Sap concentrations in gram equivalents per liter.) From Steward and Martin (215).

The content of each ion is shown as the mean sap concentration for the whole population. The mean ± twice the standard deviation sets the reasonable limits within which single sap determinations could lie even if they were withdrawn from populations not significantly different from the one in question. The mean ± twice the standard deviation of the mean sets the reasonable limits within which the mean sap composition of other populations might lie without indicating that they were essentially different from the first.

The data show that the sap compositions of the organisms as they existed in the sea were subject to more variation than had been im-

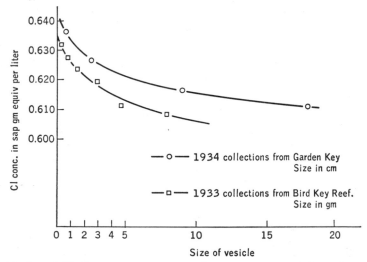

FIG. 10. (cont.) B. Sap concentration in relation to size of vesicle, *Valonia ventricosa.* From Steward and Martin (215).

plied. When represented in the form of frequency diagrams (Fig. 10) several interesting conclusions may be drawn.

The first conclusion is that many of the analyses of single sap collections made from plants collected in various geographical locations fall within the normal range of variation of the plant in question as it occurs in one locality. The second conclusion is that *V. ventricosa* and *V. macrophysa* as they occurred at the Dry Tortugas cannot be distinguished by their chloride concentration. However, the frequently cited enrichment of potassium in *V. ventricosa,* with the consequent reduction in sodium concentration, still stands; for it is quite clear that the two algal populations can definitely be distinguished by the greater content of potassium in *V. ventricosa* and the greater content of sodium in *V. macrophysa.* Even so, however, the two plant populations did show some overlap at their extremes with respect to these ions. *V. ocellata*, composed of much smaller vesicles, is distinguished from other species of *Valonia* because it has a higher chloride concentration (0.704 *M*), the extra chloride being accompanied by sodium rather than by potassium (207).

Having demonstrated the range of variation which these organisms

TABLE VIII

The Composition of the Sap of *Valonia macrophysa* in Relation to Light Conditions during Growth and to the Diffuse Light of Laboratory after Collection[a,b]

Habitat	K	Na	Cl
"High light"[c]	0.531	0.093	0.628
"Medium light"[d]	0.512	0.103	0.616
"Low light"[e]	0.450	0.162	0.621
Collected vesicles in diffuse laboratory light	0.497	0.123	0.622

[a] From Steward and Martin (215).
[b] Mean sap composition in gram equivalents per liter.
[c] All plants in this group received the direct sunlight of late morning or early afternoon.
[d] All plants in this group grew under the shadow cast by masonry and received only oblique light for a briefer period each day.
[e] Plants in this group grew in deep shade.

TABLE IX

Daily Fluctuation in Sap Composition of *Valonia macrophysa* Growing in an Isolated Pool[a,b]

Time	K	Na	Cl
9:30 A.M.	0.453	0.164	0.623
4:00 P.M.	0.471	0.155	0.644
7:00 P.M.	0.427	0.180	0.622

[a] From Steward and Martin (215).
[b] Sap composition in gram equivalents per liter.

exhibit in their habitat, the next problem relates to the environmental factors that are associated with this variation.

   c. Habitat factors which affect the sap of Valonia spp. The "normal" sap composition could be seen to vary with factors in the environment. *V. ventricosa* commonly grows completely submerged, rarely exposed at low tide, in situations where it is exposed to surf and protected from the effects of diurnal fluctuations in the composition of the surrounding medium. *V. macrophysa* on the other hand inhabits various situations in which it is protected from surf but is exposed to wide daily variations in the composition of the external medium (as, e.g., in dissolved oxygen and carbon dioxide concentrations of tide-filled pools) and tolerates varied light conditions.

In the case of *V. macrophysa* the natural sap composition seemed to be richest in potassium and poorest in sodium when the plants grew in

TABLE X

THE COMPOSITION OF THE SAP OF *Valonia ventricosa:*
THE EFFECT OF SIZE OF THE VESICLE ON COLLECTION[a,b]

Average weight (gm)	Cl	K
0.70	0.637	0.617
2.50	0.627	0.593
9.00	0.617	0.590
18.00	0.611	0.570

[a] From Steward and Martin (215).
[b] Sap composition in gram equivalents per liter.

TABLE XI

THE COMPOSITION OF THE SAP OF *Valonia ventricosa*
IN RELATION TO HABITAT FACTORS[a,b]

Habitat	Factor[c]	K	Na	Cl
In cavities on masonry and coquina rock exposed to surf	Exposed situation	0.579	0.044	0.620
On masonry in isolated body of sea water	Protected situation	0.604	0.036	0.635
Collected vesicles from exposed situations adjusted to changed sea water in the laboratory		0.599	0.044	0.638

[a] From Steward and Martin (215).
[b] Mean sap composition in gram equivalents per liter.
[c] Differences between the "exposed" and the "protected" situations and the collected vesicles in the laboratory are significant for K and Na ($P < 0.01$) and for Cl, just significant ($P = 0.05$).

situations furnishing the longest periods of daily exposure to bright light (Table VIII), and, in tide-filled pools, both potassium and chloride content of the sap tended to be greatest after periods of photosynthesis such that the pools were supersaturated with oxygen (Table IX). In the case of *V. ventricosa*, age of the vesicle, as reflected by size, was a prominent factor (Fig. 10 and Table X); and the composition of the sap also reflected the degree of protection or shelter which the natural habitat afforded by the higher content of potassium relative to sodium.

The more varied habitats of *V. macrophysa* are compatible with the wider range of variation in its sap (215). In both cases, however, when the organisms were collected, somewhat dissected to single vesicles (in the case of *V. macrophysa*), and superficially cleaned (in the case of *V. ventricosa*), their sap changed and readjusted to values which were characteristic of the diffuse light and quiet conditions of the laboratory (Table XI).

These points have been stressed to show that "normal" sap composition of *Valonia* is not fixed and definite, nor does it remain constant during growth. Although the composition of the vesicle may seem to be relatively stable, in comparison with that of many other cellular fluids, it is still subject to a range of biological variation and response to environmental factors which might well have been predicted from a general knowledge of plant nutrition.

A similar problem has been investigated with reference to *Nitella*. The species that have been investigated are *N. clavata* by Hoagland et al. (90–94), *N. flexilis*, *N. gracilis*, and *N. hyalina* by Collander (40). The sap of these organisms was analyzed and compared with the water in which they grew, as shown in Table V.

d. Variations in the sap of Nitella and Chara. Hoagland realized that the cell sap was not constant, for it varied between April and November, as shown by figures recorded in Table V. The principal cations in *Nitella* are potassium and sodium; the principal anion is chloride. The range of variation of chloride in the sap was much less than that of potassium and sodium during the period of investigation. Although present in very dilute solution externally, the potassium ion was accumulated more than the other ions.

Comparing the *Nitella* species (as in Table V), it may be seen that they came from solutions which varied very greatly in salt content, but the level of chloride in the sap varied much less than the external concentration. The accumulation ratio for chloride was as high as 3300 in the case of *N. flexilis* (which came from the most dilute external solution), and was as low as 3.6 for *N. hyalina* (which came from the most concentrated external solution). The accumulation ratios for potassium were higher than those for sodium over the whole range of organisms. The actual content of sodium and potassium in the sap, however, varied greatly, from *N. gracilis* with a potassium content some 16 times its sodium content, to *N. hyalina* with a sodium content nearly twice that of the potassium content. These figures show that in these closely related plants the internal sap concentration is much more uniform than the composition of the external solutions in which they grow. This results from the fact that when the external concentration of an ion is low, its relative accumulation is high. This is especially true with respect to chloride, the content of which was the most constant over the entire range of saps analyzed. Granting that the total salt concentration is indicated by the level of the principal anion, namely chloride, it is evident that the proportion of sodium to potassium can vary appreciably within one organism, e.g., *N. clavata*. Considering different species of *Nitella*, the ratio of sodium to potassium can also vary greatly.

Various members of the Characeae normally occur in brackish waters. Collander (39–40) found that the internal sap concentration was stabilized against wide variations in the salinity of the external solution. Data are available for *Chara ceratophylla* in waters of widely different degrees of salinity. Whereas the external solution varied over a wide range, the internal composition of the sap was much more constant (Table XII).

TABLE XII

THE SAP COMPOSITION OF *Chara ceratophylla*
AS GROWN IN MEDIA OF DIFFERENT COMPOSITION[a,b]

	Na	K	Mg	Ca	Cl
Sap	152	66	26	13	233
Medium	68	1.4	14	3.8	80
Accumulation ratio	2.2	4.7	1.8	3.4	2.9
Sap	126	61	20	11	208
Medium	31	0.6	6.5	2.0	36
Accumulation ratio	4	101	3	5.5	5.8
Sap	84	77		13	176
Medium	0.21	0.04		3.3	0.13
Accumulation ratio	400	1900		4	1350

[a] From Collander (40) and (42), see page 320 of reference (42).
[b] All sap concentrations in milliequivalents per liter.

e. The sap of angiosperms. Work in Collander's laboratory (42) showed that the general principles enunciated above apply also to aquatic angiosperms, with special reference to *Ceratophyllum* and *Lemna* (see Table XIII). The analyses were made on the pond water on the one hand, and on sap expressed from the plants on the other. It is obvious that the accumulation ratio is high when the external solution is least concentrated in the ion in question and the ratio is lower when the external ionic concentration is higher.

As early as 1919, Hoagland had realized that the total osmotic pressure of the aqueous fluids of terrestrial angiosperms, when grown in water cultures, was very much less affected by drastic changes in the total concentration of the nutrient medium than might have been supposed from the then prevalent tendency to control, fastidiously, the

TABLE XIII

Relations between the Accumulation of Ions by Aquatic Angiosperms (*Ceratophyllum demersum* and *Lemna trisulca*) and the Composition of the External Solution[a]

	Chloride		Sodium		Potassium	
Plant	Ext. solution (meq/liter)	Accum. ratio	Ext. solution (meq/liter)	Accum. ratio	Ext. solution (meq/liter)	Accum. ratio
Ceratophyllum	71	1.6	61	0.26	1.3	146
	46	2.2	40	1.2	0.84	77
	31	2.3	26	1.2	0.56	325
	33	24	3.2	3.8	0.15	520
	0.97	95	0.93	53	0.18	494
Lemna	46	1.6	40	0.90	0.84	62
	31	3.2	26	2.7	0.56	61
	5.1	15	6.5	4.0	0.19	374
	1.5	43	1.2	5.8	0.087	943
	0.08	1400	0.11	245	0.088	948

[a] From Collander (42).

TABLE XIV

Effect of Concentration of Nutrient Medium on the Expressed Sap of Barley[a]

Concentration and osmotic pressure of culture solution		Osmotic pressure (atm) of expressed sap	
ppm	O.P. (atm)	Roots	Leaves
200	0.07	3.62	8.01
4300	1.7	5.62	10.2

[a] From Hoagland (84a).

osmotic pressure of plant culture solutions. Table XIV shows that, whereas the osmotic pressure of the external solution was varied over an approximately 20-fold range, the osmotic pressure of the sap expressed from roots did not vary more than 1.5-fold, and the sap of leaves changed even less than this (1.2-fold).

In part this stability of the internal osmotic concentration, despite changes in the external osmotic concentration, was due to the operation of the "accumulation" process, as already observed for other plants. In part, however, it was due to a compensatory accumulation of organic solutes (notably sugars, and especially in roots) at the lower external salt concentrations. The latter point was to be demonstrated much more forcibly when, later, barley plants were deliberately grown with a restricted supply of nutrients as a means of developing a system which is capable of rapid absorption of salts (87).

Collander (41) investigated the ability of 21 terrestrial angiosperms to discriminate between the various alkali metals. These ions were furnished at the same equivalent concentration in nutrient solutions supplied to *all* the plants grown in a *single* container. The analyses were made on shoots, and the data for each ion were expressed as a percentage of the cation determined in the tissue. Figure 11 illustrates the results obtained. The figure shows that while the amounts of sodium absorbed are very different, the plants absorbed almost the same percentage of their total cation as potassium and the absolute amounts of potassium in the plants did not vary greatly. An interesting feature is that only a very few species (two or three) contained more sodium than potassium, and these were halophytes.

f. Previous nutrition of excised barley roots grown in water. If seedling barley plants are grown in a limited volume of a standard inorganic nutrient solution (such as Hoagland's), the excised roots are rendered much more active in subsequent absorption of salt than roots which are excised from plants grown with access to a full and renewed supply. To quote Hoagland and Broyer (87), the explanation is as follows: "When the volume and concentration of the nutrient solution does not keep pace with the translocation of salt from root to growing shoot, as a consequence the root suffers partial depletion of its salt content . . . low salt root tissues developed under these experimental conditions also have a high sugar content, a point to be discussed presently." Thus, in these cases, the roots of plants growing in water respond to a reduced concentration of total salt, not only by an increase in the relative absorption of the salt, but by an actual increase in their sugar content. Hoagland and Broyer also observed the effect of the environmental factors on the development and absorbing power of roots grown under these relatively low salt conditions. When a given strain of barley was grown at different seasons of the year, the yield of shoot was relatively constant with a controlled supply of external nutrient, but it was found that the roots differed significantly. Roots of higher fresh weight and higher absorbing capacity were produced in the summer months, and roots of lower fresh weight and lower absorbing capacity were formed during the winter months. This was ascribed to the intensity and quality of the light during the growing period.

Such effects indicate again that the mechanism by which solutes are absorbed from the external solution is an intimate function of the

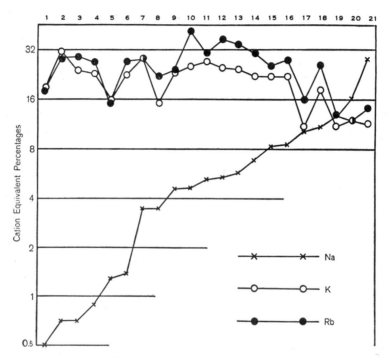

FIG. 11. Percentage of Na, K, and Rb in the cation complement of 21 angiosperms all furnished the same nutrient containing these elements. From Collander (41). Key to the plants used: 1. *Fagopyrum*; 2. *Zea*; 3. *Helianthus*; 4. *Chenopodium*; 5. *Salsola*; 6. *Pisum*; 7. *Nicotiana*; 8. *Solanum*; 9. *Spinacia*; 10. *Avena*; 11. *Aster*; 12. *Papaver*; 13. *Lactuca*; 14. *Plantago lanceolata*; 15. *Melilotus*; 16. *Vicia*; 17. *Atriplex littoralis*; 18. *Sinapis (Brassica)*; 19. *Salicornia*; 20. *Plantago maritima*; 21. *Atriplex hortensis*.

The table contains two examples of pairs of species from the same genus. Therefore, the differences to which Collander referred operate at the specific, as well as the generic, level.

general physiological status of the organism, and this is determined by a variety of conditions of which the concentration of the nutrient solution is only one. In the case of plants with aerial shoots, the effect of shoot on root is an important feature which determines, not only the natural sap composition of the root, but its ability subsequently to absorb salt under experimental conditions.

3. Experimental Variables in the Investigation of Salt Accumulation

After the choice of experimental material, the recognition and control of the variables which affect its further salt uptake will determine the type of conclusions that can be drawn. This will become apparent when the work done on selected types of experimental material is discussed, but it will be convenient first to recognize the different variables that need to be considered.

a. Nature of the solute. To be informative, the means of measurement need to be specific for the solutes in question.

The natural occurrence in quantity of many ions in the cell sap so complicates the analytical detection of small changes with respect to one of the natural constituents that attention was turned to ions which were either absent from, or inconspicuous in, the natural sap. The bromide ion as an anion and rubidium as a cation early came into prominence because analytical methods were developed for the one (Br) and spectrographic methods for the other (Rb). Steward and Harrison (214) made such a quantitative study of the uptake of Rb and of Br ions by potato tissue disks before the use of radioactive isotopes greatly simplified this kind of experiment. Even before this, the use of bromides of various cations in the study of ion accumulation by *Nitella,* as practiced by Hoagland *et al.* (90–94), set the pattern for much of the work which was to follow. Although much earlier work was done with basic dyes and their colorimetric determination, these substances often combine with acidic compounds in the cells and may even form precipitates.

The ions K^+, Na^+, Ca^{++}, Mg^{++}, Cl^-, SO_4^{--}, $H_2PO_4^-$, HCO_3^-, and possibly NO_3^- may be present in the cell sap at the outset, so the detection of small experimentally induced changes in any one of these may be difficult. Analysis of the external solution only, calculating the absorption by difference, suffers because the accumulation process is most interesting when it operates from the most dilute solutions. Since these need to be furnished in large volume to maintain an adequate total supply, the concentration differences to be established by analysis, therefore, are often small. Electrical conductivity may indicate electrolyte concentration, either internal or external, but it lacks specificity. For these reasons Hoagland turned to the bromide ion, which was not present appreciably in the tissue. Analytical methods were developed for the determination of bromide in the presence of the other halogens, chloride in particular. The bromide ion, like chloride, is not metabolized, so removing a factor which complicates the study of absorption of nitrate and phosphate. The alkali bromides have been mostly used. Chemical determinations of rubidium, lithium, and cesium in the presence of potassium and sodium have never been easy. Lithium detected spectroscopically, had already been used as an indicator of translocation. Rubidium is well fitted as an indicator of cation accumulation because of its similar behavior to potassium. Quantitative spectrographic methods which are applicable to rubidium and all the alkali metals (122, 214) made experiments with rubidium bromide feasible. In these experiments neither of the absorbed ions were originally present in the tissue in appreciable amounts. The availability of a number of radioactive isotopes of the common chemical elements has given a new degree of freedom to the choice of solutes. Isotopes of lead had been used by Hevesy at a very early date, and Devaux had early studied the binding of heavy metals by cells, but these early experiments were robbed of much physiological significance by the chemical properties of the elements in question. The most useful isotopes for the study of salt accumulation do not participate directly in metabolic reactions and have half-lives sufficiently long that decay may be either neglected or is slow enough to permit experiments of relatively long duration. The isotopes which have found the most important application are listed in Table XV.

Using isotopes, experiments can now be performed in extremely dilute solutions. Certain carrier-free isotopes may be used in solutions as dilute as 10^{-11} equivalents per liter. At these extreme dilutions care must be taken so that transient and purely physical interchanges with the tissue are not overemphasized.

b. Concentration. The variable concentration enters into salt accumulation studies, both as it relates to the concentration of salts initially present in the absorbing cells and to the concentration of the salts as supplied.

Even though the accumulation process normally involves movement against diffusion gradients, higher external concentrations promote the *total* salt absorbed, even though the relative uptake, or accumulation ratio, may be decreased thereby.

Regarding movement of the solute from a low to a high concentration as one which requires the expenditure of energy, a familiar relationship states the amount of energy to be expended per mol of solute moved from one level to the other. This relationship is that the maximum work per mol is related to log C^1/C^2, where C^1 and C^2 are the respective concentration levels between which the solute is moved. Thus, if the tissue performs osmotic work (albeit isothermally and reversibly) on each gram mol of solute moved, then the energy so expended should be approximately doubled for each tenfold increase in the accumulation, or absorption, ratio.

Since the degree of accumulation is greater at lower external concentrations, the internal concentration, after absorption, is not greatly affected by the small changes in external concentration that may result from the absorption: this is particularly true if the volumes of the

TABLE XV

A TABLE OF RADIOACTIVE ISOTOPES USEFUL IN BIOLOGICAL RESEARCH[a]

Isotopes	Half-life	Type of radiation
Br^{82}	34 hours	Beta, gamma
C^{14}	6000 years	Beta
Cs^{134}	2.3 years	Beta, gamma
Cs^{137}	33 years	Beta, gamma
Ca^{45}	152 days	Beta
Cl^{36}	4.4×10^5 years	Beta
Co^{60}	5.3 years	Beta, gamma
I^{131}	8.0 days	Beta, gamma
Fe^{59}	47 days	Beta, gamma
Mo^{99}	68 hours	Beta, gamma
P^{32}	14.3 days	Beta
K^{42}	12.4 hours	Beta, gamma
Rb^{86}	19.5 days	Beta, gamma
Na^{22}	2.6 years	Beta, gamma
Na^{24}	14.9 hours	Beta, gamma
S^{35}	87.1 days	Beta
Sr^{85}	65 days	X, gamma
Sr^{89}	53 days	Beta
Zn^{65}	250 days	Beta, gamma

[a] See British Atomic Energy Research Establishment publication "Radioactive Materials and Stable Isotopes," 1950; and Comar, C. L., "Radioactive Isotopes in Biology and Agriculture," Chapter 6. McGraw-Hill, New York, 1955. The rare earths are here omitted.

external solutions are relatively large. Nevertheless, the external concentration of the solute being absorbed is an obvious variable to be controlled. Its effect can be indicated by an experiment (204) in which potato disks accumulated the bromide ion from the following external concentrations (in milligram equivalents per liter) of potassium bromide, namely 0.468; 4.68; 46.8. After the same periods of absorption from the same volume of aerated external solution and by the same number of tissue disks, the bromide concentrations in the expressed sap were 22.78; 43.06; 88.04, respectively; thus each tenfold increment in external concentration caused the internal concentration to be doubled. These experiments were done with the tissue absorbing ions at steady rates (cf. Section I,C,2). Collander reports a similar effect of concentration on the uptake of lithium prior to the attainment of "equilibrium" in species of *Chara* [cf. page 325 of reference (42)]. At external concentrations of 0.1; 1.0; 10; 100 meq per liter the sap concentrations were 0.62; 1.9; 3.7; and 8.8, respectively; again the internal concentration is approximately doubled for a tenfold increase in external concentration.

Another effect of concentration concerns the total *initial* concentration of the electrolytes in the cells. This may be regulated in a variety of ways.

When tissue is prepared for experiment by cutting from a relatively large mass, such as a storage organ, there may be an initial loss of electrolytes, incidental to the procedure of cutting and to the preliminary washing of the disks; this loss produces a salt deficit that can later be made good by absorption during the experiment. Also, when cut into the form of thin disks, the tissue may take in water and thus be able to absorb salt to restore the internal concentration of solutes.

As already indicated (cf. page 295) the internal concentration was controlled by Hoagland and Broyer in their work with excised "low-salt" and "high-salt" roots respectively, although "low-salt roots" are also high in sugar.

c. Temperature. As in all physiological processes, temperature is a readily recognizable variable affecting salt accumulation. Where diffusion alone is the limiting factor in the absorption process, the temperature coefficient of absorption should be low. It is now familiar that the temperature coefficient of the absorption and accumulation process is high (of the order of two to three), and this is commonly cited as evidence that the process involves the chemical reactions of metabolism. By contrast, it is now a common device to carry out absorption experiments at a relatively low temperature, of the order of $2°C$, in order to distinguish that part of the absorption that may occur "passively," without metabolism, from that which is "active" and involves metabolism. While in general this distinction may be valid, the technique is not without complications. Many starch-containing tissues actually show *enhanced* metabolism (respiration) at temperatures of the order of $+1°C$, although it is true that some time must often elapse for this to become apparent. Absorption and accumulation of ions commonly declines to a low level, if not actually to zero, at temperatures below $4°C$. This was shown for the uptake of bromide by potato and Jerusalem artichoke disks [(206), cf. Table I] and for the uptake of bromide by *Elodea* (*Anacharis*) (180) and excised roots (87). It is, however, not safe to assume, in all cases, that this is because metabolism is virtually nonexistent at low but nonfreezing temperatures. Rather it may be that at this low temperature the metabolism is uncoupled from the accumulation of ions. In the case of the potato tuber, the kind of metabolism to which the tissue is diverted by long storage at +1–2°C is *not* the kind of metabolism that is directly related to salt accumulation (210). In fact, cells from tubers so treated may fail to retain their salts even against water, even though they are not killed by the temperature treatment.

Therefore, temperature may have very different effects upon the processes of salt accumulation, and these require to be evaluated.

d. Light. Light was first detected as a prominent variable in ion accumulation through the now well-known experiment of Hoagland and Davis (91) on *Nitella.* At the outset it was recognized that the effect of light was not upon such properties as permeability, but that it operated indirectly through the nutrition and metabolism of the green cells. The first belief was that exposure to light replenished the carbohydrate supplies which were not, it was thought, conspicuously stored in *Nitella*. The supply of organic substrate would then determine any metabolic reactions, dependent upon the carbohydrate. There are many other situations, however, in which light impinges upon salt accumulation. Leaf formation and enlargement are characteristically associated with light, protein synthesis in leaves is frequently stimulated by light, and protein breakdown by darkness. Light, however, also regulates oxygen production, carbon dioxide fixation and carbohydrate formation and thus fosters conditions conducive to respiration, whereas darkness has the converse effects. Other light-determined morphogenetic effects occur because, as Hoagland and Broyer pointed out, the kind of absorbing root system which is developed on seedling barley plants is, indirectly, a function of the light conditions that prevail during their growth. Light, therefore, has many indirect effects on salt accumulation.

e. Oxygen concentration: aeration. Cells and tissues used to investigate ion accumulation are usually submerged, with the notable exception of the accumulation of ions in the tissues of the shoot when they are received via the vascular system. The rate of dissolved oxygen supply to the cell surface of submerged cells, or tissues, is of a different order from the diffusion of oxygen as a gas in the air spaces of tissue which is exposed to air. Many submerged tissues, as, for example, the roots of aquatic plants, contain large air spaces, or lysigenous cavities, in which air may be present and which, in fact, communicate with the air spaces of the shoot. As Cannan (33, 34) showed, the shoot in the light may effectively aerate the submerged root and, especially in mono-cotyledons, there is a more ready transfer of oxygen from shoot to root, and even to the external solution, than is often realized. Indeed, roots of plants that normally do not grow in water differ in their structure when they are developed in water culture, and the extent to which they contain air-filled cavities may be influenced by aeration of the culture medium. For rice (*Oryza sativa*) and barley plants this was shown by Vlamis and Davis (242). Therefore, to maintain physiological activity unlimited by oxygen deficit may require more effective oxygen supply than can occur, unaided, by solution of oxygen from the air and by diffusion. When, for purposes of experimental observation, tissues that normally develop in contact with air, as for example roots, slices of storage organs, etc., are submerged, supplementary aeration may be necessary in order that their metabolic activity may be maintained. Attention was being given to the use of supplementary aeration of plants almost as soon as plants were grown in water culture [see (84), cf. page 133]. Indications that disks of storage tissue, which were totally submerged and shaken in closed bottles, suffered from lack of oxygen were implied by an observation of Stiles (224). This observation was that the conductivity of the external solution changed after each occasion on which the bottles were opened for the measurements to be made!

Plants that normally grow in water, water-grown root systems, and aquatic angiosperms like *Elodea* normally have metabolism which is only limited by dissolved oxygen at concentrations lower than that of water in equilibrium with air. By contrast, tissues which normally develop in air, as for example disks cut from storage organs, are maintained in a state of oxygen saturation only if the solutions are in equilibrium with air (209) or with a gas even richer in oxygen. For this purpose, flowing streams of gas of known oxygen partial pressure need to be, and have been, used. But even so, attention also needs to be given to the rate of the flow of gas through the solution and the dispersal of the oxygen-equilibrated solution to the tissue by mechanical stirring, or other devices. Also, to maintain the tissue in a state of oxygen saturation, the disks used need to be thin—not more than a few cells thick. For the special case of disks of storage tissue a technique for this purpose was devised and described (202) and similar procedures have been used in the investigation of the uptake of ions by excised roots (87).

Rapid aeration by carbon dioxide-free air supplies oxygen and also removes carbon dioxide. For disks of potato tissue, it has been shown that the effectiveness of *rapid* aeration with carbon dioxide-free air in promoting bromide accumulation is due not only to oxygen supply, but

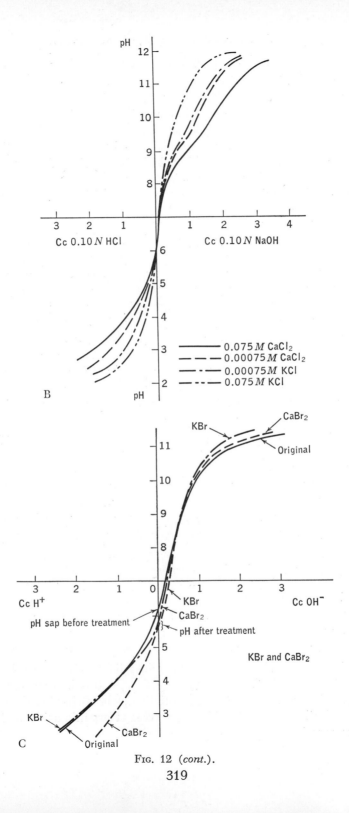

Fig. 12 (cont.).

319

of ions from the tissue which counterbalance unequal absorption of anion and cation. The absorption and accumulation of the ions is not, however, greatly affected by relatively small changes in the external pH, at least not to an extent that requires that the external solution should be rigorously buffered. There is a relatively broad zone in which changes in the accumulation process, brought about specifically by the pH of the external medium, are relatively small, and, when working within this range, it is preferable for most purposes *not* to employ buffer solutions with their complex composition and physiological effects.

An examination of the effect of pH on the uptake of ions by disks of red beet (*Beta vulgaris*) tissue in the presence of a relatively inert buffer system, namely tris(hydroxymethyl)aminomethane (tris) plus sulfuric acid has been made (98a, b). Hurd found that, while the uptake of chloride from a solution of buffered potassium chloride was only slightly greater at pH 8.5 than at pH 6.5, the absorption of potassium was approximately doubled at the more alkaline pH. Whereas the uptake of potassium and chloride was approximately equal at pH 6.5, there was a considerable excess of potassium over chloride absorbed at pH 8.5. The unequal uptake of potassium and chloride in alkaline solutions was found to be associated with the accumulation of organic acid anions in the tissue (88, 100, 239). It was further shown that at pH 8.5 the uptake of potassium and chloride was accompanied by absorption of bicarbonate ions from the medium in an amount equivalent to the difference in the quantities of potassium and chloride absorbed. Experiments with $HC^{14}O_3{}^-$ demonstrated that the bicarbonate ions were incorporated in the cells into organic acid anions—mainly malate. The difference in the rate of absorption of potassium at pH 6.5 and pH 8.5 was therefore attributed to the absence of bicarbonate ions in the acid medium and their presence, in relatively high concentration, at the alkaline pH.

Tissues from several other storage organs, including the cultivated Brassicas, turnip and swede, or yellow turnip, behaved in a similar manner to beet, but others, notably potato, showed no pH effects over the range studied. This difference in behavior was thought to be associated with the different abilities of the various tissues to assimilate bicarbonate ions and to incorporate these into organic acids. It may be noted, in passing, that when potato disks fix $C^{14}O_2$ from solution the principal product is C^{14}-glutamine (217) and this in turn can readily form protein, so that potato tissue might not be as liable as some other tissues to build up a cation-binding excess of organic acids at the more alkaline reactions.

To modify the reaction of the cell sap is more difficult than to control the external pH. This may occur in response to drastically unbalanced absorption of anion and cation, as, for example, penetration of ammonium as a weak base. However, the consequences of the penetration of the ammonium ion (marked loss, or exchange, of potassium ion) may outweigh the direct effects of the pH of the sap, and especially so because this resists change by the extent to which it is naturally buffered.

Thus, the implication is that absorption and accumulation of neutral salts can sometimes be studied from single salt solutions, without the need to employ buffer solutions to stabilize the pH of the medium. When control of the reaction is desirable, however, a convenient and physiologically appropriate buffer system is that which often operates in nature, namely the carbon dioxide-bicarbonate buffer system. Here again a complication is encountered, for it may be difficult to distinguish between effects which are due to the hydrogen ion per se, and the physiological effects which are due to the carbon dioxide as a metabolite. The contrasting effects of bicarbonate and phosphate buffer systems at the same pH may be indicative here (219).

g. Variables which determine metabolic and vital activity. As salt accumulation became identified with metabolic and vital activity, other variables became apparent. There are two obvious stages in the growth of cells. In the first, division and multiplication of self-duplicating structures are the conspicuous events—in the second, growth by cell enlargement is mainly in question. In order to carry out the latter, cells must first complete, or be capable of, the former. One can conceive of continuous growth without cell enlargement—one cannot conceive of growth completely without cell division. These two phases of growth in the cell may well be characterized by different kinds of absorption processes. This will be discussed later. However, it goes without saying that the bulk of the water and the bulk of the solutes in the cell are taken up *during the phases of extension growth.* Thus conditions which promote increased volume and increase of fresh weight are often associated with increased salt accumulation. This was found in the early attempts to determine the most appropriate conditions for the accumulation of ions by potato disks, for salt accumulation and increased fresh weight invariably proceeded concomitantly. Similarly, effects due to auxin and to other means by which extension growth is facilitated are, therefore, to be expected, and some have been found (235).

Cells or organs may be compared at different stages of their normal growth when their capacity for further growth can thus be varied. In fact one may expect to see in the rate of intake of salt to leaves at different locations on the axis a reflection of their relation to Sachs's Grand Period of Growth for that organ. Though discussed in more detail in Part III of this chapter it may here be noted that the leaf takes in, and reaches its highest concentration of an ion (bromide) when its own growth by enlargement is greatest (see Part III). Gradients along the axis of a submerged aquatic, like *Elodea* (180), bear somewhat similar interpretation. An outstanding example of a gradient of ion accumulation is that which normally occurs along the axis of un-branched roots. Here the highest concentrations are reached at points near the root tip at which cells are most actively growing, in this case by enlargement. The ability of thin disks of storage tissue to absorb salt has been shown to depend on disk thickness, or specific surface, which determines the number of metabolically active cells.

Thus the process of salt accumulation cannot be dissociated from the type of metabolic activity which, in the leaf, is made evident by its peak ability to grow. In the root this is indicated by cells whose rate of expansion is at the maximum, and in a cut disk of potato it is con-centrated in the surface cells. All these examples lead to the recogni-tion of variables which affect metabolic and vital activity in one way or another.

Cells which have previously completed their growth and development and are unable to return to a more actively growing state, as for ex-ample in the mature parenchyma of pome fruits and parenchyma of bulb scales of monocotyledons, often fail to respond to conditions other-wise favorable to salt accumulation, and they may even release to water the solutes which had previously been absorbed. Even potato tubers can be placed in this condition, so that disks cut from them will not absorb. This is done by treatments (relatively long exposure to $+1°C$ at which starch turns to sugar) under which the tissue changes so that the cells lose their ability to divide again when exposed to moist air (210). Disks cut from such tubers not only failed to accumulate salts, but they failed to retain against distilled water those salts which had previously been absorbed.

Therefore, the level of salt accumulation in cells readily responds to a number of external variables which may be identified and controlled. It is important also to recognize certain internal variables which will determine the rate and kind of growth of which the cells remain capa-ble, for this in turn determines their ability to harness the vital ma-chinery for the accumulation of ions in their sap.

The results obtained on the most important tissue systems will now be summarized in a manner that will show the views developed by different schools of thought.

Part II. Salt Accumulation in Selected Systems—Experimental Results and Their Interpretation

The results obtained by the use of typical systems will be summarized by reference to work on *Nitella*, *Valonia*, disks of storage tissue, root systems, microorganisms and certain aquatic angiosperms. Inevitably, however, work was proceeding simultaneously on all these different systems, and experiments carried out with one type of material were necessarily influenced by results being obtained with another.

A. *Nitella*

The publications on *Nitella* (*N. clavata*) from Hoagland's laboratory began in 1923 with a paper by Hoagland and Davis. This was very largely exploratory in nature, but it had some points that in retrospect are interesting. Hoagland and Davis compared *cell* sap and *expressed* sap and invariably found that the cell sap was more concentrated than the expressed sap. The *Nitella* sap was found to be weakly buffered, but yet its internal pH was remarkably stable despite a wide change of pH in the external solution. Direct penetration of ammonium from ammonium salts caused the pH of the cell sap to rise, and this was accompanied by loss of chloride which was thus recognized as a sensitive symptom of injury to the cells. Nitrate, which does not normally accumulate in this sap, was found to penetrate more readily from a slightly acid solution than from an alkaline one. Having observed that very little absorption of chloride occurred from potassium chloride by *Nitella* in the dark and that the absorption varied with the intensity and the duration of light, Hoagland and Davis proceeded to demonstrate with potassium bromide solutions that the effects of illumination were definitely on the penetration of ions into the cell sap. The most important results to 1929 were summarized by Hoagland and Davis (92) and are reproduced in Figs. 13–16. Using these data as a basis, the most important conclusions on *Nitella* may be summarized.

First the interacting effects of other ions on the one being accumulated can be mentioned. The greater penetration of anion from more acid external solutions, which had been observed with nitrate, was not observed with bromide. This ion consistently gave the greatest concentration in the sap at nearly neutral external reactions (pH 6.4 to 7.0): this may be compared with later results obtained with potato disks (cf. page 335). Addition of other slow-moving anions (SO_4^{--}) did not retard bromide uptake, in fact, if supplied as the potassium salt they increased slightly the bromide absorption due to the increase in the effective concentration of potassium. The other halides (chloride and iodide), being themselves readily absorbed, reduced the uptake of bromide when present in the solutions (see Table XVII). Prior absorption of potassium chloride by the cells reduced the subsequent uptake of bromide [cf. (93), page 481] and compelled all that uptake to be by equivalent replacement of chloride from the cell populations.

The effect of the nature of the accompanying cation on bromide uptake (Fig. 13) was described as follows. The cations fell into three groups: K and Rb > Na > Ca, Mg, Sr, Li, in order of decreasing bromide accumulation from their salts. In mixtures the effect of potassium predominated over calcium. The mobility of the cations in water and in membranes is determined both by their degree of hydration and by the weight of the ion. The smaller the ion, the greater the density of the

TABLE XVII

EFFECT OF OTHER ANIONS ON ACCUMULATION OF BR IONS IN
Nitella clavata CELL SAP FROM 0.005 *M* KBR[a,b]

Nature and concentration of added salt		Br in sap (mgm equiv. per liter)
none		32
K_2SO_4	0.005 *M*	35
KNO_3	0.005 *M*	31.8
KCl	0.005 *M*	13.8
KI	0.005 *M*	15.5

[a] From Hoagland *et al.* (93).
[b] 12-Day period, diffuse light, phosphate buffered solutions pH 5.0.

charge and the larger the hydration shell. The order K, Rb > Na > Li reflects this, and so the bromide was accumulated most readily when it was accompanied by the most mobile cation.

The most important of Hoagland's results, however, were those which showed (a) the exit of chloride as bromide entered, (b) the contrasted effects of light and darkness and (c) the effects of temperature.

The data of Fig. 14 are commonly cited to show that the accumulation of bromide in *Nitella* was by ionic exchange for previously absorbed chloride. Hoagland, however, stressed that in addition to this exchange there was some accumulation of potassium accompanied by bromide, because the final Br + Cl concentration was greater than the initial chloride concentration. The entry of bromide without equivalent replacement of chloride was recognized to be a feature of cells that had not previously acquired their full salt content and could, therefore, still take in some halide plus potassium [cf. Hoagland *et al.* (94), page 132]. However, there are alternative explanations of these data which do not necessarily indicate direct bromide-chloride exchange in cells. The most significant feature of the experiment was that *it took 40 days in continuous light.* In 40 days a miscellaneous population of *Nitella* cells must undergo change. Some already mature cells go into decline and lose their previously absorbed salt, e.g., chloride; while

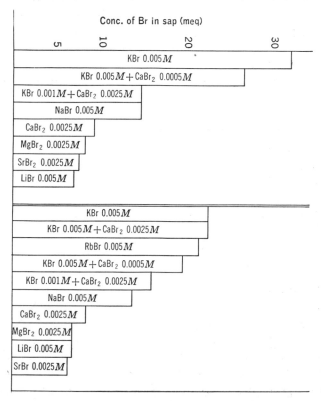

Fig. 13. Effect of the nature of the cation in the external solution on absorption of bromide in *Nitella clavata*. From Hoagland *et al.* (93).

others, in the light, may sluggishly expand and absorb the only halide present (bromide). Indeed Hoagland recognized this point when discussing another very long experiment which lasted eight months (92) but seemed not to stress it particularly in relation to the mechanics of bromide-chloride exchange. Therefore, although this celebrated experiment has been most often cited as evidence of accumulation by ion exchange, it is doubtful if any such direct exchange occurred in the cells that actually absorbed the bromide. The most significant feature of this experiment is its time course, which clearly indicates the role of growth in this process.

Earlier, Hoagland *et al.* (94) had shown that if *Nitella clavata* cells are in contact with bromide only in the light they accumulate the ion, but if they are in contact with bromide only in the dark they do not acquire a concentration greater than the external. By various experiments Hoagland and his associates showed (Figs. 15 and 16) that the

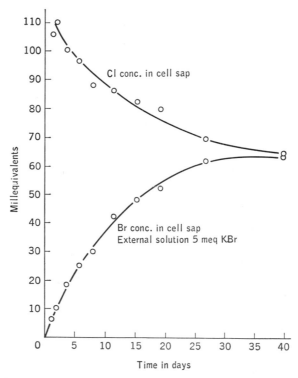

Fig. 14. Effect of time on bromide and chloride concentration of *Nitella clavata* sap. From Hoagland *et al.* (94).

accumulation of the bromide in the cell population, largely but not exclusively acquired by replacement of chloride, was a direct function of the intensity and the daily duration of light. They also showed that at low temperatures, even in the light, accumulation did not occur; and that the temperature coefficient of the process was high. A very suggestive experiment was briefly mentioned (92). In the endeavor to stimulate photosynthesis, which was believed to be limiting bromide accumulation, the external carbon dioxide concentration was increased. This treatment decreased, instead of stimulating, the accumulation.

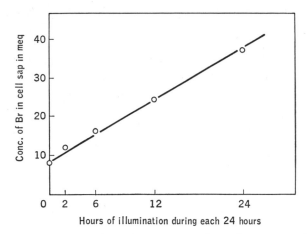

FIG. 15. Effect of daily period of illumination on bromide concentration of *Nitella clavata* sap. From Hoagland *et al.* (94).

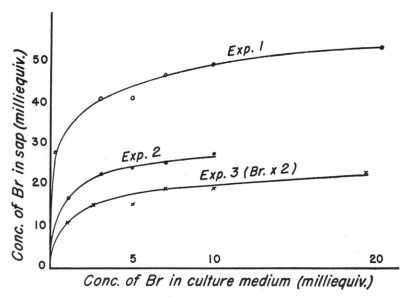

FIG. 16. Effect of external KBr concentration on Br concentration of the sap in *Nitella clavata* under various conditions: Experiment 1: 9 days; continuous light of two 300-watt lamps; 23–25°C. Experiment 2: 13 days; diffuse daylight; room temperature. Experiment 3: 27 hours; daylight plus continuous artificial light. From Hoagland *et al.* (93).

Therefore, from all these results, and in retrospect, one can now infer that:

(a) The undoubted effects of light on bromide accumulation involved morphogenetic effects on those cells which could still grow during the course of the experiment.

(b) In large measure the apparent exchange of bromide for chloride in the cell population was probably a reflection of loss of chloride from some older cells and preferential absorption of bromide by other, younger, growing cells.

(c) Additional to the morphogenetic effects of light mentioned above there were more direct physiological effects. These determined that the cells, through photosynthesis, were under oxygen and carbon dioxide conditions that also promote metabolism, whereas in the dark the converse was true. Both of these effects now seem to be more important than the direct replenishment of organic reserves upon which the chief stress was laid in the first interpretation.

In later work with *Nitella* the experiments have tended to follow interpretations from other materials. Work done both by Brooks (23) and by Hoagland and Broyer (89) stresses that ions may be accumulated first in the cytoplasm and later in the vacuole (see also page 362). Hoagland and Broyer used two isotopes: rubidium as the chloride, and bromide as its potassium salt. By separating cytoplasm and vacuole, Hoagland and Broyer obtained a ratio of the content of the isotope in the vacuole to that in the residue, including the cytoplasm. After 1 day, the isotopes were present mainly in the cytoplasmic residue, but later, after 10 days, the content in the vacuole was greater than that in the cytoplasm. Also, in these later experiments, it was recognized that maximum absorption by *Nitella* was obtained in the light *with supplementary aeration*, whereas minimum absorption was obtained, not merely in the dark, but also by *depriving the cells of oxygen* in an atmosphere of nitrogen.

B. *Valonia*

In large part the work on *Valonia* is identified with Osterhout and his collaborators. Reference has already been made to the knowledge that may be gained from the natural sap composition of these organisms; to the habitat factors involved and also to the difficulties that arise when the vesicles are interpreted simply as single cells. Therefore, the discussion of the experimental work which has involved *isolated* vesicles of *Valonia* will be restricted here to that which illustrates the more important generalizations drawn. The work on *Valonia* has been extensively summarized by Osterhout (149–154).

Valonias have been used in different types of investigation. By the use of dyes as the entering solute they have been used extensively for studies upon permeability, principally by M. Irwin and M. M. Brooks. They have also been used extensively in bioelectric studies by a number of workers associated with Osterhout, (for refs., see 151, 152) and for studies of wall structure. Conspicuously lacking, however, are detailed investigations on their metabolism and general nutrition.

Discussion of the work on *Valonia*, bearing upon permeability and membrane structure, lies outside the scope of this chapter (see Volume I, Chapter 6). Suffice it to say, however, that work in these areas led Osterhout to certain general ideas which were extended into the interpretation of the salt accumulation process. These ideas are:

(a) The functional protoplasmic membranes are regarded as fluid and nonaqueous.

(b) *Valonia* behaves as though its functional protoplasmic membranes are not freely permeable to ions but only to undissociated molecules (ion pairs) of free base or acid.

(c) The membrane surface comprises an acidic substance (HG), which can combine with entering bases and transport them across the protoplasm, where they are hydrolyzed in contact with the sap, which is more acidic (pH 5.2 to 5.5) than the sea (pH 8.2 to 8.3).

Based upon these general ideas there are extensive series of papers dealing with the kinetics of penetration in cells and in models; these cannot be discussed here in detail.

The point at (a) above is regarded as consistent with the electrical properties of the *Valonia* membrane surfaces, as these are revealed by measurements of resistance, which is high, and of electromotive force (emf).

The points (b) and (c) derived initially from observations upon the passive penetration of molecules, particularly of weak bases and acids. Notably, it has been shown that the effect of external pH upon the penetration of a weak base (ammonia or dyes) or a weak acid (H_2S, H_2CO_3) from sea water is a function, not of its total concentration, but of the concentration of the undissociated base or acid. The type of data obtained are shown in Fig. 17. However, the experiments from which these data were obtained were of short duration, and accumulation of the entering solute was not in question.

Osterhout extended the ideas gained from the study of weak acids and bases to the penetration of the salts which are normally accumulated. In doing this Osterhout combined the ideas summarized in (a) to (c) above with an ingenious exploitation of the peculiar pH relations of sap and external medium which apply to *Valonia* in the sea.

There are, therefore, two problems: (a) How far do these ideas explain the accumulation of potassium in *Valonia?* (b) How extensively may they be applied to other systems? The following analysis is directed to these pertinent questions.

The novel concept of Osterhout was that potassium and sodium do not enter cells in the form of ions, but as "undissociated free base," and the gradient in the "thermodynamic potential of free base" as measured by the ionic product $K \times OH$ determines the entry or exit of the cation. Thus, potassium could continue to enter the vesicle from the sea

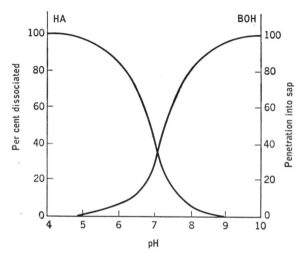

FIG. 17. Penetration of weak acid (HA) and weak base (BOH) into the sap of *Valonia macrophysa* as affected by external pH and as related to the external concentration of undissociated molecules. After Osterhout (150).

so long as the product $K_o \times OH_o$ is $> K_i \times OH_i$ (where the subscripts *o* and *i* represent outside and inside, respectively). Since the normal hydroxyl ion concentration of the sea approaches one thousand times that of the sap, this could furnish a driving force which counteracts the dilution of potassium in the sea. At the membrane surface the molecules of free base $(K \times OH)$ encounter the acidic carrier molecule HG and form salts which are hydrolyzed at the pH of the sap. Entry of the principal anion (chloride), which is but little accumulated in *Valonia*, could be exchanged as HCl for issuing H_2CO_3 or as HCO_3^- for the chloride ion.

The crucial experiments are those which describe the effect upon the distribution of potassium and sodium of the penetration of ammonium from sea water to which ammonium chloride was added. Al-

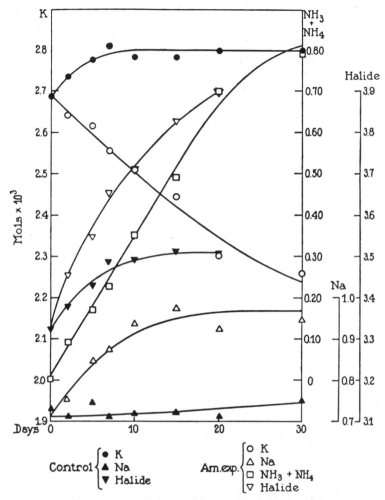

FIG. 18. Penetration of NH_4 from NH_4Cl sea water into *Valonia macrophysa* and its effect upon the distribution of other ions. Graphs show the change in moles of K, Na, NH_4, and halide in a typical lot of cells of *Valonia macrophysa* in sea water (control) and in an equivalent lot of cells in sea water containing 0.001 *M* NH_4Cl (the scales of ordinates for each substance are the same but are displaced vertically to bring the curves into one figure.)

though the interpretation should be based upon the respective *concentrations* of the solutes in the two phases, the data were in fact reported in terms of *total amounts* in the cells. These effects are shown in Fig. 18 which is reproduced from Jacques and Osterhout (103). In summary these results show: (a) The continued entry of potassium and of chloride, with but little change in sodium, into the vesicles of *Valonia macro-physa* which remained in sea water. This entry may be interpreted as due to growth (that is, sluggish expansion) or to the readjustment of the sap after the vesicles have been collected, dissected, and transferred from one environment to another. (b) The rapid penetration of ammonium chloride in sea water into the experimental vesicles and, associated with this, a loss of potassium, a gain of sodium, and a gain of chloride by the vesicles.

Under these conditions the pH of the sap was raised in the experimental vesicles, the internal ionic product $K_i \times OH_i$ increased, and the gradient of this product from the outside to the inside was reduced, until potassium could actually come out of the cells. However, the observed changes which are crucial to the interpretation of this experiment are (a) the loss of potassium, (b) the gain of sodium. Both of these are movements of ions due to the treatment, but they were from *a region of high to one of low concentration* and thus were *with, not against, the concentration gradient of ions as well as of "undissociated free base."*

A similar experiment carried out with *Valonia ventricosa* vesicles, by Steward and Martin (215), produced a different result. In response to the entry of ammonium from 0.001 *M* ammonium chloride in sea water and to the consequential increase in the pH of the sap, visible changes occurred. After 8 days when the sap pH had risen to 7.2 the potassium concentration of the sap had fallen from an initial value of 0.597 *M* to 0.462 *M*, while the sodium concentration had risen from 0.035 *M* to 0.173 *M*. During this time the chloride concentration remained steady. Thereafter, at 11 days, extreme injury due to ammonium was encountered and this was accompanied by the drastic loss of potassium and gain of sodium. In *Valonia macrophysa* similar, but less extreme, results were obtained.

An experiment made with potato disks to simulate the conditions of the *Valonia* experiment, though without the special pH relations between sap and external medium, also produced a quite different result (Fig. 19). Ammonia rapidly entered potato disks from ammonium chloride solution, the sap being buffered so that its pH does not change much, but despite this potassium came out rapidly, being replaced by ammonium. As a result of the much increased ammonium content in the cells the concentration of every previously accumulated ion in the system decreased, as also did the total water content. These experiments could not be interpreted in the manner used by Osterhout for *Valonia*, because the special pH relations of *Valonia* in the sea do not apply. On the contrary, the experiments show a direct replacement of potassium by ammonium, followed by sufficiently toxic conditions to reduce the internal concentration of all ions initially in the system.

A strong argument, however, against the concept elaborated by Osterhout is the following: The ionic product K × OH should be equally affected by the same relative change in K or OH, and therefore both

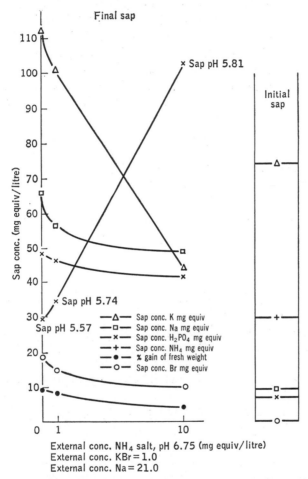

FIG. 19. Penetration of NH₄ from buffered NH₄Br solution and its effect upon the distribution of other solutes. From Steward (206).

the potassium concentration and the hydroxyl ion concentration of the medium should influence the distribution of potassium, and in a similar manner. That is, a threefold change in external potassium concentration should have the same effect as threefold change in hydroxyl concentration. This is not so. By the appropriate control of the carbon dioxide tension, the external pH of the sea water was varied over a wide range (pH 5.8 to 9.2). and the potassium concentration was changed by addition of potassium chloride to sea water to supplement the potassium concentrations by 0.025 M and 0.05 M. In 4 days the sap composition of the Valonia (*V. ventricosa* and *V. macrophysa*) exposed to these widely different external pH's did not change significantly. However, in response to the relatively small increase in total potassium and total halide the internal concentration of potassium did increase significantly. Therefore, it is quite clear that a change in external potassium concentration has a much greater effect than the same relative change in hydroxyl ion concentration. It is quite impossible, therefore, to harmonize these results with the idea that potassium enters as an undissociated free base and not in the form of ions. In a later section it will be shown (see page 366) that excised roots can accumulate ions from solutions more acid than their own sap (88).

Osterhout has devised ingenious model systems. These consist of an outer alkaline (KOH) aqueous layer, a central nonaqueous acidic layer composed of phenolic substances, and an inner layer acidified by the passage of carbon dioxide. In ways which are of physicochemical interest, the penetration of the cation from the outer to the inner layer, via the phenolic layer, has been investigated. The formation of KG (potassium phenolate), its passage across the membrane as a carrier molecule, and its entry into the aqueous phase by combination with acid (HA $=$ H$_2$CO$_3$) is implicit in this model. This type of model, however, has not yet reproduced the essential feature of the salt accumulation process—a feature which, it is true, is not well illustrated by *Valonia*—namely, the simultaneous accumulation of anions and cations in approximately equivalent amounts. For these reasons the model experiments are not discussed in detail, for they seem to be too remote from physiological reality.

Even if the basic concepts of Osterhout really satisfied the requirement of *Valonia* they could hardly be extended generally to plant cells, which could, only rarely, take advantage of the unusual circumstances that the hydroxyl concentration of the external solution was almost one thousand times that of the interior of the cell! The Osterhout model and the ideas derived from *Valonia* really represent ways in which the features of the environment are manipulated to allow accumulation to occur by movements which are essentially diffusion. Inasmuch as these do not lead to knowledge of the intimate relations between metabolism and salt uptake, upon which it seems necessary to base the ultimate interpretations of the process, we shall refrain from further discussion of this experimental system in favor of others with which more direct metabolic evidence has been obtained; these seem more suitable situations from which to draw generalizations.

C. DISKS OF STORAGE TISSUE

1. Conditions Conducive to Salt Accumulation in Potato Disks

Early knowledge derived from the use of storage tissue has already been summarized. When the conditions most conducive to salt accumulation by thin slices of potato tissue were realized, a technique to control the variables which regulate the process was devised, and the section to follow is largely the outcome from its use. This work was published in two main groups of papers, one dealing with factors affecting accumulation and the other dealing with the metabolism of the disks (see the last paper of each group for references namely (210, 219).

The technique utilized disks cut from large tubers; the disks were of known thickness (the minimum thickness 0.3 mm, the usual thickness 0.6 mm, and any desired multiple of this could be used) and of constant diameter (3.4 cm). By the use of disks of varying thickness, a wide range of specific surface (cm^2/gm fresh weight) could be covered. The disks were immersed in a relatively large volume (2 liters) of dilute solution (of the order of 0.001 gram equivalent per liter) of a neutral salt (usually potassium bromide). The solution was contained in a specially designed vessel of 4 liters' capacity, equipped with slowly rotating stirrer and devices for the rapid flow (of the order of 10–15 liters per hour) of carbon dioxide-free air through the solution. The carbon dioxide respired by the tissue during the absorption period was collected and determined by titration or by measurements of electrical conductivity in the absorbing solution. By these devices standard conditions for the investigation of the salt accumulation could be established and different variables could be investigated, one by one, under otherwise comparable conditions. Conclusions that emerged from this work, together with representative data, will now be summarized. Unless otherwise stated for a particular factor under consideration, the standard conditions that applied in these experiments were as follows: Not more than 40 disks of thickness 0.6 mm (approximately 30 gm) were used in 2 liters of solution of 0.00075 to 0.001 equivalents per liter at 23°C, aerated by carbon dioxide-free air at a controlled rate of 10–15 liters per hour.

2. Time Course of Absorption

The cut disks were prepared for experiment by washing, usually in running tap water or in changes of distilled water, and this plays an

TABLE XVIII

The Effect of Washing in Tap Water on Subsequent Bromide Intake by
Potato Disks during 48 Hours at 23°C from KBr Solution
(0.00075 mg equiv. per liter)[a]

Duration of washing (hours)	Respiration rate (mg CO_2/ gm/hr)	Bromide conc. in sap (mg equiv. per liter)
1	0.187	5.81
25	0.191	13.04
50	0.214	18.94
553	0.210	0.73

[a] From Steward *et al.* (210).

essential part in the subsequent absorption process. The tissue reaches
a relatively stable ability to absorb salt after periods of washing of the
order of 24–48 hours, and a respiration rate (of the order of 0.20 mg

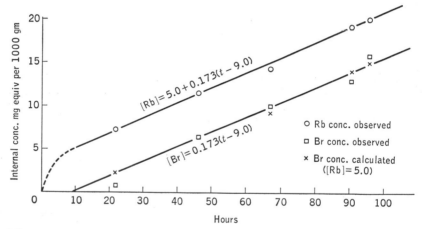

Fig. 20. The effect of time in the absorption by potato disks of rubidium and
bromide from 0.001 equivalents RbBr in aerated solution. From Steward and
Harrison (214).

carbon dioxide per gram fresh weight per hour) is established which
remains approximately constant for long periods of time. Very prolonged washing impairs salt accumulation before it retards respiration
(Table XVIII).

The washing period has various implications. Solutes from cut cells
are removed and simultaneously, no doubt, some are leached from the
living cells, producing an initial salt deficit. Metabolic inhibitors, characteristic of the storage tissue, may also be removed; but in the washing

period the tissue begins the transition from the inactive metabolic state of the tuber to the active metabolic condition which obtains in the disks under the conditions which are conducive to salt accumulation. This transition is not completed when the surface-dried disks are placed in the solution, but a brief lag period elapses thereafter (approximately 12

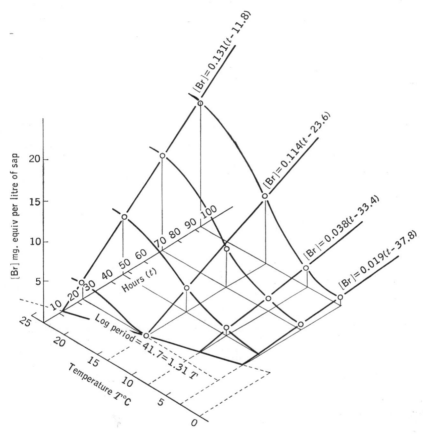

Fig. 21. Effects of time and temperature in absorption by potato disks of bromide from 0.00075 equivalents KBr in aerated solution. From Steward *et al.* (210).

hours at 23°C and of longer duration at lower temperatures). During the lag period some solutes are lost by the tissue and these may subsequently re-enter when the absorption begins (201, 225). Thereafter, however, anions (bromide) and cations (potassium or rubidium) may be absorbed at linear rates in approximately stoichiometrically equivalent amounts for relatively long periods of time (100–200 hours), during which the respiration rate is maintained steady. These effects are illustrated by Fig. 20, which shows the time course of absorption, and also illustrates the initial lag period.

During the initial period of such experiments and before the steady rate of absorption for anions has been established, the tissue shows some ability to bind cations, e.g., rubidium, unaccompanied by anions. This, however, is a separate and transient phenomenon requiring later discussion but, as Fig. 20 shows, the subsequent intake involves cation and anion simultaneously and in equal amounts. Figure 21 shows also the marked effect of temperature on salt accumulation, from 3°C to 23°C, which is characteristic of the simultaneous intake of cations and anions and indicates that this process has a basis in metabolic reactions.

3. pO_2 Effects; Surface—Volume Effects; CO_2/HCO_3 Effects

The accumulation of ions in thin disks of potato tissue at 23°C may be summarized with reference to three sets of variables. These are (a) the oxgyen tension in the flowing gas stream, with which the solutions are maintained in equilibrium, (b) the surface:volume relations of the

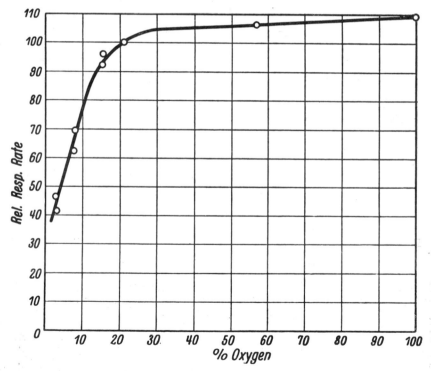

Fig. 22. Effect of pO_2 on respiration. Respiration of potato disks in equilibrium with air = 100. From Steward (204).

disks, and (c) the carbon dioxide:bicarbonate relations of the external solution at constant oxygen tension. Two of these variables, namely oxygen tension and carbon dioxide tension, bear directly on the respiration of the disks, and the other bears indirectly on this, by emphasizing the difference between the behavior of cells situated near the tissue surface and those that are deep-seated within the tissue mass. Data which cover this range of variables for experimental periods of

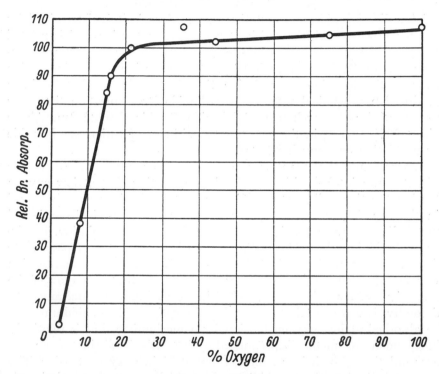

FIG. 23. Effect of pO₂ on bromide accumulation. Sap concentration in potato disks in equilibrium with air = 100. From Steward (204).

the order of 70 hours are shown, in six figures—Figs. 22–27. In these, the concentrations of absorbed ions are shown as milligram equivalents per liter of expressed sap and respiration rates are shown as milligrams of carbon dioxide evolved per gram of tissue per hour.

Figures 22 and 23 show the effect of oxygen tension on the uptake of bromide by the tissue and on its simultaneous respiration, as measured by carbon dioxide evolved. In these figures the values that relate to tissue in equilibrium with air (20.9% oxygen by volume) were used for comparative purposes and assigned the value 100. Oxygen saturation is maintained in submerged potato tuber tissue in solutions at, or near, equilibrium with air. In solutions of lower oxygen concentration bromide accumulation was limited by oxygen deficit and approached zero at zero oxygen tension. The respiration also was limited by oxygen deficit in solutions in equilibrium with gas mixtures poorer in oxygen than air but, approaching zero oxygen tension, the total respiration tended to approach a value about one-third that of the respiration of tissue in solutions in equilibrium with air.

Thus the bromide accumulation occurred as if it is determined by aerobic processes which vanish at zero oxygen tension and is related to a strongly aerobic part of the total carbon dioxide production; however, carbon dioxide produced at low oxygen tension seemed ineffective in promoting bromide accumulation. Calculations showed that the tissue produced a great deal more carbon dioxide by respiration than was necessary for the exchange of bromide for bicarbonate ion. Above about 20% oxygen in the gas stream, neither respiration nor bromide accumulation were much affected.

It was found that what was true of bromide uptake was also true of potassium intake, with the added complication that at low oxygen tension the tissue did not restrain completely the ions already stored in the sap. These results suggested the use of aerobically released metabolic energy to furnish energy for the salt accumulation. By the use of rubidium bromide as the accumulated salt it was possible to demonstrate two effects: first, that the general effect of oxygen tension was the same on cation and anion absorption, so that over most of the range ions of both sign are absorbed in approximately equal amounts. Secondly, it was possible to show that the cations (e.g., rubidium) can be absorbed at low oxygen tension by a mechanism that is relatively independent of metabolism, and also that this is distinct from the accumulation of both ions which occurs at higher oxygen tension. The nonmetabolic intake of rubidium, by which some accumulation is achieved, is to be regarded as a physicochemical binding to specific sites in the tissue which may involve one ion being displaced by another.

Furthermore, there was an effect due to the thickness of the storage tissue disks on ion absorption, as Ruhland (quoted by Stiles, 223; also see 222) had indicated long before, and also upon metabolism. The relation of the concentration of absorbed rubidium and bromide to the relative thickness and to the specific surface of the disk, is shown in Fig. 27. In this figure 0.6 mm is arbitrarily designated as thickness 1.0. The way in which respiration interacts with these variables and with temperature is shown in Fig. 24 by data obtained at 23°C and at 5°C.

The interpretation of Fig. 24 is that the carbon dioxide which issues in the flowing gas stream is produced in much greater quantity by cells which lie at the surface of the tissue disk than those which are deep-seated and which contribute to the bulk of the thicker disk. Extrapolation of the relative thickness curve to zero thickness shows the limiting value which represents the respiration rate per gram of tissue of cells that lie wholly in the surface (0.40 mg CO_2 per gram per hour at 23°C and 0.14 mg CO_2 per gram per hour at 5°C). Extrapolation of

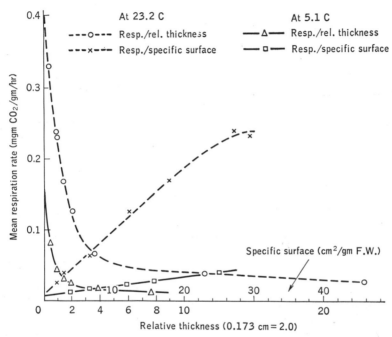

FIG. 24. Surface effects on respiration: the effect of temperature. From Steward et al. (210).

the specific surface curves to zero surface shows the low limiting values of the respiration rate (per hour per gram of tissue) of cells that lie wholly within a large bulk and are so remote from the surface that they are unaffected by the conditions that prevail. Using these extrapolated values, and recognizing that there is a gradient there of respiration falling steeply from the value at the surface to that which prevails within the disk, it is possible to calculate the effective depth of the zone of tissue which participates in the enhanced respiration rate at the disk surface (203, 214). The respiration rate at the surface has been termed for convenience the "surface respiration rate," and respiration in the center of the tissue disk, assumed uniform throughout its bulk, has been termed the "bulk respiration rate." The thickness of the outer shell or zone, of metabolically active tissue, measured in millimeters, has been designated by the symbol "z." From Fig. 24 it is obvious that the respiration at the surface is very markedly affected by temperature, and from Fig. 25 one can also conclude that it is strikingly affected by oxygen tension, requiring a relatively high oxygen concentration in the medium to achieve oxygen saturation. Whereas the general

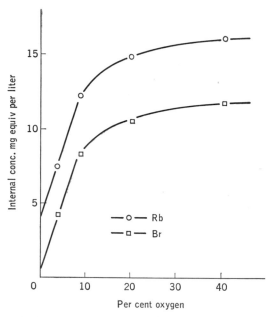

FIG. 25. Effect of pO_2 on rubidium and bromide accumulation by potato disks in 72 hours at 23°C. From Steward and Harrison (214).

order of magnitude of the so-called bulk respiration, upon which the surface effects are superimposed, is approximately the respiration rate at 23°C of whole potato tubers, the respiration of the cut disks is very much greater than this. Much of the older physiological literature made reference to what was called wound respiration [for summary see Steward et al. (222)]; this is here assumed to be a response directly to oxygen tension and to the conditions in the surface cells which, as a result of cutting and washing, maintain a much higher level of metabolic activity than prevails in the intact tuber.

Figures 26 and 27 show that similar surface:volume considerations apply to the absorption of both anion and cation. These data are taken from an experiment in which rubidium bromide was the absorbed salt; the one ion (rubidium) was determined spectrographically, and the other (bromide) was determined chemically, although the same kind of behavior had previously been worked out for the ion bromide alone. From these figures it is apparent that the rapid uptake of rubidium, which was previously shown by the time series in Fig. 20 and which was again evident in the oxygen tension experiment (Fig. 25), also entered into the experiments on the surface:volume relations of the

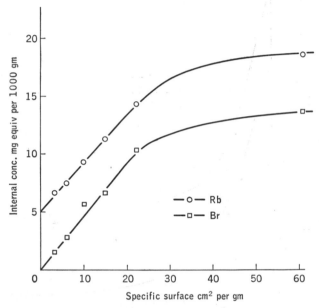

FIG. 26. Effects of disk surface on absorption of rubidium and bromide by potato disks during 72 hours at 23°C. From Steward and Harrison (214).

disks (Fig. 26). Apparently rubidium can bind to the tissue as deeply within as it can penetrate, and it reaches the same average concentration (5 micron equivalents per gram of tissue from a medium containing 1 micron equivalent). This uptake is regarded, therefore, not as a special property of the surface cells of the disk, but as one which is uniform throughout its bulk. Superimposed on this type of absorption, however, rubidium and bromide simultaneously are absorbed to a high concentration near the surface (see Fig. 28) but this type of uptake is absent from the middle of a thick disk. Calculations show that the simultaneous accumulation of anion and cation was confined to a depth of tissue (z), of the order of about three cells and, further within the disk, this type of absorption vanished simultaneously for both ions.

Thus the simultaneous uptake of anion and cation is superimposed upon a uniform uptake of rubidium throughout the disk; the latter requires neither the high level of oxygen concentration, nor the long periods of time, over which salt accumulation occurs. Figures 28 and 29 show diagrammatically the relations of ion absorption and respiration rate at the surface of the disk and throughout its bulk.

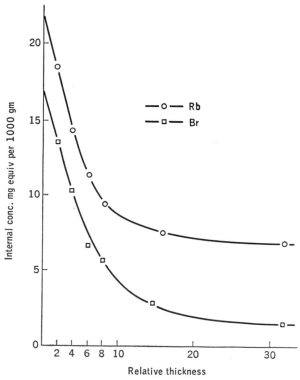

Fig. 27. Effect of disk thickness on absorption of rubidium and bromide by potato disks during 72 hours at 23°C. From Steward and Harrison (214).

The simultaneous accumulation of rubidium or potassium and bromide therefore occurs in cells which exhibit an enhanced aerobic respiration rate over that which prevailed in the intact tuber. Other experiments show that bromide will penetrate deeper into the disk when it is supplied at much higher concentrations, but it never reaches a greater concentration than the external solution in cells which are so deep within the disk that their respiration is that of the bulk of the tissue, rather than that of the surface zone (X_s).

Rapid aeration with carbon dioxide-free air not only supplies oxygen

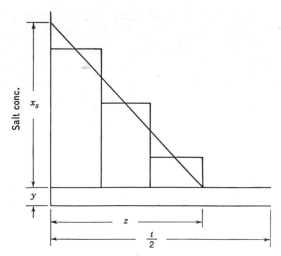

FIG. 28. Diagrammatic relation between basal uptake (y) of rubidium, accumulation in the surface cells (x_s), and depth of tissue (z) which accumulates RbBr. From Steward and Harrison (214).

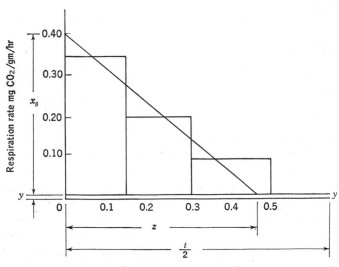

FIG. 29. Diagrammatic relation between respiration rate (y), increased respiration at the surface (x_s), and depth of tissue and showing enhanced respiration at 23°C. From Steward and Harrison (214).

to the tissue surface but it effectively removes carbon dioxide. The effects observed, therefore, may be in part due to oxygen pressure and in part due to carbon dioxide removal. Supply of carbon dioxide, at known partial pressures into the flowing gas stream, produces effects which are in part due to carbon dioxide per se and are in part due to the concomitant increase of hydrogen ion concentration; Figs. 30 and 31 show the effects obtained.

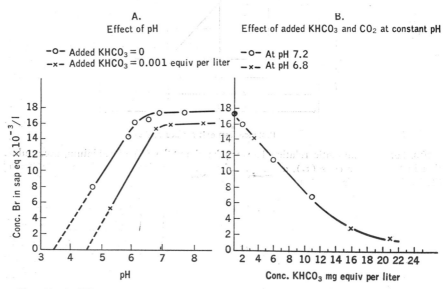

FIG. 30. A. Effect of CO_2 concentration and pH on bromide accumulation by potato disks at constant $KHCO_3$ concentration. B. Effect of $KHCO_3$ concentration on accumulation of bromide by potato disks at constant pH. From Steward and Preston (219).

In these aerated salt solutions (potassium or rubidium bromide) from which both ions are absorbed in approximately equal amount, the external pH remains at or near neutrality. As Fig. 30 shows there is a relatively broad region of pH in which the tissue is relatively little affected by pH per se. When the carbon dioxide concentration rises to a value consistent with a pH below 6, the bromide intake is depressed, although all other factors are conducive to active accumulation of this ion. With increased potassium bicarbonate concentration, higher pressures of carbon dioxide can be introduced into the gas stream at constant pH and their effects studied. At constant pH and increased bicarbonate concentration and pressure of carbon dioxide in the flowing gas stream, bromide absorption was steadily depressed until, at about 20 milligram equivalents of added bicarbonate per liter, the bromide accumulation vanished. The effect of pH on the absorption of the bromide ion is shown by a family of curves in which each shows the effect of the external concentration of carbon dioxide at a constant concentration of bicarbonate (Fig. 31).

From all these effects it is clear that the concentrations of bicarbonate ion and of carbon dioxide determine the ability of the tissue to ac-

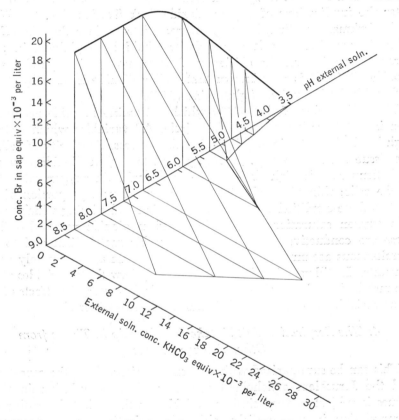

FIG. 31. Interactions of pH of solution, potassium bicarbonate concentration, and bromide absorption by potato disks. From Steward and Preston (219).

cumulate salt in quite as striking a manner as the other variables that have been indicated above. The effects observed are, however, due to a summation of those due to bicarbonate ions, hydrogen ions, and undissociated carbon dioxide. Figure 31 shows how these interactions can be represented. For any concentration of added bicarbonate or carbon dioxide in the solution, reactions near to neutrality are the most favorable to bromide uptake, though there is a relatively broad range of hydrogen ion concentration which is compatible with bromide accumulation in the tissue. At constant pH, bicarbonate ion or undissociated carbon dioxide specifically suppress bromide accumulation. This effect occurs at concentrations of these solutes that will be shown to have an effect on metabolism (cf. page 357).

When these effects were first discovered the direct use of carbon dioxide as a metabolite by dark fixation was little known. However, after the implications of carbon dioxide dark fixation, i.e., the Wood and Werkman reaction, became apparent, it was clear that the effects of rapid aeration on salt accumulation could be explained, in part, through its effect on the efficiency of the removal of respired carbon dioxide (221). It is now known that carbon dioxide is fixed by potato disks with the formation of glutamine (217).

Thus, discussion of the effects which determine bromide accumulation in potato disks has come full circle. All variables which foster a high level of metabolic activity affect the process. Proximity of cells to the tissue surface, to oxygen supply, and to carbon dioxide removal; high temperature; high oxygen tension to maintain oxygen saturation in the cells; thin disks which expose the maximum proportion of the cells to these conditions; and rates of aeration and stirring which maintain oxygen saturation and efficient carbon dioxide removal—all of these are conductive to high rates of aerobic metabolism and to the simultaneous accumulation of anion and cation in approximately equal amounts. It still remains, however, to specify how these variables affect the metabolism of the cells. Before doing this, however, the effects of the same variables on other storage tissues will be summarized.

4. The Respiration-Salt Uptake Relationship in Tissue from Other Storage Organs

This can be summarized with reference to tissue from the carrot root and the Jerusalem artichoke tuber. Bromide accumulation in these tissues is subject to very similar variables as those that determine the accumulation in potato disks, although each tissue presents special features.

In both carrot and artichoke disks, maximum accumulation of bromide from dilute solution requires a rapid stream of carbon dioxide-free air. To maintain these tissues unlimited by oxygen deficit, insofar as salt accumulation is concerned, again requires oxygen tensions of the order of those that prevail in air, and solutions in equilibrium with mixtures poorer than this cause reduced intake of ions. This is shown in Fig. 32 for the uptake of potassium and bromide ions by artichoke tuber tissue. The absorption of ions by tissue from both storage organs has a high temperature coefficient and, in both cases, the investigation of disks of varying thickness shows that the surface cells contribute more to the respiration and salt intake of the disk than those which are deep-seated. However, because of the greater anatomical complications and certain time effects now to be discussed, the precise interpretation of these surface:volume effects is more difficult.

An outstanding difference between tissue cut from the potato tuber

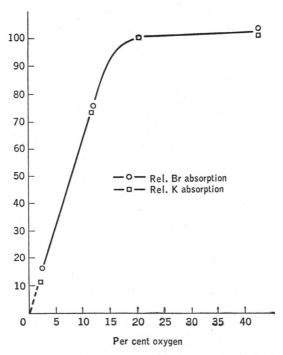

FIG. 32. Uptake of potassium and bromide ions by Jerusalem artichoke tuber tissues. From Steward *et al.* (209).

and tissue cut from the carrot root, and particularly from the artichoke tuber, is as follows: After preliminary washing and a brief lag period, thin disks of potato tubers will maintain a uniform respiration rate and rate of salt intake for a protracted period provided that the tissue is not limited by oxygen lack. This is not so for tissue cut from the carrot root or the artichoke tuber. In these cases the tissue disks reach a relatively high rate of respiration in the early periods of exposure to dilute salt solution, but this rate declines with time. It is as if a new limitation enters, progressively with time, into the behavior of the surface cells and renders them unable to maintain their maximum and initial response to external oxygen by a high rate of respiration. This declining respiration rate with time is most strikingly shown for artichoke tissue, and, after the lapse of 120 hours, the respiration rate falls from its initial value of 0.22–0.26 mg of carbon dioxide per gram fresh weight per hour to a value as low as 0.07 mg of carbon dioxide per gram fresh weight per hour. Indeed in experiments in which this procedure was very prolonged the tissue eventually reached a steady respiration level of approximately 0.022 mg CO_2 per gram fresh weight per hour. In parallel with this declining aerobic respiration rate, a decreased intake of potassium bromide from the external solution was also observed (208), but from the first observation it was recognized that there was no simple stoichiometric relationship between salt absorbed and the carbon dioxide evolved.

These observations mean that it is that part of the aerobic respiration of the disk which resides particularly in the surface cells and which is only oxygen-saturated in solutions which are in equilibrium with air that decreases in intensity with time. This declining respiration has associated with it other metabolic processes which equally become limited with the lapse of time and, it seems, these are also linked with the intake of ions. After the complete lapse of this time drift, so that the respiration rate has fallen to the low level indicated above, the cells of the disk behave at the surface very much as do those which are deeper within the disk.

It was an obvious suggestion that artichoke tissue, unlike potato tissue, is unable to maintain the flow of organic substrates for its respiration. Explanations of the time respiration drift were therefore sought through the availability of various substrates. Although transient responses to sugar, phosphate, and nitrogen compounds were, in fact, obtained, the initial respiratory rates were not permanently restored. In fact, the only way now known to restore the submerged and long-washed artichoke disks to a high and maintained level of metabolic activity, is to submit them to growth factors, such as those now known to be present in coconut milk. Under aseptic conditions, these treatments cause the artichoke tissue to proliferate and grow by cell division (212). The salt-absorbing properties of the artichoke tissue in this latter state will be referred to below.

Thus, disks from other storage organs respond much as does potato tissue; though artichoke disks differ in that they are unable to maintain, for long periods, their initial rates of absorption and metabolism. The basic causes, however, of salt accumulation in the cells of disks of carrot and artichoke are not to be thought of as different from those in potato tissue (at least during the early periods of experiments, when they still show high respiration rates) which are maintained at the oxygen concentrations of solutions in equilibrium with air.

5. *The Role of Growth in Salt Accumulation in Disks of Storage Tissue*

Ion accumulation can occur in different ways which make different demands on the growth of the accumulating cells. The simultaneous and progressive absorption of anion and cation in potato disks was recognized to be part of a recrudescence of growth in the otherwise dormant cells of the tuber; this relationship should be made clear before the detailed metabolic characteristics of the cells are described.

Mature parenchyma tissues differ widely in their ability, spontaneously, to grow again. Leaf parenchyma is often difficult to grow, though in some cases (*Bryophyllum* [= *Kalanchoe*]) this occurs spontaneously with the re-formation of buds and the regeneration of shoots. In wound healing and regeneration at a cut surface, mature cells may show a temporary return to the growing state. This occurs more frequently in parenchymatous tissues of dicotyledons than of monocotyledons, though all such parenchyma cells, even in dicotyledons, do not possess these properties; e.g., the fleshy tissue of mature pome fruits do not regenerate. Identifying the ability to accumulate bromide ions, in these previously mature parenchyma cells which first acquired their salt during growth, with renewed ability to grow, one may ask whether there is any relationship between the known growth responses of the tissues and their renewed ability to accumulate. The relationship that appeared is as follows.

Tissues which in the form of disks accumulated bromide in some degree were also able, in some degree, to undergo regeneration at a cut surface in moist air. This regeneration may vary greatly from tissue to tissue. Studying a range of such tissues Berry and Steward (14) found that, despite very varying responses in respiration, the ability to accumulate the bromide ion consistently implied the presence of some cells which had the ability to grow again, as shown by some degree of regeneration at a tissue surface exposed to moist air. When the ability to regenerate was destroyed, even in potato tissue, by prestorage of the tubers at 1°C to 2°C, the ability to accumulate bromide also disappeared (210).

Therefore, renewed ability to accumulate is associated in these storage tissues with cells which have the metabolism characteristic of cells that can grow again, as best shown by their *ultimate* ability to divide at a cut surface in moist air. This relationship, however, indicates the *kind* of metabolism of which the cells are capable; it does not necessarily mean that the submerged cells actually achieve, during the course of an experiment, the same dividing state. As already stated (page 321) it is the ability of cells to grow first by cell division that carries with it their ability to grow subsequently by the different process of cell enlargement. Thus, regeneration of tissues in moist air is a convenient visible index of their underlying capacity for growth. This relationship now requires to be expressed in more precise metabolic terms so that its significance for salt accumulation may be understood. The most characteristic metabolic property of cells which both regenerate and accumulate is not alone a high aerobic respiration rate, but respiration which is also linked to protein synthesis from soluble nitrogenous reserves.

With modern knowledge of hormones and growth-regulating substances the behavior of cells may be modified more widely than hitherto. Auxins may stimulate growth by cell enlargement and, in particular cases, factors are now known that stimulate cell division, causing cells rapidly to proliferate. Cells at the extremes of these contrasted states will also be found to have different characteristics in the absorption of ions. First, however, the metabolism which is most closely correlated with the salt accumulation will be described for potato disks.

6. Biochemical Characteristics of Accumulating Potato Cells

This may be summarized from a group of papers dealing with the biochemistry of potato disks under conditions conducive to salt accumulation; for reference see Steward and Preston (219).

Initially, washed potato disks are high in starch and low in sugar. Their total nitrogen contents are about one-third in the form of protein, and the rest is in the form of soluble compounds, such as various amino acids and the amides, asparagine and glutamine. The variety of potato will determine, to some extent, the detailed composition with respect to nitrogen, though the variety King Edward (which has been much used in the salt-accumulation work) usually contains a preponderance of amino acid nitrogen over amide nitrogen, and the latter may be largely in the form of asparagine. Titration curves of expressed sap show the buffering effect of these nitrogen compounds between pH 8 and 9, and also the relatively strong buffering at more acid reactions which is due to the organic acids. When washed disks are exposed to aerated water and to high temperature (23°C), the following changes occur. Starch turns to sugar, which accumulates in excess of that which is respired away; soluble nitrogen is converted to protein and, as its content decreases, the composition of the soluble nitrogen also changes toward an increased emphasis upon glutamic acid and glutamine. The organic acid content of the tissue also decreases as may be seen by a shift in the titration curve. A curious, and unexpected, conversion of carbohydrate to uronic acids which accumulate at the disk surface as a film of mucilage or pectin, especially in contact with potassium solutions, occurs. In response to oxygen pressures which saturate the disks, they undergo the browning reaction which is due to enhanced activity of the potato polyphenol oxidase system. In the cells certain oxidizable substances are present in the reduced state (e.g., ascorbic acid) and these substances increase in aerated solutions. The potato

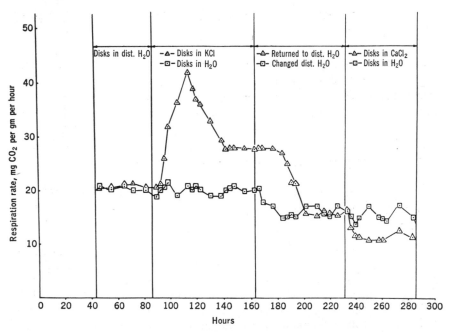

FIG. 33. Reversibility of the contrasted effects of K^+ and Ca^{++} on respiration of potato disks. From Steward (unpublished).

oxidase acts upon its natural substrates, which contain the catechol grouping, but the oxidation products (orthoquinones) may in turn, bring about metabolic oxidations which do not occur in a resting cell. As a result of all this metabolic activity, considerable heat is evolved. The total heat lost by the tissue system, most of it being absorbed by the surroundings, may be measured by the decline in heat content in a bomb calorimeter. All these diverse metabolic processes, conditioned by oxygen and affected by the salt supply, contribute to the metabolic machinery in the cells which makes salt accumulation possible; in short they represent the recrudescence of growth and metabolism to which earlier reference has been made.

The processes outlined above proceed in potato disks in distilled water —but their pace may be modified greatly by variables which also determine the accumulation of the bromide ion. These are the nature of the salt, the oxygen pressure, and the carbon dioxide-bicarbonate concentrations in the solution. The metabolic effects of these variables are summarized below.

The potassium ion causes an increased rate of metabolism of potato disks, over and above the level attained in aerated distilled water. By

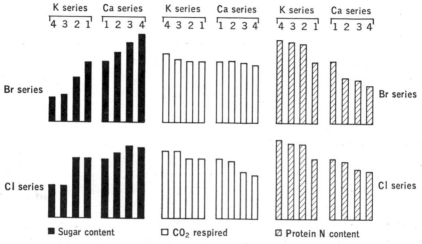

FIG. 34. Effects of salt concentration on respiration, sugar content, and protein synthesis of potato disks at 23°C during 72 hours. From Steward and Preston (218).

Data on Relative Basis—Tissue in Salt Solution 1 = 100
Salt concentration 1 = 0.0075 gm equivalents per liter
Salt concentration 2 = 0.015 gm equivalents per liter
Salt concentration 3 = 0.050 gm equivalents per liter
Salt concentration 4 = 0.075 gm equivalents per liter

contrast, the calcium ion causes a decreased rate; this is shown by the contrasted effect of potassium and calcium salts with a common anion. These effects are shown to be reversible in the hitherto unpublished data of Fig. 33 [cf. Fig. 3 of Steward and Preston (218)].

It might be supposed that the effect of K^+, which increases respiration, would be to increase the starch hydrolysis and sugar concentrations, with the converse effect due to Ca^{++}. This, however, is not so. The high respiration rates due to K^+ were established in tissue which had *lower* sugar concentration than the tissue in distilled water, and the lower respiration rates due to Ca^{++} were established in tissue with *higher* sugar concentrations. The effects of the salts on respiration were, however, similar to their effects on the protein synthesized by the tissue, since K⁺ stimulated and Ca⁺⁺ depressed this from the level established in distilled water (Figs. 34 and 35).

The original papers (218a) show in more detail the contrasted effects of K⁺ and Ca⁺⁺ concentrations on the conversion of soluble nitrogen to

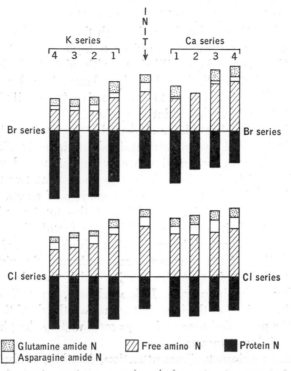

Fig. 35. Effects of salt concentration on the nitrogen metabolism of potato disks. From Steward and Preston (218).

Data on Relative Basis—Initial Total Nitrogen = 100
Salt concentration 1 = 0.00075 gm equivalents per liter
Salt concentration 2 = 0.015 gm equivalents per liter
Salt concentration 3 = 0.050 gm equivalents per liter
Salt concentration 4 = 0.075 gm equivalents per liter

protein, and it is to be noted that the synthesis was reduced, in the tissue used, virtually to zero at high Ca⁺⁺ concentrations (0.075 gram equivalents per liter).

Balance sheets of metabolites, drawn up for the recovery of the initial content of the disks in carbon, total carbohydrate, dry weight, calorific value, and organic nitrogen showed that these could be accounted for. However, the balance sheets and the effects of salts and oxygen also emphasize the following points:

(a) Salts with a common anion affected the gain of fresh weight, respiration, the activity of enzymes involved in the browning reaction, and protein synthesis by the tissue in accordance with the series $Ca^{++} < H_2O < K^+$.

(b) Combined effects of salts and oxygen affected the *residual* sugar concentrations in accordance with $Ca^{++} > H_2O > K^+$.

(c) The salts affected the tissue through that part of the metabolism which, for oxygen saturation, requires a high external oxygen concentration (i.e., water in equilibrium with air). This is more suggestive of metabolism limited by phenolases than by cytochrome, although the phenolase seems to be effective in mobilizing the substrates before they enter the oxidative cycle (221).

(d) Previous pO_2 effects have shown that approximately two-thirds of the steady level of respiration attained in tissue in distilled water was mediated through systems that are oxygen-saturated only in water in equilibrium with air. These data also show that two-thirds of the respiration is mediated by processes that vary, *pari passu*, with protein synthesis and the remaining one-third of the total respiration as it proceeds in water is independent of nitrogen metabolism. The oxygen requirement of that part of the respiration which is most closely related to the metabolism of the nitrogen compounds and the visible changes in the tissue links this to pathways in which the phenolase system seems to play a dominant role. Independently, Boswell and Whiting (17), had also arrived at the idea that some two-thirds of the respiration of potato disks proceeded over pathways mediated by the phenolase- and catechol-containing compounds.

(e) These experiments direct attention, forcibly, to the idea that at least part of the respiration of the tissue, and particularly that which is sensitive to salt entry and to oxygen, requires that the carbon of the emergent carbon dioxide be canalized through nitrogen compounds and particularly protein. Thus, in the metabolic role of K^+ which is to stimulate protein synthesis, carbon is drawn from sugar and nitrogen from the organic reserves to form protein, but the carbon of the latter is respired away.

Therefore, the cardinal feature of potato disks accumulating bromide and potassium is their synthesis of protein, which is determined by oxygen concentration, stimulated by K^+, depressed by Ca^{++}, and linked to that part of the respiration which requires an oxygen concentration of solutions in equilibrium with 20.9% oxygen in the gas phase. The nature and concentration of cations in the external solution regulate the metabolism which is conductive to salt accumulation, and the anion accumulation seems to follow the lead of the cations.

Confirmation of the causal connection between protein synthesis and anion accumulation (Br⁻) arose from two other lines of work.

Disks cut from tubers after storage at 1–2°C progressively declined in their ability to accumulate bromide at 23°C with the length of the storage period. When tissue failed to accumulate, it also failed to

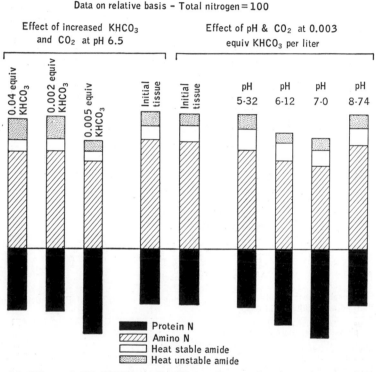

Data on relative basis – Total nitrogen = 100

FIG. 36. Effect of CO₂ KHCO₃, and pH on nitrogen fractions of potato disks after 72 hours at 23°C. From Steward and Preston (218).

synthesize protein, although, as shown by protoplasmic streaming, the cells were not dead.

Disks cut from normal potato tubers decline in their bromide uptake and in their protein synthesis concomitantly with the CO_2/HCO_3^- concentrations added to the medium at a pH of 7.0. At an appropriate bicarbonate concentration *both* protein synthesis and bromide accumulation vanish. The similarity throughout of the effects of CO_2/HCO_3^- in the external solution on metabolism and the absorption of bromide leads again to protein synthesis as the most intimate property of the tissue with which the ion accumulation is associated. Inasmuch

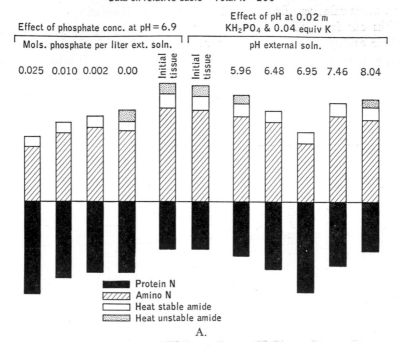

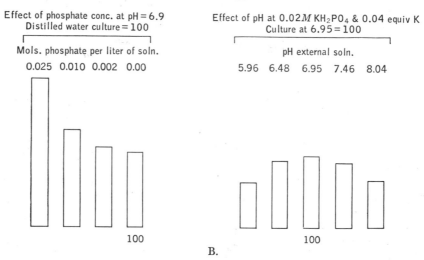

Fig. 37. A. Effect of KH₂PO₄ and pH on nitrogen fractions of potato disks after 72 hours at 23°C. B. Effect of KH₂PO₄ and pH on respiration of potato disks at 23°C. From Steward and Preston (218).

as bromide and an alkali cation (K⁺ or Rb⁺) may be accumulated together in equivalent amounts, all these results apply to the cations also. The general effects of pH, CO_2/HCO_3^- and $H_2PO_4^-$ concentrations on respiration and nitrogen metabolism of potato disks are shown in Figs. 36–37.

Thus salt accumulation in potato disks is not to be identified with some easily specified single reaction, but rather it requires a metabolic state which is associated with a definite, but prescribed, ability of the cells to grow. While the tissue uses metabolic energy to accumulate salts, the conservation of that energy seems very wasteful, the over-all metabolic turnover being far greater than that which is necessary to release energy for salt accumulation alone. The metabolism creates the over-all condition in the cell which renders it capable of salt accumulation, but, this being done, the cell then takes salt accumulation "in its stride." The focal point in the metabolism where the effects of salts and oxygen converge upon salt accumulation is that part of it where respiration and protein synthesis connect. K⁺ stimulates and Ca⁺⁺ depresses that synthesis, no doubt by affecting the release of energy for synthesis as mediated by phosphorylation and dephosphorylation. To understand salt accumulation in this system it is, therefore, necessary to visualize how energy could be donated, simultaneously, to protein synthesis and to salt accumulation and how this energy could emerge from a part of the respiration which bears characteristic relations to salts and to oxygen.

At this point the further development of these ideas impinges upon the subject matter of other chapters and they can only be mentioned here in summary form. The main points are as follows:

First, the regulatory control of carbon dioxide over metabolism is now known to be due to its role as a metabolite in dark-fixation processes. Complete starvation of cells of carbon dioxide can amount to a block, or limitation, in the mobilization of carbon for carbon dioxide evolution via the Krebs cycle, and this may bring into play alternative substrates for respiration. These alternative sources may require different oxidases. Whether this entire lack of carbon dioxide is brought about by light, or by removal by aeration, one can now visualize, under these circumstances, that the carbon which eventually enters the respiratory pathway may do so, not immediately from sugar, but from the amino acids. The nitrogenous groups so released are reconverted to protein, using carbon from sugar, when synthesis is stimulated by K⁺ and the residual carbon (keto acid) framework after deamination and deamidation is respired away (211).

Chromatographic studies show that the range of soluble nitrogen compounds present in the tissue is *not* in the expected proportions to form protein. The soluble compounds are present, not as the immediate precursors of synthesis, but as storage products and as the products of protein breakdown. The ultimate seat of protein synthesis, no doubt at a pentose nucleic acid surface, is the location in the cell at which the respiratory energy, probably mobilized in phosphate compounds, is mediated for both the processes of ion accumulation and protein synthesis.

By the use of tissue culture techniques and of substances that promote cell division in preference to cell enlargement, tissue from storage organs can be put, at will, into the continuously dividing state or into the state in which enlargement, but little division, may occur. This has been done mainly for carrot phloem and Jerusalem artichoke tuber explants. The consequential effects on ion accumulation have been investigated by the use of Cs^{137}, supplied at will in the presence, or absence, of a carrier. An outstanding case is that of the artichoke tuber disks far enough along their time drift so that the initial intake of salt and high surface respiration had declined. On the application of coconut milk, containing cell division growth factors, respiration increased but the Cs^{137} concentration adjusted to a low level of accumulation. When the tissue was deprived of the coconut milk and absorbed water, without dividing, it adjusted to a *higher* level of Cs^{137} accumulation, though with a lower level of general respiration and metabolism. The newer ideas (216) to which these results give rise are as follows:

In the dividing cell, ions (particularly cations) are bound stoichiometrically to sites which multiply while cells are in the state of growth by division in which self-duplication occurs. The metabolism which is here involved is that of the dividing cell and it differs in many interesting respects from that of cells not so actively dividing. Cells in this state are typified by a low content of soluble nitrogen and a high content of protein. In terms of concentration in the *total water* of the cell the ionic concentrations at this stage are not high, though they are greater than in the external solution. In this state of the cell, the novel feature seems to be that the accumulation ratio for ions does not increase greatly with dilution, as it more commonly does in cells which are now seen to be growing also by enlargement.

When the cells pass from the actively dividing state, or when the continued binding of ions at synthetically produced sites is completed, the ions move into the vacuole. In this second stage of growth and by this process the concentrations attainable in the total water of the cell are much greater. Normally, no doubt this process of secretion is preceded by the ionic binding, just as cell expansion cannot occur without prior division. In both types of ionic accumulation in cells the processes of growth and metabolism are involved. In one case (ionic binding to specific cytoplasmic sites) it is the metabolism of cells that are multiplying, but not conspicuously enlarging, that is involved; in the other (secretion into the vacuole) it is the metabolism of cells that grow mainly by enlargement.

In the earlier work on storage tissue and on *Nitella*, the first phase of absorption as described above was submerged in the observations on the second. For this reason the growth scheme of 1935 left open the possibility that, by extrapolation, dividing cells would have the highest

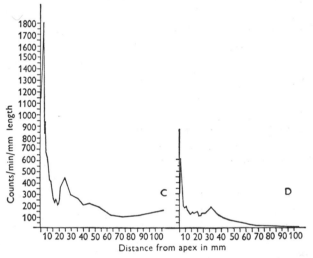

FIG. 38. Distribution of accumulation of Cs¹³⁷ along axis of *Narcissus pseudonarcissus* roots. From Steward and Millar (216).

degree of accumulation. It is now known that this is not so, but the highest degree of accumulation occurs *at the point of most rapid cell growth by enlargement*. In most ordinary situations these two stages so rapidly follow that they may well be inseparable. But it can now be shown, by the use of isotopes (Cs¹³⁷) that at the very tip of a root the concentration of absorbed ions in the total water is less than that just behind the root apex (Fig. 38).

Some early data—long inexplicable—obtained with potato tissue, may now be explained. In strong salt solutions (0.05 to 0.075 equivalents per liter) the normal gradient of absorbed bromide to the surface is interrupted in ways that suggest that the *maximum* salt content of the outermost cells is fixed at a limit below that of cells immediately

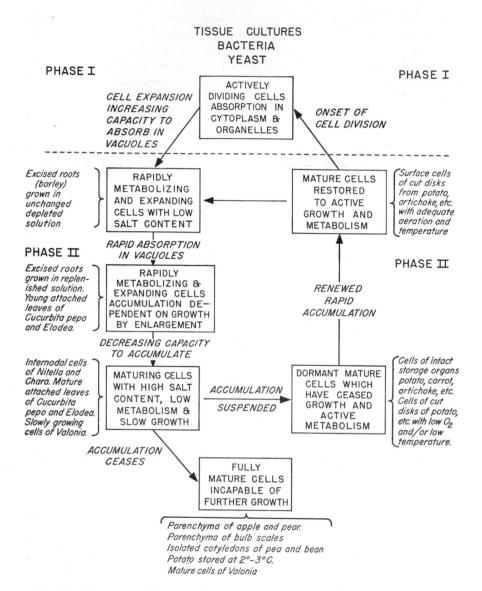

TISSUE CULTURES
BACTERIA
YEAST

PHASE I

CELL EXPANSION
INCREASING
CAPACITY TO
ABSORB IN
VACUOLES

ACTIVELY
DIVIDING CELLS
ABSORPTION IN
CYTOPLASM &
ORGANELLES

PHASE I

ONSET OF
CELL DIVISION

Excised roots
(barley)
grown in
unchanged
depleted
solution

RAPIDLY
METABOLIZING
AND EXPANDING
CELLS WITH LOW
SALT CONTENT

MATURE CELLS
RESTORED
TO ACTIVE
GROWTH AND
METABOLISM

Surface cells
of cut disks
from potato,
artichoke, etc.
with adequate
aeration and
temperature

PHASE II

RAPID ABSORPTION
IN VACUOLES

Excised roots
grown in replen-
ished solution.
Young attached
leaves of
Cucurbita pepo
and Elodea.

RAPIDLY
METABOLIZING &
EXPANDING CELLS
ACCUMULATION DE-
PENDENT ON GROWTH
BY ENLARGEMENT

PHASE II

RENEWED
RAPID
ACCUMULATION

DECREASING CAPACITY
TO ACCUMULATE

Internodal cells
of Nitella and
Chara. Mature
attached leaves
of Cucurbita
pepo and Elodea.
Slowly growing
cells of Valonia.

MATURING CELLS
WITH HIGH SALT
CONTENT, LOW
METABOLISM &
SLOW GROWTH

ACCUMULATION

SUSPENDED

DORMANT MATURE
CELLS WHICH
HAVE CEASED
GROWTH AND
ACTIVE
METABOLISM

Cells of intact
storage organs
potato, carrot,
artichoke, etc.
Cells of cut
disks of potato,
etc. with low O_2
and/or low
temperature.

ACCUMULATION
CEASES

FULLY
MATURE CELLS
INCAPABLE OF
FURTHER GROWTH

Parenchyma of apple and pear.
Parenchyma of bulb scales
Isolated cotyledons of pea and bean
Potato stored at 2°-3°C.
Mature cells of Valonia

Fig. 39. Salt accumulation in relation to growth and metabolism. In Phase I, the main emphasis is upon binding of ions of specific sites which can multiply. In Phase II, the main emphasis is upon active secretion into vacuoles. Modified from Steward (205).

within. This may be due to the operation of a similar effect as in the root apex

On these ideas, the syntheses involved in self-duplication or multi-publication of binding sites on protein account for initial cytoplasmic accumulation. The subsequent secretion to the high concentrations of the vacuole would be typical of cells that, having accomplished this phase, are now most characterized by cell enlargement and the particular kind of metabolism and protein synthesis which this entails. No doubt sites which are vacated by secretion into the vacuole may be reoccupied by entering ions.

The scheme of 1935 attempted to relate salt accumulation to growth, development, and the nutritional status of cells and organs; one may now extend this to bring it into accord with later and modern knowledge. The chief requirement is to recognize that salt accumulation reaching high concentrations in the vacuole begins with the onset of vacuolation and that prior to this, in dividing but relatively nonvacuolated cells, a different mechanism obtains. In dividing cells ions are bound, stoichiometrically, at sites which multiply [Phase I of the absorption process, after Steward and Millar (216)] and subsequently they accumulate in vacuoles as the latter develop (Phase II). These ideas are summarized in Fig. 39.

D. EXCISED ROOTS

(This section was written by the authors in collaboration with
T. C. Broyer)

With higher plants, roots are the primary organs for the absorption of mineral elements and water. It is for this reason that they have received so much attention. The morphology of the root imposes problems in the interpretation of results. The use of excised roots obviates some interrelated effects between shoot and root, although it, in turn, imposes certain limitations on the absorptive capacity of the isolated root system. The use of roots detached from the shoot came into special prominence in the investigation of salt accumulation in the 1930's, due principally to the work of Hoagland and Broyer on the one hand, and Lundegårdh and Burström on the other. Brief résumés of these investigations will now be presented. Lundegårdh preferred to use decapitated plants rather than submerged excised roots, since this obviated exposure of the cut surfaces to the solution and eliminated any artificial effects of the exudate on the culture media when root systems are completely immersed, as in most of the experiments of Hoagland and Broyer. In the latter case the net effect of absorption and exudation

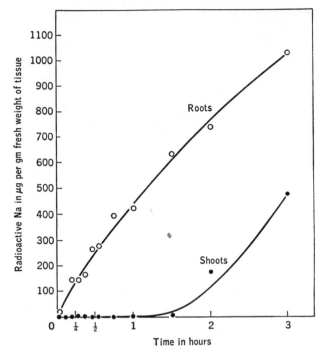

A. Rate of absorption and translocation of radioactive sodium in barley plants of low-salt, high-sugar type. Barley plants were grown from September 29 to October 18 (19 days) by standard technique, 168 plants per set. Experimental culture solution 3000 ml containing radioactive Na as sodium chloride. From Broyer and Hoagland (30).

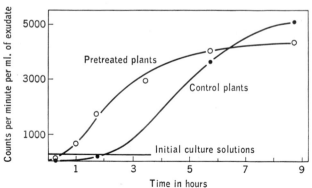

B. Exudation of radioactive K from decapitated barley plants. Pretreated plants, initial salt content increased before beginning absorption experiment. Control plants, initial salt content low. Sacramento barley plants grown from April 12 to May 1: 19 days growth, in standard nutrient solution (half strength). Exudation period 8¾ hours. Experimental culture solution of composition 0.0008 N K*Cl, aerated continuously. Temperature 25°C. From Hoagland (85).

FIG. 40. Salt absorption by excised barley roots.

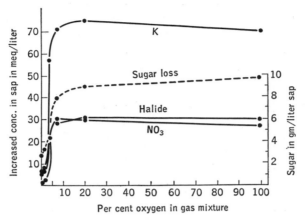

C. Relation of oxygen tension in flowing gas stream to accumulation of salt by excised barley root systems. From Hoagland and Broyer (87).

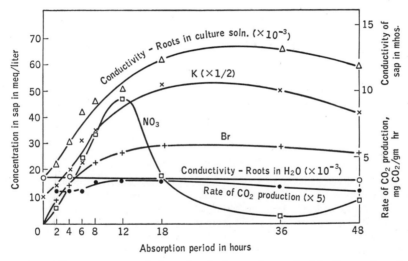

D. Time effects on accumulation of salt in sap by excised barley root systems. From Hoagland and Broyer (87).

FIG. 40 (*cont.*). Salt absorption by excised barley roots.

is measured by following the changes in the composition of the expressed sap.

With roots, ionic movements of two kinds are in question: (a) those across individual protoplasts into vacuoles of parenchyma cells, and (b) those across tissues into the stele. Though predominantly it is the inwardly directed movements of salt and water that are most in question, it should not be forgotten that some other components of the system

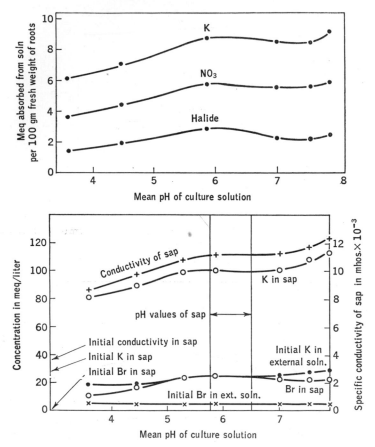

E. Effect of hydrogen ion activity of buffered culture solutions on salt absorption. Above: absorption based upon data obtained from culture solution analyses. Below: accumulations as indicated by composition of composite expressed sap. Plants grown from August 7 to August 29. Excised roots were exposed for 9 hours to culture solutions (KBr 0.005 M; KNO$_3$ 0.005 M) buffered with phosphate (approximately 0.01 M). From Hoagland and Broyer (88).

Fig. 40 (cont.). Salt absorption by excised barley roots.

are also moving across these barriers, and the resultant "flux" is directed and controlled by the growth and development of the cells and organs and of the plant as a whole. In the cell the inner phase may be regarded as a more "closed" one, whereas in the case of the stele it can be regarded as more "open" by the ready removal of solutes to the shoot (27). Thus steady-state conditions may be achieved for the root as a whole as movements of ions into the root are balanced by movements upward to the shoot. The simultaneous movement from root to

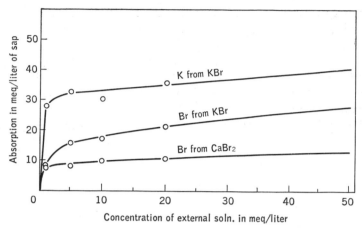

F. The effect of concentration of the culture medium on absorption of KBr and CaBr₂ solutions by excised barley roots. From Hoagland (85).

FIG. 40 (*cont.*). Salt absorption by excised barley roots.

shoot, concomitantly with absorption by the root, is particularly well illustrated by Figs. 40,A [cf. (30), Fig. 3, page 265] and 40,B [cf. (85), Fig. 8, page 191], for these show that the supply to the shoot will follow more or less promptly upon absorption by the root according to the previous salt supply.

1. Factors Which Affect Ion Uptake by Root Systems

About 1929 or 1930 Hoagland returned to the consideration of ion uptake by root systems which had engaged his attention almost ten years earlier. In the interim the outlook had entirely changed as a result of work described under heading II A, B, and C above. Hoagland now enquired how far the ideas derived from the other systems mentioned were applicable to root systems. This phase of investigation is documented in a group of papers identified principally with the names of Hoagland and Broyer and with the activities of Hoagland's laboratory in this period.

Hoagland's first preoccupation was with experimental technique to give the investigation of roots the precision desired. This was achieved by a device which not only did this but also laid stress upon a variable which is particularly important in the behavior of the root, namely its initial salt status, which is determined by the balance between internal supply during growth and removal to the shoot. Hoagland and Broyer achieved root systems which were extremely active and reproducible in their ability to accumulate ions, by growing genetically uniform seedlings of barley under conditions which furnished a minimal supply of external nutrients (particularly nitrate). These conditions were fully described in 1936 in a well-known paper (87). Excised root systems from such plants were high in organic solutes (sugars), low in salts and, when transferred to salt solutions under experimental conditions they had, for a brief period at least, a great and reproducible ability to absorb ions. Such root systems were designated "low-salt roots" in contrast to the less actively absorbing "high-salt-roots" obtained from plants fully furnished with nutrients during growth. This contrast is shown in Table XIX [modified from work of Hoagland and Broyer (87)]. Roots

TABLE XIX

Relation of Initial Salt Status of Root System to Absorption of Salt in Subsequent Period[a]

Conditions of experiment	Absorbed from solution (meq/liter)	
Experiment A	K	Br
High-salt, excised roots	−0.20	0.50
Low-salt, excised roots	2.66	1.90
Experiment B	K	NO₃
Entire plants	2.64	2.56
Excised root systems	2.53	2.43
Experiment C	K	NO₃
High-salt, entire plants	2.05	2.40
High-salt, excised root systems	0.67	0.89
Low-salt, entire plants	3.66	3.49
Low-salt, excised root systems	3.02	2.93

[a] From Hoagland and Broyer (87).

which had been depleted of their initially high sugar content, with consequential decrease in their ability to accumulate salts, could be benefited somewhat by an exogenous supply of sugar (88) as shown by Table XX quoted by Broyer (27a). An exogenous sugar supply is, however, not as effective in promoting absorption as an endogenous supply.

Profiting from the work done with bromides on *Nitella* and storage tissues, Hoagland and Broyer examined the response of such excised root systems to the principal external variables with particular reference to the concomitant effects of the external conditions upon the absorption of the bromide ion and upon the respiration of the roots, as measured by carbon dioxide output. In addition the changes in the internal composition of the roots, which accompany the rapid uptake of salt, were measured with particular reference to their sugar and organic acid content, total electrolyte content as measured by electrical conductivity, and to the constituents of the sap responsible for its buffer capacity.

TABLE XX

EFFECTS OF SUGAR SUPPLIED TO CULTURE MEDIUM ON ACCUMULATION OF POTASSIUM AND NITRATE[a]

Preliminary treatment[b]	Treatment[c] during absorption period	Absorbed from solution (meq/liter)		Sugar in sap (gm/liter)	
		K	NO₃	Initial	Final
Experiment A					
1. Aerated in tap H₂O	Aerated	0.87	1.61	2.3	0.3
2. Aerated in tap H₂O + glucose	Aerated	1.97	2.57	7.2	1.5
3. Aerated in tap H₂O	Aerated plus glucose	1.74	3.46	2.3	0.8
Experiment B					
1. No treatment	Aerated	1.71	2.05	15.1	2.5
2. Aerated in tap H₂O	Aerated	0.86	0.80	2.3	0.3
3. Aerated in tap H₂O	Aerated plus glucose	1.53	3.32	2.3	0.2

Concentration of K in Sap, Experiment B
(Initial conc. of K in sap approx. 25)
meq/liter
1.   93.6
2.   63.0
3.   89.6

[a] From Broyer (27a).
[b] Experiment A: 3 days; Experiment B: 4 days.
[c] 27 hours.

Briefly the conclusions which emerged from these experiments were as follows.

a. The effect of oxygen pressure. As in the case of the disks of storage tissue, so in the case of roots, the active uptake of bromide requires a maintained optimum level of oxygen concentration to maintain in turn a high rate of aerobic respiration. The general form of the pO₂, bromide absorption, respiration relationship was similar for roots, Fig. 40,C (87) and storage tissues (Figs. 22 and 23), but with the important difference that the limiting oxygen concentration at which the process was oxygen-saturated was lower in the case of the roots (5% instead of 20% in the gas phase). A similar result was also obtained with potato roots (209).

b. *The effect of temperature.* The temperature coefficient of bromide absorption by excised roots was high, although the interaction of time and temperature effects gave variable values for the Q_{10}. In general accumulation of bromide in the expressed sap virtually ceased at temperatures approaching 0°C. These observations were later confirmed by Ulrich (240).

c. *The effect of the accompanying ions in single salt solutions; mixed salts; competition or stimulation.* The effects on the electrical conductivity of the sap clearly showed a net increase of total electrolyte during absorption by the roots, presupposing that both anion and cation were being absorbed—a result confirmed by actual analysis of accompanying anions, Fig. 40,D (87). An outstanding result, already evident in Fig. 40,C above, was the rapid metabolism of sugar, which had accumulated during the "low-salt" type of growth, not only to carbon dioxide but also to organic acids; these furnished organic anions which played a prominent part when there were discrepancies between the uptake of anions and cations from the external solution. Changes in the organic acid content of the tissue during absorption were shown by the effects on the titration curve of the sap. These studies were extended by Ulrich (239), who furnished data on the relationships between salt absorption, respiration, and the loss of sugar and organic acid. These data showed that the excess of cation over anion absorbed was largely balanced by organic acid anions produced from sugar.

Following the general experience, univalent anions were more readily absorbed by roots along with univalent cations than with divalent cations [Fig. 40,F (85)]; similar effects of the anion on the absorption of a common anion were observed. In mixed solutions, one ion may affect the absorption of another of similar sign by competing for an ion of opposite charge: this was particularly so in cultures containing a mixture of salts with monovalent radicals.

d. *The effect of the pH of the external medium.* With the hypothesis of Osterhout in mind, and in contrast to the conditions which normally prevail in marine plants, Hoagland and Broyer showed that accumulation of bromide could readily occur from solutions as acid, or even more acid, than the cell sap; thus disposing of any idea that, in this system, absorption of the potassium ion was dependent upon a positive gradient of "undissociated free base" between solution and cell sap, Fig. 40,E (88).

Using buffered solutions over the pH range 3.5–8.0 it was found that the absorption of cations and anions did not vary appreciably. Within this range, the greatest accumulation of potassium, nitrate, and chloride occurred near the pH of the barley root sap (i.e., pH 6.0). When similar tests were made of the effect of the external pH on the composition of the exudate from roots similar results were obtained (88).

When absorption of the ions of a salt occurs from unbuffered solutions the pH of the medium which bathes the excised roots of barley may change. The direction and magnitude of the pH shift varied with time, the conditions of aeration and temperature, and with the previous salt status of the tissues, which in turn determined their content of sugar. In this general study of the effects of external pH on absorption of ions by roots the effects of unequal intake of anions and cations upon the buffer value of the sap were first appreciated. Where cations were absorbed in excess of anions the buffer capacity in regions of pH 7.8–2.6 increased, indicating that organic acid anions were formed to fill the deficit of anions; with an excess of anion uptake over cation a decrease of organic acid anions was observed to occur. Subsequent work by Ulrich (239) and Jacobson and Ordin (100) confirmed these effects and showed that malic acid was the acid most affected during the processes of absorption.

It is now convenient to refer to certain later experiments, on the effects of pH, by Fawzy and associates (63). Not surprisingly, losses of ions occurred from roots exposed to solutions more acid than pH 3.0, and these effects were attributed to displacement of other ions on binding sites by hydrogen at these low pH's. It is difficult, however, to dissociate effects at so acid a reaction from injury to the living system.

e. The effect of external solute concentration. The effect of concentration upon the absorption of ions by excised roots has been extensively investigated. Total solute concentration has little effect per se on the absorption of a salt by excised barley roots. By the use of glycerol, Broyer (28) increased the total solute concentration to levels 1½ times that of the expressed sap of the roots without greatly affecting the absorption of potassium bromide supplied at a constant concentration.

Typical data on the effect of concentration on the absorption of a single salt are given in Fig. 40f (85) and, later, experiments showing concomitant effects on other metabolic changes were made (77, 240).

The effect of higher external concentration is to reduce the gradient from external solution to cell sap and to decrease the energy required per molecule absorbed, or retained, by the cell. The effects observed can be explained on these general lines, and in this respect they resemble those observed with storage tissue.

f. The effect of respiratory inhibitors. Throughout these experiments conditions which affected the bromide intake also affected the respiration, but, as in the case of storage tissues, there was a similarity in these effects, not a direct stoichiometrical relationship. These results were interpreted in a manner consistent with the findings with *Nitella* and disks of storage tissue that the aerobic respiration was the ultimate energy source for a process that requires it to be expended. For excised roots these effects were well illustrated by Hoagland and Broyer (87) and Handley and Overstreet (77).

Reference should now be made to effects of specific inhibitors of salt absorption and respiration, as illustrated in the work of Machlis (135) and of Hoagland and Broyer (89). Numerous substances which also affect metabolism in various ways modify the absorption of ions; namely, glycolytic inhibitors, substances which uncouple phosphorylation, or which block the Krebs cycle, inhibitors of electron transfers to α-keto acids, and inhibitors which affect the oxidases containing heavy metals—all these depress salt accumulation. All these respiratory regulators except dinitrophenol (DNP) variably suppress aerobic respiration; DNP on the other hand may at some concentrations increase respiration.

It is pertinent here to point to differences, as well as to similarities, in the results obtained by similar techniques with storage tissues and with roots. Both sets of data stress the importance of oxygen and aerobic metabolism, both stress the necessity for cells with the necessary degree of metabolic activity if salt accumulation is to occur. Because of the nature of the excised monocotyledon root and the device of using "low-salt" roots, the work on barley roots placed greater emphasis on the previous salt status of the root and less upon its ability for further growth in contrast to the storage tissues. (These ideas were incorporated into the unified scheme of Fig. 39.) Whereas the limiting metabolic reactions of potato disks which are linked to respiration and salt accumulation concern nitrogen metabolism and protein synthesis, they seem to concern organic acid metabolism in the case of the nitrogen-starved, excised, roots of barley. These differences, however relate more to the detailed mechanism by which the metabolic energy is brought to bear upon salt accumulation: in both systems the energy is required for the accumulation process.

2. The Concept of Anion Respiration: Relations to Cytochrome

In 1932, Lundegårdh compiled a comprehensive work entitled "Die Nährstoffaufnahme der Pflanze," in which he reviewed much of the early work (123). He also added material from his own investigations with higher plants, and his discussion covered methods of study for solute absorption, permeability in plant systems, diffusion and ion exchange, and early observations on salt accumulation in cell vacuoles and xylem exudates. Against this background Lundegårdh devised a

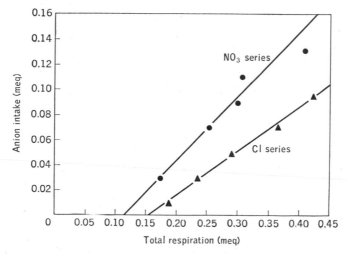

FIG. 41. Relation of anion uptake and respiration in two series of experiments with KNO₃ and KCl. From Lundegårdh and Burström (130).

technique for simultaneous measurement of ion absorption and respiration by roots of intact seedlings in water cultures. Extensive experiments with numerous single salt solutions led Lundegårdh and Burström (130) to deduce a relationship between the respiration of the roots and ion absorption. They concluded that anions are accumulated metabolically in relation to an enhanced aerobic respiration, while cation accumulation is primarily by exchange and is indirectly associated with the anion absorption process. The empirical relationship $R_t = R_g + K \times A$ was proposed in which R_t represented the total respiration (measured in equivalents of CO_2 evolved), R_g, a "ground" or fundamental respiration believed to be characteristic of roots which were not absorbing anions and which was derived by extrapolation as in Fig. 41; and K, a respiration coefficient specific for each anion, and which indicates the amount of additional respiration observed during accumulation; and A represented the equivalents of anion absorbed. From the early data with intact wheat (*Triticum* sp.) seedlings Lundegårdh and Burström concluded that the anion absorption coefficients (K) for NO_3, Cl, and SO_4, were 2, 3, and 6, respectively. The test of this empirical relationship with the original data is shown in Table XXI in which a comparison was made between the total respiration as observed

TABLE XXI
ANION UPTAKE AND RESPIRATION
TOTAL RESPIRATION (R_t) = GROUND RESPIRATION (R_g) + K ANION RESPIRATION (A)

NO₃ Series (meq.)					
Cation	Na	K	Mg	Ca	Ba
A (Anion Respiration)	0.049	0.073	0.068	0.105	0.044
KA (where $K = 2$)	0.098	0.146	0.136	0.210	0.088
R_t (observed)	0.205	0.248	0.261	0.325	0.278
R_t (calc.; $R_g = 0.115$)	0.213	0.261	0.251	0.325	0.203
Cation uptake	0.027	0.064	0.027	0.026	0.021

Cl Series (meq.)							
Cation	Li	Na	K	Rb	Cs	Mg	Ca
A (Anion Respiration)	0.036	0.026	0.042	0.036	0.031	0.055	0.070
KA (where $K = 3$)	0.108	0.078	0.126	0.108	0.093	0.165	0.180
R_t (observed)	0.235	0.231	0.264	0.262	0.239	0.314	0.323
R_t (calc.; $R_g = 0.148$)	0.256	0.226	0.274	0.256	0.241	0.313	0.357
Cation uptake	0.027	0.032	0.051	0.036	0.028	0.038	0.033

SO₄ Series (meq.)			
Cation	Na	K	Mg	Ca
A (Anion respiration)	0.030	0.034	0.042	0.013
$6A$ ($K = 6$)	0.180	0.204	0.252	0.078
R (observed)	0.298	0.294	0.349	0.201
R_t (calc.; $R_g = 0.110$)	0.290	0.314	0.362	0.188
Cation uptake	0.051	0.044	0.040	0.016

and that calculated by use of the derived values for R_g and K and the observed values for A. It was this apparent quantitative relationship which originally led Lundegårdh to propose the anion-respiration theory.

Following this early work of Lundegårdh and Burström, much experimental work has been done to substantiate the theory. A series of papers have continued to emphasize the relationship between anion absorption and the enhanced respiration. While the fundamental concept has been retained, modification and additions to the original proposal have been made in the light of more recent experiments.

a. The effects of accompanying cations. In an early work the conclusion was reached that the total respiration rises and falls with the anion uptake but is to a great extent independent of the cation absorption and of the cation-anion imbalance (130, 131). In later publications the independence of respiration and cation absorption was minimized (124, 132). However, the coefficient K was found to vary, not only with the kind of anion absorbed (increasing with decreasing mobility of the anion in the organ), but also with the kind of accompanying cation. The coefficient was generally smallest when the cation is most easily absorbed and changed according to the series of cations as follows: Na < K < Mg < Ca < Sr < Ba. The cause of the cation influence is not related to the amount of cation absorbed, but to their mere presence in the solution.

b. The effects of concentration of the salt. As others have also shown, Lundegårdh has observed that absorption by roots varies with the concentration of the salt supplied (126). The data show that absorption is a logarithmic function of concentration, but the evidence also was that the coefficient k varied with the concentration of the salt supplied; as the external concentration was raised, the value of K decreased and vice versa.

c. The effects of oxygen concentration. Lundegårdh, as indeed others have done, found that absorption by roots was not greatly modified by reduction in the oxygen tension of the solution to approximately one-fifth that of solutions in equilibrium with air; but below this, both absorption and respiration were reduced. In an atmosphere of nitrogen, salt uptake was completely inhibited and ground and anion respiration were similarly affected by reduced oxygen. Inhibitory effects of increased carbon dioxide on salt absorption did not become perceptible until the carbon dioxide concentration became relatively high; i.e., more than ten times the concentrations attributable to respiration (130).

d. The effects of hydrogen ion and the role of organic acids. In 1933 Lundegårdh and Burström indicated that there was little effect of external pH on the absorption of salt. However, the differential absorption of cations and associated anions caused the pH of the cultures to change. Thus the large absorption of NO_3^- relative to Ca^{++} from calcium nitrate led to a rise of pH, and the relatively greater absorption of K^+ than SO_4^- from potassium sulfate led to increasingly acid solutions.

The role of organic acids when absorption of cations and anions is imbalanced was recognized (31a, 124). Thus where cations are accumulated in excess of anions, increased production of organic acids balances the excess; the organic acid anions so produced and forming part of the salt accumulated contribute to the enhanced respiration. Where absorption of anions occurs in excess, Lundegårdh and Burström recognize that decreases of organic acid simultaneously occur, either through modified respiration as indicated by the respiratory quotient or in part through loss to the external solution (cf. page 370).

e. The effects of respiratory inhibitors. The effects of aeration and oxygen tension on absorption and respiration have been referred to above. Effects of specific inhibitors of particular metabolic reactions, on respiration and salt accumulation, were also studied by Lundegårdh. For example, in 1935 data were published [Table 7, page 248 of Lundegårdh and Burström (132)] on the effects of KCN, which specifically inhibits cytochrome oxidase. These data show a marked depression of K^+ and NO_3^- accumulation and of total carbon dioxide evolution. At the appropriate concentrations of cyanide "ground respiration" was not affected but "anion respiration" was inhibited in ways that were interpreted to mean that the cytochrome system is involved in the salt uptake. Subsequently Lundegårdh has confirmed the presence, and has studied the activity, of the cytochromes in wheat roots; and in 1951 he reported data, from spectrophotometric measurements, showing the concomitant decrease of salt absorption and of cytochrome activity with increasing concentration of KCN in the medium bathing the roots (127). The ground respiration, catalyzed in part by manganese (124) is characterized by insensitivity to cyanide; the "anion" respiration is independent of manganese supply, but is involved in nitrate reduction in the roots. Effects of other enzyme inhibitors, for example dinitrophenol, were also later recognized by Lundegårdh.

Lundegårdh's thinking in 1940 was extensively summarized (124) and an epitome of his views follows.

f. Successive schemes of Lundegårdh.

(1) The absorption of ions is coupled with an aerobic respiration process, the "anion respiration," the respiratory quotient of which is regarded as unity. The anion respiration is sensitive to oxygen and to KCN. Simultaneously KCN also checks the absorption of anions.

(2) A quantitative relation exists between the absorption of anions and the anion respiration, and a constant in this relationship is the coefficient k, which has specific values for different anions. It should be noted that the constancy of this relationship was first emphasized, but later it is allowed that k may increase with the increasing mobility of the anion, with increasing concentration gradient, and with increasing "negative electrical charge on the surface of the root" [(124),

see page 389]. Lundegårdh also postulated that the nature of the cation affects the value of the anion respiration by determining the value of the coefficient k. These later qualifications seem greatly to weaken the concept of the relationship between anion respiration and ion absorption as it was originally enunciated.

(3) According to Lundegårdh it is only necessary to explain the accumulation of one ion of a salt because, from electrochemical considerations, the active absorption of one ion will necessarily be followed by the absorption of another of opposite sign.

(4) By contrast with the anion respiration, the "ground" or "fundamental" respiration is insensitive to cyanide, though stimulated by manganese.

(5) The respiration of roots varies along the axis, being greatest nearest the tip where the absorption of anions is also greatest. Recognizing this, it again becomes difficult to equate the respiration of the *whole root system* quantitatively with the absorption, lacking positive evidence that the same cells are involved in both effects. In point of fact, however, this was what Lundegårdh attempted to do in the first experiments which have been described and from which the anion respiration concept emerged.

(6) As early as 1940 Lundegårdh was already implicating cytochrome directly in the ion absorption process. This was visualized as occurring in the following way:

The simplified diagram, Fig. 42, based on Lundegårdh [(124), Figs. 34 and 36], summarizes his essential ideas at that time. Across the cell membrane there is conceived to be an oxidation-reduction potential gradient, depending upon a high oxygen concentration at the outer surface and lower oxygen concentration at the inner surface. Elements of the redox system, represented by X in the diagram, are regarded as capable of becoming charged by loss of electrons, and they can then combine with anions at the outer surface. It was suggested that the anion passes from element to element traversing the membrane in this fashion. At the inner surface, the molecule X in the redox system becomes again reduced, the charged anion passing into the vacuole. Correspondingly there is a wave of electrons passing from the outer to the inner surface where the positive charges in the redox system are eliminated by reacceptance of their electrons. Already Lundegårdh was thinking that the elements of the redox system X could possibly be identified with hemin iron in the "shape of cytochromes." This was clearly the germ of the concept which has later been much elaborated.

Cations, on the other hand, were conceived to enter by direct combination with hypothetical negatively charged regions in the protoplasm and were supposed to move along "adsorption tracks" across the membrane. At this time Lundegårdh conceived that the cations reaching the inner surface were displaced by hydrogen ions from the cell sap or the vacuole, and, in consequence, he visualized a movement of hydrogen ion in opposition to the flow of entering cation. Clearly, this was not a truly workable scheme because unequal numbers of charged anions and cations were being transported from the outer to the inner surface. Anticipating later developments of these concepts, one may see that this difficulty may be removed by recognizing the hydrogen ion at the

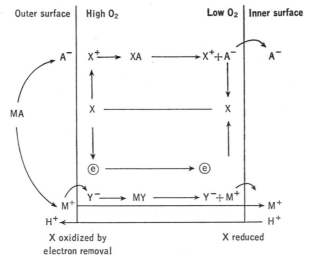

Fig. 42. Diagram summarizing Lundegårdh's concept of ion absorption in 1940. X-elements in a redox system capable of combining with the anions and possibly "heme iron in the shape of cytochrome." Y-hypothetical negatively charged carrier for cations. Based on Lundegårdh [(124), Figs. 34 and 36].

outer surface could re-form water, with the simultaneous acceptance of electrons.

The further elaboration and clarification of this scheme presupposes the contributions which were made by Robertson. These will be discussed below. However, by 1953, in the light of all these developments, Lundegårdh had modified his views substantially (128). These may now be briefly summarized as follows (see Fig. 43):

(1) The idea that there is a gradient of oxidation-reduction across the cytoplasm is still retained in this scheme, but the nature of the redox element is now positively identified with cytochrome, the metal of which can exist with three or two positive charges—that is in the ferric and ferrous state respectively.

(2) The ferric iron of cytochrome is thus regarded as able to combine with three monovalent anions, forming a complex which can physically move from the outer to the inner surface of the membrane where reduction of the cytochrome occurs, and simultaneously one of the anions is released on the inside. Movement of the cytochrome molecules is thought to occur at random by thermal agitation. Movement over only

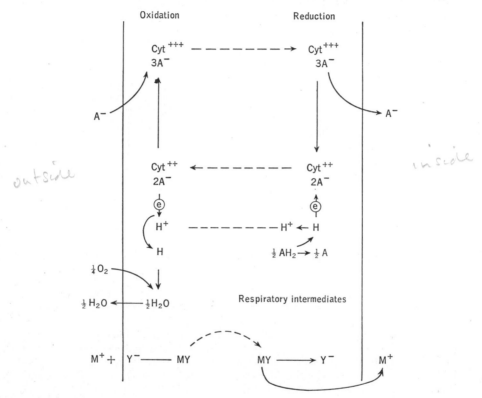

Fig. 43. Diagram summarizing Lundegårdh's concept of ion absorption in 1953. Based on Lundegårdh (128).

short physical distances is supposed to occur through the agency of a single molecule; but transport may be visualized over longer distances by a "handing on" of the anion from one cytochrome molecule to another across the barrier.

(3) To bring about the reduction of cytochrome at the inner surface one electron is required, and this is conceived to arise from reduced respiratory intermediates represented in the scheme as shown by the symbol AH_2. As the reduced substrate (AH_2) is oxidized, the atomic

hydrogen thus released becomes a hydrogen ion by release of one electron. In turn the oxidized substrate A becomes re-reduced, acting as a hydrogen acceptor presumably through the Krebs cycle. The hydrogen ions so formed move along a gradient from the inner to the outer surface, where they are available to accept an electron from reduced cytochrome, which thus becomes oxidized, and the molecular hydrogen now formed is donated to oxygen at the outer surface to produce water, this being the end product of respiration.

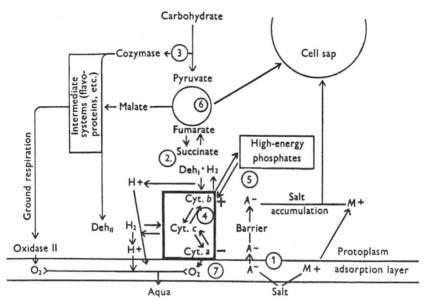

Fig. 44. Diagrammatic representation of the aerobic respiration in wheat roots and its linkage to salt accumulation. The encircled figures indicate points sensitive to inhibitors: (1) the coenzymatic effect of anions (A^-); (2) malonate and fluoride inhibiting succinic dehydrogenase (deh.$_1$); (3) inhibitors of cozymase; (4) urethane, inhibiting the oxidation of cytochrome b; (5) DNP, inhibiting phosphorylation and the reduction of b; (6) fluoride, inhibiting the production of organic acids; (7) cyanide, azide, CO, etc., inhibiting the oxidation of cytochrome oxidase. From Lundegårdh (128).

The mode of entry of cations remains as it was visualized in the earlier diagram (see Figs. 42 and 43) except that it is now no longer necessary to conceive of a direct interchange of cation for issuing hydrogen ion as long as an equal uptake of anions and cations occurs. Where the uptake of cations exceeds that of anions, Lundegårdh suggests that the excess cations may be absorbed in exchange for hydrogen ions which are derived from organic acids synthesized within the tissues.

(4) It will be seen from the scheme of Fig. 43 that corresponding to the entry of one univalent anion there is the alternate donation and removal of one electron in the cytochrome cycle; this in turn corresponds to the utilization of one-fourth of a molecule of oxygen in respiration. This is the basis of the idea that the maximum number of monovalent anions that may be transferred per molecule of oxygen used in anion respiration is 4.

(5) All the events described in Fig. 43 were conceived to occur at the outer protoplasmic surface where the primary act of ion accumulation was assumed to occur, the subsequent movement into the cell sap from the cytoplasm being along a diffusion gradient.

(6) While the full implications of this idea can not be appreciated until the work of Robertson has been discussed, Lundegårdh in 1954

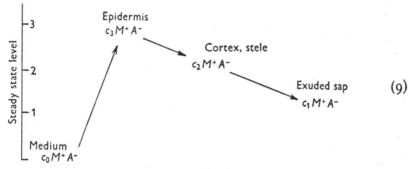

FIG. 45. The steady state of the root tissue. From Lundegårdh (128).

associated the process of anion absorption primarily with cytochrome b. In Fig. 43 all reference to the other cytochromes has been eliminated for simplicity, but they are included in Lundegårdh's detailed scheme reproduced in Fig. 44 [diagram from Lundegårdh (128)].

The comparison of Fig. 42 and Fig. 44 will therefore make clear the progress of these ideas between 1940 and the Symposium of 1953 (128).

(7) In making reference to the problem of secretion of ions into the xylem sap Lundegårdh proceeds as follows: Consistently with ideas derived from the electrochemical properties of the outer surface of the root, Lundegårdh postulated that the important region of the root in which the primary accumulation occurs is the outermost layer of cells (piliferous layer or epidermis). The highest ion concentrations are conceived to exist in these outermost cells. Thereafter there is conceived to be a diffusive spread along a downward gradient across the cortex into the stele and into the exuded sap, if the shoot is removed. This scheme is summarized in Fig. 45. It will be recognized that these ideas

are quite different from those which postulate that the movement of ions into the xylem sap is itself an activated process (see pages 280, 458).

The latest exposition of the views of Lundegårdh to which reference can be made will be found in a review published in 1955 (129). The additional features, over and above the scheme of Fig. 44, are essentially as follows:

(1) It is now recognized that the cytochromes which are conceived to be an essential part of the anion absorption mechanism are not freely distributed at the outer cell surface but are contained in particulate inclusions in the cell. Therefore, the events portrayed in Fig. 43 are now assumed to occur, not at the outermost protoplasmic surface, but at the membrane surface of the mitochondria. This implies that the primary act of ion accumulation occurs not into the cytoplasm itself, certainly not into the vacuole, but into the mitochondria. It should be recognized at this point, however, that these ideas flow mainly from the contributions which were made by Robertson on plants and even earlier by zoologists [for example Bartley and Davies (10)].

It is now recognized by Lundegårdh that metabolically active anions, like phosphate, nitrate, and sulfate, which participate in well-recognized cyclical metabolic events may well be absorbed by mechanisms that are distinct from the anion accumulation system as it has been thus far presented (128, 129). This being so, the anion respiration mechanism now becomes particularly applicable, according to Lundegårdh, to the accumulation of nonmetabolically active anions of which the halides represent probably the only significant example (although, when originally enunciated the concept was applied to nitrate and sulfate as well as to halide).

The account which has been given, necessarily abbreviated, will have shown how the ideas of anion respiration which derived originally from relatively simple and direct experimental observations have been progressively elaborated in response to the changing contemporary ideas of cellular metabolism and respiration. Nevertheless their validity rests upon their ultimate basis in experimental fact and so they should be judged.

E. Later Work on Storage Tissues

1. Disks of Carrot Root

The work of Robertson intervened in this general problem when the views of Lundegårdh were essentially at the stage which has been summarized in Fig. 42. Though obviously influenced by these views of Lundegårdh, Robertson chose to work primarily not with intact roots but with disks of storage organs, primarily carrot root tissue. It will be evident from what follows that the work of Robertson gave definition to certain ideas which were incorporated in the later Lundegårdh scheme, e.g., Fig. 43, and an attempt will now be made to show how this came about and upon what evidence these ideas rest.

In the earlier work, extending from 1940 to approximately 1948, Robertson, working with carrot tissue, was accumulating evidence on the relationship between salt uptake and respiration. The essence of these earlier conclusions may be indicated as at 1945 by the following quotations (173). "The experiments described . . . confirm Lundegårdh's theory of a cyanide sensitive respiration which is induced by salt and on which accumulation depends. . . . We are of the opinion that a distinct salt respiration does exist in some tissues, but the original suggestion of Lundegårdh—that it is an anion effect—cannot be taken as established on the evidence available." Thus while Robertson's views seemed patterned along the general lines of those of Lundegårdh he was not ready, at that time, definitely to identify the absorption with a specific component of respiration which is associated *only* with anions. It should be mentioned at this point that the general technique of Robertson involved manometric measurements of gaseous exchange and the determinations of salt absorption were derived from simultaneous observations on changes in the conductivity of the external solution. To this extent they were not specific for particular ions, and it should be noted that they did not involve, for the most part, analysis of the tissue, nor did they take account of the ions that pre-existed in that tissue.

One of the distinctive contributions which flowed from the experiment of Robertson was a clearer concept of the nature of salt respiration in the Lundegårdh hypothesis by more closely specifying its quantitative relations. The relevant experiments were by Robertson and Wilkins (174). They determined the relation between salt-induced respiration and salt uptake by carrot tissue, using solutions of various halides and a range of external concentrations. As the external concentration of salt increased, the salt respiration increased also, in the manner of Fig. 46. Some representative data, in the manner in which they were analyzed by Robertson and Wilkins are presented in Table XXII, based on their Tables 6 and 7. (Parenthetically, it may be noted that where Robertson and Wilkins speak of chloride *accumulation*, or salt accumulation, they should more appropriately have stated total salt or ion *absorbed*.) The interpretation of Robertson and Wilkins was that while the ratio of equivalents of salt absorbed to the respiration (molecules of O_2 absorbed or CO_2 evolved) induced by the salt was variable, it rarely exceeded the value of 4. In the few cases where the value did exceed 4, Robertson could explain this as due to experimental error. From these data Robertson and Wilkins concluded that the Lundegårdh hypothesis of anion respiration is a feasible one, though it should be emphasized that data of this type do not necessarily prove this mechanism. As quoted by Robertson and Wilkins the values of the ratio of accumulation to anion respiration in the work of Lundegårdh were usually quite low—often less than 1—and other values from other studies were quoted, varying up to the figure of 4 as maintained by Robertson and Wilkins. Values much less than 4 indicate that only

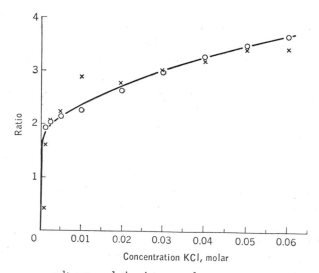

FIG. 46. The ratio $\dfrac{\text{salt accumulation in gm mol}}{\text{salt respiration in gm mol}}$ for carrot tissue in dilute solutions of potassium chloride. From Robertson and Wilkins (174).

a comparatively small part of the salt-induced respiration is specifically devoted to the intake of the salt in question. This dilemma was circumvented by the assumption that this respiration was also related to other accumulatory processes, such as movement of organic solutes in the cells or reabsorption of ions previously in the tissue which leaked out during the experimental procedures.

It should be noted that values of this ratio much higher than 4 would render the hypothesis of Lundegårdh invalid. This stands in contrast to the situation mentioned above where values turn out to be less than 4, for, while these do not help with the proof of the hypothesis, it is possible to rationalize them with it. Since this time, however, it has become quite apparent, with both the animal and plant systems, that formulations of the sort made by Robertson and Wilkins can lead to values of salt-induced respiration to the salt absorbed which greatly exceed 4. Crane and Davies (52), for example, claimed to have obtained values for HCl secreted:stimulated O_2 uptake as high as 11 during active secretion by frog gastric mucosae; while Leaf and Renshaw (116) obtained values from 2 to 13 for Na transported:molecules O_2 consumed with

TABLE XXII

RELATION BETWEEN SALT-INDUCED RESPIRATION AND
SALT UPTAKE BY CARROT TISSUE[a]

Concentration of salt (normal)	Salt respiration [gm mol/gm/hr ($\times 10^5$)]		Chloride accumulation [gm mol/gm/hr ($\times 10^5$)]		Ratio[b]	
	KCl	CaCl₂	KCl	CaCl₂	KCl	CaCl₂
0.01	0.090	0.083	0.240	0.180	2.67	1.60
0.04	0.098	0.110	0.360	0.140	3.30	1.57
0.07	0.118	0.138	0.320	0.170	3.34	1.51
0.10	—	0.122	0.390	0.230	3.36	1.48
0.13	0.114	0.126	0.390	0.180	3.36	1.48
0.16	0.114	0.131	0.380	0.200	3.36	1.48

No. of Exp.	Salt respiration [gm mol/gm/hr ($\times 10^5$)]			Salt accumulation [gm mol/gm/hr ($\times 10^5$)]			Ratios		
	KCl	NaCl	LiCl	KCl	NaCl	LiCl	KCl	NaCl	LiCl
1	0.065	0.155	0.109	0.270	0.299	0.175	4.15	1.93	1.79
2	0.106	0.100	0.057	0.203	0.255	0.136	1.91	2.55	2.39
3	0.121	0.095	0.060	0.196	0.169	0.116	1.62	1.78	1.93
4	0.057	0.080	0.054	0.336	0.310	0.264	5.90	3.88	4.89
5	0.042	0.071	0.054	0.175	0.198	0.128	4.17	2.79	2.37

[a] Reproduced from Robertson and Wilkins (174).
[b] The ratios given in this table are calculated from the curves of best fit to the salt respiration and accumulation data.

a mean value of 6.83. With excised barley roots, Handley and Overstreet (77) calculated values for salt absorbed:O_2 as equal to 4.3–7.9 in KBr solutions, 2.8–4.6 in NaBr, and 9.3–18.7 in CaBr₂ solutions. Sutcliffe (232) has consistently obtained values for anions absorbed:salt respiration > 4 during uptake of KBr, NaBr, KCl, and NaCl by slices of red beet tissue in experiments performed at low temperatures (3–5°C) after the completion of physical uptake.

Robertson then proceeded to a study of the simultaneous effects of the metabolic inhibitor 2,4-dinitrophenol on salt accumulation and on salt-induced respiration, again using carrot tissue (177). These authors show that at certain concentrations this substance stimulated respiration, but the salt absorption was concurrently reduced (Fig. 47). Moreover it was shown that this respiration, as stimulated by DNP was cyanide sensitive, mediated by the cytochrome system, and, under the Lundegårdh hypothesis, it should have resulted in *stimulated* salt absorption. The fact that the salt uptake was *inhibited* under these circumstances therefore indicates that there is no simple and direct relationship between the respiration and the salt absorbed (see also Sections

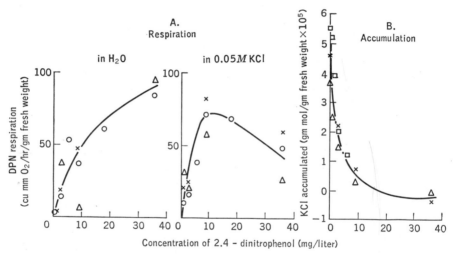

Fig. 47. Relationship between concentration of dinitrophenol (DNP) and (A) the rate of the dinitrophenol respiration (i.e. the rate in DNP over and above that in water or in salt), and (B) amount of accumulation in carrot tissue; X, O, and △ disks from one batch of carrots, respectively 144, 264, and 312 hours from cutting; □ from another batch of carrots, 96 hours from cutting. From Robertson *et al.* (177).

IIC and IID). In spite of this apparent contradiction, however, Robertson and Lundegårdh have adhered to their original hypotheses, but they explain the results obtained with DNP in different ways. It would have seemed more logical, however, to re-examine the bases of the original postulates.

Lundegårdh maintains that the part of the cytochrome system which is especially associated with ion transfer across the membrane is cytochrome b (129). It is supposed that cytochrome b *only* is inhibited by DNP and that, in the presence of DNP, electrons are transferred from the reduced substrate to oxygen by an alternative pathway through the remaining cytochromes. This permits stimulated respiration, sensitive to cyanide but unrelated to ion transfer, to occur. These relations are indicated in the Fig. 44. While Lundegårdh claims to have obtained some spectroscopic evidence in support of this view, it is not evidence which is generally acceptable to workers on electron transfer through the cytochrome system (cf. Volume IB, Chapter 5).

Robertson, on the other hand, makes the suggestion that DNP indirectly prevents the mechanism of accumulation from operating by causing some "disorganization within the cell, possibly in the mitochondria." This suggestion is linked to Robertson's more recent views according to which the mitochondria are primarily associated with the accumulation process. The ideas which implicate the mitochondria specifically in ion accumulation stem from the following observations:

(1) The cytochrome system in cells is primarily located in the mitochondria.

(2) Mitochondria tend to accumulate at surfaces where secretion is prominent, especially in certain animal systems which have been studied.

(3) It has been claimed that mitochondria isolated from animal cells are demonstrably capable of accumulating ions (10).

Robertson et al. (175, 176) have isolated mitochondria from storage tissues, and using these for the very limited duration of the activity of such preparations, have exposed them to known salt concentrations. Then after centrifuging the particles they have determined their content, particularly of chloride and alkali cations, and have attempted from these data to establish the concentrations of ions which they contain. On the necessary assumption that the anions absorbed by the mitochondria are in free solution, it is claimed that active transfer across the mitochondrial membrane occurs. It has to be recognized, however, that whereas the entire cells are capable of producing accumulation ratios which are very large, the demonstrable accumulation ratios in mitochondria are at best *very small*. Robertson's general idea is that the mitochondria may act as the primary organs of accumulation in the cytoplasm, move bodily to the inner membrane, where they tend to break down and thus release ions passively into the vacuole. This implies replenishment of the mitochondria after each active transfer, whereas the mechanism of self-duplication of mitochondria is still quite unknown. An alternative hypothesis is obviously tenable. The mitochondria, like other cellular cytoplasmic inclusions, may be able to maintain independently their own ionic gradients across their membranes, these being necessary for the normal functioning of the particles. Indeed it would be surprising if this were not so. This, however, does not necessarily mean that this phenomenon is an integral part of the transfer of ions into the vacuole across the tonoplast. It is also evident that any metabolic properties of the mitochondria may contribute to the energy requirements of absorption, but the simple view is that the mitochondria produce ATP or some other high-energy source which then, in some way, donates this energy *at the active site* to the process of accumulation at a point which may well be remote from the mitochondrial system. A parallel situation would seem to be found in the relationship between mitochondria and microsomes in protein synthesis as this is now visualized in animal cells. Here the mitochondria may furnish much of the energy, but this is transferred to the microsomal surface, for the act of protein synthesis.

2. *Disks of Beet Root*

The researches of Steward discussed above emphasized the importance of growth, and of protein synthesis as the process identified with growth, in promoting the primary uptake of mineral salts. From these investigations protein synthesis and growth appeared as concomitant manifestations of the active metabolism of cells capable of accumulating salt. It is therefore a problem to decide how far protein synthesis is directly or indirectly implicated in ion absorption in such systems and how far both processes of ion uptake and protein synthesis depend upon similar underlying energy-yielding metabolic reactions.

Sutcliffe (228) began a study of salt accumulation with disks cut from red beet root (*Beta vulgaris* var.). Prior to this Berry and Steward (14) had shown that beet disks, well washed in running tap water, showed an increased rate of respiration when they were first placed in well-aerated solutions, and, like potato disks, they also accumulated bromide. Beet slices, like potato tuber slices, were known to contain cells capable of division in moist air and were regarded in this respect as capable of further growth and development, and cut slices owed their inception of salt-accumulating powers to this recrudescence of growth in a tissue slice. Sutcliffe erected certain criteria designed to establish the absence of growth during the later absorption period used in his experiments. These were:

(1) Failure of the slices to increase in volume as shown by fresh-weight measurements, in a later period subsequent to 24–48 hours after cutting.

(2) No evidence was obtained of cell divisions occurring when cell counts were made by the method of Brown and Rickless (26) on freshly cut slices and on similar slices subjected to prolonged washing in water or salts solution.

(3) Absence of *net* protein synthesis in this later period.

It is not denied that some cell divisions may have occurred, but they were not detected by the method employed, and it is therefore likely that the μ number was small so that new cells did not contribute, as such, significantly to the accumulatory capacity of this tissue. It must be emphasized however that Steward never maintained that cell division actually needs to accompany salt absorption, but rather that the cytological and metabolic effects observed in cells capable of accumulation were those which would logically culminate in cell division under appropriate conditions (e.g., when the slices are kept in moist air).

The absence of *net* protein synthesis does not of course imply that protein synthesis did not occur. This would be manifestly untrue since it is certain that protein synthesis proceeds continuously in all living cytoplasm. What is claimed is that in beet disks protein synthesis did not exceed breakdown in these slices, and in this sense therefore the cells *were not growing in that absorption period.* Thin slices of dormant tissue from the root of the red beet when placed in aerated distilled water do increase in volume by absorption of water for 24–48 hours after cutting, and at the same time they also convert 70% alcohol-soluble nitrogen compounds to insoluble protein. During this time their ability to accumulate salts upon transference to a suitable medium remained low. Gradually however, over a period of several days, the tissue acquired a progressively increasing capacity to absorb salts (Fig. 48) without there being any concurrent measurable increase in volume or over-all synthesis of protein. Therefore, in the sense that the absorption events in the beet disks are associated with tissue which has, compared with the original tissue, undergone a "recrudescence of activity of a kind which is identifiable with growth," the beet root resembles the potato tissue. As will be shown, however, Sutcliffe addressed himself to the difference that whereas in the potato *both* the metabolic and growth effects and the ion uptake were somewhat progressive and concomitant in time, in the beet root the ionic phenomena gather pace while the growth phenomena (uptake of water and protein synthesis) are soon arrested. The somewhat converse situation in Jerusalem artichoke tuber disks in which *both* growth phenomena and ion accumulation are quickly arrested has been discussed (see page 350 et seq.). Because the situation in beet disks is instructive, the work of Sutcliffe will now be summarized.

The development of an increased absorptive ability in beet disks does, however, depend upon metabolic processes since it was retarded by low-temperature treatment and by the presence of respiratory inhibitors. The increased ability to accumulate salts is associated with enhanced metabolic activity, as indicated by measurements of oxygen uptake, and is especially associated with respiration mediated through the cytochrome oxidase system. It was concluded from these observations that the frequently observed effect of the washing of storage tissue disks in water in stimulating salt absorption is accomplished through the activation of energy-releasing metabolic processes which may be likewise responsible for increased protein synthesis and growth, under appropriate conditions.

To harmonize these newer views with the earlier experimental results two ideas need to be distinguished.

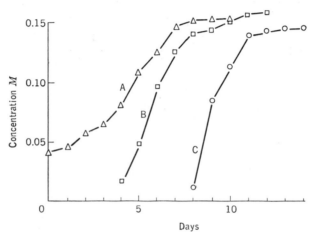

FIG. 48. Effect of washing on the absorptive capacity of red beet tissue. Internal potassium concentration of disks during aeration at 25°C in 0.02 M KCl after preliminary washing in distilled water for: A, a few hours (—△—); B, 4 days (—□—); C, 8 days (—○—). From Sutcliffe (228).

(a) The properties of cells that permit the inception of salt accumulation in previously mature tissue when cut into slices from massive storage organs.

(b) The factors that accentuate or minimize that accumulation, granted that the tissue is capable of it in the first place.

The effects of long washing of disks on their subsequent absorptive ability bear largely upon (b). The earlier experiments of Steward were primarily concerned with (a), in systems in which the washing procedure was standardized. Sutcliffe concluded that the effects of washing of dormant tissue slices on salt uptake are to be sought primarily through the activation of metabolic processes. Such possible alternatives as the removal of previously accumulated solutes or accentuated protein synthesis and growth, caused by the washing procedure, are not believed to be directly involved in the case of beet root disks.

On the basis of a "carrier hypothesis" Sutcliffe proposed that active metabolism may promote salt accumulation in storage tissue cells either by directly supplying energy to the transport mechanism or by causing synthesis of carrier molecules. Skelding and Rees (197) claimed that there is present in dormant red beet tissue a specific inhibitor of salt absorption which is lost from the cells by leaching or by metabolic degradation during washing. Dale and Sutcliffe (54) attempted to obtain evidence for such an inhibitor by comparing the inhibitory effect of extracts from dormant and washed tissue; they got a similar effect with both kinds of extract and concluded that the inhibitions observed by Skelding and Rees were attributable to the presence of various organic acids in the extracts used which interfered in a gross manner with the uptake of cations.

Subsequently Dale (53) found that the failure of dormant beet tissue to absorb salts is related to the nonfunctioning of an anion rather than a cation-accumulating system. He observed that whereas freshly cut slices of tissue will not absorb potassium ions actively from a solution of potassium chloride, they will do so from solutions containing potassium salts of various organic acids. (Parenthetically it may be recalled that Steward (201) stressed in the first experiments with bromide uptake by potato disks that organic acid anions, leached in the first 24 hours from the tissue, were reabsorbed along with potassium in preference to the halide ion.) Dale's conclusion was that a chloride-accumulating system is less active in freshly cut beet disks, so that the ability of the tissue to absorb the associated cations is indirectly curtailed. Organic acid anions are, however, quite rapidly absorbed, and concomitant accumulation of cations is, therefore, possible.

a. Effects of high internal salt concentration on uptake. As was shown by Hoagland and Broyer (87) with barley roots, the initial salt status of an organ or tissue is important in determining the rate at which ions are absorbed from an external medium. Tissues with a high internal salt concentration absorb salts more slowly than do those in which the salt content is low. This phenomenon is well exemplified in washed beet disks which absorb salts at a rate which decreases progressively with time as the internal concentration increases (Fig. 48). Nongrowing cells in general exhibit a definite limit beyond which the salt content does not increase even though external conditions are favorable for further absorption. The concentration of salt attained in the cells at saturation is independent of temperature even though at a higher temperature the equilibrium is attained more rapidly. The external concentration of salt likewise affects the rate at which salt saturation is achieved, but does not have much effect on the maximum internal concentration. (This is in fact the essence of the accumulatory mechanism which admits of uptake of *total salt* almost independently of the external dilution.) On the other hand, the nature of the salt is of considerable importance in determining the concentration of ions reached at saturation. Cells cease to absorb potassium from a solution of potassium sulfate when the potassium concentration inside is considerably lower than that attained before absorption stops from solutions of

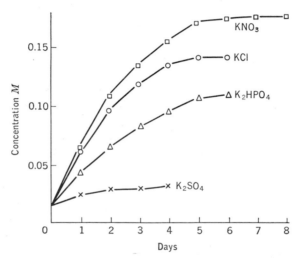

FIG. 49. Effect of the anion on the potassium content of red beet tissue at saturation. Internal potassium concentration of disks during aeration at 25°C in KNO₃ (—□—), KCl (—○—), K₂HPO₄ (—△—), K₂SO₄ (—×—) at a concentration of 0.02 M after preliminary washing for 8 days with distilled water. From Sutcliffe (228).

potassium chloride, potassium nitrate, or dipotassium hydrogen phosphate (Fig. 49).

Many studies of time effects have been made which showed the ability of a system to absorb ions from dilute solution progressively and linearly with time extending over a relatively long period (e.g., the experiments of Steward). As a biological phenomenon, however, some upper limit of salt content is inevitable unless the tissue system should become endowed with indefinite ability to grow. The failure of salt-saturated cells in which growth has subsided to increase their salt content may be attributed to one of two causes:

(1) It may be assumed that the net uptake of salt by a cell represents the resultant of influx and efflux; movement of salt out of the cell might

depend upon active or passive processes. It is possible that as the internal concentration of salt increases, the rate of efflux increases while influx is unaffected, so that eventually a steady state is established in which influx and efflux are equal and no net absorption occurs.

(2) Greater internal concentrations may inhibit the absorption mechanism directly, the rate of efflux being negligible.

These two possibilities can be distinguished by experiments with radioactively labeled ions (229). Cells were allowed to accumulate salt containing a labeled ion (e.g., K^{42} or Br^{86}), and after some time they were washed superficially and transferred to an inactive solution of the same salt. The increase in the radioactivity of the medium with time was then used as an estimate of the rate of efflux of salt. Alternatively cells were allowed to become saturated with salt containing the

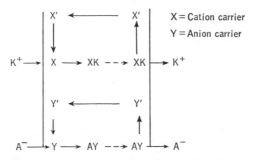

FIG. 50. An ion-carrier mechanism (for explanation see text).

stable isotopes only, and then they were placed in a comparable medium containing a radioactive isotope. The rate at which the isotope came to equilibrium between the tissue and medium was then determined. The results obtained by either method with beet root tissue showed that in healthy cells the rate of leakage of salt from vacuoles is very low even when the salt content of the tissue is high, and it was concluded that the main effect of a high internal salt concentration is to inhibit the accumulation mechanism *directly*.

A somewhat general hypothesis that accumulation of salt occurs by combination of the ions with a hypothetical carrier may here be invoked. On this hypothesis, entering substances form complexes which move, or are moved, across a membrane which is itself impermeable to the free ions. On the inner side of this membrane the complex breaks down to release the ions and to reform the carrier molecule (perhaps in a modified form), which returns to the other side of the membrane where it becomes capable of accepting another ion (Fig. 50). Although numerous suggestions have been made of suitable carrier molecules, and these are discussed elsewhere in this chapter, their specific nature is here ignored.

Energy must be supplied to move the ions across the membrane against the activity gradient. This might be involved in the initial synthesis of the carrier molecules, the combining of the ion and carrier; the physical movement of the complex, or in the final breakdown of the carrier complex.

In order to explain the effect of internal concentration on ion transport, it may be postulated that the final breakdown of the ion-carrier complex is an exergonic reaction which decreases the free energy of the system and therefore tends to proceed spontaneously. As the internal concentration of the ions increase the breakdown reaction $XK \rightleftharpoons$, $X' + K^+$ is forced increasingly toward the left, and the ion tends to remain in the combined form. This reduces the number of effective carrier molecules, and uptake eventually ceases when all the carriers remain permanently in the combined form.

In a system so visualized (Fig. 50) the uptake of salt will be prevented when *either* the cation-transporting, or the anion-transporting, system, or both, are blocked, since from electrical considerations at least the two are interdependent. The fact that cells become saturated for example in potassium sulfate when the internal concentration of K^+ is lower than that attained in, say, potassium chloride may, on this hypothesis, be attributed to blocking of the sulfate-accumulating system when a lower internal concentration of ion is attained than is necessary for the inhibition of the chloride- or potassium-accumulating system during uptake from potassium chloride.

The effects of internal concentration on ion uptake are more easily demonstrated in nongrowing cells because their volume remains constant and a steady state can be established. In growing cells, on the other hand, such an equilibrium is never attained because of the progressive uptake of water and also because there is probably a continual synthesis of new carrier molecules as the bulk of the cytoplasm increases.

b. Location of the carrier mechanism in vacuolated cells. When slices of well-washed storage tissue are placed in salt solutions, there is a rapid initial absorption of certain ions (see page 336), and it is known that this may be completed within about 30 minutes. Such uptake presumably occurs by physical processes, diffusion, exchange and absorption, since it is little affected by temperature. The amount of potassium absorbed in this way by red beet tissue at 2°C from a 0.02 M solution of potassium chloride or bromide is 8–10 μeq per gram fresh weight. On the assumption that the ion does not become concentrated in one part of the tissue, an "apparent free space" (20a) can be calculated, which in this case equals 40–50% of the tissue volume. From a 0.01 M solution of salt under the same conditions 5–6 μeq per gram fresh weight are absorbed, giving an "apparent free space" of 50–60%. Thus it is evident that the "apparent free space" is a variable quantity as far as cations are concerned because of adsorption and Donnan effects and bears no direct relation to the volume of the tissue actually penetrated passively by the salt in this manner.

It seems unlikely that significant amounts of salt enter the vacuoles of plant cells in 30 minutes at 2°C, and therefore the bulk of the salt absorbed in this time should be located in those intercellular spaces injected with liquid, or in cell walls or protoplasts. The volume occupied by intercellular spaces in this tissue is 5–10%, and even these spaces may contain air at the moment the disks are immersed in salt. Therefore, at most 1–2 μeq of potassium chloride will be absorbed as the water in the spaces comes to equilibrium with 0.02 M KCl. The remaining 6–9 μeq of K must be accommodated in the cell walls or protoplasts. It seems unlikely that this can be entirely located in the walls which occupy at most 10% of the volume in this tissue, even after allowance has been made for some adsorption onto cellulose and other cell wall constituents, and it is probable that some physical binding in the cytoplasm also occurs. [The argument is, however, complicated by the unknown degree to which cytoplasm may permeate the wall, for some authorities are now disinclined to postulate a definite line of demarkation between cytoplasm and wall (see also Volume IA, Chapter 1).]

With regard to anions the situation is even more ambiguous. Cells of red beet tissue absorb on 1–3 μeq of Br or Cl from 0.02 M potassium bromide or chloride solutions in 30 minutes at 2°C. This may easily be accommodated in the intercellular spaces and wet cell walls. It is doubtful if anions can penetrate the cytoplasm passively to an appreciable extent, since even if the outer surface is permeable, entry would be limited by the predominantly negatively charged protoplasm. Nevertheless, other workers (32, 60, 96) have obtained evidence for the rapid penetration of anions into the cytoplasm of cells in various root systems, and the solution of this problem must await further investigation.

In any case it seems probable that the tonoplast is the most important membrane which controls the movement of ions into cell vacuoles and that active transport of *both ionic species* occurs across this barrier. The mechanism for transference of ions from the outer cytoplasmic surface to the tonoplast, however, remains obscure.

c. The selective absorption of cations. Sutcliffe (230) has re-examined this phenomenon in red beet root tissue with respect to the alkali metal cations. It was found that whereas this tissue will absorb either Na+ or K+ at a similar rate from solutions of single salts under comparable conditions, a strong preference for Na+ is exhibited when the two ions are supplied together (Fig. 51). (In this respect beet tissue may represent a special case, for claims have been made that this plant requires sodium and responds to sodium application in the field.) The uptake of Na+ by the cells was little affected by the presence of K+ at the same concentration, but the uptake of K+ was strongly inhibited when sodium was present. This indicates that there is a common carrier for Na+ and K+ in these cells which is capable of transporting either ion with equal efficiency but which combines *preferentially* with Na+ when a choice is available. The total amount of cations absorbed from a mixed

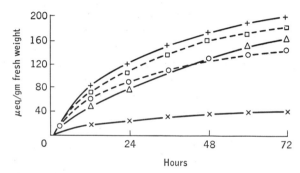

Fig. 51. Uptake of Na and K by red beet root slices in 72 hours at 25°C. From Sutcliffe (230).

△—△ Na uptake from 0.01 *M* NaCl; +—+ Na uptake from 0.02 *M* NaCl; ×—× K uptake from 0.01 *M* KCl + 0.01 *M* NaCl; ○- - -○ Na uptake from 0.01 *M* NaCl + 0.01 *M* KCl; □....□ Na + K uptake from 0.01 *M* NaCl + 0.01 *M* KCl.

solution appears to depend primarily on the external concentration of salt (probably the concentration of anions is the limiting factor) and was the same as that from the single salt solutions of comparable concentration.

Red beet root and sugar beet root tissues are the only ones, among those examined, which exhibited a preference for Na+ over K+. Others, e.g., potato tuber, swede, and turnip hypocotyl tissues show, on the contrary, a strong preference for K+. Moreover, a contrast to red beet, these tissues absorb K+ more rapidly than Na+ even from single-salt solutions under the same conditions. A few tissues, e.g., carrot root phloem and xylem, are relatively unselective with respect to these two ions. It is clear that although similar transport systems may be involved in these various tissues, the carrier molecules are sufficiently distinct to account for their different discriminatory characteristics: the problem of the selective absorption of alkali cations will be further discussed below (see pages 404–406).

d. *The effect of pH on salt uptake.* A number of investigators have examined the effect of pH on the uptake of salts by tissues and whole plants (8, 63, 88, 144, 161, 219) with variable results. In many cases little, if any, influence of pH has been observed over a range of pH values from 5 to 9. A major difficulty in the interpretation of such experiments is that buffered solutions must be employed if moderately constant pH values are to be maintained, and these may give rise to

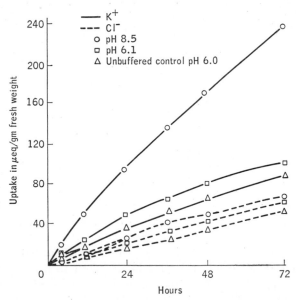

FIG. 52. Effects of pH on rate of absorption of K⁺ and Cl⁻ by red beet root tissue from 0.0024 *M* KCl. From Hurd and Sutcliffe (98b).

spurious effects either by interfering with the entering ions or by metabolic influences which indirectly affect absorption. Such effects can be minimized by use of a buffer system which does not contain easily absorbable ions likely to compete with the entering ions and which is itself metabolically inactive. The buffer system tris(hydroxymethyl) + aminomethane + sulfuric acid meets these requirements and has been used in a study of the effect of pH on salt absorption by red beet root tissue (98a).

In an experiment with red beet root tissue it was observed that the rate of absorption of K⁺ from a 0.0024 *M* solution of KCl was much greater at pH 8.5 than it was at pH 6.5 (Fig. 52). The rate of uptake of chloride was approximately the same at the two pH values. Consequently, whereas the rate of absorption of chloride was only slightly lower than that of potassium at pH 6.1, considerably more potassium than chloride was absorbed at pH 8.5. When the experimental period was extended beyond 72 hours, potassium absorption from the alkaline medium continued to exceed the uptake of chloride until the tissue became saturated with salt. At saturation the total amount of potassium in the disks at pH 8.5 was approximately twice that held by the disks in the acid solution.

The stimulation of cation absorption induced at the alkaline pH varied greatly with different tissues. White beet root tissue, Jerusalem artichoke tuber tissue, and carrot root tissue showed effects similar in magnitude to those obtained with red beet. The greatest stimulation of all was obtained with tissue from parsnip (*Pastinaca sativa*) roots. On the other hand the rate of absorption of potassium by slices of potato tuber tissue and swede tissue was not significantly different at pH 8.5 from what it was at pH 6.5 (cf. page 320).

It was concluded that the effects of pH observed in these experiments is related to the presence of bicarbonate ions in the alkaline solutions and to the ability of some of these tissues to absorb these ions. Experiments with C^{14}-labeled bicarbonate have shown that the tissues which can absorb bicarbonate from solutions in the presence of chloride ions are also those which show stimulated cation uptake at the more alkaline pH. It has further been shown that at alkaline pH the number of chloride plus bicarbonate ions absorbed is equal to the total number of cations absorbed. Bicarbonate ions are not accumulated by plant cells (86), and by some people this has been taken to mean that they are not absorbed. Hurd and Sutcliffe showed by paper chromatography of extracts of red beet tissue fed with C^{14}-labeled bicarbonate that most of the bicarbonate is converted in the tissue into various organic acids, particularly malic acid. In these experiments, therefore, the apparently unequal uptake of cations and anions observed with some tissues at pH 8.5 is really an equal uptake of positively and negatively charged particles, and the stimulated uptake of potassium at alkaline pH is attributed to a greater uptake of anions (bicarbonate plus chloride) from the alkaline solution.

The conclusion that cations may be absorbed in association with bicarbonate ions is of interest in another connection. Ulrich (239) observed that when an excess uptake of cations over anions occurs in barley roots (e.g., when absorption takes place from a solution of potassium sulfate) there is an increase in the organic acid content of the tissue. Ulrich looked upon this as a device to maintain electric neutrality following unequal uptake of oppositely charged particles. Lunde-gårdh (129) on the other hand believed that synthesis of organic acid anions was a *cause* rather than an effect of the excess absorption of cations. The above hypothesis indicates how apparent unequal uptake may be accomplished together with the production of organic acid anions by uptake of some bicarbonate ions, followed by their conversion into organic acid anions with retention of the cation. The extent to which this phenomenon is of widespread occurrence remains obscure.

Overstreet *et al.* (155) also considered the possibility that simul-taneous uptake of K^+ and HCO_3^- ions may occur from the soil into plant roots, accounting for the very rapid uptake of cations from extremely dilute soil solutions. They, however, rejected this idea in favor of the "contact hypothesis" because of an observation by Overstreet and co-workers (159) using the newly available short-lived isotope C^{11} that the uptake of bicarbonate ions by barley plants from solutions of potas-sium bicarbonate in short periods of time was only about one-seventh of the uptake of potassium during the same period. It is possible, however, that most of the potassium uptake in this short-term experiment oc-curred by physical binding and that the metabolic absorption of potas-sium was actually comparable to the observed uptake of bicarbonate. The point is worth re-examination with $HC^{14}O_3^-$, since the simultaneous absorption of potassium and bicarbonate may occur from soils at least in some plants (see also page 426).

F. Later Work with Roots

1. Binding Sites, Carriers, and Chelation Phenomena

By the 1940's it was becoming generally accepted that the first stage in ion absorption involves the binding of ions on to cytoplasmic con-stituents either by physical forces alone or by chemical combination. On account of earlier observations on the exchangeability of cations be-tween soil and roots, and vice versa (105), Jacobson and Overstreet were predisposed to the view that absorption depends initially on a reversible physical binding involving exchange of cations for H ions according to the equation $HR + M^+ \rightleftharpoons MR + H^+$.

Jacobson and Overstreet's first experiments (101, 156) were directed toward an examination of the region of maximum ion binding in ex-cised barley roots. The experimental procedure was to immerse 2-cm long sections of roots in solutions containing radioactively labeled ions at extremely low concentration (about 10^{-9} mol per liter) at either a low (0°C) or a higher (25°C) temperature. After 3 hours the root sec-tions were removed, washed, and cut into smaller sections (0.5–1 mm

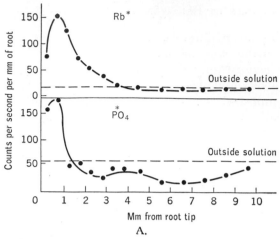

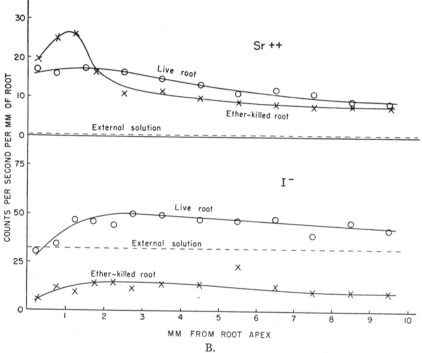

Fig. 53. A. Graphs showing the magnitude of nonmetabolic absorption of Rb[86] and P[32] (expressed in counts per second per millimeter) as a function of distance from the root apex. The dotted lines labeled "outside solution" in each case correspond to the activity of a volume of the bathing solution equal to that of 1 mm of root segment. From Overstreet and Jacobson (156).

B. Graphs showing the distribution of absorbed Sr[85] and I[131] in live and ether-killed roots as a function of distance from the root apex. The absorption and subsequent sectioning were carried out at 0°C. The dotted line labeled "external solution" corresponds to the activity of a volume of the bathing solution equal to that of 1 mm of root segment.

in length) upon which radioactivity determinations were made. Some
of the results obtained are reproduced in Fig. 53, A and B. They show
that for Rb^+ and $H_2PO_4^-$ in the experiment at $0°C$ there was a relatively
sharp maximum in the absorption at about 0.75 mm from the tip. Sr^{++}
and I^- under similar conditions showed less sharply defined maxima
somewhat further along the root. Nevertheless in all cases maximum
binding occurred very near the apex. Relatively little absorption was
found in the root hair zone which was situated about 5 mm from the
tip in these roots. In demonstrating the limited importance of the root
hair zone with respect to salt absorption these experiments confirm the
similar conclusion arrived at earlier by Scott and Priestley (193) and
by Prevot and Steward (163). Kramer and Wiebe (112) however have
obtained some evidence that this situation may not be universal in
plants. Under the conditions of their experiments a maximum uptake
of phosphorus was observed some distance from the apex.

Jacobson and Overstreet examined the binding of ions on to killed as
well as living roots, and observed a qualitatively similar but quantitatively different pattern of absorption in the two cases. Dead roots
actually took up more radioactive Sr^{++} at low temperature than did
living ones, but the ions bound by killed roots were more readily removed again, by exchange with inactive ions, than were those absorbed
by living tissue. However, a considerable proportion of the ions bound
by living tissue was exchangeable—85% of the Sr^{++} being exchanged
in 1 hour and 70% of the I^- in 4 hours. On the basis of the exchange
curves obtained, Jacobson and Overstreet suggested that the ions are
fixed in the form of chemical compounds with varying degrees of stability. Chelation was postulated as one of the phenomena involved.
Subsequent accumulation of free ions in vacuoles or in the stele is
thought to depend on breakdown of the carrier-ion complex causing a
release of the ions.

In later experiments, Jacobson, Overstreet, and their collaborators
(63, 102) studied the effect of immersing roots in hydrochloric acid at
different pH values on the amount of potassium ions lost from the
tissue. Not surprisingly, they observed that 100% of the potassium was
lost from tissue kept for 2 hours in solutions at pH values of 1 to 2.
There was only a small percentage loss into solutions at pH 6 to 7, and
intermediate amounts were lost at pH values between these extreme
values. In all cases the loss was greater in 2 hours at $26°C$ than it was
at $0°C$ (Fig. 54). Although it is difficult to dissociate these results from
a mere killing of the cells at more acid pH's with release of cell contents, Jacobson and Overstreet concluded that they had evidence supporting the reversible binding of potassium in roots as indicated by the

equation above. In view of the observation of Sutcliffe (230) that much of the potassium bound in cytoplasm is attached to sites which are inactive in ion transport, the relationship between Jacobson's and Overstreet's observations and accumulation remains obscure.

Overstreet and associates (157) have also examined the effect of the presence of other cations than hydrogen ions on the uptake of potassium by barley roots. They showed, for example, that the uptake of potassium is markedly decreased in the presence of sodium and concluded that the same binding sites are involved in the uptake of these two ions. This is in contrast to the contention of Epstein and Hagan that separate sites are involved in the uptake of Na and K by this tissue (see below).

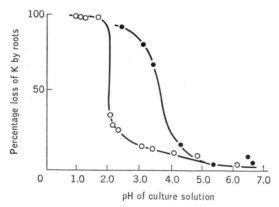

Fig. 54. Loss of potassium by barley roots during a 2 hr period of immersion in HCl solutions at 0°C (○—○) and 26°C (●—●). From Jacobson *et al.* (102).

Ca ions were shown to have a somewhat stimulatory effect on the uptake of potassium, an observation that had also been made by Viets (241). Jacobson and Overstreet concluded that calcium ions were stimulating the uptake of potassium either by acting as a cofactor in the formation of HR, or by facilitating the removal KR from the site of absorption in some way. A more likely explanation is that by adding Ca^{++} (in the form of calcium chloride) to the solution of potassium chloride from which absorption occurred, the concentration of chloride was increased. This caused an increased uptake of chloride which indirectly stimulated cation absorption, especially of the more rapidly absorbed potassium, rather than of the slowly absorbed calcium, ions. Sutcliffe (230) has explained the effect of added potassium ions in promoting the uptake of sodium by red beet root cells in this way, and much earlier Collander (41), observing a similar effect with whole plants, tentatively attributed it to the same cause. Whereas the uptake of K^+ was stimulated by the presence of calcium (except when the concentration of potassium relative to calcium was very low), the uptake of calcium was markedly inhibited by even small amounts of potassium ions. This observation suggested that calcium and potassium compete with one another for the same cation binding sites and that it is potassium which is preferentially absorbed (cf. page 395–396).

Although their experimental work has been entirely confined to a study of the absorption of cations by plant roots, Jacobson and Overstreet have implied that a similar, parallel, mechanism may be operative in the absorption of anions also. This has been represented by the following equation $A + R'OH \rightleftharpoons R'A + OH^-$. No evidence for the operation of such a mechanism in plants has ever been obtained. On the contrary, Hurd (98) observed with various storage tissues that the uptake of chloride is either unaffected or somewhat stimulated at alkaline pH, whereas an inhibition would be predicted on the above hypothesis.

2. Analogies with Enzyme Kinetics

Epstein and his collaborators (61, 62) have adopted a somewhat different approach to the study of ion-binding in excised barley roots. Accepting the view that salt transport involves attachment of the ions to a carrier, followed by subsequent breakdown of the ion-carrier complex, they have analyzed the kinetics of the reactions in order to understand the basic mechanisms involved. The analyses are based on the well-known kinetic studies of enzyme reactions developed by Lineweaver and Burk (118) in which it was proposed that the substrate combines with an enzyme to form an unstable complex which then breaks down to yield free enzyme and the product of the reaction as shown by the following equations:

$$E + S \underset{K_2}{\overset{K_1}{\rightleftharpoons}} ES$$

$$ES \underset{K_4}{\overset{K_3}{\rightleftharpoons}} P + E$$

where E is the free enzyme; S, the substrate; ES, the enzyme substrate complex; P, the products of the reaction; and K_1, K_2, K_3, and K_4 are the rate constants.

An analogous treatment can be applied to salt uptake by equating the carrier with E, and the ion with S. It must be emphasized that this treatment does not require that the carrier should be an enzyme; it can be applied whether the carrier is an enzyme or not. The value of the analysis lies in the fact that from the two equations a velocity equation can be derived as follows:

$$v = \frac{V(S)}{K_s + (S)}$$

where v represents the observed rate of absorption at ion concentration (S). V is the maximum rate of absorption at which the carrier is saturated and K_s a constant corresponding to the Michaelis constant in enzyme kinetics.

By taking the reciprocal of both sides of the last equation we get

$$\frac{1}{v} = \frac{K_s}{V(S)} + \frac{1}{V}$$

On plotting $1/v$ against $1/(S)$, a straight line results in which the ordinate intercept is $1/V$ and the slope K_s/V, thus giving a convenient method of determining both V and K_s.

The analysis can be developed to include situations in which two or more ions are competing for the same absorption sites, by an analogy with the effects of competitive inhibitors in enzyme reactions. In this way Epstein and Hagen treated the interactions between the various alkali cations observed in an absorption experiment with excised barley roots over a period of 3 hours at 30°C. The uptake of rubidium was determined in the presence or absence of the other ions, and the aim was to decide whether or not these ions were competing with rubidium for common carrier molecules. On the basis of the above analysis it is to be expected that when two ions are competing for the same site, the uptake of *either* will be reduced in the presence of the other. If different sites are involved the absorption of each ion will proceed independently of the presence of the other: some of the results obtained are presented in Fig. 55. It may be noted that straight lines were obtained by the double reciprocal plots as predicted by the theory. The slopes of the various lines indicate that whereas potassium and cesium compete with rubidium for absorption sites, lithium does not. The plots for sodium (Fig. 55B) demonstrate an intermediate situation. Below concentrations of about 10 meq per liter sodium did not apparently compete with rubidium for uptake, whereas above this concentration competition occurred (as shown by the greater slope of the line for 25 meq per liter concentration, whereas that for 10 meq per liter has a slope about the same as that obtained in the absence of sodium).

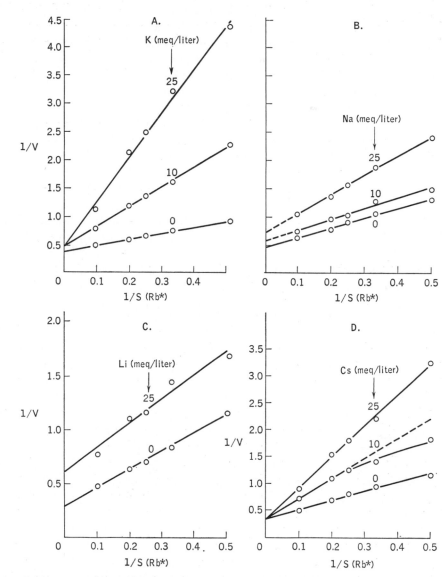

Fig. 55. A. and B. Double reciprocal plots of interference of K (A) and Na (B) with Rb uptake. Rb: 2 to 10 meq/liter. C. Double reciprocal plot of interference of Li with Rb uptake. Rb: 2–10 meq/liter. D. Double reciprocal plot of competitive interference of Cs with Rb uptake. Rb: 2–10 meq/liter. From Epstein and Hagen (61).

Thus it was concluded that K, Rb, Cs compete for the same absorption sites, whereas Li does not compete with Rb and presumably not with K or Cs. Na does not apparently compete with Rb when the concentration of both ions is below 10 meq per liter. To complicate matters, however, it was found that Na at a concentration *below* 10 meq per liter will compete with rubidium if the concentration of the latter ion is raised above 10 meq per liter.

Sutcliffe (230, 231) has criticized the conclusions of Epstein and Hagen because he finds, as have many other workers with different materials, that when two ions are competing for the same mechanism the uptake of one may be strongly inhibited, while uptake of the other (the preferred ion) is unaffected, or may even be stimulated. This contradicts the postulate of Epstein and Hagen that when two ions are competing for a site the *uptake* of *either* is inhibited in the presence of the other; this calls in question the suitability of the kinetic treatment under the experimental conditions prevailing. A study of the uptake of the preferred ion alone (Rb in the present case, considering Na as the competing ion) could lead to the conclusion that these two ions were not competing for the same sites, whereas examination of the absorption of the nonpreferred ion might indicate that they undoubtedly were.

In excised barley roots as in other plant cells the uptake of cations and anions is closely linked, and a stimulation of one invariably leads to some adjustment in the other. In some situations at any rate it appears that the rate of accumulation of *anions* is the limiting factor in determining the total number of cations absorbed, and in these circumstances the rate constants of the cation-accumulating reactions may be without importance. This situation too has its analogy in linked enzyme reactions in which one stage is rate-limiting on the others. In the experiments of Epstein and Hagen the concentration of the anion (invariably chloride) was varied between the different treatments as the total cation content of the solution was adjusted. This might lead to an increase in the *total* amount of anions (and thus cations) absorbed, as the concentration was increased, and tend to stimulate the uptake of both cations in a mixture. This stimulatory effect on the preferred ion might more than offset the inhibition induced by addition of the competing cation [see Table VII, of Sutcliffe (230)]. This possibility is easily tested by studying the uptake of a nonpreferred ion (in this case sodium) in the presence of its competitors. In this case it is observed, with both storage tissues (e.g., potato tuber and barley roots), that Na uptake is inhibited by K, Rb, and Cs over a whole range of concentrations, indicating that all four ions are competing with one another for the same common site (231).

Subsequently, Epstein and Leggett (62) studied in a similar manner the absorption and mutual interference of alkaline-earth cations Ca, Sr, Ba, and Mg. They distinguished two phases of uptake, as have other workers with storage and other tissues:

(1) A physical reaction in which the root acts as a cation exchanger; the uptake is reversible, nonspecific, and temperature-insensitive.

(2) An active transport mediated by metabolic processes which is irreversible, temperature-dependent, and highly selective. Evidence was adduced that Ca, Ba, and Sr compete with one another for identical binding sites and that Mg does not compete for these sites during active transport.

Epstein and Hagen studiously refrain from any discussion of the location of their supposed sites in the root systems with which they deal. A root system is so heterogenous, and even a single unbranched root so complex, that it seems a great shortcoming to pursue mathematical analyses along these lines without locating the phenomena in question in relation to the known structure and development of the root. This will be necessary before this method of approach can acquire much meaning.

G. ION ACCUMULATION IN MICROORGANISMS

Not until recently have the potentialities of microorganisms for the study of salt accumulation been appreciated. In these cells a large aqueous vacuole is not conspicuous and the accumulation of free ions is therefore limited. Rapid growth by cell division accompanied by great metabolic activity is, however, to be found with these organisms. It is not surprising, therefore, that the problem of ion accumulation in these cells presents different features which may well resemble the properties of dividing meristematic cells more closely than those of cells which secrete ions into their vacuolar sap.

1. Ion Transport in Yeast

Pulver and Verzár (165) were apparently the first investigators to study extensively the movement of cations into yeast cells. They found that when these organisms were placed in a medium containing K^+ ions and glucose under conditions favorable for fermentation, the K^+ disappeared from the solution during an initial period when sugar also was being absorbed. After about 10 minutes, however, when fermentation commenced, as indicated by the evolution of carbon dioxide, and the external supply of sugar was almost exhausted, the ions were returned to the external medium. When no glucose was supplied in the solution these movements of K^+ did not occur, and it was noted that there was no change in the Na$^+$ content of the medium even when glucose was present.

It was assumed that glycogen was synthesized in the cells during the initial absorption of glucose and that this was broken down again when fermentation began. Pulver and Verzár therefore concluded that the movements of K$^+$ were associated in some way with the production of phosphorylated intermediates in the polymerization and subsequent breakdown of glycogen.

Subsequently, Rothstein and Enns (181) re-examined the relationship between this reversible K$^+$ absorption and carbohydrate metabolism. They observed that during the absorption of potassium there was not an equivalent uptake of anions and concluded therefore that the bulk of the cations were absorbed in exchange for H$^+$ ions, in effect by fixation to acids produced by metabolism. This was confirmed by the observation that H$^+$ ions were released by the cells in an amount which was almost equivalent to the quantity of K$^+$ taken up. The movement of both K$^+$ and H$^+$ ions appeared to occur along their respective concentration gradients; a high concentration of H$^+$ ions being maintained within the cells by synthesis of organic acids through metabolism. Uptake of K$^+$ was prolonged if sufficient glucose was supplied and the medium became very acid. When salt was not present in the external medium both hydrogen ions and organic acid anions (mainly succinate) were excreted from the cells.

Knowledge of the movements of sodium and potassium ions, into and out of yeast cells, has been extended through the work of Conway and his collaborators, and Conway (44) has proposed the hypothesis of the "redox pump" to account for cation transport.

In fermenting yeast under anaerobic conditions it was shown that the uptake of K$^+$ was inhibited by acids and dinitrophenol at concentrations which scarcely affected fermentation. Under aerobic conditions, cyanide was also an effective inhibitor, but during anaerobic fermentation K$^+$ uptake was relatively insensitive to poisoning by cyanide.

When potassium is being exchanged for hydrogen in yeast cells, the interior of the cells becomes more alkaline (45). At the same time there is a marked fall in the pH in an outer region of the cell, which has been identified with the cell wall and which is freely permeable to ions as well as to some other solutes.

Conway *et al.* (46) demonstrated the absorption of sodium ions by cells which were suspended in $M/5$ sodium citrate. When such so-called "Na-yeasts" were washed and resuspended in distilled water, sodium was extruded together with organic anions. If potassium chloride was present in the external medium, much more Na$^+$ was lost by the cells, and most of this occurred by exchange for potassium. The excretion of sodium in these cases appears to be an active process since it was inhibited by cyanide or by anoxia. On the other hand if glucose was present in the medium, sodium extrusion would proceed anaerobically.

The effects of altering the redox potential of the medium, by addition of various dyes, on the absorption of K^+ and the extrusion of Na^+ by yeast cells have also been examined. It was found that K^+ absorption and the associated loss of hydrogen were both inhibited when the redox potential was reduced below 180 millivolts and exchange ceased entirely at 100 millivolts potential. On the other hand when the potential was artificially increased above its normal value, a marked increase in K^+ absorption was observed. Changes of redox potential had precisely the opposite effects on Na^+ excretion to those observed on K^+ uptake. There was a reduced Na^+ loss when the potential was increased and a stimulated Na^+ output when the redox potential was reduced. These observations were taken to mean that two distinct transport mechanisms are involved in the movements of K^+ into, and Na^+ out of, yeast cells.

Further evidence that there are distinct mechanisms of K^+ absorption and Na^+ extrusion is that each movement is highly specific for the ion concerned. A small amount of K^+ relative to Na^+ in the external medium is sufficient to suppress entirely the uptake of Na^+, whilst in "Na^+-rich" yeast cells, containing approximately equal amounts of Na^+ and K^+, much Na^+ and hardly any K^+ is extruded.

On the basis of these observations Conway proposed the hypothesis of the "redox pump." This is a "carrier mechanism" for cations which bears some resemblances to the Lundegårdh hypothesis for the transport of anions into plant cells (see pages 376–377). It is proposed that an unspecified respiratory enzyme in a reduced state acts as the carrier for Na^+, and by its oxidation, i.e., tranference of an electron to another system of higher redox potential, the ion is subsequently released. Alternately a similar system in the oxidized state is able to bind K^+ and acts as the carrier for this ion in the opposite direction from that in which the Na^+ is transported.

Taking Na^+ as an example of the transported ion the following sequence is visualized:

(1) The ion combines with an organic carrier molecule (M^-) within the cell. The carrier itself is produced by oxidation of a reduced respiratory intermediate (CtH_2) according to the reaction:

$$\tfrac{1}{2}CtH_2 + M = M^- + H^+ + \tfrac{1}{2}Ct$$

and combines with the ion thus:

$$M^- + Na^+ = M^-Na^+$$

(2) The complex is capable of moving across a cell membrane which is impermeable to the free cation. This may be attributed to the over-all electrical neutrality of the complex, or alternatively the complex may be lipid-soluble, whereas the ion itself is not.

(3) At the outer surface of the membrane the complex becomes oxidized and the cation is set free, possibly at a much higher chemical potential level from that at which it was taken up. Oxidation would involve the transfer of an electron which could go to another respiratory catalyst which is not itself capable of complexing with the ion. Ultimately the electron is transferred to oxygen in the presence of H^+ ions to form water under aerobic conditions, presumably via cytochrome oxidase. Under anaerobic conditions, other electron acceptors must be involved.

It appears to be necessary that the final transfer of electrons to O_2 should occur on the *inner* side of the membrane since H^+ must be taken up in amounts equivalent to those released into the cytoplasm during the synthesis of the carrier. The over-all neutrality of the membrane and cell is thus retained during transport, and according to this scheme Na^+ may be excreted simultaneously with the outward movement of a negatively charged anion or uptake of an alternative cation.

A similar scheme is visualized for the active extrusion of H^+ which is responsible for the secondary transport of K^+ *into* the cell in equivalent amount along the electrochemical gradient so created.

An advantage of the Conway hypothesis is that it does not involve substances or reactions in the cells which are not already known to occur in relation to respiration. It is only assumed that the process of liberation of H^+ in normal metabolism is separated from the terminal stages in which it is combined with oxygen, in order to allow intermediate reactions connected with ion transport to occur. It may be noted that the "redox pump" utilizes electron energy directly without the intervention of energy-rich phosphorylated respiratory intermediates. It is therefore difficult to understand why cation transport is inhibited by such poisons as DNP, which prevent phosphorylation but do not interfere with electron transfer. This is also a most cogent argument against the Lundegårdh hypothesis (see page 386). It must also be emphasized that, like the Lundegårdh hypothesis, the redox pump is limited to the transport of four ions per oxygen molecule involved. This theoretical limit has not been shown to be exceeded experimentally in aerobic yeasts, although of course when active transport occurs in yeasts under anaerobic conditions the value for the ratio must obviously approach infinity.

Duggan and Conway (57) have attempted to calculate the concentration of potassium carriers in the yeast cell in milliequivalents per kilogram of centrifuged yeast. This was determined from the amount of K^+ instantaneously adsorbed by the cells in an atmosphere of hydrogen and subsequently displaced by excess of rubidium. A value of 0.12 meq per kilogram of wet yeast was obtained. This was compared with the equivalent concentration of iron (0.60 meq) and copper (0.06 meq), and it was concluded that iron might be a constituent of the carrier, but copper probably was not. This calculation does not take into account the possibility that some of the potassium adsorbed is bound to inactive sites in the cell wall or cytoplasm.

The absorption of phosphate by yeast cells has been examined by numerous workers who have again emphasized the intimate relationship between uptake and carbohydrate metabolism. Hevesy et al. (83) showed that phosphate absorption is negligible in the absence of sugar at 20°C or with sugar at 0°C. Lawrence et al. (115) also demonstrated that sugar is essential for phosphorus uptake and that absorption is inhibited by fluoride even when glucose is present. The latter workers concluded that phosphate uptake is related more intimately to carbohydrate metabolism than it is to growth. Mullins (141) showed that the absorption of phosphorus is affected by temperature but will proceed in the absence of oxygen. It was suggested that the bulk of the phosphate absorbed is utilized in phosphorylation reactions associated with carbohydrate metabolism within the cell.

Kamen and Spiegelman (108) have discussed the evidence for the entry of phosphate into yeast cells (a) by simple diffusion and (b) by esterification at the cellular surface, and have concluded that the latter is probably the operative mechanism. Amongst the evidence cited in support of this contention is:

(1) The incorporation of radioactive phosphorus supplied as orthophosphate in the external medium has a high temperature coefficient.

(2) Exchange of radioactive phosphorus between the external and internal orthophosphate is inhibited by azide, arsenate, iodoacetate, and to a lesser extent by fluoride.

(3) Uptake is reduced in a suspension of slowly metabolizing cells.

(4) Pretreatment with carbohydrate in the absence of phosphorus accelerates the subsequent absorption of orthophosphate.

Rothstein and Meier (183) have demonstrated that a series of phosphatases are present at the yeast cell surface which can hydrolyze such compounds as inorganic and hexose phosphates. Moreover, enzymes involved in glucose phosphorylation are probably also located at the cell surface. It is considered possible that the transport of inorganic phosphate may involve the same enzyme systems as are concerned in the metabolism of glucose (182).

2. Bacteria

Bacteria, like yeasts, present favorable material for studies of ion absorption because homogeneous populations of cells, which will metabolize actively and grow rapidly under rigidly controlled conditions, can be easily obtained. In spite of these advantages, and despite the importance of these organisms in nature, relatively few investigations have been made of salt absorption by bacteria. Interest has centered mainly on the mechanisms for binding potassium and phosphate ions into cells and their relationship to metabolism. Now, however, work on bacteria is particularly appropriate inasmuch as it may serve to illuminate the absorption processes of nonvacuolated cells in contrast to the vacuolated ones that have been more studied.

The physical adsorption of various cations on to the cells of several species of bacteria was first demonstrated by McCalla (134). He found that hydrogen ions were adsorbed by *Escherichia* (*Bacterium*) *coli*, *Bacillus subtilis*, *Azotobacter* sp., and *Rhizobium* sp. from solutions of hydrochloric acid. Since the uptake was quantitatively the same for both living and dead cells it was concluded that only physical processes were involved. Ions adsorbed were more or less readily exchangeable for others in the external medium. For example, it was observed that methylene blue cations would displace various cations adsorbed by bacterial cells. The cations examined, Na^+, K^+, Ca^{++}, Ba^{++}, Hg^{++}, are here arranged in the order in which they are most readily displaced—sodium being most easily and mercury most reluctantly replaced by the dye. Adsorbed magnesium could hardly be displaced at all by sodium or potassium and was only partially exchangeable for calcium, barium, or mercurous ions. From the number of ions adsorbed under optimum conditions, it was estimated that there are about 100,000 adsorption sites per cell in *E. coli*. The adsorption phenomenon in bacteria was claimed to be comparable to physical adsorption in nonliving materials, such as proteins or inorganic colloids. It was, however, realized that if the adsorbed ions are utilized in metabolism, they are thereby removed from the absorption system and sites then become available for further uptake.

Leibowitz and Kupermintz (117) made observations on the relationship between potassium uptake and metabolism in *E. coli* and came to somewhat similar conclusions as did Pulver and Verzár (165) in a comparable study with yeast (see page 407). Leibowitz and Kupermintz observed that potassium ions were taken up by the bacterial cells for a short time while sugar was being assimilated, and the ions were later released to the external medium. In contrast to the results with yeast, however, they found that uptake ceased and release began before the fermentable sugar in the medium was exhausted, and acid production was still proceeding vigorously at this time. During the absorption phase, it was found that temperature, pH, substrate concentration, and aeration had comparable effects on the rates of fermentation and potassium uptake. It was noted that during the uptake of potassium, glucose was disappearing from the medium at a rate far exceeding its utilization in fermentation as measured by the production of acid. This discrepancy was attributed to the temporary synthesis of polysaccharide in the cells (cf. Pulver and Verzár). A close correlation was observed between the time of maximum polysaccharide accumulation and maximum potassium absorption. It was concluded that potassium has a specific function in the synthetic reactions associated with polysaccharide synthesis and that it is probably bound temporarily to intermediate substances.

Eddy, Hinshelwood, and their collaborators (58, 59) on the other hand favor the idea that the absorption of potassium by *Aerobacter aerogenes* (*Bacterium lactis aerogenes*) is associated with the binding of the ion to an essential enzyme system which is involved in the metabolism of sugar. They suggested that potassium ions are indispensable for the activity of certain enzymes and that for this reason the bacterium cannot grow in the absence of this element. It may be noted that potassium had already been shown to be essential for a number of metabolic reactions *in vitro*, e.g., the phosphorylation of fructose-6-phosphate to fructose-1,6-diphosphate (143) and the formation of ATP during the transformation of phosphopyruvic acid to pyruvic acid (18). Eddy and Hinshelwood observed that sodium cannot replace potassium, but rubidium can do so partially, as judged by the effects of sodium and rubidium ions on the growth of *Streptococcus lactis* (*Bacterium lactis*) in the absence of potassium ions.

The absorption and associated utilization of potassium by *E. coli*, and the contrasting behavior of sodium, has been extensively investigated by Cowie *et al.* (49) and Roberts *et al.* (170, 171). Their experiments indicate that the outer membrane of the bacterial cell is freely permeable to both sodium and potassium, and that the difference between these two ions lies in the fact that while potassium forms indiffusible complexes with organic substances within the membrane, sodium does not.

In the case of sodium, no metabolic influence on uptake was observed.

The amount of this ion entering the cells when they were transferred from distilled water to a solution of salt was shown to be that expected if sodium was diffusing freely until an equal concentration inside and outside the membrane was attained. This equilibrium was established in less than 4 minutes, and when the cells were replaced in water, sodium diffused out equally quickly, thus demonstrating the higher permeability of cell membrane and the ease of diffusion in either direction. It was calculated that the amount of bound sodium does not exceed about 0.05 mg/ml of cells.

On the other hand the situation with respect to potassium ions was found to be very different. Although there is a freely diffusible component comparable to that demonstrated for sodium, there is, in addition, a considerable amount of potassium bound on to cell constituents. The amount of bound potassium may reach 15 mg/ml of cells, and it is little affected by change of external concentration whereas the amount of free potassium depends directly on the concentration of this ion in the medium. The bound potassium thus represents the bulk of the potassium in the cells, except when the external concentration is high. Unbound potassium is rapidly lost when growing or resting cells are transferred to distilled water. Resting cells (but not growing cells) will also lose their bound potassium under these conditions, but at a much slower rate.

Although there is an initial binding of potassium when bacterial cells are placed in media containing this ion in the absence of glucose, Roberts et al. (171) observed that uptake is stimulated by the presence of glucose in proportion to the amount of sugar added (Fig. 56, A). Furthermore it was shown that immersion of the bacteria in glucose *prior* to the presentation of salt also stimulates the uptake of potassium when the ions are subsequently supplied alone. These observations support the view of other workers that some product of carbohydrate metabolism is responsible for the binding of potassium ions. In an attempt to examine the hypothesis of Leibowitz and Kupermintz that uptake is associated with the synthesis of polysaccharide, the effect of pH on the absorption was examined since it is known from the work of Sussman et al. (227) that, in yeast at any rate, polysaccharide synthesis proceeds much more rapidly in acid than in alkaline solutions. It was found, however, that potassium uptake is little affected by pH, and if anything it tends to be lower at pH values which favor polysaccharide synthesis (Fig. 56, B). The conclusion was drawn therefore that the binding of potassium is not intimately involved with the formation of polysaccharides in bacteria.

It seems probable that potassium can react with a number of dif-

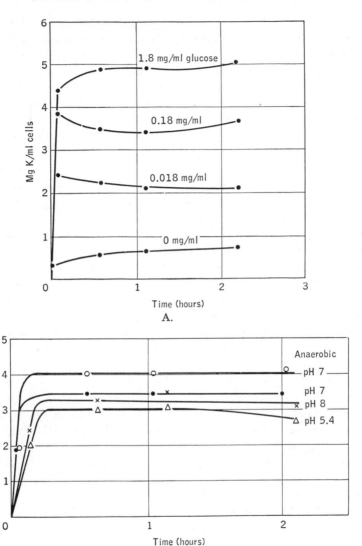

Fig. 56. A. Effects of glucose concentration on potassium uptake by *Escherichia coli*. From Roberts *et al.* (171). B. Effect of pH on potassium uptake by *Escherichia coli*. From Roberts *et al.* (171).

ferent cell constituents, but there is no evidence from exchange experiments that substances with widely different binding capacity are involved, and it is assumed that a single compound or a group of related compounds are largely responsible for fixation in *Escherichia coli*. The fact that potassium is bound to a much greater extent than sodium makes it unlikely that simple salts of organic acids such as lactic, pyruvic, or succinic acids contribute significantly to the bound potassium. It is thought that most of the binding takes place on to respiratory intermediates in the anaerobic pathway leading to the breakdown of sugar, and hexose phosphates are probably especially involved. Attempts to isolate potassium-binding substances from bacterial cells for the purposes of identification have so far proved unsuccessful.

Rorem (179) has studied the uptake of rubidium and phosphate ions by several polysaccharide-synthesizing bacteria including *Leuconostoc dextranicum, Leuconostoc mesenteroides,* and *Streptococcus salivarius.* These bacteria can be caused to produce polysaccharides in the cytoplasm and at the cell surface, or not to do so, according to the form of carbohydrate supplied to them in the external medium. Using P^{32} and Rb^{86} as tracer ions it was shown that when the cells are producing polysaccharide they absorb 2–20 times as much salt as they do when polysaccharide is not being synthesized (see Fig. 56, C). Rorem proposed that the secretion of polysaccharide at the cell surface facilitates the accumulation of ions by physical binding at the surface and enables more rapid absorption to occur from dilute medium where uptake might be limited by the low external concentration. It is proposed that the pectin in cell walls of higher plants may function in a similar manner to increase the efficiency of uptake by roots.

Several investigators have examined the mechanism of uptake of phosphate ions by bacteria, and there seems no reason to doubt that this is intimately associated with the utilization of phosphorus in metabolism. In support of this O'Kane and Umbreit (145) showed that the incorporation of both extracellular and intracellular inorganic phosphate into organic substances in *Streptococcus faecalis* is facilitated by the presence of glucose. Moreover, Eddy *et al.* (58) observed a general parallelism between the initial uptake of potassium, rubidium, and phosphate by *Aerobacter aerogenes* and concluded that the uptake of both cations and anions is related to the early stages of carbohydrate metabolism. They found that, whereas growing cells incorporate much more phosphate than resting cells, the potassium requirement of growing and resting cells was not very different. This indicated to them that while phosphorus is being incorporated into permanent cell materials (mainly nucleic acids) during growth, potassium is being continuously released and used over again.

Mitchell (139, 140) has concluded that the movements of phosphate into and out of cells of *Staphylococcus aureus* (*Micrococcus pyogenes*) occurs across a membrane that is impermeable to the free diffusion of phosphate ions. This is in contrast to the contention of Roberts and

Roberts (170) that the outer membrane of *Escherichia coli* cells allows ions to move freely across it and react with constituents in the body of the cytoplasm. Mitchell postulates the existence of a specific carrier mechanism in the membrane, which is responsible for the transport of phosphate into or out of the cell. He suggests that the carrier binds with

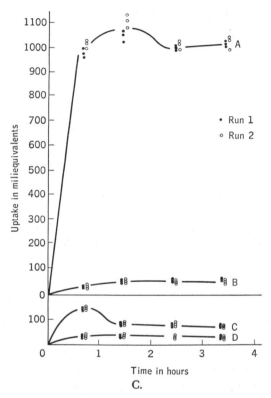

FIG. 56. C. Ion uptake per 3×10^4 cells of *Streptococcus salivarius* from solutions containing 20 meq of Rb or PO_4 per liter. Curve *A*, phosphate uptake by young living cells plus capsule or slime. Curve *B*, phosphate uptake by young living cells without capsule or slime. Curve *C*, rubidium uptake by young living cells plus capsule or slime. Curve *D*, rubidium uptake by young living cells without capsule or slime. After Rorem (179).

the ion on one side of the barrier and moves freely to and fro across the membrane by thermal agitation. Subsequent breakdown of the complex with release of the ion is thought to occur by dissociation or by enzymic hydrolysis. The transport was shown to be specifically inhibited by traces of heavy metals, by "uncouplers" of oxidative phosphorylation, and by substances reacting with thiol groups. It is claimed that parts of the osmotic barrier became separated as small lipoprotein particles after mechanical disintegration of the cells. These particles have been shown to contain more than 90% of the acid phosphatase of the cell, and an active cytochrome system is also present. The observations on the location of enzymes in the particles are taken as confirmatory evidence that a metabolically mediated active transport may occur across the membrane of bacterial cells.

The observations on ion absorption by bacterial cells, although as yet scanty, serve to emphasize the interplay of physical forces, metabolic binding, and active transport in accumulation phenomena. Although the results appear to be in some ways distinct from those obtained with vacuolated cells of higher plants, this may be related rather to the different emphasis on active transport, as opposed to metabolic binding of ions, in the two systems than to any fundamental differences. An intensive study of ion accumulation in meristematic cells of higher plants may well bridge the gap between bacterial cells on the one hand and vacuolated parenchyma cells on the other. A tissue culture system is now available which makes such a study possible (35). So far as this work has gone [cf. Steward and Millar (216)] it does indicate that absorption by stoichiometrical binding of cations is a more conspicuous feature of the absorption by dividing cells than of those which are enlarging; in this respect the meristematic cells and the bacteria may have features in common.

3. Unicellular Algae

Before leaving the subject of microorganisms, some studies of ion accumulation by unicellular algae will be mentioned briefly. Work on the uptake of inorganic materials by these plants has been largely confined to metabolically important ions, e.g., bicarbonate, phosphate, and nitrate, and these studies will be described elsewhere in relation to photosynthesis and nitrogen metabolism. Only a few investigations have been primarily concerned with the absorption, as opposed to the utilization, of mineral salts, and these will now be considered.

Scott (188, 189) has examined the cation composition of *Chlorella pyrenoidosa* under various experimental conditions. He found that the mineral content shows considerable variation when the alga is grown in media containing varying proportions of sodium, potassium, magnesium, and calcium ions. The ratio of calcium to magnesium in the cells was directly related to the proportion of calcium to magnesium in the external medium. On the other hand, potassium was absorbed preferentially to sodium, and the ratio of potassium to sodium was always higher in the cells than in the medium. In potassium-deficient media, however, this ion was replaced by an equivalent amount of sodium, indicating that sodium absorption is normally restricted by competition with potassium.

When *Chlorella* cells were transferred from a nutrient medium to distilled water, some calcium was lost from the cells, but there was no significant loss of other ions within 2 hours. It seems therefore that the bulk of the cations in the algal cells are bound relatively firmly in the protoplast by fixation to cytoplasmic constituents or by enclosure within impermeable membranes. Most of the calcium could be removed, however, by placing the cells in a calcium-free medium containing magnesium ions, the calcium being replaced by an equivalent amount of magnesium. Some, but not all, of the magnesium in a cell could be displaced by alternative ions, and it was concluded that magnesium is bound in the cells in at least two states. Only about 25% of the non-displaceable magnesium could be accounted for in chlorophyll.

The relationship between potassium uptake and metabolism in *Chlamydomonas humicola* was examined by Roberts *et al.* (171). They found that absorption was increased by the presence of glucose when experiments were conducted in the dark and that uptake was also stimulated by light. Some probable effects of light in promoting the uptake of cations by *Nitella* have been discussed above and may also be applicable here. Ketchum (109, 110) has shown that by growing a marine diatom, *Nitzschia closterium*, in the light without adding phosphate to the medium, it is possible to produce a "phosphorus debt" such that the organism will later take up phosphorus in the dark, which it would not do in normal circumstances. The uptake occurring in the dark with phosphorus-starved cells is independent of external concentration of phosphate within wide limits and probably depends on chemical combination with metabolic intermediates which are produced in the cells in the light.

Knauss and Porter (111) have recently made a study of the effect of concentration in the nutrient medium on the absorption of a number of different ions by *Chlorella pyrenoidosa*. Their results show that the amount of an element absorbed depends on the particular element involved and its concentration in the medium. Iron and manganese were absorbed much more rapidly than other elements studied at comparable concentrations, and it was concluded that adsorption of these ions on to the cell surfaces was one of the factors involved. The phosphorus and sulfur contents of the algal cells remained constant when the nutrient concentration of these elements was high. At lower levels absorption of these two elements varied with the log of the concentration according to the equation $A = K \log C + m$ where A is the amount absorbed;

C, the concentration; *K* and *m* being constants, characteristic of the particular element. In the case of other elements examined (Ca, Zn, Cu, Sr), surprisingly enough, the absorption was directly proportional to concentration; i.e., $A = KC + m$, *K* and *m* again being constants characteristically different for each element. The different rates of absorption of the various elements point to independent mechanisms of uptake and utilization in each case. The extent to which active transport across membrane, as opposed to metabolic utilization and incorporation, is involved in these processes is not yet clear.

H. The Cellular Mechanisms of Absorption and Accumulation: Summary

After this rather extensive survey of work carried out on a variety of cellular systems and after the consideration of views which have emerged from various schools of thought, one must recognize that no single, detailed and specific mechanism of ion accumulation in all cells can be regarded as experimentally and finally established. In view of the nature of the process, which is so intimately related to the organization of the living cell as a whole, this is not surprising.

Nevertheless there is somewhat general agreement on many broad aspects of the problem. It is now generally agreed that metabolism in the cell is involved because work is done to move ions from a low to a high concentration; since the free energy of the system is thus increased, energy must be donated in the process. Since ion accumulation has been investigated, for the most part in organisms which are aerobic, the importance of oxidative processes in the supply of the necessary energy has been naturally emphasized. This does not preclude, however, the possibility—largely unexplored—that an aerobic organisms bring about ion transfers by using energy sources which are independent of molecular oxygen, since it is evident that they perform many other processes which must be equally dependent upon the use of energy.

Inherent in most interpretations is the idea of a "carrier" system in which the ions are bound to some organic carrier complex in the form of which they are transported across the membrane system. Moreover the carrier system is supposed to operate in some reversible and cyclical manner; it being the contribution of metabolism to keep these cyclical processes operating. Clearly energy might be involved at several points, such as the initial synthesis of the carrier and its replenishment after breakdown. Energy might also be involved to maintain the physical movement of the complex, as for example by protoplasmic streaming or by the energy relations of contractile proteins as in the hypothesis of

Goldacre (71). Finally metabolism may be invoked in order to displace the ions from the carrier complex.

Various kinds of carrier molecules have been postulated; these range from direct chemical combination of an acid-base kind, as in the original hypotheses of Osterhout, in which the carrier for cations is regarded as an acidic substance in the nonaqueous membrane which combines with "undissociated free base." The ideas of "carriers" become more sophisticated when they assume that ions, the cations particularly, may be sequestered in chelated complexes, as in the ideas of Overstreet and Jacobson particularly. According to these views the carrier molecule would be some cytoplasmic constituent with strong chelating properties, and it is evident that many naturally occurring compounds might seem fitted for this role, although, to date, proof that any have actually assumed it, is lacking. A variety of hypotheses have been developed on the assumption that the ion binds to electrostatically charged surfaces, such as those on cytochrome containing particles, as in the hypothesis of Lundegårdh. Various suggestions have been made to utilize as a carrier molecule one of the accepted intermediates in the aerobic respiratory sequence in the electron chain and thus account for the manifest relations between ion uptake and aerobic respiration. Recognizing the importance of phosphorylation as the principal means of energy transfer, a variety of suggestions have been made according to which the carrier system might be a phosphorylated complex which embodies energy by these means and is also capable of combining with the ions transferred. This is true of such work as that on bacteria by Roberts *et al.* (171) and by Eddy and Hinshelwood (58, 59); and the suggestion was also made by Steward that phosphorylated intermediates of nitrogen and protein metabolism might function in this way (216).

In any system which invokes carriers to explain the mechanism a relatively impermeable membrane barrier is essential, since otherwise the carrier molecule, like the ions transported, would be freely diffusible through the whole system. This also implies that work in accumulating salt would be done at the barrier. There is obviously abundant evidence that the cytoplasm as a whole behaves as a relatively impermeable barrier, because it retains the substances within the vacuole, and the electrochemical properties of cells are also consistent with this idea. Nevertheless, attempts have been made to specify more precisely the location of the functional membrane within the cytoplasm. These attempts are naturally bound up with ideas on the site where the primary act of accumulation occurs. Again it should be recognized that no final and generally accepted answer to this question can be given, although it will be evident from the trend of the preceding discussion that the special importance of the tonoplast in vacuolar accumulation seems now to be inescapable.

Stemming from the importance of the tonoplast as an important barrier have come certain ideas that the ions are freely accessible to all parts of the cytoplasm external to this membrane. It seems a retrograde step, however, to assume that ions must be freely diffusible in and out of the cytoplasm. Doubtless the cytoplasm has a binding capacity for cations and some of these are freely replaceable, but this should not imply that entry and exit from the cytoplasm is not subject to control. Even were this so for cations, the presence and entry of anions into the cytoplasm are almost certain to be intimately bound up with metabolic processes.

These thoughts lead naturally to the consideration of the ionic relationships of vacuolated and nonvacuolated cells, as shown for example by meristematic cells of higher plants and cells of microorganisms, on the one hand, and vacuolated parenchyma on the other. In general terms it is now possible to visualize the meristematic cell as representing a system in which the ions may be bound to specific metabolically reproduced, sites within the cytoplasm. From these sites the ions may be subsequently removed when, in the more two-stage process as it occurs in the vacuolated cell, they are transferred into the vacuole. Thus the absorption by the meristematic cell and by the vacuolated cell are seen, not as entirely separate and distinct phenomena, but as two parts or stages of the over-all process.

Passing now from these more general views to particular interpretations, it is evident that investigators have paid the greatest attention to attempts to stipulate the exact linkage between metabolism and the transport system. The first ideas in this direction based on direct stoichiometrical combination of an acid-base kind, between entering cations and a carrier molecule (HA), carry with them rigorous restrictions on the relative pHs of sap and external solution which permit of cation accumulation; for this reason they are not of general application. Contrary to the ideas developed by Lundegård, as followed by Robertson, there is clearly no certain evidence that the energy for the accumulation arises from a special component of respiration which is *specifically* associated with anion, or even with salt, uptake. From all the evidence available it is still clear that only a very small part of the total energy of respiration is used specifically in the accumulation process, and this is true whether one considers the process in relation to the respiration of the cell as a whole, or only to that part of it which may be stimulated by salts. It is true to say, therefore, that ion accumulation is related to the respiration of the cell as a whole and is determined by factors which regulate its over-all metabolic activity. Granted this necessary level of metabolic activity, the cell seems able, in large measure, to take the process of ion accumulation "in its stride," much as a moving train might pick up mailbags while en route. The transfer of a letter in the mailbags in the analogy, or the transfer of an indicator ion in the cell, are equally properties of a dynamic system in being and in operation, and while there is an over-all relation to energy expended and consumed it may be, and is, very difficult to distinguish between that necessary to keep the whole system going and that specifically required for the ionic process in question. Though expressed as early as 1937 (206); these views still seem apt.

Undoubtedly support for the ideas of Lundegårdh seemed to be given by the claim that the relationship between anions absorbed and the associated respiration stimulated by the salt was quantitatively of a kind that could be theoretically interpreted. This depends upon the claim that for every additional oxygen molecule used in respiration four anions could be transferred. The experimental basis for this claim has, however, now largely disappeared with the observations that ion pairs accumulated per oxygen involved may be greater than 4 under some experimental conditions. Furthermore the evidence from the effect of dinitrophenol on respiration and on salt accumulation argues strongly against the cytochrome hypothesis although, on the contrary, it is compatible with more generalized views on the linkage between respiration and salt accumulation being mediated via phosphorylation or similar processes. In fact, on close examination, it is seen that the bulk of the evidence which was mobilized by Lundegårdh and by Robertson to support the anion, or salt, respiration hypotheses is equally applicable to a phosphorylation mechanism operating through cytochromes.

Thus in any specific case the ultimate explanation of ion transfer will need to show how the energy is mobilized and, through some known mechanism of energy transfer, linked *specifically* to the act of ion transfer. Indeed this point of view should be extended from the active transfers which involve inorganic ions only to active transfers in which other types of molecules may be involved.

Thus, in the final analysis, the problem involves a discussion of the mechanism by which respiration is linked with active transport; and the dilemma is that in the organized cell the respiration is also being linked to a variety of other processes in which energy is also used. In fact, as the preceding discussion has emphasized, one should regard the entire cell as a working machine in which energy is being released and utilized in a variety of ways, and often the relations of metabolism to the ion transfer process become more intelligible when they are related to the operation of the system as a whole.

In fact it is suggestive that growing cells are particularly active in the absorption of salt as well as in such other energy-requiring processes as protein synthesis. It may seem paradoxical that the explanation of one energy-requiring (endergonic) process by which the free energy of the system is increased may also need to invoke another (protein synthesis) which is equally, or even more, obscure.

Such ideas, however, become more acceptable when it is appreciated that synthesis of such molecules as protein produce order out of disorder—an essential feature of the creation of form during growth. This alone produces negative entropy changes, which allow the free energy to increase, even though the accompanying change in heat content may be small, or even negative, for $\Delta F = \Delta H - T\Delta S$. Such ideas, which invoke synthesis of complex molecules like protein have their counterpart in those of Goldacre which invoke contractility of a fibrous protein. Ultimately, however, cycles of protein synthesis and breakdown may well be related to any, or all, of the dynamic machinery of cells because it appears to be an increasingly characteristic feature of the cellular metabolism.

Since current ideas of protein synthesis invoke the pentose nucleic acids as templates at which condensation occurs and from which synthesized protein is removed, there is here the inherent possibility of linking together carrier molecules for ion transfer and carriers or donors of nitrogen for protein synthesis. According to such views, necessarily speculative at this stage, the carrier molecule may bring simultaneously to the protein-synthesizing surface the phosphate bond energy that may be used in synthesis, the nitrogen in a form in which it may be donated to form protein, and the ions which, when freed from the complex as their combining groups are lost in the act of synthesis, may be locally accumulated. Such ideas presuppose that the carrier molecules are regenerated after protein breakdown, and that the pace of the "protein cycle" is linked to the respiration rate as well as to the speed of ion accumulation. From ideas derived from its role in protein synthesis, it seemed reasonable to Steward and Millar (216) to suggest that a molecule like a phosphorylated glutamine would have the necessary qualifications for this role. The discovery of phosphorylcholine in plants (237) and the ideas of Bennet-Clark (12) which would implicate such a substance as a carrier molecule, the regeneration of which depends upon a different type of metabolic cyclic (involving phosphatides) suggests other means by which similar ends might be achieved.

However, the importance of this discussion is not to claim that such carrier molecules have been critically identified but rather to indicate that satisfying explanations of the process of ion accumulation must integrate with other attributes of actively metabolizing and growing cells, and in consequence they need to be more far-reaching in scope than would at first sight appear. The reason is that active ion transfers are intimate properties of dynamic living systems. In consequence they present the challenge and the difficulty that to explain, rather than to describe, them, it will be necessary to achieve far closer contact with the features which distinguish living cells than we are at present able to do. Recognition of this, however, marks an advance. The second quarter of the twentieth century made great strides in describing the process of active transfer of ions in cells—only the future can show how far ultimate explanations of its detailed mechanism can be achieved.

Part III. Salt Relations of Intact Plants

The foregoing discussion has been almost entirely concerned with salt accumulation as a cellular process since it is recognized that the more complex problems of absorption and transport and nutrition in the multicellular plant body can only be understood with an adequate knowledge at the cellular and subcellular level. There are, however, certain aspects of the problem of salt accumulation which are only encountered in the intact organism. Moreover it is at this level that the interests of the plant physiologist impinge most closely upon those of the agriculturalist concerned with the behavior of his crop plants. It is with some of the problems raised by the intact green vascular plant growing in soil that much of the remainder of this discussion is concerned.

A. MODES OF ENTRY, TRANSFER, AND EXIT OF SALTS

1. The Plant and the Soil

The soil as the source of mineral nutrients for the plant is the subject of a chapter in a later volume (Volume III). It is sufficient here to remember that the soil particles provide a large reservoir of mineral elements, some of which are readily available to the plant and others are in relatively unavailable forms. One of the most remarkable characteristics of soil is the extremely low concentration of ions normally present in the soil solution (cf. Table 3). Only in some saline soils, e.g., near the sea, does the soil solution have an appreciable osmotic pressure. Among the most puzzling problems presented by plants growing in soil is how the plant is able to extract adequate amounts of nutrients from such dilute soil solutions. In this respect a plant in soil appears to be much more efficient than one that is growing in water culture. Various explanations have been put forward to account for this anomaly. Comber (43) for example contemplated the possibility that sparingly soluble mineral phosphates might be absorbed, not in solution, but from colloidal suspension. In 1923, Breazeale (19), on the other hand, was emphasizing the absorption of salts in ionic rather than molecular form, and he held the idea that the plant did not require to be in physical contact with its nutrient supply. He suggested that ions are mobile in the soil and that when those near to the root surface are absorbed they are rapidly replaced by others further away.

Two alternative hypotheses exist as a means of explaining the ready availability to plants of salts in soil (a) the carbon dioxide hypothesis, and (b) the contact exchange hypothesis. According to the first idea, carbon dioxide produced in respiration is released from the root and forms carbonic acid. This diffuses to the clay surface where H^+ replaces other cations such as K^+, and the released ions diffuse back to the root surface accompanied by bicarbonate ions, where they are absorbed either by exchange for H^+ or as ion pairs with bicarbonate. The argument against the carbon dioxide hypothesis rests mainly on the observation of Overstreet et al. (159) that bicarbonate ions were not absorbed as rapidly as potassium in short-term experiments with barley plants. The inadequacy of these experiments has been discussed above, and it seems important that the question of bicarbonate uptake, followed by its conversion to organic acid in roots growing in soil, should be re-examined with the more adequate techniques now available. Another observation which has been cited against the carbon dioxide hypothesis is that of Overstreet and Jenny (158), who found when studying the uptake of Na^+ from sodium bentonite suspensions that absorption was reduced when the intermicellar fluid was saturated with carbon dioxide. There may, however, have been a reduction of respiration induced by the high concentration of CO_2, which limited metabolic absorption.

In the contact exchange hypothesis proposed by Jenny and Overstreet (104) it is visualized that ions are transferred from soil particles to the root or vice versa without passing into free solutions. It is known that ions adsorbed electrostatically to a surface are not held rigidly, but on account of thermal agitation each oscillates a certain distance around its adsorption site. The space to which the adsorbed ion is confined has been termed its "oscillation volume." When the oscillation volumes of two adsorbed ions overlap then one may exchange for the other, and this process has been called "contact exchange." Jenny and Overstreet suggested that contact exchange of ions might occur not only between one soil particle and another, but also between a soil particle and the root surface. The hypothesis is largely based on the observation of Jenny and associates (106) that transfer of ions occurs between soil particles and roots and vice versa. It was shown that clay suspensions are much more effective than single salt solutions in extracting radioactive K^+ from roots, and conversely K^+ is absorbed more rapidly from the colloid than from the nutrient solution.

A major difficulty in accepting the contact hypothesis is the extremely short distances over which contact exchange is possible. Many workers have doubted that the cytoplasm of the root cells comes into sufficiently close contact with the soil particle for such exchange to occur, since it is unlikely that the oscillation volumes of adsorbed ions exceed a few hundred Angstroms. Overstreet et al. (155) clearly realized this difficulty, and they sought to combine the carbon dioxide and contact exchange hypothesis by considering that carbonic acid is responsible for releasing the adsorbed cation from the soil particle and that diffusion then occurs to the root surface where the cations pass inside the root by exchange for hydrogen ions.

Whether or not the contact exchange mechanism, in either its original or modified form, is or is not of importance in the uptake of cations it seems very unlikely that it is of any significance in the entry of anions. Jenny and co-workers (106) in fact demonstrated that clay particles have a very low adsorption capacity for anions, as also do plant roots. In contrast to the observations with cations, Jenny and his associates found that clay suspensions removed smaller amounts of radioactive bromide from roots than did the corresponding salt solutions. It still, however, remains to be demonstrated conclusively that plants can absorb such anions as nitrate and phosphate in sufficient amounts entirely from the dilute soil solution which bathes the surface of the root.

The root systems of many plants exist in association with symbiotic fungi, the significance of which has long been debated. There is scattered evidence throughout the literature that infected plants grow more actively and have a higher ash content than do those not so infected.

Routien and Dawson (184) observed that not only did mycorhizal roots of *Pinus echinata* take up more salt than noninfected ones, but that they also had a higher rate of respiration. These workers suggested that the presence of the fungus stimulated respiration of the host, causing the plant to produce hydrogen ions for use in base exchange reactions with the soil. They found that the uptake of iron was especially stimulated in infected roots, and they proposed that the mobilization and adsorption of iron may be one of the major roles of mycorhizae in the growth and survival of *P. echinata* in the field.

Kramer and Wilbur (113) found a greater uptake of P^{32} in mycorhizal roots than in nonmycorhizal roots and also noted that there was a more rapid uptake in infected parts than in uninfected parts of single roots of *Pinus taeda* and *Pinus resinosa*.

None of these experiments demonstrate adequately whether uptake takes place into the plant roots via the fungus, or whether the fungus stimulates active uptake of salt directly into the root. Melin and Nilsson (138) however, obtained some evidence in this connection. They were able to grow mycorhizae in which the fungus (*Boletus variegatus*) was attached to the seedling root (*Pinus sylvestris*) at one end, and at the other it was immersed in a solution containing P^{32}. With this arrangement they were able to demonstrate that phosphate *can be transferred* to the host *through* the fungal hyphae.

The most extensive investigation of salt absorption by mycorhizal roots has been that of Harley and his colleagues (79). They have studied the uptake of phosphate by excised mycorhizal roots of beech (*Fagus sylvatica*). Their first experiments showed that infected mycorhizal root tips absorb phosphate from dilute solutions more rapidly than do noninfected tips with the same surface area. The absorption process apparently depends on metabolism, since it is reduced in an anaerobic atmosphere, or in the presence of phthalate at pH 5, or by addition of cyanide. The effect of varying pH over the range 5 to 7 on the rate of uptake suggested that the phosphate was being absorbed mainly as $H_2PO_4^-$. Later experiments showed that upon exposure of the mycorhizae to P^{32} for short periods of time (up to 6 hours) the bulk of the phosphate (about 90%) was located in the fungal sheath. A comparison between excised and attached mycorhizae indicated no difference in the proportion of phosphate accumulated in the sheath. The proportion was, however, reduced in excised mycorhizae when oxygen was replaced by nitrogen in the gas stream; in the presence of sodium azide; and to some extent when high concentrations of phosphate were supplied.

Evidence was obtained that the presence of the fungus in excised roots hinders uptake of phosphate *into the root*. This was attributed to the fact that the fungus forms an outer layer of tightly packed hyphae with thick walls and few intercellular spaces, which serves as a partial barrier to diffusion. In contrast to the behavior of excised roots, is that of intact seedlings which show a higher phosphorus content when infected than when free of the fungus. This apparent anomaly is as yet unresolved. Other questions that remain unanswered are:

(1) Does phosphate enter the root of the host plant by diffusion through spaces and cell walls of the sheath or by way of the living sheath cells, being accumulated there first?

(2) Does the presence of the fungus stimulate respiration of the host cells, and, if so, is this linked to increased uptake of salt?

2. Absorption by Aerial Shoots

Although nonaquatic angiosperms normally obtain the bulk of their mineral salts requirement from the soil via the root system, the possibility that ions may be absorbed through the surface of the shoot, in particular of the leaves of these plants, should not be excluded. Excluding such epiphytes as *Tillandsia usneoides*, with its peltate absorptive scales, it is probable that most terrestrial plants obtain only insignificant amounts of solutes through the leaves, because these structures are rarely brought into contact with an adequate nutrient medium. Nevertheless, a limited supply of salts is available through rain and through the deposition of dust particles from which ions may be dissolved by rainwater, dew, or guttated fluids. For some slow-growing "lower" plants, e.g., lichens, the dust particles of the air may be an important source of mineral nutrients. In favorable circumstances even moderately sized angiosperms, e.g., species of Bromeliaceae, may grow epiphytically; and presumably these plants depend mainly, or entirely, on mineral salts supplied to the leaves from the air.

Interest in the direct supply of ions to aerial shoots arises largely from its possible agricultural applications. In some circumstances, minerals may be more economically supplied as sprays onto the leaves than by addition to the soil; and a greater control of fruiting and vegetative response may be obtained.

Foliar application of nutrients has already proved successful in the diagnosis and treatment of trace element deficiencies (this will be discussed in Volume III).

Few physiological studies of ion uptake by the aerial leaves of plants have been attempted, and even these have been largely concerned with the question of transport rather than with the mechanism of absorption (15, 233); see also Chapters 5 and 6 in this volume. Uptake of salts has generally been detected only by observing the alleviation of deficiency symptoms or the development of certain characteristic toxicity effects after the application of the nutrient solution. Little quantitative work on absorption seems to have been attempted, although with the availability of radioactive and other isotopes, the determination of the uptake can now be made relatively easily.

The results of many experiments, mostly conducted in the field, show that nearly all the mineral elements are absorbed more or less readily by the leaves of various plants. The earliest and most convincing demonstrations of this were with micronutrients. Chandler et al. (37) for example successfully used sprays of zinc salts to remedy zinc deficiency in fruit trees; Atkinson (9) cured boron deficiency in apple (*Malus sylvestris*) trees by applying borax or boric acid in solution to the leaves. Deficiencies of other micronutrients, e.g., molybdenum, manganese, and copper have been successfully alleviated by supplying the appropriate chemical to the leaves.

Special interest centers around the question of iron absorption by leaves, since plants are often unable to absorb sufficient iron even when this element is plentiful in the soil (cf. Chapter 6). The result of spraying the leaves of iron-deficient plants show that the localized absorption generally occurs, but there often appears to be little or no translocation of this cation to other parts of the plant. Biddulph (15) found that in iron-deficient bean (*Phaseolus vulgaris*) plants, the translocation of iron depends to a considerable extent on the pH of the solution applied and the phosphate status of the plants. Using Fe^{55} as a tracer element, he observed that when the pH of the medium was 7 and the phosphate content of the leaves high, iron was precipitated in the veins and intervenal chlorosis was observed. On the other hand, when the pH of the medium was 4 and the phosphate content of the plant low, iron became rather uniformly distributed in the leaves, and there symptoms of iron deficiency were alleviated.

The application of organic complexes of iron, e.g., iron complexes with ethylenediaminetetraacetic acid (EDTA) seems in many cases to be more effective in curing iron deficiency than the supply of inorganic iron salts. Whether this is due to improved uptake, or translocation, or both, is not yet clear.

A major difficulty encountered in the supply of nutrients to leaves is that the high concentrations required to cause an adequate total uptake through the cuticle may cause leaf damage. An alternative procedure applicable to trees is to apply nutrient solutions of higher concentration to the dormant branches after leaf fall (238). It has been claimed however, that such uptake takes place through fissures in the outer layers of the twigs and that uptake of ions by the living cells only follows the renewal of growth activity in the spring (78). Direct applications to an exposed cambial surface [as described by Steward (216)] presents still a further unexplored possibility.

In recent years considerable interest has been shown, especially amongst agriculturalists, in the possibility of supplying macronutrients such as nitrogen, phosphorus, potassium, and magnesium as foliar sprays. If these could be applied satisfactorily, it is possible that definite economies, both in the amount of fertilizer required and the costs of application, might be achieved with certain crops. The most successful results so far have been obtained in supplying nitrogen in this way to fruit trees and to such crops as the sugar cane (*Saccharum officinale*), pineapple (*Ananas comosus*) and banana (*Musa*). Hamilton et al. (76) sprayed apple trees with solutions of sodium and potassium nitrate, ammonium sulfate, and urea but found that only the urea treatment caused an increase in the total nitrogen content of the leaves. The failure of the inorganic salts to be effective was attributed to the amount of leaf injury caused by the solutions at the concentration used, which masked any beneficial effects from the nitrogen absorbed. Fisher (64) also found that apple trees responded favorably to treatment with urea and claimed that foliar application was a little more effective than soil application of a similar amount of urea. Humbert and Hanson (97) reported a rapid increase in total leaf nitrogen and chlorophyll content when sugar cane was sprayed with urea, and the effect was observed much more rapidly than after a comparable treatment of the soil.

Numerous investigations indicate that phosphorus is absorbed by the leaves of various crop plants. Phosphorus is generally found to be absorbed more rapidly when applied as orthophosphate than in any other form. On the whole, foliar application is less effective than the supplying of phosphates to soil, and so is, as yet, of limited economic importance. However, Silberstein and Wittwer (196) reported a more rapid absorption of phosphate by several vegetable crop plants when phosphorus compounds were applied to the foliage rather than to the roots. Potassium also can apparently be absorbed in significant amounts by leaves (236), but it seems that foliar application of this element could never be adequate alone to satisfy the considerable potassium requirement of a growing plant.

a. Factors affecting foliar absorption. i. Nature of the leaf surface. A part of the nutrients absorbed by leaves appears to enter through the cuticle, and the structure and chemical composition of this layer is therefore of importance in determining the rate of absorption. Before the salt can enter, the leaf surface must be wetted, and the ease with which this is achieved depends on the composition of the cuticle and on the presence or absence of surface convolutions and appendages, as well as on the surface tension of the penetrating solutions. Fogg (66) found great differences in the wettability of leaves, depending on the species examined, the age of the leaf, and its water content. The increased wettability of leaves with age may be attributed in part to discontinuities which develop in the cuticle, for example, by weathering or the attacks by insects. Skoss (198) observed parallel changes in the wax content and wettability of leaves grown under different conditions and concluded that the amount of wax in the cuticle largely controls the movement of water into leaves. The penetration of aqueous solutions into leaves may be facilitated by the presence of nonwaxy regions or pores in the cuticle. Roberts and associates (168) have demonstrated the presence of layers of pectinaceous materials in the cuticle of apple leaves, which are continuous with similar layers in the walls of the epidermal cells. Microchemical tests indicate that solutes can move along these pathways.

The presence of hairs on the leaf surface may be expected to promote the uptake of solutes, because such emergences are generally covered by a thinner layer of cuticle than is the general leaf surface. Damage to these structures would be expected to facilitate the rate of entry of substances into the leaf. Although hairs tend to decrease the wettability of leaves, pubescence increases the amount of solution held at the leaf surface when wetting is achieved and may promote uptake in this way.

ii. The importance of stomata. There is still some controversy about the importance of stomata in relation to the entry of solutes. Crafts (50) claimed that, while the cuticle of the leaves of many plants are fairly permeable to water vapor and dissolved substances, a negligible amount of penetration takes place through stomata. In support of this contention, Weaver and DeRose (244) obtained no correlation between the presence of stomata and the rate of entry of 2,4-D into leaves of *Coleus blumei* and *Tropaeolum* sp. plants. Went and Carter (245) likewise concluded that sugar is absorbed mainly through the cuticle of tomato leaves and not through the stomata, and Rodney (178) came to the same conclusion in relation to the penetration of urea into apple leaves.

On the other hand, Ginsberg (70) found that petroleum oil sprays enter apple leaves more rapidly through the lower than through the upper surface and concluded that penetration occurred mainly through the stomata since these are more numerous on the lower surface of apple leaves. Cook and Boynton (47) found significant differences in the rate of absorption of urea by the two surfaces of apple leaves in relatively short-term experiments. In 3 days the same amount of urea was absorbed by the upper surfaces of the leaves as was taken in at the lower surfaces in 2 hours.

It must be noted that in all these experiments the presence of stomata may not have been the only factor which facilitated the penetration of substances through one surface. The upper and lower surfaces of leaves commonly differ in other respects, e.g., in the thickness of the cuticle, smoothness of the surface, and presence of hairs, which may have important influences on the rate of absorption.

Skoss (198) has shown convincingly that the cuticle of some leaves is relatively impermeable to water and salts, by separating the cuticle in an undamaged state from the leaf and using it as the membrane in an osmometer. No movement of chloride ions or 2,4-D through such a membrane was observed within 7 days. Amongst other evidence, cited by Skoss, to support the view that entry of spray materials occur mainly through stomata are the following points. (a) Internal leaf injury is often observed in the region of stomata before it occurs elsewhere. (b) The penetration of aqueous and oil drops into stomata can be directly demonstrated. (c) Greater herbicidal effects are observed when sprays are applied to leaves with open stomata.

In view of this conflicting evidence, it would appear probable that penetration of solutes may occur *both* directly through the cuticle *and* via the stomata. The relative importance of the two paths doubtless depends on the species investigated, the situation in which it is grown, the conditions under which absorption occurs, and the particular solute involved.

iii. Other factors which influence foilage uptake. Most investigators have found that, other things being equal, nutrients penetrate more rapidly into young, actively growing leaves than they do into mature leaves. Structural differences, such as the composition of the cuticle, may be important in this connection; but Oland and Oland (146) have suggested that a difference in physiological activity may also be involved. They claim that magnesium ions are taken up by apple leaves mainly in exchange for hydrogen and suggest that it is the greater production of organic acids which facilitates the more rapid absorption of ions by young leaves. A diurnal variation in organic acid production is thought to account for the more effective penetration of magnesium when salts were sprayed on to leaves in the evening than in the early afternoon.

Various external factors may be expected to influence the absorption of mineral nutrients by leaves. They include the concentration and nature of the solute; the presence of other ions or molecules in the solution, especially wetting or chelating agents, as discussed above; air movements; temperature; and humidity. These factors will influence the rate of drying of a liquid film on the surface of the leaf and therefore affect the opportunity presented to the solute for entry. When the vapor pressure gradient is reduced, thus reducing evaporation, a greater absorption is expected and has been observed (47). Fisher and Walker (65) found that the inclusion of glycerine in sprays increased the absorption of minerals applied. This effect may be due to the increased retention of water at the leaf surface or to the greater spreading qualities of the spray when glycerine was present.

3. Aquatic Angiosperms

Submerged aquatic angiosperms are distinguished by certain morphological characteristics such as the rudimentary development of a root system and the absence of a cuticle at the surface of the shoot. Therefore, such plants probably absorb most of their mineral nutrients from the surrounding water via the shoot rather than the root. Evidence is available to show that rapid absorption by submerged leaves, and to a lesser extent by stems, can occur.

a. *Elodea.* Rosenfels (180) demonstrated the absorption of bromide ions from dilute solutions of potassium bromide by pieces of shoots of *Elodea (Anacharis) canadensis.* He found that variations in oxygen supply, temperature, and the carbohydrate content of the plant had similar, but not identical, effects on the uptake of salt and production of carbon dioxide. Distal regions of the shoot absorbed more bromide and respired more rapidly per unit of dry weight than did regions situated nearer the base of the stem. This was attributed to the greater metabolic activity of the young, growing leaves in comparison with those that were more mature, since the stems showed a fairly constant absorptive ability throughout their length.

Rosenfels observed that *Elodea* shoots in light will accumulate bromide ions rapidly when aerated with carbon dioxide-free air, without the appearance of respiratory carbon dioxide in the external medium. This suggested that, contrary to the views of Briggs (20) and Brooks (20b) at that time, an outward movement of carbon dioxide is not essential for the intake of ions. The data were thought to be more consistent with the contention of Steward that rapid salt absorption is associated with an active metabolic condition as manifested by a high rate of respiration.

An effect of light on the uptake of ions by shoots of *Elodea canadensis* was studied by Ingold (99). He found, as have workers with other plants, that light stimulates the absorption of salts. It seems likely that photosynthesis is somehow involved in this effect, but the supplying of carbohydrates is apparently not the sole factor, since Ingold could obtain no effect on ion uptake by adding sugar to the medium in the dark. Alternatively, photosynthesis may function indirectly by improving aeration (cf. experiments with *Nitella*, page 326). Although the possibility that light may also have some influence on permeability cannot be excluded, there is little evidence to support this idea at present; it is more probable that the morphogenetic effects of light on the growth of the leaves are directly concerned. It should also be noted in this connection that light has effects on protein synthesis and growth other than by its direct effects through photosynthesis, and these may, in turn, affect the pattern of ion uptake.

Ingold observed that the solutions in which *Elodea* was illuminated were invariably more acid at the end of an experiment than those in

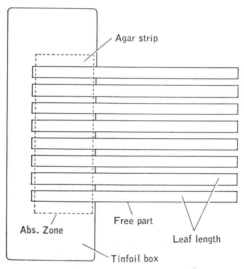

Fig. 57. The experimental arrangements for study of absorption by *Vallisneria spiralis* leaf sections. From Arisz (6).

which the plants had been in darkness. If the acidity had been due to carbonic acid, the reverse situation would be expected. This effect seemed rather to be related to a greater influence of light in promoting the uptake of cations than of anions. About equal quantities of potassium and chloride ions were absorbed from a solution of potassium chloride in the dark, whereas considerably more potassium than chloride was taken up in the light. The excess cations were presumably being absorbed mainly in exchange for hydrogen ions, thus causing acidity to develop in the medium.

b. Vallisneria. Arisz (5, 6), in the Netherlands, has made an extensive investigation of the uptake and transport of chloride ions by leaves of *Vallisneria spiralis*. Excised segments of these leaves will absorb ions at a constant rate for several days under suitable conditions. By supplying salts to one part of the leaf section, the transport to, and accumulation in, other zones can be examined. The absorption zones of the leaves were placed between two layers of agar containing the salt and generally enclosed in a tinfoil box. The free parts of the leaves were allowed to protrude through slits in the side of the box and to rest on wet filter paper in moist air (Fig. 57). The whole apparatus was maintained at a constant temperature during the experimental period. Pieces of leaves, 7.5 cm long, were used in most of the experiments; subsequently they were divided for analysis into three equal

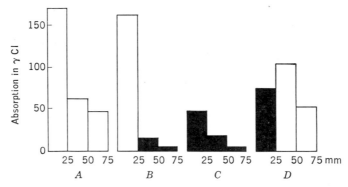

Fig. 58. The effect of light on uptake and transport of chloride in leaves of *Vallisneria*. In *A* the whole leaf is in the light, in *B* only the first zone. In *C* the whole leaf is in the dark, in *D* the first zone is in the dark and the free part in the light. The first zone of 25 mm is 24 hours in contact with a solution of 0.01 *M* KCl +CaSO₄ in 2% agar. The second and third zones are free in moist air on wet filter paper. On the ordinate the increase in Cl content; temp. 25°C. Pretreatment during 24 hours in aerated distilled water in the light. From Arisz (6).

parts; namely, the absorption zone and two others from the free part of the leaf.

An effect of light on the uptake of chloride by the absorption zone, and upon transport to other regions, was observed (Fig. 58). When the whole leaf was exposed to light, much chloride was accumulated in the absorption zone and there was little transport to other parts of the leaf (*A*). This situation was emphasized when the free part of the leaf was darkened and the absorption zone exposed to light (*B*). When the absorption zone was darkened, however, and the free part of the leaf illuminated, accumulation was reduced in the absorption zone but increased in the remainder of the leaf.

These results were interpreted on the basis of two assumptions. (a) The protoplasts of the leaf cells are continuous with one another (the symplast concept). (b) Uptake into a vacuolated cell is a two-stage process involving, first, entry into the protoplasm and subsequently a distinctive, independent movement into the vacuole. Arisz suggested, in explanation of these results, that the movement into vacuoles is stimulated by light but that the uptake into the cytoplasm is unaffected. The light effect was claimed to be independent of photosynthesis since it occurred when carbon dioxide was withheld.

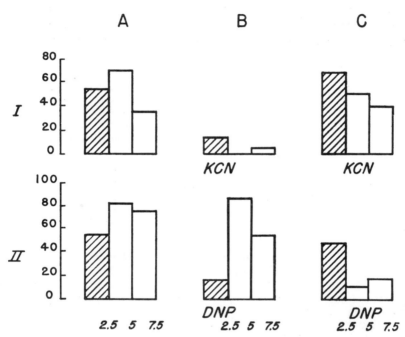

Fig. 59. The effect of inhibitors on uptake and transport of salt in leaves of *Vallisneria spiralis*. From Arisz (6a). See text for explanation.

Transport from one region of the leaf to another probably occurs mainly in the protoplasm with only negligible amounts of salt diffusing along the cell walls and intercellular spaces, since Arisz found that chloride ions are not lost in significant amounts from leaves immersed in distilled water after treatment with salt. Movement in the protoplasm was not apparently affected by light.

The distinction made between uptake in the cytoplasm and vacuolar secretion was supported by an experiment with respiratory inhibitors (Fig. 59). When potassium cyanide ($3 \times 10^{-4}M$) was applied to the absorption zone, accumulation of chloride and its transport to other parts of the leaf were both reduced. Accumulation in the absorption region was also reduced when 2,4-dinitrophenol ($10^{-4}M$) was supplied together with salt, but, in contrast to the effect of potassium cyanide, dinitrophenol did not interfere with transport and accumulation in the free part of the leaf. Application of salt to the absorption zone and potassium cyanide to the proximal section of the free zone (Fig. 59, IC) caused a reduction in the amount of chloride accumulated in the middle region, but transport through this part into the distal portion of the leaf was not curtailed. In contrast, however, dinitrophenol strongly inhibited accumulation, both in the middle and terminal zones when it was applied under the same conditions (IIC).

Most of these observations are explicable if potassium cyanide prevents uptake into the protoplasm but does not inhibit either vacuolar secretion or transport in the cytoplasm at the concentration used; and, conversely, if dinitrophenol inhibits vacuolar secretion without interfering with entry into the cytoplasm or movement through it. Arisz, however, does not offer an explanation of the observation that accumulation was reduced in the terminal zone when dinitrophenol was applied to the region between it and the absorbing zone, although this effect was as marked as others which were assumed to be significant. Dinitrophenol seems to be interfering with transport into the distal region under these conditions.

Arisz' explanation, while substantially correct, may well have oversimplified the situation. The possibility that inhibitors supplied to one zone may, like salts, be absorbed and transported to another region, where they are also active, cannot be excluded. Moreover, Arisz has assumed that the leaf of *Vallisneria* behaves as if it were a mass of undifferentiated parenchyma cells. In fact, this organ has well-developed phloem tissue, and it would be surprising if this specialized tissue does not contribute to the longitudinal transport of ions and organic materials. Experiments of Sutcliffe (unpublished), using radioautography, have indicated that the longitudinal movement of K^{24} and Rb^{86} proceeds mainly through the phloem in *Vallisneria* leaves. The transport was found to occur about twice as rapidly toward the morphological base than toward the apex of the leaf. Such polarized movement would normally serve to direct salts, absorbed from the surrounding medium, toward the growing basal region of the leaf and the shoot apex. Little is yet known about the mechanism of polar transport, of either ions or organic materials, in plants (see Chapters 5 and 6 in this volume), and *Vallisneria* leaves may prove to be a favorable material for further investigation of this problem.

   c. *Ion movements in the leaves of other submerged aquatics.* In contrast to the predominantly longitudinal transport of salts in *Vallisneria* leaves studied by Arisz and his associates are the transverse movements occurring in the leaves of other aquatic plants. These have been investigated by several workers, e.g., Arens (3), Steeman-Nielsen (198a), and Lowenhaupt (121). In the plants now to be considered, the movements of cations across the leaves appear to be closely related to the absorption and utilization of bicarbonate ions.

Arens was among the first investigators to study the movement of cations across the lamina of the leaves of water plants, including *Elodea* (*Anacharis*) *densa*, *Hydrilla verticillata*, and several species of *Potamogeton*. Single leaves were used as membranes across which movement of ions occurred from one solution to another. It was found that when solutions of various bicarbonates were supplied to the abaxial surface of an illuminated leaf, cations (e.g., K^+, Na^+, Ca^{++}) migrated across the leaf into distilled water placed on the other side. At the same time the pH of the later solution markedly increased. In the dark no ion movements or change of pH were observed. Even in the light there was no transport if Cl^- or NO_3^- were substituted for bicarbonate ions. Movement of cations across the leaf did not occur when a bicarbonate solution was applied to the adaxial surface either in the light or in darkness.

Arens concluded from these observations that bicarbonate ions are absorbed by the undersurface of the leaves if accompanied by an equivalent number of cations. Within the leaves the bicarbonate ions are utilized in photosynthesis, and the cations are then excreted from the upper surfaces together with hydroxyl or carbonate ions. The utilization of bicarbonate in photosynthesis, in contrast to carbonate or carbon dioxide, by some (but not by all) water plants has since been confirmed by several other workers (186, 198a).

The observations of Arens were to some extent confirmed by Steeman-Nielsen (198a) for *Potamogeton lucens*. However, Nielsen concluded, in contrast to Arens, that bicarbonate ions are absorbed, together with equivalent quantities of cations, by *both* the upper and lower surfaces of the leaves of this plant. He suggested that the bicarbonate ions are utilized in the chloroplasts with the production of carbohydrate and release of hydroxyl ions. The latter, he proposed, are transported exclusively to the upper surface of the leaf and released with an equivalent number of cations. The net results are thus a transference of cations from one side of the leaf to the other; uptake of bicarbonate by the abaxial surface, and an increased alkalinity of the medium in contact with the adaxial surface. Steeman-Nielsen concluded that the uptake of cations and bicarbonate ions occurs by diffusion and that active transport is involved only in the transference of OH^- out of the upper surface of the leaf.

If the uptake of bicarbonate and its associated cation are linked together, it is to be expected that the rate of photosynthesis in certain
circumstances may depend on the availability of cations in the medium. Steeman-Nielsen observed, in support of this contention, that
addition of calcium chloride to the solution increased the rate of bicarbonate assimilation in *Myriophyllum spicatum*. The rate of uptake
of carbon dioxide, on the other hand, was unaffected by addition of the
salt, indicating that the added cations were promoting bicarbonate uptake directly rather than through an influence on photosynthesis.

The transport of calcium ions across the leaves of *Potamogeton
crispus* has been intensively studied by Lowenhaupt (121). He has
observed movements of cations which are somewhat similar to those reported by Steeman-Nielsen and Arens, and seeks to explain them on the
basis of a similar hypothesis to that proposed by Conway, to account
for cation transport in yeast (see pages 409–410). Lowenhaupt suggests
that a carrier mechanism transports calcium ions across diffusion
barriers on both the adaxial and abaxial surfaces of the leaves; in the
one case the ions are moved out of the leaf, and in the other into it.
The diffusion barriers across which transport occurs are identified with
the outer protoplasmic membranes of the surface cells, because Ca^{++}
ions do not apparently penetrate into the cell vacuoles.

Lowenhaupt proposes that the carrier is a reduced metabolic intermediate synthesized during photosynthesis, and in this way light and
bicarbonate ions are indirectly implicated in the transport mechanism.
He suggests that a reservoir of carrier molecules can be built up in the
protoplasm since a limited amount of calcium was found to be fixed in
the dark, following exposure of the leaf to light. Besides facilitating the
synthesis of carrier molecules, light is thought to be involved in the
subsequent breakdown of the ion-carrier complex at both leaf surfaces.

An interesting physiological difference between the two leaf surfaces
was demonstrated convincingly by experiments in which Lowenhaupt
floated leaves on the surface of salt solutions. When the abaxial surface
was in contact with the solution in the light, calcium ions were absorbed; in the dark, on the other hand, cations were released from the
leaf. When the adaxial surface was in contact with the medium, however, in the light, ions were released, whereas in the dark they were
taken up.

Lowenhaupt proposes that within the leaf, calcium ions move from
one surface to the other via the continuous protoplasts without entering the vacuoles (cf. Arisz' suggestions). This movement he thinks
may be assisted by the association of ions with cytoplasmic constituents
and by protoplasmic streaming. Lowenhaupt could find no evidence

for comparable movements of anions in the cytoplasm or across the leaf. He showed, however, that phosphate and iodide ions can be absorbed equally by both leaf surfaces. The absorbed anions are presumably accumulated in vacuoles or, in the case of phosphate, metabolized in the cytoplasm of the leaf cells.

4. Excretion and Exclusion of Salts

a. The passive leakage of ions out of coenocytic algae, storage tissue disks and excised roots has already been mentioned above. The amount of salt lost in this way from healthy tissues, under conditions of either active or of reduced metabolism (when reabsorption is prevented) is usually very small, and most of this is probably derived from damaged or moribund cells. This, however, does not appear to be the case in the green alga Ulva lactuca, since Steeman-Nielsen (198a) found that movement of sodium chloride occurs easily into and out of these cells. He calculated that the diffusion coefficient for sodium chloride in the Ulva thallus is about 0.06 when the cells are alive, and 0.2 when they are dead, in comparison with 1.10 for the diffusion coefficient of this salt in water. Steeman-Nielsen concluded, on this evidence, that salt must diffuse freely through the thallus, via the vacuoles of the component cells, and that therefore the protoplasts do not in this plant prevent a formidable barrier to ionic exchange and diffusion.

This conclusion is supported by the work of Scott and Hayward (192), who found that sodium and potassium ions can be leached from the thallus of Ulva lactuca fairly easily, and almost completely, by repeated washing in a sucrose solution isotonic with sea water. Furthermore, almost all the sodium and potassium in the tissue were found to exchange freely with other ions from the external medium. By the use of Na^{24} as a tracer ion it was possible to show that 88% of the sodium was exchanged in 5 seconds (190).

Turning now to the spermatophyta, there is plenty of evidence that salts can be leached, albeit with some difficulty, from the aerial shoots of both angiosperms and gymnosperms. Mann and Wallace (136) may be credited with the first observation that ions, in particular potassium, can be washed from apple leaves in significant amounts, e.g., by rain. Wallace (243) later demonstrated that plants exhibit varying degrees of resistance to loss of salts in this way. An investigation by Tamm (234) showed that certain forest mosses get most of their mineral supply from the leaves of trees which are leached in situ by rain.

It is generally thought that such minerals as are released by foliage are derived mainly from aging or damaged leaves; but Long and co-workers (120) have reported losses of salts from young, healthy leaves also. These latter workers supplied potassium and rubidium salts containing radioactive tracers to the roots of high- and low-salt bean plants growing in water culture, under conditions of light and darkness. They found that only a relatively small amount (5–12%) of the salt absorbed during the preliminary treatment in the light was subsequently leached from the plants during immersion of the shoots in distilled water for 4 hours. Much greater percentage losses of potassium and slightly greater losses of rubidium were obtained under the same conditions *following absorption only in the dark*. Low-salt plants, released 42% of the previously absorbed potassium; high-salt plants, 71%; and the corresponding figures for percentage loss of rubidium were 7 and 14. Presumably, in the light, plants are able to fix potassium and, to a lesser extent, rubidium, more effectively than in the dark. Such fixation may involve transference of the ions into vacuoles, or the formation of complexes between the ions and organic protoplasmic constituents. The influence of light is probably exerted in some way through photosynthesis, but the differing rates of transpiration in light and darkness may be a contributing factor if indeed this influenced the absorption of salt or its translocation into the shoot. The nutrient status of the plants was clearly important in determining the percentage of the absorbed salt which was subsequently released. This is understandable, since salts from the xylem sap will be less quickly transferred into the vacuoles of the leaf cells when the salt content of the vacuoles is high than when it is low.

An important source of the mineral nutrients washed from leaves may be guttation fluids, which emerge from hydathodes and stomata, under conditions of reduced transpiration and evaporate on the leaf surface. Deposits of salt on the leaves of plants growing in saline or calcareous soils are sometimes large enough to be visible to the naked eye.

Root systems as well as shoots may, in suitable circumstances, release significant amounts of salt to the external medium. Achromeiko (1) showed that normal high-salt plants, when transferred to a salt-deficient medium, served as a source of nutrients to other plants growing in the same solution. Luttkus and Botticher (133) found that considerable amounts of potassium were given off by the roots of intact maize (*Zea mays*) plants growing in water culture in the dark, and Helder (81) likewise observed losses of phosphorus and nitrogen under the same conditions.

b. Active movements. Since plants generally utilize energy in the absorption of salts, it is somewhat surprising to find that sometimes the loss of ions also requires an "effort" on the part of the organism. The situation is analogous to that in some animal cells and tissues, e.g., erythrocytes, muscle, and nerve fibers, when an active extrusion of sodium ions occurs. The excretion of cations by yeast cells and by the leaves of certain aquatic plants has already been considered. Two other examples of the apparently activated movement of salts out of plants will now be described, namely the excretion of sodium ions from the thallus of *Ulva lactuca* and the functioning of salt glands in certain angiosperms.

i. Ulva lactuca. Scott and Hayward (191, 192) made an extensive investigation of the movements of cations into and out of disks cut from the thallus of *Ulva lactuca*. They found that under normal conditions the tissue accumulates potassium when immersed in sea water, and partially excludes sodium. The maintenance of this distribution of ions depends on photosynthesis, since it was shown that when the material was transferred to the dark, potassium was gradually lost and sodium taken up by the tissue as respiratory substrates became depleted. Upon subsequent illumination of the darkened thallus, sodium was excreted and potassium accumulated until finally the normal cation balance was attained (Fig. 60). If the tissue was leached in isotonic sucrose and then transferred to sea water, the cells rapidly returned to equilibrium by uptake of both potassium and sodium. The uptake mechanisms of potassium and sodium under these conditions are apparently distinct from one another, since it was observed that the rate of reaccumulation of potassium was not influenced by the presence of sodium in the external medium.

In an attempt to elucidate the mechanism of these cation movements in *Ulva*, Scott and Hayward (192) examined the effects of various respiratory inhibitors, alone or in pairs. They found that iodoacetate at a concentration of $10^{-3}M$ caused a greater reduction in the potassium content and a greater increase in the sodium content of the thallus in the dark than would normally occur, but the inhibitor did not interfere with ionic movements in the light. This observation suggests that, through photosynthesis, metabolic intermediates are produced which can supply energy for active transport by routes which are independent of reactions blocked by iodoacetate. On the other hand, active movements of ions in the dark can only occur via reactions which are sensitive to iodoacetate poisoning.

Arsenate, when supplied along at a concentration of $5 \times 10^{-3}\ M$, had no effect on the movement of cations either in the light or in the dark; but when it was supplied together with iodoacetate in the dark, it pre-

vented the iodoacetate from causing a loss of potassium. On the other
hand, the uptake of sodium stimulated by iodoacetate under these conditions was unaffected by the arsenate.

Phenylurethane at a concentration of 10^{-3} M caused a progressive loss
of potassium and gain of sodium both in the dark and in light, and, in
contrast to the effect of iodoacetate, the effect of phenylurethane was

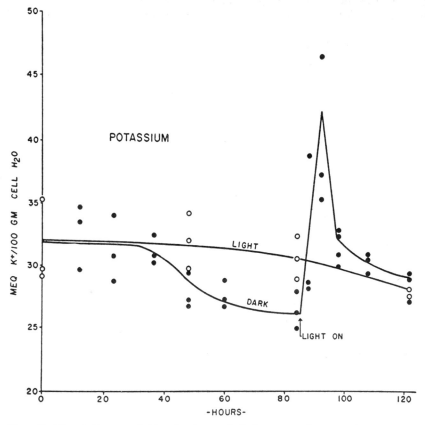

FIG. 60. The influence of illumination and darkness on the potassium content of
Ulva lactuca. From Scott and Hayward (191).

entirely reversible. When iodoacetate (10^{-3} M) and phenylurethane
(10^{-3} M) were added together, greater movements of cations were observed than with either inhibitor alone. The simultaneous effect of the
two inhibitors supplied together in promoting potassium loss was greatest during the first 10 hours, and the promoting influence on sodium
uptake reached a maximum later, suggesting independent effects on
the two mechanisms for the two ions.

p-Chloromercuribenzoic acid (an inhibitor of succinic dehydrogenase) and 2,4-dinitro-*o*-cresol (an uncoupling agent) inhibited the activated movements of both sodium and potassium in the light and in the dark. ATP (adenosine triphosphate) had a stimulating effect on the uptake of potassium but did not influence sodium excretion. The presence of pyruvate (50 mg per cent) gave complete protection to the sodium-transporting mechanism against poisoning by iodoacetate in the dark, and it also protected the potassium-absorption mechanism from

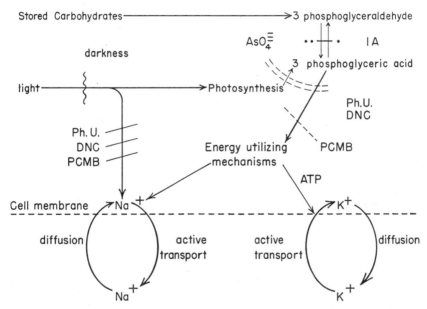

FIG. 61. Proposed scheme for the active transport of sodium and potassium in the thallus of *Ulva lactuca*. IA = iodoacetate; Ph.U. = phenylurethane; DNC = dinitro-*o*-cresol; PCMB = *p*-chloromercuribenzoate; ATP = adenosine triphosphate. From Scott and Hayward (192).

iodoacetate inhibition, but to a much smaller extent. Phosphoglyceraldehyde, on the other hand, had little effect on sodium transport but significantly reduced the loss of potassium instigated by iodoacetate.

These somewhat involved observations and the conclusions derived from them by Scott and Hayward (192) are summarized in the scheme shown in Fig. 61.

These workers suggest that 3-phosphoglyceric acid, derived either by the breakdown of stored carbohydrates or from photosynthesis, is a source of energy for metabolic transport. Iodoacetate inhibits conversion of 3-phosphoglyceraldehyde to the acid and in this way prevents active transport in the dark, but the inhibitor is ineffective in the light because phosphoglyceric acid is also produced via photosynthesis. Arsenate overcomes the inhibitory effect of iodoacetate, allowing the conversion of phosphoglyceric acid to phosphoglyceraldehyde, and this prevents the loss of potassium in presence of iodoacetate in the dark. The remaining inhibitors, effective in both light and darkness, presumably block a common pathway in the transfer of energy from 3-phosphoglyceraldehyde to the transport mechanisms. Light appears to have an independent effect on sodium extrusion (which is sensitive to phenylurethane, dinitrocresol, and p-chloromercuribenzoate), because the effects of light and of these inhibitors on sodium transport are different in some respects from those on potassium absorption. Sodium extrusion seems to be closely bound up with the metabolism of pyruvate and may depend on a supply of reduced metabolic intermediates, whereas potassium absorption apparently is identified more closely with the supply of energy in the form of phosphorylated compounds.

Salt transport in *Ulva* resembles in some respects the situation observed by Conway in yeast and differs considerably from that found in typical vacuolated plant cells. In *Ulva*, as in yeast, a highly selective active transport of cations apparently occurs in both directions across a membrane which is highly permeable to the free diffusion and exchange of ions. In these circumstances it is difficult to visualize that a significant accumulation of *free* ions within the cell occurs, because of the energy expenditure required to maintain even a modest concentration gradient. This raised the important question, not discussed by Scott and Hayward, of the location and state of the ions within these algae cells. Data on the extent of the vacuole system in *Ulva* appears to be fragmentary. The classical diagram of an *Ulva* cell by Oltmanns (147), copied by other authors, does not include a representation of the vacuole, and this may be taken to mean that if it is present the vacuolar system is small. Observations by the present authors on fresh material stained with neutral red have failed to reveal extensive vacuoles in the vast majority of the cells comprising the thallus, although a few expanded cells at the edge of the thallus do contain such vacuoles.

It must, therefore, be concluded that the bulk of the cations taken up by *Ulva* cells are held in the thick cell wall and in the protoplast, probably by binding onto cellular constituents or by establishment of Donnan equilibria. The amount of salt held in these ways may clearly be dependent on metabolism of the cell. The discrimination exhibited between sodium and potassium suggests that binding onto specific sites is of greatest importance (171), but the fact that salt is washed out with water suggests that the ions have a somewhat tenuous hold. It is doubtful whether an active carrier system operates in transporting ions across the outer membrane, as was visualized by Scott and Hayward, since a high resistance barrier is essential to the effective operation of such a system; and their results indicate that the resistance of the membrane is low.

It is probably of considerable significance that the free, two-way movement of ions across cell membranes has been observed only in *nonvacuolated* cells. Examples of this which have been discussed above include yeast and bacteria. Ruhland and also Collander (see Chapter 1) recognize the high permeability of the cell membrane in *Beggiatoa* to various substances, in accordance with their molecular size, and Collander refers to this organism as "aberrant" by comparison with other, predominantly vacuolated, cell types. He notes that diatoms, (again cells in which the cytoplasm occupies a large part of the cell) resemble *Beggiatoa* in their permeability properties.

c. Excretion of ions by salt glands. When plants are growing in highly saline or calcareous soils, deposits of salts are sometimes to be found in specific regions on the leaf surface. This phenomenon is well known in halophytic species of the Plumbaginaceae, Frankeniaceae, Tamaricaceae, and among the calcareous Saxifragaceae. In some cases the salt may be extruded passively along with guttation water, but in others active excretion seems to occur. Only in the latter should the deposit of salt on leaf surfaces be properly attributed to the activity of "salt glands." Some authors (7, 82) refer to the function of the salt glands as "secretion," but the term "excretion" is here more appropriate. If salt glands have any beneficial function in plants, it would appear to be in the removal of certain salts which might otherwise accumulate in toxic amounts in the leaf cells.

Although salt glands were described in anatomical studies during the nineteenth century, it was not until 1910 that their significance was appreciated, through the observations of Schtscherback (187). Ruhland (185) made a careful investigation of the structure and activity of salt glands on the leaves of *Limonium* (*Statice*) *gmelinii*. He showed that a single gland consists of a group of 16 cells with large nuclei, dense cytoplasm, and no chloroplasts or obvious vacuoles (Fig. 62). Four large excretory cells (*A*) are arranged centrally in a decussate manner. Closely associated with them are four small cells (*B*); and eight flattened cells (*E*), forming a two-layered cup-shaped structure, enclosing the others.

The entire gland is enclosed by a rigid cuticle of irregular thickness which is penetrated by a small pore (*P*) at the tip of each excretory cell. It is through these pores that the salt is excreted. Outside the gland proper are four collecting cells (C) communicating through the cuticle with the cup cells by means of plasmodesmata. The collecting cells serve as a link between the gland and the underlying tissues of the leaf, and it is apparently through these that salts are transferred from the mesophyll into the gland cells.

Ruhland estimated that up to 0.86 mg of liquid were excreted per hour by a square millimeter of leaf surface (i.e., by about 700 glands). He found the sodium chloride concentration of the fluid to be about equal to the concentration of salt in the leaf tissue as a whole and concluded, therefore, that the glands themselves do not perform osmotic work. His conception was that accumulation occurred in the leaf and

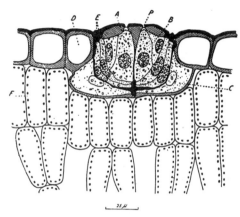

Fig. 62. Structure of salt gland in *Limonium* (*Statice*) *gmelinii* according to Ruhland, slightly altered. *A*, secretion cell; *P*, pore; *B*, adjoining cell; *C*, collecting cell; *D*, epidermal cell; *E*, cup cells; *F*, mesophyll. From Arisz *et al.* (7).

that a liquid rich in salt was squeezed from the gland in some way under pressure.

More recently Arisz and his collaborators (7) have examined the mechanism of excretion in the salt glands of another species, *Limonium latifolium*. They have made use of the earlier observations of Schtscherback and Ruhland that glands attached to isolated fragments of leaf will continue to function. Leaf disks, 3 cm in diameter, were rested on filter paper soaked in salt solutions and kept in a saturated atmosphere to prevent evaporation of the excreted fluid. At the end of an experiment the liquid excreted by the upper surface of the leaf was collected on a filter paper and analyzed. The volume of liquid was also estimated, and this enabled both the amount of salt excreted and the concentration of the excretion fluid to be determined. Since the experimental plants were grown in soils containing little chloride, it was found necessary to place excised leaves with their petioles in a concentrated sodium chloride solution for several hours before cutting the disks in order to stimulate subsequent excretion. The amount of salt excreted increased as the time of pretreatment was increased and with increased concentration of salt in the pretreatment medium up to a maximum value of 0.34 *M*. By subsequently placing the disks on a sufficiently concentrated salt solution, it was possible to demonstrate that excretion may occur when the external concentration is higher than that of the leaf tissue. Arisz concluded, therefore, that, contrary to the view of Ruhland, the glands perform osmotic work in excreting a more concentrated solution than is present in the leaf cells (Table XXIII). However, this conclusion is not unequivocal since the calculation of the concentration of salt in the leaf cells was based on the analysis of expressed sap and it is somewhat uncertain how far this will

TABLE XXIII

OSMOTIC VALUE OF LEAF SAP OF *Limonium latifolium*
AND SECRETED FLUID[a,b]

Leaf sap (mean osmotic value)	Secreted fluid (mean osmotic value)	Secreted fluid (NaCl conc.)
0.338 *M* NaCl	0.438 *M* NaCl	0.420 *M*

[a] From Arisz *et al.* (7).
[b] Secretion 24 hours at 25°C on a 0.342 *M* NaCl solution.

reflect the actual concentration of salts either in the protoplasm or in the vacuoles of the intact cells.

Excretion of salt from leaf disks continued for several days, although it decreased progressively with time, even when the tissue was supplied with a salt solution. Light and increase of temperature stimulated the amount of fluid excreted, but did not significantly influence the concentration of sodium chloride in the liquid. Various respiratory inhibitors at moderately high concentrations (10^{-3} to 10^{-4} *M*) caused a reduction in the amount of liquid excreted without affecting the concentration of salt therein. Phloridzin was apparently ineffective as an inhibitor of excretion. Very little fluid was exuded under anaerobic conditions, and no chloride ions could be detected in it.

When the osmotic pressure of the medium on which the disks were placed was increased by the addition of sucrose or magnesium sulfate, the amount of liquid excreted was reduced but the concentration of salt in it was increased, so that there was little effect on the actual amount of salt excreted.

Arisz concluded that there is an active accumulation of salt into the gland cells. This causes an uptake of water, and since the gland cells cannot swell appreciably, owing to the presence of the surrounding cuticle, fluid is exuded under pressure at the leaf surface. Since the amount of liquid excreted depends on the pressure built up inside the gland cells and this, in turn, is controlled by the accumulation of salt, it is understandable that it is reduced by metabolic inhibitors. What is less clear is why the concentration of salt in the exudation fluid remains the same in these circumstances. If the mechanism operates in the manner suggested by Arisz, reduced flow of liquid would be expected to be associated with a reduced concentration, since the pressure developed and hence the amount of liquid exuded will depend upon the concentration of salt in the gland cells. A more satisfactory explanation of glandular secretion on the basis of Arisz' observations would appear to be that salt is excreted *from* the gland cells at the pore surfaces, and this is followed by osmotic withdrawal of water from the leaf.

The influence of the increased osmotic pressure in reducing the amount of liquid excreted and causing an increase in the concentration is equally explicable on the basis of Arisz' hypothesis and the alternative suggested here. Through immersion in a medium of high osmotic pressure, the leaf tissue presumably develops a higher diffusion pressure deficit, and the osmotic withdrawal of water from it is thus achieved with greater difficulty.

B. INTERRELATIONS OF SALT ACCUMULATION WITH GROWTH AND DEVELOPMENT OF THE PLANT BODY

It is now necessary to examine the problem of salt absorption with the whole plant body and its interrelationships in mind. The underlying thought is that centers of activity in growth and salt accumulation tend to go hand in hand. The stimuli which promote the one also regulate the other.

The features of the process of salt accumulation in cells which need to be kept in mind when interpreting the whole plant are as follows: First, the process comprises events, or steps, of different kinds. In cells multiplying by division, or cells which lack conspicuous, large aqueous vacuoles, stoichiometrical binding is the outstanding feature. The first step, or phase, in salt accumulation seems to occur at specific sites, the multiplication of which is a function of metabolic activity. This process has characteristic metabolic relations, as shown by its sensitivity to enzyme inhibitors and its relations to external concentrations which are more consistent with stoichiometrical combination than is the case for accumulation in vacuoles. It is here that bacteria, the rapidly multiplying cells of tissue cultures, and the meristematic cells in the apex of the roots show similarities. As growth of the individual cell proceeds and cell multiplication gives way to cell expansion, with the enlargement of vacuoles and internal secretion of water across the tonoplast, the second, more familiar process of accumulation in vacuoles sets in. Somewhat different features now characterize absorption by the enlarging cell, for the efficiency of the process is greater from dilute solutions and the oxidative metabolism to which the process is related is more strongly cyanide sensitive (216). Arisz, too, has postulated a different sensitivity to inhibitors for entry into protoplasm and into the vacuoles of leaf cells. While both the first and the second steps in the over-all entry of ions into the vacuole have a metabolic basis, it is clear that the mere production of carbon dioxide is not alone involved, for many cells will produce carbon dioxide rapidly without appreciable salt accumulation, while other cells can accumulate salt without releasing carbon dioxide (e.g., green cells in the light). The more fundamental relationship, therefore, seems to be that the energy released by respiration is harnessed by some coupling mechanism to the process of accumulation per se; and, in this coupling mechanism, it seems that the energy can be made available simultaneously to salt accumulation and protein synthesis, for the two often go hand in hand. Active transfer of ions across cellular boundaries is related also to *activated movements of organic solutes*, because a deficit of salt may be associated with the internal secretion of sugars and the later replacement of those organic solutes when salts become available. These general ideas, unified by the broad concept that the *de novo* accumulation of salts is an integral part of the process of growth in the cell, were illustrated in Fig. 39.

The consideration of the distribution of ions within the plant body impinges upon the problems of translocation and of the ascent of sap—problems that are treated in Chapters 5 and 6 in this volume. Both these large problems also require physical explanation. Again the dilemma is that the phenomena in question occur only in the organized system, which has the inherent capacity to invoke living cells to maintain nonequilibrium states, whereas much physiological research is, and no doubt should be, directed toward explanations which rely first upon purely physical interpretations and the response to equilibrium conditions. The problem of the simultaneous movement of water and salt is a case in point. Are salts merely "swept along" by a moving water stream, or are their respective movements in a measure independent? Does physical evaporation "pull" both salt and water into plants? The relative independence of the *de novo* absorption of water and salts seemed established and well recognized even in the first quarter of the century, but it is still a somewhat disputed question how far activated movements of water occur in plants or how completely translocation of ions is merely a passive process in which, over long distances at least, they are carried along by a moving water stream. The forgoing discussion makes it clear that, at the two essential points of (a) release from cells at the source, and (b) storage and accumulation at the "sink," the transport of ions is controlled by specific mechanisms in which the cells are peculiarly and selectively involved, and to this extent at least their role in the movement of ions is a very real one.

The chastening fact is that over two centuries after Stephen Hales speculated upon the "circulation of the sap," plant physiologists have still no completely acceptable explanation of the process. That physical phenomena, like evaporation, probably operating in a somewhat unfamiliar way in the environment of the plant body, play an essential— probably an overwhelming—role, is undeniable. It is equally undeniable that the very system in which the process occurs depends for its formation and maintenance upon living cells. To dissociate the process of transpiration from essential features of the way in which the organism in which it occurs grows and develops, and without which it has no physical counterpart, would be as shortsighted a view as to attempt its explanation apart from the physics of evaporation. The fact that the modern plant physiologist can be so easily embarrassed when asked by the man in the street, "How does the sap get to the top of a tall tree?" is evidence enough that the processes are only imperfectly understood. Moreover, it still remains true to say with respect to both water and salts that, in *the first instance*, they did not "move" to the top of a tree as it were in one jump, but "they grew there" by the successive additions to the shoot of new cones of tissue laid down by each annual increment of growth. That these tissues are established in ways which make the subsequent movement of water, and salts, from the older tissue below into the new tissue above a relatively easy event is clear. It is also apparent that once the system has been so established, replenishment of the water evaporated from the shoot or salts moved to other regions can occur, rapidly and over long distances, across intervening regions of vascular tissue which are devoid of living cells. All this is quite evident. The fact remains, however, that the complete and satisfying interpretation of water and salt movement, especially in tall trees, still remains to be presented; and it is one of the challenges of modern plant physiology to do this in a manner which is as consistent with the structure, development, and growth of the organism as it is with the physical realities of the processes.

In whatever manner the longitudinal movement is effected, the solutes after entry into the shoot must pass to lateral organs, particularly leaves and buds. The mechanism by which the lateral organ will gain access to, and compete for, the solutes in question must of necessity involve the way in which its own vascular supply integrates with that of the axis. While it is improbable that strict rules for all vascular plants can be laid down, nevertheless it is to be expected that the different organization of monocotyledonous and dicotyledonous plants will be reflected in differences in their mode of access to the salts in the vascular supply and in the way they are distributed to the foliage system. In trees and in other plants with marked secondary growth, one may also expect that the way in which absorbed ions would be distributed throughout the plant body would be related to the manner of their growth. Indeed, if the claims of an individual organ in competition with others upon the solutes that are available is related to the intensity of metabolism and growth, which follows from the scheme shown in Fig. 39, one might also expect that the distribution of ions would be subject to the mechanisms which smoothly control and integrate the growth of the parts. Following the ideas already put forward by Steward and Millar (216) we may now see how these principles apply to both dicotyledonous and monocotyledonous herbaceous plants, and also to trees.

1. The Root as an Absorbing Organ for Salts

The functioning of the root as an intact organ and as part of the intact plant body is here to be considered. The morphological questions that immediately arise are as follows:

(a) Where in the root are the most active centers of salt accumulation in cells located?

(b) From which part of the root are ions most freely removed to the shoot, and how are they transferred to the stele?

(c) How is the distribution of ion absorption and removal from the root affected by, or related to, the way in which the root grows?

(d) How are the effects of the shoot on absorption by the root made manifest?

Regarding salt accumulation as a property not merely of a root but of a *growing* root, the dilemma is that the full conditions which determine the growth of the roots are imperfectly understood. Some roots, notably those of certain dicotyledons, can be easily grown in continuous culture; others, especially of monocotyledons, have not yet been grown entirely satisfactorily in this way. In other words, there are nutrient requirements or stimuli to the continued growth and activity of excised roots which are still unknown. These may emanate from the shoot or reside in certain regions of the root. With few exceptions, studies on salt accumulation in roots have concentrated on a system designed to take up salt rapidly enough for experimental purposes. They have not therefore been made, necessarily, with roots endowed with the ability to grow *continuously*.

Lateral roots arise endogenous and relatively far back from the apex, and root hairs are characteristically ephemeral structures and by no means universal. The root is therefore an organ in which accumulation of ions may be correlated with growth with the minimum of complications by the emergence of lateral organs. Since cell division and the maximum intensity of growth by cell elongation occur so close to the root tip and do so with graded intensity along the axis, it is possible to correlate these with similar gradients of salt accumulation.

Against the background of knowledge contained in papers by Scott and Priestley (193), Hoagland and Broyer (87), Steward et al. (220), and by reference to works which deal with the way in which roots grow, as for example by Goodwin and Stepka (72) and by Brown and Broadbent (25), the following general statements can now be made.

Longitudinal gradients of salt accumulation, causing high concentration near the tip, are best demonstrated with roots that either do not branch, e.g., *Narcissus*, or with relatively long unbranched regions (barley). Aggregates of roots which, on the average, show these smooth gradients do so very strikingly when the cells are absorbing their salts in accordance with their growth, and not primarily to replace an initial deficit caused by prior removal to the shoot. Here the gradient reflects that part of the absorption mechanism which is designated in Fig. 39 as Phase II, and which is associated with that part of the growth of the cell which emphasizes cell enlargement more than cell multiplication. Naturally, therefore, this type of absorption reaches its greatest intensity very close to the apex (<1 mm) where cell enlargement is most intense. This position will vary somewhat with the root in question and with the conditions under which absorption occurs. Even so, however, it is known from work with radioactive isotopes that the situation in an individual root may be more complex. The point, or points, of maximum accumulation may be located at different distances along the axis of the root, and this may vary with time (216). The full explanation of these apparently anomalous effects is not yet known, but there are strong indications that cyclical or rhythmical phenomena, which reflect the periodicity in growth, may influence the location at any one time of the maximum accumulation along the axis of the root. Rhythmical or periodic phenomena in the secretion of water from excised roots are well known (73). Generally, however, the behavior shown in Fig. 63 is encountered.

The evidence indicates that root hairs are not more fundamentally concerned with the initial process of absorption than any other living

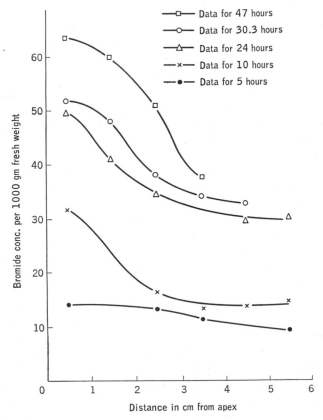

FIG. 63. Diagram showing location of maximum bromide accumulation in root. From Prevot and Steward (163).

cells of the root that are in contact with the external solution. The limited information available indicates that salts preaccumulated in an attached root are normally removed to the shoot preferentially from locations near the apex (220). This, however, does not preclude the possibility that ions may enter through the older surfaces of roots and enter the stele at gaps caused by the emergence of lateral roots, as in the observation of Kramer and Wiebe (112). While experiments may be designed to show that such processes can occur, it is nevertheless important that their role in the over-all economy of the plant body should not be overemphasized. The growing regions of roots are still the main centers of ion absorption and accumulation. The presumption is that this is due, in part, to the greater metabolic activity of the cells in this region, though it would be unwise to dismiss the possibility that there are also special features of the endodermis in this region. The apical meristem of the root can be regarded as able to bind ions by the methods indicated in Fig. 39 (Phase I), and certainly can be shown to absorb radioactive cations (e.g., Cs^{137}) although the total content of salt so absorbed is strictly limited by lack of large sap vacuoles. This absorption also has a metabolic basis since multiplication of the binding sites occurs along with that of other self-duplicating structures. The recent recognition that the central dome of apical meristem in roots may contain some larger quiescent, relatively nondividing cells, the cytogenerative region of Clowes (38), has not yet been evaluated in terms of the distribution of ions in the root tip. This quiescent region might be a source of stimuli which affect the behavior of the surrounding cells and tissues or, alternatively, it may be a vacuolated region in which ions locally accumulate.

The removal of ions from roots, which takes place so readily under the influence of the shoot, particularly in the light, causes a movement which in the *detached root* would seem to require almost that the cells should be killed, since cells in water hold on to their absorbed salts tenaciously. The nature of the stimulus which passes from shoot to root to accomplish these events is completely unknown, and it can only be described on the general principle that internal competition between accumulating centers in the plant body exists, so that those centers which are most active in growth and metabolism temporarily compete with those that are less active in these respects. Thus, shoots deplete roots, younger leaves deplete older leaves, and salts move to such local centers of growth and metabolic activity as occur in developing fruits, seeds, or storage organs in response to whatever stimuli control and regulate the growth of these organs. It is recognized, however, that this is a *description*, not an *explanation*, of what occurs. All the evidence indicates that the process of secretion into the noncellular fluids of the xylem requires the same kind of metabolic activity that is necessary for the movement of ions across the protoplasts and into the vacuole. This has been shown in such works as those of Hoagland and others, in which determinations have been made of the concentration of ions in liquids that exude from the xylem when shoots are excised. It is, however, difficult to avoid the conclusion that the living cells which border the stele—that is the endodermis and pericycle—play some major role in this connection.

a. Movement of ions across the cortex. With regard to movement of ions from the external surface of the root across the cortex into the stele there are in general two different types of hypothesis. The first involves accumulation of salt in the outer cell layer, followed by transference from vacuole to vacuole across the cortex until movement into the xylem vessels occurs. The manner in which one cell might release ions to the next is largely unknown. Lundegårdh (125) thought that ions were actively accumulated in the cells of the epidermis and leaked passively from cell to cell across the cortex into the stele.

The alternative view is that salt may move—at least as far as the endodermis—without passing through the vacuoles of the intervening cells at all. Scott and Priestley (193) suggested that salts might move freely through the water-filled cell walls up to the endodermis, where they encountered the main barrier to diffusion. In more recent years attention has been focused on the cytoplasm as an alternative pathway for movement of salts across the cortex. According to the idea of Münch (142), the protoplasts of the individual cells in roots are interconnected through plasmodesmata to form a single unit—the "symplast." On the basis of this hypothesis Crafts and Broyer (51) postulated that ions are taken up actively into the protoplasm of the cortical cells where metabolic activity is high. From there the salts diffuse inward from cell to cell until they arrive in the stele, where the metabolic activity of the cells is lower and the protoplasm is unable to retain the solute accumulated in the cortex. Consequently ions are released and move passively into the lumina of the dead conducting elements. Somewhat similar views were expressed by Wiersum (246) and by Arisz (4, 6a). Arisz suggests that ions may become bound in the protoplasts onto particles and movement of these through the symplasts may be facilitated by protoplasmic streaming, so that a rapid transference of ions across the cortex may thus be achieved.

According to the above hypotheses the transfer of ions from living cells into the dead conducting elements is a purely passive process. An alternative view to this, for which there is as yet no direct experimental evidence, is that salt is actively secreted from the living cells into the nonliving elements by processes similar to those operating in "salt glands." Such a secretory mechanism might be located at the protoplasmic surfaces adjoining the conducting elements, or the endodermis might function as a cellular secretory membrane such as those which are familiar to animal physiologists (e.g., frog skin and gastric mucosae). Supporting the idea of active transport into the stele is the observation of Sutcliffe (231) that the mechanism of ion accumulation in the root cortex of barley plants is less selective with respect to alkali cations than is the mechanism which regulates transference into the shoot. This implies that a highly selective secretory process is involved in transference of ions into the stele which is distinct from those operative in the uptake of ions into the protoplasm, or vacuoles, of the root cortex cells. (For a consideration of possible relationships between water and salt movements through the plant reference may be made to Chapter 6.)

2. Distribution of Salt in the Shoot

Turning now to the shoot apex, a different set of problems arises. Ions, instead of passing across a cortex and endodermis entering the stele from without, as they do in the root, normally reach the cells in which they are accumulated from the stele. This involves passage via the living cells of procambial strands up to the meristem and out into living cells of the bundle ends of leaves (see Fig. 4).

Two questions may now be posed. First, at what point in the development of a leaf or leaf primordium is the *intensity* of accumulation of ions at its height? Second, how are ions diverted from the vascular tissue, in which longitudinal movement occurs over long distances, to the lateral organs, whether these are leaves or buds?

Approaching these questions in the light of the limited evidence available, the following statements may be made.

(1) By studying the entry of an ion into the leaves that constitute one complete phyllotactic series (e.g., the first six leaves on the axis of seedlings of *Cucurbita pepo*) one can show that the highest concentrations are achieved in those leaves that are at the height of their expansion and that the relative accumulation in the different leaves reflects Sachs's "Grand Period of Growth" for these leaves. This was best demonstrated when the ion (Br) was absorbed by roots during a period of darkness and was transferred to shoots during ensuing periods in the light *and also when the shoot was previously fully nourished by salts,* so that the pattern of distribution was not confused by initial salt deficits in the leaves.

(2) When rapidly growing leaves of the apex compete with those below for the available salt, the vertical file of leaves which constitute one orthostichy are seen to be more closely related than the others. In this sense they constitute a nutritional unit and, as lower leaves release ions to those above, they have conferred upon them the ability to take in more ions than would be expected from their stage of development alone. This can be discerned when an indicator ion, not pre-

viously present in the tissue, is used. In *Cucurbita pepo* there are five such systems and in each the cauline bundle system seems to be the channel by which the necessary stimuli move from leaf to leaf and also the means by which the leaves at different nodes are united into a nutritional unit (lateral movement via the common anastomosing strands being demonstrably more difficult).

(3) Passage into leaves and buds implies that ions are diverted laterally from their otherwise onward movement along the axis.

The zone of living differentiating cells adjacent to the more mature xylem which, in trees, has all the metabolic activities already noted, like the cells at the surface of slices of storage tissues, or the cells of the root just behind the tip, constitutes a system which can accumulate from the older wood in which longitudinal movement over long distances may occur (see Chapter 6 in this Volume). Thus passage into the leaf or bud is associated with the presence and activity of tissue in the vascular supply to these organs which has requisite properties to accumulate ions and, in so doing, effectively to draw them off laterally. In this sense the patterns of distribution of ions in the shoot may be expected to follow those stimuli which determine the onset and activity of growth and differentiation in the procambial regions. Clear evidence of this has been seen in a study of the entry of bromide into the developing buds and leaves of *Populus nigra* throughout a growing season. It was shown (80) that: (a) bromide could move over long distances and past a ring in ways which demand that it should move up in the xylem; (b) bromide accumulation occurs in the cambial region, consisting of cambium proper and differentiating cells; (c) the presence of developing leaves, in immediate continuity with this cambial tissue on the axis, insures the prompt removal of the solutes from the main axis to the leaf; (d) techniques such as budding, defoliation, ringing, etc., which can be used to regulate the spread of the stimuli which control activity in the cambial regions, produce results which are intelligible if these ideas are accepted (see Fig. 64A). Figure 4A shows that it would be entirely reasonable to expect that the cambial cells could intervene and absorb water and ions independently, controlling the passage of ions from the main axis to the most actively accumulating centers which are situated in the most actively growing leaves.

As a consequence of this interpretation, ions should be accumulated if they were presented directly to the exposed, outer, surface of the cambial region in the spring. This is demonstrable and it can also be shown that the ions so appiled subsequently move preferentially along vertical lines to developing leaves immediately above the point

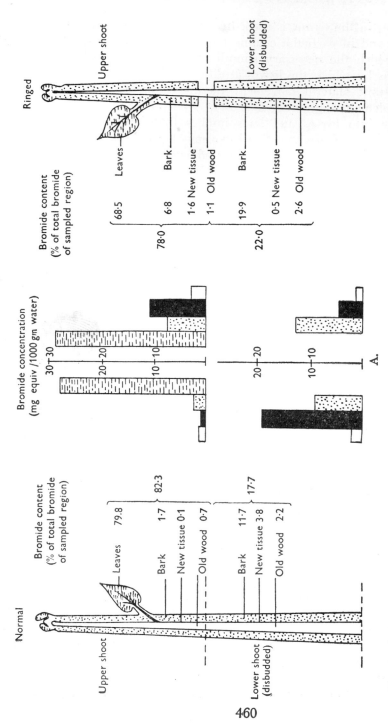

Fig. 64. A. Centers of salt absorption in the shoot of *Populus nigra*. The normal shoot is disbudded below the level shown, but the cambium is stimulated by the growing leaves above. As the buds grow, most of the bromide in the shoot goes to the leaves; the cambium region is depleted of the bromide, which it absorbs from the wood within. In the lower region of the stem—without leaves—the cambium region accumulates markedly more bromide than the old wood within. The ringed shoot is disbudded below the ring. Leaves grow less in ringed shoots. Some residual accumulation of bromide in the cambial region remains even above the ring. Below the ring, the cambium is much less active in the disbudded shoot, so that, even though there are no leaves to absorb it, there is relatively little bromide accumulated in the cambial region. From work of Harrison, see Steward and Millar (216).

of application if they are in the light (see Fig. 64B). A region of stem with its leaves in the dark does not, moreover, pass the isotope either upward or outward to leaves as readily as if it is in the light (Fig. 64B).

(4) The organization of the body of monocotyledonous plants presents these problems again in a somewhat different light. The linear leaves of monocotyledons grow in a different way, for they retain

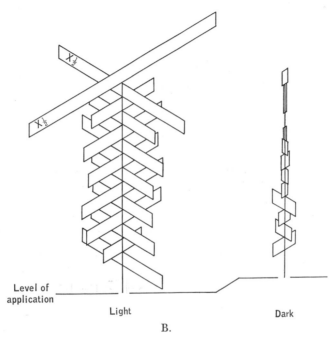

Level of
application

Light   Dark

B.

FIG. 64. B. Effect of light and dark on uptake of Cs^{137} by cambium of maple (*Acer platanoides*) and its passage into leaves. Concentrations are plotted at right angles to the axis. In view of the high concentrations in the bud, the scale was reduced by one-half for shoots in the light. From Millar (138b), based on work by Pollock and Millar in association with F. C. Steward.

active, dividing cells at their base when the leaf tips have ceased to grow. Whereas the entry of radioactive ions occurs into the lamina of net-veined leaves of dicotyledons (see Chapter 6 in this volume), it has been shown that ions (e.g., Cs^{137}) accumulate at the base of cotyledonous leaves of *Narcissus*, and even in the sheathing leaf bases where the intercalary meristems are located. The role assigned to cambial regions in the dicotyledonous plant, by which it accumulates ions and diverts them from the longitudinal path in the old xylem to the growing and accumulating lateral organs, seems to be fulfilled in the monocotyledon by the intercalary growing regions at the base of leaves.

3. *Efficiency of Salt Accumulation in Roots, Stems, and Leaves*

Within each morphological system gradients comparable with rates of growth and metabolism and with the dominance of one region over

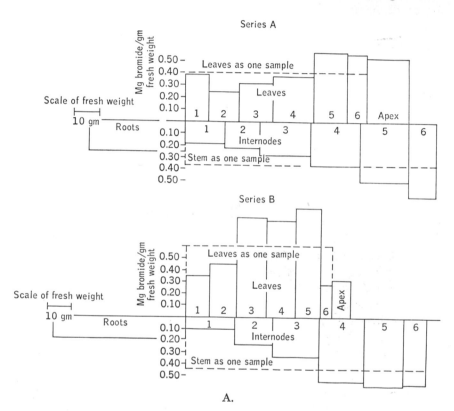

Fig. 65. A. Concentration and total amounts of salt in different regions of *Cucurbita pepo* and *Narcissus pseudo-narcissus* plants. Based on work of A. G., and F. C. Steward, see also Steward and Millar (216). Distribution of bromide in *Cucurbita pepo*. Plants of series A received KBr via their roots during periods in which the shoots were in the light; plants of series B received KBr via their roots but only during the alternating periods in which their shoots were in the dark. Light and dark periods were adjusted to 12 hours each.

another have been seen. Can any comparisons be drawn between the three broad regions of root, stem, and leaves? In *Cucurbita pepo* it was shown that the *average* concentration in the root system of the plant grown in water of a *particular ion* was lower than in each internode of the shoot and that each leaf "accumulated" from the internode which subtended it. (Within the stem, successive internodes showed higher concentrations in an acropetal succession.) Thus the order of the average concentrations of bromide in *Cucurbita* was leaf > stem

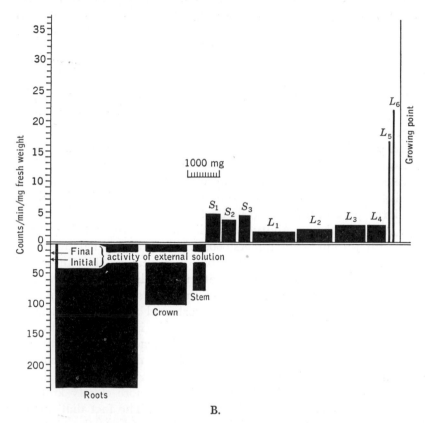

B.

FIG. 65. B. Absorption and distribution of Cs^{137} in *Narcissus*. SS, sheath leaves in acropetal succession; L, expanded leaf with sheathing base; stem, axis of lateral bud bearing S 1–3 and L 1–6; crown, flattened main axis of bulb. (Based on work of Caplin in association with Steward.)

> roots. In *Narcissus pseudo-narcissus* and for Cs^{137} this was conspicuously different, being roots > axis > sheating leaf base > growing leaf. By a graphical device which shows both concentration and total amounts in each region (Fig. 65), it may be readily seen that the anion bromide was stored principally in the shoot of *Cucurbita* while the cation Cs^{137} was stored primarily in the root of *Narcissus*.

However, when storage organs and organs of perennation form an enlarged root (as in the carrot); or stem tubers (as in potato and Jerusalem artichoke); or bulbs with scale leaves (as in the onion); it is clear that a principal part of the stimulus to their development must concern the redistribution of salts and other solutes within the plant body and then local storage and accumulation at the site in question. Remarkably little is known about the metabolic bases of these important problems, though some of the underlying stimuli to the growth in question originate in photoperiodic or thermoperiodic phenomena.

C. Epilogue

The discussion of the problem has now come full circle. The important practical question, "How do plants absorb salts?" required first a consideration of the problem in terms of cells. In turn it became necessary to invoke the metabolic machinery of these cells, but also metabolic machinery which has characteristic features according as the cells are growing, dividing, or elongating. Except in unicellular organisms, the individual cells do not behave as isolated units, for they form part of the integrated whole which constitutes the plant body. As such, their activity is coordinated and regulated by mechanisms, still incompletely understood, which preserve balance and integration in the plant body. Here one may quote: "The information that has accrued from the investigation of cells at different stages of their development tells something of the diverse ways in which cells appear to use their metabolism to absorb and accumulate their salts. Each region, however, can hardly operate in isolation, for shoots deplete roots, leaves in the same orthostichy interact with each other, and the active cambium accumulates ions from the dilute xylem fluid within and supplies them—via the current year's growth—to the buds above. Leaves in one orthostichy constitute a more closely knit nutritional unit than the leaves of the whole shoot. The fact that this complete pattern is controlled and integrated is evident; the method by which it is accomplished is, however, totally unknown. Though we now have some idea how a given cell absorbs its solutes from dilute solutions in the first place, we have no idea of the nature of the stimulus that prompts that cell to part with those solutes so that they may be directed to even more strongly accumulating cells elsewhere in the plant body." Thus the dilemma is that the problem is inherently dependent on the organization of both the living cell and the living plant. The ultimate aim is to understand and interpret these activities completely in physicochemical, metabolic, or structural terms. By recognizing active transfers in cells that utilize metabolic energy to mediate movements and to maintain conditions incompatible with true physicochemical equilibrium, much progress has been made in the present century toward such an understanding. But, in relation to the operation of the whole plant body and the integration between its parts, the problems of salt accumulation and distribution have only, as yet, been imperfectly conceived.

Thus the investigator must still stand, awed but challenged, by that "built-in capacity for growth" (to quote Sinnott) which exists in the fertilized egg and which, through the beautifully coordinated and balanced processes of organic and inorganic nutrtition, maintains the internal composition of the organism quite different from that of the environment. By the accumulation and diversification of substance in plants, one can describe their growth: it is equally true, though seemingly paradoxical, that the driving force which permits the *de novo* accumulation of salts from the very dilute external solutions and which distributes and stores them in selected regions of the plant body is incomprehensible without this ability to grow.

In short, plants grow because they can absorb and accumulate salts; they also accumulate because they can grow.

REFERENCES

1. Achromeiko, A. J. Uber die Ausscheidung mineralischer Stoffe durch Pflanzenwurzeln. *Z. Pflanzenernähr. Düng. u. Bodenk.* 42, 156–186 (1936).
2. Anderssen, F. G. Some seasonal changes in the tracheal sap of pear and apricot trees. *Plant Physiol.* 4, 459–476 (1929).
3. Arens, K. Physiologisch polarisierter Massenaustausch und Photosynthese bei submersen Wasserpflanzen. I. *Planta* 20, 621–658 (1933).
4. Arisz, W. H. Contribution to a theory on the absorption of salts by the plant and their transport in parenchymatous tissue. *Koninkl. Ned. Akad. Wetenschap. Proc.* 48, 420–446 (1945).
5. Arisz, W. H. Uptake and transport of chloride by parenchymatic tissues of leaves of *Vallisneria spiralis.* III. Discussion of the transport and the uptake vacuole secretion theory. *Koninkl. Ned. Akad. Wetenschap. Proc.* 51, 25–32 (1948).
6. Arisz, W. H. Active uptake, vacuole-secretion and plasmatic transport of chloride ions in leaves of *Vallisneria spiralis. Acta Botan. Neerl.* 1, 506–515 (1953).
6a. Arisz, W. H. Significance of the symplasm theory for transport across the root. *Protoplasma* 46, 5–62 (1956).
7. Arisz, W. H., Camphuis, I. J., Heikens, H., and van Tooren, A. J. The secretion of the salt glands of *Limonium latifolium* Ktze. *Acta Botan. Neerl.* 4, 322–338 (1955).
8. Arnon, D. I., Fratzke, W. E., and Johnson, C. M. Hydrogen-ion concentration in relation to absorption of inorganic nutrients by higher plants. *Plant Physiol.* 17, 515–524 (1942).
9. Atkinson, J. D. Progress report on the investigation of Corky-pit in apples. *New Zealand J. Sci. Technol.* B16, 316–319 (1934).

9a. Bailey, I. W. "Contributions to Plant Anatomy," 259 pp. Chronica Botanica, Waltham, Massachusetts, 1954.

10. Bartley, W., and Davies, R. E. Active transport of ions by sub-cellular particles. *Biochem. J.* **57**, 37–49 (1954).

11. Bennett, J. P., Anderssen, F. G., and Milad, Y. Methods of obtaining tracheal sap from woody plants. *New Phytologist* **26**, 316–323 (1923).

12. Bennet-Clark, T. A. Salt accumulation and mode of action of auxin—a preliminary hypothesis. *In* "The Chemistry and Mode of Action of Plant Growth Substances" (R. L. Wain and F. Wightman, eds.), 312 pp. Academic Press, New York, 1956.

13. Bennet-Clark, T. A., and Bexon, D. Expression of vacuolar sap. *Nature* **144**, 243 (1939).

14. Berry, W. E., and Steward, F. C. The absorption and accumulation of solutes by living plant cells. VI. The absorption of potassium bromide from dilute solution by tissue from various plant storage organs. *Ann. Botany (London)* **48**, 395–410 (1934).

15. Biddulph, O. The translocation of minerals in plants. *In* "Mineral Nutrition of Plants" (E. Truog, ed.), 469 pp. University of Wisconsin Press, Madison, Wisconsin, 1951.

16. Bollard, E. G. The use of tracheal sap in the study of apple tree nutrition. *J. Exptl. Botany* **4**, 363–368 (1953).

17. Boswell, J. G., and Whiting, G. C. A study of the polyphenol oxidase system in potato tubers. *Ann. Botany (London)* **2**, 847–863 (1938).

18. Boyer, P. D. The activation by K^+ and occurrence of pyruvic phosphoferase in different species. *J. Cellular Comp. Physiol.* **42**, 71–77 (1953).

19. Breazeale, J. F. Nutrition of plants considered as an electrical phenomenon. *J. Agr. Research* **24**, 41–54 (1923).

20. Briggs, G. E. The accumulation of electrolytes in plant cells. A suggested mechanism. *Proc. Roy. Soc.* **B107**, 248–269 (1930).

20a. Briggs, G. E., and Robertson, R. N. Apparent free space. *Ann. Rev. Plant Physiol.* **8**, 11–30 (1957).

20b. Brooks, S. C. The accumulation of ions in living cells—A non-equilibrium condition. *Protoplasma* **8**, 389–412 (1929).

21. Brooks, S. C. Composition of the cell sap of *Halicystis ovalis* (Lyng.) Areschoug. *Proc. Soc. Exptl. Biol. Med.* **27**, 409–412 (1930).

22. Brooks, S. C. Selective accumulation of ions in cavities incompletely surrounded by protoplasm. *Biol. Bull.* **64**, 67–69 (1933).

23. Brooks, S. C. The intake of radioactive ions by living cells. *Cold Spring Harbor Symposia Quant. Biol.* **8**, 171–180 (1940).

24. Brooks, S. C., and Brooks, M. M. "The Permeability of Living Cells," 393 pp. Borntraeger, Berlin, 1941.

25. Brown, R., and Broadbent, D. The development of cells in the growing zones of the root. *J. Exptl. Botany* **1**, 249–263 (1951).

26. Brown, R., and Rickless, P. A new method for the study of cell division and cell extension with some preliminary observations on the effect of temperature and of nutrients. *Proc. Roy. Soc.* **B136**, 110–125 (1948).

27. Broyer, T. C. The movement of materials into plants. II. The nature of solute movement into plants. *Botan. Rev.* **13**, 125–167 (1947).

27a. Broyer, T. C. The nature of the process of inorganic solute accumulation in roots. *In* "Mineral Nutrition of Plants" (E. Truog, ed.), 469 pp. University of Wisconsin Press, Madison, Wisconsin, 1951.

28. Broyer, T. C. Current views on solute movement into plant roots. *Proc. Am. Soc. Hort. Sci.* 67, 570–586 (1956).

29. Broyer, T. C., and Hoagland, D. R. Methods of sap expression from plant tissues with special reference to studies on salt accumulation by excised barley roots. *Am. J. Botany* 27, 501–511 (1940).

30. Broyer, T. C., and Hoagland, D. R. Metabolic activities of roots and their bearing on the relation of upward movement of salts and water in plants. *Am. J. Botany* 30, 261–273 (1943).

31. Burd, J. S., and Martin, J. C. Secular and seasonal changes in the soil solution. *Soil Sci.* 18, 151–167 (1924).

31a. Burström, H. The mechanism of ion absorption. *In* "Mineral Nutrition of Plants" (E. Truog, ed.), 469 pp. University of Wisconsin Press, Madison, Wisconsin, 1951.

32. Butler, G. W. Ion uptake by young wheat roots II. The "apparent free space" of wheat roots. *Physiol. Plantarum* 6, 617–635 (1953).

33. Cannan, W. A. Physiological features of roots, with special reference to the relation of roots to aeration of the soil. *Carnegie Inst. Wash. Publ.* 368, 168 pp. (1925).

34. Cannan, W. A. Absorption of oxygen by roots when the shoot is in darkness or in light. *Plant Physiol.* 7, 673–684 (1932).

35. Caplin, S. M., and Steward, F. C. A technique for the controlled growth of excised plant tissue in liquid media under aseptic conditions. *Nature* 163, 920 (1949).

36. Chambers, R., and Höfler, K. Micrurgical studies on the tonoplast of *Allium cepa*. *Protoplasma* 12, 338–355 (1931).

37. Chandler, W. H., Hoagland, D. R., and Hibbard, P. L. Little-leaf or rosette in fruit trees. *Proc. Am. Soc. Hort. Sci.* 28, 556–560 (1932).

38. Clowes, F. A. L. Nucleic acid in root apical meristems. *New Phytologist* 55, 29–34 (1956).

39. Collander, R. Permeabilitätsstudien an *Chara ceratophylla*. I. Die normale Zusammensetzung des Zellsaftes. *Acta Botan. Fennica* 6, 1–20 (1930).

40. Collander, R. Der Zellsaft der Characeen. *Protoplasma* 25, 201–210 (1936).

41. Collander, R. Selective absorption of cations by higher plants. *Plant Physiol.* 16, 691–720 (1941).

42. Collander, R. Die Elektrolyt-Permeabilität und Salz-Akkumulation pflanzlicher Zellen. *Tabulae biol.* (*Hague*) 19, 313–333 (1942).

43. Comber, N. M. The availability of mineral plant food. A modification of the present hypothesis. *J. Agr. Sci.* 12, 363–369 (1922).

44. Conway, E. J. Evidence for a redox Pump in the active transport of cations. *Intern. Rev. Cytol.* 4, 377–396 (1955).

45. Conway, E. J., and Downey, M. An outer metabolic region of the yeast cell. *Biochem. J.* 47, 347–355 (1950).

46. Conway, E. J., Ryan, H., and Carton, E. Active transport of sodium ions from the yeast cell. *Biochem. J.* 58, 158–167 (1954).

47. Cook, J. A., and Boynton, D. Some factors affecting the absorption of urea by McIntosh apple leaves. *Proc. Am. Soc. Hort. Sci.* 59, 82–90 (1952).

48. Cooper, W. C., Jr., and Blinks, L. R. The cell sap of *Valonia* and *Halicystis*. *Science* 68, 164–165 (1928).

49. Cowie, D. B., Roberts, R. B., and Roberts, I. Z. Potassium metabolism in *Escherichia coli*. I. Permeability to sodium and potassium ions. *J. Cellular Comp. Physiol.* **34**, 243–257 (1949).

50. Crafts, A. S. Sulfuric acid as a penetrating agent in arsenical sprays for weed control. *Hilgardia* **8**, 125–147 (1933).

51. Crafts, A. S., and Broyer, T. C. Migration of salts and water into xylem of the roots of higher plants. *Am. J. Botany* **25**, 529–535 (1938).

52. Crane, E. E., and Davies, R. E. Chemical and electrical energy relations for the stomach. *Biochem. J.* **49**, 169–175 (1951).

53. Dale, J. E. The effects of extracts from storage tissue on salt uptake and respiration in plants. Ph.D. Thesis, University of London, England, 1956.

54. Dale, J. E., and Sutcliffe, J. F. Inhibition of salt absorption in red beet root tissue. *Nature* **177**, 192–193 (1956).

55. Dixon, H. H. "Transpiration and the Ascent of Sap in Plants," 216 pp. Macmillan, London, 1914.

56. Doyle, W. L. Cytology of *Valonia*. Papers from Tortugas Laboratory. XXIX. *Carnegie Inst. Wash. Publ.* **452**, 13–21 (1935).

57. Duggan, P. F., and Conway, E. J. The concentration of the cation carrier in yeast cell wall as milliequivalents per kilogram of centrifuged yeast. *Biochem. J.* **64**, 40 P (1956).

58. Eddy, A. A., Carroll, T. C. N., Danby, C. J., and Hinshelwood, C. Alkali metal ions in the metabolism of *Bacterium lactis aerogenes*. I. Experiments on the uptake of radioactive potassium, rubidium and phosphorus. *Proc. Roy. Soc.* **B138**, 219–228 (1951).

59. Eddy, A. A., and Hinshelwood, C. The utilization of potassium by *Bacterium lactis aerogenes*. *Proc. Roy. Soc.* **B136**, 544–562 (1950).

60. Epstein, E. Passive permeation and active transport of ions in plant roots. *Plant Physiol.* **30**, 529–535 (1955).

61. Epstein, E., and Hagen, C. E. A kinetic study of the absorption of alkali cations by barley roots. *Plant Physiol.* **27**, 457–474 (1952).

62. Epstein, E., and Leggett, J. E. The absorption of alkaline earth cations by barley roots—kinetics and mechanism. *Am. J. Botany* **41**, 785–791 (1954).

63. Fawzy, H., Overstreet, R., and Jacobson, L. The influence of hydrogen ion concentration on cation absorption by barley roots. *Plant Physiol.* **29**, 234–237 (1954).

64. Fisher, E. G. The principles underlying foliage applications of urea for nitrogen fertilization of the McIntosh apple. *Proc. Am. Soc. Hort. Sci.* **59**, 91–98 (1952).

65. Fisher, E. G., and Walker, D. R. The apparent absorption of phosphorus and magnesium from sprays applied to the lower surface of McIntosh apple leaves. *Proc. Am. Soc. Hort. Sci.* **65**, 17–24 (1955).

66. Fogg, G. E. Quantitative studies on the wetting of leaves by water. *Proc. Roy. Soc.* **B134**, 503–522 (1947).

67. Fritsch, F. E. "Structure and Reproduction of the Algae," Vol. 2, 939 pp. Cambridge Univ. Press, London and New York, 1952.

68. Gale, E. F. The accumulation of amino acids within staphylococcal cells. *Symposia Soc. Exptl. Biol.* **8**, 242–253 (1954).

69. Gamble, J. L. "Chemical Anatomy, Physiology and Pathology of Extra-cellular Fluid." Harvard Univ. Press, Cambridge, Massachusetts, 1947.

70. Ginsberg, J. M. Penetration of petroleum oils into plant tissue. *J. Agr. Research* **43**, 469–474 (1931).

71. Goldacre, R. J. The folding and unfolding of protein molecules as a basis of osmotic work. *Intern. Rev. Cytol.* **1**, 135–164 (1952).

72. Goodwin, R. H., and Stepka, W. Growth and differentiation in the root tip of *Phleum pratense*. *Am. J. Botany* **32**, 36–46 (1945).

73. Grossenbacher, K. A. Diurnal fluctuations in root pressure. *Plant Physiol.* **13**, 669–676 (1938).

74. Guilliermond, A. "The Cytoplasm of the Plant Cell," 247 pp. Chronica Botanica, Waltham, Massachusetts, 1941.

75. Haberlandt, G. Kulturversuche mit isolierten Pflanzenzellen. *Sitzber. Akad. Wiss. Wien. Math-naturw. Kl. Abt. I* **111**, 69–92 (1902).

76. Hamilton, J. M., Palmiter, D. H., and Anderson, L. C. Preliminary tests with uramon in foilage sprays as a means of regulating the nitrogen supply of apple trees. *Proc. Am. Soc. Hort. Sci.* **42**, 123–126 (1943).

77. Handley, R., and Overstreet, R. Respiration and salt absorption by excised barley roots. *Plant Physiol.* **30**, 418–426 (1955).

78. Harley, C. P., Regeimbal, L. O., and Moon, H. H. Absorption of nutrient salts by bark and woody tissues of apple and subsequent translocation. *Proc. Am. Soc. Hort. Sci.* **67**, 47–62 (1956).

79. Harley, J. L., Brierley, J. K., and McCready, C. C. The uptake of phosphate by excised mycorrhizal roots of beech. V. The examination of the possible sources of misinterpretation of the quantities of phosphorus passing into the host. *New Phytologist* **53**, 92–98 (1954).

80. Harrison, J. A. Indicators of salt accumulation and translocation. Ph. D. Thesis, University of London, England, 1938.

81. Helder, R. J. Analysis of the process of anion uptake of intact maize plants. *Acta Botan. Neerl.* **1**, 361–434 (1952).

82. Helder, R. J. Loss of substances by cells and tissues (salt glands). *In* "Handbuch der Pflanzenphysiologie—Encyclopedia of Plant Physiology" (W. Ruhland, ed.), Vol. 2, pp. 468–488. Springer, Berlin, 1956.

83. Hevesy, G., Linderstrøm-Lang, K., and Nielsen, N. Phosphorus exchange in yeast. *Nature* **140**, 725 (1937).

84. Hewitt, E. J. Sand and water culture methods used in the study of plant nutrition. *Commonwealth Bur. Hort. Plantation Crops (Gt. Brit.) Tech. Commun.* **22**, 1–241 (1952).

84a. Hoagland, D. R. Relation of nutrient solution to composition and reaction of cell sap of barley. *Botan. Gaz.* **68**, 297–304 (1919).

85. Hoagland, D. R. Salt accumulation by plant cells, with special reference to metabolism and experiments on barley roots. *Cold Spring Harbor Symposia Quant. Biol.* **8**, 181–193 (1940).

86. Hoagland, D. R. "Lectures on the Inorganic Nutrition of Plants," 226 pp. Chronica Botanica, Waltham, Massachusetts, 1944.

87. Hoagland, D. R., and Broyer, T. C. General nature of the process of salt accumulation by roots with description of experimental methods. *Plant Physiol.* **11**, 471–507 (1936).

88. Hoagland, D. R., and Broyer, T. C. Hydrogen-ion effects and the accumulation of salt by barley roots as influenced by metabolism. *Am. J. Botany* **27**, 173–185 (1940).

89. Hoagland, D. R., and Broyer, T. C. Accumulation of salt and permeability in plant cells. *J. Gen. Physiol.* **25**, 865–880 (1942).

90. Hoagland, D. R., and Davis, A. R. The composition of the cell sap of the plant in relation to the absorption of ions. *J. Gen. Physiol.* **5,** 629–646 (1923).

91. Hoagland, D. R., and Davis, A. R. Further experiments on the absorption of ions by plants, including observations on the effects of light. *J. Gen. Physiol.* **6,** 47–62 (1923).

92. Hoagland, D. R., and Davis, A. R. The intake and accumulation of electrolytes by plant cells. *Protoplasma* **6,** 610–626 (1929).

93. Hoagland, D. R., Davis, A. R., and Hibbard, P. L. The influence of one ion on the accumulation of another by plant cells, with special reference to experiments with *Nitella. Plant Physiol.* **3,** 473–486 (1928).

94. Hoagland, D. R., Hibbard, P. L., and Davis, A. R. The influence of light, temperature and other conditions on the ability of *Nitella* cells to concentrate halogens in the cell sap. *J. Gen. Physiol.* **10,** 121–146 (1926).

95. Höber, R. "Physical Chemistry of Cells and Tissues," 676 pp. Blakiston, Philadelphia, 1945.

95a. Hollenberg, G. J. Some physical and chemical properties of the cell sap of *H. ovalis* (Lyng) Areschoug. *J. Gen. Physiol.* **15,** 651–653 (1931–1932).

95b. Holman, R. M., and Robbins, W. W. "A Textbook of General Botany for Colleges and Universities." Wiley, New York, 1924.

96. Hope, A. B., and Stevens, P. G. Electrical potential differences in bean roots and their relation to salt uptake. *Australian J. Sci. Research* **B5,** 335–343 (1952).

97. Humbert, R. P., and Hanson, N. S. *Hawaiian Sugar Planters' Assoc. Spec. Release* **64,** 1–2 (1952); cited from Boynton, D. *Ann. Rev. Plant Physiol.* **5,** 31–54 (1954).

98. Hurd, R. G. The effect of pH on salt uptake by plant tissues. Ph. D. Thesis, University of London, England, 1956.

98a. Hurd, R. G. The effect of pH and bicarbonate ions on the uptake of salts by disks of red beet. *J. Exptl. Botany* **9,** 159–174.

98b. Hurd, R. G. and Sutcliffe, J. F. An effect of pH on the uptake of salt by plant tissue. *Nature* **180,** 233–235 (1957)

99. Ingold, C. T. Effect of light on absorption of salts by *Elodea canadensis. New Phytologist* **35,** 132–141 (1936).

100. Jacobson, L., and Ordin, L. Organic acid metabolism and ion absorption in roots. *Plant Physiol.* **29,** 70–75 (1954).

101. Jacobson, L., and Overstreet, R. A study of the mechanism of ion absorption by plant roots using radioactive elements. *Am. J. Botany* **34,** 415–420 (1947).

102. Jacobson, L., Overstreet, R., King, H. M., and Handley, R. A study of potassium absorption by barley roots. *Plant Physiol.* **25,** 639–647 (1950).

103. Jacques, A. G., and Osterhout, W. J. V. The accumulation of electrolytes. III. Behavior of sodium, potassium and ammonium in *Valonia. J. Gen. Physiol.* **14,** 301–314 (1930).

104. Jenny, H., and Overstreet, R. Surface migration of ions and contact exchange. *J. Phys. Chem.* **43,** 1185–1196 (1939).

105. Jenny, H., and Overstreet, R. Cation interchange between plant roots and soil colloids. *Soil Sci.* **47,** 257–272 (1939).

106. Jenny, H., Overstreet, R., and Ayers, A. D. Contact depletion of barley roots as revealed by radioactive indicators. *Soil Sci.* **48,** 9–24 (1939).

107. Kalckar, H. M. The nature of energetic coupling in biological synthesis. *Chem. Revs.* **28,** 71–178 (1941).

108. Kamen, M. D., and Spiegelman, S. Studies on the phosphate metabolism of some unicellular organisms. *Cold Spring Harbor Symposia Quant. Biol.* **13**, 151–163 (1948).

109. Ketchum, B. H. The absorption of phosphate and nitrate by illuminated cultures of *Nitzschia closterium*. *Am. J. Botany* **26**, 399–407 (1939).

110. Ketchum, B. H. The development and restoration of deficiencies in the phosphorus and nitrogen composition of unicellular algae. *J. Cellular Comp. Physiol.* **13**, 373–381 (1939).

110a. Kirkwood, J. E., and Gies, W. J. Chemical studies of the coconut with some notes on the changes during germination. *Bull. Torrey Botan. Club* **29**, 321–359 (1902).

111. Knauss, H. J. and Porter, J. W. The absorption of inorganic ions by *Chlorella pyrenoidosa*. *Plant Physiol.* **29**, 229–234 (1954).

112. Kramer, P. J., and Wiebe, H. H. Longitudinal gradients of P^{32} absorption in roots. *Plant Physiol.* **27**, 661–674 (1952).

113. Kramer, P. J., and Wilbur, K. M. Absorption of radioactive phosphorus by mycorrhizal roots of pine. *Science* **110**, 8–9 (1949).

114. Kuckuck, P. Uber den Bau und die Fortflanzung von *Halicystis* Areschoug und *Valonia* Ginnani. *Botan. Z.* **65**, 139–185 (1907).

115. Lawrence, J. H., Erf, L. A., and Tuttle, L. W. Intracellular irradiation. *J. Appl. Phys.* **12**, 333–334 (1941).

116. Leaf, A., and Renshaw, A. Ion transport and respiration of isolated frog skin. *Biochem. J.* **65**, 82–90 (1957).

117. Leibowitz, J., and Kupermintz, N. Potassium in bacterial fermentation. *Nature* **150**, 233 (1942).

118. Lineweaver, H., and Burk, D. The determination of enzyme dissociation constants. *J. Am. Chem. Soc.* **56**, 658–666 (1934).

119. Lippmann, F. Metabolic generation and utilization of phosphate bond energy. *Advances in Enzymol.* **1**, 99–162 (1941).

120. Long, W. G., Sweet, D. V., and Tukey, H. B. Loss of nutrients from plant foliage by leaching as indicated by radioisotopes. *Science* **123**, 1039–40 (1956).

121. Lowenhaupt, B. The transport of calcium and other cations in submerged aquatics. *Biol. Revs. Cambridge Phil. Soc.* **31**, 371–395 (1956).

122. Lundegårdh, H. "Die quantitative Spektralanalyse der Elemente," 155 pp. Fischer, Jena, 1929.

123. Lundegårdh, H. "Die Nährstoffaufnahme der Pflanze," 374 pp. Fischer, Jena, 1932.

124. Lundegårdh, H. Investigations as to the absorption and accumulation of inorganic ions. *Kgl. Lantbruks-Högskol. Ann.* **8**, 234–404 (1940).

125. Lundegårdh, H. Absorption, transport and exudation of inorganic ions by the roots. *Arkiv Botan.* **A32**, 1–139 (1945).

126. Lundegårdh, H. The time course of ion absorption of wheat roots and the influence of the concentration. *Physiol. Plantarum* **2**, 388–401 (1949).

127. Lundegårdh, H. Spectroscopic evidence of the participation of the cytochrome-cytochrome oxidase system in the active transport of salts. *Arkiv. Kemi* **3**, 69–79 (1951).

128. Lundegårdh, H. Anion respiration. The experimental basis of a theory of absorption, transport and exudation of electrolytes by living cells and tissues. *Symposia Soc. Exptl. Biol.* **8**, 262–296 (1954).

129. Lundegårdh, H. Mechanisms of absorption, transport, accumulation and secretion of ions. *Ann. Rev. Plant Physiol.* **6**, 1–24 (1955).
130. Lundegårdh, H., and Burström, H. Untersuchungen über die Salzaufnahme der Pflanzen: III. Quantitative Beziehungen zwischen Atmung und Anionenaufnahme. *Biochem. Z.* **261**, 235–251 (1933).
131. Lundegårdh, H., and Burström, H. Atmung und Ionenaufnahme. *Planta* **18**, 683–699 (1933).
132. Lundegårdh, H., and Burström, H. Untersuchungen uber die Atmungsvorgänge in Pflanzenwurzeln. *Biochem. Z.* **277**, 223–249 (1935).
133. Luttkus, K., and Botticher, R. Uber die Ausscheidung von Aschenstoffen durch die Wurzeln I. *Planta* **29**, 325–340 (1939).
134. McCalla, T. M. Cation adsorption by bacteria. *J. Bacteriol.* **40**, 23–32 (1940).
135. Machlis, L. The influence of some respiratory inhibitions and intermediates on respiration and salt accumulation of excised barley roots. *Am. J. Botany* **31**, 183–192 (1944).
136. Mann, C. E. T., and Wallace, T. The effects of leaching with cold water on the foliage of the apple. *J. Pomol. Hort. Sci.* **4**, 146–161 (1925).
137. Mason, T. G., and Phillis, E. Experiments on the extraction of sap from the vacuole of the leaf of the cotton plant and their bearing on the osmotic theory of water absorption by the cell. *Ann. Botany (London)* **3**, 531–544 (1939).
138. Melin, E., and Nilsson, H. Transfer of radioactive phosphorus to pine seedlings by means of mycorrhizal hyphae. *Physiol. Plantarum* **3**, 88–92 (1950).
138a. Meyer, A. Notiz über die Zusammensetzung des Zellsaftes von *Valonia utricularis*. *Ber. deut. botan. Ges.* **9**, 77–81 (1891).
138b. Millar, F. K. Investigations upon the growth, salt accumulation and related metabolic problems of plant tissue by the use of radioactive isotopes. Ph.D. Thesis University of Rochester, 1953.
139. Mitchell, P. Transport of phosphate across the surface of *Micrococcus pyogenes*. Nature of the cell "inorganic phosphate." *J. Gen. Microbiol.* **9**, 273–287 (1953).
140. Mitchell, P. Transport of phosphate across the osmotic barrier of *Micrococcus pyogenes*—specificity and kinetics. *J. Gen. Microbiol.* **11**, 73–82 (1954).
141. Mullins, L. J. The permeability of yeast cells to radiophosphate. *Biol. Bull.* **83**, 326–333 (1942).
142. Münch, E. "Die Stoffbewegungen in der Pflanze," 234 pp. Fischer, Jena, 1930.
143. Muntz, J. A. The role of potassium and ammonium ions in alcoholic fermentation. *J. Biol. Chem.* **171**, 653–665 (1947).
144. Nielsen, T. R., and Overstreet, R. A study of the role of the hydrogen ion in the mechanism of potassium absorption by excised barely roots. *Plant Physiol.* **30**, 303–9 (1955).
145. O'Kane, D. J., and Umbreit, W. W. Transformations of phosphorus during glucose fermentation by living cells of *Streptococcus faecalis*. *J. Biol. Chem.* **142**, 25–30 (1942).
146. Oland, K., and Oland, T. B. Uptake of magnesium by apple leaves. *Physiol. Plantarum* **9**, 401–411 (1956).
147. Oltmanns, F. "Morphologie und Biologie der Algen," Vol. 1, 733 pp. Fischer, Jena, 1904.
148. Osterhout, W. J. V. Conductivity and permeability. *J. Gen. Physiol.* **4**, 1–9 (1921).
149. Osterhout, W. J. V. Is living protoplasm permeable to ions? *J. Gen. Physiol.* **8**, 131–146 (1925).

150. Osterhout, W. J. V. The kinetics of penetration. I. Equations for the entrance of electrolytes. *J. Gen. Physiol.* **13**, 261–294 (1929).

151. Osterhout, W. J. V. Physiological studies of single plant cells. *Biol. Revs. Cambridge Phil. Soc.* **6**, 369–411 (1931).

152. Osterhout, W. J. V. Permeability in large plant cells and in models. *Ergeb. Physiol. u. exptl. Pharmakol.* **35**, 967–1021 (1933).

153. Osterhout, W. J. V. The absorption of electrolytes in large plant cells. *Botan. Rev.* **2**, 283–315 (1936).

154. Osterhout, W. J. V. The mechanism of accumulation in living cells. *J. Gen. Physiol.* **35**, 579–594 (1952).

155. Overstreet, R., Broyer, T. C., Isaacs, T. L., and Delwiche, C. C. Additional studies regarding the cation absorption mechanism of plants in soil. *Am. J. Botany* **29**, 227–231 (1942).

156. Overstreet, R., and Jacobson, L. The absorption by roots of rubidium and phosphate ions at extremely small concentrations as revealed by experiments with Rb^{86} and P^{32} prepared without inert carrier. *Am. J. Botany* **33**, 107–112 (1946).

157. Overstreet, R., Jacobson, L., and Handley, R. The effect of calcium on the absorption of potassium by barley roots. *Plant Physiol.* **22**, 583–590 (1952).

158. Overstreet, R., and Jenny, H. Studies pertaining to the cation absorption mechanism of plants in soil. *Soil Sci. Soc. Am. Proc.* **4**, 125–130 (1939).

159. Overstreet, R., Ruben, S., and Broyer, T. C. The absorption of bicarbonate ions by barley plants as indicated by studies with radioactive carbon. *Proc. Natl. Acad. Sci. U.S.* **26**, 688–695 (1940).

160. Peat, S. The biological transformations of starch. *Advances in Enzymol.* **11**, 339–375 (1951).

161. Petrie, A. H. K. The intake of ions by carrot tissue at different hydrogen ion concentrations. *New Phytologist* **37**, 211–231 (1938).

162. Plowe, J. Q. Membranes in the plant cell. II. Localization of differential permeability in the plant protoplast. *Protoplasma* **12**, 221–240 (1931).

163. Prevot, P., and Steward, F. C. Salient features of the root system relative to the problem of salt absorption. *Plant Physiol.* **11**, 509–534 (1936).

164. Priestley, J. H., and Wormall, A. On the solutes exuded by root pressure from vines. *New Phytologist* **24**, 24–38 (1925).

165. Pulver, R., and Verzár, F. Connection between carbohydrate and potassium metabolism in the yeast cell. *Nature* **145**, 823–824 (1940).

166. Redfern, G. M. On the course of absorption and the position of equilibrium in the intake of dyes by disks of plant tissue. *Ann. Botany* (*London*) **36**, 511–522 (1922).

167. Rickenberg, H. V., Cohen, G., Buttin, G., and Monod, J. La galactoside perméase d'*Escherichia coli*. *Ann. inst. Pasteur* **91**, 829–857 (1956).

168. Roberts, E. A., Southwick, M. D., and Palmiter, D. H. A microchemical examination of McIntosh apple leaves showing relationship of cell wall constituents to penetration of spray solutions. *Plant Physiol.* **23**, 557–559 (1948).

169. Roberts, E. H., and Street, H. E. The continuous culture of excised rye roots. *Physiol. Plantarum* **8**, 238–262 (1955).

170. Roberts, R. B., and Roberts, I. Z. Potassium metabolism in *Escherichia coli*. III. Inter-relationship of potassium and phosphorus metabolism. *J. Cellular. Comp. Physiol.* **36**, 15–39 (1950).

171. Roberts, R. B., Roberts, I. Z., and Cowie, D. B. Potassium metabolism in

Escherichia coli. II. Metabolism in the presence of carbohydrates and their metabolic derivatives. *J. Cellular Comp. Physiol.* 34, 259–291 (1949).

172. Robertson, R. N. Mechanism of absorption and transport of inorganic nutrients in plants. *Ann. Rev. Plant Physiol.* 2, 1–24 (1951).

173. Robertson, R. N., and Turner, J. S. Studies in the metabolism of plant cells. 3. The effects of cyanide on the accumulation of potassium chloride and on respiration: The nature of salt respiration. *Australian J. Exptl. Biol. Med. Sci.* 23, 63–73 (1945).

174. Robertson, R. N., and Wilkins, M. J. Studies in the metabolism of plant cells. VII. The quantitative relation between salt accumulation and salt respiration. *Australian J. Sci. Research* B1, 17–37 (1948).

175. Robertson, R. N., Wilkins, M. J., and Hope, A. B. Plant mitochondria and salt accumulation. *Nature* 175, 640 (1955).

176. Robertson, R. N., Wilkins, M. J., Hope, A. B. and Nestel, L. Studies in the metabolism of cells. X. Respiratory activity and ionic relations of plant mitchondria. *Australian J. Biol. Sci.* 8, 164–185 (1955).

177. Robertson, R. N., Wilkins, M. J., and Weeks, D. C. Studies in the metabolism of plant cells. IX. The effects of 2,4-dinitrophenol on salt accumulation and salt respiration. *Australian J. Sci. Research* B4, 248–264 (1951).

178. Rodney, D. R. Entrance of nitrogen compounds through the epidermis of apple leaves. *Proc. Am. Soc. Hort. Sci.* 59, 99–102 (1952).

179. Rorem, E. S. Uptake of rubidium and phosphate ions by polysaccharide producing bacteria. *J. Bacteriol.* 70, 691–701 (1955).

180. Rosenfels, R. S. The absorption and accumulation of potassium bromide by *Elodea* as related to respiration. *Protoplasma* 23, 503–519 (1935).

181. Rothstein, A., and Enns, L. H. The relationship of potassium to carbohydrate metabolism in baker's yeast. *J. Cellular Comp. Physiol.* 28, 231–252 (1946).

182. Rothstein, A., and Larrabee, C. The relationship of the cell surface to metabolism. II. The cell surface of yeast as the site of inhibition of glucose metabolism by uranium. *J. Cellular Comp. Physiol.* 32, 247–259 (1948).

183. Rothstein, A., and Meier, R. The relationship of the cell surface to metabolism. IV. The role of cell surface phosphatases of yeast. *J. Cellular Comp. Physiol.* 34, 97–114 (1949).

184. Routien, J. B., and Dawson, R. F. Some interrelationships of growth, salt absorption, respiration and mycorrhizal development in *Pinus echinata*, Mill. *Am. J. Botany* 30, 440–451 (1943).

185. Ruhland, W. Untersuchungen über die Hautdrüsen der Plumbaginaceen. Ein Beitrag zur Biologie der Halophyten. *Jahrb. wiss. Botan.* 55, 409–498 (1915).

186. Ruttner, F. Die Kohlenstoffquellen für die Kohlensäureassimilation submerser Wasserpflanzen, *Scientia (Milan)* 88, 20–27 (1953).

187. Schtscherback, J. Uber die Salzausscheidung durch die Blätter von *Statice gmelini* (Vorläufige Mitteilung). *Ber. deut. botan. Ges.* 28, 30–34 (1910).

188. Scott, G. T. The mineral composition of *Chlorella pyrenoidosa* grown in culture media containing varying concentrations of calcium, magnesium, potassium and sodium. *J. Cellular Comp. Physiol.* 21, 327–338 (1943).

189. Scott, G. T. Cation exchanges in *Chlorella pyrenoidosa*. *J. Cellular Comp. Physiol.* 23, 47–58 (1944).

190. Scott, G. T., De Voe, R., Hayward, H., and Craven, G. Exchange of sodium ions in *Ulva lactuca*. *Science* 125, 160 (1957).

191. Scott, G. T., and Hayward, H. Metabolic factors influencing sodium and potassium distribution in *Ulva lactuca*. *J. Gen. Physiol.* **36**, 659–671 (1953).

192. Scott, G. T., and Hayward, H. Evidence for the presence of separate mechanisms regulating potassium and sodium distribution in *Ulva lactuca*. *J. Gen. Physiol.* **37**, 601–620 (1954).

193. Scott, L. I., and Priestley, J. H. The root as an absorbing organ. I. A reconsideration of the entry of water and salts into the absorbing region. *New Phytologist* **27**, 125–140 (1928).

194. Seifriz, W. New material for microdissection. *Protoplasma* **3**, 191–6 (1927).

195. Sharp, L. "Fundamentals of Cytology," 270 pp. McGraw-Hill, New York, 1943.

196. Silberstein, O., and Wittwer, S. H. Foliar application of phosphatic nutrients to vegetable crops. *Proc. Am. Soc. Hort. Sci.* **58**, 179–190 (1951).

197. Skelding, A. D., and Rees, W. J. An inhibitor of salt absorption in the root tissues of red beet. *Ann. Botany (London)* **16**, 513–529 (1952).

198. Skoss, J. D. Structure and composition of plant cuticle in relation to environmental factors and permeability. *Botan. Gaz.* **117**, 55–72 (1955).

198a. Steeman-Nielsen, E. Passive and active ion transport during photosynthesis in water plants. *Physiol. Plantarum* **4**, 189–198 (1951).

199. Steward, F. C. The maintenance of semi-permeability in the plant cell during leaching experiments. *Proc. Leeds Phil. Lit. Soc., Sci. Sect.* **1**, 258–270 (1928).

200. Steward, F. C. Diffusion of certain solutes through membranes of living plant cells and its bearing upon certain problems of solute movement in the plant. *Protoplasma* **11**, 521–557 (1930).

201. Steward, F. C. The absorption and accumulation of solutes by living plant cells. I. Experimental conditions which determine salt absorption by storage tissue. *Protoplasma* **15**, 29–58 (1932).

202. Steward, F. C. The absorption and accumulation of solutes by living plant cells. II. A technique for the study of respiration and salt absorption in storage tissue under controlled environmental conditions. *Protoplasma* **15**, 497–516 (1932).

203. Steward, F. C. The absorption and accumulation of solutes by living plant cells. IV. Surface effects with storage tissue. A quantitative interpretation with respect to respiration and salt absorption. *Protoplasma* **17**, 436–453 (1932).

204. Steward, F. C. The absorption and accumulation of solutes by living plant cells. V. Observations on the effects of time, oxygen and salt concentration upon absorption and respiration by storage tissue. *Protoplasma* **18**, 208–242 (1933).

205. Steward, F. C. Mineral nutrition of plants. *Ann. Rev. Biochem.* **4**, 519–544 (1935).

206. Steward, F. C. Salt accumulation by plants—the role of growth and metabolism. *Trans. Faraday Soc.* **33**, 1006–1016 (1937).

207. Steward, F. C. The growth of *Valonia ventricosa* J. Agardh, and *Valonia ocellata* Howe in culture. Papers from Tortugas Laboratory. XXXII. *Carnegie Inst. Wash. Publ.* **517**, 87–98 (1939).

208. Steward, F. C., and Berry, W. E. The absorption and accumulation of solutes by living plant cells. VII. The time factor in the respiration and salt absorption of Jerusalem artichoke tissue (*Helianthus tuberosus*) with observations on ionic interchange. *J. Exptl. Biol.* **11**, 103–119, (1934).

209. Steward, F. C., Berry, W. E., and Broyer, T. C. The absorption and accumula-

tion of solutes by living plant cells. VIII. The effect of oxygen upon respiration and salt accumulation. *Ann. Botany (London)* **50**, 345–366 (1936).

210. Steward, F. C., Berry, W. E., Preston, C., and Ramamurti, T. K. The absorption and accumulation of solutes by living plant cells. X. Time and temperature effects on salt uptake by potato discs and the influence of the storage conditions of the tubers on metabolism and other properties. *Ann. Botany (London)* **7**, 221–260 (1943).

211. Steward, F. C., Bidwell, R. G. S., and Yemm, E. W. Protein metabolism, respiration and growth. A synthesis of results from the use of C¹⁴-labeled substrates and tissue cultures. *Nature* **178**, 734–738, 789–792 (1956).

212. Steward, F. C., and Caplin, S. M. Investigations on growth and metabolism of plant cells. III. Evidence for growth inhibitors in certain mature tissues. *Ann. Botany (London)* **16**, 477–489 (1952).

213. Steward, F. C., Caplin, S. M., and Shantz, E. M. Investigations on the growth and metabolism of plant cells. V. Tumorous growth in relation to growth factors of the type found in coconut. *Ann. Botany (London)* **19**, 29–47 (1955).

214. Steward, F. C., and Harrison, J. A. The absorption and accumulation of salts by living plant cells. IX. The absorption of rubidium bromide by potato discs. *Ann. Botany (London)* **3**, 427, 454 (1939).

215. Steward, F. C., and Martin, J. C. The distribution and physiology of *Valonia* at the Dry Tortugas with special reference to the problem of salt accumulation in plants. Papers from Tortugas Laboratory. XXXI. *Carnegie Inst. Wash. Publ.* **475**, 87–170 (1937).

216. Steward, F. C., and Millar, F. K. Salt accumulation in plants. A reconsideration of the role of growth and metabolism. *Symposia Soc. Exptl. Biol.* **8**, 367–406 (1954).

217. Steward, F. C., and Pollard, J. K. Some further observations on glutamyl and related compounds in plants. *In* "Inorganic Nitrogen Metabolism" (W. A. McElroy and B. Glass, eds.), pp. 377–407. Johns Hopkins Press, Baltimore, Maryland, 1956.

218. Steward, F. C., and Preston, C. Metabolic processes of potato discs under conditions conducive to salt accumulation. *Plant Physiol.* **15**, 23–61 (1940).

218a. Steward, F. C., and Preston, C. The effect of salt concentration upon the metabolism of potato discs and the contrasted effect of potassium and calcium salts which have a common ion. *Plant Physiol.* **16**, 85–116 (1941).

219. Steward, F. C., and Preston, C. Effects of pH and the components of bicarbonant and phosphate buffered solutions on the metabolism of potato discs and their ability to absorb ions. *Plant Physiol.* **16**, 481–519 (1941).

220. Steward, F. C., Prevot, P., and Harrison, J. A. Absorption and accumulation of rubidium bromide by barley plants. Localization in the root of cation accumulation and of transfer to the shoot. *Plant Physiol.* **17**, 411–421 (1942).

221. Steward, F. C., and Street, H. E. The nitrogenous constituents of plants. *Ann. Rev. Biochem.* **16**, 471–502 (1947).

222. Steward, F. C., Wright, R., and Berry, W. E. The absorption and accumulation of solutes by living plant cells. III. The respiration of disks of potato tissue in air and immersed in salt solutions with observations on surface: volume effects and salt accumulation. *Protoplasma* **16**, 576–611 (1932).

223. Stiles, W. Permeability. *New Phytologist Reprint* **13**, 1–296 (1924).

224. Stiles, W. The exosmosis of dissolved substance from storage tissue into water. *Protoplasma* **2**, 577–601 (1927).

225. Stiles, W., and Kidd, F. The influence of external concentration of salts on the position of equilibrium attained in the intake of salts by plant cells: *Proc. Roy. Soc.* **B90**, 448–470 (1919).
226. Stiles, W., and Kidd, F. The comparative rate of absorption of various salts by plant tissue. *Proc. Roy. Soc.* **B90**, 487–504 (1919).
227. Sussman, M., Spiegelman, S., and Reiner, J. M. The dissociation of carbohydrate assimilation from catabolism. *J. Cellular Comp. Physiol.* **29**, 149–158 (1947).
228. Sutcliffe, J. F. The influence of internal ion concentration on potassium accumulation and salt respiration of red beet root tissue. *J. Exptl. Botany* **3**, 59–76 (1952).
229. Sutcliffe, J. F. The exchangeability of potassium and bromide ions in cells of red beet root tissue. *J. Exptl. Botany* **5**, 313–326 (1954).
230. Sutcliffe, J. F. The selective uptake of alkali cations by red beet root tissue. *J. Exptl. Botany* **8**, 36–49 (1957).
231. Sutcliffe, J. F. The selective absorption of alkali cations by storage tissues and intact barley plants. *Potassium Symposium Ann. Meeting Board Tech. Advisers Intern. Potash Inst. Berne*, 1–11 (1957).
232. Sutcliffe, J. F., and Hackett. D. P. The efficiency of ion transport in biological systems. *Nature* **180**, 95–96 (1957).
233. Swanson, C. A., and Whitney, J. B. Studies on the translocation of foliar applied P[32] and other radioisotopes in bean plants. *Am. J. Botany* **40**, 816–823 (1953).
234. Tamm, C. O. Removal of plant nutrients from tree crowns by rain. *Physiol. Plantarum* **4**, 184–188 (1951).
235. Thimann, K. V., and Schneider, C. L. The role of salts, hydrogen-ion concentration and agar in the response of the *Avena* coleoptile to auxins. *Am. J. Botany* **25**, 270–280 (1938).
236. Thomas, M. D., Hendricks, R. H., and Hill, G. R. Sulfur content of vegetation. *Soil Sci.* **70**, 9–18 (1950).
237. Tolbert, N. E. Phosphoryl choline in plants. *Federation Proc.* **16**, 262 (1957).
238. Tukey, H. B., Ticknor, R. L., Hinsvark, O. N., and Wittwer, S. H. Absorption of nutrients by stems and branches of woody plants. *Science* **116**, 167–168 (1952).
239. Ulrich, A. Metabolism of non-volatile organic acids in excised barley roots as related to cation-anion balance during salt accumulation. *Am. J. Botany* **28**, 526–537 (1941).
240. Ulrich, A. Metabolism of organic acids in excised barley roots as influenced by temperature, oxygen tension and salt concentration. *Am. J. Botany* **29**, 220–227 (1942).
241. Viets, F. G. Calcium and other polyvalent cations as accelerators of ion accumulation by excised barley roots. *Plant Physiol.* **19**, 466–480 (1944).
242. Vlamis, J., and Davis, A. R. Effects of oxygen tension on certain physiological responses of rice, barley and tomato. *Plant Physiol.* **19**, 33–51 (1944).
243. Wallace, T. Experiments on the effect of leaching with cold water on the foliage of fruit trees. 1. The course of leaching of dry matter, ash and potash from leaves of apple, pear, plum, black currant and gooseberry. *J. Pomol. Hort. Sci.* **8**, 44–60 (1930).
244. Weaver, R. J., and DeRose, H. R. Absorption and translocation of 2,4-dichlorophenoxyacetic acid. *Botan. Gaz.* **107**, 509–521 (1946).

245. Went, F. W., and Carter, M. Growth response of tomato plants to applied sucrose. *Am. J. Botany* **35**, 95–105 (1948).

246. Wiersum, L. K. Transfer of solutes across the young root. *Rec. trav. botan. néerl.* **41**, 1–79 (1948).

247. Wilson, J. K. The nature and reaction of water from hydathodes. *Cornell Univ. Agr. Expt. Sta. Mem.* **65**, 3–11 (1923).

247a. Woodhead, T. W. "The Study of Plants," 440 pp. Oxford Univ. Press, London and New York, 1915.

248. Zirkle, C. The plant vacuole. *Botan. Rev.* **3**, 1–30 (1937).

PREAMBLE TO CHAPTERS 5 AND 6

Chapters 5 and 6 together deal with problems of the movement of dissolved substances from organ to organ. Specialization of function creates the need to transfer solutes, organic and inorganic, from one region of the plant body to another. Plants lack a circulatory system analogous to that of the blood, but, especially in tall trees, they create obvious demands for movement of solutes over long distances. For convenience this over-all problem of movement of solutes is here presented in Volume II; Chapter 5 places the main emphasis on the movement of organic solutes with attention focused on the role of the phloem, and in Chapter 6 the attention is focused on the movement of inorganic solutes and the role of the xylem. But both aspects of the movement of solutes in plants appear as examples of movements which often appear to be physically baffling, by their rate, if not their direction, and the problem again arises how cells and tissues as they are organized in the plant body are able to bring about these movements. Answers to these questions increasingly involve many fundamental features of cells which are dealt with in the earlier chapters of this volume and Volume I, namely their permeability properties, ability to absorb water and/or solutes, ability to negotiate energy transfers which enable them to do work, etc., etc.

The movement of solutes within the plant body is here segregated for separate discussion in Chapters 5 and 6. Like the problems of solute absorption and of active transfer in cells, as discussed in Chapter 4, the movement of solutes presents problems of rate and direction which are physically difficult to duplicate in artificial systems. As in the discussion of other physiological functions which involve the organization of the living system, there is a dilemma. How far can the interpretation be pushed, as yet, in terms of known physical mechanisms? Can full understanding be reached apart from the ways in which the plant body grows and develops and the methods by which it achieves integration between its parts? The latter topics will be discussed in later sections of this work (Volumes V and VI).

CHAPTER FIVE

Translocation of Organic Solutes

C. A. Swanson

I. Concepts

Slow cell-to-cell movement of organic solutes, at maximal rates of 1–2 cm per hour but usually much slower than this, can occur in all cells or tissues of the plant. Some of the mechanics connected with this type of transport have been considered in previous chapters, as for example the process of "active transfer" across membranes (Chapter 4 in this volume). Of more immediate concern in the present chapter, however, are transport processes which are more directly associated with specific transport tissues, mainly phloem, and are characterized by a considerably higher rate (potential or capacity) as well as by a movement over greater distances, from organ to organ as opposed to cell to cell. The term *translocation* of organic solutes (as distinct from more general terms, such as *transport, transfer,* etc.) is often reserved for this type of transport. A similar application of this term is made in connection with the rapid movement of water and inorganic solutes as well (Chapters 6 and 7 in this volume).

It must be recognized, however, that in actual practice it is difficult to apply any precise definition of translocation. The rates of translocation are so variable that a separation on this basis is not always possible. Furthermore, uncertainty exists as to the extent to which the mechanism of translocation in the phloem tissue differs fundamentally from the relatively slower cell-to-cell movement of solutes in non-specialized transport tissues, such as cortex, cambium, xylem, parenchyma, etc. Observed differences may be only quantitative in character, the higher rates of translocation being simply the result of the more favorable structural relations for transport found in the vascular tissues. In practice, therefore, the terms *translocation, transport, transfer*, etc. are often used interchangeably. Despite these complications, there appear to be a number of distinctive physiological and morphological aspects which characterize translocation, especially of solutes in the phloem, and as our knowledge of the component processes involved in transport in general increases, a more consistent and discriminating terminology may be feasible.

The translocation of organic solutes from one part of the plant to another is a process of considerable magnitude, second, in quantitative terms, only to water translocation. An apple orchard, for example, will, under very favorable growing conditions, produce in excess of 60,000 pounds of fruit per acre, containing approximately 4 to 4.5 tons of organic substances. Only a small fraction of this organic increment is due to local photosynthetic production within the chlorenchyma of the fruit itself; much the greater portion is derived from compounds translocated to it from the leaves. Similar estimates can be made for many other fruit and root crops.

Even at the individual plant level, organic translocation can be of considerable magnitude. In the Zucca melon (*Lagenaria siceraria*), for example, a single fruit may attain a dry weight of 5 pounds or more within a month from the time of pollination. As in the apple fruit, this dry weight increment is principally derived from translocated compounds. An interesting calculation may be made of the molecular equivalents of this translocated material. Assuming an average molar weight of 300 gm for the translocatory compounds (on the basis that the translocate is mostly sucrose), and disregarding various second-order corrections, approximately 4.5×10^{24} molecules $(453 \times 5 \times 6 \times 10^{23}/300)$ would be translocated into such a fruit. This number of sucrose molecules, placed end to end, would extend for a distance of about $2\frac{1}{2}$ trillion miles! It should not be construed, of course, that sucrose molecules are translocated single file; the computation merely serves to emphasize the enormous number of molecules translocated.

It is evident that the mechanisms by which such amounts of organic substances are moved, the transport tissues involved, and the effect of environmental and internal factors on this process are problems of significant concern in both theoretical and applied physiology.

II. Historical Perspectives

The study of translocation of organic solutes, or in older terminology, "elaborated sap" ("*bearbeiteter Bildungssaft*" or "*suc propre*"), has a venerable history, which may be said to have begun with the ringing experiments by Malpighi and others about the mid-1600's. An interesting review of this early history through to the time of Pfeffer's work in 1877 is given by Münch [(96) pp. 212–226].

Early interest in translocation was stimulated in part by Harvey's discovery of blood circulation in 1628, and many studies were subsequently undertaken to determine to what extent a comparable system of sap circulation existed in plants. To these discussions, Stephen Hales contributed significant observations. It should be recalled that, at the time of these early studies, the Aristotelian "humus" theory of plant nutrition was generally accepted, namely, that the plants derived their nutrition wholly from the soil, although it was rather generally recognized from the time of Malpighi that the "ascending sap" from the soil was variously modified in the leaves ("*ausgekocht*," "*vergoren*," "*verdaut*," "*bearbeitet*") through the action of sunlight and thereby converted into an "elaborated sap" which moved to other parts of the plant and was used in growth. The identification of the tissues through which the "ascending sap" (or "raw sap") and the "descending sap" (or "elaborated sap") moved was a problem much discussed during these early years, but progress was slow, and as late as 1798 the problem was still so controversial that the Kaiser Academy for Natural Science offered an award for the best paper submitted on this question. The general view appeared to be that both the "ascending" and "descending" sap streams moved in the wood. The significance of the wood, therefore, as a conducting tissue was early recognized. The concept of the bark as having a similar significance, however, escaped the early investigators, despite numerous ringing experiments, many of which, during the preceding century of work, had been carried out at a high level of technical performance. Attention on the bark was centered mainly on its fibrous structures because of the commercial importance of the "bast" or phloem fibers. It was not until the extensive ringing experiments by Cotta and Knight, about the turn of the nineteenth century, and the discovery of sieve tubes by Hartig in 1837, that the concept of the bark as a conducting tissue was gradually recognized. Simultaneously with these developments in plant science, organic chemistry was established on a firm footing (the term "organic chemistry" was first applied in 1807), providing the basis for sound advance on the physiological aspects of translocation. The modern era of translocation research thus begins with the early nineteenth century.

III. Phloem: General Aspects

The primary channel for organic tranlocation is the phloem tissue. Exceptions to this generalization will be noted later, but the statement is sufficiently accurate that it is well to begin a study of translocation with a consideration of this tissue.

In most species the phloem is a highly complex tissue. Its basic components consist of *sieve elements*, several types of parenchyma cells (*companion cells, ray cells, phloem parenchyma*, etc.), *fibers*, and *sclereids*. In species possessing a laticiferous system, such as the rubber tree (*Hevea brasiliensis*), elements of this system may be found in the phloem also. The complexity of composition and the diversity of ontogenetic details regarding the various cell types and their intergrading variants are such as to preclude any concise characterization of this tissue.

From numerous phloem investigations during the past century, periodically reviewed from various points of view by different investigators [in recent years by Crafts (33); Esau (50, 51); Esau *et al.* (52); and Huber (63)], the simplifying concept has emerged that "the main, single characteristic of the phloem is the presence of highly specialized cells, the sieve elements" (51). Evidence is also accumulating that these cells are the primary conducting elements in phloem-limited translocation. Following a brief review of the systematic organization of the phloem tissue in plants, the anatomy and cytology of the sieve elements will be presented in some detail. The anatomy of the remaining constituents of the phloem does not appear to differ essentially from that of parenchyma or sclerenchyma cells in general, and no special effort will be made to characterize these cells further.

In vascular plants, the phloem is spatially associated with the xylem and has, accordingly, the same distribution pattern as the xylem, except for minor discrepancies at the ultimate terminations of the vascular system. Commonly in the shoot and root apices, the phloem extends farther than the xylem, whereas in the leaf, the situation is reversed, as will be described in more detail later.

As a general rule, the phloem occupies a position external to that of the xylem in the stem, or abaxial in the leaves (i.e., away from the axis, or stem, to which the leaf is attached). Phloem in this position is referred to as *external phloem*. In many species, however, as in certain ferns and in many families of the dicotyledons (e.g., Apocynaceae, Asclepiadaceae, Compositae, Convolvulaceae, Cucurbitaceae, Myrtaceae, Solanaceae), *internal phloem* may be present as well, that is, phloem occupying a position internal to that of the xylem in the stem, or adaxial in the leaves. Vascular bundles with both internal and external phloem are referred to as *bicollateral bundles* (Fig. 1). No instances have been discovered of plants with internal phloem only; when present, such phloem is always in addition to the external phloem. In certain species, phloem may also occur as strands or layers

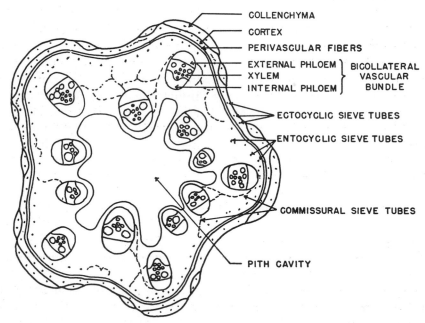

FIG. 1. Diagram of transection of *Cucurbita pepo* stem showing the distribution of various phloem systems. The ectocyclic and endocyclic sieve tubes are joined by connecting branches at the stem nodes. From Esau (51).

within the xylem tissue, and is referred to as *interxylary phloem* or *included phloem* or *phloem islands*. In cucurbitaceous plants, strands of sieve tubes and companion cells also traverse the cortex and interconnect with the internal and external phloem of the bundles by way of radially oriented sieve tubes. In view of the relatively extensive work on translocation which has been carried out with members of the Cucurbitaceae, the special names applied at times to these relatively unique phloem strands are given in Fig. 1.

Phloem tissue forms an essentially continuous unit system in the plant, from the ends of the phloem elements in the vascular bundles of the leaves to the ultimate termination of the phloem in the fruit, and in the tips of stems and roots. In some species the differentiating phloem extends to within 100–200 μ of the ultimate tips of the shoot and root axes, and the mature phloem to within 250–500 μ. According to Mason (83), the sieve element strands in certain species of the monocotyledonous genus *Dioscorea* are interrupted at each node by compact masses of thin-walled parenchyma cells (*bast glomeruli*) containing deeply staining nuclei. Roeckl (105), however, in a recent re-examination of these structures, found evidence that they may be composed of sieve elements and that phloem continuity is in actuality maintained across the node. If this view is correct, no important exceptions to the generalized picture of the continuity of the phloem system are presently known. Even during the rapid elongation phase of the gynophore of the peanut (*Arachis hypogaea*), the sieve-tube elements are continuous at all times through the active meristem (66). The same appears to be true in other intercalary meristems. Discontinuities are known to occur in the graft unions of certain monocotyledons but, although the scions may remain alive and healthy-appearing over long periods of time in these cases, the amount of their growth is greatly restricted (97).

Entry of organic solutes into the phloem system occurs primarily in the leaves. It is important, therefore, to consider the distribution and structure of the veins, especially the minor veins, and their organizational relationship to the photosynthesizing cells of the mesophyll, which constitute the major centers of synthesis of the organic translocate. The following description is drawn with reference to typical mesophytic leaves of dicotyledons. The details of organizational structure for leaf types in general are too varied for concise presentation.

The minor veins, with rare exceptions, lie in the upper layer of the spongy mesophyll, and the details are best observed, therefore, in paradermal sections. As may be noted from Fig. 2, the minor venation is closely integrated and constitutes a plexus of conducting tissues which, in some species, forms a closed system dividing the leaf into definite islets (*vein islets*). In the majority of species, the islets are invaded by vascular branches which usually fork repeatedly and then end blindly (Fig. 2). The extent of branching is such that, in many leaves, the average interveinal distance, from xylem to xylem, does not exceed about 130 μ (129). The maximal distance, therefore, separating vascular components and mesophyll cells horizontally in such species is of the order of 65 μ. Stated in other terms, the aggregate length of the vascular system in such leaves is of the order of one meter per square centimeter of leaf area. It should be recognized that in certain species, however, the extent of vascularization in the leaves is less, and the corresponding interveinal distances greater. These statistics apply specifically to the xylem. As shown below, the phloem and xylary components are not coextensive in the vein endings. Comparable data applying specifically to mesophyll-phloem distances have not been compiled, but would not be greatly different.

Fig. 2. Leaf venation in dicotyledons. Example shown: *Ulmus americana*. Average interveinal distance (D) equals 76 μ in this species. Redrawn from Wylie (130).

The number of phloem components (which are the same in the major leaf veins as in the stem) and the size of the cells diminish gradually as the veins become smaller, the very fine veins consisting, with respect to phloem constituents, of sieve elements and companion cells only. This progressive reduction in cell size is more pronounced for the sieve elements than for the companion cells, so that toward the ends of the ultimate branches the latter cells constitute the major component of the phloem. With still further reduction in vein size, the phloem constituents are frequently eliminated entirely, the final extension consisting in such cases of tracheids only.

The major veins in most leaves are surrounded by a relatively extensive development of parenchyma cells containing few chloro-plasts. The minor veins in turn are invested in a single-layered sheath of compactly arranged parenchyma cells (sclerenchyma in certain species) which is referred to as the *bundle sheath* or *border paren-chyma*. This specialized tissue extends to the ends of the vascular system and completely encloses the terminal tracheids as well as phloem cells if such are also present. Certain investigators have ascribed considerable significance to this tissue in the process of translocation, as will be discussed later.

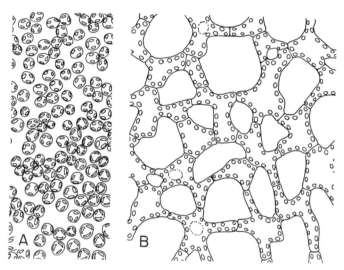

Fig. 3. Paradermal sections of leaf (*Juglans nigra*) cut parallel to leaf surface through palisade layer (A) and spongy mesophyll layer (B). Redrawn from Wylie (129).

Figure 3 presents a paradermal view of leaf structure as seen at the levels of the palisade (A) and spongy mesophyll (B) layers respec-tively. It should be noted that, in terms of translocation, the palisade constitutes essentially an anticlinal or vertical system, inasmuch as the cells are not laterally contiguous to any great degree. The junction plane with the spongy mesophyll shows the lower ends of the palisade cells closely grouped and connected to the spongy mesophyll cells. Spongy mesophyll cells, therefore, are usually intermediary between the palisade and veins. Wylie (129) has emphasized that the upper epidermis may also function as an important conducting system to and from the palisade tissue.

The functional organization of a typical leaf, therefore, is that of narrow zones of chlorenchyma developed in close proximity to the vascular network. From a more generalized standpoint, the same concept holds for the entire plant: the actively growing and "working" cells of the plant are restricted to the proximity of the vascular system.

IV. Phloem: Sieve Elements

Although the phloem tissue, in a generic sense, is frequently referred to as a "food-conducting" tissue, not all of its several components are considered to function in this process with equal facility. As previously indicated, rapid, long-distance transport is believed to occur primarily, perhaps exclusively, in the sieve elements; it appears probable that the translocational capacity of the other parenchymatous components of the phloem is of essentially the same order as that of nonspecialized transport tissues in general, as for example, the cortex of the stem, or the palisade and spongy parenchyma of the leaf. This is not to exclude the possibility, however, that the associated parenchyma cells may play an important ancillary role in translocation. In the dicotyledons, for example, the companion cells and certain of the phloem parenchyma cells are derived from the same mother cells as is the sieve element and remain alive for the same period of time as the sieve element, although other phloem parenchyma cells, not as closely related ontogenetically, may remain alive much longer. Evidently the sieve elements and the ontogenetically related parenchyma cells form a closely integrated unit physiologically, but the nature of the participating role of these cells, if any, is obscure.

In view of the dominant role of the sieve elements in phloem-limited translocation, it is important for the physiologist to be fully cognizant of the many peculiar morphological and cytological characteristics of these cells, as well as of the controversial aspects of the problem resulting from the extraordinary technical difficulties involved in the study of such cells. Function and structure, at cytological levels, cannot be disassociated, and theories relating to the mechanism of translocation in sieve elements must meet the structural requirements imposed by these cells. In a very real sense, therefore, progress in understanding the mechanism of phloem-limited translocation is closely dependent on advances in sieve-element cytology. For this reason, a relatively complete consideration of the present status of this problem will be presented. A more detailed and authoritative treatment of the problem is given in the extensive publications of Esau (see especially 50, 51, and Esau *et al.*, 52).

Sieve elements are of two types: (a) *sieve cells* and (b) *sieve-tube elements* (or *sieve-tube members*). The structure commonly referred

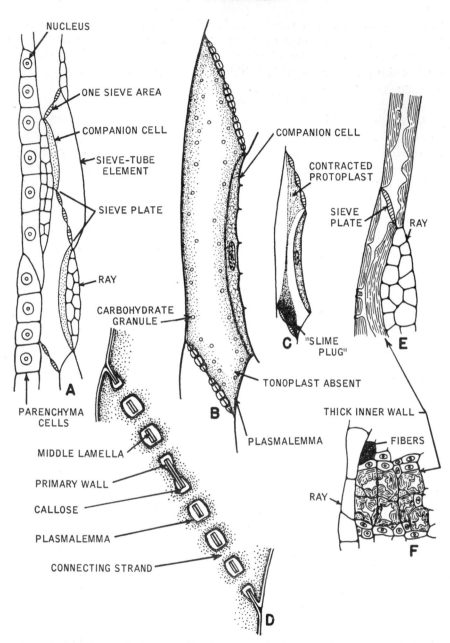

NUCLEUS

ONE SIEVE AREA

COMPANION CELL

SIEVE-TUBE ELEMENT

SIEVE PLATE

RAY

CARBOHYDRATE GRANULE

PARENCHYMA CELLS

MIDDLE LAMELLA

PRIMARY WALL

CALLOSE

PLASMALEMMA

CONNECTING STRAND

COMPANION CELL

CONTRACTED PROTOPLAST

SIEVE PLATE

RAY

"SLIME PLUG"

TONOPLAST ABSENT

PLASMALEMMA

THICK INNER WALL

FIBERS

RAY

A B C D E F

Fig. 4. Structure of phloem cells. (*A*) Tangential view of fragment of phloem of *Fraxinus americana*. Companion cells are stippled, phloem parenchyma cells indicated by nuclei. Compound sieve plates in sieve-tube elements, each with two or three sieve areas. Magnification: × 375. (*B*) Single sieve-tube element of *Fraxinus*

to as the *sieve tube* is composed of a series of sieve-tube elements joined together linearly. Prominent and highly specialized sieve areas develop on the relatively transverse walls separating the "elements" within the tube, and are known as *sieve plates*. The sieve plates are traversed by connecting strands of unknown composition which join the protoplasts of adjacent sieve-tube elements.

Sieve cells are commonly elongated elements with tapering ends. They are not arranged in as definite a longitudinal series as *sieve-tube elements*, nor are the sieve areas sufficiently specialized to be properly referred to as *sieve plates;* a series of sieve cells, therefore, should not be considered as a sieve tube. By this point of view, only plants with sieve-tube elements possess sieve tubes.

Sieve cells are considered to be phylogenetic precursors of sieve-tube elements and are found for the most part in gymnosperms and the lower orders of plants. Sieve tubes, on the other hand, are found in the phloem of angiosperms, although highly developed sieve elements which may properly be called sieve-tube elements have been described for certain species of the brown algae and other lower plants.

Although a true picture of the structure of sieve elements is extremely difficult to obtain, investigators have gradually developed the now rather generally held concept that sieve elements, at maturity, have a thin parietal layer of cytoplasm without nucleus, surrounding a large central vacuole containing varying amounts of "slime." The following features, characteristic of these cells with varying degrees of constancy depending on species, deserve special comment: the enucleate protoplast, "slime," slime bodies," and "slime plugs," the sieve areas, plasmolyzability, and nacreous wall appearance. Certain structural features which are more or less characteristic of sieve elements and of closely associated cells in the phloem tissue are presented in Fig. 4.

in tangential view. Illustration of one of the concepts of protoplast structure: enucleate, plasmalemma present, tonoplast absent; carbohydrate granules in cytoplasm, but some released into vacuolar area. Magnification: × 750. (*C*) Sieve-tube member of *Fraxinus* with protoplast contracted as a result of sectioning of the phloem. Slime, which in normal state [as in (*B*)] is dispersed throughout the protoplast, including the vacuolar area, has accumulated on the lower sieve plate and forms a "slime plug." Magnification × 375; size of connecting strands somewhat exaggerated. (*D*) Diagram illustrating the structure of the sieve plate of the sieve-tube element in (*B*). The connecting strand contains both cytoplasm and vacuolar material without a sharp separation between the two. (*E*) and (*F*) Phloem fragments of *Magnolia kobus* from tangential (*E*) and transverse (*F*) sections illustrating thick inner wall (sometimes called *nacré* wall) in sieve-tube members. The wall thickening is absent on sieve plates and the lateral sieve areas [(*E*) at left below]. In (*F*) parenchyma cells are with nuclei, companion cells stippled. Magnification: × 375. Drawings by Katherine Esau.

Enucleate protoplast. Perhaps the most outstanding feature of the sieve-element protoplast is that it lacks a nucleus when the cell is mature. This enucleate condition has been established for a very large number of species representing all major plant groups; exceptions have been reported in the literature, but it is now known that at least some of these reports were in error, and the enucleate condition is considered normal.

During its developmental stage, which is of relatively short duration, the sieve element resembles a typical parenchyma cell, with its streaming cytoplasm and clearly distinguishable nucleus and vacuole. As maturation proceeds, the nucleus swells in size, loses its chromaticity (stainability), and finally disintegrates, leaving no visible remnants, except, in certain species, the nucleoli, which are extruded from the disintegrating nucleus and retain their identity within the cell.

"Slime bodies," "slime," and "slime plugs." During the early ontogeny of the sieve elements in many dicotyledons, "slime bodies" originate in the cytoplasm. These bodies are initially of various shapes (spheroidal, spindle-shaped, drop-shaped, coiled, etc., depending somewhat on species), but they gradually, and more or less concurrently with the break-up of the nucleus, swell in size, lose their sharp boundaries, and eventually become dispersed as a colloidal "slime" in the vacuole. In some species, the slime bodies do not disperse but retain their identity in the cytoplasm for the duration of the cell; in many other species, including principally the monocotyledons, gymnosperms, and the vascular cryptogams, they are absent altogether.

Information on the chemical nature of slime bodies and slime is very fragmentary. The slime has been interpreted by some authorities as essentially cytoplasmic. On the basis of histological staining reactions, the slime bodies are considered to be mainly proteinaceous (108).

The slime content of sieve element vacuoles is highly variable from species to species. The sieve tubes of the Cucurbitaceae are notable for their high concentration of slime but are relatively atypical in this respect; more commonly, the vacuoles are quite watery, with only small amounts of slime. Although it is generally considered that the slime is derived from the dispersion of the slime bodies, a part of the slime may originate in other ways. Slime, in small amounts, has been reported as occurring in the sieve cells of gymnosperms, although slime bodies, as indicated above, have never been observed in this plant group. There is suggestive evidence, too, that slime may occur in the companion cells in certain species, despite the absence of slime bodies in these cells (except in *Vitis*).

The slime is usually considered to be colloidally dispersed in the vacuole, but it should not be inferred that it is always in the form of a structureless sol. At times it is characterized by a fibroid organization (35); in the dormant sieve tubes of grape, for example, dense masses of slime may be found stretched across the whole length of occasional sieve-tube elements and connected with similar strands in adjoining elements through the sieve plates.

The vacuolar slime of sieve tubes is often observed as relatively dense accumulations on one of the sieve plates. These slimes masses are designated as "slime plugs" (not slime bodies). It is now generally recognized that slime plugs are not normal inclusions of sieve-tube elements, but are artifacts induced by injury to the cell. Sieve tubes are normally characterized by a moderately high turgor pressure, and any sudden steepening of pressure gradients, such as would be induced by sectioning the phloem, will cause an excessive flow rate of cell sap through the sieve tubes to their cut ends, with consequent displacement of the vacuolar slime and other inclusions against the sieve plates in the "downstream" direction. The special precautions which must be exercised in order to observe the sieve-tube contents in a relatively undisturbed state have recently been described in detail by Currier et al. (35).

Much additional work on the quantitative aspects and chemical properties of the slime proteins will have to be carried out before any germane appraisal can be made of their possible role in the mechanism of translocation.

Sieve areas. Parenchyma cells in general, whether found in the mesophyll, cortex, phloem, or other tissues, have thin areas in their walls which are traversed by plasmodesmata. Such areas are referred to as *primary pit fields*. In sieve elements, these primary pit fields are considerably modified during the maturation phase of cell differentiation: first, the plasmodesmata increase in thickness and in their affinity for histological stains, becoming much more conspicuous; and second, callose, a polysaccharide, is deposited as a cylindrical jacket around each plasmodesma. When this stage is reached, the plasmodesmata are referred to as *protoplasmic connections* or *connecting strands*, and the pit fields as *sieve areas*. The composition of the *connecting strands* is imperfectly known, and, therefore, this relatively noncommittal term as regards composition is preferable to *protoplasmic connections*. The increased chromaticity of the connecting strands may indicate the intrusion of slime proteins.

The precise structure of the sieve areas or plates is the subject of much controversy, highly pertinent to translocation studies. Whether the movement of solutes and water from one sieve element to the next is channeled primarily through the connecting strands or may occur with more or less equal facility through the intervening portions of the sieve plate, or even through unmodified wall areas as well, are questions which cannot be answered at present. In the older literature, the vacuoles of adjoining sieve elements were assumed to be continuous through the connecting strands, providing an easy passageway for a pressure-actuated flow through the sieve tubes; it is now generally recognized, however, that the pores are completely filled with "cytoplasm" and callose. The connecting strands are extremely minute in diameter, seldom larger than 2 μ and frequently only a fifth to a tenth this diameter, approaching the limit in resolving power of an ordinary light microscope. Some authorities are of the opinion that the protoplasm and walls of sieve elements are both so highly permeable to water and solutes that the structure of the sieve areas is of little consequence with regard to translocation (30).

Plasmolyzability. The ability of the sieve-element protoplast to plasmolyze has received extended study because of its important bearing on the physiological state of the protoplast. The disintegration of the nucleus is unquestionably attended by a certain degree of denaturation of the cytoplasm. Certain investigators have maintained that this denaturation extends to complete inability to plasmolyze (33); others, that this capacity exists but is inherently difficult to demonstrate in sieve elements because of their sensitivity to injury and physical manipulations (35, 114).

The mature sieve-element protoplast presents the most difficult kind of material for plasmolytic studies. In young sieve elements, as in parenchyma cells in general, plasmolysis and deplasmolysis (to distinguish true plasmolysis from irreversible injury contractions of the cytoplasm) can be readily demonstrated. The situation is very different at maturity, however. At this stage, the cytoplasm occupies such a thin parietal position in the cell and loses its affinity for histological stains, as well as its inherent refractiveness, to such an extent that it can be detected in many cases only with great difficulty, even with the most modern optical equipment, including phase optics. In line with the diminished chromaticity of the cytoplasm, the tonoplast (limiting membrane between cytoplasm and vacuole) also becomes difficult to detect and may be obliterated entirely. Vital stains, such as neutral red, frequently employed in plasmolytic studies to distinguish the vacuoles more clearly and facilitate observations, cannot be used successfully in the present instance, because the mature sieve element protoplast lacks the capacity to accumulate such stains. In fresh sections, "starch" grains (red-staining with iodine) may frequently be observed in Brownian movement in the vacuole. These grains originate in plastids located in the cytoplasm; their presence in the vacuole is further evidence of the absence, or tenuous state, of the tonoplast.

Despite these profound modifications in cytoplasmic structure during maturation, it is now generally agreed that the protoplast retains the property of differential permeability and, accordingly, the inherent capacity for plasmolysis. This conclusion derives from a reinvestigation of the problem by Currier *et al.* (35), who, by careful adherence to manipulative techniques designed to avoid either hypotonic or hypertonic injury to the highly unstable protoplasts during observation, as well as pretreatments to reduce the turgor pressure of the sieve tubes prior to cutting, were able to demonstrate plasmolysis and deplasmolysis in 22 out of 23 widely different species. This work confirms the earlier conclusions reached by Schumacher (114) and Rouschal (108).

Wall structure. A widely observed characteristic of the walls is their relative thickness. As seen through the microscope, the thickened walls present a characteristic glistening aspect, described as *nacré* (a French word meaning "with pearly luster"). The thickness of the walls apparently results mainly from an unusually high degree of hydration of their constituents. Walls dehydrated by conventional histological techniques are much thinner (by as much as 50% or more) than walls in a fresh section; in many species, the walls are so extraordinarily thick that even after complete dehydration, the lumen of the cell is of negligible cross-sectional area relative to that of the walls (16).

It is evident that, in many species, the walls of the sieve elements occupy an appreciable fraction of the total phloem area. This fact, in conjunction with their highly hydrated character, has led to the suggestion that the primary channel of conduction in the phloem may be the walls per se (26). Although this hypothesis is no longer regarded as tenable, the possibility is not excluded that the high degree of wall hydration indicates a wide spacing of the cellulose polymers, resulting in a wall structure capable of accommodating rapid flow of solutions (32). It should be recognized, however, that thick, *nacré* walls are not constant characteristics of all sieve elements, but vary with the species; in the sieve tubes of grape, for example, the walls remain relatively thin and do not develop the typical *nacré* appearance.

A cell-wall substance invariably associated with the sieve areas, and recently found to be characteristic of pit fields in general, is *callose*, a substance probably of glucosan nature (36). As was described above, callose is deposited in the pores of the sieve areas, forming a collar of variable thickness around each connecting strand. Callose deposition also occurs as an overlay on wall areas between the pores. In sieve tubes which persist more than one growing season, the deposits of callose on the sieve plates and in the pores become extremely thick preceding the period of dormancy, forming a plug which is referred to at times as a *callus plug* (note the use of the word *callus* in this connotation and its distinction from *callus* tissue). During reactivation of the sieve-tube elements the following spring, the callus is reduced in amount.

Callose synthesis can occur very rapidly in injured cells—for example, within a few seconds following sectioning. The callose deposit is believed to exert a sealing effect, retarding the rate of transport across cell walls or sieve areas.

Summary. Particularly relevant to the physiologist is the general picture which has emerged regarding the nature of the sieve element protoplast at maturity. It is at this stage, which usually lasts only one season but may extend for many years, as in certain perennial monocotyledons, that the sieve element is considered actively functional in translocation. Briefly summarized, the evidence indicates that the mature protoplast has no nucleus, is incapable of accumulating vital stains, is highly labile or sensitive to injury, and is characterized by a cytoplasm which is greatly hydrated, unrefractive, without affinity for the usual protoplasmic stains, and disposed as a thin, quiescent layer around a large, usually watery, vacuole, with which the cytoplasm merges imperceptibly owing to the absence of any clearly evident tonoplast. The nature and composition of the protoplasmic connections between elements is unknown. Vacuolar continuity from element to element does not exist. The mature protoplast is often described as "denatured," "premortal," or "senescent"; the denaturation, however, does not extend to loss of differential permeability, for the protoplasts are plasmolyzable. The fragmentary data available on the histochemical and enzymological characteristics of the cytoplasm will be considered in Section VIII.

The anatomy and cytology of the sieve element define the structural requirements of any hypothesis proposed to explain the translocatory mechanism.

V. Pathways of Transport

As previously indicated, the phloem cells, and more specifically the sieve elements, comprise the major translocational system for organic solutes. We shall now consider the evidence for this generalization as well as alternative hypotheses and known exceptions.

Most of the information relative to transport pathways has been derived from ringing experiments, which have been extensively employed in translocational studies for nearly three centuries. The decade of the 1920's was an especially active period for such studies in view of the fact that a number of traditional concepts with respect to the transport pathways, of both organic and inorganic solutes, were challenged by various investigators at that time. Important reviews of this work, reflecting the major areas of controversy, are given by Curtis (39), Curtis and Clark (40), and Mason and Phillis (89).

Ringing, unless otherwise specified, refers to the removal of an annular band of tissues external to the xylem, thus completely intercepting the phloem tissue. The term is often used synonomously with *bark-girdling*, or simply *girdling*, although the term *girdling* more frequently refers to an excision around the stem which intercepts the sap wood or active water-conducting layers of the wood as well.

When bark is separated from the wood, the break usually occurs, not in the vascular cambium as is often assumed, but more often in the young xylem where the tracheary elements have their maximal diameters but are still without secondary walls. Ringing, therefore, invariably causes some injury to the xylem, resulting in a reduction of water transport to the ringed branch. The extent of this reduction has been variously disputed; it appears to differ measurably with the species and the time of the season at which the ringing is carried out (22). Ringing a branch early in the growing season, for example, prior to completion of the new xylem, causes a greater restriction to water flow than ringing later in the season. The termination of cambial activity at the ring results in a discontinuity in the xylem of the current season, causing a partial obstruction to water movement. The effect is usually more severe in ring-porous than in diffuse-porous woods. However, with adequate protection of the exposed surface of the xylem at the ring (usually by paraffining), the damaging effects of ringing are seldom sufficiently severe to vitiate the use of this experimental technique for many types of translocational studies.

A. Soluble Carbohydrates

It is well established that ringing interrupts the downward movement of soluble carbohydrates. Use has been made of this fact for many years by horticulturists for purposes of modifying the translocation pattern in various kinds of plants, notably in fruit trees and vines. Such observations, however, although in accord with the concept that downward translocation of sugar occurs in the phloem, do not establish unequivocally that this tissue is the normal pathway for such transport in intact plants.

The most careful analysis of this problem is that by Mason and Maskell (85, 86). Cotton (*Gossypium barbadense*) plants, growing under field conditions, were prepared as shown in Fig. 5. The plants were ringed in the morning, and samples of bark and wood from both above and below the ring were harvested at intervals during the following 20-hour period for chemical analyses. Samples were also taken

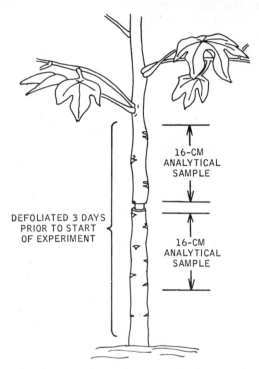

DEFOLIATED 3 DAYS
PRIOR TO START
OF EXPERIMENT

16-CM
ANALYTICAL
SAMPLE

16-CM
ANALYTICAL
SAMPLE

Fig. 5. Diagram showing experimental treatment of stem of cotton (*Gossypium barbadense*) plant. During course of experiment, defoliated stem was covered with black paper to prevent photosynthesis in the cortical chlorenchyma. Only a small portion of the foliage area of the plant is shown. Analytical results are given in Fig. 6.

of stem segments in comparable positions on the control plants (unringed). The results, condensed and generalized, are given in Fig. 6. It is evident that ringing caused a marked accumulation of sugars in both the bark and wood *above* the ring, and a marked decline in these tissues *below* the ring. These results are most easily interpreted in terms of phloem transport of sugars; however, the alternative hypothesis of xylem transport is not excluded and therefore requires consideration.

The limited cross-sectional dimensions of the phloem, especially in trees, is well known, and has led a number of investigators (10, 45) to conclude that downward transport of sugars probably occurs for the most part in the xylem. For example, the aggregate cross-sectional area of the sieve elements in the trunks of a number of trees (ranging from 30 to 50 feet in height) amounts to roughly only about 25–100 mm^2 (96). In Section VIII, we shall take up a more detailed consideration of this problem; for present purposes it is sufficient to point out that

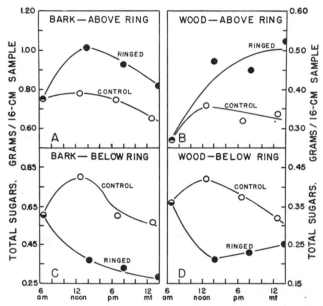

Fig. 6. Diurnal changes in the quantity of total sugars in the bark and wood of 16-cm sections of stem above and below ring. Plants were ringed at 6:30 A.M. on day of experiment. Additional details of experiment given in Fig. 5. Data of Mason and Maskell (85).

the calculated rate of translocation in the phloem tissues of a number of species required to account for observed rates of dry-weight accumulation in the roots and other organs nutritionally dependent on the translocate have been considered by some investigators to exceed the physical limitations of the phloem as a conducting tissue. Furthermore, it has frequently been noted that if a dye is injected into the xylem, its movement can be traced both upward and downward in the xylem. This observation has been taken as evidence of naturally occurring reverse streams in the xylem.

Assuming, therefore, that downward translocation of sugars actually occurs in the xylem, it may be argued that ringing indirectly blocks this transport, either by injury to the xylem or by blocking the downward transport in the phloem of certain hormones required for continued growth of the plant parts below the ring; according to the latter view, downward transport through the xylem would then cease because assimilatory utilization of the sugar is inhibited below the phloem-block. In this event, the sugar concentration in the stem below the ring should remain high.

It is evident, however, from Mason and Maskell's data that the sugar concentration fell off rapidly in both bark and wood tissues below the ring (Fig. 6, C and D), indicating strongly that ringing inhibits the downward movement of sugar mainly by interrupting its normal channel of transport, and not by preventing sugar utilization below the ring.

With respect to the possibility of injury to the xylem, Mason and Maskell reasoned that the clogging of the tracheae that is assumed, on ringing, to interrupt the transport of sugar, ought also to interrupt the transport of a dye. Accordingly, they immersed the cut ends of some of the leaves above the ring in an eosin solution, and 6 hours later examined the wood for the presence of the dye. It was found to have penetrated downward past the ring as well as upward. It is not likely, therefore, that clogging of the tracheae can explain the disruption of the transport.

The movement of water in the xylem is normally in the direction of decreasing pressures (or increasing tensions). Consequently it may be expected that when water is supplied to a plant by immersing a cut leaf or petiole attached to the plant under water, or even by partially immersing the intact leaf itself, the usual pattern of hydrostatic gradients within the plant will be modified, and movement of water out of the leaf and thence downward as well as upward from the point of leaf attachment will likely occur. Substances supplied with the water under these conditions, such as eosin (a dye) (85), systox (an insecticide) (1), or streptothricin (a bactericide) (57), will be rapidly carried throughout the plant in the xylem, but such instances of xylary distribution are undoubtedly artifacts in most cases, and cannot be used as proof that the xylem is the normal channel of conduction of photosynthates and other naturally occurring solutes originating in the leaves. Furthermore, solutes supplied exogenously to leaves in minimal quantities of water, and hence with minimal disturbance to the hydrostatic system of the plant, are translocated principally via the phloem (24). Additional information relating to various aspects of this problem is given by Crafts (31).

Not all of the present data, however, can be aligned with certainty with the above interpretation. Bondarenko (14), for example, has presented evidence, based on autoradiograms, that C^{14}-labeled amino triazole, supplied as a 0.04 ml drop of solution to Canada thistle (*Cirsium arvense*) leaves was translocated both upward and downward in the stem principally in the xylem.

Nonetheless, while recognizing the possibility of occasional participation of the xylem in the conduction of organic solutes from leaves to basal parts of the plant, we must not lose sight of the fact that downward translocation of major nutrients, such as carbohydrates, *in calorically significant amounts*, occurs only in the phloem.

The pathway of upward transport of soluble carbohydrates has been a subject of very considerable controversy. Classically, the xylem was regarded as the principal tissue involved in this translocation, the major arguments in favor of this view being: (a) the occurrence of relatively large quantities of carbohydrates (mainly starch) in the xylem, especially in the ray cells and wood parenchyma, and particularly in parenchyma cells immediately proximal to the tracheae; (b) the presence of various sugars in the "xylem sap" (that is, in the sap extracted from stem segments by centrifugation or vacuum displacement; this fraction would represent principally that present in the tracheae or vessels, and presumably, therefore, the more mobile fraction of the "ascending sap"); and (c) the observation, first published by Hartig in 1858 (59), that in ringed trees, starch stored in the wood below the ring disappeared during the period of spring growth. This disappearance was credited to hydrolysis and upward transport of the soluble products in the xylem. This classical view was generally accepted, with various qualifications, especially as regards the transport of organic solutes to the fruit, until about 1920.

None of these arguments favoring the xylary transport hypothesis is now considered to be tenable. The disappearance of starch below the ring in Hartig's experiments can be largely accounted for by the utilization of this food in growth below the ring. Many modified Hartig experiments in which a second ring is made to block off phloem transport to the roots (double-ringing experiments) demonstrate this relationship very clearly.

Furthermore, on the basis of present rather fragmentary data, the actual concentration of sugars in the xylem sap is very low in most species, varying between essentially zero values for a major portion of the summer to maximal values of only 0.02–0.05% during the winter and early spring (Fig. 7). Important exceptions to this generalization, however, occur. In *Acer saccharum*, for example, concentrations as

high as 8% have been reported (67), and in *Ilex aquifolium,* a summer
maximum comparable in magnitude to the vernal maximum has been
found (6).

Although present data do not permit a precise calculation of the
relative amounts of soluble carbohydrates transported acropetally in
the xylem and phloem, it is now generally recognized that the xylary-
translocated fraction is negligible, at least in physiological terms. This

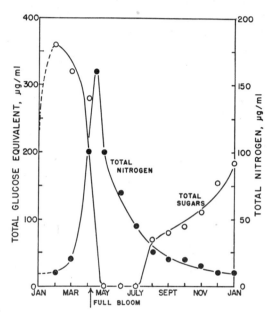

FIG. 7. Seasonal variations in the total sugar content of the xylem sap of pear
(*Pyrus communis*) tree [data of Anderssen, (2)] and the total nitrogen content
of the xylem sap of the apple (*Malus sylvestris*) tree [data of Bollard (12)]. Time
axis for apple (New Zealand experiments) advanced six months so that growing
seasons for northern and southern hemispheres coincide on abscissa.

conclusion is based on comprehensive ringing experiments by many
investigators. Typical of such studies is that by Curtis (37) on the
effect of ringing on terminal growth. Dormant branches of *Crataegus*
sp. were ringed, either partially or completely, at various distances
from the apical bud; all lateral buds were removed at the start of the
experiment. Representative data are given in Table I. It is evident from
these results that complete interception of the phloem tissue of the stem
a short distance back of the terminal bud greatly inhibited the growth
of the terminal shoot. This severe inhibition did not result from injury to the xylem in these experiments, for growth was not appreciably

checked in the partially ringed branches, even though half of the xylem was also cut away in order to exaggerate possible injury effects to the tissue. On the basis of other experiments, the restriction of growth above the ring appears to result largely from a limited supply of carbohydrates (37). Many other experiments substantiate these general relationships.

As noted above, ringing usually (perhaps invariably) reduces the rate of transpiration in leaves distal to the ring, and it may be argued that failure of soluble carbohydrates to move upward through the xylem in quantities sufficient to affect growth resulted from this reduced rate of water loss. If the transpirational factor were actually of

TABLE I

EFFECT OF RINGING AT DIFFERENT DISTANCES FROM THE TERMINAL BUD UPON
SUBSEQUENT GROWTH[a,b]

Treatment	Number of branches	Avg. shoot elongation (mm)
1. Control (not ringed)	17	26.8
2. Ringed, 2nd internode from tip	13	6.1
3. Partially ringed,[c] 2nd internode from tip	6	26.2
4. Ringed, 4th internode from tip	13	8.1
5. Partially ringed,[c] 4th internode from tip	4	28.0
6. Ringed, base of 1-year wood	8	17.0
7. Ringed, base of 4-year wood	11	22.0

[a] From Curtis (39).

[b] *Crataegus* sp., ringed April 8; measurements taken May 8.

[c] In the "partially ringed" branches, one-half of the xylem and three-fourths of the phloem were removed.

importance, however, the terminal growth in the partially ringed branches should have been reduced even more than in the conventionally ringed branches, yet, in point of fact, growth in the branches with half of the xylem removed but with a fourth of the phloem intact closely approximated to that of the controls.

It thus appears that the early growth of the terminal shoot is largely dependent on the organic nutrients, mainly carbohydrates, translocated to it by way of the phloem from reserves previously accumulated in the older parts of the stem. This is equally true in *Acer saccharum*, in which, as noted above, the xylem sap contains an unusually high concentration of sugars during the early spring. As new leaves mature, the shoots become relatively independent of the rest of the plant with respect to organic nutrients, but translocation of carbohydrates and other organic solutes from the mature leaves to the growing tips and young leaves continues to occur throughout the growth period, and this translocation is also primarily phloem-limited.

In herbaceous plants, ringing is not feasible, but physiologically equivalent phloem-blocks can be readily made by killing short segments of the stem, usually by application of hot wax or superheated steam from a fine jet. Inasmuch as only the conducting elements of the phloem, and not those of the xylem, are living, any phase of translocation interrupted by killing the cells in a stem segment can be assumed to be occurring in the phloem or in the vascular cambium. This technique has also been employed with woody species as a substitute for conventional ringing. The data from a considerable number of such experiments with respect to the distribution of carbohydrates and other carbon compounds, including C^{14}- or C^{13}-labeled derivatives accord with the general conclusions presented above, namely, that the translocation of these compounds, both acropetally and basipetally, occurs principally in the phloem (104).

B. ORGANIC NITROGENOUS COMPOUNDS

A serious difficulty in nitrogen-transport studies is the uncertainty regarding the identity of the major translocatory compounds of this element (see Section VII). Both inorganic-nitrogen and organic-nitrogen compounds (mostly amino acids and amides) have been identified in the tracheary or xylem sap as well as in phloem exudate, hence the evidence is presumptive that both forms of nitrogen are transported in each tissue. However, neither the relative amounts translocated as organic nitrogen and inorganic nitrogen, nor the relative amounts transported in phloem and xylem, have been accurately established for any species. At present the controversial data bearing on this problem can be reconciled only on the assumption that these relationships vary significantly with species and environmental conditions.

The general view has prevailed for many years that nitrogenous compounds entering the roots from the soil, principally as nitrates, are first transported in this chemical form to the leaves mainly in the xylem and then synthesized into organic compounds. The subsequent distribution of these organic compounds has been considered to occur primarily in the phloem, the basic translocation pattern from the leaves resembling closely that of carbohydrates. This generalized picture has been qualified by the recognition that, in certain species, a very considerable fraction of the entering nitrate ions are reduced to organic nitrogen in the roots (123), and that in such species the initial distribution of nitrogenous compounds to the leaves may occur principally in the form of organic compounds through both the xylem and phloem.

We shall briefly consider now some of the experiments on which this concept is based.

The most extensive studies of the nitrogenous fractions of xylem sap, collected from stems by vacuum displacement, are those of Bollard (13). Bollard found that in a wide variety of species, selected from angiosperms, gymnosperms, and ferns, most of the nitrogen in the xylem sap occurs as amino acids and amides, especially aspartic acid, asparagine, glutamine, and glutamic acid. Nitrate nitrogen was present in less than half of the species tested and usually in trace amounts only. It is now evident that the upward translocation of nitrogen-containing compounds in the xylem occurs more commonly as organic compounds than was previously anticipated.

Data showing the seasonal variations in the concentration of the total nitrogenous fraction (entirely organic nitrogen except for traces of ammonia nitrogen) in the xylem sap of apple are given in Fig. 7, along with comparative data on total sugar concentration. The nitrogen data are calculated in terms of micrograms of elemental nitrogen and represent, therefore, approximately one-tenth to one-fifth of the corresponding weight in amino acids and amides present in the sap. It should be noted that, although the sugar concentration dropped rapidly to zero during the period of most rapid transpiration (presumably because of a slow rate of supply of hydrolyzates from the carbohydrates in the parenchyma cells of the xylem), the nitrogen concentration remained moderately high throughout this period, suggesting a sustained rate of supply from the roots. The sharp increase in concentration immediately prior to anthesis, however, is perhaps mainly due to mobilization of reserves rather than to increased nitrogen uptake by the roots.

As to the major tissue involved in the upward translocation of nitrogenous compounds from the roots, the literature is in disagreement, but the weight of the evidence presently favors the xylem. Typical of the experiments supporting this view may be cited those of Mason *et al.* (87). Using defoliated stems of cotton isolated from the roots by ringing and maintained in a saturated atmosphere (to reduce transpiration), they found that the nitrogen content of the stems and roots of both ringed and control plants increased at approximately the same rate. Conversely, if the xylem rather than the bark was intercepted, little if any nitrogen ascended the stem. It is evident, therefore, that even under conditions of low water-translocation rates, the participation of the phloem in these plants in upward translocation of nitrogenous compounds from the roots was negligible. Other experiments, by various investigators under a wide variety of conditions, lead to substantially the same conclusions, e.g., 22).

Data contradictory to this view have been obtained principally by Curtis (39). For example, with shoots of *Ligustrum ovalifolium,* ringed late in August after shoot elongation and xylem formation were nearly completed for the season, he found that the gain in nitrogen content of the leaves above the ring during a 6 weeks' period was about 23%, whereas for comparable leaves on the unringed controls, the corresponding gain was about 116%, or five times as great. Although it is possible that the restricted translocation of nitrogenous compounds to the ringed shoots may have resulted from a reduced rate of transpiration in these shoots (see discussion above on effect of ringing on transpiration rate), nonetheless it should be noted that the experiment was carried out toward the end of the growing season, when damage to the xylem by ringing should be at a minimum.

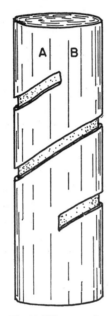

Fig. 8. Diagram showing spiral ringing of a woody stem.

Experiments of MacDaniels and Curtis (79) provide additional data confirming the importance of the phloem in the translocation of nitrogenous compounds. When the trunks of apple (*Malus sylvestris*) trees were spirally ringed (Fig. 8), it was found that nitrogen from the roots moved principally to the branches above the open end of the spiral (side *B*). Branches on the opposite side (*A*) received relatively little nitrogen. No differences in results were obtained if the spiral ring was cut to a depth of two annual rings into the xylem as well. These results are most easily interpreted on the basis that the nitrogen is translocated primarily in the phloem. Although the tracheal sap of apple trees contains a moderate concentration of nitrogenous constituents (entirely organic) throughout much of the growing season (Fig. 7), apparently the preponderant amount of nitrogen compounds transported via the phloem from the roots in the experiments of MacDaniels and Curtis was sufficient to control the major distribution pattern of this element to the branches.

A considerable amount of evidence exists that organic nitrogenous compounds synthesized in the leaf and stem tissues are redistributed to other parts of the plant principally via the phloem. The most critical and extensive of such experiments are those of Maskell, Mason, and Phillis on the cotton plant, reviewed by Mason and Phillis (89). In the experiment described earlier (Figs. 5 and 6) concerned with carbohydrate transport, analyses were also made of the nitrogenous fractions in the stem sections above and below the ring and in corresponding sections on unringed plants (82). Disregarding details, the general pattern of diurnal changes in nitrogen content in the stem as affected by a phloem block was found to be quite similar to that for carbohydrates under the same conditions. In later experiments, Phillis and Mason (103) demonstrated that nitrogenous compounds from the lower leaves of a plant may also move to the upper leaves, at least when prevented by the presence of a phloem block at the base of the stem from moving down into the roots. Ringing the stem between the upper and lower foilage regions of the plant greatly diminished, by at least two-thirds or more, the transfer of nitrogen from the lower leaves to the upper. Although in none of these experiments, including the more recent nitrogen-redistribution studies in corn by Hay *et al.* (60), and others, have the major translocatory compounds of nitrogen been unequivocally identified, the evidence is strongly presumptive that they are organic in nature. A factor complicating the interpretation of such experiments is the capacity of most tissues in the plant to reduce nitrates to organic nitrogen. For example, fruit-culture studies have shown that excised tomato (*Lycopersicon esculentum*) and gherkin (*Cucumis sativus*) ovularies can grow *in vitro* and form fruits which ripen, and occasionally even produce viable seeds, on a simple medium containing only sucrose, mineral salts, and water (76, 99). Engard (49) in a study of nitrogen transport in the raspberry (*Rubus idaeus*) cane could find no evidence of the translocation of any organic nitrogenous compounds either upward or downward in the phloem. Such nitrogen distribution as occurred appeared to be only in the form of nitrates, the distributive channel being mainly the xylem. It is evident that there are many unsolved problems with respect to nitrogen transport.

C. Fats and Fat Hydrolyzates

Information on the translocation of these compounds is very meager. A brief review is given by Miller (91); no critical information is available with respect to transport channels.

D. Miscellaneous Organic Compounds

Phosphorylcholine and glycerylphosphorylcholine have recently been identified as relatively significant translocatory compounds of phosphorus occurring in the xylem sap of a number of species (80). The translocation of auxins will be considered in a later volume.

E. Summary

Soluble carbohydrates, whether originating as primary products of photosynthesis in chlorenchymous tissues or as hydrolyzates in storage tissues, are translocated primarily in the phloem in both the acropetal and basipetal directions. Small amounts of these compounds may also move at times in the xylem, mainly in the "transpiration stream," but the amounts thus transported do not appear to be physiologically significant in plant growth.

There is less agreement as to the transport channels utilized in the distribution of nitrogenous compounds. Disregarding certain divergent data, the basic pattern appears to be as follows. The translocation of nitrogenous compounds from the roots to the leaves and other plant parts occurs primarily in the xylem. Apparently this is true of both inorganic and organic nitrogen compounds, and irrespective of whether the organic compounds were synthesized initially by soil microorganisms, in mycorrhiza or the roots proper, or entered the root system by translocation from the shoot. Subsequent retranslocation of nitrogen compounds from the leaves to other plant parts occurs primarily in the phloem.

VI. Distribution Patterns

From the preceding discussion, certain fundamental patterns of organic solute translocation are evident, as, for example, the export of carbohydrates and other organic solutes from the leaves to the stems, and thence either acropetally or basipetally to other organs and plant parts. We have now to consider some of the details of this general transport pattern, particularly the question of the relative contribution of the various leaves to the nutrition of the various parts of the plant, especially the fruit and meristems.

As will be developed in the next chapter, problems of this nature are very complex, principally because of recycling of many of the translocatory compounds, after being utilized in the synthesis of different compounds characterized by different turnover rates. In consideration of this fact, the supply pattern to or from a given leaf is a function not only of the age of the leaf but also of its relation to the ontogeny of the plant, i.e., the time of development of the leaf in relation to the age of the plant. The details of this concept have been largely worked out by the use of radioactive isotopes of iron, sulfur, phosphorus, calcium, and other mineral elements (Chapter 6 in this volume). Comparable studies with radioactive and heavy isotopes of carbon are as yet relatively incomplete, partly because of greater technical difficulties involved in the use of these isotopes and partly because of complications imposed by the respiratory losses of carbon from the plant. Hence, a less definitive treatment of this subject is presently possible for organic solutes than for the "mineral" solutes.

It is generally considered that the distribution of C^{14}-labeled translocate from a single leaf on a plant, whether derived from radioactive sugar applied to the leaf surface or from radioactive products of photosynthesis, is essentially the same. Mellor (90), for example, observed that sucrose-C^{14} applied to a single leaf on tobacco (*Nicotiana tabacum*) plants moved in relatively large quantities into the stem and younger growing leaves but only to a slight extent into the mature leaves on the plant. This distribution pattern of C^{14} accords with that observed in the mature canes of grape (*Vitis labruscana*) and red raspberry (*Rubus idaeus*), in which the C^{14} was supplied in the form of radioactive carbon dioxide. Recently, data contrary to this generalization, however, have been reported. Nelson and Gorham (98) observed that, in soybeans (*Glycine max*), the distribution pattern of C^{14} when derived from sugars exogenously supplied to one of the leaves was very different from that derived from $C^{14}O_2$. The reasons for these discrepancies are not presently known. The results indicate, however, that caution must be applied in drawing conclusions regarding the distribution patterns of photosynthetically derived sugars based on foliar applications of sugars to the plant.

The amount of C^{14}-labeled translocate from a given leaf which partitions in the stem toward the shoot and root tips respectively depends on the position of the leaf on the shoot axis, the occurrence and location of rapidly developing fruits, light intensity, and many other factors. Mellor (90) observed, using the method of foliar applications of sugars, that the movement of C^{14}-labeled compounds from tobacco leaves during the night period was predominantly toward the roots, but that at a light intensity of approximately 750 foot-candles, a major fraction of the translocate moved to the growing shoot tip. A similar reversal in the destination of the translocated products of photosynthesis has been observed in soybean plants, based on methods employing $C^{14}O_2$ (98).

A considerable number of observations indicate that the movement of phosphate from leaves is correlated to a degree with the movement of sugar from the leaves; that is, radioactive phosphate applied to a leaf is exported only under conditions when outward translocation of sugar from the leaf may also be expected (9, 23, 24, 68, 70). Furthermore, the relative rates of translocation of P^{32}- and C^{14}-labeled compounds from leaves are of the same order (50–100 cm per hour in some experiments), and the respective distribution patterns of these compounds, in short-term experiments (4–10 hours) are known to be similar in a number of details. Complete correspondence, of course, cannot be expected, because of different rates of radial transfer to the xylem and different rates of recycling in the plant, but for short translocation periods (4–8 hours) the evidence suggests that the distribution of the readily measured radioactive phosphorus can be used to indicate the distribution patterns of carbohydrate materials from the leaves.

If this premise is accepted, the detailed studies which have been made of P^{32} distribution patterns in plants are of interest in the present connection. Linck (77) has reported a comprehensive study of this type on pea (*Pisum sativum*) plants, some of the results of which are presented diagrammatically in Fig. 9. The relative amounts of the P^{32}-labeled substances which moved upward and downward from the various leaves, as a function of their position on the stem and location with respect to the developing pods, are indicated by the direction and relative size of the arrows. As shown in the figure, movement from the lower leaves was predominantly toward the roots, and from the higher leaves, predominantly toward the apex. A considerable fraction of the translocate from the leaf at the ninth node moved into the two pods, especially the pod at the eleventh node. The most striking instance of specificity of transport relationships between a given leaf and the developing pods was that observed between the leaf and pod at the same node. Usually about 70–80% of the total P^{32} content of the plant was found in the pod attached at the axil of the leaf to which the labeled phosphate was supplied. Significantly, the second pod (P 2), which normally received only 1–5% of the total labeled translocate from the leaf at the tenth node, received 50–60% of this translocate if the first pod (P 1) was removed. The results were the same whether this pod was removed many hours before, or just prior to, the application of the labeled phosphorus to the leaf. Evidently, new patterns of distribution from a given leaf may be established very rapidly in plants. This general effect was demonstrated by many other treatments (chilling the pod, removal of ovules, etc.).

A similar conclusion can be drawn from various systematic defoliation studies. Drollinger (47), working with tomatoes growing under optimal greenhouse conditions, has shown that the removal of two to four leaves on the stem, in the immediate vicinity of a cluster of young tomato fruits, did not significantly reduce the yield from that cluster, although it is very likely that normally the leaves closest to a given cluster of tomato fruits supply most of the food materials used by that cluster.

Further evidence of the polarizing effect of developing fruit on the transport patterns from leaves is provided by the data of Haller (58) on apples. For a constant ratio of 30 leaves per fruit, the leaves and fruit could be separated by distances up to 10 feet on the tree (the maximal distance varying with variety) without decreasing the size

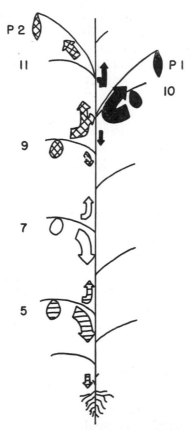

FIG. 9. Diagram of pea (*Pisum sativum*) plant showing distribution pattern of labeled translocate from individual leaves as a function of their position on the plant. Arrows from leaves 5, 7, 9, and 10, respectively, indicate the relative amounts of P³²-labeled solutes translocated from each of these leaves acropetally and basipetally. The labeled translocate from leaf 10 moves predominantly to the pod (P 1) attached at the same node; from leaf 9, more of the translocate moves to the second pod (P 2) than to P 1. From Linck (77).

and sugar content of the apples. Similar relationships have been demonstrated in other plants (8, 78, 96). It is evident that the occurrence of "acceptor" sites (such as meristems and immature fruit), and the position of these sites with respect to the supply leaves, are major determinants affecting translocation patterns from leaves. These acceptor sites, to be effective, must be characterized by a high degree of metabolic activity, probably metabolic activity of a kind peculiar to growing cells. Similar ideas have been elaborated in relation to the movement of inorganic ions (Chapter 4 in this volume). Merely establishing arbitrarily a sugar concentration gradient from the leaves by prolonged darkening of a part of the plant is ineffective in inducing translocation along the gradient (5). Skok (117) has shown that excision of the apical meristems of sunflower (*Helianthus annuus*) plants reduced by nearly 50% the amount of acropetal translocation of C^{14} labeled compounds from the basal leaves of the plants.

The translocation pattern from a given leaf or branch is predominantly linear; that is, the translocated materials are restricted to the vertical system of the stem more or less in line with the leaf or branch supplying the translocate. Dana (42) has obtained evidence with radioactive carbon techniques that this translocation "stream," in young apple trees, remains narrow even at distances of 3 or 4 feet from the source of supply. Similar conclusions may be reached from observations of growth data. It has frequently been noted, for example, that the annual rings of trees on the side directly under the larger limbs or on the open side of trees bordering on a forest are much larger than those on the opposite side. Caldwell (15) has reported that removal of all leaves from one side of the stem of the Swedish turnip (a cultivated *Brassica*) at an early stage of development results in a highly asymmetric growth of the turnips, with little expansion of the tap root occurring on the defoliated side. Many other observations of a similar nature may be cited. It is evident that lateral transport in a tangential direction during the course of translocation in the stems is greatly restricted.

Whether this impedance to tangential movement results principally from structural organization (lateral separation of sieve elements by parenchymatous cells, fewer sieve areas on the lateral walls, etc.), or is, to a greater extent, the result of a growth pattern which itself is primarily limited by a rigorously linear, basipetally polarized auxin-transport pattern, is not known. On the basis of the latter view, the asymmetric growth of the turnip root results not from any anatomical impediments to tangential transfer but from a deficiency of auxins on the defoliated side. The fact that the usual orthostichous (vertically aligned) distribution patterns from pea leaves can be readily distorted by selective removal of immature pods, as described in the studies by Linck reported above, further supports this view. Czapek's (41) modified ringing experiments, in which he found that transfer of organic materials across the ring could take place only through straight, vertical bridges of phloem, and not through oblique or stepwise connections (Fig. 10), are sometimes cited as evidence that phloem-restricted substances cannot move tangentially to any appreciable degree, but the results of all such experiments can be as adequately explained on the basis of interruption of the auxin-transport system itself [see also (25)]. It is not known to what extent spiral ringing inhibits translocation. In the experiment by MacDaniels and Curtis (described on page 506), it is possible that the nitrogenous compounds moved for the most part in the sieve tubes differentiated in the callus tissue produced along the cut edges of the spiral.

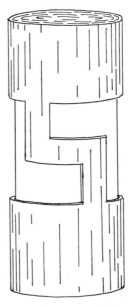

In contrast to tangential transfer, radial transfer occurs readily. This is evidenced by the fact that large quantities of carbohydrates accumulate in the parenchyma and ray cells of both phloem and xylem during the growing season. It is generally presumed that radial transfer is facilitated by the ray cells in phloem and xylem.

VII. Compounds Translocated

The discussion thus far has been principally in terms of various generic classes of organic compounds, as for example, *soluble carbohydrates* or *organic nitrogenous compounds*. A number of studies have been conducted to identify more specifically the major compounds which are translocated, and although such studies are of great theoretical interest, progress has been very slow and uncertain, principally because of the difficulty of distin-

Fig. 10. Diagram showing a partial ring with a stepwise connecting bridge.

guishing between the actual or postulated transit molecules in the main conducting cells of the phloem (presumably the sieve elements) and nontransit molecules of the same species in the nonconducting cells of the phloem and adjacent tissues. Such molecules may originate in these cells either by radial diffusion from the conducting cells in the phloem or by synthesis *in situ*.

On the basis that the empirical molecular formula for the dry-weight composition of a plant approximates to that of a carbohydrate, it may be anticipated that the major compounds translocated are themselves carbohydrates. Sucrose, glucose, and fructose are commonly present in the phloem tissue of plants, in a concentration totaling as high as 10% of the dry weight at times, and this fact is further presumptive evidence for this view.

Returning to the studies of Mason and Maskell on carbohydrate translocation in the cotton plant discussed above (page 498), analyses were also made of sucrose and hexose gradients in the stem, both longitudinally and radially, and as a function of time of day and other factors. The diurnal sucrose concentration in the bark fluctuated much more markedly than that of the reducing sugars; furthermore, the radial concentration gradient within the bark was positive for sucrose, and negative for the reducing sugars, that is, the highest concentration of sucrose occurred in the inner zone of the bark and diminished toward the cortex, whereas the opposite pattern characterized the radial distribution of hexoses. Such a sucrose gradient was found to parallel the distribution frequency of sieve tubes in the bark. On the basis that the sieve tubes constitute the primary channel of transport in the phloem, Mason and Maskell concluded that sucrose is the major translocatory sugar in cotton. Similar studies on other plants by various investigators have led to the same conclusion (48, 78). Other investigators, however, have concluded that glucose and fructose are quantitatively more important than sucrose as translocatory compounds; see literature review in Leonard (75). From these studies, therefore, the general conclusion emerged that the relative translocatability of these three sugars varied significantly with species and environmental conditions, although there has been general agreement for a long time that sucrose is the most important translocatory carbohydrate in most species.

Another approach to this problem in recent years has been chromatographic analysis of the phloem exudate. As indicated in Section IV, the sieve-tube system is characterized by a moderately high turgor pressure, and if an incision is made in the bark, a small quantity of sap representing mostly sieve-tube exudate is obtained, the volume usually ranging from about 0.05 to 0.1 ml for a cut several centimeters long (Fig. 11). Chromatographic analyses of this sap have now been reported for about 25 different species (131, 132) and reveal the singular fact that hexoses are not constituents of the sieve-tube exudate in any of the species thus far studied. Only nonreducing sugars of the raffinose family have been found: sucrose, raffinose, stachyose, and possibly verbascose. These four sugars constitute an ascending series of di-, tri-, tetra-, and pentasaccharides, differing from each other only in the number of included galactose residues. Zimmermann (132) found that sucrose was the dominant sugar in all species tested except *Fraxinus americana,* and in a considerable number it was the only sugar. The total sugar concentration of the sieve-tube sap ranges between 10 and 25%, varying with species and individual tree and fluctuating diurnally and seasonally.

If we accept the premise that the sieve-tube sap is a valid sample of the translocatory fraction in the phloem, the dominant importance of sucrose in translocation is now evident. On this basis, the common if not universal occurrence of glucose and fructose in the phloem,

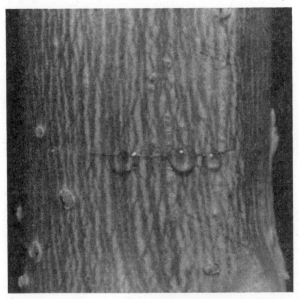

Fig. 11. Drops of sieve-tube sap exuding from an incision at $X - X$ made in the bark of box elder (*Acer negundo*). Magnification, $\times 4$. Photograph by K. L. Webb and R. H. Hodgson.

often in considerable quantities, must be attributed mainly to hydrolysis of sucrose and related sugars in the nonconducting cells of this tissue.

A similar conclusion was reached by Swanson (119) and Swanson and El Shishiny (121), based on a totally different approach. Carbon dioxide, C^{14}-labeled, was supplied to a single leaf on a grape cane (Fig. 12), and samples of bark were subsequently harvested at the positions indicated in the figure. An analysis of the labeled sugar fractions (sucrose-C^{14}, glucose-C^{14}, and fructose-C^{14}) revealed a major portion of the radioactivity present in sucrose; more significantly, the concentration of glucose-C^{14} was found to equal that of the fructose-C^{14} (if the reasonable assumption is made that both species of molecules were uniformly labeled). This 1:1 equivalence in the radiochemical glucose:fructose fraction (see Table II) is most simply interpreted as being the result of hydrolysis of uniformly labeled sucrose molecules absorbed by the nonconducting cells in the phloem from the sieve tubes. On this basis, no significant translocation of hexoses need be postulated to account for the distribution of these sugars in the stem. It should be noted that the radiochemical fractions of sugars in the

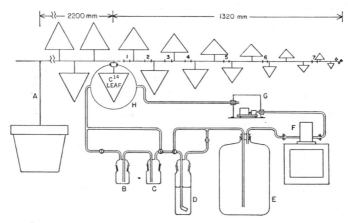

FIG. 12. Schematic drawing of apparatus used for supplying $C^{14}O_2$ to single leaf on grape (*Vitis labruscana*) cane. (*A*) Grape vine bearing two canes (one shown); (*B*) flask containing approximately 2 mc $BaC^{14}O_3$; (*C*) flask containing lactic acid; (*D*) absorption tube containing 10% NaOH; (*E*) 50-liter bottle; (*F*) ionization chamber with Brown recorder; (*G*) diaphragm pump in leak-proof housing; and (*H*) 3-liter modified Florence flask. 30-Mm sections of internodes taken for analyses at positions indicated by *1, 2,* etc. Distances given in Table II refer to distance from C^{14} leaf node to mid-point of section. From Swanson (119).

grape cane represented recent increments in the stem, with an average age for the translocatory molecules and their derivatives of approximately 1 hour. In contrast, the chemically determined fractions of glucose and fructose were not equal, indicating differential rates of utilization of these sugars in the bark. These fractions showed the effect of differential utilization because of the greater average age of their constituent molecules. The assumption is implicit in this analysis that the labeled sugars in the translocate did not rapidly equilibrate with their respective pools of native sugars in the stem.

Similar studies were also carried out by Vernon and Aronoff (124) on the translocation of carbohydrates in the soybean, leading to the

TABLE II

RELATIVE CONCENTRATIONS OF LABELED SUGARS IN THE BARK OF A GRAPE CANE AS
A FUNCTION OF TRANSLOCATION DISTANCE[a]

Distance of trans-location (mm)	Counts per minute per milligram dry weight of bark			$\dfrac{\text{Glu.}}{\text{Suc.}} \times 100$	$\dfrac{\text{Fru.}}{\text{Suc.}} \times 100$
	Sucrose	Glucose	Fructose		
88	8005	661	678	8.25	8.45
202	6268	433	481	6.90	7.68
321	5800	397	402	6.85	6.94
429	4615	220	250	4.78	5.42
652	2942	136	126	4.62	4.28
875	1749	75	69	4.30	3.95
1156	900	34	31	3.72	3.40

[a] From Swanson and El Shishiny (121).

conclusion that sucrose, glucose, and fructose were all translocatory sugars but characterized by different rates of transport. A part of the data substantiating this inference is presented in Fig. 13, which shows the change in ratio of the specific activities of sucrose, glucose, and

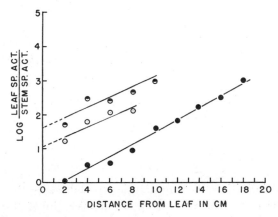

FIG. 13. Plot of leaf specific activity/stem specific activity for the translocated sugars. From Vernon and Aronoff (124). ●, sucrose; ⊙, glucose; and ◑, fructose.

fructose as a function of translocation distance, relative to the specific activities of these respective sugars in the supply leaf. According to the authors, the relative ranking of translocation rates should be inverse to the intercept values; sucrose, therefore, on this basis, had the highest, and fructose the lowest, translocation rates in these experiments.

It should be noted, however, that the magnitude of the disjunction in the specific activity values for the glucose and fructose fractions in the stem and supply leaf respectively (Fig. 13) suggests that the hexose sugars in the stem did not come from the hexose reservoirs of the leaf but were secondarily derived, to a large extent, from the translocated sucrose.

Although the evidence thus far cited indicates that the high degree of specificity of sucrose in translocation may be a phenomenon of general occurrence, it is more likely that a number of variant mechanisms exist. For example, the principal sugar present in the sieve-tube sap of white ash (*Fraxinus americana*) is stachyose, a tetrasaccharide (132). Furthermore, Dana (42) found that glucose was the primary compound in the bark containing radioactivity following a 30-min period of C^{14} photosynthesis in the terminal leaves of the branch. Although the mere fact of quantitative preponderance of a given compound cannot be used as proof of its relative importance in translocation, present indications are that the process of sugar translocation varies with species in a number of important details. It is evident that a comparative study of the biochemistry and physiology of the sugar translocation mechanism in different species is greatly needed.

Relatively extensive studies of the nitrogenous constituents of xylem sap have been reported. An inventory of these compounds now includes most of the amino acids, glutamine and asparagine, urea, and a number of cyclic ureides. In addition, peptides, as yet unidentified, have been shown to be present. The quantity of these constituents varies with species. Out of 110 species of dicotyledonous species investigated, Bollard (13) found that glutamine was a major constituent in 87 species, and asparagine in 34. In alder (*Alnus glutinosa*), and certain other species, citrulline was found to be the chief constituent, and in maple (*Acer pseudo-platanus*) and other species, allantoin and allantoic acid. Undoubtedly, many other substances remain to be identified.

Much less definitive information is available on the translocation of nitrogenous compounds in the phloem. Certain of the very serious difficulties inherent in such studies have been indicated above. In connection with the discussion of the studies of Maskell, Mason, and Phillis on nitrogen transport in the cotton plant (page 506), it was pointed out that the redistribution of nitrogenous compounds from the leaves was considered to occur primarily by way of the phloem. Further study of this problem showed that the downward movement of these compounds was against a gradient of nonprotein organic nitrogen in the bark, consisting mainly of amide nitrogen, and to a much lesser extent of amino nitrogen. The negative gradient of the amide nitrogen plus amino nitrogen fraction individually was found to exceed the negative gradient of the total organic nonprotein nitrogen fraction, hence the existence of a positive gradient of unidentified nitrogen compounds ("residual N") was inferred. On the assumption that the actual translocatory compounds move along positive gradients within the sieve-tube system (page 533), then the identity of the mobile nitrogenous compounds is to be sought in the "residual N" fraction. Such an assumption, however, is not mandatory, and other interpretations are possible. Asparagine and glutamine have been commonly accepted as important translocatory forms of nitrogen in the phloem, but the above data, while not excluding this possibility, suggest that these amides are more likely temporary storage forms of nitrogen. Maskell and Mason found that asparagine accumulated in great abundance in the phloem ray cells.

Chromatographic analyses of the nitrogenous constituents of sieve-tube exudate have been made by a few investigators (72). In general, these studies have revealed the occurrence of a considerable number of amino acids, including the amides of aspartic and glutamic acids. Although only fragmentary data are presently available, it appears that the amino acid content of the sieve-tube sap varies significantly with the developmental stage of the plant. In *Salix* spp., for example, the greatest variety and abundance of these compounds were found during the stage of rapid leaf expansion and at the termination of the growing season, during the period of leaf senescence. At other times, comprising the greater extent of the growing season, only small amounts of aspartic acid, glutamic acid, and their amides could be detected. In white ash, for example, the total concentration of amino acids in the sieve-tube exudate usually amounted to less than $0.001 \, M$ (133).

For collecting the sieve-tube sap, Kennedy and Mittler (72) devised an ingenious technique using severed aphid stylets. The feeding aphids were cut off the plants without dislodging the stylet tips from the sieve tubes into which they penetrated. The exuding sap was then harvested continuously for periods of up to 4 days by sleeving the end of a capillary collecting tube on to the stylet stump. This technique may eliminate some of the objections which have been raised to the conventional incision techniques for obtaining phloem sap reviewed by Moose (94). It is not known, of course, to what extent sieve-tube exudate is a valid sample of the total translocate in the phloem tissue, nor to what extent such exudate is a valid sample of the translocate in normal (untapped) sieve tubes. Additional consideration will be given to this problem in connection with the discussion of the mechanism of translocation.

VIII. Mechanism of Translocation

A. Basic Considerations

In view of the fact that the preponderance of organic solutes are translocated in the phloem, we shall be concerned in the present section primarily with the mechanics of transport in this tissue. The mechanism of *xylary* transport of these compounds presents no problems peculiar to organic solutes per se; the mechanism of inorganic solute transport in the xylem, as described in Chapter 6 in this volume, applies, therefore, to the transport of organic solutes as well.

As background for evaluating the various hypotheses which have been proposed to account for phloem transport, it will be helpful to review some of the pertinent facts regarding the movement of solutes in this tissue.

1. Essentiality of Living Cells

One of the important distinguishing features separating the phloem transport mechanism from that of xylem transport is the necessity of living cells. As is well known, killing the cells in a localized zone of the stem or petiole, usually by scalding the tissues with steam, boiling water, or hot wax, is an effective method of blocking the transport of phloem-limited solutes (104). Although with radioactive tracer techniques, it is possible to show that small amounts of solutes may cross the phloem-block (70), it is probable that this translocation occurs mainly through the xylem.

It should be noted, however, that the usual methods employed in killing the cells result in such severe destruction of the phloem tissue as to be tantamount to actually "ringing" this tissue. The significance, therefore, of the observed dependence of the transport mechanism on living cells in the phloem is somewhat obscured. More relevant to the basic problem are experimental procedures which involve controlled or, at least, less drastic denaturation of the phloem cells as, for example, by metabolic inhibitors, gamma radiation, and sublethal temperatures. Certain of these experiments are described below.

2. Specificity of Sieve Elements as Conducting Channels

The view that longitudinal conduction in the phloem occurs principally through the sieve elements is supported by several lines of experimental evidence:

(a) Schumacher (111) observed that immersing the blade of geranium (*Pelargonium* × *hortorum*) leaves in a dilute solution of eosin materially reduced the amount of carbohydrates and nitrogenous substances translocated from such leaves. Microscopic observation of the phloem tissues revealed that callose plugs had formed on the sieve plates at a considerable distance from the treated leaf but no injury was evident to the other phloem cells (companion cells and phloem parenchyma), as indicated by the normal streaming of their cytoplasm.

(b) Studies with radioactive C^{14} (Section VII) have led to the conclusion that sucrose is the major transport sugar. On the other hand, glucose and fructose appear to be translocated only to a minor degree, if at all. Concomitantly, the sieve-tube sap of all species thus far analyzed by chromatographic procedures has been found to be free of these hexoses.

(c) Schumacher (112) has directly observed (by fluorescence microscopy) that fluorescein supplied to leaves is translocated fastest and over the greatest distance specifically in sieve tubes.

3. Simultaneous Bidirectional Transport

It is clear from the discussion of distribution patterns in plants (Section VI) that organic solutes are translocated in opposite directions in the stem simultaneously. The major question at issue is whether this pattern of simultaneous bidirectional transport is the composite effect of oppositely directed transport in different sieve elements or whether bidirectional movement can actually occur simultaneously in the same sieve elements.

None of the experiments bearing on this problem is sufficiently critical to permit an unequivocal answer. Palmquist (101) studied the simultaneous movement of carbohydrates and fluorescein in the phloem system of bean (*Phaseolus vulgaris*) leaves and was able to show that fluorescein would move from the terminal leaflet to a lateral leaflet of a trifoliate leaf simultaneously with the movement of carbohydrates in the reverse direction. The petiolules of bean leaflets have only a single vascular bundle, and microscopic observation revealed that the fluorescein, which can be detected in minute concentrations by its yellowish fluorescence in ultraviolet light, was limited to the phloem. Although these data demonstrate that simultaneous bidirectional translocation can occur through a single vascular bundle, it is possible, of course, that different sieve tubes and phloem cells were involved in each direction. Furthermore, the translocation time required for detectable quantities of carbohydrate and fluorescein to move from one leaflet to the other was so long (1–2 days) that it is problematical whether or not transport mechanisms of major importance in the translocation process were actually measured by this technique. Chen (17) restudied the problem by means of the radioactive isotopes of carbon and phosphorus, but the results were similarly indefinite. Recently, Bauer (7) has reported an extensive series of studies which demonstrate that the movement of fluorescein is unidirectional in any given sieve tube, although not necessarily in the same direction for all sieve tubes even of the same vascular bundle. These observations were carried out on excised petioles of geranium leaves mounted as shown in Fig. 14. The petiole was notched at the center to the depth of the central vascular bundle, and fluorescein at a concentration of 0.1% in 0.8% gelatin applied at this point. The vials at either end contained water or sucrose solutions of various concentrations. Under certain conditions the fluorescein moved to both ends of the petiole simultaneously; in most such instances, it was clearly evident that the movement was not counterdirectional in the same sieve tubes. However, the transport

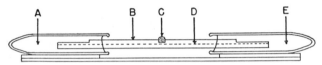

Fig. 14. Experimental method for studying the influence of certain factors on the direction of fluorescein translocation in sieve tubes. (*A*), water (or solution); (*B*), petiole; (*C*), fluorescein in gelatin; (*D*), central vascular bundle; and (*E*), solution (or water). From Bauer (7).

velocity was very slow, in the range of 5–8 mm per hour, only about one-hundredth the velocity which frequently characterizes the translocation process. Furthermore, in an open-end system, such as excised petioles, counterdirectional streaming may be partially attributable to a flux of hydrostatic tension gradients complicated by varying degrees of permeability to diffusion and flow of water in the different regions of the petiole.

In contrast to the results obtained by Bauer, Schumacher (115) has observed that bidirectional movement of fluorescein can occur in the same sieve tubes in intact stems systems. In view of the technical excellency with which the experiments by Bauer and Schumacher have been carried out, the divergence in these results attests to the extreme complexity of this problem.

4. Simultaneous Unidirectional Transport

Another approach to the mechanics of solute movement in sieve elements is to measure the respective transport velocities of various solutes moving simultaneously in the same direction.

One such study is that by Schumacher (113, 115), who studied the concurrent transport of two different fluorescent dyes, fluorescein and esculin, in sieve tubes. Schumacher observed that if a small amount of fluorescein was supplied to a leaf, the compound would advance through the sieve tubes of the leaf and into the sieve tubes of the stem for a certain distance and then stop. If the supply to the leaf was then renewed, the translocation "front" would again advance for a certain distance and again stop. This *"Schiebewirkung"* or "shoving action" could only be effected by supplying to the leaf the same kind of dye as was used to establish the original gradient; if fluorescein was used first and then esculin supplied, the fluorescein "front" would remain stationary, and vice versa. Each of these substances moved independently of the other by a mechanism analogous to an adsorption-displacement wave. Although these data suggest that different solutes may move with different velocities through the sieve tubes, it is problematical to what extent the transport of these fluorescent tracers is comparable to that of the native translocatory materials.

Zimmermann (133), taking advantage of the fact that the sieve-tube exudate of the ash (*Fraxinus americana*) contains several sugars as well as the sugar alcohol, mannitol, at fairly high concentration levels, tapped the phloem of mature trees at heights of 1, 5, and 9 meters above the ground and determined the relative change in concentration of these substances as a function of translocation distance. Although all of these substances decreased in concentration from the 9-meter to the 1-meter sampling levels, the respective rates of decrease varied independently. Furthermore, the pattern differed with individual trees. In certain trees, for example, stachyose decreased the most and sucrose the least; in others, the order was reversed. More significantly, Zimmermann observed that immediately after leaf fall, the concentration gradients for stachyose and raffinose became strongly negative (i.e., increasing in the direction from the leaves to the roots) while the gradients for sucrose and mannitol remained positive. Zimmermann attributes, in part, the independently varying gradients of these several translocatory compounds to varying degrees of hydrolysis of the sugars within the sieve tubes during transit.

If "tracer" amounts (usually 5–10 μg) of 2,4-dichlorophenoxyacetic acid are applied to leaves, the export of this compound usually occurs only under such conditions as when normal movement of foods from the leaf may also be expected (44, 92). The concurrent translocation of this compound with organic nutrients in the phloem makes it an ideal indicator in translocation studies (33).

A more extensive analysis of this general problem, using the radioactive isotopes of phosphorus, potassium, and cesium, as well as tritiated water (THO) in order to study the concurrent movement of solutes and solvent, is presented in Chapter 6 in this volume. Additional experiments are reviewed by Arisz (4).

5. Enzymology and Respiratory Level of Phloem Tissue

As previously indicated, the protoplast of the *mature* sieve element is considered by some investigators to be characterized by a condition of moribund metabolic activity. Some of the evidence for this view is given in Section IV. This conclusion is further strengthened by the recent enzymological analyses of Wanner (126) on the sieve-tube exudate of *Robinia pseudoacacia*, in which only a rudimentary glycolytic enzyme system could be shown. Wanner suggested that the sieve-element protoplast loses most of these enzymes during its maturation. Positive tests were obtained for phosphoglucomutase and a number of phosphatases, but negatives tests only for hexokinase, phosphohexoseisomerase, and aldolase. These observations reinforce the conclusions discussed in the previous section pertaining to the relative specificity of sucrose in translocation, and, according to Wanner, account for the stability of the sucrose molecule in transit.

Although Wanner was unable to find even traces of invertase in the sieve-tube exudate, this enzyme is known to occur in other cells of the phloem tissue (74). This fact, along with the consistent absence of ordinary starch in the sieve elements of plants, even in the sieve elements of the mosses (52), and its common occurrence in closely adjacent cells, suggests a high degree of enzymatic specialization between the sieve elements and the other phloem cells.

Histochemical studies of leaf tissues have revealed a significantly higher phosphatase activity (or concentration) in the bundle-sheath cells, companion cells, and young sieve elements, than in the remaining cells of the mesophyll (7, 53, 125). A possible significance of this distribution pattern with respect to the translocation mechanism will be considered below.

Several studies have been reported concerning the comparative rates of respiration in phloem and other tissues. One such study is that by Goodwin and Goddard (56). Tangential slices, 100 μ thick, were removed serially through the bark, cambium, and into the wood of ash (*Fraxinus nigra*) and maple (*Acer rubrum*) trees, and the rates of oxygen uptake in the respective sections measured. Table III presents a part of these data. The phloem section was taken approximately 100 μ from the cambium, and, considering the early time of year at which the sample was taken (May), consisted principally very likely of the previous year's cells. In this region of the phloem the respiratory activity corresponded rather closely to that of the cambium and outer sap wood; maximal activity occurred in the currently differentiating xylem. Recently, much higher rates of respiration for phloem tissue have been reported by Kursanov and Turkina [approximating to 5000 μl O_2 per gram fresh weight per hour; see reviews by Esau et al. (52) and by Willenbrink (128)].

TABLE III

OXYGEN CONSUMPTION OF TISSUES OF *Fraxinus nigra*, COLLECTED IN MAY WHEN LEAVES WERE APPROXIMATELY HALF-EXPANDED[a]

Tissue region	Rate in μl O_2 per hr[b]	
	Per gm wet wt	Per mg N
Phloem	167 ± 20	112
Cambium	220 ± 10	120
Current xylem	313 ± 24	158
Outer sapwood	78 ± 4	130
Inner sapwood	31 ± 4	76
Heartwood	15 ± 2	38[c]

[a] Data of Goodwin and Goddard (56).
[b] Each value is the mean of at least four determinations.
[c] Mostly nonenzymatic oxidation.

Thus far no estimates are available of the respiration rates of the sieve elements specifically, despite the theoretical value of such data. Admittedly, the problem entails extraordinary, perhaps insurmountable, difficulties.

6. Amount and Velocity of Translocation

In considering rate problems, one must clearly distinguish between the rate of translocation, as expressed by the amount (weight) of solute transported in unit time, and the velocity of translocation, as expressed by the lineal distance of movement per unit time. For a given concentration and rate, the velocity varies inversely with the cross-sectional area of the conducting system. The actual area, however, available for translocation cannot be identified with any specific morphological site, and cannot, therefore, be directly measured. It is generally assumed to be equivalent to the transverse area of the sieve elements, and more specifically to just the lumen of these cells. This follows from the rather general agreement that the sieve elements constitute the primary channel of organic translocation in the phloem tissue, but it is highly controversial at present whether their aggregate transverse area is uniformly available in transport. Consequently, velocity measurements computed from rate determinations involve many uncertainties.

Most rate studies have been based on calculations of the dry-weight increase in tubers, fruits, fleshy roots, and other receiving organs. Illustrative of such experiments are those of Colwell (23) on the translocation rate in pumpkins (*Cucurbita pepo*). In Fig. 15 is shown the average *hourly* increase in dry weight per fruit over a growing period

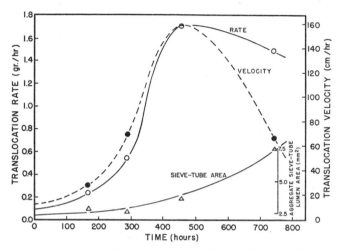

Fig. 15. Rate and velocity of translocation of dry-weight materials into pumpkin (*Cucurbita pepo*) fruit in relation to time from anthesis (time 0). Velocity calculations based on an assumed 20% solution of the translocatory solutes. Sieve-tube area refers to aggregate lumen area of the sieve elements in the fruit stalk. Data of Colwell. From Crafts and Lorenz (34).

of approximately one month. In order to equate this dry-weight increase to the rate of translocation of organic solutes into the fruit, it is necessary to correct for losses resulting from oxidation, condensation, and "backflow" or return translocation, if such occurs, and gains resulting from photosynthesis in the fruit rind and transport of solutes in the xylem. Respiratory losses probably amount to 10–20% of the dry weight, condensation losses to about 5%, and the remaining corrections are negligible with the possible exception of "backflow," for which no data are available. In the following calculations, none of these corrections has been applied, and the rates, therefore, are minimal. On this basis the maximal rate of translocation attained approximated to 1.70 gm per hour.

The calculated velocities required to account for these observed rates of translocation will depend on the values assumed for the concentration of the translocatory solutes and the cross-sectional area of the conducting system. Accepting this latter value to be equivalent to the aggregate transverse area of the sieve-tube lumina, and the concentration to a 20% solution, the velocities range from about 20 cm per hour to 155 cm per hour (Fig. 15). It should be noted that the aggregate cross-sectional area of the sieve tubes increased throughout the entire growth period (Fig. 15), accounting for the marked decline in the calculated velocity following the peak in the translocation rate.

If it is assumed that the translocate is restricted primarily to the cytoplasm during transit and that the cytoplasm occupies 5% of the transverse area of the sieve tubes, then the required velocities would be twenty times greater, or about 3100 cm per hour for the peak value $[(1.7 \text{ gm/hr} \div 1.5 \text{ gm/cm}^3 \div 0.036 \text{ cm}^2 \times 5 \times 20 = 3147 \text{ cm/hr}$, where 1.7 is the maximal hourly dry weight increase, 1.5 is the assumed average specific gravity value for the pure translocate, 0.036 cm² is the aggregate cross-sectional area of the sieve-tube lumina, 5 is the approximate dilution factor for a 20% solution, and 20 is the reciprocal of the ratio of the cytoplasmic area to that of the sieve-tube lumen]. A further twentyfold increase must be visualized during transit through the connecting strands of the sieve plates, if it is assumed that the aggregate area of these strands is one-twentieth that of the sieve plates and that the translocate is primarily restricted to these structures when traversing the sieve plates. These calculations show the many assumptions on which velocity determinations derived from rate calculations are based. Calculations based on other plant materials are given by Crafts (26–28), Clements (21), Mason and Lewin (84), and Münch (96) and indicate rates and velocities of magnitudes comparable to those above.

Attempts have also been made to measure the velocity by direct observations. Schumacher (115) and others have carried out such experiments with fluorescein and similar dyes and have obtained values in the range of about 0.5–50 cm per hour. Huber (64) measured the diurnal periodicities in sieve-tube sap of broad-leaved forest trees from tappings at various heights along the trunk and obtained evidence for a concentration "peak" which moved downward at velocities ranging up to at least 100 cm per hour. Measurements with radioactive tracers have given maximal velocities of the same order of magnitude (Chapter 6 in this volume). It should be noted, however, that because of the logarithmic decrease in the concentration of the labeled translocate along the conducting system, the time required for the translocation "front" to pass between two counter tubes spaced a known distance apart along the stem will depend on the definition of the "front," that is, whether it is considered as equivalent to an activity of 1 cpm (counts per minute), 100 cpm, 1000 cpm, etc. Hence, the velocity becomes a function of the sensitivity of the measuring instrument. It is

likely that this exponential decrease in concentration of the translocate
with distance results mainly from loss of the translocate from the conducting system during transit.

The maximal velocities which have been recorded by direct measurements are of the order of 100–200 cm per hour. On this basis, it
appears that the entire cross-sectional area of the sieve tubes must be
used in transport, inasmuch as correction to any significantly smaller
area would require higher velocities than have been measured. Much
further work, however, remains to be done on this problem.*

7. Factors Affecting Translocation

Many factors, both internal and external, are known to influence
translocation. Certain factors affecting the direction of translocation
were discussed in Section VI. In the present discussion, we shall consider primarily the factors which influence the rate or velocity of this
process.

a. Temperature. Temperature-rate curves for translocation are of
special interest in view of the generally accepted conclusion that organic solute translocation occurs primarily in mature sieve elements,
the protoplasts of which, as previously described, appear to be characterized by a relatively inactive metabolic state. Figure 16 depicts the
rate of carbohydrate translocation in bean (*Phaseolus vulgaris*) plants,
showing an optimum temperature range of 20–30°C, the rate diminish-
ing at temperatures above and below this general range. Other experiments reveal a similar optimum curve (11, 69, 112, 122). It is evident,
therefore, that translocation is influenced by temperature much in the
same manner as are most other physiological processes. The evidence
which opposes this generalization has been reviewed by Went and
Hull (127) and Hull (65).

The data of curve *A* in Fig. 16 were obtained by varying the temperature of only a restricted zone of the conducting system, specifically
the petiole of the "supply leaf," the rest of the plant being maintained
at 20 ± 1°C; translocation rates were inferred from the rates of stem
elongation over a period of 135 hours, under conditions where growth
was primarily limited by the amount of carbohydrates transported to

* According to Nelson *et al.* (*Can. J. Bot.,* in press) longitudinal transport of
photosynthetically assimilated C^{14} may comprise several components. One of these
is very rapid (order of 2000 cm per hour, *Can. J. Biochem. Physiol.* **36,** 1277–1279,
1958) and is thought to be a means of translocation through living cells which results in accumulation in the root. Another, but slower component (order of 100 cm
per hour), results in the downward spread of C^{14}-labeled material as this is accumulated in the cells of the stem. Thus it may be the manner and site of eventual
accumulation of the solute which controls the pace of its longitudinal translocation.
Ed.

the stem from the supply leaf, the blade of which was immersed in a sucrose solution. More critical experiments using labeled sugars have not as yet been carried out. The data of curve B were obtained by maintaining entire plants at the specified temperatures and measuring the respective gains and losses in dry weight in the various plant organs after an interval of 14 hours. As may be noted from the graph, the inhibitory effect of temperatures below 20°C is much greater if the

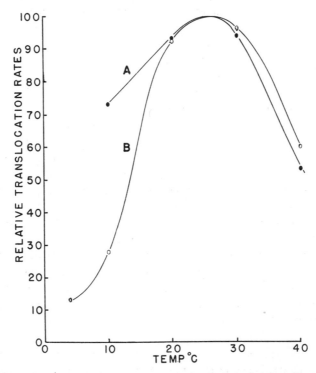

FIG. 16. Rate of carbohydrate translocation from leaves in bean plants in relation to temperature. Curve A. temperature of petiole varied. Curve B. temperature of entire plant varied [calculated from data of Hewitt and Curtis (61)]. From Swanson and Böhning (120).

entire plant is subjected to these temperatures over a short time interval than if only a restricted zone of the conducting system is subjected to the same temperatures for a much longer time period. A similar relationship was observed by Crafts (27); cooling entire cucumber (*Cucumis sativus*) plants to 17°C was found to reduce the rate of exudation from the stems considerably more than local chilling of the stems alone to temperatures as low as 1–4°C.

The agreement in the data between curves A and B in Fig. 16

at above-optimal temperatures is probably for the most part fortuitous. Experiments have shown that the inhibitory effect of high temperatures increases with time (120); hence, a 40°C petiole temperature is, with sufficient time, fully as inhibitory as maintaining the entire plant at this temperature for a much shorter time period. This time factor effect may also be observed at low temperatures, the effect, however, being in the opposite direction, that is, the inhibitory effect diminishes with time (18, 120). The divergence between curves *A* and *B* at low temperatures is perhaps partly attributable to this effect as well.

Extrapolating the curves in Fig. 16 to the zero ordinate suggests a maximal temperature for translocation of 55–60°C. The actual value would depend, of course, on the time factor.

Temperature also exerts an effect on translocation from "donor" to "acceptor" regions of the plant, even when applied outside the direct line of transport. For example, if hypocotyl regions of bean (*Phaseolus vulgaris*) plants are maintained at a series of different temperatures, the transport of sugar from the primary leaves to epicotyl is influenced in essentially the same manner as if the temperature-variable zone were in the direct line of transit (11). These data were obtained by the indirect procedure described above for the petiole-temperature studies. This "remote-control" effect has also been noted by Day (43), who showed that killing a section of the hypocotyl reduced the amount of 2,4-dichlorophenoxyacetic acid translocated from the leaves to the epicotyl region, and by Linck (77), who demonstrated that chilling the pod at the first bloom node in peas (*Pisum sativum*) (P 1 in Fig. 9) to a temperature of 8–10°C (the rest of the plant being at a temperature of 22–26°C) greatly restricted the amount of P^{32} translocated from the leaf at that node to other parts of the plant, despite the fact that because of much less translocation to the pod under these conditions, considerably more translocate should have been available for distribution to the rest of the plant. It is evident that temperature-rate relationships are highly complex, and are compounded of the separate effects of temperature on the metabolism of the supplying, conducting, and receiving tissues respectively.

b. Metabolic inhibitors. Because of the theoretical importance in identifying the specific phases of metabolism most closely associated with the transport mechanism, considerable interest attaches to studies employing selective metabolic inhibitors. The problem is complicated, however, by the consideration that such compounds, when effective in inhibiting translocation, exert this effect in two ways: (a) by directly influencing the metabolism of the conducting cells per se, and (b) by affecting the processes associated with the primary accumulation of translocatory solutes into the phloem system (mainly in the leaves) and the subsequent absorption of these solutes by tissues adjacent to the conducting cells, especially in the active receiving areas, such as growing shoot tips. Because metabolic inhibitors, such as fluorides, fluoroacetates, substituted phenols, etc., may themselves be translocated, topical application of these compounds to restricted zones of the stems or petioles of plants does not assure that the primary effect observed results from metabolic inhibition in cells of the conducting tissues. As a result, it is extremely difficult to assess separately the effects of inhibitors on these respective aspects of the translocation process. In general, however, the data appear to indicate that translocation rates may be more sensitively affected by modifying the metabolism of leaves, fruits, and growing tips (that is, the supplying and receiving tissues) than that of conducting cells. Such a conclusion is in accord with the observation that the sieve-tube protoplast (at least the exudate from these elements) is characterized by a rudimentary enzyme system.

One of the more extensive studies of these problems is that by Kendall (70, 71). Buffered solutions of various inhibitors, each at several concentrations, were injected into the central cavity of the petioles of bean (*Phaseolus vulgaris*) leaves, and the amount of translocation from the blades of the treated leaves (using the P^{32} tracer technique described in Chapter 6) was then measured after a standard period of 3 hr. Control plants were included in which the inhibitor was injected into the petiole of the primary leaf opposite to the P^{32}-supply leaf and the amount translocated from each treated leaf was measured as a percentage of the total P^{32} absorbed. Comparisons were thus available by which the direct effect of the inhibitor on the phloem cells through which translocation occurred could be evaluated. On this basis, only 2,4-dinitrophenol and sodium fluoride significantly inhibited translocation by directly affecting the metabolism of the conducting cells. Both of these compounds are considered to inactivate certain phosphoenzymes. In a highly tentative way, therefore, a possible role of adenosine triphosphate as one of the functional components in the translocation mechanism is indicated. Results with sodium fluoroacetate, idoleacetic acid, triiodobenzoic acid, and 2,4-dichlorophenoxyacetic acid were negative, and with sodium arsenite, inconclusive.

Somewhat different results were obtained by Vernon and Aronoff (124) with 2,4-dichlorophenoxyacetic acid. Although the velocity of sugar translocation from soybean leaves was not affected by this compound, the quantity translocated was reduced very considerably. In these experiments, however, the 2,4-dichlorophenoxyacetic acid was applied to the plants 24 hours prior to the start of the translocation measurements, and the effect of the inhibitor was probably systemic, therefore. By contrast, exposing a 4-cm zone of the stem to hydrogen cyanide gas at a partial pressure calculated to give a concentration of 5×10^{-3} M in cyanide within the stem tissues (assuming actual penetration of the cyanide) did not adversely affect the amount of translocation through this zone, although the evidence appears clear from earlier studies that a short-period exposure of the growing tips of tomato plants to this gas markedly depressed the translocation rate from untreated leaves (93). These observations, as well as the recent work of Bauer (7) with fluorescent tracers, thus further reinforce the view that translocation is a more sensitive function of the metabolism of supplying and receiving areas of the plant than of conducting tissues. Possibly contradictory to this view, however, are the temperature-rate relationships previously discussed, the observations of Curtis (38) and Mason and Phillis (88) that phloem translocation may be greatly retarded by restricting the supply of oxygen to a localized zone of the stem or petiole [this relationship, however, has not been confirmed in more recent experiments (128)], and the reversible and irreversible blocking of translocation from leaf blades by various enzyme inhibitors applied directly to the petioles (71, 128). On the basis of such data, some investigators consider that energy-supplying processes in the protoplasm of the sieve element participate directly in actuating the movement of materials through these elements. Such a view, of course, predicates the significance of the physiological status of the conducting cells as an important determinant affecting the rate and velocity of translocation and contrasts markedly with the view ascribing a more passive role to these cells. The critical analysis of this problem, however, is confounded by the fact that a protoplasmic structure which permits translocation can obviously be varied only within limits, and, without question, a certain amount of metabolically derived energy is required in servicing and maintaining this structure. By either view, therefore, the process of translocation should be subject to metabolic control at the level of the conducting cells.

Experiments have shown that cytoplasm is less easily damaged by ionizing radiation than is the nucleoplasm (110). It is perhaps possible that advantage can be taken of this differential effect to distinguish more clearly between the role of the sieve elements (which are enucleate) and the other cell types in the phloem in translocation. Sax (109) has reported studies in which phloem blocks were induced in trees by irradiating a stem segment with X-rays, the dosage being sufficient to suppress cell division and prevent the renewal of the phloem elements. Whether the ensuing restriction in transport resulted simply from failure of replacement of the sieve elements or from more direct effects of radiation damage is not presently known.

c. *Concentration gradients.* An extensive analysis of concentration gradients in relation to the rate and direction of translocation is given in the various papers by Mason, Maskell, and Phillis [see review by Mason and Phillis, (89)]. These studies permit the following conclusions: (a) The movement of sugars in the phloem *of the stem* (of cotton) is always in the direction of a positive gradient in sugar concentration, i.e., from a region of higher to a region of lower concentration of sugar. (b) The rate of longitudinal movement of sugars in the sieve tube is highly correlated with, and therefore, approximately proportional to, the observed gradient of sugar concentration in the bark. (c) When the stem is partially ringed so as to constrict the channel of transport to a narrow width for a short distance, the total amount of sugar translocated across the "bridge" is diminished, but the rate of movement per unit cross-sectional area of the bridge is increased. (d) The increase in rate across the constriction is proportional to the increase in the gradient of sugar concentration. (e) The movement of sucrose from the mesophyll into the phloem of the leaf is against a negative concentration gradient which exists between these respective tissues. (f) The direction of organic nitrogen transport is not correlated with the observed concentration gradients for the various nitrogenous fractions (however, it should be noted that the amount of nitrogenous compounds translocated in the phloem is very small relative to that of carbohydrates, and any positive gradient of the translocatory nitrogenous compounds, if actually present, would be readily obscured by the stationary gradients of the storage nitrogen fractions). Mason, Maskell, and Phillis concluded that the transport of sugars within—but not into—the phloem system follows a diffusion pattern, except as regards rate. The implications of these data with respect to the translocation mechanism are discussed below in connection with the hypothesis of *activated diffusion.*

Rather extensive data are also available on sugar concentration gradients within the sieve-tube systems of various species of trees (133). In general, it appears that the direction of sugar translocation within the sieve tubes is along a gradient of decreasing total sugar concentrations. This relationship also applies to the respective concentration gradients for the individual sugars during the growing season, but, as noted on page 523, the gradients of certain sugars become inverted immediately after leaf fall. We shall return to a consideration of these data in connection with the discussion of the *pressure-flow* mechanism.

d. Mineral deficiencies. Few studies have been made of the relationship of mineral deficiencies to translocation rates in the phloem. Rohrbaugh and Rice (107) observed that the translocation rate of C^{14}-labeled 2,4-dichlorophenoxyacetic acid in tomato plants was markedly reduced in phosphorus-deficient plants. A similar relationship has been noted for the rate of phloem translocation of P^{32} (73). Both P^{32} and 2,4-dichlorophenoxyacetic acid appear to function as "tracers" for organic translocation, and the observed effect of phosphorus levels on the translocation rate of these compounds probably holds, therefore, for the translocation of carbohydrates and other foods in general. The impairment of the translocation process in mineral-deficient plants probably results primarily from indirect effects; the partial effect exerted directly on the transport capacity of the conducting cells is not known and would be difficult to evaluate.

A relatively specific role of boron in the translocation of sugars is indicated by various studies (55, 116). The inclusion of 10 ppm (parts per million) of boron in a solution of C^{14}-labeled sucrose supplied to the leaves of vigorously growing, boron-sufficient, tomato plants considerably enhanced the uptake of this sugar and its subsequent distribution to other parts of the plant. Furthermore, sucrose translocation was considerably reduced in plants in an incipient stage of boron deficiency, even though no morphological symptoms of this deficiency were yet apparent. It is known that in more advanced stages of boron-deficiency the phloem tissue breaks down and becomes nonfunctional. Gauch and Dugger concluded that boron-deficiency symptoms (necrosis of meristems, abscission of flowers and fruits, high ratio of leaf:stem sugar concentrations, etc.) are in large measure simply an expression of sugar deficiency resulting from an impaired translocation of carbohydrates.

The mechanism by which boron enhances the rate of sugar translocation is not known. Various hypotheses have been proposed; for example, the known avidity with which borate ions complex with polyhydroxy compounds, such as sugars, suggests the formation of a sugar borate complex, which may be more readily transportable across membranes (55). It is more likely, however, that the relationship is purely indirect, the borate effect being mediated through the essentiality of boron to maintenance of cellular activity in apical meristems (117). Such growing regions are known to affect markedly the magnitude and direction of translocation of carbohydrates from leaves, as has been previously discussed in Section VI. On this basis, it may be expected that calcium would have similar effects on carbohydrate distribution.

e. Hormones. The possible role of auxins and other hormones in accounting for certain aspects of the observed translocation patterns in plants was briefly referred to in Section VI. Although the monopolizing effect of active meristematic regions and rapidly growing fruits on the translocation of food materials is well recognized and is considered to be mediated by growth hormones, the mechanism of this relationship is not understood. Linck (77) infiltrated the hollow interior of pea pods during an early stage of their growth with triiodobenzoic acid, an antiauxin, and noted a significant reduction in translocation to the treated pods at a concentration of 100 ppm (compared to water-infiltrated pods; water infiltration alone gave considerable reduction over untreated controls). Treatments with indoleacetic acid and phenylbutyric acid, however, gave inconclusive results due to high variability. Additional studies of certain plant hormones in relation to the translocation process have been made by Choudhri (19). The theoretical and practical importance of investigations of these relationships is obvious, and this area of research is greatly in need of more extended systematic work.

B. HYPOTHESES

Four principal hypotheses have been proposed to account for the mechanism of phloem translocation: (*1*) pressure flow (*2*) protoplasmic streaming, (*3*) interfacial flow, and (*4*) activated diffusion. These hypotheses are not mutually exclusive, and the distinctions at certain levels of interpretation are not on all points well defined. It is possible for the mechanisms visualized in these hypotheses to function simultaneously in the same cell.

1. Pressure-Flow Hypothesis. This theory of translocation, formulated principally by Münch (96) postulates a unidirectional flow of water and solutes through the sieve elements of the phloem under the driving force of a turgor-pressure gradient. The osmotic principles on which this theory is based can be most easily clarified by reference to the system illustrated in Fig. 17, in which two osmometers, *A* and *C*, are joined by the connecting tube *B*. It is assumed that the walls of the osmometers are permeable to water, but not to solutes, and that no differentially permeable membrane separates the solution in *A* from that in *C*.

The critical quantity required for flow in such a system is a *turgor pressure gradient*, and such a gradient can be established in either of two ways: (a) as a result of a difference in total solute concentration between *A* and *C*, or (b) as a result of a difference in plasticity or extensibility of the walls of the two osmometers. For example, assume first that the osmotic pressure or potentials of the solutions in A and B are respectively 20 and 15 atm and that the walls are uniformly elastic. Under these conditions, the turgor pressure in A will approach a limiting value of 20 atm if the osmometers are immersed in water, and this pressure will, by virtue of the absence of any differentially permeable membranes between A and C, be uniformly transmitted throughout the entire system. Hence at C the turgor pressure (TP) will exceed the osmotic pressure (OP) of this solution, resulting in a diffusion-pressure-deficit (DPD) value of -5 atm for the water (DPD $=$ OP $-$ TP $= 15 - 20 = -5$). That is, the water in C will have a diffusion pressure 5 atm greater than that of the water on the outside of the system, and, accordingly, will diffuse (or be forced)

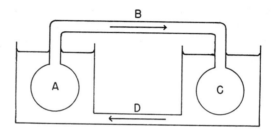

Fig. 17. Diagram of an osmotic system in which a pressure-actuated mass flow of solution will occur.

across the differentially permeable membrane which constitutes the wall of osmometer C. Thus a flow of solvent is induced from A to C through B, the solute load being carried passively in this stream. In such a system it is only required that the *total* concentration gradient be in the direction indicated; the concentration gradients for specific solute molecules may actually be negative, and such molecules will be moved against their concentration gradients so long as the total solute concentration gradient remains positive. Furthermore, the flow may, under the conditions of (b) above, be counterdirectional to the total concentration gradient as well. If the walls of A are sufficiently more extensible than those of C, the tugor pressure gradient will be positive in the direction C to A, and hence the direction of flow will be against the concentration gradient.

From one point of view, therefore, the pressure-flow mechanism may be regarded as a noncarrier active-transport system in which the movement of solutes is facilitated by a solvent flow or diffusion drag force.

In applying this concept to the transport system in plants, the physical model as described represents a system or series of sieve elements, "*A*" corresponding to the "supplying" end (mainly the sieve elements in the minor veins of the leaf) and "*C*" to the "receiving" end, or the region of removal of translocatory solutes from the transport system. The sieve element is considered to meet the structural requirements of this physical model on the basis that the connecting strands, or perhaps the sieve areas in general, are, on the basis of certain information, permeable to both solute and solvent.

In earlier considerations of this hypothesis the turgor pressure gradient was considered to extend from the chlorenchyma cells of the leaf to the sieve tubes in the veins, and thence to the receiving cells (96). Münch visualized a pressure flow of solution from the photosynthesizing cells to the sieve tubes by way of the plasmodesmata, which apparently interconnect most, if not all, living cells, and a similar mechanism for removing the solutes and water at the other end of the system. It is now generally recognized, however, that translocation from the leaf chlorenchyma into the sieve elements may occur against a concentration gradient (102, 106), and that under the conditions which prevail in the leaf, it is not likely that the turgor pressure gradient would be counterdirectional to the concentration gradient. Hence it is inferred that the movement of translocatory solutes into and out of sieve elements is principally brought about by *active metabolic transfer processes* (7, 125) (cf. Chapter 4 in this volume). As pointed out above (page 524), studies of the histochemical distribution of enzymes in the mesophyll reveal strong phosphatase activity localized in the vein sheath cells and in the companion cells. This evidence suggests that an active carrier system involving successive phosphorylation and dephosphorylation may effect the transport of sucrose across limiting membranes. The fact that phloem transport rates are especially sensitive to factors affecting the metabolism of leaves and growing parts is further presumptive evidence that metabolic transfer processes play a quantitatively significant role at these respective sites in the translocation process.

In a physical model, under certain defined conditions, solute molecules of different kinds as well as those of the solvent would move at the same velocity. Within sieve tubes, however, such a condition seldom if ever obtains. Differential and variable rates of entry of various translocatory solutes into and out of the sieve tubes, as well as varying degrees of resistance to movement of the solute molecules through the sieve plates (on the average about 3 to 5 of these must be crossed per millimeter of flow) and the dispersed "slime" proteins in the vacuole, if present, and through cytoplasm, make it most improbable that different species of solute molecules would move with the same velocities. The picture which is emerging, therefore, as pointed out by van Overbeek (100), is that of a "chromatogram," in which the solvent front moves ahead of the solutes, the different solutes having different "R_f" values. Such a view, however, is not incompatible with the pressure-flow mechanism, if we preserve the essential features of the concept of an osmotically activated flow of water which facilitates the movement of solutes unidirectionally by means of solvent drag.

In the preceding discussion, considerations of the pressure-flow hypothesis have been along essentially theoretical lines. The major tangible evidence in support of this theory derives from the well-known observation that sap, containing a high sugar content (10–30%), exudes from the sieve tubes when an incision is made into a stem (Fig. 11). This has been shown to occur in a considerable number of species of plants (29), although it is by no means a universal phenomenon. Various species of ash (*Fraxinus*) trees show this phenomenon exceptionally well; in Italy, the sieve-tube exudate of ash trees has been collected for centuries as a source of food by tapping procedures roughly comparable to harvesting latex from rubber trees (64).

There appear to be two distinct phases in the exudation of sap from sieve tubes: (a) the "expulsion" phase, which persists only for a second or less, and (b), the "bleeding" phase, in which the exudation rate is much slower and the exudate becomes progressively more and more dilute (133). The initial phase undoubtedly results from the sudden release of elastic tensions in a system which is normally characterized by a high turgor pressure. The second phase is purely osmotic; the reduction in turgor pressure which results from cutting, and the corresponding increase in the diffusion-pressure deficit of these cells, causes water to diffuse in from adjacent cells, and induces a flow of solution to the open ends. This phase of the phenomenon appears to be very similar to the mechanism involved in the flow of latex from *Hevea brasiliensis*, the rubber tree, both types of exudation showing a progressive decrease in concentration with time resulting from osmotic dilution (54). In certain species, notably of the Cucurbitaceae, the flow of sieve-tube sap is maintained only if fresh cuts are made at frequent intervals, presumably because of plugging of the sieve-tube ends by coagulation of the exudate, callose formation, or slime-plug formation (Section IV).

Further tangible evidence in support of the pressure-flow hypothesis is provided by the data on concentration gradients in trees and the gradient inversions for certain solute constituents which occur in autumn when the leaves begin to abscise (page 523). The latter phenomenon is readily explained in terms of the physical model discussed above. As previously pointed out, the direction of flow will be, under certain defined conditions, in the direction of the *total* concentration gradient, and will result, therefore, in certain species of solute molecules being moved against their respective concentration gradients, if the rate of supply of these substances to the sieve tubes in the leaves or their rate of removal "downstream" is differentially restricted (133).

In the physical model (Fig. 17), the water expressed from osmometer C returns through D (corresponding to the xylem) to compartment A', whence it may again be cycled through the system. In plants, however, such recycling of water does not occur, at least to any appreciable magnitude, and this fact has been used to reject the pressure-flow mechanism as a valid transport system in the sieve elements (21, 46). Actually, recycling is not a necessary corollary of such a system; water will not be expressed from C (Fig. 17) if this osmometer is sufficiently extensible (in plants, "extensibility" would correspond mainly to growth). The fact remains, however, that on the basis of the pressure-flow mechanism, the amount of water theoretically supplied via the phloem to rapidly growing and assimilating parts is of substantial proportions, and hence the dependency of these parts on a water supply through the xylem should be considerably reduced. For example, the water content of many kinds of immature fruit is in the range of 85–95% of their fresh weight, whereas the water content of the sieve-tube exudate is seldom less than 70–75% and may be as high as 90% at times. The finding of Döpp (46), therefore, that intercepting the water supply to fruits (by severing the xylem in the immediate vicinity of the fruits) completely inhibited their further growth is of considerable interest. A part of these results can be attributed to injury to the phloem and to a restriction in the supply of the xylem-translocated amino acids, other nitrogenous compounds, and various mineral ions, but it is also probable that, even under normal conditions, the gain of water by growing fruits through the phloem, although substantial, is never sufficient to maintain the necessary degree of turgor required for growth, especially when allowance is also made for transpirational losses.

Although the pressure-flow mechanism of transport has proved to be an effective hypothesis, which is capable of collimating many divergent facts, there are a number of serious objections to it: (a) The mechanism, to the extent that it operates in plants, is functional only in sieve tubes (or sieve-element systems). Hence, the mechanism of solute translocation to and from this system is not resolved by this theory. (b) The calculated pressure gradient required to force a sugar solution by bulk flow through the sieve tubes at a rate sufficient to ensure the observed transfer of dry matter is large, of the order of 20 atm per meter (26). A similar or greater value probably obtains for the movement of solution through the highly hydrated walls of the sieve tubes as well (118). It should be recognized, however, that many simplifying assumptions are involved in such calculations, and the computed values may be in serious error. Measurements of actual osmotic-pressure gradients in sieve tubes of trees are more of the order of 0.3 atm per meter (133), and it is most unlikely, therefore, that a turgor-pressure gradient of 20 atm per meter would be possible. (c) The flow of sap from the cut ends of the sieve tubes does not constitute proof that flow occurs normally in intact sieve tubes. Latex flows freely from severed laticifers, but present evidence indicates that there is little or no bulk transport of latex in intact systems (95). Sieve tubes and certain types of laticifers are similar in a number of anatomical and cytological features (51). (d) The sieve-element protoplast is plasmolyzable (Section IV). This fact has been demonstrated unequivocally by Currier et al. (35), Schumacher (114), and others. The possible corollary conclusion that the cytoplasm of these elements is differentially permeable, and is not materially different from that of nucleate cells, raises very serious objections to the pressure-flow hypothesis, and requires, therefore, careful consideration. The major uncertainty at present is the nature of the "differential permeability" of the cytoplasm along the end walls of the sieve-tube elements. The classical technique for measuring the relative impermeability of cytoplasm to a given solute involves measuring the time required for deplasmolysis in a hypertonic solution. Applying this technique to sieve tubes in phloem tissue sections about 15 mm in length by 50 μ in thickness, Currier et al. (35) observed no deplasmolysis in sucrose solutions over periods extending up to 2 hours. Although these observations definitely establish a high degree of impermeability of the cytoplasm along the side walls of the sieve tube to sucrose, the observations are less critical with respect to the cytoplasm along the end walls. Diffusion of sucrose longitudinally through the sieve tubes, even granting relatively permeable cytoplasmic membranes in this direction, would be a slow process at best, and coagulated slime on the sieve plates might further impede the diffusion. Failure to deplasmolyze, therefore, does not necessarily indicate a high, or even moderate, degree of impermeability to solutes of the end-wall cytoplasm. Complicating this difficult area of study is the fact that the capacity of sieve tubes in thin tissue slices to deplasmolyze in hypotonic solutions (a test for osmotic plasmolysis as opposed to injury contractions which simulate the plasmolyzed state) diminishes with time and is readily destroyed altogether by a number of factors, most of which are not presently well understood (35).

If the plasmolytic data are interpreted to indicate a degree of impermeability in the end-wall cytoplasm comparable to that along the side walls, it is difficult to reconcile such a picture with the known fact that a flow of solution can and does occur through sieve tubes, at least when they are severed.

It is evident that the pressure-flow hypothesis has been, and continues to be, a provocative and productive theory. The major defense of this theory in recent years is that of Crafts (33); a detailed critique has been given by Curtis [(39) pages 149–174].

2. Protoplasmic-Streaming Hypothesis

In contrast to the pressure-flow mechanism, which in its modern version postulates an osmotically actuated flow of water through cells, carrying the translocatory substances in solution and facilitating their transfer through more or less permeable matrices by solvent-drag forces, the protoplasmic-streaming hypothesis visualizes moving protoplasm as itself carrying the solutes. Both mechanisms involve *mass streaming*, but the mechanics actuating the flow are distinctly different; as a result, pressure-flow transport can conceivably operate through elements of the sieve tubes in which the protoplasm is at least partially "denatured" (Section IV), but the protoplasmic-streaming mechanism definitely requires metabolically active cytoplasm in each element traversed. Various experiments and observations previously described, however, although not in complete accord, indicate that sieve-element cytoplasm is not as metabolically active as that of parenchyma cells in general, and cytoplasmic streaming has never been observed in mature sieve elements. These observations constitute major weaknesses of this hypothesis, and in recent times it has not received strong support except by Curtis (39); see also Curtis and Clark (40). As is well recognized, however, the enucleate cytoplasm of the mature sieve element is delicate and unstable, and it is possible that the capacity for streaming is destroyed by the manipulative procedures required for observation. The observed velocities of streaming in other cells are, therefore, of interest as a basis for appraising the capacity of protoplasmic streaming as a system of transport. These values have been found to range from about 0.02–0.04 cm. per minute in companion cells and phloem parenchyma cells to as high as 0.5–0.8 cm per minute in the greatly elongated internodal cells of *Chara* sp. and *Nitella* sp. (39). As is apparent from the following calculations, based on an experiment by Crafts (28), these observed velocities are far slower than are required to ensure the known rates of solute transfer through the sieve tubes. For example, in a potato (*Solanum tubero-sum*) plant in which the average daily gain in dry weight of the tuber was 0.89 gm, the aggregate area of the sieve-tube lumina of the stolon amounted to 4.0×10^{-3} cm². Assuming that the volume-equivalent of the pure translocate (mainly sugars) is 0.6 ml, that the area of the sieve-tube lumen occupied by cytoplasm flowing in the proper direc-rection is 5%, and that the volume of the cytoplasm occupied by the pure translocate is 25%, then the required velocity would be: $\frac{0.6 \times 20 \times 4}{4 \times 10^{-3} \times 24} = 500$ cm per hour (not correcting for the time re-quired for diffusion across end walls, which in potato stolons number about 100 per centimeter by sieve tube). Thus the required velocity exceeds observed velocities by factors ranging from 10 to 400.

Although *a priori* one would expect protoplasmic streaming to ac-celerate solute transfer through cells, apparently this relationship does not always obtain. Arisz (3) has reported that rapid streaming in the parenchymatous cells of *Vallisneria spiralis* leaves (induced by wound-ing) did not accelerate the movement of chloride ions through the leaf. Only after prolonged submersion of the leaf in aerated water, during which time the protoplasm became more quiescent, was the transport capacity increased.

Support for the streaming hypothesis derives mainly from the ob-servation that treatments which retard the metabolic activity of stems and petioles also retard phloem translocation. This is true even though the metabolism is depressed in only a very restricted zone, a few centi-meters in length (page 528). The effect of temperature is especially marked, and the velocity of streaming is known to be temperature-dependent (39). Admittedly, however, such observations do not rule out alternative mechanisms of transport. It may be assumed, for ex-ample, that metabolic expenditure of energy is required in maintaining a specialized structure of the protoplasm necessary for the pressure-flow mechanism to operate. On this basis, pressure-flow transport would also be sensitive to metabolic conditions within the sieve elements.

3. Interfacial-Flow Hypothesis

Any substance which lowers the surface tension at the boundary phase between two immiscible liquids (or between a liquid and a gas) will be rapidly distributed at this interface, the direction of movement being from an area of low surface tension to an area of high surface tension. Van den Honert (62) demonstrated this phenomenon by layering ether over water (containing a pH indicator and adjusted to pH 5.8), and introducing at one end of the system a small quantity of potassium oleate solution with an excess of KOH. As the potassium oleate-KOH spread along the interface, an alkaline color reaction developed, providing a visual index for measuring the rate of movement. From such measurements, van der Honert calculated that the observed rate of transport in such a system was 68,000 times faster than would be expected from the actual diffusion constant of KOH in a dilute aqueous solution. Van den Honert tentatively suggested that the high rates of transport reported for phloem translocation might be accounted for in terms of such a mechanism, the phase boundary between the cytoplasm and vacuole in the sieve tubes constituting the specific pathway of transport. There would be no objection, of course, to extending this view to include the phase boundaries within the cytoplasm as well.

Lowering of the surface tension at interfaces induces an actual flow of the surface films at the boundary phase and causes a compensatory flow in the opposite direction in the system. The physical counterpart of this phenomenon in a plant cell would be protoplasmic streaming. By van den Honert's scheme, therefore, protoplasmic streaming is the consequence rather than the agent of transport.

The interfacial-flow hypothesis rests on certain demonstrable facts and may well account for the accelerated transport of certain materials. It is extremely doubtful, however, that there is sufficient surface area available in the sieve-tube cytoplasm to account for the known amounts of material transported.

4. Activated-Diffusion Hypothesis

It has been recognized for many years that known rates of phloem transport are vastly greater than can be accounted for by the process of simple diffusion. The fact remains, however, that there are certain properties of the translocation process which simulate a diffusional mechanism, as, for example, the correlation of the rate and direction of sucrose translocation with concentration gradients in the stem (page 533). Furthermore, a considerable number of measurements based on experiments with radioactive tracers reveal that a plot of radioactivity distribution in the stem as a function of distance of translocation from the source (leaf) is logarithmic (Chapter 6). This is precisely the pattern of translocation that would obtain if translocation were a diffusion process, except that the observed "diffusion constant" is of a totally different order of magnitude. Mason and Maskell (86) calculated from the observed rates of sucrose transport in cotton plants an apparent diffusion constant of 6.97×10^{-2} cm^2 per second. The actual diffusion constant for 1.97% sucrose solution in water at 25°C is 1.81×10^{-6}.* Thus the observed rate of transport in the sieve tubes is of the order of 40,000 times faster than can be accounted for on the basis of the known diffusion constant of sucrose through water, and probably even faster when compared to the slower rates of diffusion which prevail in isolated masses of cellular tissue (see Chapter 4 in this volume). Mason and Maskell also calculated the expected rate at which sucrose molecules would diffuse through the air if sucrose molecules could exist in the gaseous state; the value proved to be almost the same as the apparent diffusion constant, namely 5.77×10^{-2}. Thus the quasi-kinetics of sucrose translocation through the phloem resembles the kinetics of sucrose diffusion through a medium which offers no more resistance than that of air and led the authors to raise the question as to whether this situation could be effected through some special organization of protoplasmic structure. Further treatment of these ideas has been given by Mason *et al.* (87). The concept is sometimes referred to as *activated diffusion;* as visualized by the authors, however, this hypothesis is not committed with respect to the mechanism by which diffusion is accelerated. Its principal value has been to emphasize the fact that the translocation mechanism has certain analogies to simple physical diffusion. Important exceptions, of course, are known, as, for example, the polar transport of sucrose from mesophyll cells to the sieve elements in the vein against a concentration gradient.

In the further development of this hypothesis and related concepts, a clear distinction will be needed as to whether the energy-supplying processes function directly or indirectly in the transport process. At present, although much decisive information is lacking, the known facts regarding the cytophysiology of the sieve-element protoplasts oppose the view of an active-transport mechanism which is coupled stoichiometrically to an energy-supplying process. Active transport of this type would undoubtedly require the direct participation of enzymic surfaces, and the amount of cytoplasm in the sieve element, although imperfectly known, is probably inadequate to account for known rates of translocation, unless incredibly high "turnover" rates are assumed. The same objections apply to the somewhat related views of Mangham (81) and Clements (20). It should be recognized, however, that the dispersed proteins (slime) and perhaps cytoplasm in the vacuoles of the sieve elements may be linked into a continuous en-

* The apparent diffusion constant of sugar through membranes of tissue is of an even lower magnitude, because the effective area for that diffusion is mainly between the protoplasts. Consideration of this would even further increase the factor which measures the increased rate of translocation over physical diffusion (see Chapter 4). Ed.

zymic surface, and, if true, this would considerably increase the mechanically functional area available for a surface-controlled or surface-activated transport mechanism. It may be significant that cytoplasmic streaming has not been observed in the functional sieve elements; quiescent protoplasm would be necessary in maintaining the steric relations required for a surface-operated transport process.

Despite these conjectures, however, the present status of the problem favors the view that the sieve-element protoplast does not possess any metabolic machinery capable of driving the translocation process, except for the possibility of metabolic transfer systems which move solutes across limiting membranes. Therefore, to the extent that the physiological process of translocation requires work to be done and energy to be used, this may occur mainly in situations other than the mature sieve-tube element. It may even be that the mature sieve-tube element may be the relic of a more active system that functioned during its growth; or again it may be that the principal sources of energy which are coupled to the movement of solutes in translocation should be sought at the "donor" and "receiving" sites where "active-transport" mechanisms allow solutes to move in ways that are not restricted either in direction or rate by diffusion. Thus, certain aspects of the problems of translocation presented in this chapter, and those of active transport presented in Chapter 4, will ultimately require some common explanation.

REFERENCES

1. Ahmed, M. K., Newson, L. D., Roussel, J. S., and Emerson, R. B. Translocation of systox in the cotton plant. *J. Econ. Entomol.* 47, 684–691 (1954).
2. Anderssen, F. G. Some seasonal changes in the tracheal sap of pear and apricot trees. *Plant Physiol.* 4, 459–476 (1929).
3. Arisz, W. H. Uptake and transport of chlorine by parenchymatic tissue of leaves of *Vallisneria spiralis.* III. Discussion of the transport and uptake. Vacuole secretion theory. *Koninkl. Ned. Akad. Wetenschap. Proc.* 51, 25–33 (1948).
4. Arisz, W. H. Transport of organic compounds. *Ann. Rev. Plant Physiol.* 3, 109–130 (1952).
5. Aronoff, S. Translocation from soybean leaves. II. *Plant Physiol.* 30, 184–185 (1955).
6. Atkins, W. R. G. "Some Recent Researches in Plant Physiology." Macmillan, New York, 1916.
7. Bauer, L. Zur Frage der Stoffbewegungen in der Pflanze mit besonderer Berücksichtigung der Wanderung von Fluorochromen. *Planta* 42, 367–451 (1953).
8. Belikov, I. F. Local utilization of products of photosynthesis in soybeans. *Doklady Akad. Nauk S.S.S.R.* 102, 379–381 (1955); *Chem. Abstr.* 49, 14916 (1955).
9. Biddulph, O. Absorption and movement of radiophosphorus in bean seedlings. *Plant Physiol.* 15, 131–136 (1940).

10. Birch-Hirschfeld, L. Üntersuchungen über dio Ausbreitungsgeschwindigkeit gelöster Stoffe in der Pflanze. *Jahrb. wiss. Botan.* **59**, 171–262 (1919–20).

11. Böhning, R. H., Swanson, C. A., and Linck, A. J. The effect of hypocotyl temperature on the translocation of carbohydrates from bean leaves. *Plant Physiol.* **27**, 417–421 (1952).

12. Bollard, E. G. The use of tracheal sap in the study of apple-tree nutrition. *J. Exptl. Botan.* **4**, 363–368 (1953).

13. Bollard, E. G. Translocation of organic nitrogen in the xylem. *Australian J. Biol. Sci.* **10**, 292–301 (1957).

14. Bondarenko, D. D. 3-Amino-1,2,4,-triazol as an herbicide on Canada thistle and its effect on soil microorganisms. Ph.D. Dissertation, Ohio State University, Columbus, Ohio, 1957.

15. Caldwell, J. Studies in translocation. I. Movement of food-materials in the Swedish turnip. *Proc. Roy. Soc. Edinburgh* **50**, 130–141 (1930).

16. Cheadle, V. I. Research on xylem and phloem—Progress in fifty years. *Am. J. Botany* **43**, 719–731 (1956).

17. Chen, S. L. Simultaneous movement of P^{32} and C^{14} in opposite directions in phloem tissue. *Am. J. Botany* **38**, 203–211 (1951).

18. Child, C. M., and Bellamy, A. W. Physiological isolation by low temperature in Bryophyllum and other plants. *Science* **50**, 362–365 (1919).

19. Choudhri, R. S. Studies of the effects of certain plant hormones on growth, general behavior, and food transport of *Phaseolus vulgaris* L. Ph.D. Dissertation, Cornell University, Ithaca, New York, 1948.

20. Clements, H. F. Translocation of solutes in plants. *Northwest Sci.* **8**, No. 4, 9–21 (1934).

21. Clements, H. F. Movement of organic solutes in the sausage tree, *Kigelia Africana*. *Plant Physiol.* **15**, 689–700 (1940).

22. Clements, H. F., and Engard, C. J. Upward movement of inorganic solutes as affected by a girdle. *Plant Physiol.* **13**, 103–122 (1938).

23. Colwell, R. N. Translocation in plants with special reference to the mechanism of phloem transport as indicated by studies on phloem exudation and on the movement of radioactive phosphorus. Ph.D. Dissertation, University of California, Davis, California, 1942. Quoted in Crafts and Lorenz (34).

24. Colwell, R. N. The use of radioactive phosphorus in translocation studies. *Am. J. Botany* **29**, 798–807 (1942).

25. Cooper, W. C. Transport of root-forming hormone in woody cuttings. *Plant Physiol.* **11**, 779–794 (1936).

26. Crafts, A. S. Movement of organic materials in plants. *Plant Physiol.* **6**, 1–41 (1931).

27. Crafts, A. S. Phloem anatomy, exudation, and transport of organic nutrients in cucurbits. *Plant Physiol.* **7**, 183–225 (1932).

28. Crafts, A. S. Sieve-tube structure and translocation in the potato. *Plant Physiol.* **8**, 81–104 (1933).

29. Crafts, A. S. The relation between structure and function of the phloem. *Am. J. Botany* **26**, 172–177 (1939).

30. Crafts, A. S. The protoplasmic properties of sieve tubes. *Protoplasma* **33**, 389–398 (1939).

31. Crafts, A. S. Solute transport in plants. *Science* **90**, 337–338 (1939).

32. Crafts, A. S. Movement of materials in phloem as influenced by the porous

nature of the tissues. *In* "Interaction of water and porous materials." *Discussions Faraday Soc.* No. 3, 153–159 (1948).

33. Crafts, A. S. Movement of assimilates, viruses, growth regulators, and chemical indicators in plants. *Botan. Rev.* 17, 203–284 (1951).

34. Crafts, A. S., and Lorenz, O. A. Fruit growth and food transport in cucurbits. *Plant Physiol.* 19, 131–138 (1944).

35. Currier, H. B., Esau, K., and Cheadle, V. I. Plasmolytic studies of phloem. *Am. J. Botany* 42, 68–81 (1955).

36. Currier, H. B., and Strugger, S. Aniline blue and fluorescence microscopy of callose in bulb scales of *Allium cepa* L. *Protoplasma* 45, 552–559 (1955–56).

37. Curtis, O. F. The upward translocation of foods in woody plants. I. Tissues concerned in translocation. *Am. J. Botany* 7, 101–124 (1920).

38. Curtis, O. F. Studies on solute translocation in plants. Experiments indicating that translocation is dependent on the activity of living cells. *Am. J. Botany* 16, 154–168 (1929).

39. Curtis, O. F. "The Translocation of Solutes in Plants." McGraw-Hill, New York, 1935.

40. Curtis, O. F., and Clark, D. G. "An Introduction to Plant Physiology." McGraw-Hill, New York, 1950.

41. Czapek, F. Über die Leitungswege der organischen Baustoffe im Pflanzenkörper. *Sitzber. kais. Akad. Wiss. Wien, Math.-Naturw. Kl. Abt. I* 106, 117–170 (1897). Quoted in Curtis (39).

42. Dana, M. N. Physiology of dwarfing in apple. Ph.D. Dissertation, Iowa State College, Ames, Iowa, 1952.

43. Day, B. E. Absorption and translocation of 2,4-dichlorophenoxyacetic acid by bean plants. Ph.D. Dissertation, University of California, Davis, California, 1950.

44. Day, B. E. The absorption and translocation of 2,4-dichlorophenoxyacetic acid by bean plants. *Plant Physiol.* 27, 143–152 (1952).

45. Dixon, H. H., and Ball, N. G. Transport of organic substances in plants. *Nature* 109, 236–237 (1922).

46. Döpp, W. Beiträge zur Frage der Stoffwanderung in den Siebröhren. *Jarhb. wiss. Botan.* 87, 679–705 (1938–39).

47. Drollinger, E. L. The effects of leaf removal on the size, yield, and grade of greenhouse tomatoes. M.Sc. Thesis, Ohio State University, Columbus, Ohio, 1957.

48. Engard, C. J. Translocation of carbohydrates in the Cuthbert raspberry. *Botan. Gaz.* 100, 439–464 (1939).

49. Engard, C. J. Translocation of nitrogenous substances in the Cuthbert raspberry. *Botan. Gaz.* 101, 1–34 (1940).

50. Esau, K. Development and structure of the phloem tissue. II. *Botan. Rev.* 16, 67–114 (1950).

51. Esau, K. "Plant Anatomy." Wiley, New York, 1953.

52. Esau, K., Currier, H. B., and Cheadle, V. I. Physiology of phloem. *Ann. Rev. Plant Physiol.* 8, 349–374 (1957).

53. Frey, G. Aktivität und Lokalisation von saurer Phosphatase in den vegetativen Teilen einiger Angiospermen und in einigen Samen. *Ber. schweiz. botan. Ges.* 64, 390–452 (1954).

54. Frey-Wyssling, A. Latex flow. *In* "Deformation and Flow in Biological Systems." (A. Frey-Wyssling, ed.) Interscience, New York, 1952.

55. Gauch, H. G., and Dugger, W. M., Jr. The role of boron in the translocation of sucrose. *Plant Physiol.* **28**, 457–466 (1953).
56. Goodwin, R. H., and Goddard, D. R. The oxygen consumption of isolated woody tissues. *Am. J. Botany* **27**, 234–237 (1940).
57. Gray, R. A. The downward translocation of antibiotics in plants. *Phytopathology* **48**, 71–78 (1958).
58. Haller, M. H. The relation of the distance and direction of the fruit from the leaves to the size and composition of apples. *Proc. Am. Soc. Hort. Sci.* **27**, 63–68 (1930).
59. Hartig, Th. Ueber die Bewegung des Saftes in den Holzpflanzen. *Botan. Z.* **16**, 329–335, 338–342 (1858). Quoted in Curtis (39).
60. Hay, R. E., Earley, E. B., and DeTurk, E. E. Concentration and translocation of nitrogen compounds in the corn plant (*Zea mays*) during grain development. *Plant Physiol.* **28**, 606–621 (1953).
61. Hewitt, S. P., and Curtis, O. F. The effect of temperature on loss of dry matter and carbohydrate from leaves by respiration and translocation. *Am. J. Botany* **35**, 746–755 (1948).
62. Honert, J. H. van den. On the mechanism of transport of organic materials in plants. *Proc. Koninkl. Akad. Wetenschap. Amsterdam* **35**, 1104–1111 (1932).
63. Huber, B. Das Siebröhrensystem unserer Bäume und seine jahreszeitlichen Veränderungen. *Jahrb. wiss. Botan.* **88**, 176–242 (1939).
64. Huber, B. Gesichertes und Problematisches in der Wanderung der Assimilate. *Ber. deut. botan. Ges.* **59**, 181–194 (1941).
65. Hull, H. M. Carbohydrate translocation in tomato and sugar beet with particular reference to temperature effect. *Am. J. Botany* **39**, 661–669 (1952).
66. Jacobs, W. P. The development of the gynophore of the peanut plant, *Arachis hypogaea* L. I. The distribution of mitoses, the region of greatest elongation, and the maintenance of vascular continuity in the intercalary meristem. *Am. J. Botany* **34**, 361–370 (1947).
67. Jones, C. H., Edson, A. W., and Morse, W. J. The maple sap flow. *Vermont Agr. Expt. Sta. Bull.* **103**, 43–184 (1903).
68. Kazaryan, V. O., Avundzhyan, E. S., and Gabrielyan, G. G. Combined motion of carbohydrates and phosphorus in plants. *Doklady Akad. Nauk Armyan. S.S.R.* **20**, 197–201 (1955); *Chem. Abstr.* **49**, 14916 (1955).
69. Kendall, W. A. The effect of intermittently varied petiole temperature on carbohydrate translocation from bean leaves. *Plant Physiol.* **27**, 631–633 (1952).
70. Kendall, W. A. The effect of certain metabolic inhibitors on translocation of P[32] in bean plants. Ph.D. Dissertation, Ohio State University, Columbus, Ohio, 1954.
71. Kendall, W. A. Effect of certain metabolic inhibitors on translocation of P[32] in bean plants. *Plant Physiol.* **30**, 347–350 (1955).
72. Kennedy, J. S., and Mittler, T. E. A method of obtaining phloem sap via the mouth-parts of aphids. *Nature* **171**, 528–529 (1953).
73. Koontz, H., and Biddulph, O. Factors regulating absorption and translocation of foliar applied phosphorus. *Plant Physiol.* **32**, 463–470 (1957).
74. Kuprevic, V. F. Vnekletochnye fermenty provodīashchich tkaneĭ drevesnykh porod. [Extracellular enzymes of conducting tissues of woody species.] *Botan. Zhur.* **34**, 613–617 (1949).
75. Leonard, O. A. Translocation of carbohydrates in the sugar beet. *Plant Physiol.* **14**, 55–74 (1939).

76. Leopold, A. C., and Scott, F. J. Physiological factors in tomato fruit-set. *Am. J. Botany* **39**, 310–317 (1952).

77. Linck, A. J. Studies on the distribution of phosphorus-32 in *Pisum sativum*, in relation to fruit development. Ph.D. Dissertation, Ohio State University, Columbus, Ohio, 1955.

78. Loomis, W. E. Translocation of carbohydrates in maize. *Science* **101**, 398–400 (1945).

79. MacDaniels, L. H., and Curtis, O. F. The effect of spiral ringing on solute translocation and the structure of the regenerated tissues of the apple. *Cornell Univ. Agr. Expt. Sta. Mem.* **No. 133** (1930).

80. Maizel, J. V., Benson, A. A., and Tolbert, N. E. Identification of phosphoryl choline as an important constituent of plant saps. *Plant Physiol.* **31**, 407–408 (1956).

81. Mangham, S. On the mechanism of translocation in plant tissues. An hypothesis, with special reference to sugar conduction in sieve tubes. *Ann. Botany (London)* **31**, 293–311 (1917).

82. Maskell, E. J., and Mason, T. G. Studies on the transport of nitrogenous substances in the cotton plant. I. Preliminary observations on the downward transport of nitrogen in the stem. *Ann. Botany (London)* **43**, 205–231 (1929).

83. Mason, T. G. Preliminary note on the physiological aspect of certain undescribed structures in the phloem of the greater yam, *Dioscorea alata* L. *Sci. Proc. Roy. Dublin Soc.* **18**, 195–198 (1926).

84. Mason, T. G., and Lewin, C. T. On the rate of carbohydrate transport in the greater yam, *Dioscorea alata* L. *Sci. Proc. Roy. Dublin Soc.* **18**, 203–205 (1926).

85. Mason, T. G., and Maskell, E. J. Studies on the transport of carbohydrates in the cotton plant. I. A study of diurnal variation in the carbohydrates of leaf, bark, and wood, and the effects of ringing. *Ann. Botany (London)* **42**, 189–253 (1928).

86. Mason, T. G., and Maskell, E. J. Studies on the transport of carbohydrates in the cotton plant. II. The factors determining the rate and the direction of movement of sugars. *Ann. Botany (London)* **42**, 571–636 (1928).

87. Mason, T. G., Maskell, E. J., and Phillis, E. Further studies on transport in the cotton plant. III. Concerning the independence of solute movement in the phloem. *Ann. Botany (London)* **50**, 23–58 (1936).

88. Mason, T. G., and Phillis, E. Oxygen supply and the inactivation of diffusion. *Ann. Botany (London)* **50**, 455–499 (1936).

89. Mason, T. G., and Phillis, E. The migration of solutes. *Botan. Rev.* **3**, 47–71 (1937).

90. Mellor, J. L. Sugar absorption by tobacco plants and its application to the transplanting process. Ph.D. Dissertation, North Carolina State College, Raleigh, North Carolina, 1953.

91. Miller, E. C. "Plant Physiology." McGraw-Hill, New York, 1938.

92. Mitchell, J. W., and Brown, J. W. Movement of 2,4-dichlorophenoxyacetic acid stimulus and its relation to the translocation of organic food materials in plants. *Botan. Gaz.* **107**, 393–407 (1946).

93. Moore, W., and Williaman, J. J. Studies in greenhouse fumigation with hydrocyanic acid: physiological effects on the plant. *J. Agr. Research* **11**, 319–338 (1917).

94. Moose, C. A. Chemical and spectroscopic analysis of phloem exudate and

parenchyma sap from several species of plants. *Plant Physiol.* **13**, 365–380 (1938).

95. Moyer, L. S. Recent advances in the physiology of latex. *Botan. Rev.* **3**, 522–544 (1937).

96. Münch, E. "Die Stoffbewegungen in der Pflanze." Fischer, Jena, 1930.

97. Muzik, T. J. Role of parenchyma cells in graft unions in Vanilla orchid. *Science* **127**, 82 (1958).

98. Nelson, C. D., and Gorham, P. R. Uptake and translocation of C^{14}-labelled sugars applied to primary leaves of soybean seedlings. *Can. J. Bot.* **35**, 339–347 (1957).

99. Nitsch, J. P. Growth and development *in vitro* of excised ovaries. *Amer. J. Botany* **38**, 566–577 (1951).

100. Overbeek, J. van. Absorption and translocation of plant regulators. *Ann. Rev. Plant Physiol.* **7**, 355–372 (1956).

101. Palmquist, E. M. The simultaneous movement of carbohydrates and fluorescein in opposite direction in the phloem. *Am. J. Botany* **25**, 97–105 (1938).

102. Phillis, E., and Mason, T. G. The polar distribution of sugar in the foliage leaf. *Ann. Botany (London)* **47**, 585–634 (1933).

103. Phillis, E., and Mason, T. G. Further studies on transport in the cotton plant. IV. On the simultaneous movement of solutes in opposite directions through the phloem. *Ann. Botany (London)* **50**, 161–174 (1936).

104. Rabideau, G. S., and Burr, G. O. The use of the C^{13} isotope as a tracer for transport studies in plants. *Am. J. Botany* **32**, 349–356 (1945).

105. Roeckl, B. Nachweis eines Konzentrationshubs zwischen Palisadenzellen und Siebröhren. Inaug. Dissertation München, 1948. Quoted in Esau (50).

106. Roeckl, B. Nachweis eines Konzentrationshubs zwischen Palisadenzellen und Siebröhren. *Planta* **36**, 530–550 (1949).

107. Rohrbaugh, L. M., and Rice, E. L. Relation of phosphorus nutrition to the translocation of 2,4-D in tomato plants. *Plant Physiol.* **31**, 196–199 (1956).

108. Rouschal, E. Untersuchungen über die Protoplasmatik und Funktion der Siebröhren. *Flora (Jena)* **135**, 135–200 (1941–42).

109. Sax, K. Control of tree growth by phloem blocks. *Science* **119**, 585 (1954).

110. Schjeide, O. A., Mead, J. F., and Myers, L. S., Jr. Notions on sensitivity of cells to radiation. *Science* **123**, 1020–1021 (1956).

111. Schumacher, W. Untersuchungen über die Lokalisation der Stoffwanderung in den Leitbundeln höherer Pflanzen. *Jahrb. wiss. Botan.* **73**, 770–823 (1930).

112. Schumacher, W. Untersuchungen über die Wanderung des Fluoreszeins in den Siebröhren. *Jahrb. wiss. Botan.* **77**, 685–732 (1933).

113. Schumacher, W. Weitere Untersuchungen über die Wanderung von Farbstoffen in den Siebröhren. *Jahrb. wiss. Botan.* **85**, 422–449 (1937).

114. Schumacher, W. Über die Plasmolysierbarkeit der Siebröhren. *Jahrb. wiss. Botan.* **88**, 545–553 (1939).

115. Schumacher, W. Zur Frage nach den Stoffbewegungen im Pflanzenkörper. *Naturwissenschaften* **34**, 176–179 (1947).

116. Sisler, E. C., Dugger, W. M., Jr., and Gauch, H. G. The role of boron in the translocation of organic compounds in plants. *Am. J. Botany* **31**, 11–17 (1956).

117. Skok, J. Relationship of boron nutrition to radiosensitivity of sunflower plants. *Plant Physiol.* **32**, 648–658 (1957).

118. Steward, F. C., and Priestley, J. H. Movement of organic materials in plants: A correction. *Plant Physiol.* **8**, 482–483 (1933).

119. Swanson, C. A. Translocation of organic solutes. *In* "Atomic Energy and Agriculture." (C. L. Comar, ed.) American Association for the Advancement of Science. Washington, D.C., 1957.

120. Swanson, C. A., and Böhning, R. H. The effect of petiole temperature on the translocation of carbohydrates from bean leaves. *Plant Physiol.* **26**, 557–564 (1951).

121. Swanson, C. A., and El Shishiny, E. D. H. Translocation of sugars in grape. *Plant Physiol.* **33**, 33–37 (1958).

122. Swanson, C. A., and Whitney, J. B. Studies on the translocation of foliar-applied P32 and other radioisotopes in bean plants. *Am. J. Botany* **40**, 816–823 (1953).

123. Thomas, W. The seat of formation of amino acids in *Pyrus malus* L. *Science* **66**, 115–116 (1927).

124. Vernon, L. P., and Aronoff, S. Metabolism of soybean leaves. IV. Translocation from soybean leaves. *Arch. Biochem. Biophys.* **36**, 383–398 (1952).

125. Wanner, H. Phosphataseverteilung und Kohlenhydrattransport in der Pflanze. *Planta* **41**, 190–194 (1952).

126. Wanner, H. Enzyme der Glykolyse im Phloemsaft. *Ber schweiz. botan. Ges.* **63**, 201–212 (1953).

127. Went, F. W., and Hull, H. M. The effect of temperature upon translocation of carbohydrates in the tomato plant. *Plant Physiol.* **24**, 505–526 (1949).

128. Willenbrink, J. Über die Hemmung des Stofftransports in den Siebröhren durch lokale Inaktivierung verschiedener Atmungsenzyme. *Planta* **48**, 269–342 (1957).

129. Wylie, R. B. Relations between tissue organization and vein distribution in dicotyledon leaves. *Am. J. Botany* **26**, 219–225 (1939).

130. Wylie, R. B. Conduction in dicotyledon leaves. *Proc. Iowa Acad. Sci.* **53**, 195–202 (1946).

131. Ziegler, H. Untersuchungen über die Leitung und Sekretion der Assimilate. *Planta* **47**, 447–500 (1956).

132. Zimmermann, M. H. Translocation of organic substances in trees. I. The nature of the sugars in the sieve tube exudate of trees. *Plant Physiol.* **32**, 288–291 (1957).

133. Zimmermann, M. H. Translocation of organic substances in trees. II. On the translocation mechanism in the phloem of white ash. *Plant Physiol.* **32**, 399–404 (1957).

CHAPTER SIX

Translocation of Inorganic Solutes

O. BIDDULPH

I. Introduction

Translocation as it is herein interpreted will refer to the movement of mineral substances over distances comparable to the total length of the plant axis, and at velocities of the order of 30–100 cm per hour. This limits the discussion to translocation in the xylem and the phloem but also includes incidental treatment of transfer from one system to the other. The movement through parenchyma and similar tissues is excluded (for reference to this subject see Chapter 4).

It is very difficult to discuss translocation apart from its anatomical background. This does not necessarily mean phloem anatomy in the cellular sense, but more generally the complete vascular anatomy of the plant. There is ample evidence to show that the translocating materials follow the vascular channels. More attention has been given to the mechanics of movement in relation to cellular anatomy than to its relation with vascular anatomy and the metabolic activity which occurs at the distal ends of the translocation path. This is an area into which research efforts could be profitably channeled.

The translocation of minerals in plants, when considered in its entirety, involves the upward movement of the inorganic materials acquired by the roots, their distribution within the shoot, and any redistribution via the vascular tissue from the initial site of deposition to any other part of the plant. The redistributions which are likely to occur are divided for convenience into two categories. They are (a) the withdrawal of elements from leaves prior to their abscission, and (b) the transfer of elements initially deposited in leaves, stems, etc., to reproductive or other structures. In addition to the above transfer types of movement, there is a much more dynamic sequence of movements which appears, in some instances, to be constituted of a more or less continual circulation within the plant. There are wide differences in the degree of circulation participated in by the various nutrient elements, so no common pattern appears which is characteristic of all. They may move in the same channel and essentially by means of the same forces, but certainly they differ in the degree of mobility within the plant as a whole. This is shown by such phenomena as the selective withdrawal of nutrient elements from leaves prior to their abscission, the unequal distribution of elements in various plant parts, and the necessity of continuous root absorption of certain elements, while intermittent absorption will suffice for others. Evidence from many sources indicates that various elements exhibit markedly different patterns of movement within the tissues and organs of the plant.

One method of investigating mineral translocation involves the application of a substance to a leaf, this substance not normally being available to the plant from that site. The fact that it is absorbed and subsequently translocated from this site to other parts of the plant indicates its capacity to move but does not necessarily mean that there normally is present a supply of the substance which is free to move at all times. Additional tests are necessary to establish the degree of natural circulation which takes place within the plant under conditions of normal nutrition. Much of the data which have been reported do not lead themselves to evaluations of this kind. The degree of natural mobility within the plant apparently ranges from almost complete immobility in the case of calcium, to a condition of more or less continuous circulation in the case of phosphorus. It is relatively certain that the current status of nutrition with respect to phloem mobile elements will determine the amount of the substance which is within the pool that constitutes the mobile fraction.

The release mechanism which frees the minerals from their association with the accumulating cells of the roots is properly considered as the terminal step of the absorption process. The release appears to be conditioned by a number of environmental factors which also influence the plant in other ways. As yet no factors are known which control only this process. In addition there are superimposed on the effects of the environmental factors certain cyclic phenomena which influence the release (41). These cyclic responses are assumed to be the result of a conditioning of the protoplasm by exposure to some critical minimum number of regular variations in the environmental factors to which the process is sensitive (20). Light is the factor to which the release seems most closely related. There is conflicting evidence concerning the effect of transpiration, which is related to light through the opening and closing of stomata. The nutritional status of the root may also exert some effect on mineral release into the transpiration stream. This subject is covered in Chapter 4 in this volume.

II. Upward Movement of Solutes

A. SOME EARLIER VIEWS

Knowledge respecting the nutrition of plants became definitive in the latter half of the seventeenth century. The earlier questions as to the nature of plant foods and the manner in which they entered and moved about—questions which engaged the attention of Aristotle and Cesalpino—were to be answered to a certain degree by Malpighi and Hales. Malpighi (58) perceived that the green leaves prepared the food and that it passed to all parts of the plant to be stored or used for growth, but the nature of the substances from which the food was prepared was beyond the chemistry of the times. Mariotte believed that plants convert the food material which they derived from the ground into new chemical combinations, thereby rejecting the view of Aristotle that since the plant had no perception it could not tell good food from bad and must, therefore, gain its food from the soil in a prepared form, which was referred to as humus. At the beginning of the eighteenth century, Hales (40), by the evolution of gases in the dry distillation of plants, concluded that a considerable part of their substance was derived from the atmosphere.

In these three ideas was sufficient information on which to base a sound theory of nutrition for plants, but the ideas were not properly combined and it took a hundred years and the overthrow of the old phlogistic chemistry before carbon assimilation and mineral nutrition were fully appreciated. Ingen-Housz (46), de Saussure (32), and Senebier (83) were responsible for the initial formulation of the principals of carbon assimilation. Liebig (52) referred nitrogenous materials to ammonia and related compounds derived from the soil and claimed ash to be an essential factor in nutrition, and Boussingault (19) succeeded in growing plants in purely mineral soil free of humus and established the use of nitrates by plants. This refuted conclusively the humus theory and showed the atmosphere to be the source of carbon and the soil to be the source of nitrogen as well as minerals. A plausible theory of the nutrition of plants was a tedious development with many more participants than those mentioned above. For details to 1860, see Sachs' "History of Botany" (81).

B. Upward Movement in the Xylem

The absorption of large amounts of water, its passage through the wood, and its loss from the leaves was recognized by Malpighi and Hales, but only limited progress was made in understanding the full function of the transpiration stream until an understanding of mineral nutrition was gained. With this knowledge there was no longer a doubt about the existence of a translocation system within the plant. The questions were then concerned with the tissues involved, the substances moved, and the forces of movement.

Many uses have been made of dyes, colorless salts which can be determined spectroscopically or chemically, and colloidal substances to determine the path and circumstances of upward movement in the xylem. The general conclusion which had been reached is that water and solutes ascended the xylem together. The possibility that downward movement might also occur in the xylem was suggested by Birch-Hirschfeld (14) partly because she herself was unable to demonstrate a rapid movement of lithium salts in the phloem and perhaps partly because there had been a continuous recognition of the flow and ebb of sap through the wood since the time of Hales. The downward movement of solution in the xylem was clearly indicated by Yendo (98). Those who have shown downward movement in the xylem include Birch-Hirschfeld (14), Rumbold (79), Dixon (34), Dixon and Ball (35), Kastens (47), MacDougal (54), and Arndt (2). A number of these studies dealt with the transfer of solutions of dyes and lithium salts, which could be detected visually or spectroscopically, from cut shoot tips, petioles, leaf blades, etc., to stem tips and bases, and in some instances to roots. These materials were detected in the xylem, from which it appeared that the xylem was not the vehicle of unidirectional transport as was previously believed, but was capable of moving minerals and organic substances back down the stem. The rapid downward movement which was observed when such dyes were introduced, as described above, is now attributable to tensions existing in the xylem which, through cohesive forces in the liquid columns, draw solutions through the tissues during conditions of internal water stress. The outcome of this type of research, while furthering other ends, did little to advance our knowledge of normal translocation patterns. The findings from this type of research have been summarized by Roach (78), and the phenomenon is now generally recognized as xylem injection. Modifications have been made in order to utilize this method of introducing materials into plants to supplement nutrients normally absorbed by the roots.

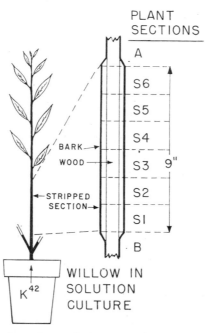

Fig. 1. Diagram to show the method of stripping the bark and the arrangement of sections of bark and wood for the purpose of measuring the movement of radioactive potassium upward in the stem. Quantitative results are shown in Table I. Adapted from Stout and Hoagland (85), page 322.

The upward transport of minerals in the wood remained unchallenged from the time of Hales until Curtis (29) emphasized the meager nature of the evidence upon which the view rested. As a result of ringing experiments, wherein it appeared that growth of the tissues of stem tips was hindered by the removal of a ring of bark, Curtis proposed the phloem as a vehicle of upward transport. There followed, in quick succession, a number of reports contrary to Curtis' interpretation. Mason and Phillis (65) give credit to Clements (24) for the first actual demonstration that soil solutes ascend in the wood, but add that he did not show that they may not also ascend in the phloem. Subsequently, it was demonstrated by Clements and Engard (26), Phillis and Mason (74), and Mason, Maskell, and Phillis (63) that ringing near the base of the stem had relatively little effect on the upward movement of minerals from the root. Many experiments involving ringing of stems have been directed at the latter problem with at least partial success in demonstrating that it is very unlikely that root-absorbed minerals normally ascend in the phloem.

TABLE I

GAIN OF POTASSIUM (FROM NUTRIENT SOLUTION) IN SECTIONS OF WILLOW (*Salix lasiandra*) AFTER ABSORPTION PERIOD OF 5 HOURS[a]

	K^* in each section[b] (parts per million)			
	Plant stripped[c]		Plant unstripped	
Section[d]	Bark	Wood	Bark	Wood
A	53	47	64	56
S6	11.6	119		
S5	0.9	122		
S4	0.7	112	87	69
S3	0.3	98		
S2	0.3	108		
S1	20	113		
B	84	58	74	67

[a] From Stout and Hoagland (85), page 322.
[b] Calculated from radioactivity.
[c] Plant stripped 1½ hours before use.
[d] See Fig. 1 for details of sections.

The fundamental conflict in the above reports perhaps can be partially resolved when the position of the ring is considered. The latter investigators employed a ring low down on the stem and were dealing primarily with root-absorbed minerals, whereas Curtis employed a girdle higher up on the stem and was, therefore, not dealing only with root-absorbed minerals but with leaf-exported minerals as well. This may prove to be the grounds upon which the two conflicting views can be at least partially reconciled, as will be shown later.

Data from the Cotton Research Station in Trinidad aided materially in reclarifying the issue as to which tissue systems were involved in upward transport, but the most definite evidence that the path of upward translocations of minerals is through the xylem was obtained by Stout and Hoagland (85) by means of radioactive tracers. These critical experimental data were obtained by following the tracers from the root environment upward through stems in which the bark and the wood had been effectively isolated from each other by an impervious layer of waxed paper. The tracers moved readily through the wood but failed to pass effectively through the bark which was isolated from it (Fig. 1, Table I, stem section A). Where the two remained in normal contact, the bark had as much tracer present as the wood. More recent work (9, 12) has revealed that an effective interchange between intact phloem, cambium, and xylem is also of normal occurrence. Stout and Hoagland stressed that the conclusion—that upward movement of minerals normally occurs through the xylem—applies to plants under normal growth conditions where leafy shoots are present. Additional data are desirable before these conclusions can be universally applied to plants in general under all circumstances.

C. MOVEMENT INTO LEAVES

As the minerals are released into the xylem, they come under the influence of the ascending transpiration stream which has the effect of sweeping them in the direction of its flow. The detailed vascular patterns differ for various plants and, as the xylem constitutes the channel through which upward movement takes place, the corresponding pattern of ascent of minerals in the stem will vary for various plants. However, it is generally assumed that minerals are carried along with the upward flow of water, and as a consequence the same channel is used for the ascension of both. It is not known, however, just what degree of freedom the ascending minerals possess in their upward movement. It is to be expected that by differential competition between the solvent (water) and an absorbent (cell-wall materials), the dissolved materials (minerals) may be differentially retarded in their progress so that different minerals ascend the stem at different rates. The closest analogy to this system of movement is with that utilized in certain types of chromatography. There is only sufficient evidence to suggest such a system, not enough to prove it (44). This suggestion applies primarily to movement of minerals along xylem channels where living cells play a minor role.

It has been shown by Auchter (4) that in straight-grained trees there is normally little cross transfer of nitrogen from the roots on one side of the tree to branches on the other, but that water readily moves from one side to the other. Failure of cross transfer of solutes from one side of herbaceous plants to the other has also been observed by Caldwell (22), and specific mineral-deficiency symptoms were even produced on halves of individual leaves of tobacco (*Nicotiana tabacum*) plants by withholding specific mineral nutrients from portions of the root system. Deficiency symptoms for nitrogen, phosphorus, potassium, sulfur, iron, and manganese were produced in this manner by McMurtrey (56). Apparently the minerals follow a very specific path from the roots which absorb them to the leaves in which they are deposited.

A study of sap movement in pear (*Pyrus communis*) and apricot (*Prunus armeniaca*) trees by Anderssen (1) revealed two important points. First, there was a wide seasonal fluctuation in the mineral content of xylem sap. The May 10 values increased over the November 10 values as follows: Ca from 16.6 to 84.7 ppm (parts per million), Mg from 0.8 to 23.5 ppm, K from 23.6 to 59.6 ppm, and PO_4 from 10.6 to 25.2 ppm. Second, the total electrolytes in the xylem sap were almost twice as concentrated in the outer annual ring as in the inner rings. This is consistent with the view that the annual rings of current origin commonly carry the burden of the transpiration stream in woody plants, but there are exceptions wherein the wood of several years' origin remains equally functional—at least in water transport (55, 72, 73).

Bollard (16) used an aspiration procedure modified from Bennett *et al.* (5) for the removal of xylem sap from apple (*Malus sylvestris*) trees. Successive studies of the mineral constituents of the sap throughout the year showed that the nitrogen, phosphorus, and magnesium concentration rapidly rose to a peak at flowering time. The magnesium concentration declined quickly after flowering, whereas the nitrogen and phosphorus concentrations declined slowly. Potassium reached a peak later, remained high, and declined with nitrogen and phosphorus. Since the maximum concentrations in xylem sap were found before the leaves were fully developed, the maximum flow may have occurred later when the concentrations were lower but the flow of xylem sap greater.

The translocation of inorganic nitrogenous materials is normally rather limited in plants. Thomas (88) showed amino acid synthesis to occur in the roots of apple trees and Eckerson (36) substantiated the work by demonstrating reducing activity in the sites of synthesis. High reducing activity was shown during fall and winter in the fine rootlets and in the buds and fine rootlets in the spring. Little activity was present in leaves at any time. Many other studies have shown that although many plants are capable of nitrate reduction in the aerial parts, most plants carry out nitrate reduction in the roots (70). However, tomato (*Lycopersicon esculentum*), sweet pea (*Lathyrus odoratus*), and soybean (*Glycine max* var. 'Bilox') may be exceptions to this rule. Since the major portion of the nitrogenous substances move as organic compounds, their movement is not considered with inorganic solutes.

Bollard (17) also showed that in apple trees nitrates are normally reduced in the roots and the upward movement of nitrogenous materials is largely a movement of organic forms of nitrogen. Aspartic and glutamic acid, asparagine, and glutamine accounted for 90% of the nitrogen in the xylem sap. The inorganic forms of nitrogen were insignificant. In an extensive study of the xylem saps of many gymnosperms, monocotyledonous and dicotyledonous plants, Bollard (18) showed that nitrates were present in less than half of the species studied and then only in trace amounts. The reduction of nitrates in the roots was then again confirmed in a much more positive manner, and it was unquestionably demonstrated that nitrogenous materials are transported largely in the organic form. This topic is discussed in Chapter 5 in this volume.

Little could be concluded as to the importance of the form of the nitrogenous materials delivered to the needs of meristematic tissues which were nourished by them (18). Whether the amides and lesser traces of specific amino acids represented only a bulk form of nitrogen from which essential constituents could be derived, or whether these particular substances each had a specific role in nutrition, could not be determined from the data obtained. The amides were shown by Mason and Phillis (64) to be important forms of nitrogen being translocated downward in the phloem of the cotton plant [a strain of sea island cotton (*Gossypium barbadense*)]. These were the forms in which nitrogen was also stored in the bark for subsequent usage in the development of reproductive structures.

Organic materials, in the form of sugars, were identified in the xylem sap of a wide variety of trees during various seasons of the year by Fischer (39) in 1888, and Wormall (97) in 1924 showed that the transport of 1.56 gm of solids per liter of exuded xylem sap occurred in the grape (*Vitis vinifera*). Two thirds of the solute was organic, of which the major part was composed of organic acids including oxalic, tartaric, malic, and succinic. The minerals present were chlorides, sulfates, nitrates, nitrites, silicates, and phosphates of sodium, potassium, calcium, iron, magnesium, and to a lesser extent manganese and aluminum. There was no implication that minerals were in any way conducted upward in chemical combination with the organic substances, and there has been rather universal agreement that the minerals which ascend the stem do so largely in the inorganic form. Recent indications by Thomas *et al.* (89) are that the translocation of sulfur occurs as sulfate ion whether it moves from stem base to stem tip or in the reverse direction. Tolbert and Wiebe (90) were unable to find organic forms of sulfur in the xylem sap of barley (*Hordeum vulgare*), spinach (*Spinacia oleracea*), and cucumber (*Cucumis sativus*) but suggested that organic phosphorus compounds might be translocated upward in the xylem sap of tomato.

Phosphorylcholine was suggested (57) as the carrier, but this serves better to exemplify the current search for carriers of nutrient ions than to prove their existence. Many difficult unanswered problems in nutrition might be explained with carriers, either as static loose adsorbers over which ions moved—or as bona fide organic carriers conducting ions.

D. Factors Influencing Distribution

It is not consistent with present views on translocation to ignore the effects which living cells have on the upward movement of minerals in the xylem. A cambium sheaths the xylem in many plants and by active accumulation may serve as a deterrent to an indiscriminate sweeping along of minerals in the transpiration stream. It has been shown previously that the xylem lying adjacent to the cambium carries the burden of the nutrient salts upward and that certain mineral nutrients are sometimes not delivered to aerial parts in proportion to the volume of transpirational water which is delivered to them. The cambium, whether it is limited to leaf traces, to cauline bundles, or to an interfascicular location, is always positionally, and usually metabolically, in the most favorable situation to exert a controlling influence over the passage of minerals in the transpiration stream (84).

The cambium, when present, together with the adjacent xylem and phloem and the rays which transcend them, forms a unit which is ideally constituted for a close regulation of the passage of minerals either upward in the xylem, downward in the phloem, or radially from one tissue to the other. Thus if the cambium within a certain region were in equilibrium with a high concentration of a given element in the phloem, it would be less likely to retard the passage of the same element ascending in the xylem than if it were low in that element and capable of active absorption.

The leaves located along the stem are connected with the vascular cylinder of the stem through their leaf traces and, by the processes of transpiration and growth, minerals are directed into them. It has not been possible to discriminate quantitatively between the influence which each of the two factors has on the delivery of minerals to the leaf when the two processes are proceeding simultaneously. There is some indication that the youngest leaves have the highest transpiration rate [(67) pp. 336–342] and consequently are aided in accumulating within themselves a sufficient quantity of minerals for their growth, but metabolic accumulation must not be minimized.

Mason and Maskell (62), in a study of the movement of solutes into the developing cotton boll, show calcium to enter the boll mainly in the xylem. The accumulation of calcium in the ovules was proportional to the calcium content of the tracheal sap, but for nitrogen, carbohydrate, phosphorus, and total ash the accumulation was vastly in excess of the delivery by the tracheal sap. In this instance, of course, the excess delivery above that in the xylem sap was assumed to be via the phloem, calcium not being phloem-mobile was restricted to only one mode of entrance, i.e., the xylem sap. The movement of phosphorus into the buds of *Carpinus* sp. and *Fagus* sp. was shown by Burström (21) to be in excess of the quantity which the xylem sap was capable of delivering at its current concentration. The water which moved into the buds was shown to come from a distance of at least 1.5–2.5 meters, for both species, and the phosphorus came a distance of 0.15–1.1 meters.

The importance of active absorption by the rapidly growing cells in immature leaves and fruits cannot be discredited as a factor in maintaining a sink into which minerals are constantly moving. The combined influence of the two factors, transpiration and metabolic accumulation, operates in such a way as to provide the continual delivery of minerals to all active plant parts. The participation of both the xylem and phloem, or intermittently one or the other, in the delivery of these minerals and organic substances is generally assumed from the mass of translocation data. Precise studies, however, on the relative importance of each tissue are lacking. The delivery of minerals to mature tissues by the transpiration stream alone may not be in proportion to the current needs. It would appear that mature leaves, which have fewer metabolic uses for certain minerals than immature ones, do export a fraction of the mineral matter delivered to them, providing the minerals in question are mobile in the phloem. For calcium, which is immobile, the precipitation of the excess as calcium oxalate may serve the purpose of disposing of the surplus.

A detailed analysis of the distribution pattern of radioactive phosphorus in the leaves at various positions on the stem of the bean (*Phaseolus vulgaris*) plant showed it to follow a rather precise pattern. There was an approximately linear relationship between the logarithm of the concentration of the element and the position of the leaves on the stem. This is shown in Fig. 2. In this instance the tracer was absorbed during the last 4 days preceding harvest (8). Figures 3, 4, and 5, which

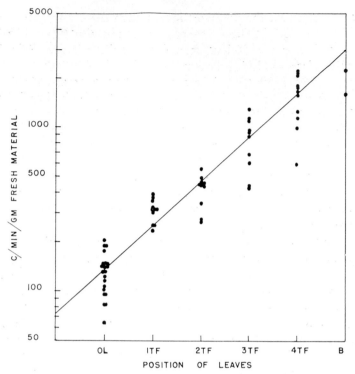

FIG. 2. The amount of P[32] moving into leaves at various positions on the stem of the red kidney bean (*Phaseolus vulgaris*) plant. The P[32] was absorbed as phosphate from the nutrient solution during a 4-day interval. The data are from the leaves of ten plants, all grown in the same tank. The oldest (opposite) leaves are to the left (OL); progression is toward younger leaves to the right (ITF, or first trifoliate leaf; to 4TF). Two very small leaves are shown at *B*. Adapted from Biddulph (8), page 266.

are discussed in Section IIIB, show the distribution patterns of aliquots of P[32], S[35], and Ca[45] which had been absorbed by the roots during a 1-hour period.

III. Mobilization Following Deposition

A. METABOLIC REUTILIZATION

The leaves constituting one orthostichy (vertical series) are more closely related nutritionally than are leaves adjacent to each other but not of the same orthostichy (84). In the cucumber the upper member of the nutritionally related pairs was able to acquire nutrients during its period of rapid growth from the lower older member. Some examples of the extent of reutilization of nitrogen and phosphorus during growth and reproductive development are given below: Gregory [see Williams (95)] has stated that 90% of the final total of nitrogen and phosphorus taken in by the developing cereal plant was absorbed by the time the dry weight constituted 25% of the final value. The nutrients were first used in the initial leaves, then withdrawn as the leaves became senescent and used in newly formed leaves, finally to be withdrawn for utilization in the inflorescence. The nitrogen and phosphorus which was absorbed early supported later growth and development, and its level determined the final yield. Williams (95) indicated that the rate of intake of nitrogen and phosphorus is governed by (a) the external supply and (b) the internal demand for growth and normal functioning of all plant parts. The internal demand in the oat (*Avena sativa*) plant was such that 93% of the inflorescence phosphorus was derived from other plant parts when an excessive external supply was available, whereas in phosphorus-deficient plants only 30% was from other plant parts. This indicates the ready availability of phosphorus within the plant for diversion into rapidly growing areas as long as an ample supply is available. In corn (*Zea mays*) plants (43), during grain formation, the vegetative parts contributed 60% of the total nitrogen moving to the grain. Only 40% was acquired from the soil and roots. Of the 60% contributed by the aerial plant parts, 60% came from leaves, 26% from stalks, 12% from husks, and 2% from shanks. Of the 60% which came from the leaves, 90% was from protein nitrogen.

In the case of phosphorus deficiency of oat plants, the whole of the phosphorus intake from the soil was retained by the root (95). It has also been shown (10) that in bean seedlings grown in a sulfur-deficient medium the roots were able to acquire from the cotyledons and withhold from the developing leaves an unusually high proportion of the cotyledonary sulfur. In both instances, i.e., that of phosphorus and sulfur deficiency, an unusually high root weight ratio was established. This appears to be characteristic of both deficiencies. The roots of apple trees, chrysanthemum (*Chrysanthemum morifolium*) plants, and bean plants have been shown to accumulate more phosphorus from foliar applications of this element when the plants were in a phosphorus-deficient condition (3, 37).

These examples indicate that internal factors such as nutrient content may operate to determine whether a root system will release minerals to the remainder of the plant or not and whether the released mineral will follow precisely the transpiration stream or be shunted into local areas of temporary intense accumulation along the channel of movement. Internal factors determine whether an element will be released from structural compounds in one part of the plant for utilization in another or be maintained in position. Little can be said about the factors which control the mobilization of minerals within the plant until more is known about the mechanisms which control the synthesis and degradation of cellular constituents. Evidence points to a complex system of controls which are under the influence of living cells at the respective ends of the plant axis as well as along the channels of movement, but the factor of greatest importance is undoubtedly the group of events associated with the reproductive process and the forces which these events set into motion.

The complete mobilization of reserves and their translocation to reproductive structures in annual plants is well known. Biennial plants, following flower induction, react similarly, and the occurrence of biennial bearing in certain varieties of perennial fruit trees also indicates that mobilization of reserves following fertilization may be excessive. Loehwing (53) has called attention to this mobilization and has furthermore shown that the permeability increases in cell membranes which allow such a complete mobilization of reserves also permit a leaching of significant quantities of minerals, especially nitrogen and potassium, from roots into the culture medium. A careful study of the stimulus which results in the freeing of reserves for translocation should prove particularly rewarding.

B. CIRCULATION PATTERNS

Hartig perhaps made the earliest suggestion that soluble products may make a circuit within the plant. He concluded that materials assimilated in the leaves passed downward in the bark and were stored in the parenchyma and rays. In the spring these materials were brought into solution, passed into the trachea and ascended with the moving current of water. Mason and Maskell (62) indicated that such a circuit might operate without the intervention of a storage period.

These investigations suggested that nitrogen, phosphorus, potassium, and other minerals ascended the stem mainly in the xylem and that any excess not used currently in the leaves was re-exported downward through the phloem. Calcium also ascended the xylem but was not re-exported in the phloem. The ratios of N, P, and K to carbohydrate moving downward in the phloem appeared to be in excess of that required for growth of the lower plant parts. It was then suggested that the excess was liberated into the xylem sap and reascended the stem.

Biddulph (7) showed the rapidity of movements of phosphorus within the plants and the possibility of a more or less continued circulation of phosphorus. A given phosphorus atom might easily make several cycles within the plants in a single day.

The methods employed in demonstrating a daily turnover of a particular element within the leafy tissue of the plant are for the most part piecemeal methods. The various experiments, to be successful, are much more easily carried out with radioactive tracers. The first step is to follow the movement of the element from the root system to the leaves, noting the pattern of distribution. This is best done in short-term experiments lasting but a few hours. The second step is to apply the element to the leaf of a previously untreated plant and determine the extent of movement to the root and to the various parts of the shoot of the plant. This is also best done in short-term experiments of a few hours. The rate of upward movement from the root is of the order of 1 meter per hour, and the downward, or outward, rate of movement from the leaf is not much less than this for such elements as phosphorus. It is evident, then, that it would take but a few hours until confusion of upward and downward movement within the plant would occur. In order to avoid such confusion, an additional experiment of a different type is carried out. This comprises the application, through the nutrient solution, of a radioactive tracer to the root system for a period as short as one hour. A group of plants so treated is then placed in a solution of normal nutrients (no radioactive isotope), and from this group, at successive intervals, a plant or more is withdrawn and analyzed or dried and prepared by one of the common methods for exposure to X-ray film. A succession of such plants then shows the fate of that particular group of atoms absorbed during the 1-hour period on a particular day. If only one experiment is to be carried out, the latter is perhaps the most useful, particularly if the half-life of the radioactive isotope permits studies over extended periods. This type of study has proved particularly helpful in interpreting the circulation patterns of various elements in the plant. The difficulties encountered in this type of work when isotopes were not used are presented by Curtis [(30) pp. 64–70].

By utilization of these methods, the translocation pattern for P^{32}, S^{35}, and Ca^{45} in the bean was fairly well worked out (see Figs. 3, 4, and 5) (11). Within 1 hour after P^{32} tracer was applied to the root, it was detected in all parts of the plant, though not uniformly. The highest concentration was in the lowermost trifoliate leaf, which in this instance was not quite fully mature.[1] The younger leaves higher on the stem had, at this instant, a lower concentration, and the bud showed

[1] To some extent this may be regarded as intake to make good the depletion during growth of the lower by the upper leaves. Ed.

still less activity. The P³² had not reached the apex of the stem in significant quantity within the first hour after application to the tracer. After 6 hours the apical bud had attained the high concentration which is characteristic of it, and subsequently as successive leaves unfolded,

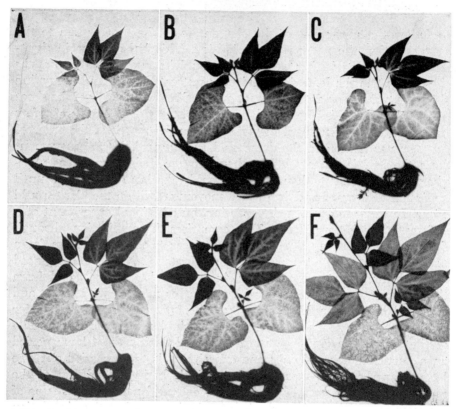

FIG. 3. A sequence of six autoradiograms showing the fate of an aliquot of P³² absorbed as $H_2P^{32}O_4^-$ during a 1-hour absorption period. The plants, after the hour in the nutrient solution containing the tracer, were removed to a normal (nonradioactive) solution where they remained for the following periods: A, 0 hours; B, 6 hours; C, 12 hours; D, 24 hours; E, 48 hours; and F, 96 hours. A fraction of the P³² remained mobile so that all parts of the plant were continuously supplied with the tracer during subsequent periods. From Biddulph et al. (11).

the apical bud continued to acquire and maintain the highest concentration of any plant part. The unifoliate leaves had their highest concentration at the end of the 6-hour period and then fell to a lower level, which was maintained throughout the 96-hour period. The trifoliate leaves in general maintained a rather uniform gradation in their concentration of the labeled phosphorus during the whole of the 96-hour period, the younger leaves always maintaining the highest concentrations. These results may be interpreted as indicating a more or less continuous circulation of phosphorus with a relatively small percentage of the total tracer being metabolically captured and incorporated into phosphorus compounds with low turnover rates.

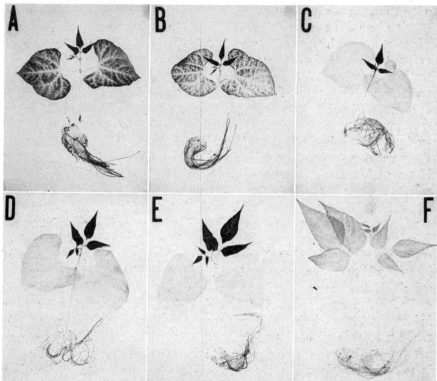

Fig. 4. A sequence of six autoradiograms showing the fate of an aliquot of S^{35} absorbed as $S^{35}O_4^{--}$ during a 1-hour absorption period. The plants, after the hour in the nutrient solution containing the tracer, were removed to a normal (nonradioactive) solution where they remained for the following period: A, 0 hours; B, 6 hours; C, 12 hours; D, 24 hours; E, 48 hours; and F, 96 hours. Most of the S^{35} which moved directly into the mature leaves was withdrawn within 12–24 hours. It moved predominantly into younger leaves near the stem apex, where it remained. From Biddulph et al. (11).

Phosphorus is known to participate in metabolic exchange cycles of short duration such as are involved in the phosphorylations accompanying glycolysis, etc. There is ample evidence that a significant amount of the phosphorus that is mobile in the phloem can be found in carbohydrate esters within a few minutes after the phosphorus enters the phloem (9). One may visualize the inorganic phosphate within the

plant as comprising a common pool from which phosphate may be
withdrawn for ester formation prior to condensation into starch, sucrose, cellulose, etc., and for participation in photosynthetic and respiratory cycles which produce the high energy phosphate bonds utilized in performing cellular work throughout the plant. Phosphorus

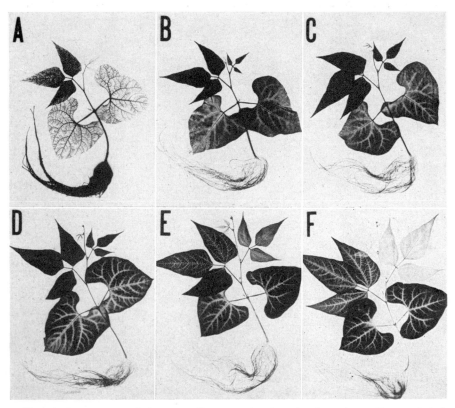

FIG. 5. A sequence of six autoradiograms showing the fate of an aliquot of Ca^{45}
absorbed as Ca^{45++} during a 1-hour absorption period. The plants, after the hour
in the nutrient solution containing the tracer, were removed to a normal (nonradioactive) solution where they remained for the following periods: A, 0 hours; B, 6
hours; C, 12 hours; D, 24 hours; E, 48 hours; and F, 96 hours. The Ca^{45} which
moved directly into the various leaves remained in place. Very little was withdrawn
to supply the younger leaves as they continued to develop. From Biddulph *et al.* (11).

may be quite unusual in the extent of its cycling, both geographically
and metabolically, and as an outcome, a usable form of it is maintained at a relatively uniform concentration throughout the plant.
    Sulfur is also freely mobile in the phloem but is rather quickly
caught up in metabolic reactions which immobilize most of it. When

absorbed by the root, it was delivered through the transpiration stream to all leaves, but within 12–24 hours the older mature leaves had lost to the young leaves at the stem apex most of the labeled sulfur which had been delivered to them. It appeared to make only one cycle within the plant body before rather complete metabolic capture occurred. Sulfur, being a component of proteins, is ultimately captured in those regions in which protein synthesis is dominant, such as stem and root tips, and in young leaves. Once deposited it appears to remain relatively immobile until proteolysis frees it for additional circulation and subsequent capture in other rapidly metabolizing regions where protein synthesis is occurring. This is a slow type of turnover which is quite different from that shown by phosphorus.

Sulfur is released by protein breakdown in grasses (*Lolium multiflorum, Lolium subulatum*), but it was stated by Wood and Barrien (96) that it did not leave the leaf as readily as nitrogen similarly released. However, Kylin (50) has shown that wheat (*Triticum aestivum* var. 'Weibull's Eroica') plants, when given radioactive sulfur followed by a period without sulfur of any kind, translocated the tracer to the sulfur-deficient roots where it then remained. It did not recirculate to newly developing leaves. In the bean plant Biddulph *et al.* (10) have shown that the sulfur acquired by the roots from the cotyledons will not be released to other parts of the plant unless there is an ample supply being acquired currently from the nutrient medium. Root growth in a medium deficient in sulfur was not significantly curtailed in either the wheat or the bean plant. Apparently the roots have the ability to capture and maintain a major portion of the sulfur that is mobile when the supply within the plant becomes limited. All data obtained with radioactive sulfur indicate that sulfur is freely mobile in the phloem and very readily captured in metabolically active regions where protein synthesis occurs. Older views (96) were that sulfur is comparatively immobile, but this must refer to the fact that little is usually free to move.

Calcium, when similarly studied, gave every evidence of being immobile in the phloem. It was very uniformly delivered through the transpiration stream to all portions of the plant, but there it remained. All new growth that was made during the 96-hour experimental period utilized the unlabeled calcium which was currently obtained from the nutrient solution by the root. The initial delivery of the tracer calcium into the leaves was the final delivery; it did not recirculate within the plant. When applied to a leaf, it also remained in place; only the slightest trace could be detected in the stem tissue below the node of the treated leaf.

FIG. 6. An Fe⁵⁵ autoradiogram of a bean (*Phaseolus vulgaris*) plant which received Fe⁵⁵ via one leaflet. The plant had been grown under nutrient conditions that promote mobility of iron within the tissues. The nutrient conditions were: pH 4.0, 0.0001 MP, and 0.002 ppm iron. All other essential elements were present. From Biddulph (8), facing page 274.

Radioactive calcium administered to the root environment of the peanut (*Arachis hypogaea*) plant resulted in a very limited Ca^{45} movement into the fruit. When Ca^{45} was administered to the fruiting zone, an active absorption occurred directly into the developing fruit. This absorption greatly exceeded the small supply obtained from the roots.

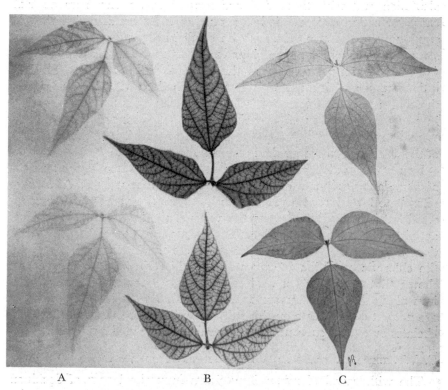

Fig. 7. An Fe^{55} autoradiogram of leaves of three separate beans plants. Fe^{55} was available from the nutrient solution to each plant. Plant (A) was grown at a pH of 7.0 and with 0.001 MP. The leaves were chlorotic throughout. Plant (B) was grown at a pH of 7.0 and with 0.0001 MP. The leaves had green veins and chlorotic interveinal tissue. Plant (C) was grown at a pH of 4.0 and with 0.0001 MP. The leaves were green throughout. The nutrient solution was otherwise complete. From Biddulph (8), facing page 270.

Because of this relatively unusual method of fruit nourishment, Bledsoe *et al.* (15) were able to obtain a very striking demonstration of calcium immobility in the phloem. The occurrence of calcium oxalate crystals in the phloem parenchyma of several plants has been reported. This is an indication of the insolubility of calcium salts at the pH of the phloem tissues. The immobility of calcium has been generally known for many years and has been confirmed a number of times by the use of Ca^{45}.

Still a different pattern of circulation is manifested by iron. It is listed among the elements which are withdrawn from leaves prior to abscission and with those that are not. This is perhaps an accurate characterization of it, for it has been shown by the sensitive methods employing radioactive tracers to be conditionally mobile. The current status of the nutrition of the plant was shown by Rediske and Biddulph (77) to determine the extent of the mobility of this element. The conditions under which mobility is maintained in the bean plant are those of a moderate to low level of phosphorus nutrition and growth in a relatively acid medium, i.e., pH 4 to 5. Iron is much more likely to move readily, following its application to a leaf, if the iron content of the tissues is low rather than adequate (see Fig. 6).

Iron can be quickly immobilized at the root surface in the presence of abundant phosphate and at a pH of approximately 7. At a lower phosphate level but at a pH of 7, iron enters the plant and moves upward but is immobilized along the veins of the leaf, producing a characteristic interveinal chlorosis. If such a plant is now shifted to a solution at pH 4, the iron is mobilized, enters the interveinal tissue, and the chlorotic condition is corrected (see Fig. 7). Weinstein (94) has shown that the uptake of chelates may also prevent the deposition of iron along the leaf veins or correct the situation if it develops. There are not yet enough data on this subject to determine the part played by chelates or related substances in the translocation of the micronutrients.

C. WITHDRAWAL FROM LEAVES PRIOR TO ABSCISSION

The withdrawal of mineral nutrients from the leaves of deciduous plants prior to their abscission has been studied by many workers [ref. (68) pp. 882–883], and with but few exceptions it has been shown that nitrogen, potassium, phosphorus, sulfur, chlorine, and, under some conditions, magnesium and iron are withdrawn, at least to some extent, prior to abscission. The amount withdrawn varies with the element and the species of plant and amounts to values of 1–90% of the maximum value present during the growing season. Some potassium has been shown to be lost completely from the plant via the root after having been first withdrawn from the leaf. Recent work by Tukey et al. (91), with radioactive tracers, indicates that leaching of minerals from leaves as a result of rain or dew may constitute a significant loss from the leaves of some plants.

Those elements which are not withdrawn or which are only slightly withdrawn prior to abscission are calcium, boron, manganese, and silicon. Iron and magnesium are perhaps intermediate in mobility since they have been listed a number of times in each category. With calcium it is quite evident that the original deposition in the leafy tissue, through the transpiration stream, is the final deposition and little, if any, recirculation occurs. This appears to be true for annual plants, but the situation may differ in perennial plants, for Ferrell (38) has shown relatively high concentrations of radioactive calcium to be present in the growing buds of pine (*Pinus monticola*) trees for each of the two years following the application of Ca^{45} to the trunk. In this instance a remobilization and accumulation in the rapidly growing buds, following a period of immobility, must coincide with a similar mobilization of carbohydrate materials when growth is resumed after winter dormancy. Curtis (30) found that the mobilized minerals utilized in early spring growth of leaves are acquired from minerals stored in the region of the buds rather than from other areas at greater distances, such as roots. He was of the opinion that such mobilization and translocation over long distances was relatively rare in deciduous plants.

IV. Translocation from Leaves

A. General Concepts

Translocation of minerals from leaves was a rather difficult subject to investigate before radioactive isotopes of the normally occurring mineral nutrients became available. The daily export of carbohydrates and their importance in the growth and development of the plant as a whole has been known and appreciated for many years. There have also been consistent reports of daily fluctuations in the mineral content of leaves. The latter consisted of over-all changes ranging from a few to a significant percentage of the total content. These changes in mineral content were based on periodic analysis of leaf tissue at frequent intervals throughout the day (75). While they indicate the total difference in mineral content at two discrete intervals, they do not provide a means of detecting and measuring the incoming minerals and the outgoing minerals when the two processes are proceeding simultaneously. It is, therefore, not surprising that a dynamic interchange of minerals within the plant was not adequately demonstrated until isotopes of the commonly occurring mineral nutrients became available. The use of these substances has made a different type of investigation possible, and from the data which have accumulated, a more precise description of the geographic wanderings and metabolic cyclings of individual elements has been possible.

Conclusive evidence as to which tissue system is involved in the export of minerals from leaves was obtained by the use of radioactive-
tracer methods. Using a technique similar to that of Stout and Hoagland previously described, it was demonstrated by Biddulph and
Markle (12) that the bark, and therefore by implication the phloem,
and not the xylem, was the pathway for the export of phosphorus from
the leaf. The phosphorus moved predominantly downward in the

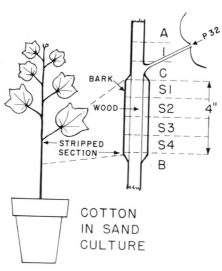

PLANT
SECTIONS

Fig. 8. Diagram to show the method of stripping the bark and the arrangement
of sections of bark and wood for the purpose of measuring the movement of P[32]
downward in the stem. Quantitative results are shown in Table II. Adapted from
Biddulph and Markle (12), page 69.

phloem of the stem, but some upward movement occurred in the same
tissue. The upward movement was erratic and did not exceed 40% of
the total mobile material; it was usually much less. A rapid transfer of
a fraction of the mobile material from the phloem to the xylem was
demonstrated, therefore indicating that both the xylem and phloem
might be responsible for upward movement in the stem when leaf
exportation rather than root exportation of minerals was involved (Fig.
8, Table II). The culminating demonstration that the phloem is the
tissue through which phosphorus and sulfur move when descending the

stem from a particular leaf was given by S. F. Biddulph (13), using microautoradiographic methods (Fig. 9).

It should not be inferred from the above statements that translocation of foods and minerals in the phloem was unknown to earlier physiologists. Studies on the downward movement of foods in the phloem began with Hartig in 1858 and in 1861 (42), although he had described the sieve tubes in 1837. Because of his observations of exudation from cut phloem, he assumed a mass flow of the sieve tube contents. However, he also observed the movement of protoplasm which he

TABLE II

DISTRIBUTION OF MIGRATORY PHOSPHATE (FROM A LEAF) IN SECTIONS OF COTTON STEM AFTER A MIGRATION PERIOD OF 1 HOUR[a]

	P^*O_4 in each section[b] (μg)			
	Plant stripped[c]		Plant unstripped	
Section[d]	Bark	Wood	Bark	Wood
A		1.11		
I	0.458[e]	0.100[e]	0.444	
C		0.610		
S1	0.544	0.064	0.160	0.055
S2	0.332	0.004	0.103	0.063
S3	0.592	0.000	0.055	0.018
S4	0.228	0.004	0.026	0.007
B		0.653	0.152	

[a] From Biddulph and Markle (12), page 69.
[b] Calculated from radioactivity.
[c] Plant stripped 16 hr before use.
[d] See Fig. 8 for details of sections.
[e] Section stripped at end of migration period.

suggested might be the moving sap. Nägeli (69) agreed with Hartig on the mass flow of sap and suggested that pressure from adjacent cells caused the flow. Sachs (80) presented the idea that proteins moved in mass through a series of sieve cells with open pores connecting one with another but that sugars moved principally in parenchyma (particularly the starch sheath) by diffusion from starch grain to starch grain.

Velten (92) had observed and described streaming in sieve tubes and many other types of cells, and when de Vries (33) also observed active streaming of both circulation and rotational types in companion cells and phloem parenchyma, he suggested that this would account for the rapid movement of substances in the phloem. At the same time

he pointed out the complete inadequacy of diffusion to do so. Lecomte
(51) observed streaming in young sieve tubes but claimed that in older
cells it ceased. He was opposed to the views of Sachs and Van Tieghem
that the protoplasm in older phloem cells was dead. He considered a
mass flow in older cells, but was also sympathetic to movement by
diffusion. Kienitz-Gerloff (48) suggested that plasmodesmata served as

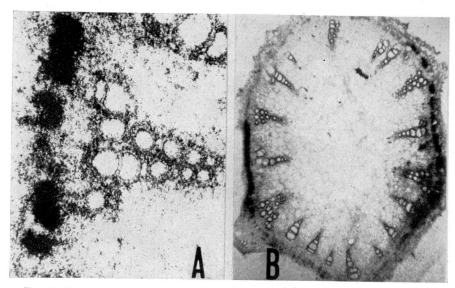

FIG. 9. Sections of the stem of the red kidney bean prepared by vacuum dehydration while frozen, with a superimposed autoradiograph due to S^{35}. The S^{35} was
placed on the surface of the leaf 1 hour prior to removal of tissue for the section.
This section was taken ⅛ inch below the node from which the leaf diverged. A
shows that the S^{35} was moving in portions of the phloem tissue. B shows that
wide areas of the phloem acquired S^{35} by tangential movement from the leaf traces.
From Biddulph (13), pages 144–145.

a path of rapid movement for protoplasm, colloidal material, and soluble substances of large molecular weight; sugars and similar substances could pass through plasmatic membranes. Strasburger (86)
thought the sieve pores in the phloem of angiosperms to be open but in
gymnosperms to be closed. He also concluded that streaming ceased in
mature sieve tubes.

Czapek (31) showed the slowness of sugar movement in parenchyma
cells and as a result considered the movement of carbohydrates and
soluble forms of nitrogen to take place through the cytoplasmic linings
of open sieve pores. He thereby rejected the dual paths of Sachs. He

also rejected protoplasmic streaming in favor of a mechanism akin to secretion. He considered that plasmolysis of sieve tubes did not stop carbohydrate movement but that narcotics slowed it up. Mangham (59) favored the mass flow of protein through sieve pores but suggested a diffusion mechanism for sugars over stationary colloids. Schmidt (82) showed that sieve pores were filled with cytoplasm and cast doubt on the ability of protein to pass from one sieve tube to another. Thus the structure of the sieve tubes contributed materially to doubts as to its ability to conduct food substances at adequate rates.

Calculations of the rate at which foods must move from leaves [Birch-Hirschfeld (14), Dixon (34), Dixon and Ball (35)] into tubers and roots [Mason and Lewin (60)] gave values between 14 and 88 cm per hour. Since rates in the order of only 1 cm per hour for movement of lithium salts, etc., in the phloem had been observed by Birch-Hirschfeld (14), these higher values cast more doubt on the adequacy of the phloem for the downward transport of foods (cf. Section IIB).

Mason and Maskell (61) showed that diurnal variations in total sugars in the leaf were correlated with those in the bark and that variations in transport were correlated with variations in concentration gradients in the same tissue. Also ringing caused an interruption in the downward translocation of carbohydrate in the bark but did not interrupt downward movement of dye in the wood. This certainly indicated again that translocation of carbohydrate took place in the bark, and there was no doubt that the translocation of minerals accompanied the carbohydrates and nitrogenous materials (62). It was at this stage in the reaffirmation of the phloem to the role of translocation of metabolites and minerals from the leaf that radioactive tracers provided the conclusive proof.

The use of tissue "blocks," such as ringing, steaming, anesthetizing, chilling, or withholding oxygen from a test portion of the stem or petiole through which translocation was anticipated, constituted the experimental procedures from which most of the early information on the pathway of movement from the leaf was obtained. These manipulations have their maximum effect on the phloem tissue since it is composed of living cells, whereas the xylem is not similarly affected as its make-up is largely limited to cells which no longer function by means of their protoplasts. The data have indicated the necessity of living cells for the normal movement of solutes from leaves to the remainder of the plant and have, therefore, indicated the phloem as the channel of movement.

As substances enter the longitudinal axis of the plant from a particular leaf, there exists an opportunity for the migratory substance to move either downward toward the root or upward toward the apex of the plant. In either event there will be zones of actively metabolizing cells at or near the ends of the channels which would benefit from the export of these materials from the leaf. It is known that the stream is effectively divided so that the nourishment of both is accomplished. The path of downward movement has been covered previously. The path of upward movement from a given leaf is not so clearly indicated.

B. ABSORPTION FROM LEAF SURFACES

One of the difficulties encountered in a study of the migration of minerals from leaves is to find a suitable method of introducing the tracer substance so that it will move readily in the phloem and yield reliable information on the pathway of movement as well as on the forces responsible for its movement. The various methods which have been used are: (a) A tracer solution is introduced by a leaf-vein injection, wherein one of the principal veins of the leaf is cut and the cut end leading to the leaf margin is dipped into a small container of the tracer. The pathway of movement is then toward the periphery of the leaf via the xylem, the channel through which the original delivery of the material from the root would be made in an uninjured plant. The tracer will then leave the leaf through the phloem, providing there is not an unusual water deficit in the remainder of the plant and providing that an additional injury is not made to the xylem system in some other part of the plant. (b) A tracer solution is introduced into the intercellular spaces of the leaf by vacuum infiltration. This is done by immersing the desired leaf, or leaflet, into a solution of tracer over which a vacuum of sufficient strength is formed to remove most of the intercellular gas from the leaf. On release of the vacuum the tracer solution penetrates the leaf tissue. (c) The tracer solution is placed in a "lake" on the leaf surface, or it may be applied as a droplet. (d) The tracer solution is applied as a fine spray to the leaf surface. The latter method is much the easiest to manipulate and is patterned after the commercial methods of application of large volumes of solution such as insecticides, herbicides, or nutrient elements to orchards, fields, lawns, gardens, etc.

A comparison of the efficiency of uptake of P^{32} when applied by the various methods listed above has been made by Koontz and Biddulph (49). Leaf-vein injection, vacuum infiltration, and spray application are approximately equally efficient methods for the application of P^{32} to the leaf surface. Absorption of tracer from a droplet is slower, and it is necessary to apply the solution at a pH of 2–3 in order to effect an efficient entrance through the cuticle.

Some of the factors which have been found to influence the absorption and subsequent translocation of phosphorus from a water solution applied to the leaf surface are: the concentration of the solution, the nature of the chemical compound and its rate of crystallization, the area and specific surface on which the application is made, the age and position of the treated leaf, and the phosphorus level of the plant. Phosphorus absorption following leaf application is independent of stomata number and degree of opening. It appears to be related to structural features such as cutinization and the proximity of plasmodesmata to the outer surface of the cutin. Wetting agents were either ineffective or decreased the uptake of P^{32}. When absorption was retarded, it appeared to be caused by the formation of a complex of the phosphorus tracer with the wetting agent (49, 87).

Phosphorus absorption from the upper and lower surfaces of the bean (49, 71) and coleus (*Coleus* sp.) (6) leaves is approximately equal. The lower surfaces of the apple and chrysanthemum (6) leaves appear most receptive, whereas in corn it may be the opposite surface (71). This characteristic obviously varies with the species.

The age of the leaf is of importance in that fully mature leaves export phosphorus freely, while immature leaves fail to export this element. The position of the leaf on the stem is also important insofar as the destination of the exported material is concerned. Those leaves nearest the root export the major part of their migratory material to the root, whereas the mature leaves higher on the stem export predominantly to the apical region (49) (see Fig. 11). Phosphorus fails to move effectively in tissues that are phosphorus deficient (49). Presumably, the deficiency must be made up within the exporting tissue and progressively along the path of movement before efficient export takes place.

The phosphorus concentration of the solution which is best for absorption appears to be that which is just short of injuring the leaf tissue. This falls in the range of 30 to possibly 50 mM for such plants as beans. Higher concentrations have been used with corresponding injury to the leaf. This may be a factor which varies with the species and with different formulations of the applied solution. The most effective pH is near 4.0, or where the ion form will be predominantly H_2PO^{4-} and the H^+ concentration will not be sufficient to injure the leaf.

The accompanying cation is important primarily through its effect on the rate of crystallization of the salt on the leaf surface. Phosphorus was absorbed most readily from those salts with the slowest rate of crystallization. If the rate of crystallization for a given salt was too rapid, a hygroscopic agent like glycerol could be added to good effect providing it did not form a complex with the phosphate ion. The compounds tested, listed in order of decreasing efficiency in promoting the absorption and translocation of phosphorus (using a 24-hour migration period) in bean plants were: NaH_2PO_4, K_2HPO_4, Na_2HPO_4, K_3PO_4, $(NH_4)_2HPO_4$, $NH_4H_2PO_4$, KH_2PO_4, and Na_3PO_4 (49). There is only general agreement in this matter as Tukey et al. (91) have shown ammonium salts to be more efficient at pH values of 2 and 3, although leaf injury was detected at these acidities.

The environmental factors which are important in leaf absorption and subsequent translocation of minerals can be deduced from the above information. Light and dark periods have little effect; that which is observable is perhaps related to the efficiency of translocation, which is higher in the light (7), rather than to absorption. There is some indication that evening, when the relative humidity of the air is highest, is the most effective part of the day for foliar application of phosphorus by spraying (91). The most effective temperature for the translocation of minerals in the bean plant appears to be between 20° and 30°C (87).

Studies on the penetration of herbicides into leaves are of much current interest, and considerable effort is being exerted to determine the factors which control entrance of these substances (71a). It appears that the penetration of minerals and herbicides have much in common, as both are dependent on many structural peculiarities of the leaf surface, on contact angles between solution and leaf surface, on the accompanying ions, the pH, the concentration of the solution, its hygroscopicity, the age of the leaf, and the nutrient status of the plant.

C. BIDIRECTIONAL MOVEMENT

The evidence presented by Curtis (30) that minerals may move upward in the phloem, at least in the region of the stem tip, introduced the possibility that movement in both directions in the phloem could occur simultaneously. There is now a large volume of literature which demonstrates the downward movement of minerals in the phloem, and a rather universal belief that upward movement of carbohydrates, nitrogenous materials, and minerals being exported from leaves also takes place in the phloem leading to developing fruits and stem tips. Observations employing tracers which show phosphorus movement to take place in both an upward and a downward direction in the stem, after application to a leaf, are given by Biddulph and Markle (12). Similar observations have been made for C^{13}-labeled photosynthate by Rabideau and Burr (76). The data of Chen (23), on studies wherein bark and wood had been separated from each other, indicated that P^{32}

and C^{14} photosynthate can move simultaneously in opposite directions in the bark. None of the reports is incontestable; either the tissues were not properly isolated or the migration periods were too long to prevent confusion of the results by interchange between tissues. The evidence, however, is strongly indicative that upward movement in the phloem actually does occur and that the presence of tracers in this tissue, at levels above the point of entrance into the stem from a given leaf, does not merely represent a lateral movement from the xylem back into the phloem at a higher level on the stem.

Traditional treatment of bidirectional movement generally infers a passage upward, and a simultaneous passage downward, of one or more substances in the same sieve tubes. Crafts (28), however, has indicated that a particular leaf may serve two "sinks," one toward the apex and the other toward the base, by a division of the solute stream between them. This implies that separate channels within the phloem must be considered, for if two leaves, one above the other, are both exporting toward the apex and the base, streams will pass each other flowing in opposite directions in the included internode. Two concepts of bidirectional movement then exist, one implying movement within a single channel, and the other implying movement in adjacent channels. It has generally been assumed that bidirectional movement in a single channel cannot occur by a pressure-flow mechanism, but is possible by a protoplasmic-streaming mechanism and by an activated-diffusion mechanism where two different substances are involved.

The vascular anatomy of the bean plant together with the patterns of translocation of P^{32} from leaves at various positions on the stem may serve as an example of the physical problems associated with the two concepts of bidirectional movement. The xylem tissues identifying the leaf traces in the stem are represented in semidiagrammatic form in Fig. 10. In considering export from a given leaf, via the phloem associated with these traces, there is only the possibility of downward movement. Upward movement to the apex must be accomplished in the secondary interfascicular phloem or the phloem of cauline bundles or young leaf traces. A tangential transfer from the phloem of the leaf trace into the adjacent phloem transcending the leaf gap must occur before ascent in the phloem is possible. The vascular anatomy of the plant in question will determine how far below the node the mobile material must descend before adjacent phloem is encountered into which movement may occur.

The initial direct exportation of tracer phosphorus from leaves at different levels on the stem (and consequently of different ages) can be summarized quite satisfactorily from Fig. 11. There are two aspects

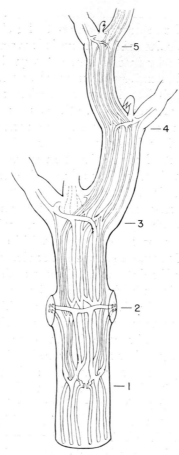

Fig. 10. A semidiagrammatic representation of the leaf traces on one side of the stem of the red kidney bean plant. This was made from the stained (basic fuchsin) and cleared (cedar oil) stems and from serial sections. In addition to these leaf traces there were from 7 to 10 cauline bundles in most specimens studied. These are not shown in the diagram. An interfascicular cambium was present to about the fourth node. The nodes are as follows: *1*, cotyledonary; *2*, opposite leaves; *3*, *4*, and *5*, trifoliate leaves.

of export which are important to a proper interpretation of the data represented in the figure. They are: (a) the total amount of phosphorus exported from each leaf, and (b) the predominant direction of its movement. The figure shows that leaves, in the age class of the plants here used, export phosphorus in proportion to their age. Exportation ranges from very pronounced in the case of the older unifoliate leaves to none at all for younger, rapidly expanding leaves. The figure

FIG. 11. P^{32} autoradiograms demonstrating the distribution of P^{32} in bean plants 24 hours after the application of 200 μl of 10 mM NaH$_2$P*O$_4$ as a spray to the following leaves: A, a unifoliate leaf; B, the terminal leaflet of the first trifoliate leaf, C, the terminal leaflet of the second trifoliate leaf; D, the terminal leaflet of the third trifoliate leaf. The predominant direction and the amount of movement from each leaf is clearly shown. From Koontz and Biddulph (49).

also shows that the predominate direction of exportation is determined by the position of the leaf on the stem. The leaves transport phosphorus to the root in proportion to their proximity to it. The combined effect of the relationship between the amount of transport and position, and the direction of transport and position, results in the lowermost leaves contributing much more phosphorus to the roots than do the upper leaves. Immature leaves fail to export phosphorus either upward or downward, but when they first become mature enough to export, they contribute phosphorus primarily to the stem apex. The exported material from leaves of an intermediate position is divided between the root and stem apex. The older basal leaves contribute only a small fraction of their exportable minerals to the apex. The net result is that the amount delivered from each leaf to the stem apex is approximately the same. These statements are applicable only to the direct translocation which occurs within 24 hr after application is made to the leaves. Longer migration periods may possibly result in freeing more P^{32} from the roots for upward transport in the transpiration stream.

Phosphorus, in moving into the stem from a leaf of the bean plant, has been shown to move tangentially from the phloem of leaf bundles into the adjacent phloem so that wide areas of the phloem become laden with the tracer (13). This is shown in part B of Fig. 9. The stem section represented in the figure was taken just below the point of divergence of the five leaf traces from the vascular cylinder of the stem into the leaf. A considerable portion of the tracer is then in a favorable position for ascending the phloem of the stem.

There are also sufficient data to indicate a significant transfer of tracer phosphorus from the phloem of the leaf traces to the xylem as the tracer descends the stem (9). The ratio of the quantity in the "bark" to that in the "wood" (of the bean stem) was found to be approximately 3:1 at several levels of the stem. The transfer of tracer from the phloem to the xylem was then quantitatively related to the amount in the phloem, and the transfer took place continuously along the bundle. As the concentration in the phloem is always greatest near the point of entrance from the leaf, the greatest transfer to the xylem occurs at this place. Ascent in the xylem to those organs and tissues near the apex of the stem has been covered previously.

The evidence presented above indicates that mature leaves, at different positions on the stem, can (Fig. 11) and probably do (Fig. 3) quite continuously export phosphorus toward both the stem apex and the stem base. This indicates, first of all, that bidirectional movement of phosphorus occurs in the stem, and if we accept the evidence that upward movement in the phloem is possible, it indicates that bidirectional movement occurs in the phloem. Second, the observation has been made that descent of tracer in leaf traces necessitates tangential transfer to adjacent bundles for ascent, thereby indicating that separate channels for upward and downward movement are a distinct possibility. The data, however, fail to provide evidence on the necessity for separate channels of upward and downward movement, so the possibility of a single functional channel with bidirectional movement cannot be dismissed.

A further inquiry into the question of channels within the phloem may be justified. Figure 11 shows a strong descent of P^{32} tracer in the phloem, regardless of which leaf is exporting it. There are indications of a transfer of a portion of this to the xylem (see Section V,A,1), but this material does not enter the general transpiration stream leading to mature leaves as can be seen from Fig. 11. If it ascends in the xylem, its ascension is predominantly in those traces leading to immature leaves, as mature leaves in general fail to acquire a significant amount of the tracer. This lends some support to the view that the phloem is definitely involved in a significant part of the upward transport of minerals from leaves.

The individual role which the phloem plays in the ascension of minerals exported by leaves has not been satisfactorily determined. In all instances in which radioactive tracers have been used in translocation experiments, a very rapid and quantitatively significant interchange of tracer between xylem and phloem has been observed. This is true regardless of whether the tracer is originally moving upward in the xylem (Table I) or downward in the phloem (Table II). This fact has been responsible for the confusion which has resulted in attempting to demonstrate upward movement of minerals in the phloem. The two tissues, with the intervening cambium, most definitely compose a unit which may be functionally as well as structurally related.

There can be little doubt that upward movement into leaf primordia occurs in the protophloem with the accomplishment of the nutrition of this highly metabolic area. As the leaf matures, the vascular tissue of the leaf matures accordingly, continuing to supply the materials for the rapid growth which is characteristic of young leaves. Importation through both the phloem and xylem then continues until the functional maturity of the leaf is attained. At this time the leaf begins the export of carbohydrates and the excess minerals delivered by the transpiration stream. The phloem of the leaf traces must then experience a reversal in the predominant direction of flow, for the leaf now exports instead of importing. As the stem apex progresses upward by its growth and the leaf in question is left farther behind, a correspondingly greater percentage of the leaf's exportable material finds its way to the root. Newer leaves then contribute to the nutrition of the stem apex. The developmental pattern of translocation is repeated as leaves progressively mature along the axis.

Downward movement in leaf traces and upward movement in cauline bundles would comprise the simplest explanation for bidirectional movement of inorganic solutes in the phloem which could be offered, if pressure flow is considered to be the mechanism of transport. If, however, the phloem of the stem as a whole is shown to function as a unit and bidirectional movement occurs within it, this explanation of movement becomes untenable. The clarification of this point would be a welcome addition to the knowledge of translocation. The bidirectional concept has been, and will continue to be, a much discussed subject until a method can be devised which can test the concept directly.

V. Mechanics of Movement in the Phloem

A. Progress in Experimentation

1. Logarithmic Distribution Patterns

Sufficient data of a quantitative nature have been accumulated from the use of radioactive tracers to permit a more exact evaluation of the translocation process than was possible by the utilization of older methods. Recently it has been shown that as a tracer moves downward in the stem, having entered from a particular leaf, it is distributed in the stem in such a way that there is a linear relationship between the logarithm of the concentration of the tracer and the distance from the point of entry into the stem (9, 12, 87). A search for the significance of this particular pattern of distribution has revealed several interesting relationships between the tracer which is moving in the phloem and that which leaves it for one reason or another.

Preliminary to an understanding of the significance of the logarithmic distribution pattern is an appreciation of the quantitative relationship between the amount of tracer in the phloem and the amount which is transferred to the xylem. In the bean the ratio between the concentration of P^{32} in the phloem to that in the xylem, when movement is originally taking place in the phloem, is approximately 3:1 (9). This holds true when tests are made at several levels of the stem. The exact ratio between the two tissues varies somewhat among individuals receiving the same treatment, but this appears to be a fairly representative figure. The transfer to the xylem represents a physical loss from the channel in which movement is taking place. It suggests that there may be other losses which effectively reduce the amount of mobile material proceeding in the phloem.

A second type of loss is encountered in the metabolic incorporation of mobile material into organic complexes. These may not necessarily be mobile, but may represent, in the case of phosphorus, a loss to metabolic processes which are taking place in stationary sites. Both of the above modes of loss from the channel of movement may be processes which proceed at rates proportional to the first power of the concentration of available substrate.

A chromatographic analysis of the alcohol-soluble constituents of the petiole of the bean through which both C^{14}-labeled metabolites and P^{32} were moving shows the extent of the metabolic incorporation of these two tracers (9). The C^{14} was applied to the upper side of the leaf as $C^{14}O_2$ and the P^{32} as a liquid spray to the under side of the leaf. Approximately 44% of the P^{32} and 90–95% of the C^{14} was extractable and represented in the alcohol extract. Identification of the constituents showed (a) inorganic phosphate, (b) fructose-1,6-diphosphate, and (c) several phosphate esters including glucose-6-phosphate and glycerol phosphate, to be present in proportions of about one-third each. Radioactive sucrose was the principal sugar being exported, but some glucose and fructose carrying the radioactive label were identified on the chromatogram. The fructose-1,6 diphosphate was not significantly C^{14}-labeled within a 1-hour migration period showing that the carbon chain was not of current photosynthetic origin. Phosphorylation must have occurred en route. Mobility tests of the fructose-1,6-diphosphate were not conducted. These data are shown in Fig. 12.

Three tracers, THO (tritiated water), P^{32}, and C^{14}, have been applied simultaneously to a leaf of the bean plant. Sections of the stem below the node of the treated leaf were analyzed for each tracer separately after migration periods of 15, 20, and 30 minutes. Typical data for a 15-minute migration period are shown in Fig. 13 (9). The data showed that there was an approximately linear relationship between the logarithm of the concentration of solute (P^{32} and C^{14}) and the distance of movement from the point of entry of tracer into the stem.

The following graphic analysis was formulated to show the means of development of the logarithmic distribution of solute. The following analogy was used: As the tracer entered the first section of a hypothetical stem, supposedly from a treated leaf, a constant proportion, e.g., 25% of the total was removed from the sieve tubes and the residue passed to the second section. Here 25% of this amount is removed and the new residue passed to the third section, etc. A second aliquot of tracer is then brought down the stem in the same manner, but is stopped one section short of the ultimate progress of the first. This is followed by a third aliquot, which stops one section short of the second, etc. The graphic analysis is shown in Fig. 14. When all values for the various sections are summarized individually and the logarithms plotted against the distance, a linear relationship results. This is true

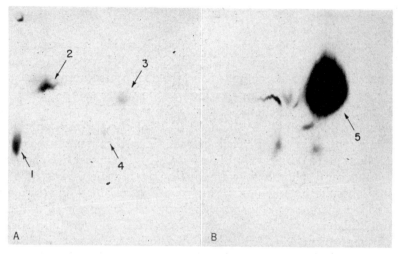

Fig. 12. (A) P^{32} autoradiogram of a filter paper chromatogram of the alcohol-soluble constituents in the petiole of an attached bean leaf. The leaf had received approximately 200 μc of P^{32} in 50 μl of solution, sprayed onto the under surface, and 100 μc of C^{14} as $C^{14}O_2$, held above the upper surface. The migration period was 1 hour. (B) The C^{14} constituents of the same chromatogram. (A) was prepared by filtering out the radiation due to C^{14} and (B) was made after the decay of the P^{32}. The exposures are calculated to be identical for both reproductions. The spots are as follows: *1*, inorganic phosphate (P^{32}); *2*, fructose-1,6-diphosphate (P^{32}); *3*, glucose-6-phosphate? (P^{32}); *4*, glycerol phosphate? (P^{32}); *5*, sucrose (C^{14}) with some glucose (C^{14}) and fructose (C^{14}). Other compounds are unknown.

for the amount remaining in the sieve tubes, or for that which escapes, or for the sum. The slope of the curve would be directly related to the amount of tracer which escaped from the sieve tubes regardless of the mechanism of escape. A more precise mathematical treatment of this aspect of translocation has been given by Horwitz (45).

The assumptions which were adopted in order to set up the mathematical model, and which fit the known facts insofar as P^{32} movement is concerned, are as follows: The tracer must be introduced at a constant rate into a moving column of liquid, but after leaving the region containing the source of tracer, the tracer must be free to escape irreversibly through the walls of the column at a rate which is proportional to the first power of the concentration and then remain in place after escaping. To be consistent with the observed facts, the "escape"

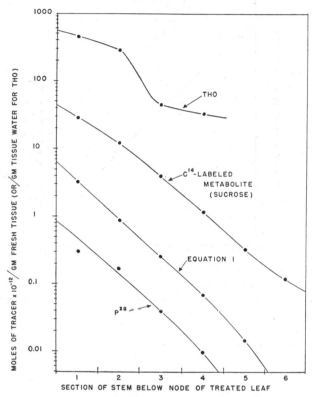

FIG. 13. The quantitative distribution of three tracers in the stem below the node of a "treated" bean leaf. The three tracers were applied to the leaf simultaneously, as $H_2P^{32}O_4^-$ (323 μc) in THO (5000 μc) (total volume 100 μl) to the underside of the leaf, and $C^{14}O_2$ (100 μc) (total CO_2 approximately $1 + \%$) to the upper side of the leaf. The migration period was 15 minutes. Each stem section was 1 inch long and the stem joined the root at the lower end of section six. The hypothetical distribution of a tracer (other than THO) in the stem, using a mathematical model based on a solution flow, is shown as a equation (1). A similar curve is obtained with diffusion phenomenon, but no mode of acceleration to this velocity is apparent.

assumed above must be considered analogous to (a) physical removal from the phloem channel into the xylem or adjacent cells, (b) metabolic exchange that results in the removal of tracer by chemical bonding with constituents along the pathway.

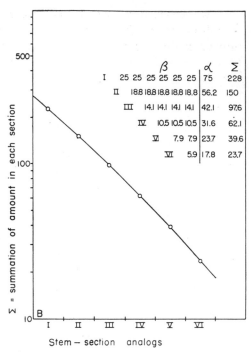

The following table appears within the figure:

	β						α	Σ
I	25	25	25	25	25	25	75	228
II	18.8	18.8	18.8	18.8	18.8		56.2	150
III	14.1	14.1	14.1	14.1			42.1	97.6
IV	10.5	10.5	10.5				31.6	62.1
V	7.9	7.9					23.7	39.6
VI	5.9						17.8	23.7

Fig. 14. The theoretical distribution of a substance in a hypothetical stem when allowed to enter at α in successive 100-unit aliquots one after another, flow downward stepwise from section to section, and move laterally to β at each level in amounts comprising 25% of the total entering the section. Each successive aliquot entering the stem is stopped one section short of its predecessor. The amounts in each section are summarized (Σ) and plotted to a log scale. From Biddulph and Cory (9).

The total amount of tracer at any distance from the point of entry of tracer into the phloem, designated as X, is given by:

$$X = C_o e - \frac{kx}{A_p v} \{A_p + k(t - x/v)\} \tag{1}$$

which is equivalent to

$$\log_e X = \log_e C_o + \log_e \{A_p + k(t - x/v)\} - kx/vA_p \tag{2}$$

in which $C_o =$ the steady state concentration of tracer in the phloem at the place where the moving column leaves the source region of tracer; $A_p =$ the effective cross-sectional area of the phloem channel; $k =$ a constant which is characteristic of the rate of escape of tracer from the stream (it is the constant for the diffusion of tracer through the walls of the phloem channel, or a first-order chemical reaction rate constant for the removal of tracer from the flowing stream into a bound form); $t =$ the time that the tracer has been flowing in the stream; $v =$ the velocity of flow of fluid in the phloem channel; $(t - x/v) =$ the distance back of the front of the tracer, in time units, to the particular point of interest. The derivations are given by Horwitz (45).

The slope of the curve relating $\log_e X$ and x is:

$$\frac{d \log_e X}{dx} = - \frac{k}{v(A_p + k(t - x/v))} - \frac{k}{vA_p} \tag{3}$$

The slope of the curve will be steep with a high rate of escape or with a low velocity of flow. If the effective cross-sectional area were to differ for different substances, this factor (A_p) would also alter the slope. In addition the slope of the curve will be steeper if the rate of entrance of tracer into the stream increases during the migration period, and the slope becomes flatter as the escape mechanism approaches saturation, as in this way the rate of escape is curtailed. A very good agreement between the actual experimental curves and the curve resulting from the mathematical treatment was obtained (Fig. 13). Removal of tracer away from the region of measurement, e.g., by upward movement in the transpiration stream, will make the relationship between the observed and the calculated distribution more perfect, for it is within the column of moving solute that the perfect logarithmic relationship exists. Deviations from the perfect pattern are due to biological irregularities in the accumulation of tracer in the tissues surrounding the moving column. The mathematical representation of the translocation phenomenon provides an explanation for the logarithmic distribution pattern, but even more important, a valid approach to a quantitative representation of the process has been formulated. This must now be tested under a variety of conditions before its usefulness can be fully established.

2. Amount and Velocity of Translocation

The total amount of labeled phosphorus moving in the phloem appears to be two to three orders of magnitude less than the amount of sucrose or water when compared on a molar basis (9). The data did not permit a valid quantitative determination of the absolute amounts of each tracer capable of moving from a whole leaf, but merely compared the amount moving from the area of application, which was approximately 1 square inch of leaf surface. The approximate observed molar concentration of tracers in the first inch of stem below the treated leaf, 20 minutes after application of the tracers, was as follows: P^{32}, $0.2 \times 10^{-12} M$; C^{14}-labeled sucrose, $75 \times 10^{-12} M$; THO, $260 \times 10^{-12} M$. The concentration of C^{14} sucrose in tritiated water would then be approximately 10 molal. This has little significance, as there is no reason to believe that the THO came to equilibrium with the tissue water within the treated area, and the $C^{14}O_2$ concentration above the leaf was enriched to the extent of 1–2% carbon dioxide.

The velocity of flow of the tracers in the phloem have been determined with considerable accuracy. The method for doing this was to determine the positions of the fronts of the three individual tracers in the stem following their simultaneous application to the leaf. A short migration period of 15 minutes was necessary in order to catch the tracers before they entered the root. The velocity of the C^{14} tracer from the instant of releasing it, in the form of $C^{14}O_2$ over the surface of the leaf, until the front had entered the root, a distance of 26.7 cm, was 107 cm/per hour. The fronts of the other two tracers were caught in sections behind those in which C^{14} was already present and gave calculated velocities of 86.4 cm/per hour (9).

When it was assumed that all tracers moved in the phloem at the same rate, i.e., 107 cm/per hour, but those with the slower apparent rate had different penetration times to the phloem, a calculated value of 2.87 minutes was obtained for the penetration time from the epidermis to the phloem for P^{32} and THO. In both of these calculations, i.e., for velociy of transport in the phloem and for penetration times to the phloem, no time has been allowed for diffusion of $C^{14}O_2$ to the plastids, for the photosynthetic reactions, and the diffusion of C^{14}-labeled sucrose to the phloem, which means that all velocities are too low. The calculated velocity of the C^{14} tracer increases to 114.6 cm/per hour and 123 cm/per hour when a 1- and a 2-minute migration time of tracer to the phloem is assumed. There was insufficient evidence to indicate whether sucrose actually had a higher velocity of movement or whether P^{32} and THO had a lower penetration time to the phloem.

3. Differential Velocities

There was some evidence, gained from a number of tests, that the P^{32} tracer characteristically lagged behind the THO tracer. The fact that sections of stem as long as 1 inch were used for the analysis of movement made it difficult to determine this point with certainty.

Swanson and Whitney (87) have presented some data which suggest that when K^{42} and P^{32}, and Cs^{137} and P^{32} in separate pairs were applied simultaneously to a leaf, the members of the pairs moved at different velocities in the phloem. The evidence is largely based on an analysis of the curves of a plot of the logarithm of the concentration of the tracer in the stem against the distance from the point of application.

Since their data were expressed in a form suitable for analysis by the scheme presented by Horwitz, it was of interest to make the analysis as an example of its usefulness. Apparently, in all cases, a plot of the logarithm of their radioactivities with distance was linear. But whereas the slopes for Cs^{137} and P^{32}, applied simultaneously, were different, those for K^{42} and P^{32} were similar. Equation 3 shows that the slope is determined by the parameters k, A_p, and v. It is not necessary, then, to suppose that a difference in slope implies a difference in velocity of the two substances, as was done, since a difference in either k or A_p can account for it. However, the data comparing K^{42} and P^{32} cannot be resolved in Horwitz's formulation without assuming different modes of transport for K^{42} and P^{32}. For although the slopes for the two substances remained the same, the rates of increase of the logarithms of their concentration with respect to time were different.

Horwitz shows that the rate of increase of $\log_e X$ with respect to time is

$$\frac{d \log_e X}{dt} = \frac{k}{A_p + k(t - x/v)} \tag{4}$$

Therefore, at least one of the parameters, k, v, or A_p must differ. But since the slopes remain the same, it is necessary from equation (3) that at least two of these parameters must differ. So either the velocity or the effective cross-sectional area of the phloem, or both, must be different for potassium and phosphorus.

This example displays the usefulness of this particular method of approach, but it must not be inferred that it establishes the case for differential rates of movement. This method of analysis requires that certain strict conditions in the mode of entrance of tracer into the phloem be met, and this assurance could not be gained from the experiments of Swanson and Whitney. There are indications, however, that the method holds some promise.

It is obvious that if different velocities of movement for the various substances in the phloem are assumed, this carries with it the implication that (a) different channels displaying different characteristics are used, or (b) different mechanics for the movement of each substance, or class of substance, is employed.

4. Physical Models

The distribution patterns of several radioactive tracers in the bean stem have been approximated in filter paper strips (9). This was done in the manner of descending filter paper chromatography, with the exceptions that the filter paper strip was moistened with water prior to the application of the tracer, and the tracer was applied continuously in a water solution to the top of the paper. The distribution of P^{32} showed a good agreement with that in the bean stem. Ca^{45}, if applied at a pH of 2.5, behaved similarly to sucrose; if applied at a pH of 5.5 it behaved similarly to P^{32}; but when the pH was 7.2, the Ca^{45} moved

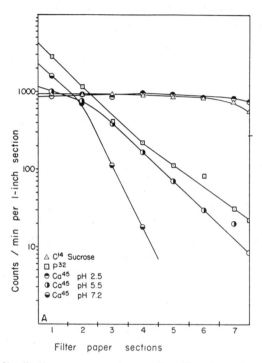

Fig. 15. The distribution patterns of several radioactive substances in separate filter paper strips (Whatman #1, 2.2 × 22 cm) following continuous application of the tracer substances, in solution, to the top of the prewetted (water) strips. Descending time approximately 90 minutes. From Biddulph and Cory (9).

poorly as it does in the phloem. C^{14} sucrose, when applied to the paper in the above manner, moved very freely; perhaps more freely than it appears to move in the phloem. The curves are shown in Fig. 15.

In these chromatographic procedures the absorptive forces in the filter papers operate to produce the same effects as the escape mechanisms mentioned in the previous section. The resultant logarithmic distribution pattern, in the filter paper, is brought about by the competition between the solvent and the absorptive forces in the filter paper for the solute.

B. Implications of Experimental Evidence to Existing Hypotheses

The data on mineral movement can provide little information on the causes of solvent movement in the phloem, so the matter will not be discussed here. Both the pressure flow and the protoplasmic-streaming hypotheses utilize the concept of solvent movement. The mass-flow mechanism assumes a solution movement by means of a turgor gradient, while the activity of the protoplast is assumed responsible for the streaming movement in protoplasmic streaming. But there is no agreement that the mass of translocation data point to a true solution movement. Clements' (25) data indicate that a smaller volume of water is involved in the movement of organic substances into fruits than one would predict from the pressure-flow mechanism. Mason and Phillis (66) also indicate that the capacity of the phloem to conduct water appears to be very limited. This has been one of the principal criticisms directed toward the pressure-flow mechanism, but an equally strong defense has been maintained by Crafts (28).

The effects which the metabolic activity of the phloem cells might have on the translocation process can be discussed equally well from the point of view of the maintenance of structure through which translocation is occurring as from the point of view of energy utilization in propulsion of solute. There are few data pertaining to the latter subject. Wanner (93) was unable to find a complete set of glycolytic enzymes in the phloem sap of *Robinia pseudoacacia* and suggested that it is difficult to understand how energy release in the phloem can be applied to the propulsion of solutes in that channel. Yet there can be little doubt that metabolic activity occurs in the phloem tissues at the time that translocation occurs, and when it stops, translocation also stops. Swanson and Whitney (87) have even shown a close dependence between translocation and temperature up to 20–30°C, with decreasing translocation at higher temperatures. This is in general the same type of temperature-dependent curve as is encountered in respiration.

Crafts (27) is of the opinion that experiments wherein translocation was modified by temperature treatments, oxygen exclusion, etc., can be of little value in deciding the issue of the dependence of translocation on metabolic energy release because of the effects of such treatments on viscosity changes and alterations of permeability to components within the channel of movement. To be realistic we must admit that we do not yet have the means for determining whether the part played by metabolic energy release within the phloem is concerned with the maintenance of the channel in a functioning condition or the propulsion of the solute or solution through the system. It would appear that we are most consistent with the accumulated data when we conclude that factors which interfere with the metabolic activity of the phloem cells also interfere with the translocation of materials in the phloem tissues.

Unfortunately the above fact need not necessarily be more damaging to one hypothesis of translocation than to another, unless it is assumed that in pressure flow the protoplast is so completely impassive that it is incapable of effecting any modification of the flow of solute. This seems a very unlikely assumption since temperature changes, oxygen exclusion, metabolic inhibitors, narcotics, etc., may increase the permeability of membranes, which in turn may affect pressure flow by altering the characteristics of movement by allowing a greater escape through the walls of the cells.

Van Overbeek (71a) has suggested that translocation may be analogous to movement of solutes in filter paper chromatography. This suggestion is unsatisfactory in its simplified form, but in pursuing the concept one might be allowed the supposition that with a protoplast as an absorbent there are more possibilities for altering the characteristics of movement to agree with observed behavior than with a filter paper. However, in reality this suggestion is only usable as an example of a possible means whereby the channel may effectively influence the flow of solute under more or less normal conditions. The motivating force must of necessity be a solvent movement. Capillarity cannot function in this instance, as the tissues are already fully hydrated.

As for activated diffusion and polar transport, no means have yet been found for treating them experimentally or hypothetically. Even though the distribution pattern of minerals in the phloem, as they move from a leaf toward the root, resembles a pattern which would result from diffusion, no inherent accelerating mechanism is known which can produce the observed velocities. And as for solvent movement, no better mechanism for propulsion has been proposed than that of the Münch system. But this does not mean that a better one will not be found, for as long as discrepancies exist between the proposed hypotheses and the observed facts, there will be a continued search for a better fit.

REFERENCES

1. Anderssen, F. G. Some seasonal changes in the tracheal sap of pear and apricot trees. *Plant Physiol.* 4, 459–477 (1929).
2. Arndt, C. H. The movement of sap in *Coffea Arabica* L. *Am. J. Botany* 16, 179–190 (1929).
3. Asen, S., Wittwer, S. H., and Teubner, F. G. Factors affecting the accumulation

of foliar applied phosphorus in roots of *Chrysanthemum morifolium. Proc. Am. Soc. Hort. Sci.* **64**, 417–422 (1954).

4. Auchter, E. C. Is there normally a cross transfer of foods, water and mineral nutrients in woody plants? *Maryland Agr. Expt. Sta. Bull.* No. **257**, 33–62 (1923).

5. Bennett, J. P., Anderssen, F. G., and Milad, Y. Methods of obtaining tracheal sap from woody plants. *New Phytologist* **26**, 316–323 (1927).

6. Bennett, S. H., and Thomas, D. E. Absorption, translocation and breakdown of Schradan applied to leaves, using P^{32} labeled material. II. Evaporation and absorption. *Ann. Appl. Biol.* **41**, 484–500 (1954).

7. Biddulph, O. Diurnal migration of injected radiophosphorus from bean leaves. *Am. J. Botany* **28**, 348–352 (1941).

8. Biddulph, O. The translocation of minerals in plants. *In* "Mineral Nutrition of Plants" (E. Truog, ed.), pp. 261–275. University of Wisconsin Press, Madison, Wisconsin, 1951.

9. Biddulph, O., and Cory, R. An analysis of translocation in the phloem of the bean plant using THO, P^{32} and $C^{14}O$. *Plant Physiol.* **32**, 608–619 (1957).

10. Biddulph, O., Cory, R., and Biddulph, S. F. The absorption and translocation of sulfur in red kidney bean. *Plant Physiol.* **31**, 28–33 (1956).

11. Biddulph, O., Biddulph, S. F., Cory, R., and Koontz, H. Circulation patterns for P^{32}, S^{35}, and Ca^{45} in the bean plant. *Plant Physiol.* **33**, 293–300 (1958).

12. Biddulph, O., and Markle, J. Translocation of radiophosphorus in the phloem of the cotton plant. *Am. J. Botany* **31**, 65–70 (1944).

13. Biddulph, S. F. Visual indications of S^{35} and P^{32} translocation in the phloem. *Am. J. Botany* **43**, 143–148 (1956).

14. Birch-Hirschfeld, L. Untersuchungen über die Ausbreitungsgeschwindigkeit gelöster Stoffe in der Pflanze. *Jahrb. wiss. Botan.* **59**, 171–262 (1919).

15. Bledsoe, R. W., Comar, C. L., and Harris, H. C. Absorption of radioactive calcium by the peanut fruit. *Science* **109**, 329–330 (1949).

16. Bollard, E. G. The use of tracheal sap in the study of apple tree nutrition. *J. Exptl. Botan.* **4**, 363–368 (1953).

17. Bollard, E. G. Nitrogen metabolism of apple trees. *Nature* **171**, 571 (1953).

18. Bollard, E. G. Nitrogenous components in plant xylem sap. *Nature* **178**, 1189–1190 (1956).

19. Boussingault, J. *Agronomie* **I** (1860); **II** (1861). [From Sach's "History of Botany" (81).]

20. Bünning, E. Endogenous rhythms in plants. *Ann. Rev. Plant Physiol.* **7**, 71–90 (1956).

21. Burström, H. The rate of nutrient transport to swelling buds of trees. *Physiol. Plantarum* **1**, 124–135 (1948).

22. Caldwell, J. Studies in translocation. II. The movement of food materials in plants. *New Phytologist* **29**, 27–43 (1930).

23. Chen, S. L. Simultaneous movement of phosphorus^{32} and carbon^{14} in opposite directions in phloem tissue. *Am. J. Botany* **38**, 203–211 (1951).

24. Clements, H. F. The upward movement of inorganic solutes in plants. *Research Studies State Coll. Wash.* **2**, 91–106 (1930).

25. Clements, H. F. Movement of organic solutes in the sausage tree *Kigelia africana. Plant Physiol.* **15**, 689–700 (1940).

26. Clements, H. F., and Engard, C. J. Upward movement of inorganic solutes as affected by a girdle. *Plant Physiol.* **13**, 103–122 (1938).

27. Crafts, A. S. Translocation in plants. *Plant Physiol.* 13, 791–814 (1938).
28. Crafts, A. S. Movement of assimilates, viruses, growth regulators, and chemical indicators in plants. *Botan. Rev.* 17, 203–284 (1951).
29. Curtis, O. F. The effect of ringing a stem on the upward transfer of nitrogen and ash constituents. *Am. J. Botany* 10, 361–382 (1923).
30. Curtis, O. F. "The Translocation of Solutes in Plants: A Critical Consideration of Evidence Bearing upon Solute Movement." McGraw-Hill, New York, 1935.
31. Czapek, F. Über die Leitungswege der organischen Baustoffe im Pflanzenkörper. *Sitzber. kais. Akad. Wiss. Wien. Abt. I* 106, 117–170 (1897); Zur Physiologie des Leptoms der Angiospermen. *Ber. deut. botan. Ges.* 15, 124–131 (1897).
32. de Saussure, N. T. Recherches chimiques sur la végétation. Nyon, Paris, 1804.
33. de Vries, H. Über die Bedeutung der Circulation und der Rotation des Protoplasmas für den Stofftransport in der Pflanze. *Botan. Z.* 43, 2–6, 18–26 (1885).
34. Dixon, H. H. Transport of organic substances in plants. *Nature* 110, 547–551 (1922).
35. Dixon, H. H., and Ball, N. G. Transport of organic substances in plants. *Nature* 109, 236–237 (1922).
36. Eckerson, S. Seasonal distribution of reducase in the various organs of an apple tree. *Contribs. Boyce Thompson Inst.* 3, 405–412 (1931).
37. Eggert, R., and Kardos, L. T. Further results on the absorption of phosphorus by apple trees. *Am. Soc. Hort. Sci. Proc.* 64, 47–51 (1954).
38. Ferrell, W. K. Mobility of calcium-45 after injection into western white pine. *Science* 124, 364–365 (1956).
39. Fischer, A. Glycose als Reservestoff der Laubhölzer. *Botan. Z.* 405, 187–195 (1888).
40. Hales, S. "Statical Essays, Containing Vegetable Staticks, or, An Account of Some Statical Experiments on the Sap in Vegetation." W. Innys, London, 1727.
41. Hanson, J. B., and Biddulph, O. The diurnal variation in the translocation of minerals across bean roots. *Plant Physiol.* 28, 356–370 (1953).
42. Hartig, Th. Ueber die Bewegung des Saftes in Holzpflanzen. *Botan. Z.* 19, 17–23 (1861).
43. Hay, R. E., Earley, E. B., and DeTurk, E. E. Concentration and translocation of nitrogen compounds in the corn plant (*Zea mays*) during grain development. *Plant Physiol.* 28, 606–621 (1953).
44. Hewitt, W. B., and Gardner, M. E. Some studies of the absorption of zinc sulfate in Thompson seedless grape canes. *Plant Physiol.* 31, 393–399 (1956).
45. Horwitz, L. Some simplified mathematical treatments of translocation in plants. *Plant Physiol.* 33, 81–93 (1958).
46. Ingen-Housz, J. An essay on the food of plants and renovation of the soil. *In* "General Report from the Board of Agriculture," appendix to Chapter 15. London, 1796.
47. Kastens, E. Beiträge zur Kenntnis der Funktion der Siebröhren. *Mitt. Inst. allgem. Botan. Hamburg* 6, 33–70 (1924).
48. Kienitz-Gerloff, F. Die Protoplasmaverbindungen zwischen benachbarten Gewebeselementen in der Pflanze. *Botan. Z.* 49, 1–10, 17–26, 33–46, 49–60, 65–68 (1891).
49. Koontz, H., and Biddulph, O. Factors regulating absorption and translocation of foliar applied phosphorus. *Plant Physiol.* 32, 463–470 (1957).
50. Kylin, A. Uptake and metabolism of sulfate by deseeded wheat plants. *Physiol. Plantarum* 6, 775–795 (1953).

51. Lecomte, H. Contribution a l'étude du liber des angiosperms. *Ann. sci. nat. Botan.* **10**, 193–324 (1889).
52. Liebig, J. "Chemistry in its Application to Agriculture and Physiology," Edited from the manuscript of the author by L. Playfair, 2nd Ed. Taylor and Walton, London, 1842.
53. Loehwing, W. F. Mineral nutrition in relation to the ontogeny of plants. *In* "Mineral Nutrition of Plants" (E. Truog, ed.). University of Wisconsin Press, Madison, Wisconsin, 1951.
54. MacDougal, D. T. Reversible variations in volume, pressure and movement of sap in trees. *Carnegie Inst. Wash. Publ.* **No. 365**, 1–90 (1925).
55. MacDougal, D. T., Overton, J. B., and Smith, G. M. The hydrostatic-pneumatic system of certain trees: Movement of liquids and gases. *Carnegie Inst. Wash. Publ.* **No. 397** (1929).
56. McMurtrey, P. E., Jr. Cross transfer of mineral nutrients in the tobacco plant. *J. Agr. Research* **55**, 475–482 (1937).
57. Maizel, J. V., Benson, A. A., and Tolbert, N. E. Identification of phosphoryl choline as an important constituent of plant sap. *Plant Physiol.* **31**, 407–408 (1956).
58. Malpighi, M. "Anatomes plantarum pars altera." Londini [From Sachs' "History of Botony" (81).] 1679.
59. Mangham, S. On the mechanism of translocation in plant tissues. An hypothesis with special reference to sugar conduction in sieve-tubes. *Ann. Botany (London)* **31**, 293–311 (1917).
60. Mason, T. G., and Lewin, C. T. On tthe rate of carbohydrate transport in the greater yam, *Dioscorea alata* L. *Sci. Proc. Roy. Dublin Soc.* **18**, 203–205 (1926).
61. Mason, T. G., and Maskell, E. J. A study of diurnal variation in the carbohydrates of leaf, bark, and wood and of the effects of ringing. *Ann. Botany (London)* **42**, 189–253 (1928).
62. Mason, T. G., and Maskell, E. J. Preliminary observations on the transport of phosphorus, potassium, and calcium. *Ann. Botany (London)* **45**, 126–173 (1931).
63. Mason, T. G., Maskell, E. J., and Phillis, E. Concerning the independence of solute movement in the phloem. *Ann. Botany (London)* **50**, 23–58 (1936).
64. Mason, T. G., and Phillis, E. Studies on the transport of nitrogenous substances in the cotton plant. VI. Concerning storage in the bark. *Ann. Botany (London)* **48**, 315–333 (1934).
65. Mason, T. G., and Phillis, E. The migration of solutes. *Botan. Rev.* **3**, 47–71 (1937).
66. Mason, T. G., and Phillis, E. Some comments on the mechanism of phloem transport. *Plant Physiol.* **16**, 399–404 (1941).
67. Maximov, N. A. "The Plant in Relation to Water." (Translation by R. H. Yapp.) pp. 336–342. Allen and Unwin, London, 1929.
68. Miller, E. C. "Plant Physiology." McGraw-Hill, New York, 1938.
69. Nägeli, C. Ueber die Siebröhren von Cucurbita. *Sitzber. Bayer. Akad. Wiss. München* **1**, 212–238 (1861).
70. Nightingale, G. T. The nitrogen nutrition of green plants. *Botan. Rev.* **3**, 85–174 (1937); **14**, 185–221 (1948).
71. Oliver, W. F. Absorption and translocation of phosphorus by foliage. *Sci. Agr.* **32**, 427–432 (1952).
71a. Overbeek, J. van. Absorption and translocation of plant regulators. *Ann. Rev. Plant Physiol.* **7**, 355–372 (1956).

72. Overton, J. B. Stem anatomy and sap conduction. *Carnegie Inst. Wash. Yearbook* **25**, 155–158 (1925).

73. Overton, J. B., and Smith, J. B. Additional observations on stem anatomy and sap conduction. *Carnegie Inst. Wash. Yearbook* **26**, 159–160 (1926).

74. Phillis, E., and Mason, T. G. The effect of ringing on the upward movement of solutes from the roots. *Ann. Botany (London)* **4**, 635–644 (1940).

75. Phillis, E., and Mason, T. G. On diurnal variation in the mineral content of the leaf of the cotton plant. *Ann. Botany (London)* **6**, 437–442 (1942).

76. Rabideau, G. S., and Burr, G. O. The use of the C^{13} isotope as a tracer for transport studies in plants. *Am. J. Botany* **32**, 349–356 (1945).

77. Rediske, J. H., and Biddulph, O. The absorption and translocation of iron. *Plant Physiol.* **28**, 576–593 (1953).

78. Roach, W. A. Plant injection as a physiological method. *Ann. Botany (London)* **3**, 155–226 (1939).

79. Rumbold, C. Injection of chemicals into chestnut trees. *Am. J. Botany* **7**, 1–20 (1920).

80. Sachs, J. Ueber de Leitung der plastischen Stoffe durch verschiedene Gewebeformen. *Flora (Jena)* **46**, 33–42, 49–58, 65–74 (1863).

81. Sachs, J. "A History of Botany." Oxford Press, London, 1890.

82. Schmidt, E. W. "Bau und Funktion der Siebröhre der Angiospermen." Fischer, Jena, 1917.

83. Senebier, J. "Mémoires physico-chimiques sur l'influence de la lumière solaire pour modifier les êtres de trois règnes, surtout ceux du règne végétal." Chirol., Geneva, 1782.

84. Steward, F. C. Salt accumulation in the plant body. *Symposia Soc. Exptl. Biol.* **8**, 393–406 (1954).

85. Stout, P. R., and Hoagland, D. R. Upward and lateral movement of salt in certain plants as indicated by radioactive isotopes of potassium, sodium, and phosphorus absorbed by roots. *Am. J. Botany* **26**, 320–324 (1939).

86. Strasburger, E. Ueber den Bau und die Verrichtungen der Leitungsbahnen in den Pflanzen. *Histologische Beiträge* **3** (1891).

87. Swanson, C. A., and Whitney, J. B. The translocation of foliar applied phosphorus-32 and other radioisotopes in bean plants. *Am. J. Botany* **40**, 816–823 (1953).

88. Thomas, W. The seat of formation of amino acids in *Pyrus malus* L. *Science* **66**, 115–116 (1927).

89. Thomas, M. D., Hendricks, R. H., Bryner, L. C., and Hill, G. R. A study of the sulphur metabolism of wheat, barley, and corn using radioactive sulphur. *Plant Physiol.* **19**, 227–244 (1944).

90. Tolbert, N. E., and Wiebe, H. Phosphorus and sulfur compounds in plant xylem sap. *Plant Physiol.* **30**, 499–504 (1955).

91. Tukey, H. B., Wittwer, S. H., Teubner, F. G., and Long, W. G. Utilization of radioactive isotopes in resolving the effectiveness of foliar absorption of plant nutrients. *Proc. Intern. Conf. Peaceful Uses Atomic Energy Geneva, 1955* **12**, 138–143 (1956).

92. Velten, W. Über die Verbreitung der Protoplasmaströmung im Pflanzenreich. *Botan. Z.* **30**, 645–653 (1872).

93. Wanner, H. Glycolitic enzymes in the phloem sap. *Ber. schweiz. botan. Ges.* **63**, 201–212 (1953).

94. Weinstein, L. H., Robbins, W. R., and Perkins, H. F. Chelating agents and plant nutrition. *Science* **120**, 41–43 (1954).
95. Williams, R. F. Redistribution of mineral elements during development. *Ann. Rev. Plant Physiol.* **6**, 25–42 (1955).
96. Wood, J. G., and Barrien, B. S. Studies on the sulphur metabolism of plants. III. On changes in amounts of protein sulphur and sulphate sulphur during starvation. *New Phytologist* **38**, 265–272 (1939).
97. Wormall, A. The constituents of the sap of the vine (*Vitis vinifera* L.) *Biochem. J.* **18**, 1187–1202 (1924).
98. Yendo, Y. Injection experiments in plants. *J. Coll. Sci. Imp. Univ. Tokyo* **38**, 1–46 (1917).

PREAMBLE TO CHAPTER 7

This chapter deals with the over-all economy of the plant in its relations to water, recognizing that the physical problems of absorption and adjustment to the immediate environment have been dealt with at the cellular level in Chapter 2 and the special mechanism of regulatory control which is exercised by the guard cells of the stomata has been considered in Chapter 3. Water movement, with its implications for solute transfer (Chapter 6) and the many adjustments that occur between plants and their habitats now properly fall to be discussed in Chapter 7. This carries the discussion of water relations beyond the topic of cell physiology, which is the main underlying theme of Volumes I and II. but, even so, the water relations of cells are so implicated in the water relations of the plant body as a whole that this is an appropriate place in this treatise for this discussion. Similarly, it was found necessary in Chapter 4 to extend the discussion of the nonnutrient relation of cells to salts, as solutes, by a treatment of the plant body as a whole (Chapter 4, Part III).

CHAPTER SEVEN

Transpiration and the Water Economy of Plants

PAUL J. KRAMER

I. Introduction

A. THE IMPORTANCE OF WATER

Water is one of the most important factors in the environment of plants, and the kind and amount of plant growth over most of the earth's surface is limited more by the availability of water than by any

607

other factor, except possibly temperature. The ecological importance of water indicates that it must be equally important physiologically because the only way in which an environmental factor can affect plant growth is through its effects on internal physiological processes and conditions.

Water is important to plants both qualitatively and quantitatively. Its qualitative importance results from the fact that it possesses the most unusual combination of properties of any known liquid. The biological importance of the unique properties of water has been appreciated by biologists for many years (22, 109, 135). Water has the highest specific heat of any known substance, except liquid ammonia. The standard used for measuring heat, the calorie, is the amount of energy required to warm 1 gm of water from zero to 1°C. The latent heat of melting, 80 calories per gram, is among the highest known, and the heat of vaporization (540 calories at 100°C) is the highest known. Evaporation of water therefore produces a strong cooling effect and condensation of water vapor, a warming effect. Water also is a good conductor of heat among liquids and nonmetallic solids, although poor as compared with metals.

Because of its peculiar internal structure water has an unusually high density and it attains maximum density at 4°C. The density of ice is less than that of water as evidenced by the fact that water expands when it freezes and ice floats. Related to its high density is its high surface tension (73 dynes per cm at 20°C) which is exceeded only by that of mercury, and its very high tensile strength which is estimated to be about 14,000 atm per square cm under ideal conditions. It also adheres firmly to surfaces possessing many oxygen atoms with which hydrogen bonds can be formed, such as clay, cellulose, and glass. This tendency to be adsorbed or bound to surfaces explains why large amounts of water occur in cell walls and protoplasm and why the imbibition of water is accompanied by swelling.

Water is very slightly dissociated, hence it is a poor conductor of electricity (i.e., has a high dielectric constant) and therefore is an excellent solvent for electrolytes because they can ionize freely in it. It also is an excellent solvent for many other substances, such as sugars, organic acids, phosphates, nitrates, and substances containing amino, carboxyl, or other groups with which it can form hydrogen bonds. Thus water is perhaps the most nearly universal solvent known.

Another property of biological importance is the relatively high transparency of water to visible radiation which permits aquatic plants to carry on photosynthesis below the surface and allows light to penetrate even into thick leaves. Finally, it should be mentioned that the boiling point and the freezing point of water are abnormally high for a substance with the formula H_2O.

It is believed that the unique properties of water can be explained by the asymmetry of the molecules and by assuming that water molecules are associated together by hydrogen bonds in an orderly structure [(260); see also (63, 92)]. In ice the water molecules are held in a widely spaced tetrahedral lattice structure by hydrogen atoms forming bonds between the oxygen atoms of adjacent molecules. As ice melts some of the hydrogen bonds break and the open lattice structure collapses, decreasing the volume and increasing the density. Above 4°C the increased thermal agitation of the molecules more than compensates for loss of structure and the volume increases progressively with temperature. The high melting point of water and high heat of melting of ice are attributed to the large amount of energy required to break some of the hydrogen bonds. To vaporize water the remainder of the hydrogen bonds must be broken, and this explains the high boiling point and high heat of vaporization.

B. THE USES OF WATER IN PLANTS

It has been stated that water has the most unusual combination of properties of any known chemical compound, and the existence of this unique combination of properties is an essential factor in the existence of life on the earth. It may equally well be said that water has the most unusual combination of functions of any substance found in plants. Its numerous functions may be grouped under four general headings.

1. Water as a Constituent

Water is an essential constituent of active protoplasm, often comprising 85–90% of the fresh weight of root and stem tips, succulent leaves and fruits, and over 50% of the fresh weight of woody structures, as shown in Table I. As will be shown later, decrease in water content much below normal is accompanied by decrease in rates of various physiological processes and if the water content falls below a certain critical value death from dehydration occurs. A few plants can be dehydrated to air-dryness or lower without being killed, but their physiological activity also is reduced to a negligible level by such treatment. Water therefore is a very important constituent of protoplasm.

2. Water as a Reagent

A second use of water in plants is as a reagent or reactant in various physiological processes, including photosynthesis and hydrolytic

TABLE I

WATER CONTENT OF VARIOUS KINDS OF PLANT TISSUE EXPRESSED
AS PERCENTAGE OF FRESH WEIGHT[a]

Organ	Type of tissue	Water content (%)	References
Roots:	Barley ((*Hordeum vulgare*), apical region	93.0	(199)
	Pinus taeda, apical region	90.2	(140)
	P. taeda, mycorrhizal roots	74.8	(140)
	Carrot (*Daucus carota* var. *sativa*), edible root	88.2	(55)
	Sunflower (*Helianthus annuus*), entire root system, 7 weeks old	71.0	(374)
Stems:	Asparagus (*A. officinalis* var. *altilis*), edible tips	88.3	(71)
	Sunflower, average of stems, 7 weeks old	87.5	(374)
	Pinus echinata, wood	50–60	(155)
Leaves:	Lettuce (*Lactuca sativa*), inner leaves	94.8	(55)
	Sunflower, average of leaves on 7-weeks-old plant	81.0	(374)
	Cabbage (*Brassica oleracea* var. *capitata*), mature	86.0	(241)
	Corn (*Zea mays*), mature	77.0	(241)
Fruits:	Tomato (*Lycopersicon esculentum*)	94.1	(55)
	Watermelon (*Citrullus vulgaris*)	92.1	(55)
	Strawberry (*Fragaria chiloensis* var. *ananassa*)	89.1	(71)
	Apple (*Malus sylvestris*)	84.0	(71)
Seeds:	Sweet corn, edible	84.8	(71)
	Field corn, dry	11.0	(55)
	Barley	10.2	(55)
	Peanut (*Arachis hypogaea*), raw	5.2	(55)

[a] Water content is highly variable, but these values are fairly representative.

processes such as the digestion of starch to sugar. It is just as essential in this role as carbon dioxide or nitrogen.

3. *Water as a Solvent*

Water also acts as the solvent in which minerals, gases, and other solutes enter plants and move from cell to cell and organ to organ within the plant. The water occurring in the cell walls is quite as important as that in the protoplasm in connection with translocation of solutes.

4. *Water for Maintenance of Turgidity*

Another essential role of water is in maintenance of turgidity of plant tissue. This is essential for growth and for the maintenance of the form of leaves, new shoots, and other slightly lignified structures. It also is important in connection with the opening of stomata and movements of leaves, flower parts, and other plant structures which are controlled by changes in turgor. The most evident and important over-all effect of an internal water deficit is reduction in vegetative growth, because maintenance of a sufficiently high water content for a certain minimum turgor seems essential for cell enlargement.

In view of these facts the claim of Walter (362) that water is the most important environmental factor in relation to plant distribution seems justified. It also seems clear that the water relations of plants are just as important in relation to their growth as are such biochemical processes as photosynthesis and respiration.

C. AMOUNT OF WATER USED

The volume of water required for its various uses in plants is quite small, ordinarily being less than 5% of all the water which passes through a plant. Miller (241) estimated the water budget of an average Kansas corn plant (*Zea mays*) as follows:

Water occurring as a constituent	1,872 gm
Water used as a reagent	250 gm
Water lost in transpiration	202,106 gm
Total water used during growing season	204,228 gm

In this instance less than 1% of the water absorbed by the plants was used in them, and the remainder simply passed through. Assuming that water was used at this rate by plants in the field, Kansas corn would lose the equivalent of 11 inches of water per acre by transpiration during the growing season. Generally stands of plants do not transpire as rapidly as isolated individuals, but the actual transpiration of an Illinois corn field was estimated to be 8.2 inches from June 1 to September 15 by Reinmann et al. (276). It is estimated that a tree 30–35 ft in height will lose at least 1000 gallons of water a month during the summer, and it has been shown experimentally that a hardwood forest in the southern Appalachian Mountains transpires 17–22 in. of water per year (146). More detailed data on water loss will be given in the section on transpiration.

Unless these heavy water losses are replaced by the absorption of equivalent quantities, a water deficit develops in the plants, loss of turgidity occurs, numerous plant processes are interfered with, growth is reduced or ceases, and death from desiccation finally occurs.

D. THE SCOPE OF PLANT WATER RELATIONS

The water relations of plants can be considered as falling into two categories, water relations of cells and tissues and water relations of entire plants. The water relations of cells and tissues involve consideration of permeability, osmotic pressure, turgor pressure, and diffusion-pressure deficit of individual cells and tissues. These relationships have been considered in Chapter 2. The water relations of an entire plant involve a number of interrelated processes, including water absorption, the ascent of sap, and transpiration. These processes are affected by numerous environmental and plant factors, and in turn they affect various plant conditions such as the osmotic pressure, diffusion-pressure deficit, and the turgidity of plant tissues. A complete understanding of plant water relations also requires a knowledge of soil moisture relationships and of the factors which govern the availability of soil moisture.

Some knowledge of the structure of plants also is necessary for an understanding of plant water relations. Absorption of water is related to root structure, translocation to the structure of the xylem, and water loss is controlled to a large extent by leaf structures.

The most essential requirement in plant water relations is maintenance of a sufficiently high degree of turgidity for the occurrence of normal rates of growth and other processes. The *Wasserzustand* or condition of a plant with respect to its water content can be described in terms of its saturation deficit, or relative turgidity, which expresses the water content relative to that which would prevail if the plant were saturated. The water balance also can be expressed in terms of diffusion-pressure deficit or osmotic pressure. The concept of water balance recognizes that the important factor in plant water relations is the amount of water actually present in the plant, relative to the amount necessary for optimum growth, rather than the absolute water content.

Water absorption, water movement in the plant, and transpiration all are important to plant growth and survival because their interaction controls the internal water balance of plants and thereby affects many other essential processes. There is room for debate concerning the most logical order in which to take up these various processes, but because water loss usually seems to dominate this group of processes it will be discussed first, the ascent of sap next, and then water absorption. Internal water balance will be discussed last because it is controlled by the interaction of the other processes.

II. The Loss of Water from Plants

Water is lost from plants in the form of vapor by transpiration and as a liquid by guttation. Although the volume of water lost as a liquid is very small compared to that lost as vapor, the puzzling nature of the process of guttation has resulted in it receiving more attention than it would otherwise deserve.

A. GUTTATION

Everyone has observed the presence of droplets of water along the margins of dicotyledon leaves or at the tips of blades of grass and cereals early in the morning. This phenomenon was termed guttation by Burgerstein (47). Most guttation occurs through hydathodes, which are pores in the epidermis over intercellular spaces which usually are separated from the ends of xylem elements by a mass of thin-walled parenchyma cells called epithem. Most hydathodes appear to be modified stomata. Guttation can also occur through ordinary stomata and lenticels, and in general guttated water appears to escape by the path of least resistance. The process of guttation has recently been discussed in detail by Stocking (331).

1. Cause of Guttation

Guttation in vascular plants seems to be related to root pressure, a phenomenon which will be discussed later in this chapter, although the process has its counterpart in nonvascular plants, e.g., fungi. Root pressure occurs when the absorption of water exceeds the loss of water, resulting in development of hydrostatic pressure in the xylem which causes water to exude wherever it can escape. Guttation can be demonstrated easily by placing a healthy plant growing in moist, warm soil in a humid atmosphere where transpiration is reduced.

Haberlandt (117) and a few other writers have distinguished between "epithem hydathodes," from which water is forced by root pressure, and "active hydathodes," from which water is secreted by forces developed in the cells themselves. It seems probable that any structure from which water is secreted by locally developed forces should be termed a gland. Stocking (332) discusses the exudation of various types of materials from plant hairs and other specialized glandular structures, but the mechanism by which exudation or secretion occurs from these is not known as yet.

Various factors which interfere with root pressure also reduce guttation, such as cold, dry, or poorly aerated soil. Mineral deficiencies also reduce or prevent guttation (275, 358).

2. Species Which Exhibit Guttation

Burgerstein (48) reported the occurrence of guttation in species from 333 genera belonging to 115 plant families, and Frey-Wyssling (101) noted 12 additional genera. Although most often observed in herbaceous species, guttation occurs in some woody species, especially in the tropics. No examples of guttation in conifers are known to the writer. Guttation is not restricted to leaves, but sometimes occurs from the stems, usually through lenticels or leaf scars. Raber (274), for example, reported sap flow from leaf scars of deciduous trees in Louisiana after the autumn leaf fall, and Friesner (102) observed exudation of sap from uninjured stems of stump sprouts of *Acer rubrum* in Indiana in February. The sap flow from wounds in maple (*Acer*) trees seems to be caused by locally developed stem pressure, but sap flow from birch (*Betula*) and grape (*Vitis*) is caused by root pressure (186).

3. Volume and Composition of Guttated Liquid

The volume of water escaping by guttation varies from a few drops on a grass blade to 10–100 ml or more per day from leaves of heavily guttating plants, such as *Colocasia antiquorum*. Some tropical species are said to guttate so vigorously that water literally drips from their leaves at night. The volume and the composition of the liquid are exceedingly variable, even among plants of similar past history, ranging from almost pure water to a solution with an osmotic pressure of 1 atm or more.

TABLE II

APPROXIMATE COMPOSITION[a] OF THE GUTTATION FLUID FROM THE LEAVES OF PLANTS[b]

	Squash (*Cucurbita* sp.)	Tomato (*Lycopersicon esculentum*)	Cucumber (*Cucumis sativus*)	Cabbage (*Brassica oleracea* var. *capitata*)
pH	6.50	6.75	6.00	6.08
Nitrite nitrogen	—	1.5	1.0	—
Nitrate nitrogen	250	1.0	3.0	4.0
Ammonia nitrogen	5	7.5	4.0	2.0
Phosphorus	75	2.0	8.0	4.0
Potassium	75	40.0	15.0	15.0
Calcium	750	125.0	100.0	100.0
Magnesium	50	8.0	6.0	5.0
Aluminum	—	0.3	0.3	0.3
Manganese	—	1.0	trace	1.0
Chlorine	—	25.0	50.0	35.0
Sulfate sulfur	—	30.0	25.0	25.0
Sodium	—	trace	trace	—
Zinc	—	—	—	—
Copper	—	—	—	—
Total solids	2500	600		600
Organic matter	1100	275		250

[a] Parts per million.
[b] From Curtis (66).

The composition of the liquid guttated by the leaves of four herbaceous species is shown in Table II. Unfortunately, the composition of the organic matter which made up nearly half of the total solids was not determined, but sucrose, glucose, fructose, and maltose have been found in the xylem sap of various species and probably occur in the guttation fluid also. (Reference to Chapter 4 may be made for the composition of xylem or tracheal sap.) Wilson (377) reported the presence of catalase and peroxidase in the guttation fluid from corn, oats (*Avena sativa*), and timothy (*Phleum pratense*) and of reductase in the fluid from timothy. Meeuse (233) found amylase in maple sap. It is probable that a variety of organic compounds might be demonstrated in guttation liquids by chromatographic methods, such as those used by Pollard and

TABLE III

Osmotic Pressure (in Atmospheres) of Guttated Solution and Solution Exuding from Stumps of Tomatoes with Roots Immersed in Solutions of Various Osmotic Pressures[a]

	Normal-salt plants			Low-salt plants
	May 24	May 25	May 26	June 5
Culture solution	0.49	0.54	0.60	0.82
Exudate from stumps	2.40	1.50	2.18	2.21
Guttated liquid, leaves:	0.76			0.13
Younger leaves		0.68	0.90	
Older leaves		0.51	0.80	

[a] From Eaton (84).

Sproston (265a) on maple sap, by Bollard (34a) on apple sap, by Tolbert and Wiebe (346) to isolate phosphorus- and sulfur-containing compounds from xylem sap and by Maizel *et al.* (229) to identify the phosphorus-containing compound as phosphoryl choline.

The kind and concentration of solutes can be varied by changing the root environment. For example, Curtis (65) reported that following heavy fertilization of a lawn the grass blades were coated with a deposit of glutamine left by evaporation of guttated sap. Eaton (84) found the osmotic pressure of guttation liquid from tomato (*Lycopersicon esculentum*) leaves was 0.13 atm for plants growing in a dilute nutrient solution and 0.5–0.9 atm for plants growing in a more concentrated solution. Both Eaton and Hohn (142) found the salt concentration of the guttated liquid to be much lower than that of the exudate from the stumps of similar decapitated plants (see Table III). This is attributed to removal of salt from the xylem sap by the living cells of the stem and leaves.

4. Periodicity of Guttation

Root pressure often shows definite periodicity with a maximum in the daytime and minimum at night (114), but guttation rarely is observed during the day under field conditions because water loss tends to exceed water absorption. Rhythms in guttation have been observed by various investigators in plants grown in a very humid environment. These seem to be related to the period of illumination (331), and there also is said to be an endogenous rhythm in darkness (89, 133). It would seem that in a constant, humid environment there ought to be a close correlation between root pressure and guttation, but it is claimed that in at least some instances a negative correlation has been found. For example, Engel and Heimann (90) found water absorption at a maximum when guttation was at a minimum in cereal seedlings grown in a saturated atmosphere. They accounted for the difference between water absorbed and water lost by guttation as water used in growth. Schmidt (295) also observed a cycle in guttation of several species of *Saxifraga* related to the stage of development, with a maximum about the time of flowering.

5. Importance of Guttation

Guttation probably plays no significant role in the life of plants, although Curtis (66) regarded it as a sort of safety valve which regulates turgor and Schmidt (295) regarded it as a means of eliminating surplus calcium from certain species of saxifrage. The water-absorption mechanism of plants is of such a nature that dangerously high turgor pressures seldom occur as a result of excessive absorption.

According to Curtis (64) and Ivanoff (163), salts left on the surfaces of leaves by the evaporation of guttated water are sometimes subsequently redissolved and drawn back into the interior, causing injury. They suggest that many instances of tip burn to succulent leaves such as lettuce (*Lactuca sativa*) are caused in this manner. Curtis (66) also believed that spray materials and other toxic substances sometimes are dissolved in guttation water and are then drawn back into the interior where they produce injury. Bald (19) and Johnson (169) suggest that guttation water provides a convenient path for the entrance of pathogenic microorganisms which otherwise would have great difficulty in penetrating leaves.

It seems probable that guttation occurs as the result of the relatively high root pressure developed in many species of plants, but it has not been shown to play any essential role in their physiology.

B. TRANSPIRATION

By far the largest amount of water lost by plants escapes in the form of vapor by the process of transpiration. Transpiration is essentially an evaporation process but differs from evaporation in physical systems because it is modified by plant structure and the behavior of stomata. Most transpiration occurs in two stages, the evaporation of water from moist cell walls into intercellular spaces and its diffusion from the intercellular spaces into the outside air. Most of the water vapor escapes through stomata, but some diffuses out through the epidermal cells and their covering of cuticle, and in woody species through lenticels of the twigs and branches.

1. Lenticular Transpiration

Loss of water from woody stems occurs through the bark in general, but particularly through the lenticels or small breaks in the corky tissue covering the twigs. The amount of water lost through the bark is quite small, being estimated by Huber (152) as amounting to about 0.1% of the water loss from the crown of a tree. According to Huber, plugging up the lenticels reduced bark transpiration at least 20%, although they occupied only 2% of the surface. The bare branches of deciduous trees lose considerable water through their bark. Kozlowski (180), for example, found a winter rate of transpiration from yellow poplar (*Liriodendron tulipifera*) twigs of 2 gm per 100 sq cm per week, compared to 80 gm per 100 sq cm of leaf surface in August. In a cold climate the rate of winter transpiration from bare twigs is approximately as great per unit of surface as the winter transpiration rate of conifers (365). Under some conditions winter transpiration produces large water deficits and even causes injury from desiccation. According to Geurten (105) transpiration from the bark of woody plants follows the same daily and seasonal course as transpiration from leaves, but the rate is much lower, the maximum rate being less than 100 mg per square decimeter per hour.

2. Cuticular or Epidermal Transpiration

Loss of water from the surfaces of the epidermal cells of leaves and herbaceous stems often is termed cuticular transpiration because such surfaces usually are covered with a waxlike layer of cutin of varying thickness. This layer, termed the cuticle, greatly reduces the loss of water, although some water escapes through cracks and pores and it is far from being completely impermeable. The thickness of the cuticle varies widely, usually being much thicker in sun than in shade leaves and in plants of dry habitats than in those of moist habitats. There also are genetic differences, the cutin of *Ilex opaca* for example being 7–10 μ in thickness, that of coleus (*Coleus blumei*) and tobacco (*Nicotiana tabacum*) only 1 μ (341). Furthermore, the cuticle varies considerably in permeability with age and other factors. Leaves of two species with widely differing thickness of cuticle are shown in Fig. 1.

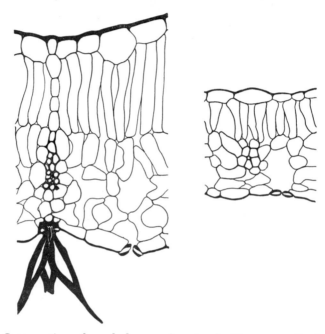

Fig. 1. Cross sections through leaves of post oak (*Quercus stellata*) (left) and beech (*Fagus grandifolia*) (right). Post oak represents an xeromorphic type of leaf with thick cutin, double layer of palisade cells, and bundle sheath extensions. Beech is a mesomorphic type with thinner cutin, a single layer of palisade cells, and no bundle sheath extensions. The beech type has much less internal surface exposed to intercellular spaces than the post oak type. Courtesy of Dr. J. Philpott.

Because of the differences in thickness and efficiency of the cutin layer, cuticular transpiration varies widely, as shown in Table IV. Plants with a thin cutin and high rate of cuticular transpiration often suffer severe water deficits and injury when exposed to condition favorable for heavy transpiration. This is particularly noticeable when plants grown in the shade are transferred to full sun. On the other hand a segment of heavily cutinized *Opuntia* stem can be exposed to sun and air for weeks or even months without losing enough water to be injured. Stålfelt (319) has recently summarized the literature on cuticular transpiration.

TABLE IV

CUTICULAR TRANSPIRATION UNDER STANDARD EVAPORATING CONDITIONS[a]

Species	Loss of water[b]
Impatiens noli-tangere	130.0
Caltha palustris	47.0
Fagus sylvatica	25.0
Quercus robur	24.0
Sedum maximum	5.0
Pinus sylvestris	1.53
Opuntia camanchica (= O. phaeacantha)	0.12

[a] From Pisek and Berger (263).
[b] In milligrams per hour per gram fresh weight.

3. Stomatal Transpiration

Most of the water lost from plants escapes through the stomata because they are the path of least resistance to diffusion of water vapor and other gases. Stomata are tiny pores in the epidermis surrounded by a pair of specialized epidermal cells, called guard cells, which control the size of their apertures. When fully open the elliptically shaped pores range from 3 or 4 to 12 μ across and from 10 to 40 μ in length. Views of two common types of stomata are shown in Fig. 2. Stomata are exceedingly numerous, varying from 1000 to 2000 per square centimeter in some cereals to over 100,000 per square centimeter in *Quercus coccinea* [see Meyer and Anderson (239), page 143] with an average of about 10,000 per square centimeter, but they are so small that even when fully open they occupy only 1–2% of the leaf surface. The number of stomata per unit of leaf area not only varies among species but also depends on the environment, usually being larger on leaves grown in sunny dry habitats than on leaves grown in shady, humid habitats (290). Stomata occur only on the lower surface of leaves of many species, including most woody plants, but in many other species they occur on both surfaces. The relative importance of stomatal and cuticular transpiration will be discussed later, but it can be said that the loss of water from land plants is controlled to a greater extent by stomatal opening than by any other single factor. It is because of this control that the daily course of transpiration differs so much from the daily course of evaporation.

Stomata are among the most interesting structures in plants, not only because they control transpiration and carbon dioxide absorption, but also because they are so responsive to light and to the internal water balance. They have deservedly received much attention from plant physiologists; only a few important items can be discussed here. Stålfelt (320) has recently summarized the literature on stomatal transpiration and Miller [(241) pages 417–447], summarized a large amount of literature on stomata and their behavior.

In Fig. 2 are shown surface views of two typical forms of stomata, that typical of dicotyledons and that typical of some grasses. In some plants the stomata are at the base of pits or grooves which are supposed

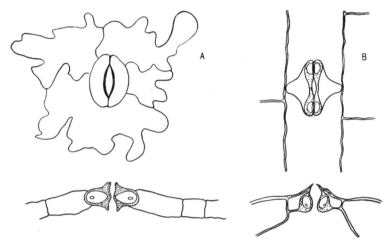

Fig. 2. Diagrams of two common types of stomata and guard cells, as seen in surface view and in cross section. Type *A*, from potato (*Solanum tuberosum*), is typical of many dicotyledons and some monocotyledons. The thickened inner walls of the guard cells cause them to become curved when turgid, producing the stomatal aperture. Type *B* is from corn (*Zea mays*) but occurs in some other monocotyledons. In this type the ends of the dumbbell-shaped guard cells swell when turgid, pushing the guard cells apart. Adapted from Eames and MacDaniels (82) and Loftfield (215).

to decrease water loss by increasing the length of the diffusion gradient. Haberlandt (117) described a wide variety of stomatal structures, and several types are described by Esau (93) and by Eames and Mac-Daniels (82). For a detailed discussion of stomatal behavior reference may be made to Chapter 3.

a. Diffusive capacity of stomata. First in importance is the great efficiency of stomata as pathways for diffusion. With a surface area of only 1% or less of the leaf surface, the movement of water vapor through them often exceeds 50% of that from a free water surface and absorption of carbon dioxide by leaves can approach that of a free surface of NaOH solution. Thus the rate of diffusion through individual stomata sometimes is at least 50 times as great as diffusion from an equal area of free water surface. This is possible because, as Brown and Escombe (44) have shown, diffusion through small openings is more nearly proportional to their diameter or perimeter than to their area. This has been verified by a number of investigators, including Sayre (291), Sierp and Seybold (305), Huber (150), and Verduin (355). The relationship between rate of diffusion and area and perimeter of openings is shown in Table V. Inspection of this table shows that as pore size is decreased water loss is decreased in proportion to reduction in perimeter rather than to reduction in area of pores. The smallest pore used had an area of only 1% of the largest pore, but its perimeter was

TABLE V

RELATION OF DIFFUSION OF WATER VAPOR THROUGH SMALL OPENINGS
IN MEMBRANES TO AREA AND PERIMETER OF OPENINGS[a]

Diameter of pores (mm)	Loss of water vapor (gm)	Relative amounts of water lost	Relative areas of pores	Relative perimeters of pores
2.64	2.655	1.00	1.00	1.00
1.60	1.583	0.59	0.37	0.61
0.95	0.928	0.35	0.13	0.36
0.81	0.762	0.29	0.09	0.31
0.56	0.482	0.18	0.05	0.21
0.35	0.364	0.14	0.01	0.13

[a] From Sayre (291).

13% of that of the largest pore and water loss was 14% of that through the largest pore.

Small pores permit a higher rate of diffusion per unit of area than large ones because substances diffusing through pores in membranes do not move in straight lines but tend to spread out in all directions from the perimeter of the openings, forming diffusion shells or vapor caps as shown in Fig. 3. Thus the rate of diffusion through the openings is related more closely to the circumference than to the area. The location of regions of equal concentration of water vapor immediately over a pore is indicated by lines which form flattened spheroids over the pore, and the diffusion lines form hyperbolas. At a distance from the pores the diffusion lines tend to straighten out, forming a thick layer of vapor over the entire surface of the membrane or leaf. This situation exists only in still air.

If the pores are 10 or more diameters apart, the vapor caps or diffusion shells over individual pores do not interfere seriously, but if closer together they tend to interfere significantly. According to Verduin

(355) diffusion becomes proportional to pore circumference only when the diffusion shells are large enough to be practically spherical, which is not until their long axis is five times the diameter of the pores. At a greater distance the diffusion lines tend to become straight and form a vapor layer over the entire membrane or leaf, which is shown in Fig. 3. Of course these conditions exist only in still air, as wind tends to sweep the vapor caps away. According to Sierp and Noack (304), in a breeze the rate of evaporation through pores tends to shift from dependence

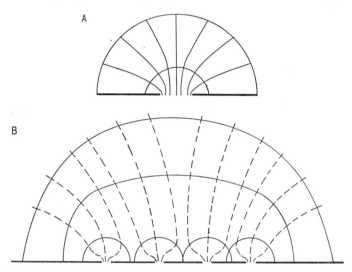

FIG. 3. A. Diagram of a so-called diffusion shell over a stoma. The curved lines show the path of diffusion through the pore, and the flattened semicircular lines inclose areas of equal concentration of water vapor above the pore. Ideally, diffusion shells are formed on both sides of the membrane, as shown by Verduin (355).

B. Diagram showing how diffusion shells tend to overlap as the distance between the pores or stomata decreases. The diffusion paths tend to straighten out at a distance from the pores and form a uniform layer of water vapor over the entire leaf surface. Adapted from Bange (20).

on the diameter of the pore to dependence upon its area. This is not important because removal of the layer of vapor over a leaf more than compensates for decreased stomatal efficiency by producing a steeper vapor pressure gradient from the intercellular spaces to the outside air. In general stomata are spaced far enough apart so that there is little interference between them. As they begin to close, their diameters decrease and the effective distance between them increases, resulting in higher rates of diffusion from partly open stomata than would be predicted from their diameters alone (355).

The application to stomata of the principles governing diffusion through small openings was first made by Brown and Escombe (44). Among the more recent contributions are those of Verduin (355) and Penman and Schofield (261a) and Bange (20). Verduin developed an equation to express interference between adjacent pores in a membrane. Various authors of whom Renner (277) was the first and Bange (20) the latest have added to the equation for diffusion the effect of the water vapor layer which lies beyond the individual diffusion shells. When this is included as a resistance to diffusion, the calculated rates of transpiration are in much better agreement with the actual rates than when the equation of Brown and Escombe is used, because they omitted the latter factor.

b. *Effect of stomatal aperture on transpiration.* Considerable difference of opinion has existed concerning the relation between stomatal aperture and rate of transpiration. Lloyd (213) observed considerable changes in rate of transpiration without corresponding changes in stomatal aperture, and Jeffreys (168) believed that stomata exerted little effect on transpiration until they were 98% closed. Loftfield (215) concluded that stomatal aperture does not exert much effect on transpiration until the stomata are more than 50% closed. This problem has been investigated extensively by Stålfelt (315, 316), and some of his results are shown in Fig 4. It is apparent that transpiration increases rapidly with increase in aperture as the stomata begin to open, but at low rates of evaporation there is little further increase in rate of transpiration over a wide range of increase in aperture. In contrast, with a high rate of evaporation the rate of transpiration increases up to the widest apertures obtained.

Bange (20) has recently reviewed the work on diffusive capacity of stomata and suggests that the following resistances to escape of water vapor exist. First is the resistance of the stomatal aperture itself; second, the water vapor shell over each stoma; and third, the water vapor shell over the entire leaf. With small apertures the resistance of the aperture controls the rate, but as this resistance decreases with increasing aperture the resistance offered by the water vapor in the air over the leaf becomes the controlling factor. Thus stomatal aperture becomes increasingly important as the air becomes drier and the potential rate is limited less and less by atmospheric conditions and more and more by the rate of diffusion through the stomata.

Midday closure of stomata occurs in many species and causes reduction in both transpiration and photosynthesis. The relation between stomatal aperture and transpiration of the oil palm (*Elaeis guineensis*) is shown in Fig. 11. Midday closure of stomata with consequent reduction in transpiration was observed in coffee trees (*Coffea arabica*) in

Africa by Nutman (249) and in Central America by Alvim and Havis (5). It has been observed in a variety of plants in the temperate zone, including several species of trees (266). It usually is attributed to dehydration of the leaf, but Nutman observed it to occur immediately when leaves were exposed to full sun while adjacent leaves were shaded and no evidence of a water deficit existed. He therefore attributed stomatal closure in coffee to the direct effects of bright light.

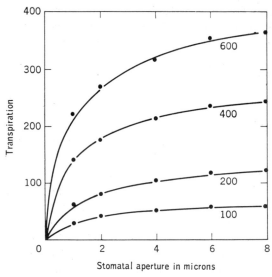

FIG. 4. The relation between rate of transpiration and stomatal aperture of leaves of *Betula pubescens* at various rates of evaporation. The curves show rates of transpiration with various amounts of stomatal opening. The number on each curve indicates the rate of evaporation from a blotting-paper atmometer, expressed in milligrams of water per hour per 25 sq cm of surface. Transpiration is expressed in milligrams of water per hour per 25 sq cm of leaf surface. From Stålfelt (315).

c. Measurements of stomatal opening. Methods of measuring stomatal opening have received considerable attention. This subject is dealt with more fully in Chapter 3 because of the interest aroused by the mechanism of stomatal movement in the water relations of guard cells: it can therefore be omitted here. Suffice it to say that the methods adopted have ranged from the use of epidermal strips as developed by Lloyd (213) and Loftfield (215); the infiltration technique as used by Molisch (244) and later by Alvim and Havis (5) to the use of various forms of porometers. According to Darwin and Pertz (70), porometers were first proposed by Dutrochet, but have later been improved successively by Knight (176), Laidlaw and Knight (203), Bolas and Selman (34),

Gregory and Pearse (113), Gregory and Armstrong (111a), and their use has been extensively discussed by Penman (261), by Heath (130), by Heath and Williams (132), and Williams (371). Wilson (372) made an extensive study of the effects of environmental factors on stomatal opening by the use of recording porometers.

d. Control of stomatal opening. The aperture of the stomatal pores is controlled by movement of the guard cells, which depends in general on their turgor. There are many types of guard cells in the plant kingdom, but according to Stålfelt (320) they all operate on the same principle— increased turgor of the guard cells causing opening movements by change in shape of guard cells and decrease in turgor resulting in closure by elastic contraction of their cell walls. Stålfelt summarizes the steps as follows:

Change in composition→turgor change→volume change→guard-cell movement

The cause of the changes in turgor is not fully understood, but it usually is attributed to changes in osmotic pressure of the guard cells resulting from changes in relative amounts of starch and sugar. Other factors may be involved, however (see Chapter 3). In the most familiar type of guard cells, the walls next to the pores are thicker than the outer walls. The latter stretch more as the turgidity increases, causing the guard cells to bend outward and so produce an opening between them. In another type the ends of the guard cells swell or contract with changes in turgor, pushing the guard cells apart or allowing them to come together (see Fig. 2).

In general stomata are open in the light and closed in darkness, but many exceptions to this generalization occur because stomatal behavior is very sensitive to water content and temperature, as well as to light. According to Loftfield (215), light intensity must be reduced to less than 50% of full sun to affect the stomata of plants growing in the open, and he reported that stomata of some plants open at night in moonlight or artificial light. According to Sayre (291), stomata of *Rumex patientia* continue to open at the usual time for 1 or 2 days in a dark chamber. This indicates an internal rhythm which is at least partly independent of environment, as does the observation of Loftfield that stomata will open more readily with artificial light toward morning than before midnight.

Loftfield studied opening and closing of stomata of about 60 species and found that most of them could be classified in one of the following three types.

Alfalfa type. Stomata of this type open during the day and close at night. They open gradually during a period of 2–6 hours after sunrise, remain fully open 3–6 hours, and then gradually close. Under unfavorable moisture conditions they may close partially during the middle of the day. Night opening often accompanies daytime closure. Most thin-leaved mesophytes including many woody species belong in this group. The midday closure of stomata in *Coffea arabica* was attributed by Nutman (249) to bright light rather than to dehydration.

Potato type. The stomata of this group are open continuously except for about 3 hours after sunset, and daytime closure occurs only when leaves are badly wilted. Tulip (*Tulipa gesneriana*), onion (*Allium cepa*), leek (*A. porrum*), cabbage (*Brassica oleracea* var. *capitata*), pumpkin (*Cucurbita pepo*), and squash (*C. maxima*) are among the species belonging in this group.

Barley type. Daytime opening and closing occurs rather rapidly and stomata are seldom fully open for more than an hour or two. They never are open at night and are sensitive to environmental conditions during the day, some plants such as corn and sorghum (*Sorghum vulgare*) opening best on bright, warm days, while others such as barley (*Hordeum vulgare*), oats, and wheat (*Triticum aestivum*) open widest on cool humid days with low light intensity.

The stomata of *Scirpus validus* and some species of *Equisetum* are continuously open, according to Loftfield.

In view of the diversity of behavior encountered among different species it is not surprising that no satisfactory explanation of the cause of stomatal opening and closing has yet been developed. It has been shown that when illuminated the pH of guard cells increases (291), probably because of use of carbon dioxide in photosynthesis, and the increased pH favors the change of starch to sugar. This increases the diffusion pressure deficit of the guard cells relative to other epidermal cells, causing water absorption and increase in turgidity. In darkness, accumulation of carbon dioxide is supposed to cause a decrease in pH, favoring conversion of glucose to starch, decrease in water content, loss of turgidity by guard cells, and closure of stomata. It is difficult, however, to distinguish the feature of the stomata which is rate-limiting in the movement of the guard cells (see Chapter 3).

Stålfelt (318, 320) states that water content has an even greater effect on stomatal opening than light. In extremely turgid leaves pressure of surrounding epidermal cells sometimes interferes with normal opening. If heavy transpiration produces an internal water deficit, stomata may close by midday, or even earlier, and early closure of stomata is a more sensitive indicator of water deficit in most species than wilting, because it occurs long before wilting is visible. Large water deficits decrease the sensitivity of guard cells to light for several hours after they have occurred, and Stålfelt regards this as a safety device which prevents too rapid response of stomata to light on bright, sunny days. Occasionally stomata open in wilting leaves. At temperatures near freezing stomata usually remain closed, and according to Wilson (372) stomatal aperture increased in several species with increasing temperature up to 25° or 30°C. There also is some rather contradictory evidence that mineral nutrition, especially nitrogen supply, affects the responsiveness of stomata. This deserves further investigation.

This brief discussion has omitted many details and uncertainties concerning the behavior of stomata and the reader is referred to Chapter 3 and to papers by Heath and his co-workers; see (131, 293, 318, 320) for more detailed, but controversial, discussions of these interesting problems. Wilson (372) presents considerable data on the effects of light, temperature, and humidity on stomatal behavior.

4. Factors Which Affect the Rate of Transpiration

The rate of transpiration is highly variable because it is affected by so many internal and external factors. Stocker (327) recently summarized the literature on environmental factors in relation to transpiration. The more important internal or plant factors will be discussed first, followed by a discussion of environmental factors.

a. Plant factors which affect transpiration. During the early part of this century great emphasis was placed on structural modifications which appeared to reduce the rate of water loss, and plants were classified as xerophytic if they possessed structural characteristics such as thick, heavily cutinized leaves, which presumably reduced water loss, or mesophytic if they did not possess these characteristics. Unfortunately this generalization proved untrue, because Maximov (232) and others [also see (241), pages 449–451] reported many instances where plants with so-called xeromorphic types of leaves transpired more rapidly than mesomorphic types, if well supplied with water. As a result the role of plant anatomy in plant water relations has been underestimated in recent years. Actually leaf structure does have important effects on the rate of water loss and the survival of plants in dry habitats. Adaptation to xerophytic conditions is now known to be due, in many cases, to structural features which prevent wilting or collapse of the leaf, rather than elimination of water loss. Among such features are thick cell walls and the presence of supporting sclereids in the mesophyll. Furthermore the characteristics of root systems as water-absorbing surfaces also plays an important role in water relations. This will be discussed in more detail in the section on water absorption. The role of stomata has already been considered and will not be discussed further.

Leaf area. If most of the water loss occurs from leaves it would be expected that plants with a large leaf area would lose more total water than plants with a small leaf area. This is true, although the rate usually is not proportional to the leaf area, smaller plants often tending to transpire more rapidly per unit of leaf area than larger plants of the same species. Miller [(241) p. 454] reported that a large corn plant transpired 629 gm per square meter and one with 89% as much leaf surface transpired 784 gm per square centimeter in the same period of time. Kelley (173) reported that removing half of the leaves from shoots of trees increased the transpiration of the remaining leaves from 20 to 90%. Reduction of the leaf surface by pruning reduces the total transpiration per plant, but usually increases the rate per unit of remaining surface, especially when environmental conditions favor a high rate of transpiration. This occurs because, if the ratio of leaves to roots is reduced, more water can be supplied to each leaf (257).

In some species development of a severe internal water deficit results in the shedding of leaves and reduction in water loss. This is very noticeable in some species native to dry habitats such as *Larrea tridentata* in the southwestern United States and species of *Euphorbia*. The leaves of some grasses roll when they begin to wilt, materially decreasing the transpiration surface. Dropping of leaves and curling are indications that injurious water deficits already exist, but they may enable plants to survive drought much longer than if they did not occur. Even more effective is the reduction of leaves to vestigial structures as occurs in the cacti and a few other groups. Killian and Lemée (175) have recently summarized the literature on morphological characteristics affecting transpiration.

Leaf structure. There is no doubt that leaf structure greatly affects the rate of water loss. For example, if a collection of detached leaves of various species are allowed to dry it will be found that some become air-dry in a few hours but others survive for several days, or even for weeks. This difference usually can be correlated with thickness and efficiency of the cutin layer, plants with a thick cuticle surviving much longer than those with a thin cuticle. On the other hand it has been shown that plants bearing thick, heavily cutinized leaves with several compact layers of palisade cells and small intercellular spaces often transpire as rapidly or more rapidly than thin, lightly cutinized leaves with a loose mesophyll structure and large intercellular spaces, if supplied with adequate water. *Gordonia lasianthus* and *Ilex glabra* have sclerophyllous leaves, but when well supplied with water they transpire more rapidly per unit of leaf surface than the thinner leaves of *Liriodendron tulipifera* (53). Swanson (341) found that American holly

(*Ilex opaca*) transpired more rapidly than *Coleus blumei*, lilac (*Syringa vulgaris*), or tobacco on a sunny day, but more slowly on a cloudy day. Huber (149) attached considerable importance to the ratio of surface to volume as a factor affecting rate of water loss.

These results are related to differences in the internal surfaces exposed to the intercellular spaces. Turrell (347, 348) has shown that xeromorphic or sclerophyllous leaves usually have more cell-wall surface exposed to the internal atmosphere than do mesomorphic leaves. The larger evaporating surface provided by the larger internal exposed cell surfaces of sclerophyllous leaves results in higher transpiration rates when the stomata are open and if the plants have an adequate supply of water, but the thicker layer of cutin results in lower cuticular transpiration when the stomata are closed. Kamp (170) reported that cuticular transpiration was over 15% of the total in certain woody mesophytes, but only about 4% in *Olea lancea* and 2% in *Laurus nobilis*, which have very heavily cutinized leaves. A heavy layer of cutin obviously only becomes effective in reducing water loss when the stomata are closed, hence responsiveness of stomata often is an important factor (95, 159) in controlling water loss. Two types of leaves with quite different structures are shown in Fig. 6.1.

As mentioned earlier stomatal frequency is highly variable, but it averages about 10,000 per square centimeter in mesophytes. Killian and Lemée (175) cite reports of less than 100 stomata per square centimeter in various desert plants. Stomata are located in pits or furrows in leaves of many species, and this reduces transpiration by reducing the steepness of the vapor pressure gradient from intercellular spaces to outside air. The presence of a thick coating of hairs on leaves was once supposed to reduce transpiration for the same reason, but Sayre (291) found that careful removal of the hairs from the leaf surface of *Verbascum thapsus* had little effect on transpiration, and the importance of leaf hairs seems doubtful. Indeed if the hairs contain water rather than air they may even increase evaporation.

Size and form of leaves may become an important factor in rate of transpiration, as pointed out by Stålfelt (321), but several opposing factors are involved. The rate of water loss usually is greater per unit of surface from a small, than from a large, evaporating surface. Also the cells from which evaporation occurs are closer to water-conducting elements in small leaves and less likely to suffer from water deficits than in large leaves where the interveinal distances are greater (96). On the other hand large, thick leaves with small specific surface will tend to be warmer than the surrounding air while small, thinner leaves (5 mm or less across) are more nearly at air temperature (300). A steeper water vapor pressure gradient from leaves to air exists in leaves which are warmer than the air, resulting in a tendency toward higher transpiration, and this may compensate, in part, for those factors which favor higher transpiration rates from small leaves. It should be remembered that differences in rate of transpiration per unit of leaf surface often are compensated by differences in total leaf area. For example, the writer found the transpiration rate per unit of leaf area was twice as high for yellow poplar as for loblolly pine (*Pinus taeda*), but loblolly pine seedlings had about three times the leaf area of yellow poplar seedlings of similar size, and the rate per seedling for seedlings of similar crown volume was greater for pine than for yellow poplar.

Orientation. Leaves generally are oriented more or less perpendicular to the average incident radiation, hence they are affected to the maximum possible extent by the heating effects of the sun. In a few plants, including certain species of *Lactuca* and *Silphium*, and *Quercus marilandica*, the leaves are oriented approximately parallel to the average incident radiation. So definite is the north and south orientation of leaves in some of these species that they are called "compass plants." Although such an orientation probably reduces leaf temperature measurably it is doubtful if the reduction in water loss is of much significance.

The rolling or curling of leaves, mentioned previously in connection with leaf area, also changes their transpiration rate. Stålfelt [(320) page 336] has recently summarized the literature on this phenomenon. He quotes work of Lemée showing that rolling of leaves causes a reduction in transpiration of about 35% in plants of moist habitats, 55% in Mediterranean species, and 75% in desert xerophytes.

Root-shoot ratio. The ratio of water-absorbing surface to transpiring surface probably is more important than the actual leaf or transpiring surface. Unless water absorption keeps pace with water loss, an internal water deficit develops which reduces transpiration. It was mentioned earlier that removal of part of the leaves usually results in an increased rate of water loss per unit of surface from the remaining leaves. Reduction in the extent of root systems by disease, insects, mechanical injury, flooding, during transplanting, or by other causes often produces severe water deficits in the attached shoots because of lack of sufficient absorbing surface. Both Bialoglowski (25) and Parker (257) found that transpiration per unit of leaf area increased as the ratio of root surface to leaf surface increased (see Fig. 5). Sorghums typically possess nearly twice as many small branch roots per unit of length of first-order roots as corn, or as many other comparable plants, and this is believed by Miller [(241) page 455] to be the principal reason for its higher transpiration rate per unit of leaf surface, especially in dry soil. This is supported by data of Slatyer (308), who found that sorghum had better-developed root systems and maintained higher turgidity than cotton (*Gossypium hirsutum*) or peanuts (*Arachis hypogaea*) under drought conditions.

Water content of leaves. Some differences in opinion exist as to whether reduction in water content reduces transpiration directly, or only indirectly, by bringing about stomatal closure. It was suggested

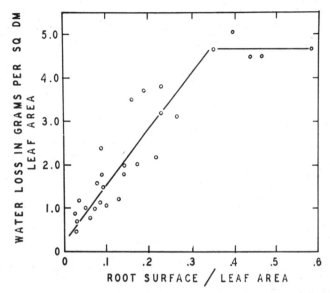

Fig. 5. The effects of variations in ratio of root surface to leaf surface on rate of transpiration of rooted lemon (*Citrus limonia*) cuttings. Root surface was not limiting until the ratio of root surface to leaf surface was redudced below 0.3. From Bialoglowski (25).

by Renner (277), Livingston (211), and Livingston and Brown (212) that rapid transpiration reduces the water content of the cell walls bordering the intercellular spaces and thereby reduces the rate of evaporation. This "incipient drying" has been claimed to explain the fact that the rate of transpiration often lags behind the rate of evaporation on bright days. On the other hand, Gregory et al. (112) found that the leaf water content must be reduced very greatly before it directly reduces transpiration. Stålfelt [(320) pp. 391–395] reviewed the literature on this subject and concluded that incipient drying of internal cell walls has no important direct effects on stomatal transpiration. He attributes most of the reduction in transpiration during dehydration to stomatal closure. On the other hand Hygen (157) points out that drying effects are complex, but reduction in water content appears to reduce transpiration independently of stomatal closure. It is certain that, when dehydration of cell walls is severe enough to reduce the vapor pressure of water in them, the rate of evaporation into the intercellular spaces will be reduced.

There is no doubt that severe dehydration causes a marked reduction in transpiration, as shown by the behavior of wilted plants (Table VI). In such instances there is probably a considerable reduction of evaporation into the intercellular spaces because of reduced vapor pressure in the cell walls and also because large reduction in water content probably reduces the permeability of the protoplasmic membranes of the

TABLE VI

TRANSPIRATION RATES AND TEMPERATURES OF WILTED AND UNWILTED
LEAVES OF CROP PLANTS[a]

Species	Time of day	Transpiration (gm/dm²/hr) Turgid	Wilted	Ratio transpiration turgid : wilted leaves	Temperature difference (°C)
Corn (Zea mays)	1–3 P.M.	1.20	0.34	3.5	2.8
	3–5 P.M.	0.80	0.20	4.0	2.5
Cowpea (Vigna sinensis)	1–2 P.M.	1.03	0.37	2.8	6.1
	3–5 P.M.	0.71	0.06	11.5	4.9
Sorghum (Sorghum vulgare)	1–2 P.M.	1.71	0.47	3.6	1.8

[a] From Miller and Saunders (242).

mesophyll cells (17, 115, 209). Dehydration also decreases the permeability of the cuticle to water (319). Recent work suggests that cutinized layers of leaves have a complex structure containing submicroscopic pores and possibly plasmodesmata, and that they vary considerably in water content.

a. Environmental factors which affect transpiration. Transpiration basically is the evaporation of water and its diffusion out of plant tissue, hence it is affected by those factors which affect evaporation from the other moist surfaces. The most important of these are humidity of the air, temperature of the plant and the air, and wind. These factors are important because they affect the steepness of the water-vapor pressure gradient from the evaporating surfaces of the plant to the outside air. Although visible light has no direct effect on evaporation it greatly modifies the rate of transpiration through its effects on stomatal opening. The amount of available soil moisture also is an important factor because of its effects on the internal water balance. It will be assumed that readers are familiar with such terms as vapor pressure, relative and absolute humidity, and vapor pressure gradient.

Humidity of the air. Atmospheric moisture conditions can be expressed in terms of absolute humidity, relative humidity, saturation deficit, vapor pressure, or vapor-pressure deficit. Assuming that the stomata are open and temperature is constant, the rate of transpiration depends on the steepness of the vapor pressure gradient from plant tissue to air, hence the water vapor pressure of the air is the most useful value. If the relative humidity is known, the vapor pressure can be calculated easily. For example, the saturation vapor pressure of water at 20°C is 17.54 mm of mercury and at 70% relative humidity the vapor pressure is 12.27 mm. If it is assumed that the intercellular

TABLE VII

Vapor Pressure Difference from Moist Surface to Air at 60% Relative Humidity (R. H.) at Various Temperatures

Temperature of air and evaporating surface (°C)	Vapor pressure of tissue (mm Hg)	Vapor pressure of air at 60% R. H. (mm Hg)	Vapor pressure difference
10	9.21	5.53	3.68
20	17.54	10.52	7.02
30	31.82	19.09	12.73

spaces of plant tissue are saturated, then at 20°C and 70% relative humidity of the outside air the vapor pressure difference is 5.27, but at 50% relative humidity it is 8.77 mm and transpiration will be correspondingly more rapid.

The rate of transpiration or evaporation depends on the steepness of the vapor pressure gradient from the evaporating surfaces to the air, and use of relative humidity alone to evaluate atmospheric conditions with respect to evaporation or transpiration can be very misleading. Experimental material sometimes is placed at various temperatures, but the same relative humidity is maintained at all temperatures. This actually produces very different environments at different temperatures, as shown in Table VII. The rate of water loss will be much higher at the higher temperatures because the vapor pressure gradient is much steeper. If material is to be stored at different temperatures under uniform moisture conditions, the relative humidities must be adjusted to produce the same vapor pressure gradient between the material and the air at all temperatures. This applies to both animal and plant tissue, and in fact to any material which loses water by evaporation.

It has been assumed thus far that the relative humidity of the intercellular spaces is 100%, but in transpiring plants this is seldom true. If the air in the intercellular spaces actually was saturated no evaporation could occur from the moist cell walls. Thut (345) has shown that the relative humidity in the intercellular spaces of transpiring plants often is far below saturation. He found a relative humidity in the intercellular spaces of unwilted leaves of about 91% and values as low as 65% in wilted leaves, when the relative humidity of the outside air was 40–48%. Transpiration can occur into a saturated atmosphere (69) if the leaf temperature is higher than the air temperature.

Temperature. An increase in temperature almost always increases the rate of transpiration because it increases the steepness of the water vapor pressure gradient from plant tissue to air. This can be shown by the following examples, in which for convenience it will be assumed that leaf and air temperatures are identical and the relative humidity in the intercellular spaces is 100%. Neither assumption is strictly true, and the reader can calculate the effects of varying leaf and air temperatures and relative humidities.

Temperature of air and leaf	10°C	20°C	30°C
Vapor pressure in leaf*	9.21	17.54	31.82
Vapor pressure of air at 60% relative humidity	5.53	10.52	19.09
Vapor pressure difference from leaf to air	3.68	7.02	12.73

* Water vapor pressures are expressed in millimeters of mercury.

In these examples an increase of 10°C almost doubled the steepness of the vapor-pressure gradient from the leaf to the outside air, and this is responsible for increased water loss as the temperature rises. It sometimes is supposed that increased temperatures cause increased transpiration because of the reduction in relative humidity which occurs when the temperature of an air mass is increased, but this is not true. If the absolute humidity remains unchanged, increase in temperature increases the vapor pressure of the atmosphere slightly, in accordance with Charles's law, in spite of the fact that the relative humidity decreases.

It was assumed at the beginning of this section that leaf and air temperature are identical, but this is not true. Temperatures of leaves, twigs, and other plant parts fluctuate over a considerable range and may be higher or lower than air temperatures. Fruits, thick leaves, stems, tree trunks, and other relatively massive plant structures often become much warmer than the air when exposed to the sun, and this increases the vapor pressure difference between them and the outside air. The evaporation of water has a cooling effect, and it sometimes has been claimed that the cooling effects of transpiration are important. In Table VI, it can be seen that wilted leaves which are transpiring relatively slowly are only 2–6°C warmer than unwilted, rapidly transpiring leaves. Curtis and Clark (68) summarized the literature on this much debated subject and concluded that the cooling effect of transpiration is not of any practical significance because it rarely amounts to more than 2–5°C in leaves. At night, leaves usually become cooler than the air, resulting in the condensation of dew on their surfaces if the air is humid.

Wind. Wind usually causes increased transpiration because it removes water vapor from the vicinity of transpiring surfaces and produces a steeper vapor pressure gradient from plant tissue to air, but it also tends to cool leaves, thereby decreasing the steepness of the vapor pressure gradient. If leaves are either cooler or warmer than the air, convection currents are set up which tend to prevent water vapor from accumulating around transpiring surfaces. Koriba (178) published a study of convection in relation to transpiration, and Gaumann and Jaag (103) made a study of the effect of wind on transpiration. Martin and Clements (230) and Wrenger (382) also published on this problem.

Stålfelt (319) recently summarized the literature on the effects of wind on transpiration. Apparently cuticular transpiration is increased about 20% and stomatal transpiration is increased 100–200% by wind of relatively low velocity. As shown in Fig. 6, most of the increase occurs at very low velocities. Also the effect of air movement is much greater during the first few minutes than after the passage of time. Wrenger (382) attributed this to dehydration of the cell walls (incipient drying), but Stålfelt thinks stomatal closure is more important. The decrease in transpiration after the initial increase undoubtedly is caused by dehydration resulting from rapid transpiration, and it probably operates through the drying of internal evaporating surfaces, closure of stomata, and dehydration of the cuticle. Readers are referred to Gaumann and Jaag (103), Hygen (157, 159), and Stocker (328) for further discussion of this subject.

Light. Visible light has important effects on transpiration through its control of stomatal opening. At night stomata are mostly closed, hence the night rate of transpiration is reduced more below the day rate than is evaporation from a water surface or an atmometer. For example, in one series of measurements in midsummer the day rate of evaporation from a white Livingston atmometer was 4.7 times the night rate, while the day rate of transpiration from yellow poplar was 12.4 times the night rate (183).

There are wide variations among species in their responses to light, as shown in Fig. 7 from work by Gaumann and Jaag (103). Some species differences appear to be related to differences in habitat. For example, Polster (266) found that on days with a cool, humid forenoon and warm, dry afternoon the shade species, spruce (*Picea abies*), had a higher rate of transpiration in the morning than in the afternoon, while the reverse was true of Scots pine (*Pinus sylvestris*), which normally

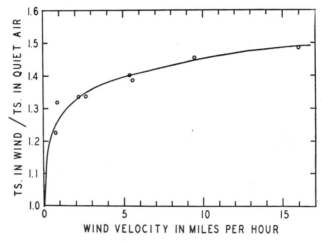

Fig. 6. The effect of wind on transpiration rate of sunflower (*Helianthus annuus*) plants. Note that most of the increase in transpiration (TS.) occurs at very low velocities where the accumulation of water vapor over the leaf surface is a limiting factor. At higher velocities transpiration probably is limited chiefly by the rate of diffusion through the stomata. From Martin and Clements (230).

grows in the sun. On sunny days the stomata of some species tend to close in the middle of the day (215, 266), and although this usually is attributed to dehydration it sometimes appears to be the direct consequence of bright light. Nutman (249) found marked stomatal closure and reduction in transpiration of *Coffea arabica* exposed to full sun. As the stomata reopened promptly when individual leaves were shaded it seems unlikely that closure resulted from dehydration of the leaves.

Cloudy weather causes decreased stomatal opening of most species native to sunny habitats, and this effect is particularly noticeable at low temperatures (373). The combination of low temperature and cloudy weather often existing in the winter is particularly unfavorable for stomatal opening.

Vital activity. It has been assumed throughout this discussion that transpiration can be treated as a physical process. Some investigators have observed transpiration into saturated atmospheres and introduced special explanations for this. Darwin (69) estimated that transpiration in a saturated atmosphere in certain experiments was equivalent to that expected if the leaf was 0.8°C warmer than the air. He attributed this

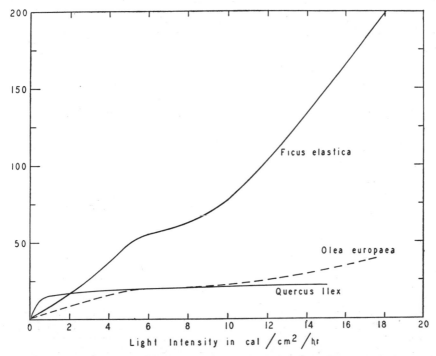

Fig. 7. The effect of increasing light intensity on the transpiration rate of leaves maintained at a constant temperature and relative humidity. The increase in transpiration results from increased stomatal aperture. Combined from various figures of Gaumann and Jaag (103).

temperature difference to the heating effect of respiration, but, as mentioned earlier, leaves exposed to sun often are warmer than the air because heat is not dissipated as rapidly as it is absorbed.

Dixon (76) suggested that the energy of respiration is used to secrete water from the mesophyll cells of the leaves when evaporation is slow. His evidence for this view was the observation that detached branches immersed in water continued to absorb eosin although there obviously could be no evaporation from the submerged leaves. Smith and associates (311) investigated this situation and concluded that the absorption observed by Dixon arose from failure to satisfy completely the saturation deficit of the branches before starting the experiment. They also point out that temperature decrease can cause contraction of gas in the stems and bring about absorption through the base of the stem.

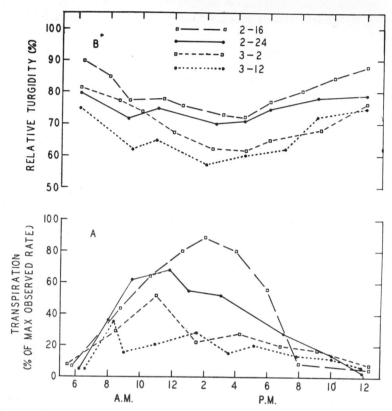

FIG. 8. The effects of decreasing soil moisture on the rate and daily course of transpiration of cotton (*Gossypium hirsutum*) (below) and on the relative turgidity of the leaves (above). These data were obtained during a period of decreasing soil moisture from Feb. 16, when the soil was near field capacity, to March 12, when there was little available water remaining in the soil. As the available water content of the soil decreased, transpiration reached its maximum earlier in the day, the relative turgidity became lower, and there was less recovery at night. Adapted from Slatyer (308).

Even so, while stressing the role of physical factors in the determination of transpiration the reality of more biological ones needs to be kept in mind, an example of this being the well-known effects of age and development on the loss of water by isolated leaves to the atmosphere.

Effects of sprays and dusts. The accumulation of dust, spray materials, or other more or less nontoxic materials may affect transpiration by modifying leaf temperatures or the permeability of the cuticle. Miller (241) summarized the older literature. Bordeaux mixture and probably all other light-colored coatings reduce leaf temperatures and thereby tend to decrease transpiration, but Bordeaux mixture seems to increase the permeability of the cuticle (148). It might, therefore, decrease stomatal transpiration during the day but increase cuticular transpiration at night and in cloudy weather. Beasley (23) reported that only those dusts containing particles smaller than 5 μ in diameter reduced transpiration, and those only at night. She thought this was because the small particles lodged in the stomata and prevented them from closing. Various kinds of waxes and latex compounds sometimes are applied to plants following transplanting to reduce transpiration, and some of these promise to be beneficial (3).

Water supply. Lack of sufficient water can easily become the most important environmental factor because of the various effects of an internal water deficit. This is shown in the rapid decrease in transpiration of plants in drying soil (Fig. 8). The decrease accompanying wilting also is shown in Table VI. The reasons for the marked effect of a water deficit have already been discussed and the factors responsible for water deficits will be discussed in a later section.

5. Measurement of Transpiration

The earliest measurement of transpiration known to the author were those of Hales, published in 1727. He measured the rate of water loss from a variety of potted plants by weighing the containers and also proved that water was lost as vapor by collecting it in a glass container enclosing a branch. The often cited studies of von Höhnel (144) on woody species also were made with potted plants. Numerous studies of agricultural plants, mostly grown in large containers, made in this country during the early part of the century were summarized by Miller (241). Methods of measuring transpiration will be discussed first, followed by a few examples of actual rates.

a. Methods of measuring transpiration. The rate of transpiration can be determined by periodically weighing potted plants or detached plant parts, by measuring the volume of water absorbed from a potometer (with the tacit assumption that it equals the loss in the same period), or by measuring directly the loss of water vapor to the air. Miller (241) and Stocker (328) discuss the various methods in some detail.

The oldest and simplest method is to grow plants in containers of soil and weigh them at frequent intervals. The containers must be waterproof to prevent loss of water from them by evaporation, and the soil moisture must be maintained in long-term experiments so that soil moisture does not become a limiting factor. This method gives very accurate results if properly carried out, but it is normally limited to plants which can be grown in at most a few hundred pounds of soil. Other problems include maintenance of good aeration and a tendency of the soil in containers to become much warmer than normal if the containers are exposed to the sun.

Because of the limitations of the pot or phytometer method many ecological studies have been made by the quick-weighing method developed by Huber and Stocker. In this method a leaf or branch is cut off and weighed at intervals of 1 or 2 minutes while it is kept in its

TABLE VIII

RELATIVE RATES OF TRANSPIRATION OF TREE SPECIES AS MEASURED BY DIFFERENT
INVESTIGATORS WITH THE RATE FOR SPRUCE AS 100[a]

Species	Weighing potted seedlings (Eidmann)	Quick weighing of detached twigs	
		Pisek and Cartellieri	Polster
Spruce [*Picea abies* (*P. excelsa*)]	100	100	100
Scots pine (*Pinus sylvestris*)	181	133	139
Larch (*Larix europaea*)	310	212	212
Beech (*Fagus sylvatica*)	268	377	372
Oak [*Quercus robur* (*Q. pedunculata*)]	282	512	460
Birch (*Betula verrucosa*)	618	541	740

[a] From Huber (151).

natural environment. Detaching a leaf or branch cuts off the water supply and eventually results in wilting, but during the first minute or two after detachment there may be an increase in transpiration if a high tension in the xylem was released by cutting. Stocker (328) claims that if proper precautions are observed the quick-weighing method gives satisfactory results, at least for ecological purposes, and it has been used frequently.

Measurements made by the quick-weighing method have some validity because, as shown in Table VIII, with one exception all of the species are ranked in the same order by the three investigators, using potted seedlings and the quick-weighing methods. Ringoet (284) found that transpiration rates obtained by weighing detached leaves, or leaflets, of several species for 2-minute intervals gave transpiration rates much higher than the results obtained with intact, potted plants. These differences were greatest with species which grow in the sun and least with shade species. In spite of this he was able to use the method for certain studies, including the daily course of transpiration.

The potometer method of measuring transpiration from detached plant parts is to place the base of the stem of a cut shoot in a closed container of water equipped with a graduated tube or other measuring device to permit careful measurement of the volume of water absorbed. A simple potometer is shown in Fig. 9. This method is useful to demonstrate effects of various environmental factors on transpiration, but it

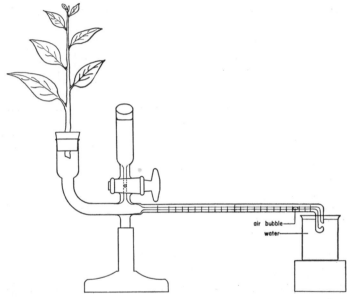

air bubble
water

Fig. 9. The potometer method of measuring transpiration of cut branches. Another type of potometer has a chamber large enough to admit root systems of small plants grown in nutrient solution.

does not give a reliable indication of the transpiration rate of the branch or leaf which is attached to a plant because the resistance to water movement of the root and stem is eliminated and the effect of any tension in the water-conducting system also is eliminated. Furthermore, absorption often is reduced by air bubbles in the open xylem vessels. More reliable results can be obtained if the roots of small plants are placed in the potometer, but even then an error occurs because absorption tends to lag behind transpiration (183, 284).

Measurement of the transpired water vapor can be made in several ways. In the method devised by Freeman (100), the transpiring object, usually an attached branch or leaf, is enclosed in a transparent container through which an air stream is passed. The difference in water content of the air before and after passing over the plant material is measured in various ways. In the early work the water vapor was absorbed in a substance such as calcium chloride or phosphorus pentoxide, but Scarth *et al.* (292), Huber and Miller (154), and others recently have used an infrared absorption apparatus to measure the content of water vapor of the air. Huber and Miller also developed a device called a "Thermoflux" in which the water vapor in the air is absorbed in sulfuric acid and the amount is measured by the heat released. Glover (108) used the difference between a wet and a dry thermocouple to measure the water vapor content of the air, and Andersson *et al.* (8) measured it by its effects on the corona of a high-tension spark discharge.

The most important problem in connection with these methods is the tendency of enclosed branches to become much warmer than they would be in the open. Undesirable wind effects also may occur if air movement is too rapid. These difficulties usually can be minimized by enclosing branches for very short periods and moving the air stream through them at a low velocity.

Cobalt chloride paper has long been used to measure the rate of escape of water vapor from leaves. Paper impregnated with cobalt chloride is blue when dry and pink when wet. If such paper is held in contact with leaves and protected from air on the outer side by a transparent covering, the time required for a color change to occur is a measure of the rate of loss of water vapor. By calibrating the paper it is even possible to calculate the amount of water being lost from the time required for change in color of the paper (236). Unfortunately, this gives no reliable indication of the rate of transpiration in the open, because it measures the rate of diffusion of water vapor in a closed system to an efficient absorbing surface, namely the anhydrous paper. The method does measure the diffusive capacity of stomata, however, if the paper is not left in position long enough to cause stomatal closure (243).

b. Rates of transpiration. Transpiration rates can be considered in terms of individual plants and plant parts or in terms of units of vegetation, as square meters of grass or crop plants, or stands of trees. Transpiration of individual plants can be expressed in terms of the plant, per unit of plant surface or per unit of fresh or dry weight of plant tissue. Furthermore, the rates can be related to the rate of evaporation by expressing them as the ratio of transpiration to evaporation from an evaporimeter of some sort such as the Livingston atmometer. This gives what Livingston (211) termed relative transpiration and serves as a method of reducing transpiration rates to a common basis so that measurements made at different times and places can be compared more satisfactorily. The most recent discussion of transpiration measurements is that by Stocker (328). Figure 10 [from Pisek and Cartellieri (264)] shows that the relative rates of transpiration of different species may be quite different when expressed per unit of surface and when expressed per unit of fresh weight, if the ratio of surface to weight is different.

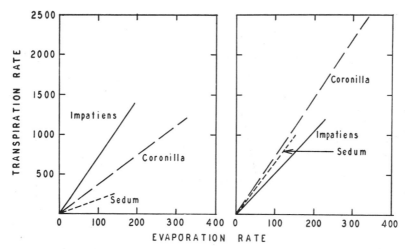

Fig. 10. Transpiration rates of three species of plants (*Impatiens noli-tangere*, *Coronilla varia*, and *Sedum maximum*) expressed in milligrams per gram of fresh weight (left), and in milligrams per square decimeter of surface (right). Transpiration is plotted over rate of evaporation from a filter-paper atmometer. The differences between species increased with increasing rate of transpiration. The rate of transpiration of *S. maximum* relative to the other species was affected most by the method used to express it because it has a lower ratio of surface to mass. Adapted from Pisek and Cartellieri (264).

Transpiration of individual plants. Stocker (328) has brought together data illustrating the transpiration behavior of plants of various species and types under a wide range of environmental conditions. Under ideal conditions the daily course of transpiration should follow the daily trend of evaporation, but often failure of water to reach the leaves as rapidly as it is lost results in development of a water deficit, closure of stomata, and a midday reduction in transpiration below the expected rate, as shown in Fig. 11. As the soil dries out and water deficits develop earlier in the day, transpiration falls off earlier on succeeding days, as shown in Fig. 8.

The rate of transpiration per unit of plant tissue varies greatly among species in the same environment, among individual plants, and even among different leaves on the same plant. Polster (266) found the rate of transpiration of *Betula verrucosa* (= *B. pendula*) to average 8.5 gm per square decimeter of leaf surface per day, but the rate was only 3.1 gm for *Fagus sylvatica*. Beech has more leaves than birch, however, hence the transpiration rate for stands of these trees was not so different, being estimated at 4.7 mm per hectare per day for birch and 3.8 mm per hectare for beech.

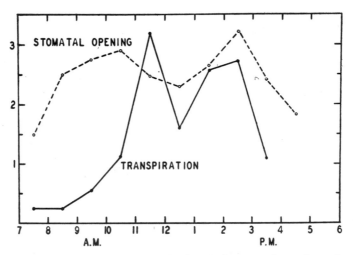

Fig. 11. The daily course of transpiration and stomatal opening of oil palm (*Elaeis guineensis*) in the Belgian Congo. The midday reduction in stomatal opening was accompanied by a reduction in transpiration. Transpiration was determined by rapid weighing of parts of leaves and is expressed in milligrams per gram fresh weight of leaves per hour. Stomatal opening was studied by the infiltration method, and the results are expressed in arbitrary units. Redrawn from Ringoet (284).

The seasonal course of transpiration is affected by both plant and environmental factors. Aging and loss of leaves in evergreens and loss of all of the leaves by deciduous species would result in large changes in rate of transpiration even if no changes occurred in the environment. As a matter of fact, large changes in soil and atmospheric moisture conditions occur even in the tropics, and there are seasonal variations, mostly related to rainfall and soil moisture. Ringoet (284) for example, found that in the Belgian Congo the oil palm (*Elaeis guineensis*) transpired more during the rainy season, when humidity was high, than during the dry season, because soil moisture is a limiting factor during the less humid season. Stocker (328) presents data from Franco and Inforzato (98), showing that coffee trees in Brazil also transpire more during the months of high rainfall than during the dry season. In contrast, bananas (*Musa paradisiaca* var. *sapientum*)* in a region of Brazil where the soil moisture-storage capacity is high, showed an increase in transpiration during the season of least rainfall [Morello, 1953, cited by Stocker (328)]. Data of this type illustrate the paramount importance of an adequate supply of soil water to maintain a high rate of transpiration.

In temperate climates with good distribution of rainfall, the annual curve for transpiration of evergreen species should follow fairly well that for evaporation of water, as Stocker (324) found for *Erica tetralix* in Germany. In areas with pronounced wet and dry seasons soil moisture is likely to become the limiting factor during at least part of the year. Differences in structure and plant behavior also become important in dry areas, as Oppenheimer (251) and others have pointed out. Some species show stomatal closure as soon as a water deficit develops and therefore have relatively low rates during the entire year, for example the evergreen oak, *Quercus calliprinos*, in Palestine. In this species the transpiration rate varied from 4 to 6 mg per gram per minute in the spring to 2 to 2.5 mg in the summer, and sometimes no measurable water loss occurred in the summer because the stomata were closed tightly. *Phillyrea media* (*P. latifolia* var. *media*), in contrast, had high transpiration rates in the spring when soil moisture was available, and these high rates persisted during the summer. The stomata remained open and the heavy water loss produced a large water deficit and high osmotic pressure in the leaves.

Loss of leaves from deciduous species abruptly reduces their transpiring surfaces, and not surprisingly, rates of water loss drop off very rapidly in the autumn when leaves are killed by frost, as shown in Fig. 12.

Evapotranspiration of stands. Agriculturists and foresters are more interested in the rate of water use per unit of land area than in the rate per plant or per unit of leaf area. This raises the question whether or not data obtained from potted plants or detached twigs and leaves can be used to calculate the rate of water loss for large areas of vegetation, such as fields or forests. It would seem unlikely that this can be done safely, unless unusual precautions are used to surround the potted plants by other similar plants and in every way maintain all conditions

---

* Recent research has indicated that the common edible bananas are properly referable to *M. acuminata* or to the hybrid *M.* × *paradisiaca* (*M. acuminata* × *M. balbisiana*).

as near as those in the field as possible. Nevertheless, the rates estimated in this manner seem to be at least of the correct order of magnitude, because Kiesselbach (174) estimated that corn transpires about 9 acre-inches during the growing season in Nebraska, and Reimann and co-workers (276) calculated from field studies that corn in Illinois transpired 8.2 acre-inches from June 1 to September 15.

The most reliable measurements of water loss from vegetation are made by measuring the differences between precipitation and runoff from lysimeters or small watersheds. Examples of this type of measurement are described by Hoover (146) and Harrold and Dreibelbis (125).

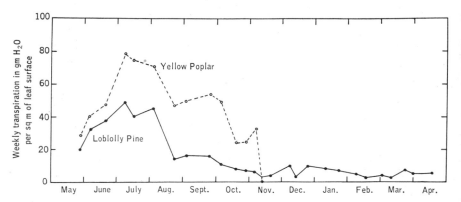

Fig. 12. Seasonal course of transpiration of an evergreen conifer, loblolly pine (*Pinus taeda*) and a deciduous tree, yellow poplar (*Liriodendron tulipifera*), at Durham, North Carolina. The trees were two years old and growing in pots of soil kept near field capacity. The pots were inclosed to prevent water loss from evaporation, and transpiration was measured by weighing the entire container. Yellow poplar continued to transpire rather heavily until a freeze killed the leaves.

Such data give a measure of evapotranspiration, which includes transpiration, evaporation of intercepted precipitation, and evaporation from the soil surface. Measurements made at Coweeta in southwestern North Carolina indicate that the hardwood forest in this mild, humid climate transpires 17–22 acre-inches of water per year, or about 30% of the annual precipitation.

A number of investigators have attempted to estimate evapotranspiration from land areas by means of equations using the amount of incident solar energy available to evaporate water. Among those who have taken this approach are Albrecht and Geiger in Europe, Penman and Schofield in England, and Thornthwaite and Mather in the United States. In this concept it is assumed that, if the soil surface is covered with vegetation and if the soil is moist, then the same amount of water will be lost by evapotranspiration, regardless of the species present. This would be true if the different kinds of vegetation were rooted to the same depth and if all species behaved in the same manner under a water deficit. These requirements seldom are met in nature, because grasses are seldom as deep-rooted as trees and different species show quite different rates of water loss when subjected to water deficits. Nevertheless, it is true that various species of trees seem to lose about the same amount of water per day during the growing season (151), and there is little doubt that there is much less difference in the evapotranspiration from different kinds of plant cover with equal depths of rooting than was formerly supposed to exist.

c. *Transpiration ratio or water requirement.* There has been considerable interest in the amount of water required to grow crops of various kinds. This usually is expressed in terms of pounds of water used per pound of dry matter produced, and it is often termed the "water requirement." Water requirement is an unfortunate term because actually there is no specific "requirement" but only a ratio which varies widely from place to place, from species to species and variety to variety, and from year to year for the same species in the same location. The term transpiration ratio is more satisfactory than water requirement. Miller [(241) pages 496–502] summarized considerable data on the transpiration ratio. Shantz and Piemeisel (301) and their colleagues studied the transpiration ratio of about 150 different species and varieties of plants at Akron, Colorado, and found it to vary from 216 for Kursk millet (*Setaria [Chaetochloa] italica*) to 1131 for *Franseria tenuifolia*, a native weed. Over a period of seven years it varied from 657 to 1068 for alfalfa (*Medicago sativa*). All of these measurements were made on plants in well-watered soil. Polster (266) estimated that the amount of water required to produce a gram of dry matter in various species of trees in Central Europe varied from about 170 gm in beech (*Fagus sylvatica*) and Douglas fir (*Pseudotsuga menziesii [P. taxifolia]* to 300 in Scots pine, 317 in birch (*Betula verrucosa*) and 344 in *Quercus robur*.

Of course the transpiration ratio varies not only with the factors which affect transpiration, but also with availability of soil moisture, soil fertility, and other factors which affect growth.

d. *Significance of transpiration.* One may well ask of what importance transpiration is to plants and whether or not the passage through a plant of 100 times or more its weight in water during a growing season is of any special significance. In general it may be said that transpiration per se contributes very little directly to plants and can there-

fore, best be described as an "unavoidable evil." It is unavoidable because of the structure of plants. The interior of leaves consists of water-
permeable cell walls bordering on intercellular spaces which are connected to the outside air through stomata, and, under the conditions
necessary for photosynthesis, it is therefore inevitable that water evaporates from the cell walls and diffuses out through the stomata whenever they are open and whenever the vapor pressure of the outside air
is lower than that inside the leaves. Transpiration is an evil because it
often produces water deficits in plants, which check photosynthesis, reduce growth, and, if too severe, may cause death from desiccation. It is
probable that more plants are injured and killed by excessive transpiration than by any other cause.

It may be asked why plants evolved with structures which make
transpiration unavoidable, if it is so often injurious. This must have
occurred because in the long run the advantages of a structure favorable
to the entrance of the carbon dioxide required for photosynthesis far
counter-balanced the disadvantages of the accompanying high rate of
water loss. Plants which have very low rates of transpiration usually
also have low rates of photosynthesis and grow slowly. For example,
Polster (266) measured photosynthesis and transpiration rates of 7 tree
species and found that those with high transpiration rates also had high
rates of photosynthesis and those with low transpiration rates had low
rates of photosynthesis per unit of leaf fresh weight. These data support
the suggestion of Hygen (158) that the transpiration rate under a given
set of conditions is an indication of the gas exchange under those conditions and the maximum rate of transpiration ought to be related to
the maximum rate of photosynthesis.

Not everyone agrees with the view that transpiration is intrinsically
unnecessary and often detrimental. It has been argued that transpiration plays an essential role in the absorption and translocation of minerals in plants and that the cooling effect on leaves keeps them from
being overheated in full sun (58). Such arguments can be refuted by
the fact that most plants can be grown just as successfully in a humid
greenhouse with a low rate of transpiration as in a dry atmosphere with
a much higher rate of transpiration. It has already been mentioned
that the cooling effects are usually not very large, and they certainly
are not essential because transpiration decreases in partly wilted leaves
just at that time of day when cooling is most needed. Those who wish
a more detailed discussion of this topic are referred to Clum (59),
Curtis (67), and Eaton and Belden (85). The relation of water absorption to salt absorption is not a simple and direct one: it will be discussed in this and also in Chapter 4.

III. The Ascent of Sap

Both botanists and nonbotanists have long been puzzled by the rise of water to the tops of trees. Several papers on this subject appeared in the Proceedings of the Royal Society of London between 1668 and 1671. Most of these were concerned with the possibility that there might be a circulation of sap in plants analogous to the circulation of blood in animals. This possibility was discussed as late as the beginning of the nineteenth century, although Hales (120) had stated early in the eighteenth century that there was no evidence of any real circulation of sap. Hales also observed that although root pressure occurred in some species of plants early in the spring, it ceased "as soon as the young leaves begin to expand enough, to perspire plentifully, and to draw off the redundant sap." This suggests that he believed that transpiration itself caused the rise of sap. It was over a century before any important additions were made toward an explanation of the problem, although many were offered.

A. MECHANISMS FOR ASCENT OF SAP

The numerous explanations of sap ascent can be grouped together and considered under a few headings.

1. Root Pressure

It has long been known that in the spring sap exudes from wounds in grape vines and stems of some other woody species and from the stumps of many herbaceous plants. Pressures of 1 or 2 atm are often developed in the spring (235), and, according to White (365a), pressures probably in excess of 10 atm were developed by growing, excised roots of tomato. Root pressure often has been assumed to be an important factor in the ascent of sap, but in the writer's opinion it scarcely can be a significant factor in the rise of sap in rapidly transpiring plants or tall trees. As soon as plants begin to transpire the pressure in the xylem begins to decrease below atmospheric pressure and tensions usually develop. It may be argued, however, that these observations are not conclusive because they necessarily are made on decapitated root systems and it is not certain that such root systems can develop as large pressures as might be developed in attached root systems. Hales observed long ago that root pressure is not detectable in rapidly transpiring plants, and furthermore root pressure is rarely observed in roots of conifers.

Some writers have suggested that root pressure plays an important role in trees by causing refilling with water of xylem vessels, which contain water under tension or which have become filled with gas bubbles during periods of rapid transpiration, even though it contributes nothing to transpiring plants. There is some increase in water content of the wood of conifers during the winter in spite of the absence of root pressure. Root pressure appears not to be essential, because trees in which it has not been shown to occur grow just as well in a wide variety of habitats as the species in which it is known to occur.

2. Vital Activity in Stems

In the 1880's several German investigators, including Westermaier, Godlewski, and Janse, concluded that the living cells of stems played an essential role in the ascent of sap. Many botanists supported this view even during the first quarter of this century. The best-known vitalist of recent times was the Indian, Bose, who claimed that alternate contraction and expansion of cells caused pulsations in the stems which brought about the ascent of sap (36). Molisch (245) supported this theory, but it now has been abandoned. Most physiologists abandoned the various vital theories as a result of experiments which showed that water will rise through segments of stems in which the living cells have been killed by heat or various poisons. Among those who have demonstrated the rise of sap through stems in which segments had been killed by various methods are Strasburger (337) and Overton (256). Miller [(241) pages 861–863] has reviewed most of these experiments.

Some investigators believe that the importance of living cells in the stem has been underestimated because the leaves usually wilt and die within a few days after the stems on which they are borne are killed. This usually is attributed to blockage of a large proportion of the xylem elements in the boundary between dead and living cells, although complete blocking rarely occurs. Priestley (272) emphasized the importance of the onset of growth and differentiation in the cambium in connection with the diversion of water and solutes to young leaves. (For a discussion of cambial tissue in supplying solutes to the leaf and bud see Chapter 4,III,B.) According to Priestley the old wood in a stem serves as a reservoir from which water is first drawn off laterally via the cambium and differentiating vascular tissue and then is moved in the leaf traces, by processes not yet fully understood, to the leaves. Handley (122) proposed that water might move upward through living cells rather than through nonliving xylem elements because chilling of stems causes decrease in ascent of sap, but he admitted that this seemed quantitatively improbable.

It also has been claimed that the ascent of sap reported to occur in aquatic plants is caused by some vital activity in the stems. Wilson

(378) reviewed the literature and made observations which led him to conclude that the uptake of water by submerged cut stems of aquatic plants probably is caused by readjustment of the internal water balance rather than by any secretory mechanisms. He attributed the effect of light and respiration inhibitors on water movement to osmotic disturbance of the sensitive water balance or to the effects of changes in gas content in the intercellular spaces, which were caused by photosynthesis. Rooted aquatics apparently develop root pressure which causes sap movement through their xylem (333, 343).

It is possible that the presence of living cells is essential to maintain conditions favorable for the ascent of sap. Death of all living cells probably allows air to infiltrate and break the continuity of the water columns. The fact that the presence of living cells is essential for the maintenance of unbroken water columns in the xylem does not constitute evidence, however, that they play a direct role in the ascent of sap.

3. Capillarity and Imbibition

It might be supposed that capillary rise would be important in the vessels of the xylem, but water can rise by capillarity only about a meter in the smaller xylem elements. Sachs claimed that the lumina of the xylem vessels were filled with air and the water moved up through the walls by imbibition, but this was disproved by experiments in which blocking the lumina of the vessels resulted in wilting of the shoots (79, 87, 337). Strugger (340) has long claimed that considerable movement of water occurs in cell walls, and this may be true in regions where no paths of lesser resistance exist, as across the cortex of roots or in the parenchyma of leaves. It certainly is not true in the xylem, however, where plugging the lumina of the xylem vessels results in immediate wilting of the leaves because of the slow rate of movement of water through the walls. Hulsbruch (156) recently summarized the literature on water movement outside of the vascular system.

4. The Cohesion Theory

Toward the end of the nineteenth century Sachs and Strasburger both concluded that transpiration produces the pull causing the ascent of sap, but they did not understand how it did this. As early as 1892 Boehm (31) had caused transpiring branches to raise mercury above barometric pressure and had suggested that water is pulled to the tops of trees in cohering columns.

Dixon and Joly (78) and Askenasy (14) almost simultaneously demonstrated that water has considerable tensile strength when con-

fined to small, completely wettable, tubes such as the xylem elements
and proposed that water is pulled to the tops of trees in such cohering
columns. Dixon found that water has a tensile strength of many atmospheres even when it contains dissolved air. He also demonstrated that
leaves could transpire into air compressed to three times normal pressure, proving that water is so firmly anchored in the walls of the mesophyll cells that they could support the pull of a water column about
30 meters in height. Dixon published two books (76, 77) and several
papers on the relation between transpiration and the ascent of sap.

Renner (278–280) also studied this problem. He estimated the suction pressure of transpiring shoots to be 10–20 atm, basing this principally on studies in which he measured the uptake of water by transpiring branches, then removed the leaves and measured the amount of
water moved through the stems under a given pressure gradient. Renner carried out some experiments on the force required to cause the
opening of fern sporangia which led him to estimate that the water in
the cells of the annulus of a fern sporangium is subjected to a tension
of about 350 atm immediately before it ruptures. Similar results were
obtained by Ursprung (349). Ursprung (350) also demonstrated the
rise of mercury to the height of 1.4 meters in a column attached to a
transpiring branch of *Clematis vitalba*. More recently Thut (344)
demonstrated the rise of mercury to a height of more than 100 cm in
tubes attached to transpiring branches.

As other theories were being disproved and abandoned, the cohesion
theory gained ground and it now has become the most generally accepted explanation for the ascent of sap, although it has its critics and
some evident weaknesses.

The cohesion theory is based on the fact that water molecules have
very high attractive forces for each other and, when confined in small
tubes such as the xylem elements of a tree, water can be subjected to a
pull or tension of many atmospheres before the cohering columns rupture. Furthermore, all of the water in a plant is conceived to be connected together through the water-saturated cell walls to form a continuous system—the hydrostatic system of MacDougal. Therefore, when
water evaporates from any part of this system, such as the walls of the
mesophyll cells of the leaves, the resulting imbibitional force developed
in the walls causes movement of water out of the xylem of the nearest
leaf veins into these cells (see Chapter 4, Fig. 4). This reduces the
pressure in the water contained in the xylem, and this reduction in
pressure is transmitted down through the water in the xylem of the
stem to the roots, where it brings about water absorption from the soil.
If the rate of water loss tends to exceed the rate of absorption, the pres-

sure on the water in the xylem falls below 1 atm and finally becomes a tension or pull. There is some question concerning the extent to which the pull developed by the imbibitional force of the cell walls from which evaporation occurs is transmitted all the way through the plant to the surfaces of the roots and the extent to which it is converted into osmotic forces by which water is moved across the parenchyma cells of the leaves and the root cortex by diffusion. It is generally assumed that although water is pulled up through the xylem, it moves by diffusion along a diffusion pressure deficit gradient from the xylem elements to the evaporating surfaces of the leaf cells. Smith *et al.* (311), for example, state that the principal function of living cells is to convert imbibitional forces into osmotic forces, which are necessary to pull on water in the xylem elements. Levitt (206) objected to this idea because, according to his calculations, movement of liquid water by diffusion is much too slow to replace the water lost by transpiration. In his opinion the imbibitional forces developed at the evaporating surfaces are transmitted as a pull causing the mass movement of water through the leaf cells, the water in the xylem, and the cells of the roots from the medium surrounding the roots. This view seems to return toward that originally postulated by Dixon (76), who wrote of the water columns in trees as if they were hanging, suspended, from the evaporating surfaces of the leaf cells and who regarded the intervening parenchyma as able to "take the strain." Renner (279) also emphasized the importance of an imbibitional gradient in the cell walls, caused by tension in the vessels.

It has been objected that mass movement of water cannot readily occur across roots because the radial and inner tangential walls of the endodermal cells usually become thickened and impermeable. It has been shown, however, that large amounts of water can be caused to move into the xylem of root systems of various species of herbaceous plants by attaching a vacuum pump to the stump and applying a pressure drop of less than an atmosphere from the exterior of the root to the xylem (182, 183). If water can be moved into the xylem by such a low pressure difference, it seems that it could also be pulled in by a reduced pressure or tension on the water in the xylem of intact transpiring plants.

The essential feature of the cohesion theory is that evaporation of water sets up imbibitional forces in the cell walls which are transmitted through the hydrostatic system, causing the absorption of water and the ascent of sap. In order for this to occur it must be assumed that water has sufficient cohesion to withstand the tension necessary to pull it to the tops of the tallest trees. On this view, the water-conducting system must also have a structure which prevents the entrance of air bubbles that might rupture the water columns, and its walls must be rigid enough to prevent collapse under tension. Presumably the cells of the leaves at the top of a plant must have an osmotic pressure in excess of the tension required to pull water to the top in order that they may obtain and retain water against the tension in the xylem.

a. Objections to the cohesion theory. Although the cohesion theory is accepted today by most plant physiologists it has certain weaknesses which have resulted in continued attempts to modify it. It has to be recognized that the content of the xylem in a tree is not the simple problem that it may seem to be because it may consist variously of (a) liquid water under positive pressure, (b) water under tension, (c) gas, even water vapor at low pressure, (d) air. Certainly between the first three of these states ready transitions may occur.

Most troublesome is the instability of water columns under tension. If the water columns under tension in the xylem of a plant stem break as readily as those in a glass tube, then swaying of stems in a wind probably would break them. In most plants the xylem elements are relatively short and of very small diameter. Furthermore, the water in them is very closely bound to the water in their walls, and this may reduce the probability of rupture. The contents of vessels can be seen to rupture under mechanical stress, but Crafts and associates (63) think that in many instances the liquid is replaced by bubbles of water vapor rather than air and the vessels later become refilled with liquid. Dickson and Blackman (74) found that bubbles of air introduced into the xylem vessels often dissolved and thought this might explain the disappearance of air from cut stems.

The possibility of a large proportion of the vessels becoming plugged with bubbles of gas also exists. During periods of rapid transpiration a considerable fraction of the xylem in some trees becomes filled with gas, especially the larger vessels of ring-porous species. The xylem is composed of many units, however, separated by walls through which air bubbles cannot move or at least move with difficulty, hence the presence of air in one part of the system does not interfere with water movement through other parts. The average plant has so much more conductive capacity in its xylem than is necessary for survival that plugging up a considerable fraction produces no serious injury.

Scholander *et al.* (298) questioned the operation of the cohesion theory because water continued to move up grapevine stems after the large xylem vessels presumably had been plugged with air. Scholander (298a) later concluded that although the vessels were plugged with air, the tracheids continued to function as the path of water movement.

Preston (267) found that when the young wood of a ring-porous species of tree was cut under dye, the new vessels were injected to a considerable distance. He concluded from this that many of the vessels were filled with gas under reduced pressure. There is little doubt that an increasing fraction of the xylem elements of trees becomes filled with gas during the course of a summer and, as noted later, in many species only a small part of the total xylem is available for the ascent of sap. Preston (268) and Greenidge (110, 111) have questioned the existence of cohering water columns under tension because horizontal cuts made more than halfway through tree trunks from opposite sides, one above the other, did not prevent the ascent of sap. Greenidge (111) assumed that such cuts must interrupt all of the water columns, yet there was no wholesale evacuation of sap from the opened vessels. Although ring-porous species have long vessels which initially may extend from the top to the base of the trunk, the effective length of these vessels often, but by no means always, is reduced by tyloses. According to Liming (210), in elm (*Ulmus americana*) the vessels in 1-year-old wood were only 5–10% as long as vessels in current wood. It seems likely that not all elements are blocked by air and that water continues to be pulled around the cuts. The water-conducting capacity of the xylem of the average tree trunk is much greater than is needed for survival, as shown by the fact that injury to one-half or more of the circumference seldom results in damage to the top. Thus plugging of over half of the xylem by air is unlikely to have any serious effect on the supply of water to the transpiring shoot.

Another question concerns the cohesive properties of water. Theoretically the intermolecular attractive forces of water are very high, and Fisher (97) estimated that rupture of stretched water columns would occur at a tension of about 1300 atm. Dixon (76), Renner (280), and Ursprung (349) thought they had demonstrated tensions of 200–350 atm, but their methods have been questioned and Greenidge (110) claims that the water in the xylem of trees probably does not support tensions much greater than 30 atm. On the other hand MacDougal *et al.* (225) estimated that tensions of up to 200 atm might occur in transpiring trees under severe conditions. Part of the diurnal shrinkage in diameter observed by MacDougal (224) in rapidly transpiring trees (see Fig. 13) is attributed to tension in the hydrostatic system, and Bode (30) observed under the microscope actual contraction of xylem vessels in stems of wilting *Impatiens sultanii* plants.

It is difficult to measure the diffusion-pressure deficit or tension in the xylem directly, but Stocking (330) found the maximum values in wilted leaves of squash to be about 9 atm. Arcichovskij and Ossipov (10) observed diffusion pressure deficits of up to 142.9 atm in desert shrubs and assumed that similar tensions must occur in the water of the xylem. Slatyer (310) observed diffusion pressure deficits of 41–77 atm in leaves of wilting plants, and it seems that equally high tensions must exist in the xylem of such plants.

The actual tension required to lift water to the top of a tree is not very large. Dixon calculated that the force required to overcome resistance to flow at a low rate is approximately equal to that required to support a column of water of the same height. Thus water could be moved to the top of a tree 100 feet high by a tension of 6 atm although,

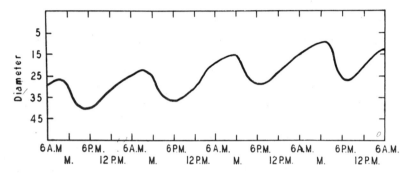

Fig. 13. Daily variation in diameter of the trunk of an Arizona ash (*Fraxinus "arizonica"*) during a period of rapid growth and high transpiration. Each division on the ordinate represents 1 mm of actual change in diameter. The afternoon shrinkage which occurred each day is the result of an internal water deficit and high tension in the water of the xylem elements. Redrawn from MacDougal (224).

if transpiration were rapid, much higher tensions doubtless would develop. Even the minimum value of 30 atm proposed by Greenidge would suffice to pull water to the top of the tallest known trees.

5. Present Status of the Problem

It seems certain that water absorption is linked to, and usually controlled by, the rate of water loss, because in moist soil the rate of water absorption follows fairly closely the rate of transpiration, as shown in Fig. 16. It also seems certain that the pressure in the contents of the xylem elements not only drops below atmospheric pressure, but that so much tension develops that xylem vessels and even tree trunks shrink in diameter. There is no doubt that during the summer an appreciable part of the xylem in tree trunks becomes occupied by air, but a large proportion of the xylem elements in at least the newest annual rings remain air-free and can function in water conduction [(153) pages

563–564]. The ease with which water columns in glass tubes are broken raises questions concerning the stability of water columns in trees when subjected to tension, but the dimensions of the xylem elements and the hydrophilic nature of their walls probably makes their contents much less subject to rupture by shock than is water in glass tubes.

There is no doubt that the cohesion theory has weaknesses, but it is the only physical theory in existence which can explain the rise of large amounts of water to the top of tall trees. It also is the only existing theory which explains how absorption is linked to transpiration. In spite of its weaknesses the cohesion theory of the ascent of sap seems to be the most satisfactory explanation of the ascent of sap.

B. The Path of Water Movement

It has been agreed at least ever since the time of Hales that water moves from the roots to the leaves through the xylem. Xylem consists of a variety of kinds of cells, including fibers and parenchyma cells of various sizes and shapes, but water movement occurs chiefly through tracheids and vessels (see Fig. 14). Tracheids are somewhat spindle-shaped cells which lose their protoplasm as they mature and are seldom more than 5 mm in length and 30 μ in diameter and which have lignified walls containing pores or pits. Tracheids are the water-conducting elements of conifers, but in angiosperms vessels form the principal path for water conduction. Vessels are formed by breakdown of the end walls of long rows of cells and disappearance of the protoplasm, forming tube-like structures with diameters of 20–400 μ and lengths varying from a few centimeters to several meters. In ring-porous trees it is believed that single vessels are differentiated from top to bottom of the trees, but in a number of species including elm (*Ulmus americana*), white oak (*Quercus alba*), black locust (*Robinia pseudoacacia*), and walnut (*Juglans nigra*) their effective length probably is quickly reduced by tyloses. Vines commonly have vessels many meters in length which offer little resistance to water movement (298).

Water movement is not entirely restricted to the xylem because water must pass through other tissues to enter the xylem of the roots and again when it passes out of the xylem in the leaves (see Chapter 4, Fig. 4). Hulsbruch (156) has recently discussed movement of water in tissues other than xylem. In young roots water enters through the epidermis, crosses the cortical parenchyma, endodermis, and pericycle, and then enters the root xylem (see Fig. 15). In older roots the cortex has disappeared and water must here enter through breaks in the corky layer, pass across the phloem, through the cambium, and into the xylem. The xylem strands of the numerous branch roots coalesce and connect with the stem xylem, from which branches separate off and pass into the stem branches, where further subdivisions supply each leaf with water. In the leaves the xylem again branches and rebranches and finally terminates in the individual xylem elements of the numer-

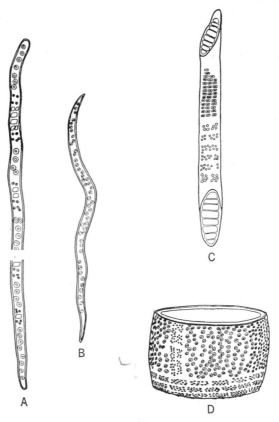

FIG. 14. Tracheids and vessel segments. A. Tracheid from white pine (*Pinus strobus*). Tracheids in conifers usually range from 25 to 80 μ in diameter and from 2.1 to 7.4 mm in length. B. Tracheid from white oak (*Quercus alba*). C. A vessel segment from yellow poplar (*Liriodendron tulipifera*). D. A vessel segment from white oak. Vessel segments usually range from 0.2 to 1.0 mm in length, and they are arranged end to end, with simple or compound perforations between the segments producing vessels which sometimes form continuous passages many feet in length. Adapted from Eames and MacDaniels (82).

ous small veins. Thus the xylem forms a continuous water-conducting system from near the tips of the roots to the mesophyll of the leaves (cf. Chapter 4, Fig. 4).

It should be emphasized that with few exceptions this system is not continuous in the same manner as a pipe, but is merely a continuous system of overlapping vessels and/or tracheids, along which water must pass through numerous cross walls. Readers who are unfamiliar with plant structure should consult a plant anatomy text, such as Eames and MacDaniels (82) or Esau (93), because it is difficult to visualize the

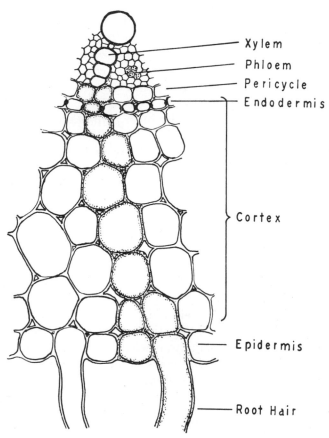

FIG. 15. Diagram of cross section through a young root, showing the path followed by water moving from the epidermis to the xylem.

path of water without more knowledge of the anatomy of the conducting system than can be presented in this chapter.

There has been considerable discussion concerning the path of water which moves outside of the xylem elements. For example, does water move from vacuole to vacuole or does at least part of it move through the cell walls, or within cytoplasm, in passing through the parenchyma cells of roots and leaves? Strugger (338–340) has presented evidence that considerable water movement occurs in cell walls, but Hulsbruch (156) points out certain flaws in the work done with fluorescent dyes. It may be assumed that water usually follows the path of least resistance, which normally is the xylem, but some lateral movement also occurs through the parenchyma cells of roots, stems, and leaves. It is uncertain whether the walls, which consist largely of water bound to the micelles and fibrils and existing free in the intermicellar spaces, offer more or less resistance than the cytoplasm. It has been shown that killing roots permits a large increase in rate of water movement through them, indicating that the protoplasm offers considerable resistance to water movement, but killing the plasmodesmata in the walls may greatly increase the permeability of cell walls to water. If the walls offer less resistance than is offered by other paths, then water movement will tend to occur through the walls.

Huber (153) recently summarized the work in the field of sap-flow dynamics, to which he has been an important contributor. He found that the relative conducting surface, expressed as square millimeters of cross section of the water-conducting system per gram of leaf (fresh weight) supplied by it varied from 0.1 in desert succulents to 0.5 in trees and 3.4 in nonsucculent desert plants, but was only 0.02 in an aquatic plant. He also found the relative conducting surface to be about ten times as great near the top as toward the base of a young tree. The leader has a much greater relative conductive surface than side branches, and the upper branches have higher relative conductivities than the lower branches. The relative conducting surface is a measure of the amount of water-supplying tissue compared to the transpiring surface but tells nothing about its efficiency. The efficiency of a conducting system is measured in terms of the amount of water transmitted per unit of cross section per unit of length per unit of pressure per unit of time ($cm^3/cm^2/m./$atmosphere of pressure applied/hr). This is known as the specific conductivity of the stem. Relative values of specific conductivity according to Huber (153) range from 20 in conifers to 65–128 in deciduous trees, 236–1273 in lianas, and 292–5388 in roots of deciduous trees.

Huber stated that the suction pressure or diffusion pressure deficit at any height in a tree depends on the soil diffusion pressure deficit, the height of the water column in the tree, rate of transpiration, and resistance to water movement through the xylem. From these values it is possible to calculate the diffusion pressure deficits in various parts of a tree or other plant.

C. Rate of Ascent of Sap

The rate of upward movement of sap varies from essentially zero for a plant in darkness in a saturated atmosphere to rates as high as 100 meters or more per hour in rapidly transpiring plants. The early measurements made by introducing dyes through cut surfaces are not very reliable because the rates of absorption and movement are quite different in cut stems from the rates in intact plants. Use of radioactive tracers (247) and fluorescing dyes, is more reliable because they can be introduced into the xylem without cutting it, but the most reliable measurements probably are those made by the thermoelectric method devised by Huber and his students (153). In this method heat is applied to the stem and the time required for the warmed sap to reach a thermocouple placed above the heating unit indicates the rate of sap rise. The velocity in diffuse-porous trees ranged from 1 to 6 meters per hour, but in ring-porous species it was often 20–40 meters. Thus in diffuse-porous species water moves slowly through a large cross-sectional area made up of several annual rings, but in ring-porous species it moves more rapidly through the more limited area of the newest annual ring. Movement usually is much slower in conifers than in deciduous species, and the most rapid rates occur in vines, which have numerous large, long vessels. Water movement in herbaceous species seems to occur at about the same rate as in ring-porous trees. Well-defined diurnal variations in rate also occur with the minimum rate at night and the maximum during the day.

IV. The Absorption of Water

A. Introduction

For a discussion of the problems of absorption of water by cells reference should be made to Chapter 2.

In discussing water absorption a distinction will be made between the so-called passive absorption of water by rapidly transpiring plants and active water absorption by slowly transpiring plants. Such a distinction was at least tacitly made by some earlier physiologists such as Sachs and Strasburger, and even Hales distinguished between plants developing root pressure and transpiring plants in which no root pressure was evident. Water absorption by rapidly transpiring plants was called passive absorption by Renner (281) because he assumed that the forces responsible for water intake are developed in the shoots and transmitted to the roots, which act as absorbing surfaces through which water enters. Active absorption depends on forces developed in the roots and is evident only in plants with low rates of transpiration which are growing in warm, moist, well-aerated soil. It produces pressure in the xylem sap, the so-called root pressure, which is responsible for guttation and "bleeding" or exudation of sap from wounds, as occurs when grapevines are pruned in the spring.

It has been suggested that active and passive water absorption can occur simultaneously, and the absence of root pressure in transpiring plants should not be taken as evidence that active absorption is not occurring. It is doubtless true that, at a rate of transpiration which just exceeds the rate at which water moves into the xylem by the active absorption mechanism, the two processes are operating simultaneously. As soon, however, as the rate of transpiration materially exceeds the rate at which water is delivered to the xylem, a diffusion pressure deficit develops in the living cells of the roots.

B. WATER-ABSORPTION MECHANISMS

Numerous explanations of water absorption have been offered, but all of them can be classified as active or passive.

1. Passive Absorption

In transpiring plants the water in the xylem usually is under reduced pressure or tension, as shown by the fact that if a stem is cut under dye, the dye instantly rushes in and spreads upward and downward in the xylem. The reduced pressure or tension on the water in the xylem exists because evaporation of water from the transpiring surfaces of the shoot develops forces which cause movement of water out of the xylem, as explained in connection with the ascent of sap (III). Reduced pressure or tension on the water in the root xylem reduces its diffusion-pressure deficit, producing a gradient along which water moves from the soil into the roots. Levitt (207) claims that diffusion of liquid water is much too slow to supply losses by transpiration and suggests that the tension in the xylem is transmitted across the cortical tissues of roots, causing mass movement of water from the soil across the intervening cells into the root xylem. According to this view the roots act merely as absorbing surfaces and water is absorbed *through* them rather than *by* them, hence the term passive absorption which was first applied by Renner (281).

That passive absorption can occur independently of any activity of the roots is indicated by the fact that transpiring shoots can absorb water through anesthetized or dead roots, or even without any roots at all. Even better evidence is the fact that the rate of absorption tends to parallel the rate of water loss in transpiring plants. This has been demonstrated by a number of workers, including Lachenmeier (202) and the writer. Some typical rates of absorption and transpiration are shown in Fig. 16.

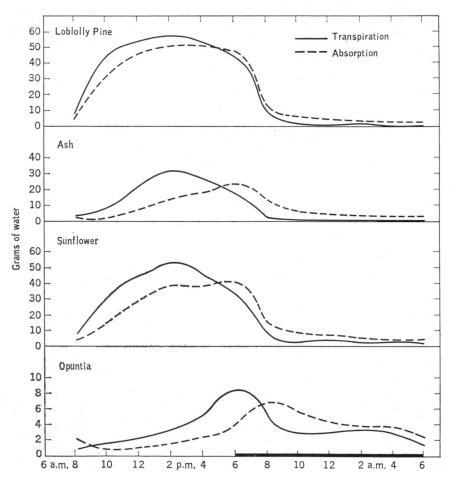

FIG. 16. Rates of transpiration and water absorption in grams per plant of several species of plants (*Pinus taeda, Fraxinus pennsylvanica, Helianthus annuus, Opuntia* sp.) on a clear, hot, summer day. From Kramer (183).

It is evident that absorption tends to lag somewhat behind transpiration in these experiments. This indicates that the diffusion pressure deficit developed in transpiring shoots is not transmitted instantly to the surfaces of the roots. The lag of absorption behind transpiration occurs because of resistance to water movement through the plant. Apparently most of this resistance occurs in the living cells of the roots because water intake through roots attached to a vacuum line is greatly increased when the roots are killed or removed (181, 184, 282). Brouwer (42) also believes that most of the resistance to water movement occurs in the roots. Certainly it is much greater than the resistance to longitudinal movement through the xylem in the stems of plants (376). The probability is that resistance to water movement in the leaves also is rather small because most of the cells are quite close to xylem elements. Figure 15 above and Fig. 4 of Chapter 4 show the path of water across a young root and into the shoot.

2. Active Absorption

The process of active absorption has received much more attention than passive absorption, probably because its mechanism is more puzzling and its results more striking. Two general types of explanation have been offered for root absorption. Osmotic explanations assume that roots behave as osmometers in which water moves from the dilute soil across a differentially permeable membrane formed by the cortex to the more concentrated solution in the xylem. Nonosmotic explanations assume that water is secreted into the xylem by the surrounding cells by a process dependent on energy released by respiration.

a. Osmotic theories. Although Dutrochet (81) seems to have first offered an osmotic explanation of water absorption, he was very vague about its exact operation. It was difficult to formulate a satisfactory osmotic explanation so long as it was believed that water moved along gradients of increasing osmotic pressure. It seemed impossible for water to move from the cortical cells with an osmotic pressure of several atmospheres into the xylem vessels whose contents usually have an osmotic pressure of less than 2 atm. This difficulty was eliminated after Ursprung and Blum (352) and others showed that movement actually occurs along gradients of diffusion pressure deficit (the suction pressure or suction tension of some authors) which are not identical with the osmotic pressure within cells. This concept is discussed in Chapter 2, hence it should suffice to remind the reader that the diffusion-pressure deficit is a measure of the pressure with which a cell or tissue can absorb water. Diffusion-pressure deficit varies from zero in fully turgid cells to a value equal to its osmotic pressure at incipient plasmolysis and may even exceed the osmotic pressure if the cell is under tension (237–239).

If the osmotic pressure of the xylem sap is higher than that of the soil solution, then water can move from the soil into the root xylem by osmosis. According to Atkins (16), the entire root cortex functions as a differentially permeable membrane, but Priestley (270, 271), Arnold (13), and many other investigators have regarded the endodermis as the effective differentially permeable membrane in this system. Neglecting all possibilities of active secretion, this osmotic system presupposes that the diffusion-pressure deficit of the root cortical cells is less than the osmotic pressure of the xylem sap.

The principal problem in connection with an osmotic explanation of root pressure is to explain how a sufficient concentration of solutes is maintained in the root xylem to produce an osmotic gradient from soil to xylem. In many herbaceous plants the solutes seem to be largely inorganic, and there seems to be a definite relationship between salt accumulation in the cortex and the occurrence of root pressure. Active absorption and root pressure occur only when healthy roots are well supplied with carbohydrates and are growing in a well-aerated medium containing the essential mineral nutrients. These conditions also are essential for the absorption of mineral nutrients, and several investigators have stressed the apparent relationship between these two processes (11, 62, 138, 222).

Broyer (45), Crafts and Broyer (62), Lundegårdh (222), Eaton (84), Arisz et al. (11), and others propose that active water absorption depends on accumulation of salt in the xylem and that water then moves in because of the resulting diffusion pressure deficit gradient. Arisz found that when root systems were transferred to a more concentrated medium the rate of exudation decreased at first and the osmotic pressure of the exudate increased, but if root systems were moved to a less concentrated medium the reverse occurred. Van Andel (6) also concluded that rate of entrance of water is correlated with the concentration of salt in the xylem vessels. The difficult problem of explaining how salt is accumulated in the xylem sap of roots is discussed in Chapter 4.

b. Nonosmotic theories. Osmotic theories assume that water moves by diffusion along a gradient of decreasing free energy or diffusion pressure, usually termed a diffusion pressure deficit or suction-force gradient. Nonosmotic movement of water is assumed to occur against a diffusion pressure gradient or at accelerated rates, and such movements presumably require the expenditure of energy released by respiration.

It was assumed originally that active absorption involved "secretion" of water into the xylem, although no satisfactory explanation of water secretion has ever been offered. Some early theories attributed it to differences in permeability between the inner and outer surfaces of the cells bordering the xylem elements. As these cells absorbed water through their outer surfaces, the increasing turgor pressure was supposed to force water out through the more permeable inner surface into the xylem vessels. Ursprung (351) reported that the diffusion-pressure deficit was higher on the inner than on the outer side of the endodermal cells, and he regarded the endodermis as a pumping mechanism which forced water into the xylem.

Interest in nonosmotic uptake of water revived in the period from 1935 to 1955; see reference (195). Among the reasons was the claim that the osmotic pressure of the sap exuding from the xylem of decapitated root systems was lower than the osmotic pressure of the solution required to stop exudation. Van Overbeek (255) regarded this difference as a measure of the nonosmotic component of active water absorption, but it more probably represents errors inherent in the methods used. It also has been shown that respiration inhibitors such as KCN partially inhibit water absorption and exudation from cut stems (255, 288). Several investigators have observed well-defined diurnal variations in water uptake and root pressure (114, 118, 313). It further has been shown that a low concentration of auxin increases exudation (221, 307). It has been claimed that these observations are difficult to reconcile with a purely osmotic theory of active absorption because they indicate that there is some relationship between water uptake and root metabolism. It is possible, however, that this relationship is an indirect one in which respiration is linked to water uptake through metabolic control of root growth and of salt uptake rather than through direct effects on water absorption.

Electroosmosis has been claimed to play a part in active water uptake by a number of investigators, including Keller (171), Lund (219), and Heyl (137). The stele appears to be electrically negative to the outer surface of the root, producing a situation favorable to the inward movement of water, but Lundegårdh (220) found the electrical-potential difference between the surface and the interior of wheat roots to be only about 100 millivolts. This small difference would produce negligible water movement (29). It therefore seems improbable that electroösmosis causes any significant amount of water absorption.

3. Relative Importance of Active and Passive Absorption

Water absorption was first regarded as a simple equilibrium process; later, considerations of active secretion came to the fore, but it now appears to the author that most water absorption should be classified as passive absorption. Among the reasons that contributed to this view are the following: (a) Root pressure rarely if ever is observed in conifers. (b) No root pressure or active absorption can be demonstrated in roots of rapidly transpiring plants. (c) The volume of xylem sap exuded from stumps as a result of active absorption usually is less than 5% of the volume lost by rapidly transpiring plants (185). (d) Intact transpiring plants can absorb water against larger diffusion pressure deficits than can excised root systems.

It has been demonstrated repeatedly that the amount of water exuding from stumps of plants is very small compared to the amount required to replace that lost by transpiration, as shown in Table IX. Furthermore, the root-pressure mechanism probably is not in operation in rapidly transpiring plants, as shown by the fact that dye and water are sucked into stems when cuts are made in them and sap usually does not begin to exude until after some time has passed. Rather typical

TABLE IX

A Comparison of the Rate of Exudation from Detopped Root Systems with Rate of Transpiration of the Same Plants Prior to Removal of Tops[a],[b]

Species	Transpiration		Exudation[c]		Exudation as percentage of transpiration
	First hour	Second hour	First hour	Second hour	
Coleus blumei	8.6	8.7	0.30	0.28	3.2
Helianthus annuus	4.3	5.0	0.02	0.02	0.4
Hibiscus moscheutos	5.8	6.7	−0.01	0.05	0.7
Impatiens sultanii	2.1	1.9	−0.22	−0.06	—
Tomato (*Lycopersicon esculentum*):					
Ser. 1	10.0	11.0	−0.62	0.07	0.6
Ser. 2	7.5	8.7	0.14	0.27	3.1

[a] From Kramer (185).
[b] Rates are in milliliters of water per half hour and are averages of 5–8 plants.
[c] Tops removed.

behavior is shown in Fig. 17, which also shows the relative amounts of water supplied by root pressure and required by transpiring shoots.

It has been argued by some investigators that absence of root pressure when the tops are removed is not proof that active absorption does not occur in intact transpiring plants (177). There seems to be little doubt, however, that a considerable water deficit generally exists in the root systems of rapidly transpiring plants, as a considerable volume of water usually is absorbed through the stumps after the shoots are removed (see Fig. 17, for example). In the writer's opinion the existence of a water deficit in the root cells will hinder or even prevent active absorption of water.

The force with which water is absorbed can be measured either directly or indirectly. An example of an indirect measurement follows.

Certain tomato plants transpired 20 ml of water per plant per hour, but when the excised root systems were attached to a vacuum pump, only 0.6 ml was moved through them with a pressure gradient of 0.85 atm. From these data it was calculated that a pressure difference of 28 atm would be required to bring about absorption of enough water to re-place that lost in transpiration. Renner (282) estimated from similar measurements that certain plants developed absorbing pressures of 4–11 atm and Köhnlein (177) estimated that pressures of 20–73 atm were

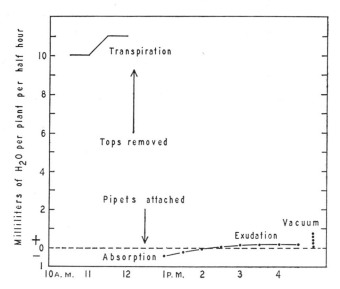

Fig. 17. Comparison of rate of transpiration of intact tomato plants (*Lycopersicon esculentum*) with rate of exudation from stumps after removal of shoots. The root systems absorbed water through their stumps for over an hour after removal of the tops and only began to show exudation after their water deficit was satisfied. These root systems were in soil near field capacity. From Kramer (192). By permission of McGraw-Hill Book Co.

developed. These estimates may be too high because, as the tension in the xylem increases, the area of the roots through which absorption occurs probably increases and, according to Brewig (39) and Brouwer (41), the permeability of the roots also increases.

Direct measurements indicate that transpiring plants can absorb water against much larger forces than can excised root systems (282). Tagawa (342) found that intact bean plants (*Phaseolus vulgaris*) could absorb water from solutions with an osmotic pressure of up to 14.6 atm but absorption by excised root systems was stopped by an osmotic pres-sure of 1.9 atm, and similar results were obtained by Army and Kozlowski (12) (see Fig. 21). It also has been found that one-half or more of the soil moisture available to intact plants is unavailable to excised root systems (12, 189, 223). This is another indication that excised root systems cannot develop as great diffusion pressure deficits as can intact plants.

The available evidence indicates to the author that passive absorption accounts for most of the water absorbed by plants. Active absorption will thus only become important in very slowly transpiring plants growing in soil near field capacity, and (as demonstrated by the phenomenon of root pressure by excised roots) has never been observed in many plants, such as the conifers. Active absorption as thus defined is probably the result of the peculiar root structure and tendency to accumulate salt in the xylem which occurs in certain species, and its role in the over-all water relations between the plant and its environment should be minimized. It is realized that this rather simple concept of water absorption eventually may prove inadequate in the light of further research, but it seems more consistent with the evidence available at present than any other explanation.

C. Factors Affecting Water Absorption by Intact Plants

The absorption of water is affected by a number of plant and environmental factors which can be discussed only briefly in the available space. The reader is referred to the article by Richards and Wadleigh (283) and to a book and recent papers by the author (192, 196, 197) for more detailed discussions of these factors.

1. Environmental Factors

Atmospheric factors affect water absorption indirectly through their effects on transpiration, but soil factors such as aeration, temperature, concentration of the soil solution, and water content affect absorption more directly and often become limiting factors for plant growth.

a. Available water content of soil. The water content of soil often becomes a limiting factor for plant growth because as the water content decreases the remaining water is held more firmly by the soil particles and is less available to plants. The water content of a soil expressed as a percentage of its oven-dry weight tells little about the availability of water for plant growth unless the water content at field capacity and permanent wilting also are known.

The field capacity of a soil is its water content after it has been thoroughly wetted and then allowed to drain until little further change in water content occurs. The remaining water, often termed capillary water, is held by surface forces in films surrounding the soil particles and in the smaller pore spaces. The permanent wilting percentage is the water content at the time leaves of plants first remain wilted overnight in a humid chamber. The water content between field capacity and permanent wilting often is termed readily available water because it can be absorbed readily by plants. Water in the range from field capacity to saturation also is readily available, but water contents above field capacity displace so much of the soil air that the plant roots usually suffer from inadequate aeration. Although plants usually continue to absorb water in soil drier than permanent wilting, absorption is too slow to replace water losses, and the resulting internal water deficit causes cessation of growth and eventually results in death from dehydration.

The availability of soil water to plants depends primarily on its diffusion pressure deficit, often termed the soil-moisture stress, and the relationship between water content and diffusion pressure deficit is shown in Fig. 18. The diffusion-pressure deficit of a soil depends on the soil-moisture tension, determined by gravitational, hydrostatic, and capillary forces, and the osmotic pressure of the soil solution. In humid regions the osmotic pressure of the soil solution usually is less than 1 atm and can be neglected, but in arid regions it often becomes a limiting factor for water absorption, as will be shown below.

As indicated in Fig. 18, the diffusion pressure deficit of a soil at field capacity is one-third of an atmosphere or less, while at the permanent wilting percentage it usually is assumed to be about 15 atm. According to Richards and Wadleigh (283), the moisture tension at permanent wilting (approximately equal to the diffusion pressure deficit in these soils) ranged from 4 to 40 atm in a long series of measurements made in their laboratories. There is increasing evidence that the diffusion pressure deficit at permanent wilting is affected more by plant factors than is generally supposed and is by no means fixed at a diffusion-pressure deficit of 15 atm (Slatyer, 310, 310a).

It has been claimed that water is equally available to plants over the entire range from field capacity down to permanent wilting (353, 354), but this scarcely can be true. The effect of decreasing soil moisture on stem elongation of sunflower plants (*Helianthus annuus*) is shown in Fig. 19, and there is considerable evidence that growth and other processes are checked at soil diffusion pressure deficits far below permanent wilting. The problem of availability of water for growth over the range from field capacity to permanent wilting was discussed in detail by Richards and Wadleigh (283).

It was mentioned earlier that excised root systems usually cannot absorb water from a medium with a diffusion pressure deficit of more than 1.5–2.0 atm, although intact transpiring plants can absorb from soil with deficits of many atmospheres. As a result of this situation, much of the soil water which is available to intact plants is unavailable to root systems from which the tops have been removed. Jantti (166) suggested that this might explain the slow resumption of growth by

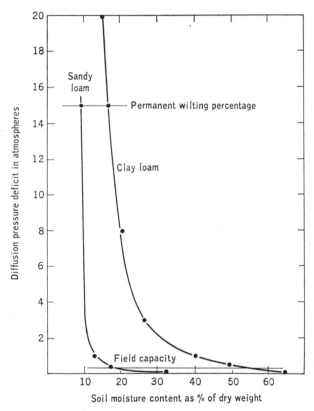

Fig. 18. The relationship between diffusion pressure deficit and moisture content in a sandy soil and a clay soil. In those soils which are low in salt content the diffusion pressure deficit is approximately equal to the soil-moisture tension, but in saline soils the osmotic pressure of the soil solution must be added to the soil-moisture tension to give the diffusion-pressure deficit of the soil. Adapted from Kramer (192).

pasture or hay crops which are defoliated when the soil is drier than field capacity. This idea was supported by further research which showed that grass and clover (*Trifolium repens, T. pratense*) make much less regrowth after cutting in soil drier than field capacity than in soil near field capacity (167).

672   PAUL J. KRAMER

b. Concentration of soil solution. Occasionally the osmotic pressure of cultivated soils, especially in greenhouses, is increased to an injurious level, but in humid climates where precipitation equals or exceeds evaporation, the osmotic pressure of the soil solution rarely becomes a limiting factor for water absorption. In regions of low rainfall, however, accumulation of salt in the soil often raises the osmotic pressure to a point where water absorption is hindered and growth is reduced. According to Magistad and Reitemeier (227), an osmotic pressure of 2 atm at permanent wilting reduces growth and an osmotic pressure of 4 atm causes injury to most crop plants.

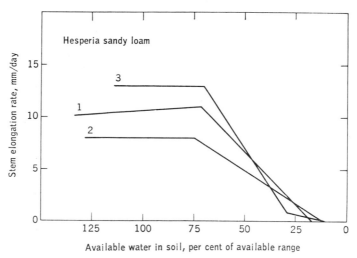

FIG. 19. The relation between available water content of soil and stem elongation of sunflower (*Helianthus annuus*) plants. Each curve is for a different soil. The permanent wilting percentage is taken as zero, field capacity as 100%. Notice that growth began to decrease while nearly 75% of the so-called readily available water was present and ceased completely before it was all removed. From Blair *et al.* (28).

The early work in this field stressed differences between ions with respect to toxicity and their effects on permeability. It is true that chloride ions appear to be more toxic to some plants than sulfate ions, and van Eijk (86) attributes the succulence of halophytes to their high chloride content. Data of Biebl (27), Boyce (38), Schmied (296), and others indicate that certain ions do have specific effects on the water relations of plants. It appears, however, from the work of the United States Regional Salinity Laboratory at Riverside, California, that the osmotic effects usually are more important than the ionic effects. For example, it was found that solutions of sucrose, mannitol, sodium sulfate, sodium chloride, and calcium chloride of the same osmotic pressure reduced water absorption by corn plants to the same extent (129). As shown in Fig. 20, growth of bean plants (*Phaseolus vulgaris*) is reduced to the same extent by a given diffusion-pressure deficit in the soil whether this is produced by low soil moisture, high salt content, or a combination of the two factors.

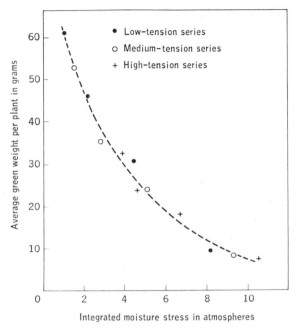

FIG. 20. Effects of combined soil-moisture tension and osmotic pressure, expressed as integrated moisture stress, on growth of bean plants (*Phaseolus vulgaris*). Low-tension plants were watered when 40–50% of the available water was removed; medium-tension plants were watered when 60–65% of the available water was removed; and high-tension plants were watered only after 90–100% of the available water was removed. Each moisture series was subdivided into four groups which received no salt, 0.1%, 0.2%, and 0.3% of sodium chloride. Reduction in amount of growth was proportional to the average diffusion pressure deficit or moisture stress to which they were subjected, whether this was increased by low soil moisture, high salt, or a combination of the two. From Kramer (192). By permission of McGraw-Hill Book Co.

These results are in accord with the view that water absorption depends on the existence of a diffusion pressure difference from soil to roots. However, a small increase in the osmotic pressure of the medium surrounding the roots produces a surprisingly large decrease in water absorption, as shown in Fig. 21; transpiration was reduced about 50% when the roots were transferred from a solution with an osmotic pressure of 0.5 atm to one of 2 atm. It is a little surprising to find that this small increase in osmotic pressure of the medium caused such a large decrease in water absorption, because as a rule a small increase in the osmotic pressure of the medium causes an equivalent increase in osmotic pressure of the sap of plants growing in it (83). Nevertheless, water absorption and growth decrease if the osmotic pressure of the medium is increased by only a few atmospheres.

Apparently the indirect effects on the plant of a high osmotic pressure in the medium are more important than the direct effects on water

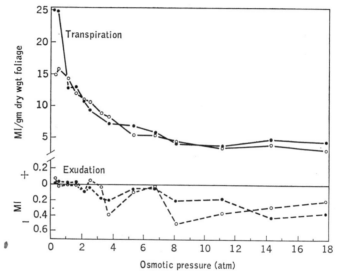

FIG. 21. The effect of the osmotic pressure of the medium surrounding the roots on the absorption of water by intact, transpiring plants and by detopped root systems. From Army and Kozlowski (12).

absorption. The cells of plants with roots in a solution even of low concentration are subjected to a continual diffusion-pressure deficit equal to that of the solution, which reduces their turgidity and the rate of cell enlargement. Root growth is reduced and the permeability of roots to water also is reduced by increased suberization (127). Thus, although a favorable gradient for water absorption exists, the tissues are subjected continuously to a diffusion pressure deficit which reduces growth and water absorption (363).

  c. *Soil temperature and water absorption.* Early in the eighteenth century Stephen Hales reported that low soil temperature reduced water absorption. In the middle of the last century Sachs observed that tobacco and gourd (*Cucurbita pepo*) plants wilted more severely than cabbage and turnip (*Brassica rapa*) plants in cold soil, and several more recent investigators have found wide differences among species in their reactions to cold soil (43, 80, 96). The difference in behavior between the cool-season collards (*B. oleracea* var. *acephala*) and the warm-season cotton (*Gossypium hirsutum*) and watermelon (*Citrullus vulgaris*) plants, when cooled, is shown in Fig. 22. Kozlowski (180) also

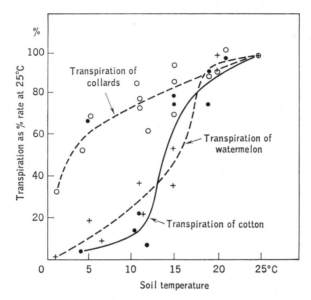

Fig. 22. The effect of decreasing the soil temperature on water absorption, as measured by the rate of transpiration. Collards (*Brassica oleracea* var. *acephala*), which is a cool weather crop, are much less affected than cotton (*Gossypium hirsutum*) and watermelon (*Citrullus vulgaris*), which thrive only in warm soil. From Kramer (190).

found much greater reduction in water absorption of the southern loblolly pine (*Pinus taeda*) than of the northern white pine (*P. strobus*), and Cameron (51) reported that cold soil causes wilting of orange trees (*Citrus sinensis*). This suggests that root permeability[1] may be affected differently in various species, possibly because of differences in reaction of the protoplasm in different species. (Permeability usually refers to the resistance offered by a membrane to passage of a substance, but in this chapter it is used in a more general sense to

[1] A descriptive term to describe the ability of water to be transmitted across a tissue, such as the root cortex, is here needed to avoid confusion with the term permeability as used in relation to cells (see Chapter 1). The "porosity" of the tissue or its "water transmissivity" might be suggested. These properties would be influenced by both physical and physiological properties of the tissue in question. Ed.

refer to the over-all resistance offered by root tissues to the passage of water, for which there is no other convenient and accepted term).

Slow cooling of root systems seems to cause less reduction in water absorption than rapid cooling (33), and if root systems are kept cool the rate of absorption often tends to rise. This suggests that changes in permeability of the protoplasm may occur during prolonged exposure to low temperature, perhaps similar to those reported by Levitt and Scarth (208) to occur during cold hardening.

Soil temperatures seldom are high enough to limit water absorption, but absorption is reduced in a number of species at 30–35°C (25, 116). Active absorption, as measured by exudation from stumps, is at a maximum at about 25°C in tomato and sunflower and decreases at both higher and lower temperatures.

Water absorption is decreased by low temperature because the viscosity of water itself increases, the permeability of the protoplasm decreases (323), root growth slows down, and respiration decreases. The latter is important insofar as it affects active water absorption, which at least quantitatively seems to be a negligible factor in water absorption at low temperatures. When root systems are attached to a vacuum pump the amount of water moved through them under constant pressure decreases rapidly as the temperature decreases. In dead roots the decrease almost exactly parallels the increase in viscosity of water (187). This suggests that most of the reduction in water absorption at low temperature is caused by the physical effects of increased resistance to water movement through the roots. A considerable fraction of this resistance resides in the protoplasm because when it is killed the rate of water movement is greatly increased, as shown in Fig. 23. That killing the protoplasm increases root "permeability" also was demonstrated by Ordin and Kramer (254). Decreased rate of water movement from soil to roots, decreased root extension, and lowered metabolic activity at low temperature appear to be of much less importance than the increased resistance to water movement through the roots.

d. Aeration and water absorption. Deficient aeration often interferes seriously with absorption of water and mineral nutrients. This was recognized for water at least as long as the beginning of the nineteenth century when de Saussure observed that pea plants (*Pisum sativum*) wilted if carbon dioxide was passed through the medium in which they were growing. Near the end of the nineteenth century considerable work was done on the effects of aeration on water absorption. The early work was summarized by Clements (57), the more recent work by Kramer (192) and by Russell (289).

There is a wide difference in tolerance of poor aeration among various species. Obviously plants such as cattails (*Typha*), bald cypress (*Taxodium distichum*), and rice (*Oryza sativa*) which normally grow in saturated soil must have root systems which can function with very little oxygen in their environment. On the other hand, tomato and

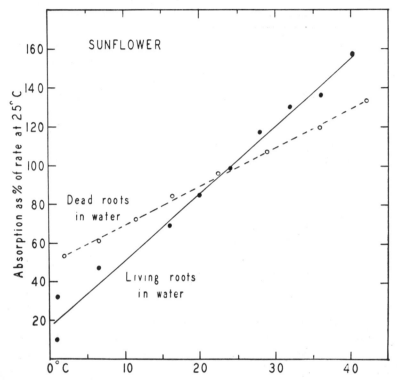

FIG. 23. Effect of temperature on rate of water movement through living and dead sunflower (*Helianthus annuus*) roots under a pressure gradient of 0.8 atm. Temperature affected movement through living roots more than through dead roots, indicating that living protoplasm offers considerable resistance to water movement at low temperatures. From Kramer (187).

tobacco usually wilt within an hour or two if the air in the soil is displaced by water, carbon dioxide, or nitrogen. Water absorption usually is reduced more rapidly and more drastically by a high concentration of carbon dioxide than by a low concentration of oxygen (54, 139, 188, 198). Apparently a high concentration of carbon dioxide definitely reduces the "permeability" of roots to water, as defined on page 675, probably through specific effects on the structure of protoplasm (54), and this is the principal reason for the rapid wilting of plants when the soil is saturated with water or carbon dioxide (42, 198). Prolonged exposure to poor aeration kills the roots of many species. An instance of the effects of flooding on rate of water movement through root systems is shown in Fig. 24. It is probable that under field conditions oxygen becomes too low more often than carbon dioxide becomes too high.

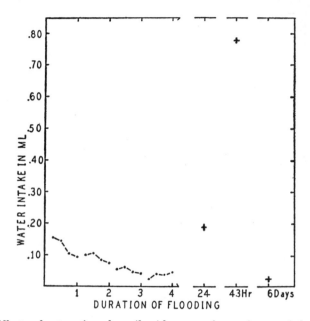

FIG. 24. Effects of saturating the soil with water, for various periods of time, on rate of water intake through tobacco (*Nicotiana tabacum*) root systems subjected to a pressure gradient of 0.2 atm. Each point is the average of 6 root systems for a 15-minute period at 25–30°C. Resistance increased rapidly at first, decreased due to root injury after 24 hours, then the rate fell off because of decay of the root systems. From Kramer and Jackson (198).

Active absorption is decreased or stopped in most plants by deficient aeration, as would be expected for a process dependent in some manner on metabolic activity of the root cells. This may occur because of reduced transport of solutes into the xylem, increased resistance to movement of water across the root, or interference in some other manner not yet understood. Some aquatic plants appear to have an active absorption mechanism which functions in a poorly aerated medium, as indicated by the fact that they exhibit root pressure (343).

Although the loss of turgor and wilting which usually occur as a result of deficient aeration are clearly the result of inadequate water absorption, the deterioration and death which result from prolonged flooding of the soil cannot be attributed entirely to inadequate water absorption (165, 194). Under some circumstances, toxic substances such as sulfides and nitrites are formed in saturated soil, and there also is a possibility that toxic substances carried upward from the roots may injure the shoots (194). Jackson (164) believes that deficient aeration prevents the roots from supplying some substance which is essential to the shoot and that this is a major cause for deterioration and death of the tops of plants when the soil is kept flooded for several days.

Deficient aeration therefore has several injurious effects on plants. First is a direct reduction in both active and passive absorption caused by decreased metabolic activity and decreased permeability of the roots, then after a day or two both water and mineral absorption are reduced by injury to the root system. After a few days of deficient aeration serious injury to the shoots develops for reasons which are not fully understood but which are more probably related to biochemical effects of injury to root systems than to reduced water absorption.

2. Plant Factors Affecting Water Absorption

If soil conditions are favorable the absorption of water by intact plants is controlled primarily by the rate of transpiration, the efficiency of their root systems, and the steepness of the diffusion-pressure gradient from soil to roots. Isolated roots are capable of absorbing water, but as indicated in the discussion on absorption mechanisms (IVB), the amount of water absorbed in this manner alone is too small to be of much significance in the over-all water economy of plants.

a. Transpiration in relation to absorption. It has been demonstrated experimentally that the rate of water absorption usually is closely correlated with the rate of water loss. This is shown for several species in Fig. 16. This would be expected because according to the cohesion theory of the ascent of sap, loss of water from leaves produces a tension or pull which is transmitted through the hydrostatic system of the plant to the roots and produces conditions favorable for the entrance of water. According to this view water absorption is necessarily closely linked to water loss.

If the two processes are linked together, it may be asked why absorption tends to lag somewhat behind transpiration, as shown in Fig. 16. This occurs because the pull developed by transpiration is not transmitted instantly to the roots but is delayed by resistance to water movement in various parts of its path. The resistance to movement through the xylem usually is rather low compared to resistance to movement across masses of living cells, such as must occur in roots and leaves (376). Resistance to flow in roots seems to be much higher than in leaves because removal of roots greatly decreases the lag of absorption behind transpiration. This is probably because water must cross a larger number of cells in roots than in leaves. Killing roots greatly decreases the resistance to water movement by either diffusion or mass flow (42, 182, 282). Ordin and Kramer (254) found that *Vicia faba* roots killed by hot water were six times as permeable to deuterium hydroxide as living roots. All of these data indicate that most of the resistance to water movement through roots occurs in the cytoplasm rather than the cell walls. On the other hand, work of Ordin and Bonner (253) indicates that most of the resistance to water movement in *Avena sativa* coleoptiles occurs in the walls, because killing them produces little increase in permeability. It would be worth-while to investigate the location of the resistance in leaves, also.

The resistance to water movement through the xylem was discussed in relation to ascent of sap (IIIB). It is much higher in conifers than in ring-porous species where long vessels often occur and a minimum of cross walls exist. Wind (379) made some estimates of rate of water flow through vessels of various diameters and concluded that there is considerable resistance to water movement through xylem elements less than 20 μ in diameter. This, he believes, definitely limits water movement through the roots of some species of grasses which have very small vessels.

Because of the resistance to water movement and the consequent lag of water absorption behind transpiration the water content of transpiring plants almost always decreases during the day and then rises again at night when absorption exceeds water loss. These relationships are shown in Fig. 31 and their effects on the internal water balance are discussed in the section on variation in water content (VIB).

b. Diffusion pressure gradient from soil to plant. Water enters roots only when there is a positive diffusion pressure or free-energy gradient from the soil or other medium surrounding the roots to the root xylem. In very slowly transpiring plants in moist soil this gradient may be caused by the accumulation of solutes in the xylem sap, resulting in a higher diffusion pressure deficit in the xylem sap than in the soil solution. Alternatively, this gradient might be produced by expenditure of metabolic energy causing secretion of water into the xylem. However, most of the intake of water occurs because of the reduced pressure or tension produced in the hydrostatic system of the plant by loss of water in transpiration.

There is no very precise information concerning the magnitude of the tensions developed in the xylem of transpiring plants. MacDougal and co-workers (225) estimated that tensions as high as 200 atm might be developed in trees, but Greenidge (110) thinks that tensions in plants seldom exceed 30 atm. Army and Kozlowski (12) found that transpiring tomato plants absorbed some water from a sucrose solution with an osmotic pressure of 17.8 atm. Stocker (325) reported osmotic pressures and diffusion-pressure deficits of 35–48 atm in plants growing in the Egyptian desert, and Slatyer (310) found diffusion pressure deficits in excess of 40 atm in wilted cotton and tomato plants. The diffusion pressure deficit of soil increases so rapidly with a small decrease in moisture content in the vicinity of the movement wilting percentage (see Fig. 18) that it quickly exceeds the maximum diffusion pressure that can be developed by plants, and water absorption ceases, even in those plants capable of developing high diffusion pressure deficits in their tissues.

c. *Efficiency of the absorbing system.* In a moist soil, with a given diffusion pressure gradient from soil to roots the amount of water absorbed depends on the extent of the root system and its "permeability" to water. There are wide variations in the "permeability" of roots by water, even on the same plant, because this varies with the age and degree of differentiation and suberization. Unsuberized root hairs are much more permeable than older roots in which suberization and lignification have occurred. The efficiency of roots as absorbing surfaces has been discussed in detail elsewhere by this author (192, 197).

There are well-defined longitudinal gradients in permeability to water which are related to root structure as shown in Fig. 25. Little water enters through the meristematic region because of the high resistance to water movement through the compact mass of cytoplasm-filled cells and the absence of xylem vessels to conduct the water (273). Maximum water intake occurs through the region where the xylem is differentiated sufficiently to provide an effective path for translocation but the epidermis and endodermis have not become sufficiently suberized to seriously decrease permeability. In onion roots maximum water absorption occurred 4–6 cm behind the tip in roots over 7 cm long (286, 287), and absorption by corn roots increased to a maximum about 10 cm behind the tip and then decreased toward the base in roots more than 10 cm in length (128).

According to Sierp and Brewig (303), maximum water absorption through *Vicia faba* roots at a low rate of transpiration occurred 1.5–8 cm behind the tip, but the zone of most rapid absorption shifted toward the base at higher rates of transpiration. Brouwer (42) also observed this shift in roots of *Vicia faba* subjected to increased tension and attributed it to increased permeability caused by decreased turgidity. Although a small decrease in turgidity may cause an increase in

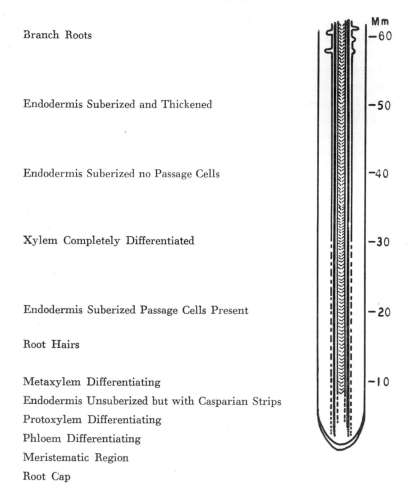

Branch Roots

Endodermis Suberized and Thickened

Endodermis Suberized no Passage Cells

Xylem Completely Differentiated

Endodermis Suberized Passage Cells Present

Root Hairs

Metaxylem Differentiating
Endodermis Unsuberized but with Casparian Strips
Protoxylem Differentiating
Phloem Differentiating
Meristematic Region
Root Cap

Mm
−60
−50
−40
−30
−20
−10

FIG. 25. Diagram of longitudinal section of a barley (*Hordeum vulgare*) root, showing relative amounts of differentiation at various distances from the apex. Although the spatial relationships vary with species and environment this order of differentiation is typical of many roots. The diagram is based on observations by Prevot and Steward (269), Heimsch (134), and Kramer and Wiebe (199). Most rapid absorption of water occurs in the region where xylem is well differentiated but suberization has not progressed far enough seriously to reduce permeability. According to Wiebe and Kramer (366) most rapid intake of salt also occurs through this region.

permeability, prolonged exposure to a more severe water deficit causes decrease in permeability (115, 193, 209).

The region of maximum water absorption generally is in the zone where root hairs are most abundant, but it probably is the differentiation of xylem and lack of suberization which determines its location, rather than the presence of root hairs. Hohn (141) probably was correct in his conclusion that root hairs have little effect on absorption from water or very moist soil, because under these conditions absorption is not limited by external surface but by internal factors such as permeability and the conductive capacity of the xylem. In soil with a water content lower than field capacity, water movement is quite slow and the amount of root surface in contact with the soil may affect the amounts of water and solutes absorbed.

It has been stated that root hairs increase the root surface 6–12 times, but Evans (94) reported that they increased the root surface of sugar cane (*Saccharum officinarum*) growing in soil only 3.2 times, and Dittmer (75) found that the root hair surface of barley roots was 1.6 times the remainder of the root surface. Root hairs seem to be rather uncommon on many woody plants, but perhaps the mycorhizae which are so common on roots of certain woody species compensate for the lack of root hairs.

There is considerable variation, even among individual roots of the same plant, with respect to the location of the absorbing zone. In rapidly growing roots it extends farther back from the tip than in slowly growing roots where suberization often occurs practically to the meristematic region. Suberization is particularly noticeable in roots growing in dry soil or concentrated solutions.

Suberized roots. It often is assumed that absorption of water and minerals occurs only through the unsuberized regions of roots, but if this is to be true, it must be considered in relation to the fact that such roots often comprise such a very small percentage of the root system that they can scarcely supply the volume of water required. In most dicotyledon species the epidermis and root hairs are destroyed within a few weeks by the activity of a cork cambium and the cortex is soon split off by cambial activity. The corky layer which develops over the outside of older roots usually contains many lenticels and cracks through which water can enter (2). In the winter very few growing roots can be found on evergreen trees, yet they often lose considerable water and must absorb corresponding quantities. Even in the summer the total surface of unsuberized roots is so small that it seems probable that a considerable part of the water and mineral absorption must occur through suberized roots. Hayward and co-workers (128) showed that considerable water absorption occurs through the suberized roots of sour orange (*Citrus aurantium*) seedlings, and Kramer (191) demonstrated absorption through suberized roots of *Pinus echinata, Cornus florida,* and *Liriodendron tulipifera.* More research ought to be done on the absorption of water and minerals through the older roots of plants.

Mycorhizal roots. The absorbing surfaces of most woody plants and some herbaceous species are modified by the development of mycorhizae. These are produced as a result of the association of a fungus with the root, and the extension of fungal hyphae into the soil appears to increase the absorbing surface. It has been demonstrated that mycorhizal roots of trees are more efficient in mineral absorption than nonmycorhizal roots (123, 126, 200, 234, 369), but their role in water absorption has not been investigated. Further work is needed on the role of mycorhizal roots, especially those on herbaceous species Apparently they occur on roots of many crop plants, but Sievers (306) and Winter (380) regard them as without benefit and possibly detrimental to crop plants, in contrast to their beneficial role on trees.

Extent of root systems. The depth and spread of root systems is an important factor in determining their efficiency as absorbing surfaces. Water moves very slowly in soils drier than field capacity, hence water and mineral nutrients are available only if the soil mass in which they occur is penetrated by roots. Although most plants have larger root systems than are essential for survival, removal of a large fraction of the root system by mechanical injury such as deep cultivation or transplanting, or reduction in efficiency by flooding the soil or by attacks of fungi and other organisms, often results in serious injury to the top. An excellent example is the little-leaf disease of *Pinus echinata* which is prevalent on poorly aerated soils in the southeastern United States. Soil conditions unfavorable for root growth and increased activity of a fungus reduce the absorbing surface so much that the trees become unthrifty and eventually die (52).

Numerous studies of the development of root systems have been made, but those of Weaver are particularly well known [see (192) for references] and these all show that roots of both annual and perennial plants penetrate to great depths where soil conditions are favorable. In well-aerated soil, corn and sorghum roots penetrate to a depth of at least 2 meters, alfalfa roots have been found at a depth of 10 meters, and roots of 18-year-old apple trees (*Malus sylvestris*) penetrated to a depth of 10 meters and completely occupied the soil between the rows (368). In heavy, poorly aerated soils, or where the water table is near the surface, roots usually are much more shallow (60). Deep-rooted plants usually are more drought resistant than shallow-rooted plants because they can absorb water from a larger soil mass. For example, corn was observed to make good growth when all of the readily available water in the upper meter of soil was exhausted, because it was absorbing from the second meter (276).

Information concerning the depth and extent of root systems originally was obtained by excavating them, but it is also possible to obtain information by measuring changes in water content at various points in the soil. Radioactive isotopes are now being used to measure mineral absorption at various depths and distances from plants, and it is safe to assume that if minerals are being absorbed at a certain point, water also is being absorbed. The reader is referred to Lott *et al.* (216) for details of the methods used.

d. Metabolism and absorption. Respiration apparently does not supply energy directly to bring about the absorption of water and neither passive nor active absorption[2] are closely related to the rate of respiration (218, 307, 375). Nevertheless, factors which interfere with metabolic activity, such as respiration inhibitors and poor aeration, are likely to reduce both active and passive absorption of water if they reduce the permeability of the roots. Active absorption also is reduced because of decreased accumulation of minerals in the xylem. Various species react differently to unfavorable environmental factors and laboratory treatments, probably because their metabolic processes and protoplasmic structure are affected to different extents.

The most important contribution of metabolic activity to water absorption is the production of a constantly expanding root system which comes into contact with a sufficiently large volume of soil to provide the water required to replace transpirational losses. Water moves rather slowly in soil drier than field capacity, and it is probable that if root extension were to cease for a day or two a serious water deficit would develop in rapidly transpiring plants growing in soil drier than field capacity. Thus, although for the most part roots act simply as absorbing surfaces through which water is absorbed, their continued growth and healthy condition is essential to their proper functioning.

[2] Those who emphasize that metabolic energy derived from respiration is concerned in movements of water or of salts would be the first to recognize the difficulty of establishing simple and direct relations between them—especially in systems like roots, in which much of the respiration measured may derive from cells which may play a limited role in the water absorption. The author allows metabolic factors to influence both "passive" and "active" movements. Most authorities would distinguish between "passive" and "active" movements of solutes and water because the former is in response to diffusion gradients and is uninfluenced by factors which affect metabolism, while the latter defies diffusion gradients but is responsive to factors which influence metabolism. By allowing that water absorption may be affected through the accumulation of salts in the xylem, which is metabolically controlled, and also through the "permeability" or "transmissivity" of the root cortex, which it is believed is affected by factors that affect the protoplasm, the author postulates conditions in which decisive distinctions between the active and passive movements are virtually impossible to obtain. Ed.

D. ABSORPTION OF WATER THROUGH LEAVES

It was reported by Stephen Hales that plants absorbed a measurable amount of water through their leaves on nights when there was a heavy dew or light rain. According to Gessner (104), who recently reviewed this subject, Mariotte had observed water absorption through leaves even earlier. Sachs doubted that water was absorbed through leaves, but other contemporary investigators observed absorption, although there was considerable uncertainty concerning its importance. This uncertainty still exists.

There is no doubt that some plants can absorb considerable water through their leaves if the leaves are sprinkled or dipped in the liquid. Brierley (40) observed that if one of a pair of wilted raspberry (*Rubus idaeus*) canes was exposed to a continuous spray of water, the other one recovered. Williams (370) was able to keep plants of several species alive and growing for some weeks with the roots and most of the leaves in the air while one or two leaves were kept immersed in water. Williams reported that a survey of the literature showed that leaves of plants of over 100 different genera had been shown to absorb water. The only question is how important such absorption is to the water economy of plants. Most of the water enters through the epidermal cells, although in some species hairs and specialized epidermal cells provide regions of high permeability.

According to Gessner (104), the rate at which water is absorbed through leaves depends on the permeability of the cutin and the diffusion pressure deficit of the leaf cells. In general water absorption is more rapid in young leaves than in old leaves of the same plant. There are wide differences among individual leaves as well as among leaves of different species. Krause (201) classified a number of species according to the rate of water absorption; beans, oats, and wheat absorbing water rapidly; *Euonymus fimbriatus, Arbutus unedo,* and *Rhododendron × hybridum* absorbing very slowly. This was related principally to thickness of cutin, although the permeability of cutin varies with its hydration and also can be modified by foreign substances such as spray materials (148, 241).

There has been a lively discussion concerning the extent to which plants absorb water from dew and fog and whether or not such water can move back through the plant and even escape from the roots into the soil. The conflicting evidence on this subject was summarized recently by Stone *et al.* (335), and only a few papers will be cited here. It was claimed by E. L. Breazeale and his colleagues that if the shoots of plants are exposed to a fine spray, water is absorbed and translocated back through the roots into the soil or even secreted into an empty flask enclosing them. Haines (119) found that such movement occurred only if a vapor-pressure gradient was produced by temperature differences between tops and roots. Hohn (143) and Janes (165a) both found that plants could decrease the moisture content of soil while their shoots were exposed to saturated air, indicating that under certain conditions plants lose water in a saturated atmosphere.

Stone et al. (336) reported that pine seedlings (*Pinus ponderosa*) in dry soil can absorb water from the air, and Stone and Fowells (334) found that artificial dew prolongs the life of pine seedlings in very dry soil. In a later experiment Stone et al. (335) found that water did not move back from the shoot through the roots into soil at the wilting percentage in measurable quantity. They concluded that no movement would be expected because no significant vapor pressure gradient exists from plant to soil, but it also is possible that poor contact between roots and soil develops as the soil dries. Slatyer (309) demonstrated that water can be caused to move in either direction through pine seedlings if the proper vapor pressure gradient is provided.

The ecological and physiological importance of dew to plants is reviewed by Gessner (104), who points out that it is a very complex problem. There seems to be little doubt, however, that dew is an important source of water under some conditions. In the experiments of Stone, absorption of dew definitely prolonged life in dry soil because it reduced the saturation deficit of the leaves. Slatyer (310) found that badly wilted cotton, tomato, and privet (*Ligustrum lucidum*) plants absorbed sufficient water overnight from the humid air in a greenhouse to reduce their diffusion-pressure deficit. Apparently little or none of the water absorbed moved back from the plant to the soil. It has been claimed (61) that plants in the "fog belt" of the Pacific Coast benefit considerably from the atmospheric moisture, although it is uncertain to what extent because of reduced transpiration and to what extent because of actual water absorption.

V. Relation between Water Absorption and Salt Absorption

One of the often debated questions of plant physiology is the relationship between water absorption and salt absorption. Most early botanists assumed that because mineral salts occur in the soil solution they must be absorbed with the water. Studies of the relation between salt and water uptake of slowly and rapidly transpiring plants by Kiesselbach (174), Muenscher (248), and others indicated that there was no consistent relationship between water and mineral absorption. In most of these experiments plants were grown under conditions favorable for high and low transpiration and the amounts of water and salt absorbed were measured. These experiments were criticized because plants grown for long periods in sun and shade or in low and high humidity differ morphologically and physiologically and are not really comparable with respect to salt absorption. Freeland (99) and Wright (383) avoided this objection by growing plants under identical conditions, then separating them into two groups each of which was subjected to either high or low humidity for only 3 or 4 days. In both series the plants which absorbed the most water also absorbed the most salt, although the relative increase in salt uptake was not as great as the relative increase in water uptake in rapidly transpiring plants.

More recently, Hylmo (160) found that increasing transpiration caused increased absorption of Ca and Cl ions and Petritschek (262) found that the concentration of the xylem sap did not decrease in proportion to the daily increase in rate of water movement, i.e., the more rapid the movement of the transpiration stream, the more salt was moved in it, although the amounts of salt and water absorbed were not proportional. Brouwer (42) found that increased water uptake caused increased salt uptake by *Vicia faba* roots, but the two processes could be separated because salt uptake was reduced by respiration inhibitors which did not reduce water uptake. Smith (312) observed that increasing the rate of transpiration of corn and pea seedlings increased the absorption and translocation of P^{32} to the shoots (see Fig. 26). In contrast, van den Honert *et al.* (145) found no relation between salt and water absorption of corn.

Hoagland and his colleagues [Hoagland (138)] emphasized the importance of salt accumulation in the root cells as the first step in salt absorption and claimed that although increased transpiration may sometimes speed up transport of salt from roots to shoots it has no direct effect on movement of salt from the external medium into the root xylem. They assumed that any increase in salt uptake in rapidly transpiring plants occurs because of conditions which dilute the xylem sap, resulting in more active transport of salt into the xylem.

Earlier, Scott and Priestley (299) had suggested that the soil solution diffuses into roots along cell walls as far as the endodermis and that ions are accumulated from it by the protoplasts of the surrounding cortical cells, but, having been so accumulated, it was not evident how the ions could be removed. They also regarded the endodermis as a barrier across which ions can move only by active transport across the protoplasts because the Casparian strips on the radial walls presumably rendered them impermeable to water. In slightly older roots the inner tangential walls of the endodermal cells also usually become suberized, but the cells opposite the xylem vessels often remain unsuberized, forming the so-called passage cells. Plasmodesmata are said to occur in the endodermis, and it also often is pierced by branch-root initials. Wiebe and Kramer (366) found considerable movement of ions into the xylem of barley roots in regions where the endodermis is supposed to be completely suberized, and salt transport into roots has been observed several

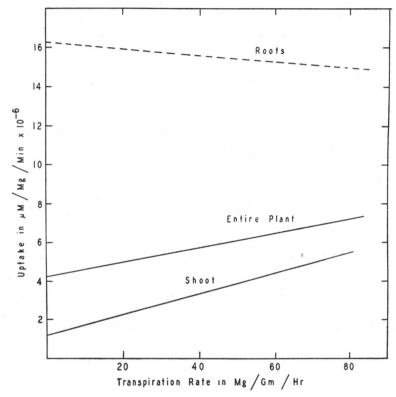

FIG. 26. The relation between rate of transpiration and rate of absorption of P^{32} by pea (*Pisum sativum*) seedlings. With rapid transpiration less P^{32} remained in the roots and more moved to the shoots. From Smith (312).

centimeters behind the tip by other investigators (42, 367, 381). It therefore seems possible that the endodermis is less of a barrier to water and solutes, especially in older roots, than is generally supposed.

It should also be remembered that secondary growth greatly modifies root structure by causing the endodermis and cortical parenchyma to slough off. In the winter, and even at times in the summer, there are few growing root tips on trees and most of the water and salt absorption must occur through older roots from which the cortex has disappeared, and a periderm may replace the outer primary surface. In such roots water and solutes must pass through the living cells of the phloem and cambial regions before reaching the xylem, and possibly these cells exercise some control over the amount and kind of solutes which pass through. If the so-called outer or free space includes part of the cytoplasm this opens up a pathway for the movement of solutes by diffusion and even by mass movement[3] in the transpiration stream.

In view of the complexity of the processes involved it is not surprising that differences of opinion have arisen concerning the relation between water and salt absorption. The effects of transpiration on salt absorption are probably modified by the age, metabolic condition, and salt content of the plant, and especially of the roots. In young, actively growing roots, low in salt, accumulation in root cells and active transport into the xylem are the dominant processes and increased water absorption has little effect on salt absorption, except possibly indirectly by reducing the concentration in the xylem sap. In older roots, especially if high in salt, less accumulation occurs and there probably is less active transport into the xylem and it is possible that some salt is carried into the xylem in the transpiration stream independently of the active transport mechanisms. The possibility of such passive movement in older roots is increased by the concept that 10–30% of root volume is "free space" into and out of which ions can move by diffusion (50, 91, 147). If ions can move by diffusion through the walls and part of the cytoplasm of root cells, then they might also be carried by mass movement in the transpiration stream through the same regions. Obviously, more work is needed on the nature and extent of free space in roots and on the relation of root structure to salt uptake by roots of various ages, before the relations between water and salt absorption can be explained fully. Mees and Weatherley (232a) demonstrated the occurrence of mass flow of water into roots under a pressure gradient and discussed its implications with respect to salt absorption.

---

[3] An essential and inescapable feature of a root system is its ability, whether attached or detached, to discriminate between the different ions of the solution and between ions and water. This degree of discrimination is incompatible with ideas of mass movement of solution or of free entry of the constituents of the outer medium into the so-called "free space" of the tissues. Judgment should be reserved upon any conclusions or arguments based upon the supposed role of the "free space," which seems to undermine the well-established classic conclusions, compatible with the known properties of membranes, that the several ions of the solution are separately absorbed and that water and solutes move under their own gradients. Were it not so, the characteristic features of the plant body, which preserves an internal composition different from that of its environment (see Chapter 4), would tend to disappear. Ed.

VI. Internal Water Relations of Plants

A. Introduction

The most important aspect of plant water relations with respect to physiological processes and conditions is that dealing with the internal water conditions of plants. Such European plant physiologists as Maximov and Montfort used the terms "water balance," "water deficit," and "water economy" (*Wasserhaushalt*) in discussing plant water relations. This is a logical terminology because we regard the water content of a plant as the water balance which fluctuates according to the relative rates of water absorption and water loss. The relative rates of transpiration and absorption are much more important than the absolute rate of either process, because rapid transpiration is not harmful if it is accompanied by sufficiently rapid absorption to prevent development of an injurious water deficit.

There is some uncertainty concerning the best way of characterizing the internal water condition (*Wasserzustand*) of a plant. Mere water content alone is unsatisfactory because wide variations in water content occur among different kinds of plant tissue, as shown in Table I, and a water content which would be normal in one kind of tissue is seriously limiting in another kind. If we assume that plant tissue normally is turgid, then a useful basis for comparing the water conditions of plant tissues is the "relative turgidity" (364) or "saturation deficit" (326) because these values compare the actual water content of plant tissues under field conditions with the water content when tissues are allowed to attain maximum turgidity. Walter (361–363) expresses the water condition as the *Hydratur* and measures it in terms of the osmotic pressure of plant sap. Diffusion pressure deficit also describes the condition of water in a plant because it is a measure of the difference in free energy, expressed as diffusion pressure in atmospheres, between free water and water as it occurs in cells, tissues, the soil, or other system. The terms "suction tension" and "suction pressure" refer to the same osmotic quantity as "diffusion pressure deficit."

B. Variations in Water Content

The water content of plants varies with age, kind of tissue, season, time of day, and other factors. There also appear to be autonomic or endogenous variations in the water content of at least some kinds of tissues.

Walter (363) and some other European physiologists separate the plant kingdom into homoiohydrous and poikilohydrous plants. Those organisms whose moisture content and diffusion pressure deficit are controlled directly by the environment are called poikilohydrous. Among these are bacteria, fungi, lichens, and mosses, and these organisms for the most part can grow only in water, dilute solutions, or moist habitats. Their growth ceases in dry air although many of them can survive for long periods in an air-dry condition. Most seed plants, in contrast, can grow in relatively dry air because they are fairly well protected against water loss by layers of cork and cutin. Such plants have a diffusion pressure deficit which is more or less independent of their environment and therefore are termed homoiohydrous. We are chiefly concerned with homoiohydrous plants in the following discussion.

Some examples of the variations in water content commonly found in plant tissue will be discussed under the various factors which are chiefly responsible for the variations.

1. Age and Kind of Tissue

Table I shows the water content of a variety of plant tissues. In general, young tissue is relatively high in water content because it consists of highly hydrated protoplasm inclosed in very thin, highly hydrated

TABLE X
WATER CONTENT OF VARIOUS REGIONS OF GROWING ROOTS[a]

Distance from root tip (mm)	Water content as percentage of fresh weight	
	Barley	Pinus taeda
0–2	89.4	82.4
2–4	93.4	89.9
4–6	94.1	91.3
6–8	94.0	92.2
8–10	94.2	91.6
18–20	93.5	90.2
48–50	93.2	87.0

[a] Data for barley (*Hordeum vulgare*) from Kramer and Wiebe (199); figures for *Pinus taeda* are from unpublished data of the same authors.

cell walls. The water content of meristematic tissue seems to be slightly lower than that of tissue in which cell enlargement is occurring, as shown in Table X. In the meristematic tissue vacuoles are very small and the protoplasts consist largely of cytoplasm, but as the cells enlarge the vacuoles enlarge rapidly, causing rapid increase in water content. As the cells mature, the walls become thicker and the proportion of dry matter increases, causing a decrease in percentage of water.

The change in water content accompanying maturation often is quite striking. In some instances this is largely the result of increase in thickness of cell walls rather than because of decrease in water content. For example, Ackley (1) found the water content of pear (*Pyrus communis*) leaves decreased from 73% of their fresh weight in May to

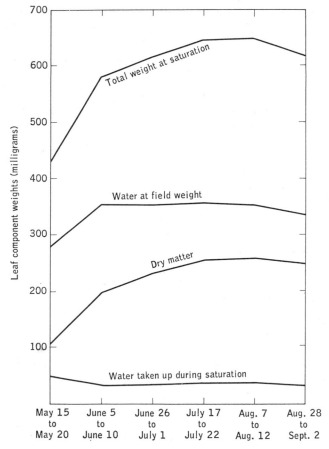

FIG. 27. Seasonal changes in water content, water deficit, and dry weight of pear (*Pyrus communis*) leaves. Note that although the actual water content remained nearly constant from June to August, the large increase in dry matter caused an increase in total weight of leaves and a decrease in water content on a percentage basis. The average water deficit underwent little change during the season. From Ackley (1).

59% in August, although the total water content increased over this period (see Fig. 27). Herbaceous species tend to become more or less dry and woody toward the end of the growing season and the decrease in water content of seeds as they mature is particularly notable, as shown in Fig. 32.

A noticeable decrease in water content at the time of, or soon after, flower-bud initiation is said to occur in at least some annual plants (214). Burns (49) and Dennison (72) both observed this in tobacco, and Hall (121) observed that there was relatively low uptake of water and salt in the period between initiation of flower primordia and syngamy in gherkins (*Cucumis sativus*). Biddulph and Brown (26) observed changes in daily increments of water uptake by developing cotton flowers and fruits, associated with time after microsporogenesis and macrosporogenesis. Changes in water content in relation to ontogeny seem to deserve more attention than they have received.

2. Seasonal Changes in Water Content

Seasonal changes in water content caused by maturation and aging of tissue have already been mentioned, but well-defined seasonal changes

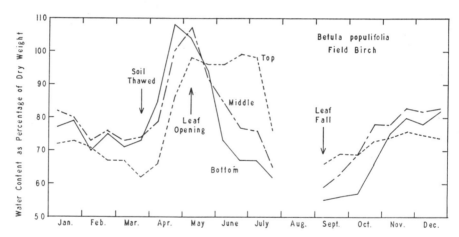

FIG. 28. Seasonal cycle in the water content of the wood from the bottom, middle, and tops of stems of *Betula populifolia*, growing near Montreal, Canada. Trees were 15–20 feet in height and 2–4 inches in diameter at breast height. From Gibbs (107).

also occur because of seasonal differences in the balance between absorption and transpiration. These changes have been studied extensively in trees by Gibbs (106, 107). An example of the seasonal cycle in trees is shown in Fig. 28. In eastern Canada, where these studies were made, tree trunks usually attain their maximum water content in the spring, just before the leaves appear. The water content decreases during the summer to a minimum just before leaf fall, because transpiration tends to exceed absorption. It increases after leaf fall, but most species show a decrease during the winter, presumably because cold soil interferes with water absorption. Conifers appear to show smaller fluctuations in water content than do nonconiferous species. It is probable that the fluctuations observed by Gibbs are more or less typical of those found in trees of the temperate zone all over the world.

The seasonal fluctuations in water content are of economic as well as physiological interest because they affect the rate of drying, cost of transport, and flotation of logs. Logs of some species are so heavy at the season when they contain their maximum water content that they tend to sink and cannot be floated to mills.

3. Diurnal Variations in Water Content

Well-defined diurnal variations in water content occur in the tissues of all transpiring plants, the water content decreasing during the day

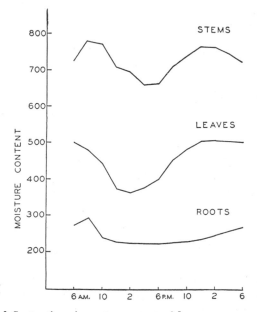

Fig. 29. Diurnal fluctuations in water content of leaves, stems, and roots of young field-grown sunflower (*Helianthus annuus*) plants, expressed as percentages of dry weight. The high water content of the stems results from the large amount of water present in their pith cells. From Wilson *et al.* (374).

and increasing at night. What appear to be typical curves for an herbaceous plant are shown in Fig. 29. The high water content of the stems results from the large amount of pith parenchyma, whereas roots have much less parenchyma and a much lower water content. They also showed less variation in water content than other parts of the plants.

The only puzzling thing is why the maximum water content of leaves occurred soon after midnight and then decreased toward morning. Wilson and associates (374) observed this in both sunflower and *Amaranthus spinosus,* and Stanescu (322) observed it in leaves of Boston ivy (*Parthenocissus tricuspidata*). No reason has been found for this decrease in leaf water content before daylight. Perhaps measurements of the total change in water content of entire plants would help to explain this puzzling phenomenon.

Diurnal rhythms in water content and turgor have been reported. Shreve (302) found diurnal variations in water-holding capacity of cactus (*Opuntia versicolor*) tissue, with the maximum during the day and the minimum at night. Periodicity in transpiration occurs in plants kept in a uniform environment (24, 246), and detopped root systems show diurnal periodicity in amount of exudation (89, 114, 118, 133).

4. Autonomic Changes in Water Content

In addition to seasonal and diurnal variations in water content there appear to be more or less rhythmic changes which are not directly related to environmental conditions. Enderle (88) observed a morning maximum and a nightly minimum in growth and turgor of carrot (*Daucus carota* var. *sativa*) tissue cultures which persisted after four months in darkness. Stomatal opening and closing continues for several days in darkness because of changes in turgor of the guard cells (291), but this probably represents the continuation of the effect of a diurnal cycle, as presumably is the case with diurnal variations in root pressure.

There are said to be definite rhythms in the protoplasmic streaming of slime molds and root hairs, and rhythmic fluctuations in viscosity of the protoplasm of *Elodea* (*Anacharis*) cells persist for about 3 days in darkness (317, 357). According to Bünning (46), various turgor movements in plants also show a rhythmic periodicity. It is probable that these rhythms in water content and other properties are related to rhythms in metabolic processes which affect such cell properties as permeability and imbibitional capacity.

C. Measurement of Water Content and Water Condition

1. Basis for Calculating Water Content

The water content of plant tissue is expressed either on a fresh- or a dry-weight basis, and both methods have certain disadvantages. Fresh weight is an unsatisfactory basis because it changes from day to day and even from hour to hour. Furthermore, as pointed out by Curtis and

Clark [(68) page 259], large changes in water content per unit of tissue result in deceptively small changes in percentage of fresh weight. Unfortunately, dry weight does not constitute a stable base either, because of changes caused by photosynthesis, respiration, and translocation. Over longer periods of time large changes can occur as a result of increase in dry matter such as increased thickness of cell walls (1, 364).

Mason and Maskell (231) attempted to reduce fluctuations in the base weight by extracting tissue in dilute HCl to remove the easily hydrolyzable materials. This left the cell walls and other resistant materials, which they termed the "residual dry weight." Denny (73) found that residual dry weight and total nitrogen were satisfactory bases for calculating changes in chemical constitutents of leaves over short periods of time. He also was able to make satisfactory comparisons between members of a pair of opposite leaves or leaflets. These methods probably could also be used for water. Miller (240) and others have calculated the water content and other constituents of leaves on a leaf-area basis, but the leaf area usually decreases with decreasing water content. There is no universally satisfactory basis for calculating water content, and investigators should be aware of the shortcomings of the method which they use.

2. Measurement of Internal Water Condition

a. *Measurement of osmotic quantities.* Numerous measurements of osmotic pressure of plant saps have been made, mostly by plasmolytic or cryoscopic methods. Care must be taken to obtain uniform samples of sap, and this requires freezing of many kinds of tissue. The methods have been discussed by Crafts *et al.* (63). Harris (124) and Walter (362) collected a large amount of data on the osmotic pressure of plants growing in various habitats, but in recent years interest has shifted somewhat from osmotic pressure to diffusion pressure deficit.

Diffusion pressure deficit is a better indicator of the internal water condition than osmotic pressure because it is a measure of the pressure with which the tissue can absorb water if placed in pure water. Turgid tissue has zero diffusion pressure deficit and the diffusion pressure deficit increases as the water deficit increases. An example of diurnal variation in osmotic pressure and diffusion-pressure deficit is shown in Fig. 30. It is very difficult to make reliable measurements of diffusion pressure deficit in some types of tissue, however. The reader is referred to Crafts *et al.* (63) for a discussion of methods. The refractive index method described by Lemée (204) seems promising for some types of tissue. Slatyer (310) determined the diffusion pressure deficit by placing samples of tissue over solutions with various diffusion pressure def-

698   PAUL J. KRAMER

icits in order to find at what diffusion pressure deficit no gain or loss of water occurred. This method requires very careful control of temperatures.

b. *Saturation deficit.* Stocker (326) proposed that the best indication of the internal water balance was the amount of water taken up by leaves or twigs when allowed to absorb water until they reach equilibrium. The leaves or twigs are weighed and their bases set in water in a

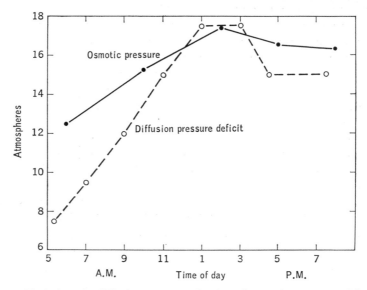

Fig. 30. Variations in diffusion pressure deficit and osmotic pressure of leaves of great ragweed, *Ambrosia trifida*, during a hot summer day. From Herrick (136).

saturated atmosphere until equilibrium is attained, which usually requires 1 or 2 days, then reweighed and oven-dried. This procedure has been used quite extensively, especially in Europe.

The saturation deficit is then calculated as follows:

$$\text{Saturation deficit} = \frac{\text{Saturated weight} - \text{original fresh wt.}}{\text{Saturated weight} - \text{oven-dry weight}} \times 100$$

The chief difficulty with this method is the long period of time required to attain equilibrium. During this period an appreciable loss in dry weight may occur as the result of the use of food in respiration.

c. *Relative turgidity.* Weatherley (364) proposed "relative turgidity" as a slightly different term for the difference in water content before and after saturation. Weatherley cuts out disks of tissue, weighs them, floats them on water until they attain equilibrium, and then obtains their fresh weight and their oven-dry weight. The relative turgidity can be calculated as follows:

$$\text{Relative turgidity} = \frac{\text{Original fresh weight} - \text{dry weight}}{\text{Saturated weight} - \text{dry weight}} \times 100$$

By this method of calculation, the original water content is expressed as a percentage of the water content at saturation and the smaller the change, the larger the figure representing relative turgidity. In the method generally used for calculating saturation deficit, the change in water content is expressed as a percentage of the water content at saturation and the smaller the change, the smaller the figure representing saturation deficit. Thus relative turgidity decreases and saturation deficit increases toward midday in transpiring plant tissue. Curves showing the daily course of relative turgidity of leaves of cotton in drying soil are shown in Fig. 8.

Equilibrium usually is attained by leaf disks in a few hours, hence no appreciable loss of dry matter occurs. Sometimes the disks become waterlogged, although this usually is not a serious problem in chlorophyll-containing tissue if they are kept illuminated. According to Oppenheimer (252), both methods will give good results if used judiciously, but preliminary studies should be made to learn the best method of handling any particular kind of tissue.

D. Causes of Variation in Water Content

It has already been pointed out that changes in proportions of cell-wall material associated with differentiation and maturation are responsible for some of the changes in water content during maturation of plants and plant organs. For the most part diurnal changes in water content occur because absorption and transpiration occur at different rates, bringing about changes in the water balance.

It is not difficult to find the explanation for the daily reduction in water content which occurs in transpiring plants. It occurs because during the day the rate of water loss tends to exceed the rate of absorption, as shown in Fig. 31, and the difference is made up by removal of water from the tissues of the plants themselves (see Figs. 29 and 31). This absorption lag occurs even in plants growing in moist soil or water culture, where water supply certainly is not limiting, hence it must be caused chiefly by resistance to water movement through the plant itself.

Under certain conditions the absorption lag and consequent water deficit develops because of an inadequate absorbing system. Injury to root systems by transplanting often results in wilting, and transplanted trees and shrubs are often pruned severely to produce a more favorable ratio of transpiring surface to roots. Sometimes injury to root systems by excavation, flooding, or disease so reduces the absorbing surface that severe injury or even death of the top results. Plants grown in shade often develop severe water deficits when exposed to full sun because they have a low ratio of roots to leaves and thin cutin and a poorly developed vascular system in their leaves. It will be repeated, however,

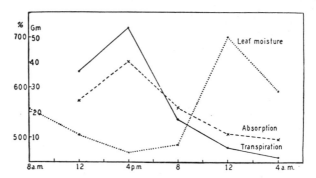

FIG. 31. The relationship between water absorption, transpiration, and leaf moisture of autoirrigated sunflower (*Helianthus annuus*) plants. Leaf moisture is expressed as percentage of dry weight; absorption and transpiration, as grams per plant per 4-hour period. From Kramer (183).

that a midday water deficit often develops in transpiring plants which have completely normal root systems and an adequate supply of water.

E. EFFECTS OF INTERNAL WATER DEFICITS

The effects of internal water deficits are shown in many ways, some easily visible such as wilting and reduced growth, others equally important but quite invisible, such as decreased photosynthesis and changes in composition.

1. Loss of Turgor and Wilting

The most immediate and direct effect of decrease in water content is loss of turgor and wilting. Three degrees of wilting often are distinguished—incipient, temporary, and permanent wilting. Incipient wilting refers to a slight loss of turgor, too small to produce visible drooping of leaves. It probably occurs daily when conditions are favorable for transpiration, and it often develops into temporary wilting where visible drooping of leaves and succulent shoots occurs. Although the loss of turgor occurs first in leaves it spreads throughout plants, as indicated by the reduction in water content of stems and roots shown in Fig. 29, causing wilting of young shoots, shrinkage of fruits, and reduction or cessation of growth. Because of the large amount of lignified tissue in many stems and leaves, severe loss of turgor often occurs without visible wilting, but its effects are just as injurious as when visible wilting occurs.

If water loss continues to exceed absorption, permanent wilting occurs. Temporarily wilted plants regain their turgidity over night, but permanently wilted plants cannot regain turgidity unless water is added to the soil. A few plants can survive long periods of dehydration, but in most species prolonged permanent wilting results in death. The occurrence of permanent wilting is used as an indication that the soil moisture content has reached a critically low level, the permanent-wilting percentage.

Loss of turgor and wilting of leaves is accompanied by measurable reduction in leaf area and thickness. Closure of stomata also occurs, and this often interferes with photosynthesis. Midday stomatal closure accompanied by a reduction in transpiration is shown in Fig. 11.

2. Growth

Everyone knows that water deficits sufficiently severe to cause wilting also reduce growth. The dwarfing effect of drought on plants is largely due to decreased cell enlargement and earlier cell maturation. The rapid maturation of tissues in plants subjected to severe water deficits reduces the possibility of recovery, so if growth of plants is reduced by drought early in the season it seldom can be made up later in the season by an adequate supply of water. The thicker cuticle, more compact structure, thicker cell walls, and better-developed vascular system of sun leaves probably results from the more frequent and severe water deficits to which they are subjected.

Occasionally, moderate water deficits are beneficial. For example, an abundance of water often results in large, succulent seedlings with a low ratio of roots to leaves which do not survive transplanting as well as seedlings which are less succulent and have a higher root-shoot ratio. Careful control of the water supply enables skillful growers to produce seedlings with the desired characteristics. Clark and Levitt (56) found that soybean (*Glycine max*) plants subjected to water deficits developed larger amounts of lipids in their leaves, which reduced cuticular transpiration and caused the plants to be more drought resistant than those not subjected to water deficits.

3. Osmotic Pressure and Diffusion Pressure Deficit

Changes in water content produce larger variations in osmotic properties. Osmotic pressure tends to increase during the day, both because decreasing water content increases the concentration of cell sap and because photosynthesis adds soluble carbohydrates. (These relations do not invariably apply as, for example, to "succulents.") The osmotic pressure tends to increase during droughts and has often been used as an indicator of the moisture conditions under which plants are growing (124, 179, 362). Harris and Walter have published extensive collections of data on osmotic pressure of plant tissue.

In recent years more attention has been devoted to variations in diffusion pressure deficit or suction tension because it is a more sensitive indicator of the internal water condition of a plant than the osmotic pressure. In a fully turgid plant the diffusion pressure deficit approaches zero, but as a water deficit develops, wall and turgor pressure decrease and the diffusion pressure deficit increases rapidly. If a severe water deficit develops, the tension in the hydrostatic system sometimes exceeds the osmotic pressure of the cell sap and under these conditions the diffusion pressure deficit exceeds the osmotic pressure. Thus the diffusion pressure deficit can vary over a range from zero to values in excess of the osmotic pressure of the cell sap. An example of diurnal variation in diffusion pressure deficit and osmotic pressure is shown in Fig. 30.

4. Physiological Processes

The physiological activity of plant tissue is closely related to its water content. The relation between water content and respiration in maturing seeds is shown in Fig. 32. The respiration rate of dry seeds is quite

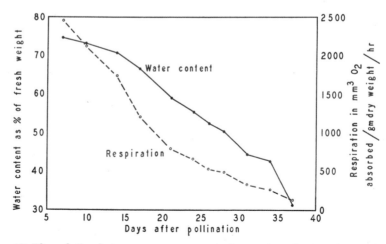

FIG. 32. The relation between water content and rate of respiration of maturing rye (*Secale cereale*) seeds. A similar relationship also was found in wheat (*Triticum aestivum*) seeds. From Shirk (301a).

low and increases very slowly with increasing water content up to a critical content which is about 15% in corn, wheat, and oats, beyond which further increase in water content causes very rapid increase in respiration rate (see Fig. 33). Appleman and Brown (9) found that a 5% increase in water content caused a fivefold increase in aerobic respiration of corn seeds. This is of considerable practical importance because if the moisture content of stored grain exceeds the critical moisture content, excessive respiration causes "heating" and spoilage.

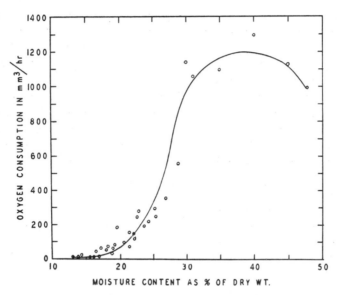

FIG. 33. The relation between water content and respiration rate of dry seeds. Notice that at a low water content a very small increase in water produces a very large increase in respiration. From Bakke and Noecker (18).

Although the direct effect of reduction in water content is decreased in respiration, indirect effects such as conversion of starch to sugar may cause increase in respiration. Parker (258) observed that as the needles of Austrian pine (*Pinus nigra*) dried, the rate of respiration decreased then increased temporarily before ceasing, and others have observed similar phenomena. Schneider and Childers (297) found that as soil moisture decreased the rate of photosynthesis decreased, but the rate of respiration often increased. The increased use of carbohydrates combined with decrease in their synthesis caused serious depletion of reserve food.

Reduction in water content usually reduces the rate of photosynthesis both because dehydrated protoplasm is less effective and because stomatal closure reduces the supply of carbon dioxide. Schneider and Childers (297) found that photosynthesis of apple trees began to decrease before wilting occurred and was reduced over 80% by the time that wilting was visible. Various investigators have reported midday reduction in photosynthesis which usually is attributed to stomatal closure resulting from internal water deficit (249, 266), but others have failed to observe it. Possibly this depends on the severity of the midday leaf water deficit, the extent of stomatal closure, and the degree to which carbon dioxide can enter through the epidermis. A slight decrease in water content sometimes causes an increase in photosynthesis. Stålfelt (320) attributes this to wider opening of stomata following a slight reduction in turgor of the cells surrounding the guard cells.

Severe water deficits often produce changes in chemical composition. Best known is the conversion of starch to sugar which has been observed in both woody and herbaceous species by several investigators (228, 314). Water deficits cause accumulation of hemicelluloses and pentosans in some species and nitrogen metabolism sometimes is affected, but these effects are not very consistent. A deficient water supply decreased the total alkaloid and quinine sulfate content of cinchona (*Cinchona ledgeriana*) seedlings (217). A generous water supply produced large bushes of guayule (*Parthenium argentatum*), but plants subjected to water deficits contained a higher percentage of rubber, resin, and lignin (172, 360). Water deficits appear to affect such protoplasmic properties as viscosity and permeability, but there is considerable disagreement concerning the nature of these effects (205, 207).

F. INTERNAL COMPETITION FOR WATER

Whenever water deficits occur in plants, competition for water develops among the various organs and tissues. Movement of water in plants occurs along gradients of increasing diffusion pressure deficits. These deficits usually are attributed to increased concentration of cell sap and decreased wall pressure, but they also are produced by the imbibitional forces resulting from evaporation of water from cell walls. If a plant in moist soil were surrounded by a saturated atmosphere it might become completely turgid and internal water movement would practically cease, but such a static condition rarely exists. Transpiration, changes in concentration of sap and turgor, and growth produce changes in the relative diffusion pressure deficits of various tissues which cause more or less continuous redistribution of water.

When large water deficits develop, as in dry soil, competition becomes severe and those tissues which can develop the highest diffusion pressure deficits obtain water at the expense of tissues having lower diffusion pressure deficits. Young leaves usually can obtain water at the expense of older leaves and the latter usually die first during drought. Shaded leaves often suffer from water deficits, and even die, because they cannot develop as high diffusion pressure deficits as unshaded leaves which are carrying on more photosynthesis. Growing stem tips often continue to elongate even when the remainder of the stem is shrinking from water deficit because they can absorb water from the older regions (373).

The competition between leaves and fruits has been studied by several investigators. Fruits usually shrink during the day when transpiration produces high water deficits in leaves and water moves from leaves to fruits. They enlarge again at night when the direction of water movement is reversed because of decreasing deficits in the leaves. These fluctuations are more severe in fruits on trees in dry soil (21). Rokach (285) found that water transfer from fruits to leaves does not occur in young orange fruits and Anderson and Kerr (7) found that cotton leaves did not remove water from young cotton bolls. As soon as bolls ceased to enlarge, however, daily changes in diameter occurred similar to those reported for fruits. Thus young fruits appear to resemble stem tips in their ability to compete with older tissues for water but lose this ability as they mature.

VII. Drought Effects and Drought Resistance

A. Introduction

Drought usually is thought of as a deficiency of available soil moisture which produces internal water deficits in plants severe enough to reduce plant growth. Although drought injury results primarily from deficient soil moisture, its injurious effects often are aggravated by atmospheric factors, such as high temperature, low humidity, and wind, which increase the rate of transpiration and hasten development of internal water deficit. Schimper (294) introduced the concept of "physiological drought" for special instances where water deficits develop in plants because cold soil or high osmotic pressure of the soil solution interferes with water absorption.

Drought is an environmental condition, and its effects on psysiological processes have already been discussed. Different species show great differences in their ability to survive drought. Although the causes of drought resistance have been studied for over half a century there still are wide differences in opinion and terminology (205, 207, 259, 329). The earliest view seems to have been that drought-resistant plants have low rates of water loss, but Maximov (232) and others attacked this view because they found that many xeromorphic species transpire rapidly if supplied with water. As a result emphasis was shifted from structural characteristics tending to reduce water loss to ability of the protoplasm to endure dehydration as the primary factor in drought resistance (161, 162, 232, 329). At present there is a tendency toward a broader viewpoint which admits that numerous factors are involved, including those which postpone dehydration as well as those which increase ability to endure dehydration.

B. CLASSIFICATION

Perhaps we can differentiate between various types and degrees of drought resistance as follows:

(1) Plants which cannot endure drought. Some plants are severely injured or die as soon as soil moisture becomes deficient because they are very quickly dehydrated. Many species which characteristically grow in the shade belong in this category.

(2) Plants such as the cacti and other succulents that store large amounts of water and have very low rates of water loss because of their small ratio of surface to volume, thick cutin, and small number of stomata. These plants often have relatively low resistance to dehydration.

(3) Plants which are truly drought-enduring because their protoplasm can be dehydrated without permanent injury. In this group are many mosses, lichens, and ferns and a few seed plants.

(4) Plants which have only limited or moderate ability to endure dehydration, combined with structural characteristics which decrease the rate of water loss or improve absorption and thereby postpone the development of critical internal water deficits. Most of the so-called drought resistance of crop plants falls in this category.

C. STRUCTURAL BASIS OF DROUGHT RESISTANCE

Among the structural characteristics which postpone the development of internal water deficits are deep and wide-spreading root systems, efficient conducting systems, a small leaf area, thick cutin, and few stomata or stomata which close promptly when a water deficit occurs. A deep, well-branched, and wide-spreading root system is excellent protection against drought injury, because the more extensive the root system, the larger the reservoir of soil water on which the plant can draw. Some plants of dry regions grow only where their roots can penetrate deeply and reach underground supplies of water. The superior drought resistance of sorghum as compared to corn has been attributed to the more extensive branching of sorghum roots (241), which results in more surface in contact with the soil and more rapid absorption of water. Slatyer (308) found sorghum was affected more slowly by decreasing soil moisture than were cotton or peanuts and attributed this to better root systems and better internal control of transpiration in sorghum.

The importance of leaf structure probably has not been adequately recognized in recent years, because there is considerable evidence that leaf structure does affect drought resistance. A thick layer of cutin which retards cuticular transpiration combined with stomata which close promptly when leaf moisture begins to decrease can greatly reduce water loss and prolong survival. The effectiveness of this kind of control of water loss is shown in the differences in rate of drying of detached leaves. Reduction in transpiring surface by leaf abscission reduces water loss of some species. Some plants normally shed their leaves as soon as a water deficit develops, and this greatly reduces the rate of further water loss. Absorption of dew and fog through the leaves may prolong the life of some plants (104, 310, 335, 336). This probably is most important where heavy fogs occur, as in the fog belt along the Pacific coast of the United States.

D. Physiological Basis of Drought Resistance

The structural characteristics just discussed merely postpone the evil day when water content falls to a critical level. Ultimately drought resistance of plants depends on the degree of dehydration which their protoplasm can endure. Species and even individuals of a species differ widely with respect to this characteristic. Olive (*Olea europaea*) leaves could be dried to a saturation deficit of 70% before serious injury occurred, but fig (*Ficus carica*) leaves were injured at a deficit of 25% (250). According to Bourdeau (37) the more drought-resistant post oak (*Quercus stellata*) and blackjack oak (*Q. marilandica*) can endure a higher saturation deficit than white oak (*Q. alba*) or eastern red oak (*Q. borealis*). Seasonal differences in resistance to drought also occur. According to Pisek and Larcher (265), drought resistance increased during the autumn to a high level in the winter, then decreased again during the spring, much like cold resistance.

Attempts have been made to relate drought resistance to osmotic pressure and bound water, and it is true that osmotic pressure and bound-water content often are higher in plants in dry habitats than in well-watered plants, but these increases are largely the result of dehydration rather than protective changes. A high osmotic pressure does confer some advantages because of reduced transpiration (35) and because the higher diffusion pressure deficit provides a more favorable gradient for water absorption from dry soil, but water bound so firmly that it cannot be removed by evaporation or freezing probably is not available for physiological activity, either.

There obviously is a variety of factors which increase drought resistance, and some are more important in one species, others in other species (259). Efficient root systems, thick cutin, and stomata which close promptly when water deficits develop serve to postpone the occurrence of critical water deficits (308), but eventually the ability of

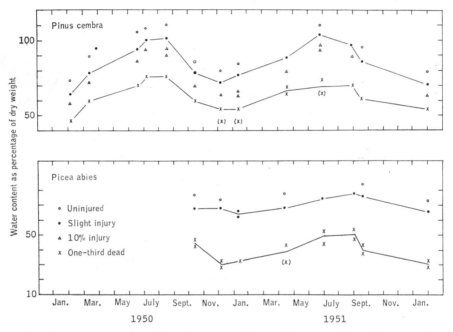

FIG. 34. Seasonal changes in resistance to dehydration of needles of *Pinus cembra* and *Picea abies* (*P. excelsa*). From Pisek and Larcher (265).

the protoplasm to endure dehydration becomes the most important factor.

Iljin (162), who has studied this problem for many years, claims that tissues consisting of small, elongated cells with a large ratio of surface to volume are most resistant to dehydration. Slowly dehydrated tissue is said to endure more desiccation than that which is rapidly dehydrated or rehydrated. Although cell size and shape appear to be important, they cannot explain the seasonal variations in dehydration endurance of the same leaves reported by Pisek and Larcher (265) (see Fig. 34). These changes in resistance must involve changes in the protoplasm which make it less susceptible to injury from dehydration, just as exposure to cold makes many kinds of tissue more resistant to freezing. The nature of these changes remains debatable (205, 207, 329), and it is obvious that more research must be done before we have a satisfactory explanation of differences in ability of different kinds of protoplasm to endure dehydration.

REFERENCES

1. Ackley, W. B. Seasonal and diurnal changes in the water contents and water deficits of Bartlett pear leaves. *Plant Physiol.* 29, 445–448 (1954).
2. Addoms, R. M. Entrance of water into suberized roots of trees. *Plant Physiol.* 21, 109–111 (1946).
3. Allen, R. M. Foliage treatments improve survival of longleaf pine plantings. *J. Forestry* 53, 724–727 (1955).
4. Alvim, P. de T. Studies on the mechanism of stomatal behavior. *Am. J. Botany* 36, 781–791 (1949).
5. Alvim, P. de T., and Havis, J. R. An improved infiltration series for studying stomatal opening as illustrated with coffee. *Plant Physiol.* 29, 97–98 (1954).
6. Andel, O. M. van. The influence of salts on the exudation of tomato plants. *Acta Botan. Neerl.* 2, 445–521 (1953).
7. Anderson, D. B., and Kerr, T. A note on the growth behavior of cotton bolls. *Plant Physiol.* 18, 261–269 (1943).
8. Andersson, N. E., Hertz, C. H., and Rufelt, H. A new fast recording hygrometer for plant transpiration measurements. *Physiol. Plantarum* 7, 753–767 (1954).
9. Appleman, C. O., and Brown, R. G. Relation of anaerobic to aerobic respiration in some storage organs with special reference to the Pasteur effect in higher plants. *Am. J. Botany* 33, 170–181 (1946).
10. Arcichovskij, V., and Ossipov, A. Die Saugkraft der baumartigen Pflanzen der zentralasiastischen Wüsten nebst Transpirationsmessungen am Saxaul (*Arthrophytum haloxylon* Litw.). *Planta* 14, 552–565 (1931).
11. Arisz, W. H., Helder, R. J., and van Nie, R. Analysis of the exudation process in tomato plants. *J. Exptl. Botany* 2, 257–297 (1951).
12. Army, T. J., and Kozlowski, T. T. Availability of soil moisture for active absorption in drying soil. *Plant Physiol.* 26, 353–362 (1951).
13. Arnold, A. Über den Functionsmechanismus der Endodermiszellen der Wurzeln. *Protoplasma* 41, 189–211 (1952).
14. Askenasy, E. Ueber das Saftsteigen. *Botan. Centr.* 62, 237–238 (1895).
15. Askenasy, E. Beiträge zur Erklärung des Saftsteigens *Verh. naturh. med. Ver.* (*Heidelberg*) 5, 429–448 (1896).
16. Atkins, W. R. G. "Some Recent Researches in Plant Physiology." Whitaker, London, 1916.
17. Aykin, S. The relations between water permeability and suction potential in living and non-living osmotic systems. *Rev. fac. sci. univ. Istanbul, Sér. B* 11, 271–295 (1946).
18. Bakke, A. L., and Noecker, N. L. The relation of moisture to respiration and heating in stored oats. *Iowa. Agr. Expt. Sta. Research Bull.* 165 (1933).
19. Bald, J. G. Stomatal droplets and the penetration of leaves by plant pathogens. *Am. J. Botany* 39, 97–99 (1952).

20. Bange, G. G. J. On the quantitative explanation of stomatal transpiration. *Acta Botan. Neerl.* **2**, 255–297 (1953).

21. Bartholomew, E. T. Internal decline of lemons. III. Water deficit in lemon fruits caused by excessive leaf evaporation. *Am. J. Botany* **13**, 102–117 (1926).

22. Bayliss, W. M. "Principles of General Physiology," 4th edition, Chapter VIII. Longmans, Green, London, 1924.

23. Beasley, E. W. Effects of some chemically inert dusts upon the transpiration rate of yellow coleus plants. *Plant Physiol.* **17**, 101–108 (1942).

24. Biale, J. B. Periodicity in transpiration of lemon cuttings under constant environmental conditions. *Proc. Am. Soc. Hort. Sci.* **38**, 70–74 (1941).

25. Bialoglowski, J. Effect of extent and temperature of roots on transpiration of rooted lemon cuttings. *Proc. Am. Soc. Hort. Sci.* **34**, 96–102 (1936).

26. Biddulph, O., and Brown, D. H. Growth and phosphorus accumulation in cotton flowers as affected by meiosis and fertilization. *Am. J. Botany* **32**, 182–187 (1945).

27. Biebl, R. Borwirkungen auf *Pisum sativum*. *Jahrb. wiss. Botan.* **90**, 731–749 (1942).

28. Blair, G. Y., Richards, L. A., and Campbell, R. B. The rate of elongation of sunflower plants and the freezing point of soil moisture in relation to permanent wilt. *Soil Sci.* **70**, 431–439 (1950).

29. Blinks, L. R., and Airth, R. L. The role of electroosmosis in living cells. *Science* **113**, 474–475 (1951).

30. Bode, H. R. Beiträge zur Dynamik der Wasserbewegung in den Gefässpflanzen. *Jahrb. wiss. Botan.* **62**, 92–127 (1923).

31. Boehm, J. Ueber einen eigenthümlichen Stammdruck. *Ber. deut. botan. Ges.* **10**, 539–544 (1892).

32. Boehm, J. Capillarität und Saftsteigen. *Ber. deut. botan. Ges.* **11**, 203–212 (1893).

33. Böhning, R. H., and Lusanandana, B. A comparative study of gradual and abrupt changes in root temperature on water absorption. *Plant Physiol.* **27**, 475–488 (1952).

34. Bolas, B. D., and Selman, I. W. An inexpensive recording porometer. *Ann. Botany (London)* **99**, 803–807 (1935).

34a. Bollard, E. G. Nitrogenous compounds in plant xylem sap. *Nature* **178**, 1189–1190 (1956).

35. Boon-Long, T. S. Transpiration as influenced by osmotic concentration and cell permeability. *Am. J. Botany* **28**, 333–343 (1941).

36. Bose, J. C. "Plant Autographs and Their Revelations." Macmillan, New York, 1927.

37. Bourdeau, P. Oak seedling ecology determining segregation of species in Piedmont oak-hickory forests. *Ecol. Monographs* **24**, 297–320 (1954).

38. Boyce, S. G. The salt spray community. *Ecol. Monographs* **24**, 29–67 (1954).

39. Brewig, A. Permeabilitätsänderungen der Wurzelgewebe, die von den Sprosz en aus beeinflusst werden. *Z. Botan.* **33**, 481–540 (1937).

40. Brierley, W. G. Absorption of water by the foliage of some common fruit species. *Proc. Am. Soc. Hort. Sci.* **32**, 277–283 (1934).

41. Brouwer, R. Water absorption by the roots of *Vicia faba* at various transpiration strengths. I. Analysis of the uptake and the factors determining it. *Koninkl. Ned. Akad. Wetenschap., Proc. Ser. C* **56**, 106–115 (1953).

42. Brouwer, R. The regulating influence of transpiration and suction tension on

the water and salt uptake by the roots of intact *Vicia faba* plants. *Acta Botan. Neerl.* **3**, 264–312 (1954).

43. Brown, E. M. Some effects of temperature on the growth and chemical composition of certain pasture grasses. *Missouri Agr. Expt. Sta. Research Bull.* **299** (1939).
44. Brown, H. T., and Escombe, F. Static diffusion of gases and liquids in relation to the assimilation of carbon and translocation of plants. *Phil. Trans. Roy. Soc. London, Ser. B* **193**, 223–291 (1900).
45. Broyer, T. C. Experiments on imbibition and other factors concerned in the water relations of plant tissues. *Am. J. Botany* **38**, 485–495 (1951).
46. Bünning, E. *In* "Lehrbuch der Pflanzenphysiologie," Vols. 2 and 3, Entwicklungs und Bewegungsphysiologie der Pflanze. Springer, Berlin, 1953.
47. Burgerstein, A. Materialien zu einer Monographie betreffend die Erscheinungen der Transpiration der Pflanzen. *Zool.-Botan. Ges. Wien* **37**, 691–782 (1887).
48. Burgerstein, A., "Die Transpiration der Pflanzen," Part 2. Fischer, Jena, 1920.
49. Burns, R. E. Composition, structure, and ontogeny of cortex and pith of tobacco stem in relation to potassium and nitrogen deficiency. *Am. J. Botany* **38**, 310–317 (1951).
50. Butler, G. W. Ion uptake by young wheat plants. II. The "apparent free space" of wheat roots. *Physiol. Plantarum* **6**, 617–635 (1953).
51. Cameron, S. H. The influence of soil temperature on the rate of transpiration of young orange trees. *Proc. Am. Soc. Hort. Sci.* **38**, 75–79 (1941).
52. Campbell, W. A., and Copeland, O. L. Little leaf disease of shortleaf and loblolly pines. *U.S. Dept. Agr. Circ.* **940** (1954).
53. Caughey, M. G. Water relations of pocosin or bog shrubs. *Plant Physiol.* **20**, 671–689 (1945).
54. Chang, H. T., and Loomis, W. E. Effect of carbon dioxide on absorption of water and nutrients by roots. *Plant Physiol.* **20**, 221–232 (1945).
55. Chatfield, C., and Adams, G. Proximate composition of American food materials. *U.S. Dept. Agr. Circ.* **549**, 1–191 (1940).
56. Clark, J. A., and Levitt, J. The basis of drought resistance in the soybean plant. *Physiol. Plantarum* **9**, 598–606 (1956).
57. Clements, F. E. Aeration and air content. *Carnegie Inst. Wash. Publ.* **315** (1921).
58. Clements, H. F. Significance of transpiration. *Plant Physiol.* **9**, 165–171 (1934).
59. Clum, H. H. The effect of transpiration and environmental factors on leaf temperatures. I, II. *Am. J. Botany* **13**, 194–231 (1926).
60. Coile, T. S. Distribution of forest tree roots in North Carolina Piedmont soils. *J. Forestry* **35**, 247–257 (1937).
61. Cooper, W. C. Redwoods, rainfall and fog. *Plant World* **20**, 179–189 (1917).
62. Crafts, A. S., and Broyer, T. C. Migration of salts and water into xylem of the roots of higher plants. *Am. J. Botany* **25**, 529–535 (1938).
63. Crafts, A. S., Currier, H. B., and Stocking, C. R. "Water in the Physiology of Plants." Chronica Botanica, Waltham, Massachusetts, 1949.
64. Curtis, L. C. Deleterious effects of guttated fluid on foliage. *Am. J. Botany* **30**, 778–781 (1943).
65. Curtis, L. C. The exudation of glutamine from lawn grass. *Plant Physiol.* **19**, 1–5 (1944).
66. Curtis, L. C. The influence of guttation fluids on pesticides. *Phytopathology* **34**, 196–205 (1944).

712 PAUL J. KRAMER

67. Curtis, O. F. Leaf temperatures and the cooling of leaves by radiation. *Plant Physiol.* **11**, 343–364 (1936).
68. Curtis, O. F., and Clark, D. G. "An Introduction to Plant Physiology." McGraw-Hill, New York, 1950.
69. Darwin, F. The effect of light on the transpiration of leaves. *Proc. Roy. Soc.* **B87**, 281–299 (1914).
70. Darwin, F., and Pertz, D. F. M. On a new method of estimating the aperture of stomata. *Proc. Roy. Soc.* **B84**, 136–154 (1912).
71. Daughters, M. R., and Glenn, D. S. The role of water in freezing foods. *Refrig. Eng.* **52**, 137–140 (1946).
72. Dennison, R. A. Growth and nutrition responses of Little Turkish tobacco to long and short photoperiods. *Plant Physiol.* **20**, 183–189 (1945).
73. Denny, F. E. Bases for calculations in measuring changes in leaves during the night. *Contrib. Boyce Thompson Inst.* **5**, 181–194 (1933).
74. Dickson, H., and Blackman, V. H. The absorption of gas bubbles present in xylem vessels. *Ann. Botany (London)* [N. S.] **2**, 293–299 (1938).
75. Dittmer, H. J. A quantitative study of the roots and root hairs of a winter rye plant (*Secale cereale*). *Am. J. Botany* **24**, 417–420 (1937).
76. Dixon, H. H. "Transpiration and the Ascent of Sap in Plants." Macmillan, London, 1914.
77. Dixon, H. H. "The Transpiration Stream." Univ. London Press, London, 1924.
78. Dixon, H. H., and Joly, J. On the ascent of sap. *Phil. Trans. Roy. Soc. London Ser. B* **186**, 563–576 (1895).
79. Dixon, H. H., and Joly, J. The path of the transpiration current. *Ann. Botany (London)* **9**, 416–419 (1895).
80. Döring, B. Die Temperaturabhängigkeit der Wasseraufnahme und ihre ökologische Bedeutung. *Z. Botan.* **28**, 305–383 (1935).
81. Dutrochet, R. H. J. "Mémoires pour servir à l'histoire anatomique et physiologique des végétaux et des animaux." Baillière, Paris, 1837.
82. Eames, A. J., and MacDaniels, L. H. "An Introduction to Plant Anatomy," 2nd edition. McGraw-Hill, New York, 1947.
83. Eaton, F. M. Toxicity and accumulation of chloride and sulfate salts in plants. *J. Agr. Research* **64**, 357–399 (1942).
84. Eaton, F. M. The osmotic and vitalistic interpretations of exudation. *Am. J. Botany* **30**, 663–674 (1943).
85. Eaton, F. M., and Belden, G. O. Leaf temperatures of cotton and their relation to transpiration and varietal differences and yields. *U.S. Dept. Agr. Tech. Bull.* **91** (1929).
86. Eijk, M. van. Analyse der Wirkung des NaCl auf die Entwicklung, Sukkulenz und Transpiration bei *Salicornia herbacea*, sowie Untersuchungen über den Einfluss der Salzaufnahme auf die Wurzelatmung bei *Aster Tripolium*. *Rec. trav. botan. Néerl.* **36**, 559–657 (1939).
87. Elving, F. Ueber die Wasserleitung im Holz. *Botan. Z.* **40**, 706–724 (1882).
88. Enderle, W. Tagesperiodische Wachstums- und Turgor-Schwankungen an Gewebekulturen. *Planta* **39**, 570–588 (1951).
89. Engel, H., and Friederichsen, I. Weitere Untersuchungen über periodische Guttation etiolierter Haferkeimlinge. *Planta* **40**, 529–549 (1952).
90. Engel, H., and Heimann, M. Weitere Untersuchungen über periodische Guttation. *Planta* **37**, 437–450 (1949).

91. Epstein, E. Passive permeation and active transport of ions in plant roots. *Plant Physiol.* **30**, 529–535 (1955).

92. Erichsen, L. von. Das Wasser, seine physikalischen und chemischen Eigenschaften unter besonderer Berücksichtigung seiner physiologischen Bedeutung. *In* "Handbuch der Pflanzenphysiologie—Encyclopedia of Plant Physiology" (W. Ruhland, ed.), Vol. 1, pp. 168–193. Springer, Berlin, 1955.

93. Esau, K. "Plant Anatomy." Wiley, New York, 1953.

94. Evans, H. Studies on the absorbing surface of sugar-cane root systems. I. Method of study with some preliminary results. *Ann. Botany (London)* **2**, 159–182 (1938).

95. Ferri, M. G., and Labouriau, L. G. Water balance of plants from the "Caatinga." *Rev. brasil. biol.* **12**, 301–312 (1952).

96. Firbas, F. Untersuchungen über den Wasserhaushalt der Hochmoorpflanzen. *Jahrb. wiss. Botan.* **74**, 457–696 (1931).

97. Fisher, J. C. The fracture of liquids. *J. Appl. Physics* **19**, 1062–1067 (1948).

98. Franco, C. M., and Inforzato, R. Quantidade de aqua transpirada pelo cafeeiro cultivado ao sol. *Bragantia* **10**, 247–258 (1950).

99. Freeland, R. O. Effect of transpiration upon the absorption of mineral salts. *Am. J. Botany* **24**, 373–374 (1937).

100. Freeman, G. F. A method for the quantitative determination of transpiration in plants. *Botan. Gaz.* **46**, 118–129 (1908).

101. Frey-Wyssling, A. Die Guttation als allgemeine Erscheinung. *Ber. schweiz. botan. Ges.* **51**, 321–325 (1941).

102. Friesner, R. C. An observation on the effectiveness of root pressure in the ascent of sap. *Butler Univ. Botan. Studies* **4**, 226–227 (1940).

103. Gaumann, E., and Jaag, O. Der Einfluss des Windes auf die pflanzliche Transpiration I, II. *Ber. schweiz. botan. Ges.* **49**, 178–238; 555–626 (1939).

104. Gessner, F. Die Wasseraufnahme durch Blätter und Samen. *In* "Handbuch der Pflanzenphysiologie—Encyclopedia of Plant Physiology" (W. Ruhland, ed.), Springer, Berlin, 1956.

105. Geurten, I. Untersuchungen über den Gaswechsel von Baumrinden. *Forstwiss. Centr.* **69**, 704–743 (1950).

106. Gibbs, R. D. Studies of Wood II. The water content of certain Canadian trees, and changes in the water-gas system during seasoning and flotation. *Canadian J. Research* **12**, 727–760 (1935).

107. Gibbs, R. D. Studies in tree physiology. I. General introduction. Water contents of certain Canadian trees. *Canadian J. Research* **C17**, 460–482 (1939).

108. Glover, J. A method for the continuous measurement of transpiration of single leaves under natural conditions. *Ann. Botany (London)* **5**, 25–34 (1941).

109. Gortner, R. A. "Outlines of Biochemistry." Wiley, New York, 1929.

110. Greenidge, K. N. H. Studies in the physiology of forest trees. I. Physical factors affecting the movement of moisture. *Am. J. Botany* **41**, 807–811 (1954).

111. Greenidge, K. N. H. Studies in the physiology of forest trees. II. Experimental studies of fracture of stretched water columns in transpiring trees. *Am. J. Botany* **42**, 28–37 (1955).

111a. Gregory, F. G., and J. I. Armstrong. The diffusion porometer. *Proc. Roy. Soc.* **B121**, 27–42. 1936.

112. Gregory, F. G., Milthorpe, F. L., Pearse, H. L., and Spencer, H. J. Experimental studies of the factors controlling transpiration. II. The relation between transpiration rate and leaf water content. *J. Exptl. Botany* **1**, 15–28 (1950).

113. Gregory, F. G., and Pearse, H. L. The resistance porometer and its application to the study of stomatal movement. *Proc. Roy. Soc.* **B114**, 477–493 (1934).

114. Grossenbacher, K. A. Autonomic cycle of rate of exudation of plants. *Am. J. Botany* **26**, 107–109 (1939).

115. Haan, I. de. Protoplasmaquellung und Wasserpermeabilität. *Rec. trav. botan. Néerl.* **30**, 234–335 (1933).

116. Haas, A. R. C. Growth and water losses in citrus as affected by soil temperature. *Calif. Citrograph* **21**, 467, 469 (1936).

117. Haberlandt, G. "Physiologische Pflanzenanatomie," 6th ed. Engelmann, Leipzig, 1924.

118. Hagan, R. M. Autonomic diurnal cycles in the water relations of nonexuding detopped root systems. *Plant Physiol.* **24**, 441–454 (1949).

119. Haines, F. M. The absorption of water by leaves in fogged air. *J. Exptl. Botany* **4**, 106–107 (1953).

120. Hales, S. "Vegetable Staticks." Innys and Woodward, London, 1727.

121. Hall, W. C. Effects of photoperiod and nitrogen supply on growth and reproduction in the gherkin. *Plant Physiol.* **24**, 753–769 (1949).

122. Handley, W. R. C. The effect of prolonged chilling on water movement and radial growth in trees. *Ann. Botany (London)* **3**, 803–813 (1939).

123. Harley, J. L. The mycorrhiza of forest trees. *Endeavour* **15**, 43–48 (1956).

124. Harris, J. A. "The Physico-chemical Properties of Plant Saps in Relation to Phytogeography." Univ. Minnesota Press, Minneapolis, Minnesota, 1934.

125. Harrold, L. L., and Dreibelbis, F. R. Agricultural hydrology as evaluated by monolith lysimeters. *U.S. Dept. Agr. Tech. Bull.* **1050** (1951).

126. Hatch, A. B. The physical basis of mycotrophy in Pinus. *Black Rock Forest Bull.* **6** (1937).

127. Hayward, H. E., and Blair, W. M. Some responses of Valencia orange seedlings to varying concentrations of chloride and hydrogen ions. *Am. J. Botany* **29**, 148–155 (1942).

128. Hayward, H. E., Blair, W. M., and Skaling, P. E. Device for measuring entry of water into roots. *Botan. Gaz.* **104**, 152–160 (1942).

129. Hayward, H. E., and Spurr, W. Effects of isosmotic concentrations of inorganic and organic substrates on entry of water into corn roots. *Botan. Gaz.* **106**, 131–139 (1944).

130. Heath, O. V. S. Studies in stomatal behavior. V. The role of carbon dioxide in the light response of stomata. *J. Exptl. Botany* **1**, 29–62 (1950).

131. Heath, O. V. S., and Russell, J. Studies in stomatal behavior. VI. An investigation of the light responses of wheat stomata with the attempted elimination of control by the mesophyll. 2. Interactions with external carbon dioxide and general discussion. *J. Exptl. Botany* **5**, 1–15 (1954).

132. Heath, O. V. S., and Williams, W. T. Studies in stomatal action—Adequacy of the porometer in investigations of stomatal aperture. *Nature* **161**, 178–179 (1948).

133. Heimann, M. Einfluss periodischer Beleuchtung auf die Guttationsrhythmik. *Planta* **38**, 157–195 (1950).

134. Heimsch, C. Development of vascular tissues in barley roots. *Am. J. Botany* **38**, 523–537 (1951).

135. Henderson, L. J. "The Fitness of the Environment." Macmillan, New York, 1913.

136. Herrick, E. M. Seasonal and diurnal variations in the osmotic values and suction

tension values in the aerial portions of *Ambrosia trifida. Am. J. Botany* **20**, 18–34 (1933).

137. Heyl, J. G. Der Einfluss von Aussenfaktoren auf das Bluten der Pflanzen. *Planta* **20**, 294–353 (1933).

138. Hoagland, D. R. "The Inorganic Nutrition of Plants." Chronica Botanica, Waltham, Massachusetts, 1944.

139. Hoagland, D. R., and Broyer, T. C. Accumulation of salt and permeability in plant cells. *J. Gen. Physiol.* **25**, 865–880 (1942).

140. Hodgson, R. H. A study of the physiology of mycorrhizal roots on Pinus taeda L. M.A. Thesis, Duke University, Durham, North Carolina, 1954.

141. Hohn, K. Die Bedeutung der Wurzelhaare für Wasseraufnahme der Pflanzen. *Z. Botan.* **27**, 529–564 (1934).

142. Hohn, K. Beziehungen zwischen Blutung und Guttation bei *Zea mays. Planta* **39**, 65–74 (1951).

143. Hohn, K. Untersuchungen über das Wasserdampfaufnahme und WasserdampfabgabeVermögen höherer Landpflanzen. *Beitr. Biol. Pflanz.* **30**, 159–178 (1954).

144. Höhnel, F. von. Über das Wasserbedürfniss der Wälder. *Centr. ges. Forstw.* **10**, 384–409 (1884).

145. Honert, J. H. van den, Hooymans, J. J. M., and Volkers, W. S. Experiments on the relation between water absorption and mineral uptake by plant roots. *Acta Botan. Neerl.* **4**, 139–155 (1955).

146. Hoover, M. D. Effect of removal of forest vegetation upon water yields. *Trans. Am. Geophys. Union.* **25**, 969–977 (1944).

147. Hope, A. B., and Stevens, P. G. Electrical potential differences in bean roots and their relation to salt uptake. *Australian J. Sci. Research Ser. B* **5**, 335–343 (1952).

148. Horsfall, J. G., and Harrison, A. L. Effect of bordeaux mixture and its various elements on transpiration. *J. Agr. Research* **58**, 423–443 (1939).

149. Huber, B. Die Beurteilung des Wasserhaushaltes der Pflanze. *Jahrb. wiss. Botan.* **64**, 1–120 (1924).

150. Huber, B. Untersuchungen über die Gesetze der Porenverdunstung. *Z. Botan.* **23**, 839–891 (1930).

151. Huber, B. Was wissen wir vom Wasserverbrauch des Waldes. *Forstwiss. Centr.* **72**, 257–264 (1953).

152. Huber, B. Die Transpiration von Sprossachsen und anderen nicht foliosen Organen. *In* "Handbuch der Pflanzenphysiologie—Encyclopedia of Plant Physiology" (W. Ruhland, ed.), Vol. 3, pp. 427–435. Springer, Berlin, 1956.

153. Huber, B. Die Gefässleitung. *In* "Handbuch der Pflanzenphysiologie—Encyclopedia of Plant Physiology" (W. Ruhland, ed.), Vol. 3, pp. 541–582. Springer, Berlin, 1956.

154. Huber, B., and Miller, R. Methoden zur Wasserdampf-und Transpirationsregistrierung im laufenden Luftstrom. *Ber. deut. botan. Ges.* **67**, 223–234 (1954).

155. Huckenpahler, B. J. Amount and distribution of moisture in a living shortleaf pine. *J. Forestry* **34**, 399–401 (1936).

156. Hulsbruch, M. Die Wasserleitung in Parenchymen. *In* "Handbuch der Pflanzenphysiologie—Encyclopedia of Plant Physiology" (W. Ruhland, ed.), Vol. 3, pp. 522–540. Springer, Berlin, 1956.

157. Hygen, G. On the transpiration decline in excised plant samples. *Skrifter I Norske Videnskaps-Akad. Oslo Matmatisk-Naturvidenskapelig Klass* **No. 1** (1953).

158. Hygen, G. Studies in plant transpiration. II. *Physiol. Plantarum* **6**, 106–133 (1953).
159. Hygen, G. The effect of wind on stomatal and cuticular transpiration. *Nytt Mag. Botan. Oslo* **3**, 83–94 (1954).
160. Hylmo, B. Transpiration and ion absorption. *Physiol. Plantarum* **6**, 333–405 (1953).
161. Iljin, W. S. Die Ursachen der Resistenz von Pflanzenzellen gegen Austrocknen. *Protoplasma* **10**, 379–414 (1930).
162. Iljin, W. S. Causes of death of plants as a consequence of loss of water: conservation of life in desiccated tissues. *Bull. Torrey Botan. Club* **80**, 166–177 (1953).
163. Ivanoff, S. S. Guttation-salt injury on leaves of cantaloupe, pepper, and onion. *Phytopathogy* **34**, 436–437 (1944).
164. Jackson, W. T. The role of adventitious roots in recovery of shoots following flooding of the original root systems. *Am. J. Botany* **42**, 816–819 (1955).
165. Jackson, W. T. The relative importance of factors causing injury to shoots of flooded tomato plants. *Am. J. Botany* **43**, 637–639 (1956).
165a. Janes, B. E. Absorption and loss of water by tomato leaves in a saturated atmosphere. *Soil Sci.* **78**, 189–197 (1954).
166. Jantti, A. Grassland practices in relation to soil water in Central, West, and North European countries. *Acta Agral. Fennica* (1953).
167. Jantti, A., and Kramer, P. J. Regrowth of pastures in relation to soil moisture and defoliation. *Intern. Grassland Congr. 7th Congr.* Palmerston, New Zealand *1956, Proc.*
168. Jeffreys, H. Some problems of evaporation. *Phil. Mag.* [6] **35**, 270–280 (1918).
169. Johnson, J. Relation of root pressure to plant disease. *Science* **84**, 135–136 (1936).
170. Kamp, H. Untersuchungen über Kutikularbau und kutikuläre Transpiration von Blättern. *Jahrb. wiss. Botan.* **72**, 403–465 (1930).
171. Keller, R. Der elektrische Faktor des Wassertransports in Lichte der Vitalfärbung. *Ergebn. Physiol.* **30**, 294–407 (1930).
172. Kelley, O. J., Hunter, A. S., and Hobbs, C. H. The effect of moisture stress on nursery-grown guayule with respect to the amount and type of growth and growth response on transplanting. *J. Am. Soc. Agron.* **37**, 194–216 (1945).
173. Kelley, V. W. The effect of pruning of excised shoots on the transpiration rate of some deciduous fruit species. *Proc. Am. Soc. Hort. Sci.* **29**, 71–73 (1932).
174. Kiesselbach, T. A. Transpiration as a factor in crop production. *Nebraska Agr. Expt. Sta. Research Bull.* **6** (1916).
175. Killian, C., and Lemée, G. Les xérophytes: leur économie d'eau. *In* "Handbuch der Pflanzenphysiologie—Encyclopedia of Plant Physiology" (W. Ruhland, ed.), Vol. 3, pp. 787–824. Springer, Berlin, 1956.
176. Knight, R. C. A convenient modification of the porometer. *New Phytologist* **14**, 212–216 (1915).
177. Köhnlein, E. Untersuchungen über die Höhe des Wurzelwiderstandes und die Bedeutung aktiver Wurzeltätigkeit für die Wasserversorgung der Pflanzen. *Planta* **10**, 381–423 (1930).
178. Koriba, K. Über die Konvektion und Verdunstung als physikalische Grundlage der Transpiration. *Japan. J. Botany* **13**, 1–242 (1943).
179. Korstian, C. F. Density of cell sap in relation to environmental conditions in the Wasatch Mountains of Utah. *J. Agr. Research* **28**, 845–907 (1924).

180. Kozlowski, T. T. Transpiration rates of some forest tree species during the dormant season. *Plant Physiol.* **18**, 252–260 (1943).

181. Kramer, P. J. The absorption of water by root systems of plants. *Am. J. Botany* **19**, 148–164 (1932).

182. Kramer, P. J. The intake of water through dead root systems and its relation to the problem of absorption by transpiring plants. *Am. J. Botany* **20**, 481–492 (1933).

183. Kramer, P. J. The relation between rate of transpiration and rate of absorption of water in plants. *Am. J. Botany* **24**, 10–15 (1937).

184. Kramer, P. J. Root resistance as a cause of the absorption lag. *Am. J. Botany* **25**, 110–113 (1938).

185. Kramer, P. J. The forces concerned in the intake of water by transpiring plants. *Am. J. Botany* **26**, 784–791 (1939).

186. Kramer, P. J. Sap pressure and exudation. *Am. J. Botany* **27**, 929–931 (1940).

187. Kramer, P. J. Root resistance as a cause of decreased water absorption by plants at low temperatures. *Plant Physiol.* **15**, 63–79 (1940).

188. Kramer, P. J. Causes of decreased absorption of water by plants in a poorly aerated media. *Am. J. Botany* **27**, 216–220 (1940).

189. Kramer, P. J. Soil moisture as a limiting factor for active absorption and root pressure. *Am. J. Botany* **28**, 446–451 (1941).

190. Kramer, P. J. Species differences with respect to water absorption at low soil temperatures. *Am. J. Botany* **29**, 828–832 (1942).

191. Kramer, P. J. Absorption of water through suberized roots of trees. *Plant Physiol.* **21**, 37–41 (1946).

192. Kramer, P. J. "Plant and Soil Water Relationships." McGraw-Hill, New York, 1949.

193. Kramer, P. J. Effects of wilting on the subsequent intake of water by plants. *Am. J. Botany* **37**, 280–284 (1950).

194. Kramer, P. J. Causes of injury to plants resulting from flooding of the soil. *Plant Physiol.* **26**, 722–736 (1951).

195. Kramer, P. J. Water relations of plant cells and tissues. *Ann. Rev. Plant Physiol.* **6**, 253–272 (1955).

196. Kramer, P. J. Physical and physiological aspects of water absorption. *In* "Handbuch der Pflanzenphysiologie—Encyclopedia of Plant Physiology" (W. Ruhland, ed.), Vol. 3, pp. 124–159. Springer, Berlin, 1956.

197. Kramer, P. J. Roots as absorbing organs. *In* "Handbuch der Pflanzenphysiologie—Encyclopedia of Plant Physiology" (W. Ruhland, ed.), Vol. 3, pp. 188–214. Springer, Berlin, 1956.

198. Kramer, P. J., and Jackson, W. T. Causes of injury to flooded tobacco plants. *Plant Physiol.* **29**, 241–245 (1954).

199. Kramer, P. J., and Wiebe, H. H. Longitudinal gradients of P^{32} absorption in roots. *Plant Physiol.* **27**, 661–674 (1952).

200. Kramer, P. J., and Wilbur, K. M. Absorption of radioactive phosphorus by mycorrhizal roots of pine. *Science* **110**, 8–9 (1949).

201. Krause, H. Beiträge zur Kenntnis der Wasseraufnahme durch oberirdische Pflanzenorgane. *Österr. Botan. Z.* **84**, 241–270 (1935).

202. Lachenmeier, J. Transpiration und Wasserabsorption intakter Pflanzen nach vorausgegangener Verdunkelung bei Konstanz der Lichtientensität und der übrigen Aussenfaktoren. *Jahrb. wiss. Botan.* **76**, 765–827 (1932).

203. Laidlaw, C. G. P., and Knight, R. C. A description of a recording porometer and a note on stomatal behavior during wilting. *Ann. Botany (London)* **30**, 47–57 (1916).

204. Lemée, G., and Laisné, G. La méthode réfractométrique de mesure de la succion. *Rev. gén. botan.* **58**, 336–347 (1951).

205. Levitt, J. Frost, drought, and heat resistance. *Ann. Rev. Plant Physiol.* **2**, 245–268 (1951).

206. Levitt, J. The physical nature of transpirational pull. *Plant Physiol.* **31**, 248–251 (1956).

207. Levitt, J. "The Hardiness of Plants." Academic Press, New York, 1956.

208. Levitt, J., and Scarth, G. W. Frost-hardening studies with living cells. II. Permeability in relation to frost resistance and the seasonal cycle. *Can. J. Research* **C14**, 285–305 (1936).

209. Levitt, J., Scarth, G. W., and Gibbs, R. D. Water permeability of isolated protoplasts in relation to volume change. *Protoplasma* **26**, 237–248 (1936).

210. Liming, F. G. A preliminary study of the lengths of the open vessels in the branches of the American elm. *Ohio J. Sci.* **34**, 415–419 (1934).

211. Livingston, B. E. The relation of desert plants to soil moisture and to evaporation. *Carnegie Inst. Wash. Publ.* **50** (1906).

212. Livingston, B. E., and Brown, W. H. Relation of the daily march of transpiration to variations in the water content of foliage leaves. *Botan. Gaz.* **53**, 309–330 (1912).

213. Lloyd, F. E. The physiology of stomata. *Carnegie Inst. Wash. Publ.* **82** (1908).

214. Loehwing, W. F. Nutritional factors in plant growth and development. *Proc. Iowa Acad. Sci.* **49**, 61–112 (1942).

215. Loftfield, J. V. G. The behavior of stomata. *Carnegie Inst. Wash. Publ.* **314**, 104 pp. (1921).

216. Lott, W. L., Satchell, D. P., and Hall, N. S. A tracer-element technique in the study of root extension. *Proc. Am. Soc. Hort. Sci.* **55**, 27–34 (1950).

217. Loustalot, A. J., Winters, H. F., and Childers, N. F. Influence of high, medium, and low soil moisture on growth and alkaloid content of *Cinchona ledgeriana*. *Plant Physiol.* **22**, 613–619 (1947).

218. Loweneck, M. Untersuchungen über Wurzelatmung. *Planta* **10**, 185–228 (1930).

219. Lund, E. J. Electric correlation between living cells in cortex and wood in the Douglas fir. *Plant Physiol.* **6**, 631–652 (1931).

220. Lundegårdh, H. Bleeding and sap movement. *Arkiv Botan.* **A31**, 1–56 (1944).

221. Lundegårdh, H. The effect of indol acetic acid on the bleeding of wheat roots. *Arkiv Botan.* [n.s.] **1**, 295–299 (1949).

222. Lundegårdh, H. The translocation of salts and water through wheat roots. *Physiol. Plantarum* **3**, 103–151 (1950).

223. McDermott, J. J. The effect of the moisture content of the soil upon the rate of exudation. *Am. J. Botany* **32**, 570–574 (1945).

224. MacDougal, D. T. Growth in trees. *Carnegie Inst. Wash. Publ.* **307** (1921).

225. MacDougal, D. T., Overton, J. B., and Smith, G. M. The hydrostatic-pneumatic system of certain trees: movements of liquids and gases. *Carnegie Inst. Wash. Publ.* **397** (1929).

226. Machlis, L. The respiratory gradient in barley roots. *Am. J. Botany* **31**, 281–282 (1944).

227. Magistad, O. C., and Reitemeier, R. F. Soil solution concentrations at the wilting point and their correlation with plant growth. *Soil Sci.* **55**, 351–360 (1943).

228. Magness, J. R., Regeimbal, L. O., and Degman, E. S. Accumulation of carbohydrates in apple foliage, bark, and wood as influenced by moisture supply. *Am. Soc. Hort. Sci. Proc.* **29**, 246–252 (1933).
229. Maizel, J. V., Benson, A. A., and Tolbert, N. E. Identification of phosphoryl choline as an important constituent of plant sap. *Plant Physiol.* **31**, 407–408 (1956).
230. Martin, E. V., and F. E. Clements. Studies of the effect of artificial wind on growth and transpiration in *Helianthus annuus*. *Plant Physiol.* **10**, 613–636 (1935).
231. Mason, T. G., and E. J. Maskell. A study of diurnal variations in the carbohydrates of leaf, bark and wood and of the effects of ringing. *Ann. Botany (London)* **42**, 188–253 (1928).
232. Maximov, N. A. "The Plant in Relation to Water." Allen and Unwin, London, 1929.
232a. Mees, G. C., and P. E. Weatherley. The mechanism of water absorption by roots II. The role of hydrostatic pressure gradients across the cortex. *Proc. Roy. Soc.* **B147**, 381–391 (1957).
233. Meeuse, B. J. D. Observations on enzymatic action of maple and birch saps. *New Phytologist* **48**, 125–142 (1949).
234. Melin, E., and Nilsson, H. Transfer of radioactive phosphorus to pine seedlings by means of mycorrhizal hyphae. *Physiol. Plantarum* **3**, 88–92 (1950).
235. Merwin, H. E., and Lyon, H. Sap pressure in the birch stem. *Botan. Gaz.* **48**, 442–458 (1909).
236. Meyer, B. S. Studies on the physical properties of leaves and leaf sap. *Ohio J. Sci.* **27**, 263–288 (1927).
237. Meyer, B. S. A critical evaluation of the terminology of diffusion phenomena. *Plant Physiol.* **20**, 142–164 (1945).
238. Meyer, B. S. The hydrodynamic system. *In* "Handbuch der Pflanzenphysiologie—Encyclopedia of Plant Physiology" (W. Ruhland, ed.), Vol. 3, pp. 596–614. Springer, Berlin, 1956.
239. Meyer, B. S., and Anderson, D. B. "Plant Physiology," 2nd ed. Van Nostrand, New York, 1952.
240. Miller, E. C. Daily variation of water and dry matter in the leaves of corn and the sorghums. *J. Agr. Research* **10**, 11–46 (1917).
241. Miller, E. C. "Plant Physiology," 2nd ed. McGraw-Hill, New York, 1938.
242. Miller, E. C., and Saunders, A. R. Some observations on the temperature of the leaves of crop plants. *J. Agr. Research* **26**, 15–43 (1923).
243. Milthorpe, F. L. The significance of the measurement made by the cobalt chloride paper method. *J. Exptl. Botan.* **6**, 17–19 (1955).
244. Molisch, H. Das Offen-und Geschlossensein der Spaltöffnungen veranschaulicht durch eine neue Methode (Infiltrations methode) *Z. Botan.* **4**, 106–122 (1912).
245. Molisch, H. The movement of sap in plants. *Science* **69**, 217–218 (1929).
246. Montermoso, J. C., and Davis, A. R. Preliminary investigation of the rhythmic fluctuations in transpiration under constant environmental conditions. *Plant Physiol.* **17**, 473–480 (1942).
247. Moreland, D. F. A study of the translocation of radioactive phosphorus in loblolly pine (*Pinus taeda* L.). *J. Elisha Mitchell Sci. Soc.* **66**, 175–181 (1950).
248. Muenscher, W. C. Effect of transpiration on the absorption of salts by plants. *Am. J. Botany* **9**, 311–330 (1922).
249. Nutman, F. J. Studies of the physiology of *Coffea arabica*. II. Stomatal move-

ments in relation to photosynthesis under natural conditions. *Ann. Botany (London)* **1**, 681–693 (1937).

250. Oppenheimer, H. R. Zur Kenntnis der hochsommerlichen Wasserbilanz mediterraner Gehölze. *Ber. deut. botan. Ges.* **50**, 185–243, (1932).

251. Oppenheimer, H. R. An experimental study on ecological relationships and water expenses of Mediterranean forest vegetation. *Palestine J. Botany, Rehovot Ser.* **8**, 103–124 (1953).

252. Oppenheimer, H. R. Critique expérimentale de deux méthodes employées en vue d'établir déficit de saturation hydrique (DSH) des feuilles. *Congr. intern. botan., 8ᵉ Congr., Paris, 1954*, Sections 11 and 12, pp. 218–220 (1954).

253. Ordin, L., and Bonner, J. Permeability of *Avena* coleoptile sections to water measured by diffusion of deuterium hydroxide. *Plant Physiol.* **31**, 53–57 (1956).

254. Ordin, L., and Kramer, P. J. Permeability of *Vicia faba* root segments to water as measured by diffusion of deuterium hydroxide. *Plant Physiol.* **31**, 468–471 (1956).

255. Overbeek, J. van. Water uptake by excised root systems of the tomato due to non-osmotic forces. *Am. J. Botany* **29**, 677–683 (1942).

256. Overton, J. B. Studies on the relation of living cells to transpiration and sap flow in *Cyperus. Botan. Gaz.* **51**, 28–63, 102–120 (1911).

257. Parker, J. Effects of variations in the root-leaf ratio on transpiration rate. *Plant Physiol.* **24**, 739–743 (1949).

258. Parker, J. Desiccation in conifer leaves: anatomical changes and determination of the lethal level. *Botan. Gaz.* **114**, 189–198 (1952).

259. Parker, J. Drought resistance in woody plants. *Botan. Rev.* **22**, 241–289 (1956).

260. Pauling, L. C. "The Nature of the Chemical Bond and the Structure of Molecules and Crystals," 2nd ed. Cornell Univ. Press, Ithaca, New York, 1940.

261. Penman, H. L. Theory of porometers used in the study of stomatal movements in leaves. *Proc. Roy. Soc.* **B130**, 416–434 (1942).

261a. Penman, H. L., and Schofield, R. K. Some physical aspects of assimilation and transpiration. *Symp. of Soc. Exptl. Biol.* **5**, 115–129 (1952).

262. Petritschek, K. Über die Beziehungen zwischen Geschwindigkeit und Elektrolytgehalt des aufsteigenden Saftstromes. *Flora (Jena)* **140**, 345–385, 1953).

263. Pisek, A., and Berger, E. Kutikuläre Transpiration und Trockenresistenz isolierter Blätter und Sprosse. *Planta* **28**, 124–155 (1938).

264. Pisek, A., and Cartellieri, E. Zur Kenntnis des Wasserhaushaltes der Pflanzen. I. Sonnenpflanzen. *Jahrb. wiss. Botan.* **75**, 195–251 (1932).

265. Pisek, A., and Larcher, W. Zusammenhang zwischen Austrocknungsresistenz und Frosthärte bei Immergrünen. *Protoplasma* **44**, 30–46 (1954).

265a. Pollard, J. K., and Sproston, T. Nitrogenous constituents of sap exuded from the sapwood of *Acer saccharum. Plant Physiol.* **29**, 360–364 (1954).

266. Polster, H. "Die physiologishen Grundlagen der Stofferzeugung im Walde." Bayerischer Landwirtschaftsverlag, Munich, 1950.

267. Preston, R. D. The contents of the vessels of *Fraxinus americana* L. with respect to the ascent of sap. *Ann. Botany (London)* **2**, 1–21 (1938).

268. Preston, R. D. Movement of water in higher plants. *In* "Deformation and Flow in Biological Systems," pp. 257–321. North Holland, Amsterdam, 1952.

269. Prevot, P., and Steward, F. C. Salient features of the root system relative to the problem of salt absorption. *Plant Physiol.* **11**, 509–534 (1936).

270. Priestley, J. H. The mechanism of root pressure. *New Phytologist* **19**, 189–200 (1920).

271. Priestley, J. H. Further observations upon the mechanism of root pressure. *New Phytologist* **21**, 41–48 (1922).
272. Priestley, J. H. Sap ascent in the tree. *Science Prog.* **117**, 42–56 (1935).
273. Priestley, J. H., and Tupper-Carey, R. M. The water relations of the plant growing point. *New Phytologist* **21**, 210–229 (1922).
274. Raber, O. Water utilization by trees, with special reference to the economic forest species of the north temperature zone. *U.S. Dept. Agr. Misc. Publ.* **257** (1937).
275. Raleigh, G. J. The effect of various ions on guttation of the tomato. *Plant Physiol.* **21**, 194–200 (1946).
276. Reimann, E. G., Van Doren, C. A., and Stauffer, R. S. Soil moisture relationships during crop production. *Soil Sci. Soc. Am. Proc.* **10**, 41–46 (1946).
277. Renner, O. Beiträge zur Physik der Transpiration. *Flora (Jena)* **100**, 451–547 (1910).
278. Renner, O. Experimentelle Beiträge zur Kenntnis der Wasserbewegung. *Flora (Jena)* **103**, 175–247 (1911).
279. Renner, O. Versuche zur Mechanik der Wasserversorgung. *Ber. deut. botan. Ges.* **30**, 576–580, 642–648 (1912).
280. Renner, O. Theoretisches und Experimentelles zur Kohäsionstheorie der Wasserbewegung in der Pflanze. *Jahrb. wiss. Botan.* **46**, 617–667 (1915).
281. Renner, O. Die Wasserversorgung der Pflanzen. *Handwörterbuch Naturwissenschaften* **10**, 538–557 (1915).
282. Renner, O. Versuche zur Bestimmung des Filtrationswiderstandes der Wurzeln. *Jahrb. wiss. Bot.* **70**, 805–838 (1929).
283. Richards, L. A., and Wadleigh, C. H. Soil water and plant growth. *Agronomy* **2**, 73–251 (1952).
284. Ringoet, A. Recherches sur la transpiration et le bilan d'eau de quelques plantes tropicales. *Publs. inst. natl. étude agron. Congo belge, Sér. sci.* **56** (1952).
285. Rokach, A. Water transfer from fruits to leaves in the Shamouti orange tree and related topics. *Palestine J. Botany, Rehovot Ser.* **8**, 146–151 (1953).
286. Rosene, H. F. Distribution of the velocities of absorption of water in the onion root. *Plant Physiol.* **12**, 1–9 (1937).
287. Rosene, H. F. Comparison of rates of water intake in contiguous regions of intact and isolated roots. *Plant Physiol.* **16**, 19–38 (1941).
288. Rosene, H. F. Effect of cyanide on rate of exudation in excised onion roots. *Am. J. Botany* **31**, 172–174 (1944).
289. Russell, M. B. Soil aeration and plant growth. *Agronomy* **2**, 254–301 (1952).
290. Salisbury, E. J. On the causes and ecological significance of stomatal frequency, with special reference to the woodland flora. *Phil. Trans. Roy. Soc. London Ser.* **B216**, 1–65 (1927).
291. Sayre, J. D. Physiology of the stomata of *Rumex patientia*. *Ohio J. Sci.* **26**, 233–267 (1926).
292. Scarth, G. W., Loewy, A., and Shaw, M. Use of the infrared total absorption method for estimating the time course of photosynthesis and transpiration. *Can. J. Research* **C26**, 94–107 (1948).
293. Scarth, G. W., and Shaw, M. Stomatal movement and photosynthesis in *Pelargonium*, I. Effects of light and carbon dioxide. *Plant Physiol.* **26**, 207–225 (1951).

294. Schimper, A. F. W. "Pflanzengeographie auf physiologischer Grundlage." Fischer, Jena, 1898.

295. Schmidt, H. Zur Funktion der Hydathoden von Saxifraga. *Planta* 10, 314–344 (1930).

296. Schmied, E. Spurenelementdüngung und Wasserhaushalt einiger Kulturpflanzen. *Österr. Botan. Z.* 100, 552–578 (1953).

297. Schneider, G. W., and Childers, N. F. Influence of soil moisture on photosynthesis, respiration, and transpiration of apple leaves. *Plant Physiol.* 16, 565–583 (1941).

298. Scholander, P. F., Love, W. E., and Kanwisher, J. W. The rise of sap in tall grapevines. *Plant Physiol.* 30, 93–104 (1955).

298a. Scholander, P. F., Ruud, B., and Leivested, H. The rise of sap in a tropical liana. *Plant Physiol.* 32, 1–6 (1957).

299. Scott, L. I., and Priestley, J. H. The root as an absorbing organ. I. A reconsideration of the entry of water and salts into the absorbing region. *New Phytologist* 27, 125–141 (1928).

300. Seybold, A., and Wey, H. G. van der. Untersuchungen über iso- und heterokalorische Laubblätter. *Rec. trav. botan. néerl.* 26, 97–127 (1929).

301. Shantz, H. L., and Piemeisel, L. N. The water requirement of plants at Akron, Colorado. *J. Agr. Research* 34, 1093–1190 (1927).

301a. Shirk, H. G. Freezable water content and the oxygen respiration in wheat and rye grain at different stages of ripening. *Am. J. Botany* 29, 105–109 (1942).

302. Shreve, E. B. An analysis of the causes of variations in the transpiring power of cacti. *Physiol. Researches* 2, 73–127 (1916).

303. Sierp, H., and Brewig, A. Quantitative Untersuchungen über die Absorptionszone der Wurzeln. *Jahrb. wiss. Botan.* 82, 99–122 (1935).

304. Sierp, H., and Noack, A. Studien zur Physik der Transpiration. *Jahrb. wiss. Botan.* 60, 459–498 (1921).

305. Sierp, H., and Seybold, A. Weitere Untersuchungen über die Verdunstung aus multiperforaten Folien mit kleinsten Poren. *Planta* 9, 246–269 (1929).

306. Sievers, A. Untersuchungen uber die Mycorrhizen von Allium- und Solanum-Arten. *Arch. Mikrobiol.* 18, 289–321 (1953).

307. Skoog, F., Broyer, T. C., and Grossenbacher, K. A. Effects of auxin on rates, periodicity, and osmotic relations in exudation. *Am. J. Botany* 25, 749–759 (1938).

308. Slatyer, R. O. Studies of the water relations of crop plants grown under natural rainfall in northern Australia. *Australian J. Agr. Research* 6, 365–377 (1955).

309. Slatyer, R. O. Absorption of water from atmospheres of different humidity and its transport through plants. *Australian J. Biol. Sci.* 9, 552–558 (1956).

310. Slatyer, R. O. The influence of progressive increases in total soil moisture stress on transpiration, growth, and internal water relationships of plants. *Australian J. Biol. Sci* 10, 320–336 (1957).

310a. Slatyer, R. O. The significance of the permanent wilting percentage in studies of plant and soil water relations. *Bot. Rev.* 23, 585–636 (1957).

311. Smith, F., Dustman, R. B., and Shull, C. A. Ascent of sap in plants. *Botan. Gaz.* 91, 395–410 (1931).

312. Smith, R. C. Studies on the relation of ion absorption to water uptake in plants. Ph.D. Dissertation, Duke University, Durham, North Carolina, 1957.

313. Speidel, B. Untersuchungen zur Physiologie des Blutens bei höheren Pflanzen. *Planta* 30, 67–112 (1939).

314. Spoehr, H. A., and Milner, H. W. Starch dissolution and amylolytic activity of leaves. *Proc. Am. Phil. Soc.* **81**, 37–78 (1939).
315. Stålfelt, M. G. Der stomatäre Regulator in der pflanzlichen Transpiration. *Planta* **17**, 22–85 (1932).
316. Stålfelt, M. G. Die Spaltöffnungsweite als Assimilationsfaktor. *Planta* **23**, 715–759 (1935).
317. Stålfelt, M. G. The influence of light upon the viscosity of protoplasm. *Arkiv Botan.* **A33**, 1–17 (1946).
318. Stålfelt, M. G. The stomata as a hydrophotic regulator of the water deficit of the plant. *Physiol. Plantarum* **8**, 572–593 (1955).
319. Stålfelt, M. G. Die cuticuläre Transpiration. *In* "Handbuch der Pflanzenphysiologie—Encyclopedia of Plant Physiology" (W. Ruhland, ed.), Vol. 3, pp. 342–350. Springer, Berlin, 1956.
320. Stålfelt, M. G. "Die stomatäre Transpiration und die Physiologie der Spaltöffnungen." Encyclopedia of Plant Physiology (W. Ruhland, ed.), Vol. 3, pp. 351–426. Springer, Berlin, 1956.
321. Stålfelt, M. G. Morphologie und Anatomie des Blattes als Transpirationsorgan. *In* "Handbuch der Pflanzenphysiologie—Encyclopedia of Plant Physiology" (W. Ruhland, ed.), Vol. 3, pp. 324–341. Springer, Berlin, 1956.
322. Stanescu, P. P. Daily variations in products of photosynthesis, water content, and acidity of leaves toward the end of vegetative period. *Am. J. Botany* **23**, 374–379 (1936).
323. Stiles, W. "Permeability." Wheldon and Wesley, London, 1924.
324. Stocker, O. Die Transpiration und Wasserökologie nordwestdeutscher Heide- und Moorpflanzen am Standort. *Z. Botan.* **15**, 1–41 (1923).
325. Stocker, O. Der Wasserhaushalt ägyptischer Wüsten- und Salzpflanzen. *Botan. Abhandl.* (*Jena*) **13** (1928).
326. Stocker, O. Wasserdeficit von Gefässpflanzen in verschiedenen Klimazonen. *Planta* **7**, 382–387 (1929).
327. Stocker, O. Die Abhängigkeit der Transpiration von den Umweltfaktoren. *In* "Handbuch der Pflanzenphysiologie—Encyclopedia of Plant Physiology" (W. Ruhland, ed.), Vol. 3, pp. 436–488. Springer, Berlin, 1956.
328. Stocker, O. Messmethoden der Transpiration. *In* "Handbuch der Pflanzenphysiologie—Encyclopedia of Plant Physiology" (W. Ruhland, ed.), Vol. 3, pp. 293–311. Springer, Berlin, 1956.
329. Stocker, O. Die Dürreresistenz. *In* "Handbuch der Pflanzenphysiologie—Encyclopedia of Plant Physiology" (W. Ruhland, ed.), Vol. 3, pp. 696–741. Springer, Berlin, 1956.
330. Stocking, C. R. The calculation of tensions in *Cucurbita pepo*. *Am. J. Botany* **32**, 126–134 (1945).
331. Stocking, C. R. Guttation and bleeding. *In* "Handbuch der Pflanzenphysiologie—Encyclopedia of Plant Physiology" (W. Ruhland, ed.), Vol. 3, pp. 489–502. Springer, Berlin, 1956.
332. Stocking, C. R. Excretion by glandular organs. *In* "Handbuch der Pflanzenphysiologie—Encyclopedia of Plant Physiology" (W. Ruhland, ed.), Vol. 3, pp. 503–510. Springer, Berlin, 1956.
333. Stocking, C. R. Vascular conduction in submerged plants. *In* "Handbuch der Pflanzenphysiologie—Encyclopedia of Plant Physiology" (W. Ruhland, ed.), Vol. 3, pp. 587–595. Springer, Berlin, 1956.

334. Stone, E. C., and Fowells, H. A. The survival value of dew as determined under laboratory conditions. I. *Pinus ponderosa. Forest Sci.* **1**, 183–188 (1955).

335. Stone, E. C., Shachori, A. Y., and Stanley, R. G. Water absorption by needles of ponderosa pine seedlings and its internal redistribution. *Plant Physiol.* **31**, 120–126 (1956).

336. Stone, E. C., Went, F. W., and Young, C. L. Water absorption from the atmosphere by plants growing in dry soil. *Science* **111**, 546–548 (1950).

337. Strasburger, E. Ueber den Bau und die Verrichtungen der Leitungsbahnen in den Pflanzen. *Histol. Beitr. (Jena)* **3**, 849–877 (1891).

338. Strugger, S. Studien über den Transpirationsstrom im Blatt von *Secale cereale* und *Triticum vulgare. Flora (Jena)* **35**, 97–113 (1940).

339. Strugger, S. Der aufsteigende Saftstrom in der Pflanze. *Naturwissenschaften* **31**, 181–194 (1943).

340. Strugger, S. "Praktikum der Zell- und Gewebephysiologie der Pflanzen," 2nd ed. Springer, Berlin, 1949.

341. Swanson, C. A. Transpiration in American holly in relation to leaf structure. *Ohio J. Sci.* **43**, 43–46 (1943).

342. Tagawa, T. The relation between the absorption of water by plant root and the concentration and nature of the surrounding solution. *Japan. J. Botany* **7**, 33–60 (1934).

343. Thut, H. F. The movement of water through some submerged plants. *Am. J. Botany* **19**, 693–709 (1932).

344. Thut, H. F. Demonstrating the lifting power of transpiration. *Am. J. Botany* **19**, 358–364 (1932).

345. Thut, H. F. The relative humidity gradient of stomatal transpiration. *Am. J. Botany* **26**, 315–319 (1939).

346. Tolbert, N. E., and Wiebe, H. H. Phosphorus and sulfur compounds in plant xylem sap. *Plant Physiol.* **30**, 499–504 (1955).

347. Turrell, F. M. The area of the internal exposed surface of dicotyledon leaves. *Am. J. Botany* **23**, 255–264 (1936).

348. Turrell, F. M. Correlation between internal surface and transpiration rate in mesomorphic and xeromorphic leaves grown under artificial light. *Botan. Gaz.* **105**, 413–425 (1944).

349. Ursprung, A. Über die Kohäsion des Wassers im Farnanulus. *Ber. deut. botan. Ges.* **33**, 153–162 (1915).

350. Ursprung, A. Dritter Beitrag zur Demonstration der Flüssigkeits-Kohäsion. *Ber. deut. botan. Ges.* **34**, 475–488 (1916).

351. Ursprung, A. The osmotic quantities of the plant cell. *Proc. Intern. Corngr. Plant Sci. 1st Congr., Ithaca, 1926*, pp. 1081–1094 (1929).

352. Ursprung, A., and Blum, G. Zur Methode der Saugkraftmessung. *Ber. deut. botan. Ges.* **34**, 525–539 (1916).

353. Veihmeyer, F. J. Soil Moisture. *In* "Handbuch der Pflanzenphysiologie—Encyclopedia of Plant Physiology" (W. Ruhland, ed.), Vol. 3, pp. 64–123. Springer, Berlin, 1956.

354. Veihmeyer, F. J., and Hendrickson, A. H. Soil moisture in relation to plant growth. *Ann. Rev. Plant Physiol.* **1**, 285–304 (1950).

355. Verduin, J. Diffusion through multiperforate septa. *In* "Photosynthesis in Plants" (J. Franck and W. E. Loomis, eds.). Iowa State College Press, Ames, Iowa, 1949.

356. Villaca, H., and Ferri, M. G. On the morphology of the stomata of *Eucalyptus*

tereticornis, Ouratea spectabilis and *Cedrela fissilis. Univ. São Paulo Bol. fac. filosof., ciénc. letras, Botan.* **11** (Bol. No. 173), 31–52 (1954).

357. Virgin, H. I. The effect of light on the protoplasmic viscosity. *Physiol. Plantarum.* **4**, 255–357 (1951).

358. Volk, A. Einfluss der Ernährung auf Wurzeldruck, Blutung und Guttation. *Bodenk. u. Pflanzenernähr.* **35**, 190–204 (1945).

359. Wadleigh, C. H., and Ayers, A. D. Growth and biochemical composition of bean plants as conditioned by soil moisture tension and salt concentration. *Plant Physiol.* **20**, 106–132 (1945).

360. Wadleigh, C. H., Gauch, H. G., and Magistad, O. C. Growth and rubber accumulation in guayule as conditioned by soil salinity and irrigation regime. *U.S. Dept. Agr. Tech. Bull.* **925** (1946).

361. Walter, H. "Die Hydratur der Pflanze." Fischer, Jena, 1931.

362. Walter, H. "Grundlagen der Pflanzenverbreitung," Part 1. Ulmer, Stuttgart, 1951.

363. Walter, H. The water economy and the hydrature of plants. *Ann. Rev. Plant Physiol.* **6**, 239–252 (1955).

364. Weatherley, P. E. Studies in the water relations of the cotton plant. I. The field measurement of water deficits in leaves. *New Phytologist* **49**, 81–97 (1950); *Biol. Abst.* E24, 34240 (1950).

365. Weaver, J. E., and Mogensen, A. Relative transpiration of conifers and broad-leaved trees in autumn and winter. *Botan. Gaz.* **68**, 393–424 (1919).

365a. White, P. R. Root pressure, an unappreciated force in sap movement. *Am. J. Botany* **25**, 223–227 (1938).

366. Wiebe, H. H., and Kramer, P. J. Translocation of radioactive isotopes from various regions of roots of barley seedlings. *Plant Physiol.* **29**, 342–348 (1954).

367. Wiersum, L. K. Transfer of solutes across the young root. *Rec. trav. bot. néerl.* **61**, 1–79 (1948).

368. Wiggans, C. C. The effect of orchard plants on subsoil moisture. *Am. Soc. Hort. Sci. Proc.* **33**, 103–107 (1936).

369. Wilde, S. A. Mycorrhizal fungi: their distribution and effect on tree growth. *Soil Sci.* **78**, 23–31 (1954).

370. Williams, H. F. Absorption of water by the leaves of common mesophytes. *J. Elisha Mitchell Sci. Soc.* **48**, 83–100 (1932).

371. Williams, W. T. The continuity of intercellular spaces in the leaf of *Pelargonium zonale*, and its bearing on recent stomatal investigations. *Ann. Botany (London)* **12**, 411 (1948).

372. Wilson, C. C. The effect of some environmental factors on the movements of guard cells. *Plant Physiol.* **23**, 5–37 (1948).

373. Wilson, C. C. Diurnal fluctuations in growth in length of tomato stems. *Plant Physiol.* **23**, 156–157 (1948).

374. Wilson, C. C., Boggess, W. R., and Kramer, P. J. Diurnal fluctuations in the moisture content of some herbaceous plants. *Am. J. Botany* **40**, 97–100 (1953).

375. Wilson, C. C., and Kramer, P. J. Relation between root resipration and absorption. *Plant Physiol.* **24**, 55–59 (1949).

376. Wilson, J. D., and Livingston, B. E. Lag in water absorption by plants in water culture with respect to changes in wind. *Plant Physiol.* **12**, 135–150 (1937).

377. Wilson, J. K. The nature and reaction of water from hydathodes. *Cornell Univ. Agr. Expt. Sta. Mem.* **65** (1923).

378. Wilson, K. Water movement in submerged aquatic plants, with special reference to cut shoots of *Ranunculus fluitans*. *Ann. Botany* 11, 91–122 (1947).

379. Wind, G. P. Flow of water through plant roots. *Neth. J. Agr. Sci.* 3, 259–264 (1956).

380. Winter, A. G. Zum Problem der Mykorrhiza bei landwirtschaftlichen Kulturpflanzen. *Z. Pflanzenernähr. Düng. Bodenk.* 60, 221–243 (1953).

381. Woodford, E. K., and Gregory, F. G. Preliminary results obtained with an apparatus for the study of salt uptake and root respiration of whole plants. *Ann. Botany* (*London*) 12, 335–370 (1948).

382. Wrenger, M. Über den Einfluss des Windes auf die Transpiration der Pflanzen. *Z. Botan.* 29, 257–320 (1935).

383. Wright, K. E. Transpiration and the absorption of mineral salts. *Plant Physiol.* 14, 171–174 (1939).

Midday Closure of Stomata and the Minimum Concentration of Carbon Dioxide in Intercellular Spaces

The noncommittal symbol Γ is used for this steady-state concentration, observed when leaves are illuminated in a closed system, to avoid at present implications as to the precise relations of photosynthesis and respiration that decide its value; it is *effectively* the "carbon dioxide compensation point" (Heath, 26) and this name has been used by Egle and Schenk (9), Rabinowitch (102), and others. It should be made clear that this concentration can only persist in the intercellular spaces under the above conditions, when there is no diffusion gradient into (or out of) the leaf. If in fact escape of respiratory carbon dioxide directly through the cuticle to the exterior (instead of into the intercellular spaces) is not negligible, at Γ there will be a gradient into the leaf through the stomata and the minimum intercellular space carbon dioxide concentration will be less than the measured Γ. With a gradient out of the leaf (e.g., in carbon dioxide-free air) the intercellular space concentration must of course be below Γ and if these somewhat unnatural conditions are to be considered, Γ is not a minimum concentration. Theoretical considerations, as well as some very indirect experimentation (101) suggest that with leaves in ordinary air of 0.03% carbon dioxide content the concentration in the intercellular spaces seldom falls much below 0.025%; on the other hand, experiments with air forced through the stomata (26) suggest that once inside the leaf the carbon dioxide can be absorbed with great efficiency, which implies a value much nearer to Γ. Be this as it may, any factor found to increase Γ, in a closed system, will tend to raise the concentration of carbon dioxide in the substomatal cavities and so tend to cause closure in the open air.

Rees (103) found that midday closure of oil palm (*Elaeis guineensis* Jacq.) stomata occurred at temperatures above 30°C, and plotting "closure" (based on infiltration) against air temperature gave a relation reminiscent of the log Γ versus temperature curves for onion and coffee in Fig. 32. He found, however, that water strain was also concerned, for an unwatered palm showed more complete midday closure, accompanied by higher air temperatures, than one which had been watered. Heath and Meidner (99) found that in wheat leaves Γ itself was affected by water-strain, increasing when the leaf sheath was placed in 0.2 M or 0.4 M mannitol and decreasing again when the solution was replaced by water once more (Table I). These changes of Γ were accompanied by appropriate stomatal movements i.e., closure with mannitol and reopening with water. (Similar changes in carbon dioxide concentration could be produced by merely removing and replacing the water supply, indicating that the effects need not be attributed to the mannitol being metabolized.) It thus seems that closure due to water-strain is in part attributable to a carbon dioxide effect. It is of interest to note that the preliminary stomatal opening observed when mannitol was substituted for water was in all cases accompanied by a

TABLE I[a]

MEAN Γ VALUES (MAINTAINED FOR 20 MINUTES) IN PARTS PER MILLION OF CARBON DIOXIDE, FROM EIGHT WHEAT LEAVES[b]

Successive treatments	0.2 M mannitol	0.4 M mannitol
Water	78	77
Mannitol	94	116
Water	83	82

Significant difference for P 0.05 = 5.9
Significant difference for P 0.001 = 10.6

[a] From Heath and Meidner (99).
[b] Leaf temperature, 25°C; light intensity, 900 foot candles; relative humidity, 50%.

small temporary fall in carbon dioxide concentration; such opening (pp. 212, 229) therefore, may also in part be operated through carbon dioxide changes.

The Necessity for Chlorophyll and (or) Carotenoids in Light Responses of Stomata

In contrast to Wilson's findings with apparently chlorophyll-free etiolated *Ipomoea* leaves (see p. 224), both Virgin (106) and Shaw (104) failed to obtain any evidence of stomatal opening in leaves of albino mutant barleys. Virgin reported that the plants he used contained no traces of chlorophyll and very small amounts of carotenoids. The strain used by Shaw was found to have a total chlorophyll content amounting to 0.16%, and a carotenoid content to 0.75%, of the values for a normal barley—the guard cells were apparently virtually chlorophyll-free; no evidence of fixation of $C^{14}O_2$ by the guard cells could be obtained and the stomata failed to open in (external) carbon-dioxide-free air. These results clearly give no support for the suggested light effects independent of carbon dioxide (pp. 220 and 224), but do not necessarily conflict with them. Such effects might be energized by light absorbed by carotenoids (p. 224), or by chlorophyll, both of which were virtually absent from the albino barleys. The finding that the grana of chloroplasts (which must contain the carotenoids as well as chlorophyll) can photochemically produce $TPNH_2$ and ATP from TPN and ADP, while carbon dioxide fixation is an entirely dark reaction carried out by the enzymes of the stroma (105) strongly suggests that the absorption of light energy by the grana may operate other systems besides that of carbon assimilation; this might especially be expected in guard cells, which have in many ways abnormal metabolism and whose chloroplasts not only differ in appearance from those of the mesophyll (p. 215) but are much less efficient in fixing carbon dioxide (p. 222). Arnon et al. (98) have in fact speculated that the photochemical production of ATP by chloroplasts may represent a pattern evolved by photosynthetic cells to use light energy for accomplishing cellular work independently of carbon dioxide assimilation.

Mouravieff (100) obtained results with *Veronica beccabunga* L. which show red light less effective than blue in causing stomatal opening, in the ratio 100:1700 (p. 245), which would seem to support the suggestion of carotenoids playing an important role. Mouravieff observed a reduction of guard cell starch in blue light and an increase in red light. In a further experiment he found that 0.3% hydroxylamine sulfate, used as an inhibitor of photosynthesis, prevented all light opening (in red, green, or blue) in ordinary air but allowed an opening response to carbon dioxide-free air. It is not clear why the blue light did not cause opening by hydrolysis of starch under these conditions, but if this effect was inhibited by the hydroxylamine the same might be true of the postulated light effects independent of carbon dioxide.

REFERENCES

98. Arnon, D. I., Whatley, F. R., and Allen, M. B. Assimilatory power in photosynthesis. *Science* **127**, 1026–1034 (1958).
99. Heath, O. V. S., and Meidner, H. Effects of water-strain on the minimum intercellular space carbon dioxide concentration Γ in wheat leaves. *Nature* **182**, 1524–1525 (1958).
100. Mouravieff, I. Action de la lumière sur la cellule végétale. I. Production du mouvement d'ouverture stomatique par la lumière des diverses regions du spectre. *Bull. de la Soc. Bot. Fr.* **105**, 467–475 (1958).
101. Penman, H. L., and Schofield, R. K. Some physical aspects of assimilation and transpiration. *Symposia Soc. Exptl. Biol.* **5**, 115–129 (1951).
102. Rabinowitch, E. I. Photosynthesis and related processes. Vol. II parts 1 and 2. Inter. Science, New York (1951 and 1956).
103. Rees, A. R. Field observations of midday closure of stomata in the oil palm, *Elaeis guineensis* Jacq. *Nature* **182**, 735–736 (1958).

104. Shaw, M. The physiology of stomata. II. The apparent absence of chlorophyll, photosynthesis and a normal response to light in the stomatal cells of an albino barley. *Can. J. Botany.* **36,** 575–579 (1958).
105. Trebst, A. V., Tsujimoto, H. Y., and Arnon, D. I. Separation of light and dark phases in the photosynthesis of isolated chloroplasts. *Nature* **182,** 351–355 (1958).
106. Virgin, H. I. Stomatal transpiration of some variegated plants and of chlorophyll-deficient mutants of barley. *Physiol. Plantarum* **10,** 170–186 (1957).

AUTHOR INDEX

Numbers in boldface refer to pages on which the complete reference is listed at the end of a chapter. Numbers in lightface indicate the pages in the text on which the references are cited. Numbers in parentheses are reference numbers and are included to assist in locating the reference in cases where the authors' names are not mentioned in the text.

A

Abelson, P. H., 57(139), **100**
Achromeiko, A. J., 442, **465**
Ackley, W. B., 693, 697(1), **709**
Adams, G., 610(55), **711**
Addoms, R. M., 683(2), **709**
Äyräpää, T., 60, 61, 74(6), 84, **94**
Ahmed, M. K., 500(1), **545**
Airth, R. L., 666(29), **710**
Allen, M. B., 729(98), **729**
Allen, R. M., 639(3), **709**
Allsopp, A., 112(1), **188**
Alvim, P. de T., 197, 216(1), 218, 222(2), 246, 624, **709**
Andel, O. M. van, 665, **709**
Anderson, D. B., 619, 664(239), 705, **709, 719**
Anderson, L. C., 431(76), **469**
Anderssen, F. G., 280(2, 11), **465, 466**, 502, **545**, 560(5), **598, 599**
Andersson, N. E., 208(4), 225, 246, 642, **709**
Appleman, C. O., 703, **709**
Archer, R. J., 24(1), **93**
Arcichovskij, V., 655, **709**
Arens, K., 439, **465**
Arisz, W. H., 15, 34, 93, 94, 146, **188**, 270, 293, 294, 435, 436, 437, 447(7), 448, 449, 457, **465**, 524, 542, **545**, 665(11), **709**
Armstrong, J. I., 203, 205(14), 246, 625, **713**
Army, T. J., 669(12), 674, 681, **709**
Arndt, C. H., 556, **598**
Arnold, A., 665, **709**
Arnon, D. I., 397(8), **465**, 729(105), **729, 730**
Aronoff, S., 29(5), **94**, 512(5), 516, 517, 531, **545, 551**

Asen, S., 565(3), **598**
Askenasy, E., 107, **188**, 651, **709**
Atkins, W. R. G., 502(6), **545**, 664, **709**
Atkinson, J. D., 430, **465**
Auchter, E. C., 559, **599**
Audus, L. J., 238, **246**
Avundzhyan, E. S., 509(68), **548**
Ayers, A. D., 427(106), 470, **725**
Aykin, S., 632(17), **709**

B

Bärlund, H., 30, 41(35), 45, 46, 47, 81(8, 35), **94, 95**
Bailey, I. W., 286, **466**
Baker, H., 163, **190**
Bakke, A. L., 703, **709**
Bald, J. G., 616, **709**
Ball, N. G., 499(45), **547**, 556, 579, **600**
Ballentine, R., 88(129), **94, 100**
Bandurski, R. S., 183(10), 184(10), **189**
Banga, I., 141, **188**
Bange, G. G. J., 622, 623, **710**
Barger, G., 118, **188**
Barker, J. W., 122(7), 140(7), **188**
Barlee, J. S., 188, **189**
Barrer, M. R., 22(9), **94**
Barrett, F. A., 216(91), 231, **250**
Barrien, B. S., 571, **603**
Bartels, P., 62(10), 64, **94**
Bartholomew, E. T., 705(21), **710**
Bartley, W., 382, 387(10), **466**
Bauer, L., 522, 524(7), 532, 537(7), **545**
Bayliss, W. M., 608(22), **710**
Beasley, E. W., 639, **710**
Beevers, H., 58(151), **101**
Beischer, D., 24(11), **94**
Belden, G. O., 648, **712**
Belikov, I. F., 511(8), **545**
Bellamy, A. W., 530(18), **546**

731

Mason, T. G., 298, **472**, 486, 497, 498, 499, 500(85), 505, 506, 507(82), 527, 532, 533, 537(102), 543, 544, **549, 550**, 557, 558, 561, 563, 566, 575(75), 579, 597, **601, 602**, 697, **719**

Maximov, N. A., 563(67), **601**, 627, 705, 706(232), **719**

Maxwell, K. M., 216(73), 217(73), **249**

Mead, J. F., 532(110), **550**

Mees, G. C., 690, **719**

Meeuse, B. J. D., 615, **719**

Meidner, H., 202(52), 207, 235, **247, 248**, 727, 728, **729**

Meier, R., 411, **474**

Melin, E., 428, **472**, 684(234), **719**

Mellor, J. L., 509, **549**

Mervin, H. E., 649(235), **719**

Meyer, A., 277(138a), **472**

Meyer, B. S., **190**, 619, 642(236), 664 (237, 238, 239), **719**

Milad, Y., 280(11), **466**, 560(5), **599**

Millar, F. K., 55, **101**, 360(216), 361, 363, 418, 421(216), 424, 430(216), 451(216), 453, 454(216), 460, 461, 462, **472, 476**

Miller, E. C., 507, **549**, 574(68), **601**, 610(241), 611, 620, 627(241), 628, 630, 632, 639, 647, 650, 686(241), 697, 706(241), **719**

Miller, E. S., 234, **248**

Miller, R., 642, **715**

Millerd, A., 183(10), 184(10), **189**

Milner, H. W., 704(314), **723**

Milthorpe, F. L., 212(54), 219, 222, 224, 228, 229, 233(30), **247, 248**, 631 (112), 642(243), **713, 719**

Misra, P., 112(1), **188**

Mitchell, J. W., 523(92), **549**

Mitchell, P., 16(112a), 57, **99**, 416, **472**

Mittler, T. E., 519(72), **548**

Mogensen, A., 617(365), **725**

Mohl, H. von, 209, 213, **248**

Molisch, H., 196(56), **249**, 624, 650, **719**

Monod, J., **473**

Montermoso, J. C., 696(246), **719**

Moon, H. H., 430(78), **469**

Moore, W., 532(93), **549**

Moose, C. A., 519, **549**

Moreland, D. F., 661(247), **719**

Morse, W. J., 502(67), **548**

Mothes, K., 88, **99**

Mouravieff, I., 216(57), 241, **249**, 729, **729**

Moyer, L. S., 540(95), **550**

Moyle, J., 57, **99**

Münch, E., 457, **472**, 483, 499(96), 511 (96), 527, 535, 537, **550**

Muenscher, W. C., 687, **719**

Mullins, L. J., 411, **472**

Muntz, J. A., 413(143), **472**

Muzik, T. J., 486(97), **550**

Myers, G. M. P., 150, 155(45), 162, **190**

Myers, L. S., Jr., 532(110), **550**

N

Nadel, M., 211, **249**

Nägeli, C., 5, **99**, 577, **601**

Nathansohn, A., 10, **99**

Nelson, C. D., 509, **550**

Nestel, L., 263(176), 387(176), **474**

Newson, L. D., 500(1), **545**

Nielsen, N., 411(83), **469**

Nielsen, T. R., 397(144), **472**

Nightingale, G. T., 560(70), **601**

Nilsson, H., 428, **472**, 684(234), **719**

Nirenstein, E., 73(116), **99**

Nitsch, J. P., 507(99), **550**

Noack, A., 622(304), **722**

Noeker, N. L., 703, **709**

Nold, R. H., 187, **190**

Nollet, M., 106, **190**

Nutman, F. J., 234, **249**, 624, 626, 636, 704(249), **719**

O

Obmüller, E., 215(37), **248**

Oechsel, G., 24(11), **94**

O'Kane, D. J., 416, **472**

Oland, K., 433, **472**

Oland, T. B., 433, **472**

Oliver, W. F., 581(71), **601**

Oltmanns, F., 289, 446, **472**

Oppenheimer, H. R., 172, **190**, 645, 699, 707(250), **720**

Orchard, B., 209, 216, 218, 235, **247**

Ordin, L., 69(117), **99**, 184, **190**, 320 (100), 371, **470**, 676, 680, **720**

INDEX TO PLANT NAMES

Numbers in this index designate the pages on which reference is made, in the text, to the plant in question. No reference is made in the index to plant names included in the titles that appear in the reference lists. In general, where a plant has been referred to in the text sometimes by common name, sometimes by its scientific name, all such references are listed in the index after the scientific name; cross-reference is made, under the common name, to this scientific name. However, in a few instances when a common name as used cannot be referred with certainty to a particular species, the page numbers follow the common name.

A

Acer (maple), 461, 518, 524, 614, 615
Acer platanoides (maple), 461
Acer pseudo-platanus, 518
Acer rubrum (maple), 524, 614
Acer saccharum, 501, 503
Aerobacter aerogenes (*Bacterium lactis aerogenes*), 413, 416
Aesculus hippocastanum, 165
Alder (see *Alnus*)
Alfalfa (see *Medicago*)
Algae, 88, 129
Algae, blue-green, 46, 48
Algae, brown, 45, 48, 57
Algae, coenocytic, 260, 281, 441
Algae, fresh-water, 9
Algae, green, 45, 48, 441
Algae, marine, 28, 45
Algae, red, 33, 48
Algae, unicellular, 29, 418–420
Allium, 242
Allium cepa (onion), 33, 51, 86, 122, 136, 150, 155, 158, 163, 184, 217, 223, 234, 235, 244, 464, 626, 727
Allium porrum (leek), 626
Alnus glutinosa (alder), 518
Amaranthus spinosus, 696
Amaryllis formosissima (see *Sprekelia*)
Ambrosia trifida, 698
Anacharis (see *Elodea*)
Ananas comosus (pineapple), 431
Apocynaceae, 484
Apple (see *Malus sylvestris*)
Apricot (see *Prunus armeniaca*)
Arachis hypogaea (peanut), 165, 486, 573, 610, 631, 707
Arbacia, 155

Arbutus unedo, 686
Artichoke, Jerusalem (see *Helianthus tuberosus*)
Asclepiadaceae, 485
Ash (see *Fraxinus*)
Asparagus officinalis var. *altilis*, 610
Aspergillus niger, 144
Aster, 310
Atriplex hortensis, 310
Atriplex littoralis, 310
Avena sativa (oats), 184, 310, 565, 615, 626, 680, 686, 703
Avicennia marina (mangrove), 179
Azotobacter, 412

B

Bacillus subtilis, 412
Bacteria, 10, 29, 46, 48, 52, 57, 281, 283, 362, 412–418, 421, 447, 451, 692
Bacterium coli (see *Escherichia*)
Bacterium lactis (see *Streptococcus lactis*)
Banana (see *Musa*)
Barley (see *Hordeum*)
Bean, 442, 581, 686 (see also *Phaseolus* and *Vicia*)
Beech (see *Fagus*)
Beet (see *Beta*)
Beggiatoa, 46, 49, 73, 74, 77, 78, 82
Beggiatoa mirabilis, 10, 46, 57
Begonia "Lucerna," 317
Begonia sempervirens, 140
Bergenia, 140
Beta vulgaris (beet), 5, 6, 29, 122, 129–132, 139–142, 150, 151, 155, 256, 257, 270, 291, 320, 385, 388–399, 402

745

Nitella, 28, 35, 38, 42, 51, 54–56, 73, 75, 77, 79–81, 84, 86, 113, 127, 135, 151, 152, 154, 155, 158, 164, 166, 167, 171–173, 180, 277, 287, 290, 292, 306, 307, 315, 323–328, 361, 362, 368, 372, 419, 541
Nitella clavata, 260, 278, 306, 307, 323–327
Nitella flexilis, 51, 278, 289, 306
Nitella gracilis, 278, 306
Nitella hyalina, 278, 306
Nitella mucronata, 37, 41, 43, 51, 67, 75, 76
Nitellopsis, 44, 56, 64
Nitellopsis obtusulus, 42, 51, 67
Nitzschia closterium, 419

O

Oak (see *Quercus* spp.)
Oats (see *Avena*)
Oedogonium, 48
Olea europaea (olive), 637, 707
Olea lancea, 629
Olive (see *Olea europaea*)
Onion (see *Allium cepa*)
Opuntia, 618, 663
Opuntia camanchica (see *O. phaeacantha*)
Opuntia phaeacantha (*O. camanchica*), 619
Opuntia versicolor, 696
Orange (see *Citrus sinensis*)
Orange, sour (see *Citrus aurantium*)
Orchids, 283
Oryza sativa (rice), 316, 677
Oscillatoria, 46, 49, 77
Oscillatoriales, 46
Oscillatoria princeps, 47, 48
Oxalis, 317

P

Palm, coconut (see *Cocos*)
Palm, oil (see *Elaeis*)
Papaver, 310
Parsnip (see *Pastinaca*)
Parthenium argentatum (guayule), 704
Parthenocissus tricuspidata (Boston ivy), 696

Pastinaca sativa (parsnip), 398
Pea (see *Pisum*)
Peanut (see *Arachis*)
Pear (see *Pyrus communis*)
Pelargonium × *hortorum* (geranium), 195, 196, 216, 217, 231–233, 235, 236, 238–242, 520, 522
Pelargonium zonale (geranium), 196
Phaseolus vulgaris (bean, kidney bean), 362, 430, 521, 528–531, 563–565, 567, 571–574, 578, 583–586, 589–591, 595, 596, 668, 673
Phillyrea media, 645
Phleum pratense (timothy), 280, 615
Picea abies (*P. excelsa;* spruce), 174, 636, 640, 708
Picea excelsa (see *P. abies*)
Pine (see *Pinus* spp.)
Pineapple (see *Ananas*)
Pinus cembra, 708
Pinus echinata, 427, 428, 610, 683, 684
Pinus monticola, 575
Pinus nigra (Austrian pine), 703
Pinus ponderosa, 687
Pinus resinosa, 428
Pinus strobus (white pine), 658, 675
Pinus sylvestris (Scots pine), 428, 619, 636, 640, 647
Pinus taeda (loblolly pine), 428, 610, 630, 646, 663, 675, 692
Pisum sativum (pea), 310, 362, 510–512, 530, 535, 676, 688, 689
Plagiothecium denticulatum, 48
Plantago lanceolata, 310
Plantago maritima, 310
Plumbaginaceae, 447
Poplar, yellow (see *Liriodendron*)
Populus nigra, 459, 460
Potamogeton, 439
Potamogeton crispus, 441
Potamogeton lucens, 440
Potato (see *Solanum tuberosum*)
Potato, sweet (see *Ipomoea*)
Privet (see *Ligustrum*)
Prunus armeniaca (apricot), 560
Prunus laurocerasus, 236
Pseudotsuga menziesii (*P. taxifolia;* Douglas fir), 647
Pseudotsuga taxifolia (see *P. menziesii*)
Pumpkin (see *Cucurbita pepo*)

SUBJECT INDEX

A

Absorption ratio, 254–259
  definition, 254
  for dyes, 257
Accumulation ratio, 254
  definition, 254
Acids,
  penetration into yeasts, 36
  permeability to, 57–60
Active transport, 9, 11–16
  of amino acids, 15–16
  definition, 9
  enzymic inhibition of, 13–14
  of sugars, 14
"Adenoid activity" *see* Active transport
"Adhesion pressure"
  of cytoplasm to cell wall, 129–130
Ammonia,
  accumulation of, 58–60
  penetration of, 332
Anaesthetics,
  influence on permeability, 69
Anion respiration, 373–376
  cytochromes and, 373
  effect of cations on, 375
  oxygen concentration and, 375
  role of organic acids in, 375–376
Anomalous osmosis,
  and water movement, 168–170
"Apparent free space," 131, 133, 293–294, 395
  barriers to diffusion and, 293–294
  calculation of, 293–294
  definition, 35
  nature and extent of, 131, 133
Arsenate, *see* Metabolic inhibitors
Ascent of sap,
  capillarity and, 651
  cohesion theory, 651–656
  imbibition and, 651
  rate of, 661
  root pressure and, 649–650
  vitalistic theories of, 650–651
Auxins,
  in water uptake, 69–70, 187

B

Bases,
  penetration into yeast, 60
  permeability to, 60–61
Bicarbonate uptake, 399, 426, 438–440
  and cation transport, 438–440
Boron,
  role in sugar transport, 534
Bound water, 143–144
  drought resistance and, 144
  estimates of, 143–144
Bromide uptake, 323–327, 336–342, 346–352, 367–370
  by barley roots, 367
  CO_2 concentration and, 346–347, 349
  effect of cations on, 324
  and ion exchange, 324–325
  and light intensity, 326–327
  oxygen tension and, 339–340, 342, 348, 369
  pH effects on, 346–347
  by potato disks, 336–337
  and protein synthesis, 357
  in relation to growth, 351–352
  temperature effects on, 370

C

Calcium,
  immobility in phloem, 571, 573–574
  see also Respiration
Cell permeability,
  of bacteria, 57, 447
  of diatoms, 447
  effect of enzymes on, 88
  effect of pH on, 68
  influence of oxygen on, 69
  and lipid solubility, 8–11, 39, 72–77
  methods of measurement, 27–32
  and molecular size, 20–22, 74–77, 80–82
  mosaic hypothesis, 10
  Overton's rules of, 38–39
  to oxygen, 52
  "retention pressure" theory, 9
  studies with radioactive isotopes, 29, 55–56
  temperature coefficients of, 66

752

W

Water,
  competition among organs for, 704–705
  extraxylary movement of, 659–660
  permeability to, 49–52
  physical properties of, 608
  uses in plants, 609–611
Water absorption,
  active, 664–669
  aeration and, 676–679
  concentration of soil solution and, 672–674
  diffusion pressure gradient of soil and, 680–681
  effect of flooding the soil on, 678–679
  by individual root hairs, 150
  metabolism and, 685
  regions of, 681–682
  and salt uptake, 665, 687–690
  and soil temperature, 674–677
  soil water content and, 669–671
Water conductivity,
  influence of metabolic inhibitors on, 161–162, 164
  of root systems, 159
Water content of plants,
  causes of variation in, 696, 699–700
  diurnal variations in, 695–696
  effects of age and nature of tissue, 692–694
  measurement of, 696–697
  and photosynthesis, 703–704
  relation to respiration, 702–703
  seasonal changes in, 694–695
Water glands,
  secretion by, 187
Water permeability,
  estimation by osmotic methods, 147–155
  of isolated protoplasts, 150
  isotope exchange and, 155–157
  see also Auxins
Water requirement, *see* Transpiration ratio
Wilting, 177, 700–701
  incipient, temporary, and permanent, 700–701

X

Xylem,
  movement of solutes in, 556
  as path of water movement, 657–660
Xylem sap,
  composition of, 279–280
  mineral content of, 560–561
  nitrogenous constituents of, 518–519
  osmotic pressure of, 144
  secretion of ions into, 381–382
  sugars in, 501–502

Z

Zeolites,
  intracrystalline diffusion into, 22